Turbulence from Earth to Planets, Stars and Galaxies—Commemorative Issue Dedicated to the Memory of Jackson Rea Herring

Turbulence from Earth to Planets, Stars and Galaxies—Commemorative Issue Dedicated to the Memory of Jackson Rea Herring

Guest Editors

Boris Galperin
Annick Pouquet
Peter Sullivan

MDPI

Basel • Beijing • Wuhan • Barcelona • Belgrade • Novi Sad • Cluj • Manchester

Guest Editors
Boris Galperin
University of South Florida
St. Petersburg, FL
USA

Annick Pouquet
National Center for
Atmospheric Research
Boulder, CO
USA

Peter Sullivan
National Center for
Atmospheric Research
Boulder, CO
USA

Editorial Office
MDPI AG
Grosspeteranlage 5
4052 Basel, Switzerland

This is a reprint of the Special Issue, published open access by the journal *Atmosphere* (ISSN 2073-4433), freely accessible at: https://www.mdpi.com/journal/atmosphere/special_issues/0ZB2T055OT.

For citation purposes, cite each article independently as indicated on the article page online and as indicated below:

Lastname, A.A.; Lastname, B.B. Article Title. *Journal Name* **Year**, *Volume Number*, Page Range.

ISBN 978-3-7258-2739-8 (Hbk)
ISBN 978-3-7258-2740-4 (PDF)
https://doi.org/10.3390/books978-3-7258-2740-4

Cover image courtesy of UCAR
Jack Herring outside the National Center for Atmospheric Research Mesa Laboratory circa 1980-07-10.

Contents

About the Editors

Boris Galperin

Boris Galperin works in the areas of geophysical and planetary fluid dynamics, in areas related to large-scale and small-scale turbulence and wave–turbulence interactions. He employs theoretical, numerical and laboratory tools in an attempt to understand and explain the observations.

Annick Pouquet

Annick Pouquet is an Emeritus scientist at CISL/NCAR, working on the dynamics of nonlinear systems and turbulence, with an emphasis on rotating stratified flows and MHD, and with applications to solar wind, the solar photosphere, as well as the atmosphere and oceans. In particular, she focuses on the origin and dynamics of magnetic fields, and the intense dissipation in systems dominated mostly by waves. She has supervised a number of students and postdoctoral fellows, and she has also held several administrative positions in scientific laboratories.

Peter Sullivan

Peter Sullivan, Ph.D., is Senior Scientist in the Mesoscale and Microscale Meteorology Laboratory at the National Center for Atmospheric Research. Dr. Sullivan's research focuses on turbulence in the atmospheric and oceanic boundary layers, air–sea interaction, and subgrid-scale modeling. He uses large-eddy and direct numerical simulations on supercomputer platforms to conduct fundamental process studies, and in the past was part of a team carrying out outdoor surface layer measurements to quantify subgrid-scale variables that appear in large-eddy simulation.

Preface

Jackson Rae Herring (1931–2022) was very much an appreciated colleague by those of us who knew and worked with him. He was a scholar of turbulence theory, influenced in particular by his close friend, Robert Kraichnan. He lived a gentle and simple life, at least as far as he let on. It would not surprise me if he holds the world record for the number of bicycle ascents of NCAR's Mesa road, if not overall, at least in the senior competitors category.

He was born in Louisville, KT, was a student in Physics at Wake Forest College and the University of North Carolina, worked for NASA at the Goddard Space Science Center and the Institute for Space Studies, and then came to the National Center for Atmospheric Research through the Advanced Study Program in 1972. Anyone meeting him or his wife Betty would immediately know they were southern. Nor would they be surprised if told they were dedicated Quakers.

We worked alongside each other for more than twenty years at NCAR, coming and leaving around the same times. He was usually cheerful, but occasionally irritated by his bosses. He also had a mischievous side: he taught me the useful practice of "dynamic incompetence": when asked to do something unnecessary or unappealing, do it badly enough that they won't ask again. In contrast, when we co-organized the summer-long Geophysical Turbulence Conference in 1987 and then edited the book of its principal lectures, he worked long and enthusiastically.

This Special Issue of *Atmosphere* is an eclectic collection of papers by some of his many colleagues, respectfully submitted in honor of his long career and his genial, generous personality.

By James C. McWilliams
Department of Atmospheric and Oceanic Sciences, University of California, Los Angeles, CA 90095, USA.

Boris Galperin, Annick Pouquet, and Peter Sullivan
Guest Editors

atmosphere

MDPI

Editorial

Personal Recollections on Jack Herring and Developments in Theory of Turbulence, Atmospheric Sciences, and Computational Fluid Dynamics

Boris Galperin [1,*] and Semion Sukoriansky [2,*]

1 College of Marine Science, University of South Florida, St. Petersburg, FL 33701, USA
2 Department of Mechanical Engineering, Ben-Gurion University of the Negev, Beer-Sheva 84105, Israel
* Correspondence: bgalperin@usf.edu (B.G.); semion@bgu.ac.il (S.S.)

The majority of articles in this Special Issue illuminate various aspects of Jack Herring's contributions to the theory of turbulence, atmospheric sciences, and computational fluid dynamics (CFD), be it through his publications, presentations, collaborations, work with colleagues and students, or personal contacts. Our (BG and SS) interactions with Jack took place mostly at meetings, conferences, and during visits to NCAR, particularly those organized under the auspices of the Geophysical Turbulence Program (GTP). These interactions were also both professional and personal. Jack was deeply interested in our progress with the development of the quasi-normal scale elimination (QNSE) theory, particularly after it expanded into anisotropic turbulence with dispersive waves excited under stable stratification or system rotation. These issues were discussed at length during Jack's visit to the Abdus Salam International Center for Theoretical Physics (ICTP) in Trieste, Italy, in August 2011, which was one of the last scientific conferences he attended. We shall come back to that conference later on.

To set up the background, we start with a recollection of our time at Princeton University, where BG was affiliated with the Program in Atmospheric and Oceanic Sciences, AOS/GFDL, and then with the Program in Applied and Computational Mathematics, PACM (1983–1989), while SS was affiliated with PACM during his entire stay (1989–1991). The PACM during that time was led by Steve Orszag and Andrew Majda.

Both of us were working on the development and implementation of theoretical and semi-empirical models of turbulence in different systems. BG's focus was on oceanic and atmospheric flows, while SS was concerned with engineering flows and magnetohydrodynamics (MHD). Our collaboration was significantly enhanced by contributions from Ilya Staroselsky, who joined Orszag's group, coming from the Landau Institute of Theoretical Physics.

During that time, Steve Orszag and Victor Yakhot, who have just relocated to Princeton from MIT, were working on the renormalization group approach to turbulence theory, in particular its application to modeling and simulation of turbulent flows [1], and we joined that effort.

It should be noted that the application of the field-theoretical methods to turbulence started in Princeton much earlier. One of the most significant and far-reaching breakthroughs was the application of the quantum field theory methods, initiated by Robert Kraichnan around the 1950s. His interest in turbulence was stimulated by the search for highly nonlinear, particle-like solutions to unified field equations.

To put the narration in a proper frame, recall that Kraichnan's earliest studies, undertaken on his own starting at age 13, were in general relativity [2]. Reaching the age of 18, he wrote an insightful undergraduate thesis at MIT, *Quantum Theory of the Linear Gravitational Field*. In 1949, he earned a PhD in physics from MIT for his thesis, *Relativistic Scattering of Pseudoscalar Mesons by Nucleons*. His thesis advisor was Herman Feshbach. Upon graduation from MIT, he was accepted as an assistant to Albert Einstein at the Institute for Advanced Study in Princeton. The IAS has kindly provided to us Kraichnan's

Citation: Galperin, B.; Sukoriansky, S. Personal Recollections on Jack Herring and Developments in Theory of Turbulence, Atmospheric Sciences, and Computational Fluid Dynamics. *Atmosphere* **2024**, *15*, 968. https://doi.org/10.3390/atmos15080968

Received: 17 June 2024
Accepted: 1 August 2024
Published: 13 August 2024

original application and two files (one from the Director's Office and one from the School of Mathematics/Natural Sciences of the IAS) that include the original correspondence from the Director of IAS Robert Oppenheimer and IAS staff confirming Robert Kraichnan's application, and then, later, referring to his work as a Research Assistant to Professor Einstein. The letter of acceptance, signed by Robert Oppenheimer, is shown in Figure 1. Kraichnan's work from that period has not lost of its breadth and significance, e.g., [3]. As a background reflecting these historic years of the interdisciplinary, breakthrough research in the quantum field theory, gravitation, and fluid turbulence, the photo in Figure 2 captures a moment of a vivid discussion between Albert Einstein and Robert Oppenheimer, the interaction that was one of the high points in the book [4] and the following up movie *Oppenheimer*.

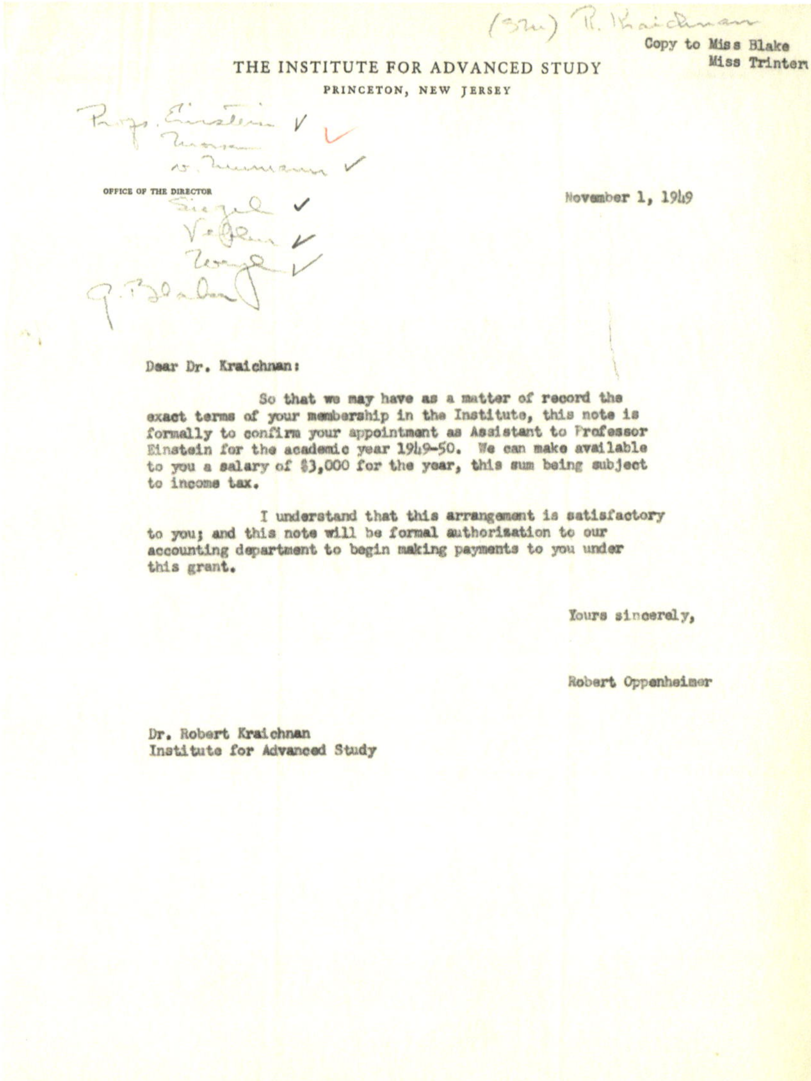

Copy to Miss Blake
Miss Trinten

THE INSTITUTE FOR ADVANCED STUDY
PRINCETON, NEW JERSEY

OFFICE OF THE DIRECTOR

November 1, 1949

Dear Dr. Kraichnan:

So that we may have as a matter of record the exact terms of your membership in the Institute, this note is formally to confirm your appointment as Assistant to Professor Einstein for the academic year 1949-50. We can make available to you a salary of $3,000 for the year, this sum being subject to income tax.

I understand that this arrangement is satisfactory to you; and this note will be formal authorization to our accounting department to begin making payments to you under this grant.

Yours sincerely,

Robert Oppenheimer

Dr. Robert Kraichnan
Institute for Advanced Study

Figure 1. A letter of appointment of Robert H. Kraichnan as Assistant to Professor Einstein at the Institute for Advanced Study in Princeton.

Figure 2. Robert Oppenheimer with Albert Einstein in LIFE Magazine, 1947. Photograph by Alfred Eisenstaedt. Courtesy of *LIFE* Magazine, 1947, Time Inc. Copied from J. Robert Oppenheimer, *Uncommon Sense* (Birkhäuser Boston, 1984).

As documented in [2], in the early 1990s, Kraichnan wrote to Jack Herring, who was his close friend and colleague for many years, that he "realized I had no real idea of what I was doing and turned to Navier–Stokes as a nonlinear field problem where experiment could confront speculation. After initial surprise that turbulence did not succumb rapidly to field-theoretic attack, I have been trapped ever since. My overall research theme regarding turbulence has been to understand what aspects of turbulence can and cannot be described by statistical mechanics; that is, what characteristics follow from invariances, symmetries, and simple statistical measures, and what, in contrast, can be known only from experiment or detailed solution of the equations of motion. The principal tool I have used is the construction of a variety of stochastic dynamical systems that incorporate certain invariances and other dynamical properties of actual turbulence but whose statistics are exactly soluble. The similarities and differences between these model solutions and real turbulence help illuminate what is essential in turbulence dynamics. My interests have centered on isotropic Navier–Stokes and magnetohydrodynamic turbulence in two and three dimensions".

The quest to find an analytical solution of Navier–Stokes equations and, thus, fully solve the problem of turbulence has remained one of the oldest and most famous challenges of classical physics for centuries. When Kraichnan started looking at turbulence, the community was trying to turn A.N. Kolmogorov's 1941 ideas into a quantitative theory. However, it withstood attacks employing advanced statistical and perturbation methods borrowed from the quantum field theory and statistical mechanics [5]. In fact, Kraichnan was the first one to really make a difference in this endeavor, based on pioneering applications of the field theory methods of theoretical physics.

Further developments and Jack Herring's contributions in this area are accounted for in two articles in this Special Issue [6,7]. Jack had many other interests in turbulence. One was turbulent convection, where he produced some pioneering results in the weakly nonlinear regime, both analytical and theoretical. These contributions are discussed in [8]. Jack had an abiding interest in the universality of small-scale turbulence. Professor Katepalli Sreenivasan of New York University fondly recalls Jack's invitation in 1978, when he was still a post-doc at Johns Hopkins University, to attend a meeting at NCAR on small-scale universality. He is grateful to Jack for providing the opportunity to interact with a number of stalwarts, such as Stan Corrsin, Bob Kraichnan, Uriel Frisch, Carl Gibson, Steve Orszag, Eric Siggia, and others, at that meeting, which also explored the implications of Kolmogorov's work to geophysical problems.

During approximately the same time period, 1946–1953, the Meteorology Group was established within the IAS. As cited in [9], "A project whose ultimate effects on weather forecasting may be revolutionary has been quietly under way during the past year (1946) in the academic surroundings of the Institute for Advanced Study, Princeton, New Jersey... In August 1946, a conference of meteorologists met in Princeton to discuss the project... Since last summer, work has gone forward in promising fashion, though it is still far too early to expect immediate, tangible results... The immediate aims of this group are the selection and mathematical formulation of meteorological problems to be solved by the electronic computer... the most interesting feature of the project is the effort being made to link the theory behind atmospheric processes with future weather". Among the early major achievements of the group were the first numerical weather forecasts on the Electronic Numerical Integrator and Computer (ENIAC).

Short and long stays with the IAS of such prominent mathematicians and atmospheric scientists as John von Neumann, Julius Charney, Arnt Eliassen, Joseph Smagorinsky, Carl Gustav Rossby, Norman A. Phillips, and many others facilitated the IAS becoming the world leading center for the numerical weather prediction (NWP). In 1955, the U.S. Weather Bureau created a General Circulation Research Section (GCRS) under the direction of Joseph Smagorinsky, who continued the von Neumann/Charney computer modeling program based upon a 3D, global, primitive-equation model of the general circulation of the atmosphere. The GCRS was initially located in Suitland, Maryland, but in 1959, it was moved to Washington, D.C., and its name was changed to the General Circulation Research Laboratory. In 1963, the GCRL was renamed again, to become the Geophysical Fluid Dynamics Laboratory. In 1968, the GFDL settled at Princeton University's James Forrestal Campus under the National Oceanic and Atmospheric Administration (NOAA), where it remains. The first GFDL Director was Joseph Smagorinsky.

While recruiting the personnel, Smagorinsky followed several strict principles that he developed over the years [10]; among them were

- Personnel decisions are the most important actions taken.
- It is better to not fill a vacancy than to compromise quality; patience always pays off.

An example of these rules' effectiveness was revealed in 1959, when he invited Syukuro Manabe of the Tokyo NWP Group to join the GCRL. As described in [10], Smagorinsky recalled: "I had been reading some papers about Japanese scientists, came across two names that seemed to crop up time after time. One was Manabe, the other was Kikuro Miyakoda. The thing that intrigued me about Manabe's name is that it didn't so much appear on papers that he had written, but on papers that his colleagues had written where they were crediting him with some of the basic ideas... So I made him an offer... he came as a visitor... I couldn't offer him a permanent job, he was an alien. I needed again very special permission to hire a foreigner. It really probably had never been done, but that started a precedent".

Manabe was assigned the task of coding and developing the General Circulation Model (GCM) of the atmosphere. Manabe's group had become one of the most vigorous and prominent GCM programs in the world. He retired in 1998. In 2021, he won a Nobel Prize in Physics for his work in climate modeling.

In the late 1950s to early 1960s, it was noticed that numerical simulations with the primitive equations of the atmospheric circulation produce numerical instabilities. To ameliorate the problem, Smagorinsky introduced a nonlinear viscosity [11]. This viscosity was adopted in many other areas of CFD, and became known as Smagorinsky's viscosity.

BG left Princeton in 1989 to assume a position at the University of South Florida (USF) in St. Petersburg, Florida. In 1990, he organized a conference entitled "Large Eddy Simulation (LES)—Where Do We Stand?" that intended to summarize developments in various areas of CFD and outline directions of future research. At that time, LES was still to gain maturity, and this conference was one of the first in this area. One of the central issues were discussions and comparisons of various formulations of the nonlinear viscosity that were used in various LES schemes, and so the first presenter was Joseph Smagorinsky. He provided a historical perspective of his approach. By that time, he had retired from GFDL, and did not know about the strong impact his formulation had on the development of LES and CFD in general. After his talk, presenter after presenter was reporting on LES results with Smagorinsky's viscosity, and Smag (the name Joseph Smagorinsky was called at GFDL) was simply overwhelmed.

Jack Herring was one of the invited presenters at the conference and his lecture, co-authored with Robert Kerr, was on the contributions of two-point closures to LES, the topic that was going back to his collaborative research with Robert Kraichnan. The conference featured numerous exchanges between scientists working in different fields, and of course, Joe Smagorinsky and Jack Herring did not miss their chances to converse. Presentations and panel discussions at the conference were summarized in the book [12].

Over the following decades, our group has developed a spectral closure based upon [1] that was named Quasi-Normal Scale Elimination (QNSE). The basic technique of this closure was described in [13]. It was extended to flows with stable stratification [14], rotation [15], and magnetic field [16]. In 2008, SS visited NCAR to implement the QNSE scheme in the atmospheric model Weather Research and Forecasting (WRF). During the visit, he had many discussions with Jack Herring about the QNSE model and its performance for stably stratified turbulence.

During the following several years, our and Jack Herring's ways crossed with IAS again in a somewhat symbolic way. In 1964, largely due to the efforts of the Pakistani Nobel Prize winner theoretical physicist Abdus Salam, the International Center for Theoretical Physics (ICTP) was founded in Trieste, Italy. Robert Oppenheimer was the first Chairman of ICTP's Scientific Council. The first Council's meeting, chaired by Robert Oppenheimer, took place in May 1964 at the International Atomic Energy Agency (IAEA) Headquarters in Vienna. The meeting is pictured in Figure 3. From that meeting arose ICTP's scientific reputation and steadfast support for scientists from developing countries. In 1965, Robert Oppenheimer wrote: "It seems to me that the Centre has been successful in these eight or nine months of operation in three important ways. It has cultivated and produced admirable theoretical physics, making it one of the great foci for the development of fundamental understanding of the nature of matter. The Centre has obviously encouraged, stimulated and helped talented visitors from developing countries who, after rather long periods of silence, have begun to write and publish during their visit to the Centre in Trieste. This is true of physicists whom I know from Latin America, from the Middle East, from Eastern Europe, and from Asia. It is doubtless true of others. The Centre has become a focus for the most fruitful and serious collaboration between experts from the United States and those from the Soviet Union on the fundamental problems of the instability of plasmas, and of means for controlling it. Without the Centre in Trieste, it seems to me doubtful that this collaboration would have been initiated or continued. In all the work at the Centre of which I know, very high standards prevail. In less than a year it has become one of the leading institutions in an important, difficult and fundamental field" [17].

The ICTP building still features the Oppenheimer Meeting Room with a large portrait of Robert Oppenheimer on the wall.

Years later ICTP was headed by Katepalli Sreenivasan, who also developed close collaboration with both Robert Kraichnan and Jack Herring. He worked with Victor Yakhot on further development of the statistical theory of turbulence. The ICTP hosted many conferences in various areas of Physics, including Fluid Dynamics. Jack Herring, SS, and BG took part in one of these conferences in 2011. During one of the break days at the conference, he and BG planned to visit several historic landmarks around Trieste including the ruins of a Roman Amphitheater (Figure 4) and a beautiful synagogue built in Italian architectural style. Jack was very curious during the visit, asked many questions about the history of Trieste and events during the WWII. After the visit, he said that he had learned a lot and was humbled by what he learned.

Figure 3. Abdus Salam, left, Paolo Budinich, second left, and Robert Oppenheimer, right, attend the first meeting of the ICTP Scientific Council at the IAEA Headquarters in Vienna in May 1964.

Figure 4. Jack Herring and Boris Galperin at the ruins of a Roman Amphitheater in Trieste while attending a conference at ICTP in August 2011.

Upon return to the US, he sent BG a message that spoke volumes about his human side and high moral standards:

9/4/2011, 7:18 PM
Dear Boris.

Thanks for the web site about the Jews in Trieste during the NAZI era. It is indeed a sad story. I looked up the quote from Rousseau about the Jews, and have copied it here:

The Jews, said Rousseau afford an astonishing speckle. The laws of Salon, Numa, and Lycurgus are dead those of Moses, much more ancient, continue to live. Athens, Sparta, and Rome have perished and left no offspring on the earth. But Zion destroyed, has not lost her children, they are preserved, they multiply, they spread through out the world. They mingle with all peoples yet are not confused with them; they have no rulers, yet they are always a people... What must have been the force of a legislator capable of effecting such a marvel! Of all the systems of legislation now known to us, this one has undergone tests, has been steadfast.

In Masson, La religion de Rousseau, II, 240.

I did enjoy the conference, and thank you for including me.

Best, Jack

Author Contributions: Writing—original draft preparation, B.G., writing—review and editing, B.G. and S.S. All authors have read and agreed to the published version of the manuscript.

Funding: This research received no external funding.

Conflicts of Interest: The authors declare no conflicts of interest.

References

1. Yakhot, V.; Orszag, S. Renormalization group analysis of turbulence. I. Basic theory. *J. Sci. Comput.* **1986**, *1*, 3–51. [CrossRef]
2. Eyink, G.; Frisch, U. Robert H. Kraichnan. In *A Voyage through Turbulence*; Davidson, P.A., Kaneda, Y., Moffatt, K., Sreenivasan, K.R., Eds.; Cambridge University Press: Cambridge, UK, 2011; pp. 329–372.
3. Maia, M.D.; Bezerra, V.B. The Higgs impact on Einstein's gravity: The Einstein–Kraichnan quantum gravity. *Int. J. Mod. Phys. A* **2020**, *35*, 2040012. [CrossRef]
4. Bird, K.; Sherwin, M.J. *American Prometheus*; Atlantic Books: London, UK, 2023.
5. Kraichnan, R. Inertial-Range Spectrum of Turbulence. *Phys. Rev. Lett.* **1973**, *31*, 744–746. [CrossRef]
6. McComb, D. Jackson R. Herring and the Statistical Closure Problem of Turbulence: A Review of Renormalized Perturbation Theories. *Atmosphere* **2023**, *14*, 827. [CrossRef]
7. Frederiksen, J.S.; O'Kane, T.J. Realizable Eddy Damped Markovian Anisotropic Closure for Turbulence and Rossby Wave Interactions. *Atmosphere* **2023**, *14*, 1098. [CrossRef]
8. Sreenivasan, K.R.; Niemela, J.J. Turbulent Convection at Very High Rayleigh Numbers and the Weakly Nonlinear Theory. *Atmosphere* **2023**, *14*, 826. [CrossRef]
9. Smagorinsky, J. The Beginnings of Numerical Weather Prediction and General Circulation Modeling: Early Recollections. In *Advances in Geophysics*; Elsevier: Amsterdam, The Netherlands, 1983; Volume 25, pp. 3–37.
10. Lewis, J.M. SMAGORINSKY'S GFDL: Building the Team. *Bull. Amer. Meteorol. Soc.* **2008**, *89*, 1339–1353. [CrossRef]
11. Smagorinsky, J. General circulation experiments with the primitive equations, Part 1: The basic experiment. *Mon. Weather Rev.* **1963**, *91*, 99–152. [CrossRef]
12. Galperin, B.; Orszag, S. (Eds.) *Large Eddy Simulation of Complex Engineering and Geophysical Flows*; Cambridge University Press: Cambridge, UK, 1993.
13. Sukoriansky, S.; Galperin, B.; Staroselsky, I. Cross-terms and ϵ-expansion in RNG theory of turbulence. *Fluid Dyn. Res.* **2003**, *33*, 319–331. [CrossRef]
14. Sukoriansky, S.; Galperin, B.; Staroselsky, I. A quasinormal scale elimination model of turbulent flows with stable stratification. *Phys. Fluids* **2005**, *17*, 085107. [CrossRef]
15. Sukoriansky, S.; Galperin, B. QNSE theory of turbulence anisotropization and onset of the inverse energy cascade by solid body rotation. *J. Fluid Mech.* **2016**, *805*, 384–421. [CrossRef]
16. Sukoriansky, S.; Zemach, E. Theoretical study of anisotropic MHD turbulence with low magnetic Reynolds number. *Phys. Scr.* **2016**, *91*, 034001. [CrossRef]
17. Hamende, A. ICTP in Trieste: Creative research for global development. *IAEA Bull.* **1987**, *61*, 60–64.

atmosphere

Review

Beyond Scale-by-Scale Equilibrium

John C. Vassilicos

Univ. Lille, CNRS, ONERA, Arts et Metiers Institute of Technology, Centrale Lille, UMR 9014-LMFL-Laboratoire de Mécanique des Fluides de Lille-Kampé de Fériet, F-59000 Lille, France; j.c.vassilicos@imperial.ac.uk

Abstract: Homogeneous turbulence and turbulence in scale-by-scale equilibrium, played a leading role in the turbulence research of the second half of the twentieth century, and Jack Herring was an important contributor to these developments. The research activity which has followed these developments over the past ten to fifteen years concerns turbulence, which is out of scale-by-scale equilibrium either because it is non-stationary or because it is non-homogeneous or both. This paper is a short review of recent progress in this relatively new direction of turbulence research.

Keywords: turbulence; turbulent flows; non-equilibrium; non-stationarity; non-homogeneity

Citation: Vassilicos, J.C. Beyond Scale-by-Scale Equilibrium. *Atmosphere* **2023**, *14*, 736. https://doi.org/10.3390/atmos14040736

Academic Editors: Boris Galperin, Annick Pouquet and Peter Sullivan

Received: 16 March 2023
Revised: 4 April 2023
Accepted: 5 April 2023
Published: 19 April 2023

1. Introduction

The last time I met Jack Herring was in April 2019 at NCAR in Boulder, Colorado. I was there to give a seminar and was very touched by the fact that he took the time to come to my seminar and even contributed to the discussion at the end. I had already met Jack Herring a few times over the years between 1999 and 2010 at various turbulence events in the UK and in France. I had come out of every single one of my interactions with him very much struck by his gentle manner, kindness and quiet sense of humor. I remember him, in particular, smiling when hearing mention of the Craya–Herring frame and expressing bemusement at his name being attached to a frame of reference. However, his work, which spanned five decades from the 1960s to the 2010s, was wide enough and deep enough to definitely deserve some such way of lasting recognition. One remembers, for example, his self-consistent field theory of turbulence [1,2] his theory of quasi-geostrophic turbulence [3], his contribution to confirming the inverse cascade in two-dimensional turbulence [4] (arguably one of the most important results of homogeneous turbulence research in the second half of the twentieth century), and his observations of three-dimensional small-scale structure in strongly stratified turbulence [5].

Most if not all of Jack Herring's work (that I know of, at least) is concerned with homogeneous turbulence. Statistical homogeneity is a necessary prerequisite for scale-by-scale equilibrium between the average turbulence dissipation rate and average interscale transfer rate at scales within an inertial scale range. The past ten to fifteen years have seen significant research activity on turbulence and turbulent flows beyond such equilibrium (e.g., [6–36]) In particular, in the four years since I last saw Jack Herring, there has been progress on scale-by-scale non-equilibrium arising from non-homogeneity. In this paper, I summarize some of these recent developments, as they are a natural next step after the turbulence research era of Jack Herring, which was, to a large extent, dominated by scale-by-scale equilibrium and homogeneity.

2. The Turbulence Problem

The turbulence problem is to reliably reduce the number of degrees of freedom as much as possible in as general/universal of a way as possible or within well-defined universality classes. This requires to understand the physics of turbulence in a variety of turbulent flows. In general, the rate of change of the turbulent kinetic energy (following a mean flow if it is meaningful to define one) is balanced by the turbulence production rate, turbulence transport rate, work performed by pressure gradient–velocity correlations, viscous diffusion rate and turbulence dissipation rate. Of these, only turbulence production and viscous diffusion are linear mechanisms. Research since the 1950s on the linearized Navier–Stokes equation, starting perhaps with [37], effectively addresses these two terms. However, the other three mechanisms, turbulence transport, pressure gradient–velocity correlations and turbulence dissipation, are fully non-linear. Turbulence dissipation requires fully non-linear interscale transfer physics (e.g., cascade) to be understood, modeled and scaled.

3. Turbulence Dissipation and Equilibrium Cascade

The Kolmogorov equilibrium turbulence cascade (see [38–41]) has been an essential cornerstone of turbulence theory and modeling. It leads to the turbulence dissipation scaling $\varepsilon \sim K^{3/2}/L$ (where ε stands for average turbulence dissipation rate, K for average turbulent kinetic energy and L for a measure of the average size of the largest turbulent eddies), which is, in turn, central for the following:

(i) Estimating the Reynolds number (Re) scaling of the number of turbulent degrees of freedom $(L/\eta)^3 \sim Re^{9/4}$ (where η is the Kolmogorov length scale) (e.g., [39,42,43]);
(ii) Establishing a formula for the turbulent viscosity in one-point turbulence models, such as the $K - \epsilon$ model (e.g., [43,44]);
(iii) Predicting the power law decay of homogeneous isotropic turbulence (e.g., [39]);
(iv) Establishing the spreading rates and mean flow evolutions of self-preserving boundary-free turbulent shear flows, such as wakes and jets (e.g., [45–48]);
(v) Estimating the entrainement and the motion of turbulent/non-turbulent interfaces in self-preserving turbulent shear flows (e.g., [48,49]);
(vi) Providing, with the related Kolmogorov form of the turbulent energy spectrum, a physical justification of subgrid scale models for large eddy simulations (e.g., [42,43]).

The Kolmogorov equilibrium turbulence cascade has its foundation on a two-point turbulent kinetic energy equation (see [50,51]), where the rate of change of the turbulent kinetic energy at scales smaller than r (integrated over all two-point directions) is, in general, balanced by the two-point turbulence production rate, two-point interspace turbulence transport rate, interscale turbulent transfer rate at scale r, two-point pressure gradient–velocity correlations, two-point viscous diffusion rate and the sum of turbulence dissipation rates at two points in the flow. The Kolmogorov scale-by-scale equilibrium in an inertial range of scales r is achieved contingent on the existence of the following:

(a) Local homogeneity and therefore zero average two-point turbulence production rate, zero average two-point interspace turbulence transport rate, and zero average two-point pressure gradient-velocity correlations at scale r;
(b) Local stationarity and, therefore, zero average rate of change of the turbulent kinetic energy at scales smaller than r.

In that case, the average interscale transfer rate balances the average turbulence dissipation rate at all scales small enough for (a) and (b) to hold and large enough to be able to neglect a two-point viscous diffusion rate (see Appendix B of [52] and pages 86–87 in [39]). This balance is the Kolmogorov scale-by-scale equilibrium, from which one may deduce the scaling $\varepsilon = C_\varepsilon K^{3/2}/L$ with C_ε being a dimensionless constant independent of Reynolds number if the inertial range can be assumed to extend all the way to scales comparable to L. The scaling $\varepsilon \sim K^{3/2}/L$ is the Taylor–Kolmogorov equilibrium dissipation law first proposed by [53] and then given a theoretical underpinning by Kolmogorov in 1941 (Kolmogorov's three turbulence papers of 1941 can all be found in [40]).

4. Non-Equilibrium Turbulence Dissipation Laws

A different turbulence dissipation law has been observed in a variety of turbulent flows, including decaying and forced periodic turbulence [12,13,54], turbulence generated by fractal and regular grids [9], turbulent bluff body wakes [14], turbulent jets [48], atmospheric turbulence [34,35], and even wall turbulence [20,55]. This is a non-equilibrium dissipation scaling such that $C_\varepsilon \sim \sqrt{Re_G}/Re_\lambda$, where Re_G is some global Reynolds number characterizing the inlet/initial/boundary conditions and Re_λ is a local Taylor length-based Reynolds number. Given its clear difference from the Taylor–Kolmogorov equilibrium scaling $C_\varepsilon = Const$, the underlying interscale transfer and cascade must be out of scale-by-scale equilibrium. There are currently two different attempts at explaining this non-equilibrium dissipation scaling: one by [56], who obtain it by setting to zero all turbulent kinetic energy at scales larger than L and whose calculation leads to different dissipation scalings if different energy spectral shapes are assumed at wavenumbers below L^{-1}; and one by [18], who obtain it by making the hypothesis that the turbulence dissipation rate at scales larger than L evolves in proportion to the turbulence dissipation rate at scales smaller than L in the presence of large scale coherent structures. Even if validated in the near wake of a square prism by [57], this hypothesis of [18] is as puzzling as the non-equilibrium turbulence dissipation scaling that it attempts to explain. If correct, this hypothesis could perhaps imply that qualitatively different large-scale coherent structures may lead to different turbulence dissipation laws. Indeed, Ref. [30] found a different non-equilibrium dissipation scaling in the high Reynolds number wake of a slender (rather than bluff) body, namely $C_\varepsilon \sim (\sqrt{Re_G}/Re_\lambda)^{4/3}$, and they made the point that the large scale coherent structures are helical in the wake of a slender body rather than dominated by vortex shedding as in bluff body wakes. While their observation supports the hypothesis of [18] in some qualitative sense, it invalidates it quantitatively because the argument of [18] leads to $C_\varepsilon \sim \sqrt{Re_G}/Re_\lambda$ but does not lead to $C_\varepsilon \sim (\sqrt{Re_G}/Re_\lambda)^{4/3}$.

A satisfactory theoretical explanation of the new non-equilibrium turbulence dissipation laws discovered in various turbulent flows over the past 15 years remains elusive. Nevertheless, these new laws make it clear that non-equilibrium interscale transfer and cascade physics common to various classes of turbulent flows must exist. In the following section, we present a classification of scale-by-scale non-equilibria and some of their properties as they seem to currently appear from research over the past few years.

5. Classification of Scale-by-Scale Non-Equilibria

Given that stationarity and homogeneity are the pillars of Kolmogorov scale-by-scale equilibrium, the attempt at a classification of scale-by-scale non-equilibria summarized here is in terms of different cases of non-stationarity and non-homogeneity. This section is no more than a synthesizing summary intended to help interested readers direct themselves toward the cited papers, where detailed expositions of each category of turbulent flow can be found, and also help them see the broader picture, which results from all of these papers together but is in none of them individually. Note that the classification into categories A to F below is not finalized yet. It is still evolving, hopefully toward a complete classification of physically meaningful universality classes.

5.1. Category A: Statistically Stationary and Homogeneous Turbulence

This type of turbulence is the case where, on average, Kolmogorov scale-by-scale equilibrium holds. However, there is fluctuating non-stationarity and non-homogeneity, and resulting non-zero fluctuations of the rate of change of the turbulent kinetic energy at scales smaller than r, two-point interspace turbulence transport rate and the two-point pressure gradient–velocity term. Even though these processes average to zero in statistically stationary homogeneous turbulence, their fluctuations are so intense that they dwarf the non-zero average interscale transfer and turbulent dissipation rates. Even though these fluctuations are not within the scale-by-scale equilibrium, they nevertheless obey certain rules. The interscale transfer rate can be decomposed into a solenoidal part resulting from the integrated vorticity equation and an irrotational part resulting from the integrated Poisson equation for pressure. Only the solenoidal part is involved in the average scale-by-scale Kolmogorov equilibrium. The irrotational part of the interscale transfer rate, the irrotational part of the two-point interspace transport rate and half the two-point pressure gradient–velocity term average to zero but are exactly equal to each other everywhere in space and at any time. The fluctuating rate of change of the turbulent kinetic energy at scales smaller than r is typically balanced by the solenoidal fluctuations of the interscale transfer rate and the two-point interspace transport rate. At intense interscale transfer rate events, the interscale transfer rate and the two-point interspace transport rate tend to have opposite signs and partially cancel each other. Further aspects of a detailed picture of fluctuating scale-by-scale non-equilibrium in statistically stationary homogeneous turbulence can be found in [58].

5.2. Category B: Non-Stationary Statistically Homogeneous Turbulence

This is the case of freely decaying statistically homogeneous and isotropic turbulence. Ref. [59] by mathematical treatment of the Navier–Stokes equations, Ref. [60] by analysis of experimental data from grid-generated turbulence, and [61] by EDQNM simulations showed that average (over space or realizations) scale-by-scale equilibrium is achieved asymptotically, as the Reynolds number tends to infinity only in the vicinity of the Taylor length. They also showed that there are systematic increasing departures from average scale-by-scale equilibrium both as the scale increases toward L because of increasing non-stationarity and as the scale decreases toward η because of increasing viscous diffusion. Given that the Taylor length λ depends on kinematic viscosity and on the kinetic energy of the turbulence, the asympotic balance between average interscale transfer rate and turbulence dissipation rate at the vicinity of λ is not an inertial range balance.

5.3. Category C: Non Homogeneous Statistically Stationary Turbulence with Two-Point Turbulence Production and without Two-Point Interspace Transport

This is the case of the intermediate layer in fully developed turbulent channel flow (FD TCF), which is known to be in approximate one-point equilibrium between the one-point turbulence production rate and turbulence dissipation rate locally in distance y from the wall. Ref. [62] showed theoretically and confirmed by using direct numerical simulation (DNS) data that scale-by-scale equilibrium between the interscale transfer rate and turbulence dissipation rate (both averaged over a sphere of radius r) occurs only at the vicinity of $r = \lambda$ in the limit of the infinite global Reynolds number δ/δ_ν and very large local Reynolds number y/δ_ν (where δ_ν is the wall unit length, δ is the channel's half width, and $y \ll \delta$). As the scale r grows above λ, there is a systematic departure from scale-by-scale equilibrium because of the increasing two-point turbulence production rate (averaged over a sphere of radius r), which is negligible only at scales close to η in the intermediate layer $\delta_\nu \ll y \ll \delta$.

The DNS data analysis of [62] has also shown that the two-point turbulence production rate is positive (scales smaller than r gain energy) mainly because one-point turbulence production is positive. However, two-point correlations conditioned on more or less aligned fluctuating velocities act to reduce this positivity. The positivity of two-point turbulence production is enhanced by two-point correlations conditioned on more or less anti-aligned fluctuating velocities, particularly at larger two-point separations r.

Finally, their DNS data analysis revealed that the approach to scale-by-scale equilibrium is at length scales (around λ), where aligned fluctuating velocities (which reduce two-point production) are stretching with their difference maximally aligned with the separation vector between the two points and where anti-aligned fluctuations (which enhance two-point production) are maximally skewed toward fast compression. The word "maximally" is used to mean that as the separation distance r grows above λ and the departure from scale-by-scale equilibrium increases, the alignment between fluctuating velocities and the skewness of anti-aligned fluctuations progressively diminish.

5.4. Category D: Non-Homogeneous Statistically Stationary Turbulence with Negligible Two-Point Turbulence Production

A case of turbulent flow, which falls under this category, is the turbulent flow under the rotating blades in a baffled container (mixer), where the baffles break the rotation in the flow to enhance mixing. In the absence of turbulence production, Refs. [50,51]

make the hypothesis that every term in the non-homogeneous but statistically stationary scale-by-scale (two-point) energy balance is self-similar in the sense that it has the same dependence on two-point separation r at different positions in space if rescaled by local (in space) velocity and length scales. The resulting intermediate asymptotic theory concludes that the interscale transfer rate, the two-point interspace turbulence transport rate and the two-point pressure gradient velocity correlation term are all proportional to the turbulence dissipation rate over a wide range of length scales r larger than λ. Two-dimensional two-component highly space-resolved particle image velocimetry (PIV) confirms two of these three conclusions [52], the one involving the pressure field remaining inaccessible to the measurement technique. The experimental measurements also confirm the absence of two-point turbulence production over the range of scales concerned. Category D is different from categories B and C in one essential way: there is no balance between interscale transfer rate and turbulence dissipation rate at any scale, not even asymptotically at scale λ, just a proportionality between them over an inertial range of scales r. The accompanying proportionality confirmed by the PIV between two-point interspace turbulence transport rate and turbulence dissipation rate over the same range of scales reflects the non-homogeneity and related absence of scale-by-scale equilibrium.

5.5. Category E: Streamwise-Decaying Non-Homogeneous Turbulence with Negligible Two-Point Turbulence Production

The term "streamwise decaying" here reflects the presence of non-zero streamwise mean advection, which, in the absence of sufficient turbulence production, can result in a streamwise decay of the turbulence. This is the case of boundary-free turbulent shear flows such as wakes and jets along their streamwise centerline. Ref. [50] developed a theory along the lines described under category D (see above) and applied it to the centerline of the turbulent wake of two side-by-side square prisms. By varying the distance between the two prisms, the flow can change qualitatively, which gives the possibility to easily and meaningfully test generality of results in the wind tunnel. Ref. [50] considered the variation along the cross-stream direction of second order structure functions and their dependence on two-point separation r and showed both theoretically and experimentally that they collapse as $K(r/L)^{2/3}$ rather than $(\varepsilon r)^{2/3}$ over a cross-stream area that is small enough for two-point turbulence production to be negligible. The difference comes from the fact that ε does not scale as $K^{3/2}/L$ in the cross-stream direction of such flows as demonstrated experimentally by ref. [63]. The $K(r/L)^{2/3}$ rather than $(\varepsilon r)^{2/3}$ scaling reflects non-equilibrium emanating from non-homogeneity via non-homogeneity's presence in the scale-by-scale energy balance in terms of the non-zero two-point interspace turbulence transport rate and/or non-zero two-point pressure gradient velocity correlation.

5.6. Category F: External Intermittency: A Case of Extreme Fluctuating Non-Homogeneity and Non-Stationarity

The turbulent/non-turbulent (TNTI) interface is an extreme case of fluctuating non-homogeneity and non-stationarity. Ref. [64] reported non-equilibrium turbulence dissipation scalings at the vicinity of the TNTI. Refs. [65,66] developed tools to study scale-by-scale energy transfers at and near the TNTI. These three studies used DNS of boundary-free turbulent shear flows. A significant conclusion concerns interscale transfers at the TNTI. In the direction more or less aligned with the local normal to the TNTI, the interscale transfer is down-scale (from large to small scales) and dominated by compressive motions, whereas in the direction more or less aligned with the local tangent to the TNTI, the interscale transfer is up-scale (from small to large scales) and dominated by stretching motions. There is an in-between region in terms of orientations relative to the TNTI where stretching motions dominate on average, but the interscale transfer is nevertheless down-scale because of extreme fluctuating compression events, which can be significantly more likely than extreme fluctuating stretching events. Furthermore, interscale energy transfer rates are typically half the magnitude of two-point interspace energy transport rates, and the two types of transfer/transport are anti-correlated. This anti-correlation is reminiscent of an anti-correlation in the fluctuations of interscale and two-point interspace turbulent energy transfer/transport in statistically stationary and homogeneous turbulence (category A). It must, of course, be noted that the interscale and interspace transfer/transport mechanisms are related by the obvious fact that they are both direct consequences of the turbulence nonlinearity.

6. Conclusions

Jack Herring was a leading contributor to the post-war era of turbulence research, where homogeneity and scale-by-scale equilibrium were central reference concepts. The Taylor–Kolmogorov turbulence dissipation scaling, which results from considerations of scale-by-scale equilibrium, was used in those years to obtain theoretical predictions even for non-homogeneous turbulent flows, for example, mean flow properties of boundary-free turbulent shear flows [45,46].

Following this era, turbulence research over the past ten to fifteen years includes a strand concerned with scale-by-scale non-equilibrium and the different ways that such non-equilibrium may result from different types of non-stationarity and non-homogeneity. A starting attempt at a classification of different non-equilibrium categories is summarized in the previous section. For example, the non-homogeneity of the producing but non-transporting type, as in the intermediate layer of FD TCF, leads to asymptotic scale-by-scale equilibrium only around the Taylor length with increasing deviations caused by two-point turbulence production in the inertial range. This is similar to decaying homogeneous turbulence, where the increasing deviations in the inertial range are caused by non-stationarity.

On the other hand, the non-homogeneity of the transporting but non-producing type, as in a baffled mixer under the rotating blades, leads to inertial range scale-independent interscale transfer rate, two-point interspace transport rate and two-point pressure gradient velocity correlation term, all of which scale with turbulence dissipation.

Funding: This research was funded by Chair of Excellence CoPreFlo funded by I-SITE-ULNE (grant number R-TALENT-19-001-VASSILICOS), MEL (grant number CONVENTION_219_ESR_06) and Region Hauts de France (grant number 20003862).

Informed Consent Statement: Not applicable.

Data Availability Statement: Not applicable.

Conflicts of Interest: The author declares no conflict of interest.

References

1. Herring, J.R. Self-consistent-field approach to turbulence theory. *Phys. FLuids* **1965**, *8*, 2219. [CrossRef]
2. Herring, J.R. Self-consistent-field approach to nonstationary turbulence. *Phys. FLuids* **1966**, *9*, 2106. [CrossRef]
3. Herring, J.R. Statistical theory of quasi-geostrophic turbulence. *J. Atmos. Sci.* **1980**, *37*, 969–977. [CrossRef]
4. Herring, J.R.; McWilliams, J.C. Comparison of direct numerical simulation of two-dimensional turbulence with two-point closure: The effects of intermittency. *J. Fluid Mech.* **1985**, *153*, 229–242. [CrossRef]
5. Herring, J.R.; Métais, O. Numerical experiments in forced stably stratified turbulence. *J. Fluid Mech.* **1989**, *202*, 97–115. [CrossRef]
6. Valente, P.; Vassilicos, J.C. Universal dissipation scaling for non-equilibrium turbulence. *Phys. Rev. Lett.* **2012**, *108*, 214503. [CrossRef]
7. Isaza, J.C.; Salazar, R.; Warhaft, Z. On grid-generated turbulence in the near- and far field regions. *J. Fluid Mech.* **2014**, *753*, 402–426. [CrossRef]
8. Meldi, M.; Lejemble, H.; Sagaut, P. On the emergence of non-classical decay regimes in multiscale/fractal generated isotropic turbulence. *J. Fluid Mech.* **2014**, *756*, 816–843. [CrossRef]
9. Vassilicos, J.C. Dissipation in turbulent flows. *Ann. Rev. Fluid Mech.* **2015**, *47*, 95–114. [CrossRef]
10. Hearst, R.J.; Lavoie, P. Decay of turbulence generated by a square-fractal-element grid. *J. Fluid Mech.* **2014**, *741*, 567–584. [CrossRef]
11. Hearst, R.J.; Lavoie, P. Velocity derivative skewness in fractal-generated, non-equilibrium grid turbulence. *Phys. Fluids* **2015**, *27*, 071701. [CrossRef]
12. Hearst, R.J.; Lavoie, P. Effects of multi-scale and regular grid geometries on decaying turbulence. *J. Fluid Mech.* **2016**, *803*, 528–555. [CrossRef]
13. Nedić, J.; Tavoularis, S. Energy dissipation scaling in uniformly sheared turbulence. *Phys. Rev. E* **2016**, *93*, 033115. [CrossRef] [PubMed]

14. Castro, I. Dissipative distinctions. *J. Fluid Mech.* **2016**, *788*, 1–4. [CrossRef]
15. Horiuti, K.; Yanagihara, S.; Tamaki, T. Nonequilibrium state in energy spectra and transfer with implications for topological transitions and SGS modeling. *Fluid Dyn. Res.* **2016**, *48*, 021409. [CrossRef]
16. Keylock, C.; Kida, S.; Peters, N. JSPS Supported Symposium on Interscale Transfers and Flow Topology in Equilibrium and Non-equilibrium Turbulence (Sheffield, UK, September 2014). *Fluid Dyn. Res.* **2016**, *48*, 020001. [CrossRef]
17. Goto, S.; Vassilicos, J.C. Local equilibrium hypothesis and Taylor's dissipation law. *Fluid Dyn. Res.* **2016**, *48*, 021402. [CrossRef]
18. Goto, S.; Vassilicos, J.C. Unsteady turbulence cascades. *Phys. Rev. E* **2016**, *94*, 053108. [CrossRef]
19. Nagata, K.; Saiki, T.; Sakai, Y.; Ito, Y.; Iwano, K. Effects of grid geometry on non-equilibrium dissipation in grid turbulence. *Phys. Fluids* **2017**, *29*, 015102. [CrossRef]
20. Nedić, J.; Tavoularis, S.; Marusic, I. Dissipation scaling in constant-pressure turbulent boundary layers. *Phys. Rev. Fluids* **2017**, *2*, 032601. [CrossRef]
21. Rubinstein, R.; Clark, T.T. 'Equilibrium' and 'non-equilibrium' turbulence. *Theor. Appl. Mech. Lett.* **2017**, *7*, 301–305. [CrossRef]
22. Breda, M.; Buxton, O. Influence of coherent structures on the evolution of an axisymmetric turbulent jet. *Phys. Fluids* **2018**, *30*, 035109. [CrossRef]
23. Meldi, M.; Sagaut, P. Investigation of anomalous very fast decay regimes in homogeneous isotropic turbulence. *J. Turbulence* **2018**, *19*, 390–413. [CrossRef]
24. Sunita; Layek, G.C. Nonequilibrium turbulent dissipation in buoyant axisymmetric plume. *Phys. Rev. FLuids* **2021**, *6*, 104602. [CrossRef]
25. Mora, D.O.; Pladellorens, E.M.; Turró, P.R.; Obligado, M.L. Energy cascades in active-grid-generated turbulent flows. *Phys. Rev. Fluids* **2019**, *4*, 104601. [CrossRef]
26. Liu, F.; Lu, L.P.; Bos, W.J.T.; Fang, L. Assessing the nonequilibrium of decaying turbulence with reversed initial fields. *Phys. Rev. Fluids* **2019**, *4*, 084603. [CrossRef]
27. Stein, V.P.; Kaltenbach, H.-J. Non-equilibrium scaling applied to the wake evolution of a model scale wind turbine. *Energies* **2019**, *12*, 2763. [CrossRef]
28. Thiesset, F.; Danaila, L. The illusion of a Kolmogorov cascade. *J. Fluid Mech.* **2020**, *902*, F1. [CrossRef]
29. Chongsiripinyo, K.; Sarkar, S. Decay of turbulent wakes behind a disk in homogeneous and stratified fluids. *J. Fluid Mech.* **2020**, *885*, A31. [CrossRef]
30. Ortiz-Tarin, J.L.; Nidhan, S.; Sarkar, S. High-Reynolds-number wake of a slender body. *J. Fluid Mech.* **2021**, *918*, A30. [CrossRef]
31. Saunders, D.C.; Britt, J.A.; Wunsch, S. Decay of the drag wake of a sphere at Reynolds number 10^5. *Exp. Fluids* **2022**, *63*, 71. [CrossRef]
32. Steiros, K. Balanced nonstationary turbulence. *Phys. Rev. E* **2022**, *105*, 035109. [CrossRef] [PubMed]
33. Steiros, K. Turbulence near initial conditions. *Phys. Rev. Fluids* **2022**, *7*, 104607. [CrossRef]
34. Waclawczyk, M.; Nowak, J.L.; Malinowski, S. Nonequilibrium dissipation scaling in atmospheric turbulence. *J. Phys. Conf. Series* **2022**, *2367*, 012032. [CrossRef]
35. Waclawczyk, M.; Nowak, J.L.; Siebert, H.; Malinowski, S. Detecting nonequilibrium states in atmospheric turbulence. *J. Atmos. Sci.* **2022**, *79*, 2757–2772. [CrossRef]
36. Xiong, X.-L.; Laima, S.; Lui, H. Novel scaling laws in the nonequilibrium turbulent wake of a rotor and a fractal plate. *Phys. Fluids* **2022**, *34*, 065130. [CrossRef]
37. Batchelor, G.K.; Proudman, I. The effect of rapid distortion of a fluid in turbulent motion. *Q. J. Mech. Appl. Math.* **1954**, *7*, 83–103. [CrossRef]
38. Tennekes, H.; Lumley, J.L. *A First Course in Turbulence*; MIT Press: Cambridge, MA, USA, 1972.
39. Frisch, U. *Turbulence: The Legacy of A.N. Kolmogorov*; Cambridge University Press: Cambridge, UK, 1995.
40. Hunt, J.C.R.; Philips, O.M.; Williams, D. *Turbulence and Stochastic Processes: Kolmogorov's Ideas 50 Years on*; The Royal Society: London, UK, 1991.
41. Mathieu, J.; Scott, J. *An Introduction to Turbulent Flows*; Cambridge University Press: Cambridge, UK, 2000.
42. Lesieur, M. *Turbulence in Fluids*; Kluwer: Alphen aan den Rijn, The Netherlands, 1997.
43. Pope, S.B. *Turbulent Flows*; Cambridge University Press: Cambridge, UK, 2000.
44. Leschziner, M.A. *Statistical Turbulence Modelling for Fluid Dynamics—Demystified: An Introductory Text for Graduate Engineering Students*; Imperial College Press: London, UK, 2016.
45. Townsend, A.A. *The Structure of Turbulent Shear Flow*; Cambridge University Press: Cambridge, UK, 1976.
46. George, W.K. The self-preservation of turbulent flows and its relation to initial conditions and coherent structures. In *Advances in turbulence*; Cambridge University Press: Cambridge, UK, 1989; pp. 39–73.
47. Vassilicos, J.C. From Tennekes & Lumley to Townsend and to George: A slow march to freedom. In *Whither Turbulence and Big Data in the 21st Century*; Pollard, A., Castillo, L., Danaila, L., Glauser, M., Eds.; Springer International Publishing: Cham, Switzerland, 2016.
48. Cafiero, G.; Vassilicos, J.C. Non-equilibrium turbulence scalings and self-similarity in turbulent planar jets. *Proc. R. Soc. Lond. A* **2019**, *475*, 20190038. [CrossRef]
49. Cafiero, G.; Vassilicos, J.C. Non-equilibrium scalings of the turbulent/non-turbulent interface speed in planar jets. *Phys. Rev. Lett.* **2020**, *125*, 174501. [CrossRef]

50. Chen, J.; Vassilicos, J.C. Scalings of scale-by-scale turbulence energy in non-homogeneous turbulence. *J. Fluid Mech.* **2022**, *938*, A7. [CrossRef]
51. Beaumard, P.; Braganca, P.; Cuvier, C.; Steiros, K.; Vassilicos, J.C. Scale-by-scale non-equilibrium with Kolmogorov-like scalings in non-homogeneous stationary turbulence. Preprint 2023.
52. Valente, P.; Vassilicos, J.C. The energy cascade in grid-generated non-equilibrium decaying turbulence. *Phys. Fluids* **2015**, *27*, 045103. [CrossRef]
53. Taylor, G.I. Statistical theory of turbulence. *Proc. R. Soc. Lond. A* **1935**, *151*, 421–444. [CrossRef]
54. Goto, S.; Vassilicos, J.C. Energy dissipation and flux laws for unsteady turbulence. *Phys. Lett. A* **2015**, *379*, 1144–1148. [CrossRef]
55. Apostolidis, A.; Laval, J.-P.; Vassilicos, J.C. Scalings of turbulence dissipation in space and time for turbulent channel flow. *J. Fluid Mech.* **2022**, *946*, A41. [CrossRef]
56. Bos, W.J.T.; Rubinstein, R. Dissipation in unsteady turbulence. *Phys. Rev. Fluids* **2017**, *2*, 022601. [CrossRef]
57. Alves-Portela, F.; Papadakis, G.; Vassilicos, J.C. Turbulence dissipation and the role of coherent structures in the near wake of a square prism. *Phys. Rev. Fluids* **2018**, *3*, 124609. [CrossRef]
58. Larssen, H.S.; Vassilicos, J.C. Spatio-temporal fluctuations of interscale and interspace energy transfer dynamics in homogeneous turbulence. *J. Fluid Mech.* **2023**. [CrossRef]
59. Lundgren, T.S. Kolmogorov two-thirds law by matched asymptotic expansion. *Phys. Fluids* **2002**, *14*, 638. [CrossRef]
60. Obligado, M.; Vassilicos, J.C. The non-equilibrium part of the inertial range in decaying homogeneous turbulence. *Europhys. Lett.* **2019**, *127*, 64004. [CrossRef]
61. Meldi, M.; Vassilicos, J.C. Analysis of Lundgren's matched asymptotic expansion approach to the Karman-Howarth equation using the EDQNM turbulence closure. *Phys. Rev. Fluids* **2021**, *6*, 064602. [CrossRef]
62. Apostolidis, A.; Laval, J.-P.; Vassilicos, J.C. Turbulent cascade in fully developed turbulent channel flow. *J. Fluid Mech.* **2023**.
63. Chen, J.; Cuvier, C.; Foucaut, J.-M.; Ostovan, Y.; Vassilicos, J.C. A turbulence dissipation inhomogeneity scaling in the wake of two side-by-side square prisms. *J. Fluid Mech.* **2021**, *924*, A4. [CrossRef]
64. Watanabe, T.; da Silva, C.B.; Nagata, K. Non-dimensional energy dissipation rate near the turbulent/non-turbulent interfacial layer in free shear flows and shear free turbulence. *J. Fluid Mech.* **2019**, *875*, 321–344. [CrossRef]
65. Watanabe, T.; da Silva, C.B.; Nagata, K. Scale-by-scale kinetic energy budget near the turbulent/non-turbulent interface. *Phys. Rev. Fluids* **2020**, *5*, 124610. [CrossRef]
66. Zhou, Y.; Vassilicos, J.C. The energy cascade at the turbulent/non-turbulent interface. *Phys. Rev. Fluids* **2020**, *5*, 064604. [CrossRef]

atmosphere

Article

Two-Dimensional Flow on the Sphere

Rick Salmon *,† and Nick Pizzo †

Scripps Institution of Oceanography, UCSD, La Jolla, CA 92093-0213, USA
* Correspondence: rsalmon@ucsd.edu
† These authors contributed equally to this work.

Abstract: Equilibrium statistical mechanics predicts that inviscid, two-dimensional, incompressible flow on the sphere eventually reaches a state in which spherical harmonic modes of degrees $n = 1$ and $n = 2$ hold all the energy. By a separate theory, such flow is static in a reference frame rotating at angular speed $2\Omega/3$ with respect to the inertial frame. The vorticity field in the static frame is an accident of the initial conditions, but, once established, it lasts forever under the stated assumptions. We investigate the possibility of such behavior with a stereographic-coordinate model that conserves energy and enstrophy when the viscosity vanishes.

Keywords: turbulence theory and modeling; atmospheric and oceanic turbulence; planetary and astrophysical turbulence

Citation: Salmon, R.; Pizzo, N. Two-Dimensional Flow on the Sphere. *Atmosphere* **2023**, *14*, 747. https://doi.org/10.3390/atmos14040747

Academic Editors: Boris Galperin, Annick Pouquet and Peter Sullivan

Received: 19 March 2023
Revised: 13 April 2023
Accepted: 15 April 2023
Published: 20 April 2023

1. Introduction

Jack Herring made fundamental contributions to the theory of two-dimensional turbulence and to quasigeostrophic turbulence [1–3]. The former is a prototype of the latter, which adds the key dynamical ingredients—rotation, density stratification, and topography—required to make the theory relevant to the Earth's atmosphere and oceans. Much of the theory is based upon planar geometry and the hope that the beta-plane approximation captures the most significant effects of curvature and coordinate system rotation. A primary purpose of this paper is to show that this is not the case: Two-dimensional incompressible flow on the sphere differs significantly from two-dimensional flow on the plane, even in the case of vanishing rotation, or, as we prefer to say, even in the case of vanishing angular momentum.

A second primary purpose is to illustrate the advantages of stereographic coordinates for both analytic and numerical work on the sphere. In stereographic coordinates, the equations governing two-dimensional turbulence on the sphere take a form, (31) or (37)–(39), that is very similar to the corresponding formulation in Cartesian coordinates on the plane. Only the appearance of the smoothly varying metric coefficient h distinguishes the two dynamics. From a mathematical point of view, h is responsible for all of the differences between flow in the two geometries.

To quickly appreciate the effects of spherical geometry, first consider inviscid, incompressible, two-dimensional flow on the plane. Unlike on the sphere, which has no boundary, the prescribed boundary on the plane is determinative. Let it be the rectangle shown in Figure 1. Let the flow within this rectangle be truncated to exclude all the Fourier modes with wavenumbers greater than some prescribed cutoff k_c. Let the initial energy be concentrated at a wavenumber k_0 that lies between k_c and the lowest wavenumber as determined by the boundary of the system.

Equilibrium statistical mechanics [4] predicts the statistically steady state attained by this system after an undetermined period of adjustment. Interest attaches to the sequence of equilibrium states that occur as $k_c \to \infty$. In this limit, the theory predicts that *all* of the energy ends up in the lowest-wavenumber mode, and that all of the enstrophy not contained in that mode appears in modes near k_c. As $k_c \to \infty$, this excess enstrophy is expelled to infinity, disappearing in much the same way as if it had been removed by viscosity. The tendency for all the energy to crowd into a single lowest mode is often referred to as 'Bose–Einstein condensation', after a closely analogous result in quantum mechanics.

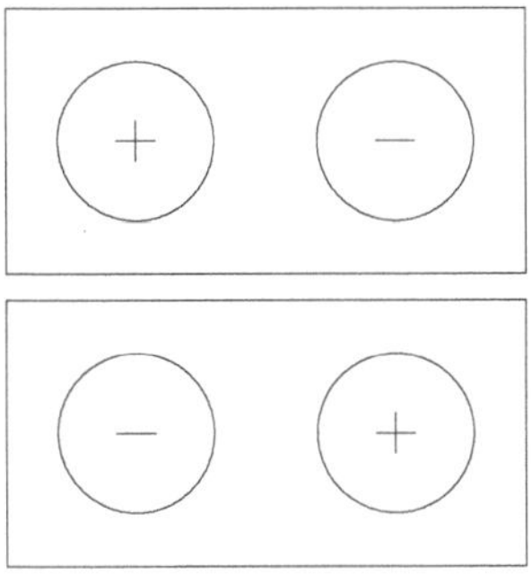

Figure 1. If all the energy ends up in the lowest wavenumber mode of the rectangular domain, it must occupy one of the two states depicted.

Suppose that the initial conditions are such that the conserved circulation around the rectangular boundary vanishes. Then the lowest-wavenumber mode, which has a single extremum within the rectangle, remains unexcited. All of the energy flows into the next-lowest mode, the one depicted in Figure 1. Because the boundary is a rectangle (and not, say, a square or a circle), there is only one next-lowest mode. If all of the energy ends up in this mode, it must appear in one of the two states depicted in Figure 1. Moreover, once either of these states is established, the system cannot flip suddenly to the opposite state because that would require the energy in one of these states to flow temporarily into higher-wavenumber modes, and that, according to statistical mechanics, is extremely unlikely. Thus, the two states depicted in Figure 1 act as a single-register binary memory. The state that actually occurs is an accident of the initial conditions, but once established, it lasts for a very long time. If the conserved circulation around the perimeter of the rectangle does not vanish, the amplitude of the lowest mode remains fixed at a nonzero value, but the Bose–Einstein condensation would still proceed only as far as the next-lowest mode depicted in Figure 1.

Still considering flow on the plane, suppose that there are *two* lowest modes available to receive all the energy. For the statistical mechanics, we take the microcanonical ensemble as seems appropriate. Let the energy be $E = x^2 + y^2$, where x and y are the amplitudes of the two modes. If x is the only mode, then, as already noted, its probability distribution would be

$$P(x) = \delta(x - \sqrt{E}) + \delta(x + \sqrt{E}) \tag{1}$$

where $\delta(\cdot)$ is Dirac's delta function and, here and below, we ignore normalization constants. However, if there are two modes, then the probability distribution of either mode is

$$P(x) = \int_{-\infty}^{+\infty} dy\ \delta(x^2 + y^2 - E) \propto \frac{1}{\sqrt{E - x^2}} \tag{2}$$

if $x^2 < E$ and zero otherwise. Like (1), the distribution (2) is sharply peaked at $\pm\sqrt{E}$, but now there is significant probability of finding x anywhere in between. If there are three lowest modes, then the analogous calculation predicts that x is uniformly distributed between $\pm\sqrt{E}$. The implication of these calculations is clear: if the lowest available mode is degenerate with two or more members, then each member sees the other members as a reservoir from which energy may be borrowed or loaned, allowing leakage between states and a breakdown of memory. In all of these examples, the average flow predicted by statistical mechanics is misleading and beside the point. For example, if the ensemble consists of the two equal-probability states depicted in Figure 1, then the average flow vanishes, but a vanishing velocity field is very unrepresentative of either state. Such considerations demonstrate how even the simplest calculations with probability distributions require careful interpretation.

Now consider the corresponding equilibrium state on the sphere [5]. Spherical harmonics replace Fourier modes, and the harmonic degree n with cutoff n_c replaces the wavenumber k with cutoff k_c. The lowest mode Y_0^0 is irrelevant because the stream function is arbitrary by a constant and because the vorticity integrates to zero over the sphere (Gauss's constraint). The next lowest modes are three: $Y_1^0, Y_1^{\pm 1}$. However, the amplitudes of these three modes are determined by the three components of angular momentum, which is conserved by our dynamics, even with nonzero viscosity. Thus, the lowest-degree modes into which the energy can pile up are those of degree $n = 2$, namely, Y_2^0, $Y_2^{\pm 1}$, $Y_2^{\pm 2}$. By the reasoning of the previous paragraph, the Bose–Einstein condensate would be one in which these five modes freely exchange energy. Instead, it turns out, the five modes of degree two lock together in a pattern that is perfectly static in a uniformly rotating reference frame. This behavior is dramatically different from flow on the plane, and is solely a consequence of spherical geometry.

Let the sphere be centered at $x = y = z = 0$, and suppose, without loss of generality, that the angular momentum (if nonzero) points toward positive z. Then, the amplitude of the Y_1^0-mode is a nonzero constant, and the amplitudes of $Y_1^{\pm 1}$ vanish. The Y_1^0-mode corresponds to a solid-body flow about the z-axis with constant angular speed Ω, and the corresponding vorticity is called the Coriolis parameter. As shown in Section 3, the flow consisting of the Y_1^0-mode and the 5 Y_2-modes with arbitrary amplitudes represents an exact solution to the inviscid vorticity equation on the sphere. If $\Omega = 0$, this flow field is steady in the inertial frame. If $\Omega \neq 0$, the flow field is steady in a reference frame rotating at speed $2\Omega/3$ with respect to the inertial frame. The amplitude Ω of the Y_1^0-mode is determined by the conserved angular momentum. The amplitudes of the Y_2-modes are accidents of the initial conditions in the same way that the initial conditions determine which of the two states in Figure 1 eventually emerge. However, once these amplitudes are determined via Bose–Einstein condensation, they persist indefinitely under the stated assumptions. Thus, Bose–Einstein condensation into Y_2-modes represents a potential memory storage that does not, as in planar geometry, require a 'probability well' for its preservation. For the plane, as for the sphere, triads with the same wavenumber magnitude (the analog of spherical harmonic degree) do not interact, either with each other or with modes of different wavenumber magnitude. However, typical planar geometries forbid the occurrence of many modes with the same wavenumber magnitude. For instance, in the case of an infinitely periodic box, in which the wave vector corresponds to integer pairs (n, m), modes with the same $n^2 + m^2$ are rare. We come back to this in Section 6.

More generally, flow consisting only of Y_1 modes and modes of a single spherical harmonic degree n is perfectly static in a reference frame rotating at angular speed

$$\Omega - \frac{2\Omega}{n(n+1)} \tag{3}$$

with respect to the inertial frame. According to [6], this result was known to Ertel. Section 3 supplies details. There are $2n + 1$ spherical harmonics with the same degree n. These facts invite us to think of the stream function for flow on the sphere as the sum

$$\psi(\xi, \eta, t) = \sum_{n=1}^{\infty} \psi_n(\xi, \eta, t) \tag{4}$$

of contributions ψ_n of degree n, where each ψ_n is itself the sum of $2n + 1$ spherical harmonics. The coordinates (ξ, η) are arbitrary, but we strongly prefer them to be stereographic coordinates. Again, if only ψ_1 and one ψ_n with $n > 1$ are present, then ψ is steady in the frame rotating at angular speed (3). In the general case where all ψ_n are present, the various

ψ_n interact to produce a turbulent flow. However, the interaction between ψ_n with very different autorotation rates (3)—very different n—is likely to be weak. Additionally, since the difference between autorotation rates is greatest and varies most rapidly at small n, we expect the energy transfer to large spatial scales to slow dramatically as the energy reaches low n for sufficiently large Ω. This is essentially Rhines's 'beta-arrest' phenomenon for flow on the beta plane [7] but now taking a form particular to the sphere. See also [8].

This study was begun with the hope that the special static solutions described above might be a factor in intermediate-range weather prediction. Although flickers of that hope remain, there are significant differences between the Earth's atmosphere and the very idealized dynamics considered in this paper. First and foremost, angular momentum conservation applies only to the entire Earth–atmosphere–ocean system. Mountain torque provides the main coupling between the atmosphere and the solid planet. The latter, with its enormously greater moment of inertia, experiences only small changes in its rotation rate, but fluctuations in the Y_1^0 component of atmospheric flow are very significant and lead to a peak in the atmospheric energy spectrum at $n = 1$ [9]. Second, because the Rhines scale corresponding to the Earth's atmosphere is smaller than the planetary radius, beta-arrest prevents the leftward movement of energy through the spectrum from reaching $n = 2$. As a result of these two factors, the atmospheric energy spectrum actually shows a minimum at $n = 2$. See [10] for the statistical mechanics including mountain torque and [11] for a thorough review of the atmospheric angular momentum.

The plan of the paper is as follows. In Section 2 and Appendix A, we derive the Navier–Stokes equations for flow on the sphere in stereographic coordinates. The form of the viscosity for flow on curved surfaces has been the subject of disagreement. For instance, the viscosity given by [12] does not conserve angular momentum. We follow the recommendation of [13] that the viscosity conserves angular momentum, and in Appendix A, we show that their recommended viscosity takes an especially simple form in stereographic coordinates.

In Section 3, we prove the special solutions described briefly above. Our proof using stereographic coordinates is substantially simpler than the proofs given by [14] or [15] using spherical coordinates.

In Section 4 and Appendix B, we use the method of [16] to construct an Arakawa-type model in stereographic coordinates. The model conserves energy and enstrophy when the viscosity vanishes.

Section 5 demonstrates the accuracy of the model and tests our speculation that Bose–Einstein condensation on the sphere produces a static flow. Our results suggest that the time required to achieve the static state is probably infinite but that the flow rather quickly becomes quasi-static.

It is a pleasure to dedicate this paper to the memory of Jack Herring. Rick Salmon acknowledges the importance of Jack's mentorship at an early stage of his career and holds the happy memories of a lifelong friendship. Rick cannot think of Jack without also remembering his wife Betty, who shared Jack's kind and generous spirit, and greatly enjoyed the foibles and eccentricities of his scientific friends.

2. Stereographic Coordinates

We solve the vorticity equation for incompressible flow on the unit sphere,

$$x^2 + y^2 + z^2 = 1. \tag{5}$$

We refer to the point $(x, y, z) = (0, 0, 1)$ as the 'north pole', and $(x, y, z) = (0, 0, -1)$ as the 'south pole', even when the angular momentum vanishes, the case commonly referred to as 'non-rotating'. The 'equator' is the intersection of the plane $z = 0$ with (5). The conserved angular momentum is a vector in the direction of $(0, 0, 1)$.

Besides the Cartesian embedding coordinates (x, y, z), we shall refer to three other coordinate systems. The first of these are the usual spherical coordinates, defined by

$$\begin{aligned} x &= \cos\theta\cos\lambda \\ y &= \cos\theta\sin\lambda \\ z &= \sin\theta \end{aligned} \tag{6}$$

where λ is the longitude and θ the latitude. The (λ, θ) coordinates cover the sphere, except for singularities at the two poles. We also use the stereographic coordinates

$$\xi = \frac{x}{1-z}, \quad \eta = \frac{y}{1-z} \tag{7}$$

defined by a line, emanating from the north pole, that intersects the equatorial plane at $(\xi, \eta, 0)$ and the sphere at (x, y, z). (ξ, η) cover the sphere except for a singularity at infinity, corresponding to the north pole. We also use

$$\hat{\xi} = \frac{x}{1+z}, \quad \hat{\eta} = \frac{y}{1+z} \tag{8}$$

defined by the intersection with the equatorial plane of a line emanating from the south pole. $(\hat{\xi}, \hat{\eta})$ cover the sphere, except for singularity at infinity corresponding to the south pole. The stereographic systems (7) and (8) prove to be more useful than the spherical coordinates; the primary challenge is to match them together. Our strategy is to solve the southern hemisphere dynamics in (ξ, η) within the unit circle

$$\xi^2 + \eta^2 < 1, \tag{9}$$

to solve the northern hemisphere dynamics in $(\hat{\xi}, \hat{\eta})$ within the unit circle

$$\hat{\xi}^2 + \hat{\eta}^2 < 1, \tag{10}$$

and to match the two solutions together at the equator,

$$\xi^2 + \eta^2 = \hat{\xi}^2 + \hat{\eta}^2 = 1. \tag{11}$$

The transformation equations between the two stereographic systems are

$$(\xi, \eta) = \frac{1}{\hat{r}^2}(\hat{\xi}, \hat{\eta}), \quad (\hat{\xi}, \hat{\eta}) = \frac{1}{r^2}(\xi, \eta) \tag{12}$$

where

$$r^2 \equiv \xi^2 + \eta^2, \quad \hat{r}^2 \equiv (\hat{\xi})^2 + (\hat{\eta})^2. \tag{13}$$

A useful relation is

$$r\hat{r} = 1. \tag{14}$$

Points within the unit circle on the $(\hat{\xi}, \hat{\eta})$ plane transform to points outside the unit circle on the (ξ, η) plane, and vice versa. On the equator itself, $(\xi, \eta) = (\hat{\xi}, \hat{\eta})$. It is also useful to record the inverse transformations from either set of stereographic coordinates to the Cartesian embedding coordinates,

$$x = \frac{2\xi}{\xi^2 + \eta^2 + 1} = \frac{2\hat{\xi}}{\hat{\xi}^2 + \hat{\eta}^2 + 1} \tag{15}$$

$$y = \frac{2\eta}{\xi^2 + \eta^2 + 1} = \frac{2\hat{\eta}}{\hat{\xi}^2 + \hat{\eta}^2 + 1} \tag{16}$$

$$z = \frac{\xi^2 + \eta^2 - 1}{\xi^2 + \eta^2 + 1} = \frac{1 - \hat{\xi}^2 - \hat{\eta}^2}{\hat{\xi}^2 + \hat{\eta}^2 + 1}. \tag{17}$$

For a thorough introduction to stereographhic coordinates, see [17,18].

Infinitesimal displacements on the surface of the sphere satisfy

$$\begin{aligned} ds^2 = dx^2 + dy^2 + dz^2 &= \cos^2\theta d\lambda^2 + d\theta^2 \\ &= \frac{4}{(1+\xi^2+\eta^2)^2}\left(d\xi^2 + d\eta^2\right) \\ &= \frac{4}{(1+\hat{\xi}^2+\hat{\eta}^2)^2}\left(d\hat{\xi}^2 + d\hat{\eta}^2\right) \end{aligned} \tag{18}$$

where ds is the infinitesimal distance tangent to the surface of the sphere. Thus, all of our systems fit the general form

$$ds^2 = h_1(\xi_1,\xi_2)^2(d\xi_1)^2 + h_2(\xi_1,\xi_2)^2(d\xi_2)^2 \tag{19}$$

where (ξ_1, ξ_2) are general orthogonal coordinates, and $\mathrm{diag}(h_1^2, h_2^2)$ is the metric tensor. In arbitrary orthogonal coordinates, the incompressibility condition takes the form

$$\frac{\partial}{\partial\xi_1}(h_1 h_2 \dot{\xi}_1) + \frac{\partial}{\partial\xi_2}(h_1 h_2 \dot{\xi}_2) = 0 \tag{20}$$

where the overdot denotes the time derivative following a fluid particle. Since the velocity components tangent to the sphere in the directions of ξ_1 and ξ_2 are, respectively,

$$U_1 = h_1\dot{\xi}_1, \quad U_2 = h_2\dot{\xi}_2 \tag{21}$$

(20) is equivalent to

$$\frac{\partial}{\partial\xi_1}(h_2 U_1) + \frac{\partial}{\partial\xi_2}(h_1 U_2) = 0 \tag{22}$$

which implies

$$U_1 = -\frac{1}{h_2}\frac{\partial\psi}{\partial\xi^2}, \quad U_2 = +\frac{1}{h_1}\frac{\partial\psi}{\partial\xi^1} \tag{23}$$

where $\psi(\xi_1, \xi_2, t)$ is the stream function. The vorticity equation takes the form

$$q_t + \frac{1}{h_1 h_2}\frac{\partial(\psi, q)}{\partial(\xi_1, \xi_2)} = 0 \tag{24}$$

where

$$q = \nabla^2_{LB}\psi \tag{25}$$

is the vorticity, and

$$\nabla^2_{LB}\psi \equiv \frac{1}{h_1 h_2}\left(\frac{\partial}{\partial\xi_1}\left(\frac{h_2}{h_1}\frac{\partial\psi}{\partial\xi_1}\right) + \frac{\partial}{\partial\xi_2}\left(\frac{h_1}{h_2}\frac{\partial\psi}{\partial\xi_2}\right)\right) \tag{26}$$

is the Laplace–Beltrami operator.

In the case of spherical coordinates, $(\xi_1, \xi_2) = (\lambda, \theta)$, $(h_1, h_2) = (\cos\theta, 1)$, and (23)–(25) take the familiar forms

$$U_\lambda = -\frac{\partial\psi}{\partial\theta}, \quad U_\theta = +\frac{1}{\cos\theta}\frac{\partial\psi}{\partial\lambda} \tag{27}$$

$$q_t + \frac{1}{\cos\theta}\frac{\partial(\psi, q)}{\partial(\lambda, \theta)} = 0 \tag{28}$$

and

$$q = \frac{1}{\cos^2\theta}\frac{\partial^2\psi}{\partial\lambda^2} + \frac{1}{\cos\theta}\frac{\partial}{\partial\theta}\left(\cos\theta\frac{\partial\psi}{\partial\theta}\right) \equiv \nabla^2_{LB}\psi. \tag{29}$$

To obtain the dynamics in stereographic coordinates, we proceed from the general Equations (24) and (25). Some care must be taken concerning the handedness of the coordinate systems. Viewed from the exterior of the unit sphere, $(\hat{\xi}, \hat{\eta})$ and (λ, θ) constitute right-handed systems, whereas (ξ, η) constitutes a left-handed system. That is,

$$\frac{\partial(\hat{\xi}, \hat{\eta})}{\partial(\lambda, \theta)} > 0, \quad \frac{\partial(\xi, \eta)}{\partial(\lambda, \theta)} < 0. \tag{30}$$

The general covariant point of view cares nothing about handedness, but we want our computed vorticity and stream function to be the same as would be obtained by solving the entire problem in spherical coordinates, and, in particular, we want these quantities to be continuous at the equator; it would be a great nuisance to keep track of differently defined stream functions in the two hemispheres. With this in mind, the northern hemisphere equations take the forms

$$\hat{h}^2 q_t + \frac{\partial(\psi, q)}{\partial(\hat{\xi}, \hat{\eta})} = 0, \quad \hat{h}^2 q = \hat{\nabla}^2 \psi, \quad \hat{h} = \frac{2}{1 + \hat{\xi}^2 + \hat{\eta}^2} \tag{31}$$

within the unit circle (10); the southern hemisphere equations take the forms

$$h^2 q_t - \frac{\partial(\psi, q)}{\partial(\xi, \eta)} = 0, \quad h^2 q = \nabla^2 \psi, \quad h = \frac{2}{1 + \xi^2 + \eta^2} \tag{32}$$

within the unit circle (9). Here,

$$\nabla^2 \psi \equiv \frac{\partial^2 \psi}{\partial \xi^2} + \frac{\partial^2 \psi}{\partial \eta^2}, \quad \hat{\nabla}^2 \psi \equiv \frac{\partial^2 \psi}{\partial \hat{\xi}^2} + \frac{\partial^2 \psi}{\partial \hat{\eta}^2}. \tag{33}$$

The Laplace–Beltrami operator in (26) and (29) is equivalent to $\hat{h}^{-2}\hat{\nabla}^2$ and $h^{-2}\nabla^2$. Apart from the factors of h and $\hat{h}$, (31)–(33) have the same form as the corresponding equations for planar geometry in Cartesian coordinates. This greatly facilitates numerical solution. The sign change between (31) and (32) reflects the change in handedness. To see that this makes sense, recall that the first equation in (31) and (32) also applies to a passive scalar advected by the velocity field represented by ψ. The sign change between (31) and (32) ensures that the (cross-equatorial) flux of q out of the unit circle (9) matches the flux of q into the unit circle (10), with no need to redefine or re-interpret ψ or q.

Alternatively, by replacing the vorticity q in (31) and (32) by

$$q + f \equiv q + 2\Omega \sin\theta = q + 2\Omega z = q + 2\Omega \frac{\xi^2 + \eta^2 - 1}{\xi^2 + \eta^2 + 1} = q + 2\Omega \frac{1 - \hat{\xi}^2 - \hat{\eta}^2}{\hat{\xi}^2 + \hat{\eta}^2 + 1} \tag{34}$$

we obtain the equations of motion in a frame rotating at angular speed Ω.

We solve (31) on (10), and (32) on (9), with the matching conditions that ψ and q are continuous at the equator (11). If the initial condition is given by prescribing q, then q must satisfy the Gauss constraint,

$$\iint d\xi d\eta \; h^2 q + \iint d\hat{\xi} d\hat{\eta} \; \hat{h}^2 q = 0 \tag{35}$$

where the integrals are over the two unit circles (i.e., the whole sphere), but the constraint (35) is automatically maintained thereafter. If the initial condition is given by prescribing ψ, the Gauss constraint is automatically satisfied.

Although we work primarily with (31) and (32), it is useful to record the corresponding forms of the momentum equations. Introducing the notation,

$$(u, v) \equiv h^2(\dot{\xi}, \dot{\eta}) \tag{36}$$

the southern hemisphere equations take the forms

$$\frac{\partial u}{\partial t} - qv = -\frac{\partial p}{\partial \xi} \tag{37}$$

$$\frac{\partial v}{\partial t} + qu = -\frac{\partial p}{\partial \eta} \tag{38}$$

$$\frac{\partial u}{\partial \xi} + \frac{\partial v}{\partial \eta} = 0 \tag{39}$$

$$q = \frac{1}{h^2}\left(\frac{\partial v}{\partial \xi} - \frac{\partial u}{\partial \eta}\right). \tag{40}$$

The northern hemisphere equations take the same form as (37)–(40) but with hats applied to all the variables except q. Again, if $h^2 = 1$, (37)–(40) are formally identical to the equations for two-dimensional incompressible motion in planar geometry. However, no choice of variables can make h^2 equal unity on the unit sphere; instead it must obey the requirement

$$\frac{1}{h^4}\left[\frac{\partial h}{\partial \xi_1}\frac{\partial h}{\partial \xi_1} + \frac{\partial h}{\partial \xi_2}\frac{\partial h}{\partial \xi_2} - h\left(\frac{\partial^2 h}{\partial \xi_1^2} + \frac{\partial^2 h}{\partial \xi_2^2}\right)\right] = 1 \tag{41}$$

that the Gaussian curvature be uniform and equal to unity. The effects of curvature can never be transformed away, but it is interesting that they can be confined to a single factor in (40). This is what motivates the definition (36). However, it must be emphasized that the symbols (u, v) invite misinterpretation: the velocity of fluid particles tangent to the sphere is

$$(U, V) = h(\dot{\xi}, \dot{\eta}) = \frac{1}{h}(u, v) \tag{42}$$

and *not* (u, v).

The inviscid dynamics (31) and (32) or (37)–(40) conserve energy in the form,

$$E = \frac{1}{2}\iint d\xi d\eta\ \nabla\psi \cdot \nabla\psi \tag{43}$$

where $\nabla\psi \equiv (\partial\psi/\partial\xi, \partial\psi/\partial\eta)$, total vorticity in the form

$$\iint d\xi d\eta\ h^2 q = \iint d\xi d\eta\ \nabla^2\psi \equiv 0 \tag{44}$$

and enstrophy in the form

$$Z = \frac{1}{2}\iint d\xi d\eta\ h^2 q^2 = \frac{1}{2}\iint d\xi d\eta\ h^{-2}(\nabla^2\psi)^2. \tag{45}$$

The integrations in (43)–(45) are over the whole sphere, with the understanding that they are actually computed as the sum of integrations over the two unit circles. The vanishing of (44) corresponds to the Gauss constraint. More generally, the inviscid dynamics conserves Casimirs of the form

$$\iint d\xi d\eta\ h^2 F(q) \tag{46}$$

where $F(q)$ is any function of the vorticity, for which the integral in (46) converges; (44) and (45) are particularly important cases of (46).

The inviscid dynamics also conserves angular momentum. Angular momentum is a vector with three components, each of which is conserved. However, since the location of the axis that determines the origin of our stereographic system has no particular significance, we may assume, without loss of generality, that the angular momentum vector points in the z-direction. Then, the sole nonvanishing component of angular momentum is

$$M = \iint d\xi d\eta\ h^2(\xi v - \eta u) = \iint d\xi d\eta\ h^3 q = \iint d\xi d\eta\ h\nabla^2\psi. \tag{47}$$

It remains to discuss the viscosity. In [13], Gilbert et al. examine three proposed definitions of viscosity for incompressible flow on a curved surface. All three are covariant, and none of the three can be rejected on purely fundamental grounds. However, only one of the three forms of viscosity conserves angular momentum. We agree with [13] that the angular-momentum-conserving form is the proper choice for applications in which the viscosity is actually an eddy viscosity representing the mixing—but not the destruction—of angular momentum by unresolved motions. In Appendix A, we show that the viscosity suggested by [13] takes the surprisingly simple form

$$\frac{\partial u}{\partial t} + \cdots = -\nu \frac{\partial q}{\partial \eta} \quad (48)$$

$$\frac{\partial v}{\partial t} + \cdots = +\nu \frac{\partial q}{\partial \xi} \quad (49)$$

in the dynamics (37)–(40), and

$$\hat{h}^2 q_t + \cdots = \nu \hat{\nabla}^2 q, \quad h^2 q_t + \cdots = \nu \nabla^2 q \quad (50)$$

in the dynamics (31) and (32). The ellipses represent the conservative terms already discussed. When the viscosity in (50) is added to the dynamics (31) and (32), we find that

$$\frac{dE}{dt} = -\nu \iint d\xi d\eta \; h^2 q^2 \quad (51)$$

$$\frac{dZ}{dt} = -\nu \iint d\xi d\eta \; \nabla q \cdot \nabla q \quad (52)$$

and

$$\frac{dM}{dt} = 0. \quad (53)$$

Thus, viscosity dissipates energy and enstrophy, but not angular momentum. The importance of angular momentum conservation by viscosity seems not to be widely appreciated, but see [19] and references therein.

Many modelers prefer a more scale-selective hyperviscosity to any form of Navier–Stokes viscosity. However, hyperviscosity introduces non-physical effects. For example, hyperviscosity creates artificial local extrema in vorticity [20]. For this reason, we believe that Navier–Stokes viscosity is well worth the extra cost in spatial resolution.

We emphasize that, in our dynamics, the sphere plays no role except to shape the flow. Without drag or mountain torque, only curvature and angular momentum conservation distinguish our dynamics from the flow in a box on the plane.

3. Exact Solutions

Let Y_n be any solution of

$$\nabla^2_{LB} Y_n = -n(n+1) Y_n \quad (54)$$

where n is a positive integer. The coordinate system used to represent Y_n is arbitrary. In a spherical coordinate system, the general solution of (54) is

$$Y_n(\theta, \lambda) = \sum_{m=0}^{n} A_m P_n^m(\sin\theta) \cos(m\lambda + \gamma_m) \quad (55)$$

where P_n^m is a Legendre function of the first kind, and A_m and γ_m are arbitrary constants. We recognize (55) as the sum of all spherical harmonics of degree n with arbitrary amplitudes A_m and phases γ_m. The customary normalization coefficients have been lumped into the A_m. The representation (55) applies to any system of spherical coordinates. That is, if we transform (55) to an alternative set of spherical coordinates—obtained, for example, by changing the polar axis with respect to which the spherical harmonics are defined—we obtain an expression identical to (55) but with a different set of A_m's and γ_m's. We may in fact use any system of coordinates, including the more convenient stereographic coordinates, because the following proof relies only on the covariant property (54).

Let $\psi^{\Omega}_{(l,m,n)}$ be the streamfunction corresponding to solid-body flow at angular speed Ω about an axis in the direction of the unit vector (l, m, n). Let Y_n be a solution of (54), and let Y^{ω}_n be the pattern Y_n rotating at the angular speed

$$\omega = \Omega - \frac{2\Omega}{n(n+1)} \tag{56}$$

about the axis (l, m, n). Then

$$\psi = \psi^{\Omega}_{(l,m,n)} + Y^{\omega}_n \tag{57}$$

is a solution of our dynamics. For example, if $(l, m, n) = (0, 0, 1)$, and if the spherical harmonics are defined with respect to this same axis, then the solution (57) takes the form

$$\psi(\theta, \lambda, t) = -\Omega \sin\theta + \sum_{m=1}^{n} A_m P^m_n(\sin\theta) \cos(m(\lambda - \omega t) + \gamma_m) \tag{58}$$

in spherical coordinates. However, the axis (l, m, n) is arbitrary, and, more importantly, the spherical harmonics can be defined with respect to any axis whatsoever; it only matters that they are all of degree n. In fact, the solution (57), properly interpreted, holds in any system of coordinates. Viewed from a reference frame rotating with the solid-body flow—that is, viewed in rotating coordinates—the solution (57) corresponds to a retrograde, 'pseudo-westward', propagation of the Y_n-pattern at an angular velocity equal to the last term in (56), which differs from the frequency of Rossby–Haurwitz waves in its lack of a factor m in the numerator. This m factor is missing because we here define ω as angular frequency rather than wave frequency. That is, in (58) we write $m(\lambda - \omega t)$ rather than $m\lambda - \omega t$.

In [6] Kochin et al. attribute the solution (57) to Hans Ertel in 1945, but they do not give a reference. See also [21]. It was re-discovered by Herring's colleague Phil Thompson [14] for the special case in which (l, m, n) coincides with the z-axis, as in (58), and Y_n consists of a *single* spherical harmonic whose axis of symmetry is *inclined* with respect to the z-axis by a prescribed angle. Thompson proposed this special exact solution as a test of numerical codes based on spherical coordinates. In [15], Verkley matched Thompson's solution in one region of the sphere with an analogous solution in another region in which Y_n is replaced by the solutions of $\nabla^2_{LB} Y_n = dY_n$, with d a *positive* constant. The matching conditions are that the two solutions precess about the z-axis at the same angular speed and have continuous derivatives at the boundary between regions. The family of solutions discovered by Ertel, Thompson, and Verkley is noteworthy in that they seem to be the only analytic solutions of the vorticity equation on the sphere (besides 'zonal' flows in which $\psi = F(\theta)$ with F arbitrary) that do not include point vortices. See also [22] and the references therein.

We prove (57) using stereographic coordinates. Because rotations are involved, it is best to use the complex notation $\zeta = \xi + i\eta$ and to regard $\psi(\xi, \eta, t)$, $q(\xi, \eta, t)$ as $\psi(\zeta, \bar{\zeta}, t)$, $q(\zeta, \bar{\zeta}, t)$, where $\bar{\zeta} = \xi - i\eta$ denotes the complex conjugate. In this notation,

$$\nabla^2_{LB} = (1 + \zeta\bar{\zeta})^2 \frac{\partial^2}{\partial\zeta\partial\bar{\zeta}} \tag{59}$$

and the dynamics (32) takes the form

$$h^2 q_t + 2i\frac{\partial(\psi, q)}{\partial(\zeta, \bar{\zeta})} = 0, \quad q = \nabla^2_{LB}\psi, \quad h = \frac{2}{1 + \zeta\bar{\zeta}} \tag{60}$$

given by [23]. Suppose, without loss of generality, that the axis of the solid body flow is $(0,0,1)$. Then

$$\psi^{\Omega}_{(0,0,1)} = -\Omega \sin\theta = -\Omega z = -\Omega \frac{(\xi^2+\eta^2-1)}{(\xi^2+\eta^2+1)} = -\Omega \frac{(\zeta\bar{\zeta}-1)}{(\zeta\bar{\zeta}+1)} = \Omega h - \Omega. \tag{61}$$

The last term in (61) is an irrelevant constant. Let $Y_n(\zeta,\bar{\zeta})$ be a solution of (54). To demonstrate Thompson's solution, we must show that

$$\psi = Y_n(s,\bar{s}) + \Omega h \tag{62}$$

satisfies (60), where

$$s \equiv \zeta e^{-i\omega t}, \quad \bar{s} \equiv \bar{\zeta} e^{+i\omega t} \tag{63}$$

and ω is given by (56). In the new coordinates $s,\bar{s}$,

$$\nabla^2_{LB} = (1+s\bar{s})^2 \frac{\partial^2}{\partial s \partial \bar{s}}, \quad h = \frac{2}{1+s\bar{s}}. \tag{64}$$

The vorticity corresponding to (62) is

$$q = \nabla^2_{LB}(Y_n(s,\bar{s}) + \Omega h) = -n(n+1)Y_n(s,\bar{s}) - 2\Omega h + 2\Omega. \tag{65}$$

The last term in (65) is needed to satisfy the Gauss constraint, but it does not contribute to the vorticity equation, in which only derivatives of q appear. By the chain rule,

$$\begin{aligned} h^2 q_t = -n(n+1)h^2 \frac{\partial}{\partial t} Y_n(s,\bar{s}) &= -n(n+1) i\omega h^2 \left(-s\frac{\partial Y_n}{\partial s} + \bar{s}\frac{\partial Y_n}{\partial \bar{s}} \right) \\ &= -2n(n+1) i\omega \frac{\partial(Y_n,h)}{\partial(s,\bar{s})} \end{aligned} \tag{66}$$

where we used the identities

$$\frac{\partial h}{\partial s} = -\frac{1}{2}h^2 \bar{s}, \quad \frac{\partial h}{\partial \bar{s}} = -\frac{1}{2}h^2 s. \tag{67}$$

For the second term in the vorticity equation, we have

$$2i\frac{\partial(\psi,q)}{\partial(\zeta,\bar{\zeta})} = 2i\frac{\partial(Y_n+\Omega h, -n(n+1)Y_n - 2\Omega h)}{\partial(s,\bar{s})} = 2\Omega i(n(n+1)-2)\frac{\partial(Y_n,h)}{\partial(s,\bar{s})}. \tag{68}$$

The sum of (66) and (68) vanishes for ω given by (56). Thus, (62) satisfies our dynamics for any positive integer n. This proof is simpler than the proofs given by [14] or [15] using spherical coordinates.

4. Numerical Model

We solve (31) and (32) by the method of [16], a generalization of Arakawa's method [24] that conserves discrete analogues of the energy (43), total vorticity (44), and enstrophy (45) when $\nu = 0$. This method was further generalized to the shallow water equations; see [25] and references therein.

We begin by noting that the vorticity equation in (32) is equivalent to the statement that

$$\iint d\xi d\eta\, \alpha h^2 q_t = \iint d\xi d\eta\, \alpha \frac{\partial(\psi,q)}{\partial(\xi,\eta)} \tag{69}$$

for any function $\alpha(\xi,\eta)$. The integration is over the entire sphere. By parts integrations, (69) takes the more useful form

$$\iint d\xi d\eta\, \alpha h^2 q_t = \frac{1}{3}\iint d\xi d\eta \left[\alpha\frac{\partial(\psi,q)}{\partial(\xi,\eta)} + \psi\frac{\partial(q,\alpha)}{\partial(\xi,\eta)} + q\frac{\partial(\alpha,\psi)}{\partial(\xi,\eta)} \right]. \tag{70}$$

Our strategy is to discretize all the variables in (70) in such a way that the conservation of energy, total vorticity, and enstrophy correspond to purely algebraic cancellations between the terms. The conservation of enstrophy (45) corresponds to the choice $\alpha = q$ in (70). This requires that the discrete form of

$$\iint d\xi d\eta \left[q\frac{\partial(\psi,q)}{\partial(\xi,\eta)} + \psi\frac{\partial(q,q)}{\partial(\xi,\eta)} + q\frac{\partial(q,\psi)}{\partial(\xi,\eta)} \right] \tag{71}$$

vanishes, and this occurs automatically if the discrete Jacobian is antisymmetric. The conservation of total vorticity corresponds to $\alpha = 1$, and requires that the discrete form of

$$\iint d\xi d\eta \, \frac{\partial(\psi,q)}{\partial(\xi,\eta)} \tag{72}$$

vanishes. Again, this occurs automatically. Energy conservation corresponds to $\alpha = -\psi$, and we discuss it further below.

We discretize (70) by covering the interior of both stereographic circles with quadrilateral finite elements. The placement of the elements is the same in both circles. Most of the elements are perfect squares, as illustrated in Figure 2 for a very low-resolution case, but near the unit circle itself, the quadrilaterals are deformed, with one or more nodes lying on the circle.

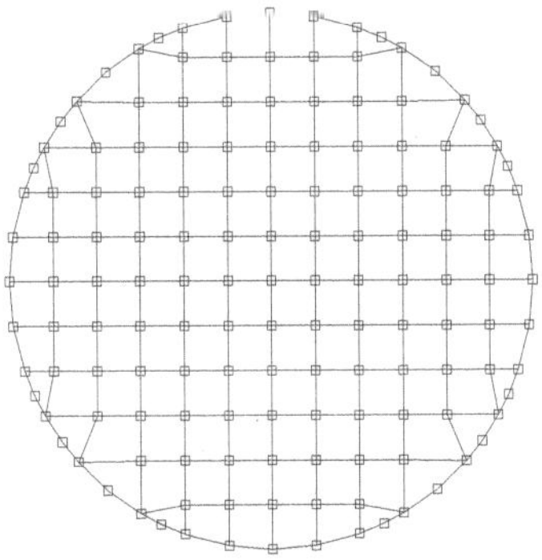

Figure 2. Quadrilateral elements cover the interior of the unit circle in the stereographic plane. The circle itself corresponds to the equator. Most of the interior elements are perfect squares. Elements with 1, 2, or 3 equatorial nodes are deformed quadrilaterals that conform to the circle.

The nodal values of ψ and q are the dependent variables of the model. Each interior node corresponds to a point on one of the hemispheres, and is shared by four elements. Each equatorial node is shared by up to six elements, counting elements on both sides of the equator. At every time, the state of the system is determined by the values of ψ or q at $N = 2N_i + N_{eq}$ nodes, where N_i is the number of interior nodes within each of the two unit circles, and N_{eq} is the number of (shared) equatorial nodes. Appendix B offers further details.

In claiming conservation properties, we regard time derivatives as exact. That is, we ignore errors that result from the finite time step. Experience shows these errors to be small in comparison to those that result from space discretization. This issue arises when we consider the finite-element discretization of the left-hand side of (70), namely

$$\sum_i A_i \, \alpha_i h_i^2 \frac{dq_i}{dt} \tag{73}$$

where the summation is over all N nodes, A_i is the area of the $\xi\eta$-plane associated with node i (in a sense to be made precise), and α_i, h_i, q_i are the nodal values of α, q and h. Note that the metric factor

$$h_i = \frac{2}{1+\xi_i^2+\eta_i^2} \tag{74}$$

is determined by the arrangement of the quadrilateral elements, and is independent of time. Replacing the left-hand side of (70) with (73) and the right-hand side of (70) with the finite-element representation described briefly above and more thoroughly in Appendix B, we obtain the model dynamics by requiring that the discrete form of (70) vanishes for arbitrary nodal values α_i. Setting $\alpha_i = q_i$ and treating the time derivative as exact, we conclude that the discrete enstrophy

$$Z \equiv \frac{1}{2}\sum_i A_i\, h_i^2 q_i^2 \tag{75}$$

is conserved.

Energy conservation corresponds to the choice $\alpha_i = -\psi_i$. Again, the right-hand side of (70) vanishes if the discrete Jacobian is antisymmetric, and hence

$$-\sum_i A_i\, \psi_i h_i^2 \frac{dq_i}{dt} = 0. \tag{76}$$

To make (76) the time derivative of a discrete energy, we must consider the Poisson equation in (32) relating ψ and q. It is equivalent to the statement that

$$-\iint d\xi d\eta\ \beta h^2 q = \iint d\xi d\eta\ \nabla\beta\cdot\nabla\psi \tag{77}$$

for any function $\beta(\xi,\eta)$. The arbitrary function $\beta(\xi,\eta)$ is analogous to $\alpha(\xi,\eta)$. Let

$$-\sum_i A_i\, \beta_i h_i^2 q_i = \sum_{ij} w_{ij}\beta_i\psi_j \tag{78}$$

be the discrete approximation to (77), where w_{ij} is the set of weights that arise by discretizing the right side of (77). We require (78) to vanish for any set of β_i. The β_i are completely arbitrary and can take different values at different times. By first regarding the β as time independent, we obtain the general result

$$-\sum_i A_i\, \beta_i h_i^2 \frac{dq_i}{dt} = \sum_{ij} w_{ij}\beta_i \frac{d\psi_j}{dt}. \tag{79}$$

Then, setting $\beta_i = \psi_i(t)$, treating time derivatives as exact, and assuming that the weights w_{ij} are symmetric ($w_{ij} = w_{ji}$), we obtain

$$-\sum_i A_i\, \psi_i h_i^2 \frac{dq_i}{dt} = \sum_{ij} w_{ij}\psi_i \frac{d\psi_j}{dt} = \frac{1}{2}\frac{d}{dt}\sum_{ij} w_{ij}\psi_i\psi_j. \tag{80}$$

Since by (76) this must vanish, we conclude that the discrete energy

$$E \equiv \frac{1}{2}\sum_{ij} w_{ij}\psi_i\psi_j \tag{81}$$

is conserved.

In overall summary, enstrophy is conserved if (70) is discretized in such a way that its Jacobians are antisymmetric, and energy is conserved if (77) is discretized in such a way that the dot product is symmetric. These two steps may be interpreted more generally in terms of the Poisson bracket and the Hamiltonian of the system [25]. The model does not conserve other Casimirs of the form (46). Appendix B gives further details of the discretizations.

5. Numerical Solutions

For flow with rms velocity U on a sphere of radius a with angular momentum corresponding to the solid body rotation rate Ω, the Rossby number $Ro = U/\Omega a$ and the Rhines scale $Rh = Ro^{1/2}a$. For values typical of the Earth's atmosphere—$U = 10$ m/s, $a = 6400$ km, and $\Omega = 2\pi/\text{day}$—we have $Ro = 0.02$ and $Rh = 0.15a$. For Earth's ocean and the giant planets, the ratio of the Rhines scale Rh to the planetary radius a is even smaller. This regime, which features zonal jets with widths on the order of Rh, has attracted the greatest interest from modelers, but the non-rotating case $\Omega = 0$ is also interesting because it reveals the effects of curvature by itself.

In our experiments, $U = a = 1$, where U is the rms velocity at the initial time in the rotating frame, that is, the rms velocity excluding solid body rotation. The Rossby number $Ro = 0.02$, corresponding to the Earth's atmosphere, then corresponds to $\Omega = 50$. We present solutions of the two cases $\Omega = 0$ and $\Omega = 50$. In the solutions with $\Omega = 50$, the displayed stream function and vorticity do not include the $n = 1$ modes associated with the angular momentum and therefore depict the fields in the rotating frame. In all of our experiments, the viscosity, if nonzero, has the value $\nu = q_{rms}\Delta^2$, where q_{rms} is the rms vorticity and Δ^2 is the area of the square gridboxes on the stereographic plane. The number $N = 241{,}000$ of nodes corresponds to the cutoff $n_c = 490$ in spherical harmonic degree n.

Figure 3 shows the streamfunction and vorticity in three views of the sphere at time $t = 2.0$ in an experiment with $\Omega = 0$ (the non-rotating case). The initial energy is confined to spherical harmonics of degree $n = 16, 17, 18$ with the amplitudes randomly assigned. Figure 4 shows the same fields in an experiment beginning from the same initial conditions but with $\Omega = 50$.

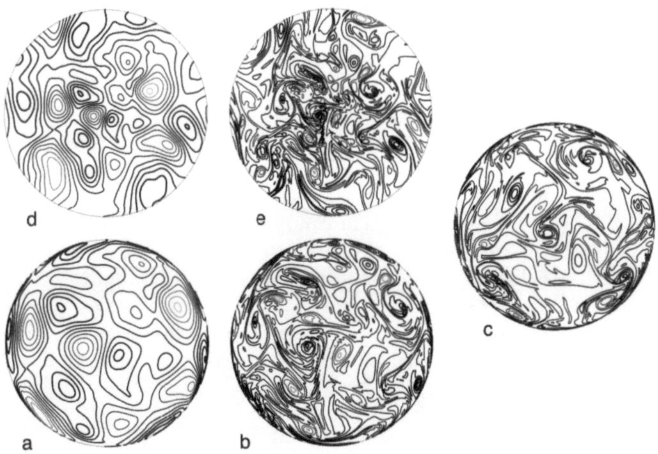

Figure 3. Stream function, (**a**,**d**), and vorticity, (**b**,**c**,**e**), at time $t = 2.0$ in a non-rotating, viscous experiment in which the initial energy is confined to spherical harmonic degrees $n = 16, 17, 18$. (**a**,**b**) are polar views, as seen by an observer at $z = \infty$, above the north pole. (**d**,**e**) are northern-hemisphere stereographic views that depict the same field as the polar views directly below. Compared to the polar views (**a**,**b**), the stereographic views (**d**,**e**) are relatively free of distortion. Panel (**c**) is an equatorial view, as seen by an observer at $x = \infty$. In all views, darker contour lines correspond to larger values.

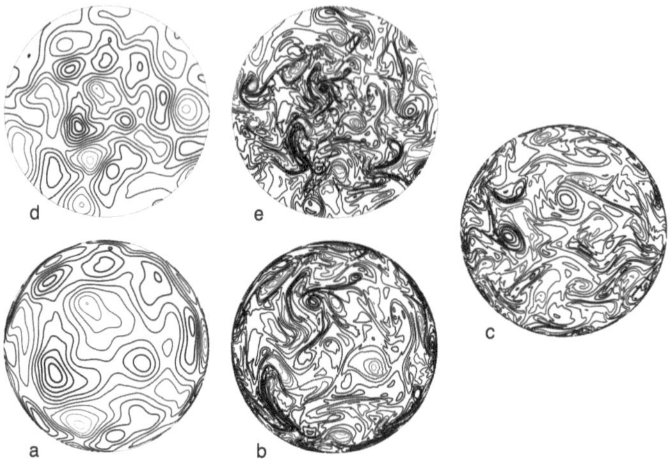

Figure 4. The same as Figure 3 but for the case $\Omega = 50$ of Earth-like rotation. This solution differs from that of Figure 3 in that isolated coherent vortices are less prominent, and that elongated, jet-like structures are beginning to appear, especially near the equator in (c), where the β-effect is greatest.

In Figures 3 and 4, (a) is a polar view of the stream function in the northern hemisphere, as seen by an observer at $z = \infty$ above the north pole, and (d) is the same field within the stereographic cirle $\hat{\xi}^2 + \hat{\eta}^2 < 1$ corresponding to the northern hemisphere. Although both views cover the entire northern hemisphere, the polar view (a) hides features near the equator: the ratio between true distance and apparent distance in (a) varies between unity at the pole and infinity at the equator. In contrast, the stereographic view (d) introduces a scale distortion that varies only between 1 and 2. It offers the superior view of the entire northern hemisphere field.

Figure 4 shows the same 5 views as in Figure 3 but for the case $\Omega = 50$ of non-vanishing angular momentum. Figure 4 differs from Figure 3 in that its isolated coherent vortices are less prominent than in Figure 3, and elongated, jet-like structures are beginning to appear, especially near the equator (Figure 4c), where the β-effect is greatest.

Figures 3 and 4 illustrate the advantage of stereographic projection. Whereas at least six conventional views of the sphere are needed to give a fair representation of a single field, two stereographic projections offer a complete and relatively undistorted depiction.

Next, we consider two experiments, in which the initial state is sharply confined to spherical harmonics of degree $n = 6$. Angular momentum corresponding to a solid body rotation rate $\Omega = 50$ is present, but now it points in the direction of the y-axis in the Cartesian embedding space. This is achieved by replacing the Coriolis parameter f in (34) by

$$2\Omega y = 2\Omega \frac{2\eta}{\xi^2 + \eta^2 + 1} = 2\Omega \frac{2\hat{\eta}}{\hat{\xi}^2 + \hat{\eta}^2 + 1}. \tag{82}$$

According to Section 3, the $n = 6$ initial state should rotate rigidly, in a retrograde manner, about the y-axis at the angular rate $\omega \equiv 2\Omega/(6(6+1))$. Figure 5 shows this motion as seen by an observer at $x = \infty$. From its perspective, the pattern rotates upward in the figure, around the sphere, completing a full cycle of revolution in the recurrence time $2\pi/\omega = 2.639$ shown on the far right of the figure. The solutions depicted in Figure 5 are both a check on the theory and a strong test of the numerical code because the rotation carries the pattern from the interior of one stereographic circle to the other and then back again, twice crossing the irregularly shaped elements near the bounding circles (11).

The inviscid experiment (bottom row of Figure 5) exhibits small, grid-scale oscillations in the vorticity field, which do not appear in the viscous experiment (top row).

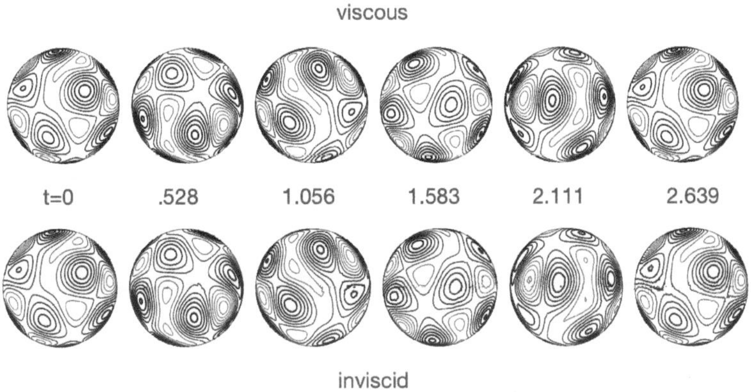

Figure 5. Vorticity as seen from $x = \infty$ at 6 times in two solutions in which the angular momentum vector points in the y-direction. z increases upward in the figure. The initial state at $t = 0$ consists only of harmonics of degree $n = 6$. The pattern rotates upward in the figure, around the sphere, to return to its original location at the predicted time $t = 2.639$. The inviscid solution shows small, grid-scale oscillations in the vorticity that do not appear in the viscous solution.

Finally, we explore the suggestion in Section 1 that solutions lock into a static state as Bose–Einstein condensation occurs. Theory predicts Bose–Einstein condensation only as $n_c \to \infty$ and $t \to \infty$—two limits that are impossible to achieve numerically. The interesting question is whether quasi-static states appear with finite resolution and in finite time. According to equilibrium statistical mechanics [5], the equilibrium energy in spherical harmonic degree n is

$$E_n = \frac{2n+1}{\alpha + \beta n(n+1)}, \quad n \geq 2 \tag{83}$$

where the constants α and β are determined by the energy and enstrophy. The equilibrium spectrum (83) is the analogue of Kraichnan's spectrum

$$E(k) = \frac{k}{\alpha + \beta k^2} \tag{84}$$

for flow on the plane [4]. The spectra (83) and (84) represent the final states attained by the inviscid, spectrally truncated dynamics after an indefinite period of adjustment. The numerator in (83) is the number of modes of degree n. Bose–Einstein condensation corresponds to the limit $\alpha \to -2(2+1)\beta$ as $n_c \to \infty$.

Several authors investigate freely decaying turbulence on the sphere and its resemblance to the predictions of equilibrium statistical mechanics, including the more recondite Miller–Sommeria–Robert–Montgomery (MRSM) theory ([26–29]). For a summary of the MRSM theory, see [30]. The requirement that $t \to \infty$ has proved formidable. MRSM theory claims to predict the structure of coherent vortices by the principle of maximum entropy. Ref. [31] offers a different picture, in which coherent vortices are *low* entropy sites that exist in order to maximize the *rate* of entropy production in the area outside the vortices.

Here, we consider the case $\Omega = \nu = 0$ of non-rotating, inviscid flow with a modest resolution corresponding to $n_c = 240$. The initial condition corresponds to energy distributed equally among the three degrees $n = 4, 5, 6$, that is, $E_4 = E_5 = E_6$ with all other $E_n = 0$. The amplitudes of the spherical harmonics within each band are randomly assigned, and the initial state is normalized such that $u_{rms} = 1$. A calculation based upon (83) predicts that 99.6% of the energy will end up in $n = 2$, the lowest available mode. Again,

$E_1 = 0$ at all time by the conservation of angular momentum. Figure 6 summarizes the evolution of the flow from $t = 0$ to $t = 60$. The time $t = 60$ is the time required for a fluid particle to circumnavigate a great circle $60/2\pi = 9.5$ times at the rms velocity of the flow. The dashed curve in Figure 6 is the fraction of energy in $n = 2$. Beginning from zero, it only gradually approaches its predicted value of 0.996.

The solid curves in Figure 6 measure the approach to a static configuration. Let $A_{2,m}$ be the amplitude of Y_2^m in the spherical harmonic representation of ψ, and define

$$\mathbf{p} = \frac{(A_{2,0},\ A_{2,1},\ A_{2,-1},\ A_{2,2},\ A_{2,-2})}{[(A_{2,0})^2 + (A_{2,1})^2 + (A_{2,-1})^2 + (A_{2,2})^2 + (A_{2,-2})^2]^{1/2}}. \tag{85}$$

Thus, $\mathbf{p}$ is a unit vector in the 5-dimensional space spanned by the amplitudes of the five spherical harmonics comprising $n = 2$. If the flow were indeed to become static, then the five components of $\mathbf{p}$ would become time independent at values that are accidents of the initial conditions. Our choice of conventionally defined spherical harmonics as amplitudes to be tested is both arbitrary and irrelevant; any other choice of mutually orthogonal functions that span ψ_2 would serve as well.

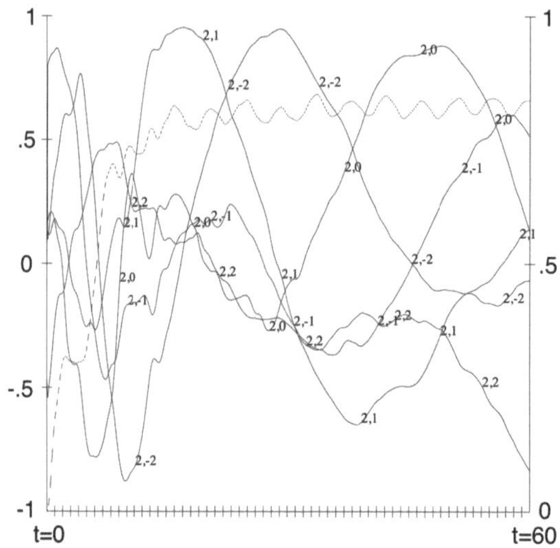

Figure 6. Evolution of a non-rotating, inviscid flow, in which the initial energy is confined to spherical harmonic degrees $n = 4, 5, 6$. The dashed curve (right scale) is the fraction of energy in $n = 2$, which vanishes initially, but, according to theory, will eventually reach 0.996. The other curves, labeled $2, m$, represent the normalized coefficient of Y_2^m in the spherical harmonic representation of the flow, which vary between ± 1 (left scale).

Figure 6 shows that the approach to the putative equilibrium state is extremely slow. Although E_2 generally increases, its increase is not monotonic but instead exhibits persistent, relatively rapid oscillations. The components of $\mathbf{p}$, labeled $2, m$ in the figure oscillate over their full range of ± 1, but the periods of their oscillations appear to lengthen as time increases. It is conceivable that this flow eventually reaches a static state, but this evidently requires a very long time. On the other hand, since the time unit is the time required for a fluid particle to move a distance equal to the radius of the sphere, the flow at $t = 60$ is already quasi-static in the sense that it requires at least several such time units for the components of $\mathbf{p}$ to change appreciably.

6. Discussion

Spectral methods based upon the conventionally defined spherical harmonics in spherical coordinates are still probably the method of choice in atmosphere modeling. Norman Phillips [32] proposed the use of two stereographic coordinate systems to overcome the pole problem, but there remained the problem of matching the two stereographic systems together. Subsequent workers sandwiched a band of spherical or Mercator coordinates between separate stereographic regions at the poles ([33–36]), or omitted the band and interpolated between the two stereographic systems in an overlapping region near the equator [37]. In the present paper, we solve the stereographic equations within two non-overlapping circles and match the solutions on the circles themselves.

Our strategy requires the use of irregular quadrilaterals that conform to the circles, as shown in Figure 2. The existence of many irregular shaped equatorial elements is cause for concern, but none of our numerical experiments, including those depicted in Figures 3–5, show evidence of spurious behavior near the equator. This good behavior is typical of Arakawa-type models that conserve energy and enstrophy when the viscosity vanishes.

Gridpoint methods outperform spectral methods when sharp features, such as boundaries, are present. Suppose that the spherical flow domain excluded an irregularly shaped island, or consisted of an ocean with an irregular boundary. It is very likely that our method of using quadrilaterals that conform to the equator would work equally as well for the boundary of the excluded region.

Compared to the plane, there are relatively few exact solutions for incompressible flow on the sphere. In fact, zonal flows and the solutions discussed in Section 3 seem to be the only exact analytical solutions for the sphere that do not contain point vortices or other singularities.

On the other hand, the solutions of Section 3 have no analogue on the plane because, as noted in Section 1, typical planar geometries do not correspond to many modes with the same wavenumber magnitude. In the commonly considered case of an infinitely periodic box, the modes correspond to integer pairs (n, m), but relatively few of these occupy the same circle in the wavenumber plane. Circles with integral radius l contain the modes $(\pm l, 0)$ and $(0, \pm l)$, but the circle contains the additional mode (n, m) only if

$$n^2 + m^2 = l^2. \tag{86}$$

Integers n, m, l that satisfy (86) are called Pythagorean Triples and are relatively sparse, the best known example being $3, 4, 5$. In contrast, there are $2n + 1$ spherical harmonics with the same value of $n(n+1)$.

We associate the exact solutions of Section 3 with the rotational symmetry of the sphere, which corresponds by Noether's theorem to angular momentum conservation. This symmetry is broken by all the usual planar geometries, including the infinitely periodic box, exceptions being the infinite plane and a perfectly circular domain. These facts only serve to underline the importance of angular momentum conservation in models, including by viscosity.

This paper advocates strongly in favor of stereographic coordinates for analysis, modeling, and flow visualization, but it is only fair to point out that different coordinate systems could be used for each of these purposes: one could use the cubed-sphere technique to perform the calculations, and still view the results in stereographic coordinates. We do admit a strong prejudice against spherical coordinates and the usual association between spherical coordinates and spherical harmonics. The latter are defined by the covariant property (54) and can be written in any system of coordinates.

Author Contributions: Both authors contributed equally to this work. All authors have read and agreed to the published version of the manuscript.

Funding: This research received no external funding.

Data Availability Statement: No data was used.

Acknowledgments: We thank David Randall and three anonymous referees for important suggestions.

Conflicts of Interest: The authors declare no conflict of interest.

Appendix A. Viscosity in Stereographic Coordinates

The angular-momentum-conserving viscosity suggested by [13] is

$$\frac{\partial \dot{\xi}^i}{\partial t} + \cdots = \nu\left(g^{kl}\nabla_k\nabla_l\dot{\xi}^i - K\dot{\xi}^i\right) \tag{A1}$$

where (ξ^1, ξ^2) are general coordinates covering the sphere, $\dot{\xi}^i = d\xi^i/dt$ following a fluid particle, g^{kl} is the inverse of the metric tensor g_{kl}, ∇_k is the covariant derivative, and K is the Gaussian curvature. In this appendix, we use raised indices to denote contravariant objects and lowered indices to denote covariant objects, as is customary in tensor analysis. Repeated indices are summed. For the unit sphere, $K = 1$. The ellipses denote the non-viscous terms that are not an issue here. In the stereographic coordinates (ξ, η), this viscosity law takes the form

$$\frac{\partial u^i}{\partial t} + \cdots = \nabla_\xi\nabla_\xi(\gamma^2 u^i) + \nabla_\eta\nabla_\eta(\gamma^2 u^i) - u^i \tag{A2}$$

where $(u^1, u^2) = (u, v) = h^2(\dot{\xi}, \dot{\eta})$ as previously defined, and $\gamma \equiv 1/h$. To ease the notation, we set $\nu = 1$.

For a general contravariant vector w^i

$$\nabla_k w^i = \frac{\partial w^i}{\partial x^k} + \Gamma^i_{jk} w^j \tag{A3}$$

where

$$\Gamma^i_{jk} = \frac{1}{2} g^{il}\left(\frac{\partial g_{lj}}{\partial x^k} + \frac{\partial g_{lk}}{\partial x^j} - \frac{\partial g_{kj}}{\partial x^l}\right). \tag{A4}$$

For mixed tensors T^i_k, such as $\nabla_k w^i$

$$\nabla_l T^i_k = \frac{\partial T^i_k}{\partial x^l} + \Gamma^i_{k\mu} T^\mu_l - \Gamma^\mu_{kl} T^i_\mu. \tag{A5}$$

In stereographic coordinates, the non-vanishing metric coefficients are

$$g_{\xi\xi} = g_{\eta\eta} = g, \quad g^{\xi\xi} = g^{\eta\eta} = \frac{1}{g} \tag{A6}$$

where

$$g \equiv h^2 = \frac{4}{(1 + \xi^2 + \eta^2)^2} \tag{A7}$$

and the corresponding Christoffel symbols are

$$\Gamma^\xi_{\xi\xi} = -\Gamma^\xi_{\eta\eta} = \Gamma^\eta_{\xi\eta} = -h\xi \tag{A8}$$

and

$$\Gamma^\eta_{\eta\eta} = -\Gamma^\eta_{\xi\xi} = \Gamma^\xi_{\eta\xi} = -h\eta. \tag{A9}$$

The needed second covariant derivatives are of the form

$$\nabla_\xi\nabla_\xi w^i = \frac{\partial}{\partial \xi}\nabla_\xi w^i + \Gamma^i_{\xi\mu}\nabla_\xi w^\mu - \Gamma^\mu_{\xi\xi}\nabla_\mu w^i \tag{A10}$$

and

$$\nabla_\eta \nabla_\eta w^i = \frac{\partial}{\partial \eta} \nabla_\eta w^i + \Gamma^i_{\eta\mu} \nabla_\eta w^\mu - \Gamma^\mu_{\eta\eta} \nabla_\mu w^i \tag{A11}$$

where $(w^\xi, w^\eta) \equiv \gamma^2(u,v) = (\dot{\xi}, \dot{\eta})$. Using (A6)–(A9), we obtain, we obtain

$$\nabla_\xi \nabla_\xi w^\xi = \frac{\partial}{\partial \xi} \nabla_\xi w^\xi - h\eta \left(\nabla_\xi w^\eta + \nabla_\eta w^\xi \right) \tag{A12}$$

$$\nabla_\xi \nabla_\xi w^\eta = \frac{\partial}{\partial \xi} \nabla_\xi w^\eta + h\eta \left(\nabla_\xi w^\xi - \nabla_\eta w^\eta \right) \tag{A13}$$

$$\nabla_\eta \nabla_\eta w^\xi = \frac{\partial}{\partial \eta} \nabla_\eta w^\xi + h\xi \left(\nabla_\eta w^\eta - \nabla_\xi w^\xi \right) \tag{A14}$$

$$\nabla_\eta \nabla_\eta w^\eta = \frac{\partial}{\partial \eta} \nabla_\eta w^\eta - h\xi \left(\nabla_\eta w^\xi + \nabla_\xi w^\eta \right) \tag{A15}$$

for the second covariant derivatives. For the first covariant derivatives, we obtain

$$\nabla_\xi w^\xi = \frac{\partial w^\xi}{\partial \xi} - h\xi w^\xi - h\eta w^\eta \tag{A16}$$

$$\nabla_\xi w^\eta = \frac{\partial w^\eta}{\partial \xi} + h\eta w^\xi - h\xi w^\eta \tag{A17}$$

$$\nabla_\eta w^\xi = \frac{\partial w^\xi}{\partial \eta} - h\eta w^\xi + h\xi w^\eta \tag{A18}$$

$$\nabla_\eta w^\eta = \frac{\partial w^\eta}{\partial \eta} - h\xi w^\xi - h\eta w^\eta. \tag{A19}$$

By (A17) and (A18)

$$\nabla_\xi w^\eta + \nabla_\eta w^\xi = \frac{\partial w^\eta}{\partial \xi} + \frac{\partial w^\xi}{\partial \eta}. \tag{A20}$$

The incompressibility condition

$$\nabla_\xi w^\xi + \nabla_\eta w^\eta = 0 \tag{A21}$$

implies that

$$h\xi w^\xi + h\eta w^\eta = \frac{1}{2} \left(\frac{\partial w^\xi}{\partial \xi} + \frac{\partial w^\eta}{\partial \eta} \right) \tag{A22}$$

so that (A16) and (A19) can be written as

$$\nabla_\xi w^\xi = \frac{1}{2} \left(\frac{\partial w^\xi}{\partial \xi} - \frac{\partial w^\eta}{\partial \eta} \right) \tag{A23}$$

$$\nabla_\eta w^\eta = \frac{1}{2} \left(\frac{\partial w^\eta}{\partial \eta} - \frac{\partial w^\xi}{\partial \xi} \right). \tag{A24}$$

Thus,

$$\nabla_\xi w^\xi - \nabla_\eta w^\eta = \left(\frac{\partial w^\xi}{\partial \xi} - \frac{\partial w^\eta}{\partial \eta} \right). \tag{A25}$$

Using these equations to simplify the needed second covariant derivatives (A12)–(A15), we obtain

$$\nabla_\xi \nabla_\xi w^\xi = \frac{1}{2}\frac{\partial}{\partial \xi}\left(\frac{\partial w^\xi}{\partial \xi} - \frac{\partial w^\eta}{\partial \eta}\right) - h\eta\left(\frac{\partial w^\eta}{\partial \xi} + \frac{\partial v^\xi}{\partial \eta}\right) \quad \text{(A26)}$$

$$\nabla_\xi \nabla_\xi w^\eta = \frac{\partial}{\partial \xi}\left(\frac{\partial w^\eta}{\partial \xi} + h\eta w^\xi - h\xi w^\eta\right) + h\eta\left(\frac{\partial w^\xi}{\partial \xi} - \frac{\partial w^\eta}{\partial \eta}\right) \quad \text{(A27)}$$

$$\nabla_\eta \nabla_\eta w^\xi = \frac{\partial}{\partial \eta}\left(\frac{\partial w^\xi}{\partial \eta} - h\eta w^\xi + h\xi w^\eta\right) + h\xi\left(\frac{\partial w^\eta}{\partial \eta} - \frac{\partial w^\xi}{\partial \xi}\right) \quad \text{(A28)}$$

$$\nabla_\eta \nabla_\eta w^\eta = \frac{1}{2}\frac{\partial}{\partial \eta}\left(\frac{\partial w^\eta}{\partial \eta} - \frac{\partial w^\xi}{\partial \xi}\right) - h\xi\left(\frac{\partial w^\eta}{\partial \xi} + \frac{\partial w^\xi}{\partial \eta}\right). \quad \text{(A29)}$$

Finally, setting $(w^\xi, w^\eta) = \gamma^2(u, v)$, substituting (A26)–(A29) into (A2), and simplifying the resulting expressions, we obtain the result (48) and (49).

Appendix B. Finite-Element Discretization

Every finite element is a quadrilateral. Most of the quadrilaterals are perfect squares, but those near the unit circles are deformed. Figure A1 shows a representative element with nodes numbered 1 to 4.

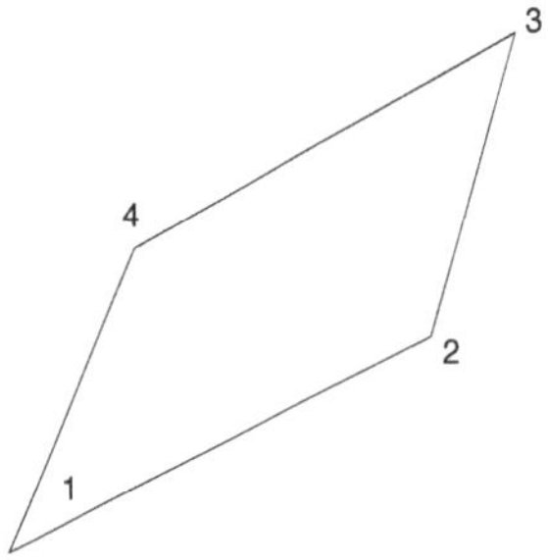

Figure A1. A representative quadrilateral element with nodes labeled 1 to 4.

Let $\{\gamma_i,\ i = 1,4\}$ be the nodal values of $\gamma(\xi, \eta)$, which could be $\beta(\xi, \eta)$, $\psi(\xi, \eta)$, or either of the coordinates ξ or η itself. Within each element,

$$\gamma(\xi, \eta) = \sum_{i=1}^{4} \gamma_i N_i(s, t) \quad \text{(A30)}$$

where (s, t) are 'labeling' coordinates that cover the element in the domain $-1 < s, t < 1$;

$$N_i(s, t) = \frac{1}{4}(1 + ss_i)(1 + tt_i), \quad i = 1 \ldots 4 \quad \text{(A31)}$$

is the standard shape function for four-node quadrilateral finite elements, and

$$(s_1, t_1) = (-1, -1), \quad (s_2, t_2) = (1, -1), \quad (s_3, t_3) = (1, 1), \quad (s_4, t_4) = (-1, 1) \quad \text{(A32)}$$

correspond, respectively, to the corners numbered 1, 2, 3, 4 in Figure A1. Thus, $N_i = 1$ at node i and $N_i = 0$ at the other three nodes. (In this appendix, the symbol t has nothing to do with time.)

The contribution of a single element to the right-hand side of (77) is

$$\iint d\xi d\eta\, \nabla\beta \cdot \nabla\psi = \int_{-1}^{1}\int_{-1}^{1} dsdt\, \frac{\partial(s,t)}{\partial(\xi,\eta)}\left[\frac{\partial(\beta,\eta)}{\partial(s,t)}\frac{\partial(\psi,\eta)}{\partial(s,t)} + \frac{\partial(\xi,\beta)}{\partial(s,t)}\frac{\partial(\xi,\psi)}{\partial(s,t)}\right]. \quad \text{(A33)}$$

The first factor in the integrand is the ratio of the quadrilateral area on the st-plane, namely 4, to the corresponding area A_{1234} on the $\xi\eta$-plane, which is conveniently calculated by bisecting the deformed quadrilateral into two triangles. Thus,

$$\iint d\xi d\eta\ \nabla\beta\cdot\nabla\psi = \frac{4}{A_{1234}}\iint dsdt\left[\frac{\partial(\beta,\eta)}{\partial(s,t)}\frac{\partial(\psi,\eta)}{\partial(s,t)} + \frac{\partial(\xi,\beta)}{\partial(s,t)}\frac{\partial(\xi,\psi)}{\partial(s,t)}\right]. \tag{A34}$$

For the first Jacobian in (A34), we obtain

$$\frac{\partial(\beta,\eta)}{\partial(s,t)} = \sum_{i,j=1,4}\beta_i\eta_j F_{ij}(s,t) \tag{A35}$$

where

$$F_{ij}(s,t) = \frac{1}{16}\left[s_i t_j(1+tt_i)(1+ss_j) - s_j t_i(1+tt_j)(1+ss_i)\right]. \tag{A36}$$

Substituting (A35), (A36) and the corresponding expressions for the other Jacobians back into (A34), we obtain

$$\iint d\xi d\eta\ \nabla\beta\cdot\nabla\psi = \frac{4}{A_{1234}}\sum_{i,j,n,m=1}^{4}\Gamma_{ijnm}\left[\beta_i\eta_j\psi_n\eta_m + \xi_i\beta_j\xi_n\psi_m\right] = \sum_{i,j=1}^{4}\beta_i C_{ij}\psi_j \tag{A37}$$

where

$$\Gamma_{ijnm} = \int_{-1}^{1}\int_{-1}^{1} dsdt\ F_{ij}(s,t)F_{nm}(s,t) \tag{A38}$$

and

$$C_{ij} = \frac{4}{A_{1234}}\sum_{pq}\left[\eta_p\eta_q\Gamma_{ipjq} + \xi_p\xi_q\Gamma_{piqj}\right]. \tag{A39}$$

The C_{ij} may be calculated analytically, computed and stored.

The final sum in (A37) represents the contribution of a single element. To discretize the integral (A33) over the entire sphere, we sum over all the elements to obtain (78). The weight w_{ij} vanishes unless node i and node j share at least one element. Thus, up to six elements contribute to each w_{ij}. However, since C_{ij} is symmetric for each contributing element, w_{ij} is symmetric as well, and therefore the discrete energy (81) is conserved.

The right-hand side of (70) is

$$\frac{1}{3}\iint d\xi d\eta\left[\alpha\frac{\partial(\psi,q)}{\partial(\xi,\eta)} + cyc\{\alpha,\psi,q\}\right] \tag{A40}$$

where the integral is over the whole sphere. We discretize it in a similar manner to (A37). For the integral over the single element depicted in Figure A1, we approximate

$$\iint d\xi d\eta\ \alpha\frac{\partial(\psi,q)}{\partial(\xi,\eta)} = \left(\frac{1}{4}\sum_{i=1}^{4}\alpha_i\right)\iint d\xi d\eta\ \frac{\partial(\psi,q)}{\partial(\xi,\eta)} \tag{A41}$$

and evaluate

$$\iint d\xi d\eta\ \frac{\partial(\psi,q)}{\partial(\xi,\eta)} = \oint q\frac{\partial\psi}{\partial s}ds = \frac{1}{2}\sum_{j=1}^{4}q_j(\psi_{j+1}-\psi_{j-1}) \tag{A42}$$

where s is distance on the $\xi\eta$-plane, measured counter-clockwise around the perimeter of the element. In the sum over j, if $j=1$ then $j-1=4$, and if $j=4$, then $j+1=1$. Thus, the contribution of each element to (70) is

$$\frac{1}{24}\sum_{i=1}^{4}\alpha_i\sum_{j=1}^{4}q_j(\psi_{j+1}-\psi_{j-1}) + cyc\{\alpha,\psi,q\}. \tag{A43}$$

Summing over all the elements gives the discrete approximation to (70).

The left-hand sides of (69) and (77) take the discrete form

$$\sum_i A_i \, \alpha_i h_i^2 f_i \tag{A44}$$

where $f_i = dq_i/dt$ in the case of (69) and $f_i = q_i$ in the case of (77). In (A44), the summation is over nodes, and A_i is the area of the $\xi\eta$-plane that is 'assigned' to, i.e., closest to, node i. The precise rule of assignment is arbitrary, but it must obey the normalization requirement that the sum of the A_i's equals the area within a unit circle, or that the sum of the $h_i^2 A_i$'s on both unit circles equals the area of the unit sphere.

By requiring the coefficient of each β_i in (78) to vanish, we obtain N equations representing a logical discrete approximation to $h^2 q = \psi_{\xi\xi} + \psi_{\eta\eta}$. By requiring the coefficient of each α_i in the above-described discrete analogue of (32) to vanish, we obtain N equations representing a logical discrete approximation to (32).

References

1. Herring, J.; Orszag, S.; Kraichnan, R.; Fox, D. Decay of two-dimensional homogeneous turbulence. *J. Fluid Mech.* **1974**, *66*, 417–444. [CrossRef]
2. Herring, J.R. On the Statistical Theory of Two-Dimensional Topographic Turbulence. *J. Atmos. Sci.* **1977**, *34*, 1731–1750. [CrossRef]
3. Herring, J.R. Statistical theory of quasi-geostrophic turbulence. *J. Atmos. Sci.* **1980**, *37*, 969–977. [CrossRef]
4. Kraichnan, R.H. Inertial ranges in two-dimensional turbulence. *Phys. Fluids* **1967**, *10*, 1417–1423. [CrossRef]
5. Frederiksen, J.; Sawford, B. Statistical dynamics of two-dimensional inviscid flow on a sphere. *J. Atmos. Sci.* **1980**, *37*, 717–732. [CrossRef]
6. Kochin, N.E.; Kibeĺ, I.A.; Roze, N.V. *Theoretical Hydromechanics*; Interscience Publishers: New York, NY, USA, 1964.
7. Rhines, P.B. Waves and turbulence on a beta-plane. *J. Fluid Mech.* **1975**, *69*, 417–443. [CrossRef]
8. Vallis, G.K.; Maltrud, M.E. Generation of mean flows and jets on a beta plane and over topography. *J. Phys. Oceanogr.* **1993**, *23*, 1346–1362. [CrossRef]
9. Boer, G.J.; Shepherd, T. Large-scale two-dimensional turbulence in the atmosphere. *J. Atmos. Sci.* **1983**, *40*, 164–184. [CrossRef]
10. Sawford, B.; Frederiksen, J. Mountain torque and angular momentum in barotropic planetary flows: Equilibrium solutions. *Q. J. R. Meteorol. Soc.* **1983**, *109*, 309–324. [CrossRef]
11. Egger, J.; Weickmann, K.; Hoinka, K.P. Angular momentum in the global atmospheric circulation. *Rev. Geophys.* **2007**, *45*, RG4007. [CrossRef]
12. Gill, A.E. *Atmosphere-Ocean Dynamics*; Academic Press: New York, NY, USA, 1982.
13. Gilbert, A.D.; Riedinger, X.; Thuburn, J. On the form of the viscous term for two dimensional Navier–Stokes flows. *Q. J. Mech. Appl. Math.* **2014**, *67*, 205–228. [CrossRef]
14. Thompson, P.D. A generalized class of exact time-dependent solutions of the vorticity equation for nondivergent barotropic flow. *Mon. Weather Rev.* **1982**, *110*, 1321–1324. [CrossRef]
15. Verkley, W. The construction of barotropic modons on a sphere. *J. Atmos. Sci.* **1984**, *41*, 2492–2504. [CrossRef]
16. Salmon, R.; Talley, L.D. Generalizations of Arakawa's Jacobian. *J. Comput. Phys.* **1989**, *83*, 247–259. [CrossRef]
17. Needham, T. *Visual Complex Analysis*; Oxford University Press: Oxford, UK, 1997.
18. Needham, T. *Visual Differential Geometry and Forms: A Mathematical Drama in Five Acts*; Princeton University Press: Princeton, NJ, USA, 2021.
19. Becker, E.; Burkhardt, U. Nonlinear horizontal diffusion for GCMs. *Mon. Weather Rev.* **2007**, *135*, 1439–1454. [CrossRef]
20. Jiménez, J. Hyperviscous vortices. *J. Fluid Mech.* **1994**, *279*, 169–176. [CrossRef]
21. Rochas, M. Comments on "A Generalized Class of Time-Dependent Solutions of the Vorticity Equation for Nondivergent Barotropic Flow". *Mon. Weather Rev.* **1984**, *112*, 390. [CrossRef]
22. Neven, E. Quadrupole modons on a sphere. *Geophys. Astrophys. Fluid Dyn.* **1992**, *65*, 105–126. [CrossRef]
23. Crowdy, D.G. Stuart vortices on a sphere. *J. Fluid Mech.* **2004**, *498*, 381–402. [CrossRef]
24. Arakawa, A. Computational design for long-term numerical integration of the equations of fluid motion: Two-dimensional incompressible flow. Part I. *J. Comput. Phys.* **1966**, *135*, 103–114. [CrossRef]
25. Salmon, R. A shallow water model conserving energy and potential enstrophy in the presence of boundaries. *J. Mar. Res.* **2009**, *67*, 779–814. [CrossRef]
26. Qi, W.; Marston, J. Hyperviscosity and statistical equilibria of Euler turbulence on the torus and the sphere. *J. Stat. Mech. Theory Exp.* **2014**, *2014*, P07020. [CrossRef]
27. Dritschel, D.G.; Qi, W.; Marston, J. On the late-time behaviour of a bounded, inviscid two-dimensional flow. *J. Fluid Mech.* **2015**, *783*, 1–22. [CrossRef]

28. Modin, K.; Viviani, M. A Casimir preserving scheme for long-time simulation of spherical ideal hydrodynamics. *J. Fluid Mech.* **2020**, *884*, A22. [CrossRef]
29. Jagad, P.; Samtaney, R. Effects of rotation on vorticity dynamics on a sphere with discrete exterior calculus. *Phys. Fluids* **2021**, *33*, 107117. [CrossRef]
30. Eyink, G.L.; Sreenivasan, K.R. Onsager and the theory of hydrodynamic turbulence. *Rev. Mod. Phys.* **2006**, *78*, 87. [CrossRef]
31. Salmon, R. Entropy budget and coherent structures associated with a spectral closure model of turbulence. *J. Fluid Mech.* **2018**, *857*, 806–822. [CrossRef]
32. Phillips, N. A map projection system suitable for large-scale numerical weather prediction. *J. Meteor. Soc. Jpn.* **1957**, *75*, 262–267. [CrossRef]
33. Phillips, N.A. Numerical integration of the primitive equations on the hemisphere. *Mon. Weather Rev.* **1959**, *87*, 333–345. [CrossRef]
34. Chen, Y.; Kuo, H. A global model with overlapping mercator and stereographic grids. *Adv. Atmos. Sci.* **1986**, *3*, 302–313. [CrossRef]
35. Browning, G.L.; Hack, J.J.; Swarztrauber, P.N. A comparison of three numerical methods for solving differential equations on the sphere. *Mon. Weather Rev.* **1989**, *117*, 1058–1075. [CrossRef]
36. Lanser, D.; Blom, J.; Verwer, J. Spatial discretization of the shallow water equations in spherical geometry using Osher's scheme. *J. Comput. Phys.* **2000**, *165*, 542–565. [CrossRef]
37. Dudhia, J.; Bresch, J.F. A global version of the PSU–NCAR Mesoscale Model. *Mon. Weather Rev.* **2002**, *130*, 2989–3007. [CrossRef]

atmosphere

Article

Turbulent Convection at Very High Rayleigh Numbers and the Weakly Nonlinear Theory

Katepalli R. Sreenivasan [1,*,†] and Joseph J. Niemela [2,†]

1 Department of Physics, Department of Mechanical and Aerospace Engineering, Courant Institute of Mathematical Sciences, New York University, New York City, NY 10012, USA
2 The Abdus Salam International Center for Theoretical Physics, 34151 Trieste, Italy; niemela@ictp.it
* Correspondence: krs3@nyu.edu; Tel.: +1-347-410-4509
† These authors contributed equally to this work.

Abstract: To provide insights into the challenging problem of turbulent convection, Jack Herring used a greatly truncated version of the complete Boussinesq equations containing only one horizontal wavenumber. In light of later observations of a robust large-scale circulation sweeping through convecting enclosures at high Rayleigh numbers, it is perhaps not an implausible point of view from which to reexamine high-Rayleigh-number data. Here we compare past experimental data on convective heat transport at high Rayleigh numbers with predictions from Herring's model and, in fact, find excellent agreement. The model has only one unknown parameter compared to the two free parameters present in the lowest-order least-squares power-law fit. We discuss why the underlying simplistic physical picture, meant to work at Rayleigh numbers slightly past the critical value of a few thousand, is consistent with the data when the single free parameter in it is revised, over some eleven decades of the Rayleigh number—stretching from about a million to about 10^{17}.

Keywords: turbulent convection; heat transport; high-Rayleigh-number asymptote; ultimate state of convection

Citation: Sreenivasan, K.R.; Niemela, J.J. Turbulent Convection at Very High Rayleigh Numbers and the Weakly Nonlinear Theory. *Atmosphere* **2023**, *14*, 826. https://doi.org/10.3390/atmos14050826

Academic Editors: Boris Galperin, Annick Pouquet and Peter Sullivan

Received: 11 March 2023
Revised: 27 April 2023
Accepted: 27 April 2023
Published: 4 May 2023

1. Introduction

Turbulent thermal convection is a grand problem because of its importance in astrophysical and planetary contexts and ubiquity in engineering applications from the cooling of nuclear reactors to the cooling of computer chips. The scientific paradigm of thermal convection is the so-called Rayleigh–Bénard convection (RBC), in which a layer of viscous fluid between two smooth horizontal plates is heated at the bottom wall and cooled at the top. In practice, the fluid is constrained by side walls that are non-conducting, so the heat input to the bottom plate is communicated to the top plate entirely through the fluid layer. A theoretical analysis of the resulting fluid flow usually incorporates the so-called Boussinesq approximation [1]. Allowing non-Boussinesq effects opens up the problem to many incompletely understood features [1–4].

We had earlier set out to measure the heat transfer due to turbulent convection in a 1 m tall sample space with a diameter-to-height aspect ratio $\Gamma = 1/2$, using cryogenic helium gas as the working fluid [5]. The purpose of using low-temperature helium gas was to enable extremely high Rayleigh numbers (up to 10^{17}) to be attained while keeping conditions nearly Boussinesq (see [2] for a detailed assessment), and to take advantage of the thermal isolation of the apparatus afforded by the cryogenic environment. The details of the experiment and its execution are provided in [2,5] and summarized in the following section. Our goal in this short paper is to compare the analysis of Herring [6,7] with experimental results on heat transport measured in high-Rayleigh-number turbulent convection. While Herring's formula was derived for the immediate supercritical state of convection where nonlinearities are weak, we find that it agrees with experimental data over 11 decades of Rayleigh number—all in the turbulent state. For an introduction to turbulent convection and its scaling, we refer the reader to reference [8].

A few background comments are in order on the traditional framework of RBC. The thermal driving of the flow is the temperature difference between the top and bottom plates, which is measured in terms of the Rayleigh number, defined as

$$Ra = \frac{g\alpha \Delta T H^3}{\nu\kappa}, \tag{1}$$

where g is the acceleration due to gravity, α is the isobaric thermal expansion coefficient, ΔT is the temperature difference across a vertical fluid layer of height H, and ν and κ are the kinematic viscosity and thermal diffusivity of the fluid, respectively. The response of the flow is the heat transport across the fluid height, measured in terms of the so-called Nusselt number, Nu, which is the actual amount of heat transport effected by convection to that possible (for the same ΔT) by thermal conduction alone. One could also include the dependence of Nu on the Prandtl number, $Pr = \nu/\kappa$, so a fundamental problem of thermal convection is to determine the functional dependence of Nu on Ra and Pr. Another response of the flow is the Reynolds number of the convective motion, but we shall not consider it here.

As in most other turbulence problems (and for all many-body problems in 3D), RBC also cannot be solved fully from a theoretical point of view, so there are only scaling theories. There are two schools of thought. That due to Malkus [9] and Spiegel [10] says that

$$Nu = 0.073 Ra^{1/3}, \tag{2}$$

with no dependence on Pr. The underlying physics is the so-called marginal stability of the top and bottom boundary layers [11]. On the other hand, Kraichnan [12] argued that the boundary layers will become irrelevant at *very* high Ra and obtained the explicit formula

$$Nu = CPr^{-1/4}\{Ra/log(Ra)^3\}^{1/2} \tag{3}$$

on the basis of an analysis of the top and bottom boundary layers (which were particularly incompletely understood then). Even though Kraichnan made valiant efforts to obtain the constant C, the details are tenuous and so nothing is lost, at least for our purposes, in regarding it as an unknown constant. See also [13,14] for similar predictions of the Rayleigh number dependence, but they do not go into any boundary layer details.

Because Kraichnan's formula demands very high Ra, the $\frac{1}{2}$-power dependence on Ra (with logarithmic corrections), including an explicit Prandtl number dependence, is thought to represent the "ultimate state" or the "asymptotic state" of RBC. As Spiegel [13] remarked, the difference between formulae (2) and (3) needs to be resolved because it would then suggest the correct physics that operates at high Ra. We will take up this thread shortly.

From a different perspective, the weakly nonlinear theory of Herring, as summarized by Busse [15], with a long pedigree involving Lou Howard, Willem Malkus, Paul Roberts, and Fritz Busse, gives

$$Nu = D\{Ra^{3/2} ln(Ra)^{3/2}\}^{1/5}, \tag{4}$$

where the constant $D = 0.24$ was theoretically calculated by maximizing the heat transport accomplished by a single wavenumber. We now examine the relevance of each formula, (2)–(4), using experimental data described below. We emphasize that the data of [5] have been repeated in part [16], but a completely independent effort would be desirable.

2. Brief Comments on the Experimental Data

As stated above, our goals in the experiments of Ref. [5] were to maximize the Rayleigh number attainable and to place all high Rayleigh numbers within the turbulent regime, for obtaining robust scaling relations. As already stated, cryogenic helium gas was used as the working fluid. The apparatus has been described in the above reference and is briefly repeated here: The sample space was cylindrical in shape and 1 m tall, formed by a thin tubular stainless steel wall 0.267 cm thick, with top and bottom heated plates 3.8 cm in thickness and made of copper annealed under oxygen-free conditions. The thermal conductivity of the copper was on the order of 2 kW m^{-1} K^{-1} at a temperature of roughly 5K. Serpentine thin film heaters were attached to both plates. The top plate was connected to a helium reservoir through an adjustable thermal link and held at constant temperature by means of a resistance bridge and servo. A constant heating was applied to the bottom plate, and constant temperature conditions at the plate were attained after an adequate waiting time. The sample space was insulated using radiation shields held at helium and nitrogen temperatures within a common cryo-pumped vacuum space. Corrections were made for a small adiabatic temperature gradient across the fluid due to the relatively large height. An important parameter is the Biot number, comparing the thermal resistance of the plates to that of the fluid, which remains negligibly small up to Rayleigh numbers of 10^{16} where it is still under 1%. Consequently, the nature of plume generation is not artificially limited by the thermal recovery time of the plates; it is a particular advantage of low-temperature systems. Cryogenic helium gas has other advantages: (1) it has the lowest kinematic viscosity of all known substances and (2) by operating the experiment close to the critical point, the divergence of the specific heat C_P means that $Ra \sim \alpha\rho^2 C_P$, where α is thermodynamically related to C_P, reaches extremely large values. Quoting from [17], where these factors were discussed in a bit more detail, we have: "For non-interacting gases, $\alpha = 1/T$, and so, low temperatures themselves have a particular advantage for buoyancy-driven flows. ...In summary, it is the combination $\alpha/\nu\kappa$ that determines the Rayleigh number...". There are two further advantages in using cryogenic helium: it is possible to stay closer to the Boussinesq approximation than in other fluids while attaining very high values of Ra, and one can achieve excellent thermal isolation as indicated above.

3. Power Laws and Herring's Formula

The experimental data of [5] suggest a power law with a scaling exponent slightly less than 1/3. We showed in [5] that the least squares fit to the data over the entire turbulent range ($10^6 < Ra < 10^{17}$) is

$$Nu = 0.124 Ra^{0.309}. \quad (5)$$

This is indeed the simplest fit possible with two parameters: the amplitude and the slope of a linear fit of $\log(Nu)$ vs. $\log(Ra)$. The fit (see the red dashed line in Figure 1) is excellent for the entire data. Note that there is no ambiguity in the data that the small difference of the exponent from 1/3 is real, perhaps to be regarded as some kind of intermittency correction (see, for example, Ref. [17]), but there could also be other valid reasons for this slight departure (which could be accounted for by introducing a subleading term). Indeed, numerical simulations at high Ra in a slender convection cell [18] suggest an exponent somewhat closer to 1/3, but it certainly appears far from the half-power in Ra.

In Figure 1 we also show as the blue dashed line Herring's prediction for turbulent convection between rigid boundaries, given by Equation (4), with the prefactor $D = 0.06$. The prefactor, the only free parameter in the expression, was obtained by fitting the function to the data. We note that the exponent 0.3 in Equation (4) is also applied to the $\ln(Ra)$ term so that the effective exponent is slightly larger and, in fact, Herring's formula fits the data as well as the power-law fit, Equation (5), as can be seen better in Figure 2. We emphasize that, instead of two constants of the power law, only the amplitude is an adjustable parameter in Herring's formula.

In fact, Figure 2 shows the raw data normalized by both fits and it is clear that each of them is equally good and satisfactory overall. We also note that a considerable substructure to the data exists, which could suggest various changes in the flow, as hinted in the caption to Figure 2, while preserving the same global trend. We shall give in Section 5 a brief interpretation of the empirically determined prefactor in Equation (4).

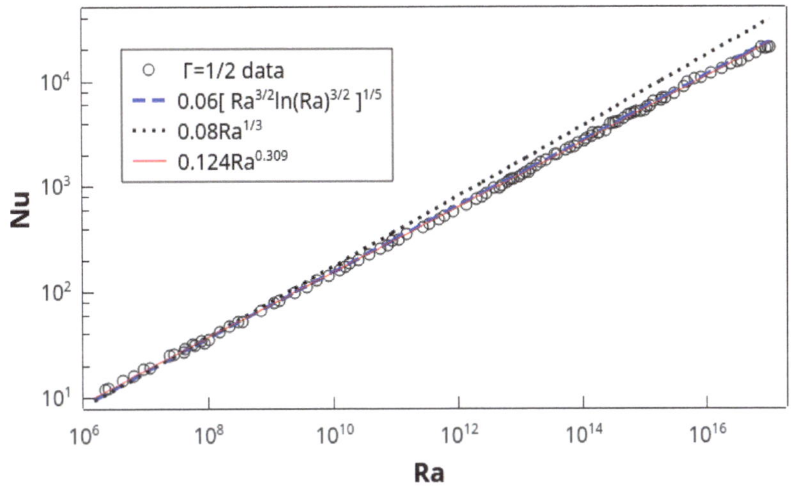

Figure 1. Circles: data obtained with the 1 m tall, aspect ratio 1/2 experiment. Only a subset of the data is plotted to make the lines visible, which would be completely masked by the data. Red solid line: Least squares fit to the raw data (Equation (5)). Dotted line: $Nu \sim Ra^{1/3}$ for comparison. Blue dashed line: Herring's model, Equation (4) with D = 0.06, this being the one free parameter in the theory (see related text, especially Section 5). The blue and red lines are essentially indistinguishable from each other because of their closeness.

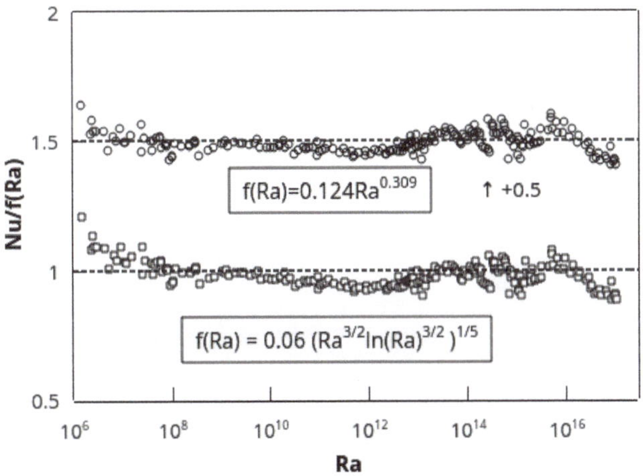

Figure 2. Nu normalized according to Equations (4) and (5). Squares, Nu normalized by Herring's formula (Equation (4)). Circles, Nu normalized by the least-squares fit to the raw data (Equation (5)) with a vertical shift of +0.5 for clarity. For low $Ra < 10^{10}$, say, the slope is smaller than 0.309, reminiscent of the 2/7-th slope discussed in [19]. Our estimate is that the boundary layers in the apparatus of [5] undergo transition at around $Ra = 10^{13}$. The last decade may have been influenced by a non-constant Prandtl number (see next section) and/or moderate non-Boussinesq effects. These features are discussed at some length in [2].

4. Remarks on the Ultimate State

A brief historical account of the "ultimate" state is provided in the Appendix A. Here we are content to compare the performance of the Kraichnan formula with those of the other two; see Figure 3. To bring them all onto one plot, we set $C = 0.04$ in the Kraichnan formula and compressed the vertical scale by a factor of about 5. On this scale the bottom two fits mask the substructure apparent in Figure 2 and are almost perfectly straight and horizontal lines, showing that both Equations (4) and (5) work very well from a global perspective. The Kraichnan formula is obviously far from being successful in reducing the data to a horizontal line, so it is clear that it has no global relevance, unlike the other two. However, the curve appears to flatten for high *Ra*, tantalizingly suggesting a possible approach to the ultimate state. We shall consider this proposition below.

Figure 4 shows an enlarged plot of the Nusselt number against *Ra* for the last decade and a half or so. The best fit to those data is a power-law exponent of 0.317, which is only minutely larger (by about 2.5%) than the global exponent of 0.309 (and quite close to the finding in [18]). No reasonable person would think that the slope is approaching a value of half. From a slightly different perspective, Figure 5 shows the high-*Ra* end of the data of Figure 3; it is clear that the tendency to flatten, which might have been inferred from Figure 3, is an illusion caused by the compressed scale. It is not clear what functional form the normalized data in Figure 5 should take, but, if we fit a power law to the last two decades of *Ra*, it yields a power law with an exponent of −0.033.

An unstated argument sometimes adduced by the adherents of the ultimate state is to point out that the Prandtl number was not constant in the measurements of [5] over the last two or so decades of *Ra*. This behavior was discussed at length in [2] and is reproduced in Figure 6. However, the variation of the Nusselt number on the Prandtl number is very weak for moderate Prandtl numbers in the range encountered here (see, e.g., [20]). Thus, one cannot argue that the rise in Prandtl number is the reason why the data do not approach the half-power. Our conclusion could be different if the interpretation of the data has the benefit of a precise theory for how the heat transport depends on the Prandtl number.

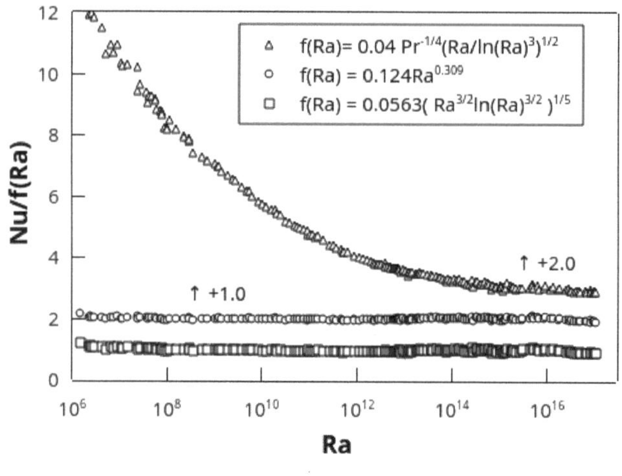

Figure 3. *Nu* normalized according to Equations (4) and (5) as in Figure 2 but with compressed vertical scale to allow for normalization by Kraichnan's formula to appear on the same plot. Squares, Nu normalized by Herring's formula (Equation (4)). Circles, *Nu* normalized by the least-squares fit to the raw data, Equation (5), with a vertical shift of +1.0 for clarity. Triangles, *Nu* normalized according to Kraichnan's formula for high *Ra* and moderate Prandtl number (Equation (3) with $C = 0.04$) with a vertical shift of +2.0 for clarity (see text). Here and in Figure 5, the dependence on the Prandtl number is left implicit in the notation f(*Ra*) as applied to Equation (3).

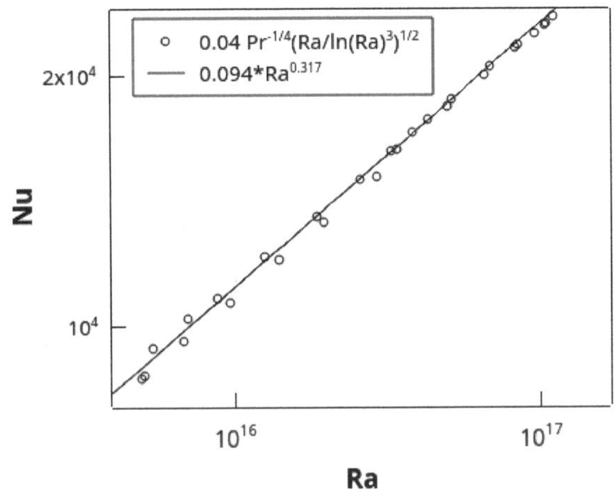

Figure 4. Nu vs. Ra for the highest decade and a half of Ra. Symbols: Nu evaluated using Kraichnan's formula (Equation (3) with $C = 0.04$). Line: Least-squares fit to the data giving a slope of 0.317.

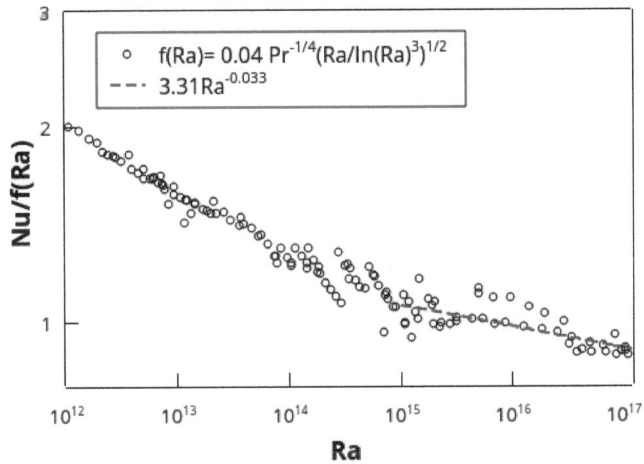

Figure 5. $Nu/\mathrm{f}(Ra)$ vs. Ra. The data are normalized using f(Ra) from Kraichnan's formula (Equation (3) with C = 0.04). Dashed line: least-squares fit to the normalized data for $Ra > 10^{15}$. The slope over the last two decades is -0.033, indicating that the experimental data do not follow Kraichnan's formula. The prefactor 3.31 is dependent on the use of $C = 0.04$ in Equation (3).

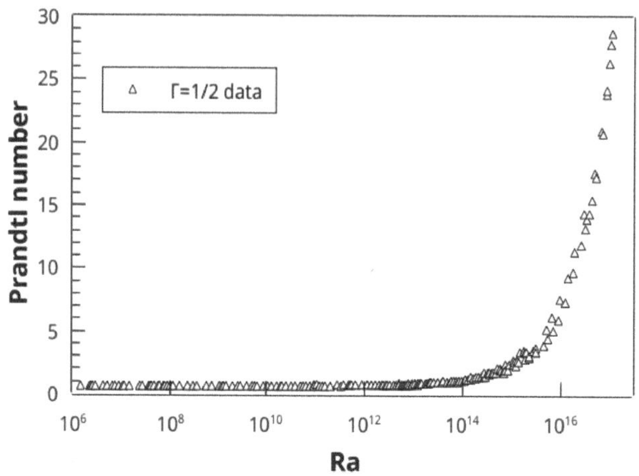

Figure 6. Prandtl number vs. Rayleigh number for the experiment of Ref. [5], showing a considerable variation at high Ra. This variation is an artifact of approaching the critical point of cryogenic helium gas to attain higher and higher Rayleigh numbers. This effect will be even more pronounced if one pushes for high Ra in a smaller apparatus.

5. Discussion

The discussion here is mainly about two questions. The first question is why Herring's formula, derived by optimizing the heat transport by single horizontal wavenumber in the slightly nonlinear supercritical regime, should work well for Rayleigh numbers up to 10^{17}. Note that the critical Rayleigh number is 1708 for RBC, and probably higher for this aspect ratio, say a few thousand. Yet the formula seems to apply for an extraordinary range of Rayleigh numbers within which the flow is decidedly turbulent and the range of wavenumbers excited is continuous.

One can perhaps say that this particular success of the Herring formula is a coincidence but that would be an unimaginative stance (given the large number of decades of Ra over which the agreement occurs). A possible reason for this behavior can perhaps be found in Herring's 1966 paper [7], where he states (for free boundaries) that "The physical picture of free boundary convective process predicted by the model is that of a large-scale motion dominating the central region between the conducting plates. This large-scale motion sweeps with it the temperature fluctuation field whose main variations occur in a thin boundary layer of vertical extent 1/Nu. The horizontal scale of both the dominant motion and the temperature fluctuation field is comparable to the distance between the conducting plates." In actuality, this describes rather well the flow observed between rigid boundaries at high Ra. Indeed, maintaining the same type of structures at the largest-scale circulation that is present at lower values of supercriticality is not, in fact, unusual (see, for example, ref. [21] for the case of turbulent wakes). Perhaps not unrelated is the later observation by Krishnamurti and Howard [22] of a large-scale flow developing and persisting at high Ra, a phenomenon that has been observed widely in turbulent convection. All of this suggests that the same functional form of the relation appropriate to the weakly nonlinear stage remains valid for the turbulent state, except that the prefactor will be different in the two states. This observation is in the spirit of renormalization. While these ideas are not precise, they have been implemented quantitatively in the isotropic and homogeneous flow generated by Gaussian large-scale forcing [23]. If this same program can be extended to convection, it would be possible formally to determine the effective transport coefficient. Until that difficult step is completed, the present observation must be regarded as merely suggestive.

The second question concerns the ultimate state. We are aware that half-power occurs in the presence of roughness [24], when convection occurs in an open-ended tube [25], when the fluid is heated by a body force such as radiation [26], etc. For a more complete enumeration of the possible states of convection, see [27]. As long as the continuum equations hold and the boundary layers on smooth surfaces are intact, it appears to us that their importance to convection will not vanish, and that theories that have no place for viscosity and thermal diffusivity will miss an essential ingredient of the flow. We call attention also to the recent paper by Lindborg [28], which, with some modifications of Kraichnan's theory, arrived at the one-third-power. We repeat that, to our knowledge, there is no compelling evidence to date, experimental or numerical, in favor of Kraichnan's formula. We have made that point explicit with respect to the present data.

6. Conclusions

Jack Herring used a greatly truncated version of the complete Boussinesq equations containing only one horizontal wavenumber and obtained a formula for calculating the Nusselt number in the weakly nonlinear supercritical state slightly past the critical Rayleigh number. The same formula works in the fully turbulent state extending over eleven orders of magnitude, if the numerical prefactor is suitably replaced. In this sense, the formula is less empirical in content than a single power law, which requires two constants to be determined from experiment. We have discussed why such a simplistic physical picture could work. We have also considered Kraichnan's asymptotic formula briefly and presented our reasoning why it does not hold for our experiment. We readily acknowledge that the half-power law exists under different conditions of convection outside the standard RBC.

Author Contributions: Conceptualization, K.R.S.; Methodology, K.R.S. and J.J.N.; Formal analysis, K.R.S. and J.J.N.; Data curation, J.J.N.; Writing—original draft, K.R.S. and J.J.N.; Writing—review & editing, K.R.S. and J.J.N. All authors have read and agreed to the published version of the manuscript.

Funding: This research was was funded by the National Science Foundation grant number DMR-95-29609.

Data Availability Statement: The data used in the paper can be obtained in tabulated form by writing to one of the authors.

Acknowledgments: We had the opportunity for interesting discussions with Jack Herring over the years and in various contexts, and one of the highlights was his model for high-Rayleigh-number convection. Jack was always graceful in his interactions. KRS remembers him as being the same from the very first time he met him in 1977 at NCAR to the last contact. Both of us are grateful that we knew him and have much pleasure in dedicating this short article to his memory. We wish to acknowledge the important role played by Ladislav Skrbek in making the measurements reported in [5].

Conflicts of Interest: The authors declare no conflict of interest.

Appendix A. A Brief History of the Half-Power Law until about the Year 2000

Here, we record the history of the half-power law as best as we know from direct conversations with Ed Spiegel (S henceforth), Charlie Doering [29] and David Goluskin (private communication). When KRS tried to draw K into a discussion on some details in the late 1990's, he was uncharacteristically adamant in his refusal to discuss any part of the paper, even in passing. We have truncated this brief history around the year 2000, but the relevant references of later origin are mentioned in the main body of the paper.

The originator of the half-power is thought to be Kraichnan [12]—to be referred to as K. However, Ed Spiegel, who was K's postdoc at the time, also developed similar ideas, though from the somewhat different perspective of astrophysical flows. A reading of [30] gives one the impression that S was well into his research on the topic by 1960, though there is no mention of the half-power law in [30]. S has said that Malkus persuaded him to omit the mention of the half-power in that report. It is amply clear, however, that S had on hand the insight into the physical ideas that led to the specific result he published [31] a year or so after K; we know from S directly that the delay was caused by issues connected to the health of one of his family members.

It is clear that K and S were influenced by each other's work. K refers to S's work and acknowledges discussions with him but records that the inspiration for his work was Malkus, who had published his one-third-scaling some years earlier [9] (K also cites C.H.B. Priestley and E. Böhm-Vitense). It may be recalled that the physical essence of Malkus' result is that the heat flux is both produced by the thermal boundary layer dynamics and limited by it (thus, H should be irrelevant). As Howard [32] clarified soon after, Malkus' arguments can be cast more elegantly in terms of the marginal instability of the thermal boundary layer. K thought that at very high Rayleigh numbers (which he estimated to be on the order of 10^{24}, essentially unattainable in practice), shear effects at the wall will overwhelm the thermal effects to yield the half-power with corrections of the form $((\log Ra)^{-3/2})$. This log correction is important to K's theory, in deriving which he used a great deal of debatable detail. K was well aware of this shortcoming and honest to state so more than once in his paper.

S's physics was somewhat different. Contrasting with K's work, he is quoted in [30] as saying: "...the best representation is contained in the recent work of K, but at the moment this is a more difficult representation than we are prepared to cope with". S had no explicit use for the walls and the boundary layers (true to his preoccupation with astrophysical flows) and regarded that thermal plumes in the bulk of the flow transported heat; and, whatever their origin, their dynamics in the bulk was the rate-controlling mechanism for transport, thus eliminating ν and κ from consideration.

The primary results of K and S are also different in detail. K's paper is contained in Equation (3) whereas S's result was that

$$Nu = C(Ra.Pr)^{1/2}, \tag{A1}$$

where the coefficient C remained unknown. Aside from the absence of numerical coefficient and log corrections in the above equation, the major difference is the Prandtl number dependence of Nu.

It should be clear from this discussion that K and S influenced each other but followed their separate paths and agenda. Moreover, as discussed above, their results are substantially different as well (despite the commonality of the half-power). Our opinion is that they each deserve the credit that came their way. To our knowledge, S always gave credit to K; only when pressed would he acknowledge his unhappiness that K's result was published before his.

The word "ultimate" to describe the half-power was first coined in the title of their paper by Chavanne et al. [14], who constructed an argument for log correction in the form $(\log Ra)^{-3}$. It suggests that the details in K's theory leading to his particular form of log correction is not sacrosanct.

Another major development is the upper-bound theory [33], whose result is

$$Nu \leq 0.167 Ra^{1/2}. \tag{A2}$$

It is clear that Equation (A1) is at variance with the upper bound if Pr increases indefinitely. Thus, one should regard Equation (A1) as having been meant only for $Pr \leq 1$.

References

1. Tritton, D.J. *Physical Fluid Dynamics (Oxford Science Publications)*; Clarendon Press: Oxford, UK, 1988.
2. Niemela, J.J.; Sreenivasan, K.R. Confined turbulent convection. *J. Fluid Mech.* **2003**, *481*, 355–384. [CrossRef]
3. Pandey, A.; Schumacher, J.; Sreenivasan, K.R. Non-Boussinesq low-Prandtl-number convection with temperature-dependent thermal diffusivity. *Astrophys. J.* **2021**, *907*, 56. [CrossRef]
4. Pandey, A.; Schumacher, J.; Sreenivasan, K.R. Non-Boussinesq convection at low Prandtl numbers relevant to the Sun. *Phys. Rev. Fluids* **2021**, *6*, 100503. [CrossRef]

5. Niemela, J.J.; Skrbek, L.; Sreenivasan, K.R.; Donnelly, R.J. Turbulent convection at very high Rayleigh numbers. *Nature* **2000**, *404*, 837–840. [CrossRef]
6. Herring, J.R. Investigation of problems in thermal convection. *J. Atmos. Sci.* **1962**, *20*, 325–338. [CrossRef]
7. Herring, J.R. Some analytical results in the theory of thermal convection. *J. Atmos. Sci.* **1966**, *23*, 672–677. [CrossRef]
8. Ahlers, G. Turbulent convection. *Physics* **2009**, *2*, 74. [CrossRef]
9. Malkus, W.V.R. The heat transport and spectrum of thermal turbulence. *Proc. R. Soc. Lond. A* **1954**, *225*, 196–212.
10. Spiegel, E.A. On the Malkus theory of turbulence. In Proceedings of the 1961 Marseille Meeting on Turbulence, Marseille, France, 28 August–2 September 1961; pp. 181–201.
11. Howard, L. Convection at high Rayleigh number. In Proceedings of the 11th International Congress on Applied Mechanics, Munich, Germany, 30 August–5 September 1964; pp. 1109–1115.
12. Kraichnan, R.H. Turbulent thermal convection at arbitrary Prandtl number. *Phys. Fluids.* **1962**, *5*, 1374–1389. [CrossRef]
13. Spiegel, E.A. Convection in stars, I. Basic Boussinesq convection. *Annu. Rev. Astron. Astrophys.* **1971**, *9*, 323–352. [CrossRef]
14. Chavanne, A.X.; Chilla, F.; Castaing, B.; Hébral, B.; Chabaud, B.; Chaussy, J. Observation of the ultimate regime in Rayleigh-Bénard convection. *Phys. Rev. Lett.* **1997**, *79*, 3648–3651. [CrossRef]
15. Busse, F.H. Nonlinear properties of thermal convection. *Rep. Prog. Phys.* **1978**, *41*, 1929–1966. [CrossRef]
16. Niemela, J.J.; Babuin, S.; Sreenivasan, K.R. Turbulent rotating convection at high Rayleigh and Taylor numbers. *J. Fluid Mech.* **2010**, *649*, 509–522. [CrossRef]
17. Niemela, J.J.; Sreenivasan, K.R. The use of cryogenic helium for classical turbulence: Promises and hurdles. *J. Low Temp. Phys.* **2006**, *143*, 163–212. [CrossRef]
18. Iyer, K.P.; Scheel, J.D.; Scumacher, J.; Sreenivasan, K.R. Classical 1/3 scaling of convection holds up to $Ra = 10^{15}$. *Proc. Natl. Acad. Sci. USA* **2020**, *117*, 7594–7598. [CrossRef]
19. Castaing, B.; Gunaratne, G.; Heslot, F.; Kadanoff, L.; Libchaber, A.; Thomae, S.; Wu, X.-Z.; Zaleski, S.; Zanetti, G. Scaling of hard thermal turbulence in Rayleigh-Bénard convection. *J. Fluid Mech.* **1989**, *204*, 1–30. [CrossRef]
20. Grossmann, L.; Lohse, D. Thermal convection for large Prandtl numbers. *Phys. Rev. Lett.* **2001**, *86*, 3316–3319. [CrossRef]
21. Roshko, A. Experiments on the flow past a circular cylinder at very high Reynolds number. *J. Fluid Mech.* **1961**, *10*, 345–356. [CrossRef]
22. Krishnamurti, R.; Howard, L.N. Large scale flow generation in turbulent convection. *Proc. Natl. Acad. Sci. USA* **1981**, *78*, 1981–1985. [CrossRef]
23. Sreenivasan, K.R.; Yakhot, V. Dynamics of three-dimensional turbulence from Navier-Stokes equations. *Phys. Rev. Fluids* **2021**, *6*, 104604. [CrossRef]
24. Toppaladoddi, S.; Succi, S.; Wettlaufer, J.S. Roughness as a route to the ultimate regime of thermal convection. *Phys. Rev. Lett.* **2017**, *118*, 074503. [CrossRef]
25. Arakeri, J.H.; Avilla, F.E.; Dada, J.M.; Tovar, R.O. Convection in a long vertical tube due to unstable stratification—A new type of turbulent flow? *Curr. Sci.* **2000**, *79*, 859–866.
26. Lepot, S.; Aumaitre, S.; Gallet, B. Radiative heating achieves the ultimate regime of thermal convection. *Phys. Rev. Lett.* **2018**, *115*, 8937–8941. [CrossRef]
27. Grossmann, S.; Lohse, D. Scaling in thermal convection: A unifying theory. *J. Fluid Mech.* **2000**, *407*, 22–56. [CrossRef]
28. Lindborg, E. Scaling in Rayleigh-Bénard convection. *J. Fluid Mech.* **2023**, *956*, A34. [CrossRef]
29. Doering, C.R. Turning up the heat in turbulent theermal convection. *Proc. Natl. Acad. Sci. USA* **2020**, *117*, 9671–9673. [CrossRef]
30. Batchelor, G.K. Considerations of convective instability from the viewpoint of physics. In *Aerodynamics Phenomena in Stellar Atmospheres, Proceedings of the 4th Symposium on Cosmical Gas Dynamics, Varenna, Italy, 18–30 August 1960*; Thomas, R.N., Ed.; Supplemento del Nuovo Cimento XXII; Nicola Zanichelli: Bologna, Italy, 1960; pp. 385–402.
31. Spiegel, E.A. A generalization of the mixing-length theory of thermal convection. *Astrophys. J.* **1963**, *138*, 216–225. [CrossRef]
32. Howard, L.N. Heat transport by turbulent convection. *J. Fluid Mech.* **1963**, *17*, 405–432. [CrossRef]
33. Doering, C.R.; Constantin, P. Variational bounds on energy dissiparion in incompressible flows. III Convection. *Phys. Rev. E* **1996**, *53*, 5957–5981. [CrossRef]

atmosphere

MDPI

Review

Jackson R. Herring and the Statistical Closure Problem of Turbulence: A Review of Renormalized Perturbation Theories

David McComb

SUPA School of Physics and Astronomy, University of Edinburgh, Peter Guthrie Tait Road, Edinburgh EH9 3JZ, UK; wdm@ph.ed.ac.uk

Abstract: The pioneering applications of the methods of theoretical physics to the turbulence statistical closure problem are summarised. These are: the direct-interaction approximation (DIA) of Kraichnan, the self-consistent-field theory of Edwards, and the self-consistent-field theory of Herring. Particular attention is given to the latter, in terms of its elegance and its pedagogical value. We then concentrate on the assessment of these theories and take the historical route of Kraichnan's diagnosis of the failure of DIA, followed by Edwards's analysis of the failure of his self-consistent theory, when compared to the Kolmogorov spectrum. As all three theories are closely related, these analyses also shed light on Herring's theory. The second-generation theories that grew out of this assessment are then discussed. First, there were the Lagrangian theories, initially stemming from the work of Kraichnan and Herring, and later the purely Eulerian local energy-transfer (LET) theory. The latter is significant because its development exposes the underlying problems with the pioneering theories in terms of the basic physics of the inertial energy transfer. In particular, later work allows us to assign a unified explanation of the incompatibility of all three pioneering theories with the Kolmororov spectrum, in that they are all Markovian approximations (in wavenumber) to the non-Markovian phenomenon of fluid turbulence. In the interests of completeness, we briefly review the formalisms of Wyld and Martin, Siggia, and Rose. More recent developments are also discussed, in order to bring the subject up to the present day.

Keywords: isotropic turbulence; statistical closure approximations; renormalization

Citation: McComb, D. Jackson R. Herring and the Statistical Closure Problem of Turbulence: A Review of Renormalized Perturbation Theories. *Atmosphere* **2023**, *14*, 827. https://doi.org/10.3390/atmos14050827

Academic Editors: Boris Galperin, Annick Pouquet, Peter Sullivan and Paul D. Williams

Received: 15 March 2023
Accepted: 28 April 2023
Published: 4 May 2023

1. Introductory Remarks

In the 1960s, Jack Herring was one of the giants of turbulence theory. The other two were Bob Kraichnan and Sam Edwards. Their revolutionary contributions transformed the study of turbulence, taking it into the realm of theoretical physics. In chronological order, Kraichnan published his direct-interaction approximation (DIA) in 1959 [1] and followed this in 1964 with a numerical computation of the DIA equations for the two-point, two-time covariance and the renormalised response function, for freely decaying turbulence [2]. Edwards derived a form of the Liouville equation for the turbulence probability distribution functional (pdf) and obtained a self-consistent-field solution for the stationary, single-time convariance in 1964 [3] and followed this with some further remarks about his theory [4]. His approach was strongly influenced by a consideration of Brownian motion, but in 1965 Herring, using a much more abstract form of self-consistent-field theory, solved the Edwards–Liouville equation for the probability distribution for the stationary case [5], and extended this to the non-stationary case the following year [6].

When I began my postgraduate research, under the supervision of Sam Edwards in October 1966, the initial excitement about the new theories was tending to die down. Important points had been established about the relationships between the three approaches, but there were still some unanswered questions. It was clear that all three theories led to the same energy-balance equation, but there were differences between their results for the renormalized response function. Attempting to understand these differences was one of the active topics of the time. However, the dominant problem was to understand why none of the three theories was compatible with the Kolmogorov–Obukhov'$-5/3$' wavenumber spectrum [7]. This aspect dominated the subject to such an extent that the 1964 computation of the DIA, which suggested that these theories were capable of making good qualitative and quantitative predictions about turbulence, tended to be overlooked. Indeed, my own first project was to rectify the Edwards theory by choosing the renormalized response such that the turbulent entropy was maximised. This did lead to a result which was compatible with the Kolmogorov–Obukhov spectrum, but the qualitative aspects were not satisfactory [8].

In the rest of this article, we shall review the subject in detail, discussing its failures and successes, and indicate how it has developed up until the present day. However, before finishing these preliminary remarks, I should point out two other contributions by Jack Herring. First, he introduced the use of negative damping to force the turbulence, rather than using stirring forces. Many years on, this is an important method of forcing numerical simulations of the Navier–Stokes equations. Secondly, he invented a simplified notation, combining wavenumber and tensor index labels, which has been used by others working in this field. For instance, Leslie made extensive use of it in his book [9], Qian used it in his work on closures [10] (see also the article by Shi [11]), and it is also used in Section 5.3 of the present work.

2. A Concise Overview of Statistical Closures

The statistical closure problem was first formulated for turbulent shear flows by Reynolds in the 1890s. He showed that the equation for the mean velocity $\bar{U}$ contains the unknown covariance of two fluctuating velocities $\langle uu \rangle$: the Reynolds stress. It was later formulated for isotropic turbulence by Taylor in the 1930s. In this case, the mean velocity is taken to be zero and the equation for the correlation of two velocities $\langle uu \rangle$ is found to contain the unknown correlation of three velocities $\langle uuu \rangle$. The equation for $\langle uuu \rangle$ contains the unknown fourth-order moment $\langle uuuu \rangle$, and so on. The problem is therefore seen as an open hierarchy of statistical equations (one-point for Reynolds, two-point for Taylor) which requires some means of closure. Theories used in engineering applications, such as the eddy-viscosity, mixing-length, and n-equation models, are all effectively statistical closure approximations for the single-point problem. In contrast, the Heisenberg eddy-viscosity and the quasi-normality theory were effectively closure approximations for the two-point problem. Further reading on this topic can be found in the following books [9,12–14].

The first *formal* treatment of the closure problem was the quasi-normality hypothesis (Proudman and Reid [15], Tatsumi [16]). The basic idea was to solve the next equation in the hierarchy for $\langle uuu \rangle$ by factorizing $\langle uuuu \rangle$ in terms of $\langle uu \rangle \times \langle uu \rangle$. This then gives a closed set of equations for the covariance. However, quasi-normality failed when computed numerically in the 1960s, as it predicted negative spectra: it was not *physically realizable* [17,18].

The attempt at quasi-normality illustrates an important aspect of all statistical theories of turbulence and indeed of all statistical field theories. In *all* theories, the averages are evaluated using the tractable properties of the Gaussian or normal distribution. In particular, even-order moments can be factorised into products of pair-correlations while odd-order moments cannot be evaluated that way, because they vanish identically by symmetry. In quasi-normality the fourth-order moment is evaluated in terms of pair correlations, whereas the triple moment is calculated from the resulting closed equation by inverting the linear operator of the Navier–Stokes equation. *Some such manoeuvre is needed in any theory if it is to be successful, with the degree of success depending on the choice of manoeuvre.*

2.1. The Pioneering Renormalized Perturbation Theories of Turbulence

The beginning of the modern age of turbulence theory was signalled by the publication in 1959 of Kraichnan's direct-interaction approximation (DIA) [1]. This was both a mean-field theory and a new kind of perturbation theory (it is an example of a renormalized perturbation theory, such as occurs in quantum field theory or many-body problems).

At the time, its most important feature was seen as being its physical realizability. It did not have the catastrophic behaviour of quasi-normality. Shortly afterwards, Wyld [19] showed that the DIA could be recovered from a conventional renormalized perturbation theory, thus establishing the validity of the mean-field assumption by Kraichnan. This work was couched in the language (and used the techniques) of quantum field theory, including the introduction of diagrams analogous to Feynmann diagrams. Although of considerable pedagogical interest, it contained some procedural errors. A discussion of this aspect may be found in Appendix A.

In 1964, Edwards published his self-consistent-field (SCF) theory [3], which he also described as a *random phase approximation*, and which was based on the probability distribution functional (pdf) of the fluctuating velocities. He began by introducing a model system with a Gaussian pdf and the same velocity covariance as the turbulence system. Then, he introduced a perturbation expansion in terms of the difference between the model system and the real system. The result was a set of equations for the single-time covariance and a renormalized response function. This result was later shown by Kraichnan to be cognate to the DIA theory and, like it, maintained the properties of the inertial term, such as conservation of energy [20].

Another attempt at a self-consistent-field theory was made by Herring [5] in 1965. This was a more abstract approach than that of Edwards, although it also started from the Edwards form of the Liouville equation. In strategy it lay somewhere between the Edwards SCF and the Kraichnan DIA, in that it formally renormalized a 'bare' operator. Herring's basic approach involved the introduction of single-mode projectors and single-mode distributions. Then, the perturbation expansion was in terms of the difference between the approximate single-mode forms and the exact coupled forms. There are similarities and differences between the two variants of SCF, and they both lead to the same form of covariance equation. However, the response function in Herring's theory was the same as the time-independent DIA form, and hence differed from that of Edwards. In the following year, Herring was able to extend his SCF theory to the two-time case [6], the result being a theory which was closely related to Kraichnan's DIA, rather than the Edwards theory.

2.2. The Later Renormalized Perturbation Theories of Turbulence

As early as 1961, Kraichnan realized that the experimental results for turbulence spectra (later published as [21]) supported the Obhukov–Kolmogorov prediction of $k^{-5/3}$, rather than his DIA prediction of $k^{-3/2}$. He began work on this and ultimately concluded that the DIA type of theory could not predict the time dependence of two-time correlations due to a violation of what he called *random Galilean invariance*. This led Kraichnan to formulate his theories in a mixed Eulerian–Lagrangian coordinate system [22]. This was not entirely satisfactory, and Herring and Kraichnan extended it to a theory based on the rate of strain, rather than on the velocity field [23]. This inspired a variety of Lagrangian renormalized perturbation theories, continuing up to the present time: [24–26]. We discuss these theories further in Section 6.1.

This was not seen as a problem for single-time correlations by Kraichnan, as such theories were invariant under the postulated random Galilean transformation. However, it should be recalled that single-time theories, such as those of Edwards [3] and Herring [5], are also incompatible with the Kolmogorov spectrum, as indeed also is the cognate single-time form of the DIA. The resolution of this problem was found in 1974, in the form of the local energy-transfer (LET) theory, and this is something that we discuss in greater detail in Section 6.2.

General background reading can be found in the books by Beran [12], Leslie [9], Monin and Yaglom [13], McComb [14], Sagaut and Cambon [27], McComb [28], Verma [29], and the review by McComb [30].

3. Basic Equations and Notation

In this review we follow a modern unified scheme of notation and employ the letter C for correlations and covariances, along with the letter R for response functions.

We consider the solenoidal Navier–Stokes equation (NSE) for the velocity field $u_\alpha(\mathbf{k},t)$ in wavenumber (k) space (see either [14] or [28]) as:

$$\left(\frac{\partial}{\partial t}+\nu k^2\right)u_\alpha(\mathbf{k},t)=M_{\alpha\beta\gamma}(\mathbf{k})\int d^3j u_\beta(\mathbf{j},t)\, ug(\mathbf{k}-\mathbf{j},t)+f_\alpha(\mathbf{k},t), \tag{1}$$

where ν is the fluid kinematic viscosity, $f_\alpha(\mathbf{k},t)$ is an arbitrarily chosen stirring force which we have to specify. Note that we follow Edwards [3] (and indeed Kolmogorov before him [31]) and use Greek letters to denote the usual Cartesian tensor indices, where these take the values $1,2$, or 3, as appropriate for a three-dimensional space. There should be no confusion with the conventional use of Greek indices in Minkowski four-space. The inertial transfer operator $M_{\alpha\beta\gamma}(\mathbf{k})$ is given by

$$M_{\alpha\beta\gamma}(\mathbf{k})=(2i)^{-1}\left[k_\beta P_{\alpha\gamma}(\mathbf{k})+k_\gamma P_{\alpha\beta}(\mathbf{k})\right], \tag{2}$$

where $i=\sqrt{-1}$, while the projector $P_{\alpha\beta}(\mathbf{k})$ is expressed in terms of the Kronecker delta as

$$P_{\alpha\beta}(\mathbf{k})=\delta_{\alpha\beta}-\frac{k_\alpha k_\beta}{k^2}. \tag{3}$$

Note that the use of the projector ensures that the velocity field remains solenoidal.

The covariance of the fluctuating velocity field may be introduced as

$$C_{\alpha\beta}(\mathbf{k},\mathbf{k}';t,t')=\langle u_\alpha(\mathbf{k},t)u_\beta(\mathbf{k}',t')\rangle, \tag{4}$$

and for isotropic, homogeneous turbulence we may write this as

$$\langle u_\alpha(\mathbf{k},t)u_\beta(-\mathbf{k},t')\rangle=P_{\alpha\beta}(\mathbf{k})\delta(\mathbf{k}+\mathbf{k}')C(k;t,t'), \tag{5}$$

where we anticipate the effects of homogeneity in writing the left-hand side. As is usual, the angle brackets $\langle\ldots\rangle$ denote the operation of taking an average. We shall discuss this further in the next sub-section, in connection with the introduction of the probability distribution functional.

If we consider the case $t=t'$, then we may introduce the spectral density function:

$$C(k,t)\equiv C(k;t,t), \tag{6}$$

and the energy spectrum:

$$E(k,t)=4\pi k^2 C(k,t). \tag{7}$$

We will also find it convenient to introduce the contracted notation:

$$C_k\equiv C(k), \tag{8}$$

for use with later time-independent theories.

4. The Statistical Formulation

4.1. The Stirring Forces

Edwards [3] introduced stirring forces in order to define the ensemble (and, when required, to study stationary turbulence). These are denoted by $f_\alpha(\mathbf{k},t)$ and can be added to the right-hand side of the equation of motion, as we have done here in Equation (1). These forces are chosen to be isotropic, homogeneous and (in order to maintain incompressibility) solenoidal. He considered random forces with a multivariate normal probability distribution, such that the associated functional integrals were analytically tractable.

He assumed that the autocorrelation of the forces was instantaneous in time and we represent this by choosing the time autocorrelation to be a delta function. A form of correlation which satisfies all these requirements is:

$$\langle f_\alpha(\mathbf{k},t) f_\beta(\mathbf{k}',t') \rangle = P_{\alpha\beta}(\mathbf{k}) F(k) \delta(\mathbf{k}+\mathbf{k}') \delta(t-t'). \tag{9}$$

Here, $F(k)$ is a spectral energy density which is related to the rate at which the force does work on the fluid. The rate at which the stirring forces do work on the fluid can be shown to take the form [3,32]:

$$\langle f_\alpha(\mathbf{k},t) u_\sigma(\mathbf{k}',t) \rangle = P_{\alpha\sigma}(\mathbf{k}) F(k) \delta(\mathbf{k}+\mathbf{k}'). \tag{10}$$

We may also introduce the energy-injection spectrum of the stirring forces $W(k)$ as:

$$W(k) = 4\pi k^2 F(k), \tag{11}$$

with associated rate of doing work ε_W, where:

$$\varepsilon_W = \int_0^\infty W(k) dk. \tag{12}$$

For stationary flows ε_W must be equal to the dissipation rate ε.

4.2. The Closure Problem

An equation for the covariance $C(k;t,t')$ may be obtained in the usual way. Multiply each term in Equation (1) by $u_\alpha(-\mathbf{k},t')$ and take the average, to obtain:

$$\left(\frac{\partial}{\partial t} + \nu k^2\right) C(k;t,t') = \tfrac{1}{2} M_{\alpha\beta\gamma}(\mathbf{k}) \int d^3 j \langle u_\beta(\mathbf{j},t) u_\gamma(\mathbf{k}-\mathbf{j},t) u_\alpha(-\mathbf{k},t') \rangle, \tag{13}$$

where we have used Equations (3) and (5), cancelled the factor $\delta(\mathbf{k}+\mathbf{k}')$ across, and invoked isotropy, along with the property $Tr\, P_{\alpha\beta}(\mathbf{k}) = 2$. We can also write this in the compact form:

$$\left(\frac{\partial}{\partial t} + \nu k^2\right) C(k;t,t') = P(k;t,t'), \tag{14}$$

where $P(k;t,t')$ is just the right-hand side of Equation (13). The problem of expressing this in terms of the covariance is the well-known statistical closure problem.

Further, using Equation (1), we can obtain an equation describing the energy balance. On the time diagonal, the covariance equation takes the form for isotropic turbulence:

$$\begin{aligned} \left(\frac{\partial}{\partial t} + 2\nu k^2\right) C(k,t) &= Re\left[M_{\alpha\beta\gamma}(\mathbf{k}) \int \mathrm{d}^3 j \langle u_\beta(\mathbf{j},t) u_\gamma(\mathbf{k}-\mathbf{j},t) u_\alpha(-\mathbf{k},t) \rangle \right] \\ &+ F(k), \end{aligned} \tag{15}$$

where $F(k)$ is the energy-injection spectral density, as defined in terms of the stirring forces by Equation (10). Again, we can write this in a more compact form as:

$$\left(\frac{\partial}{\partial t} + 2\nu k^2\right) C(k;t,t) = Q(k;t,t) + F(k), \tag{16}$$

where $Q(k;t,t)$ can be determined by comparison with the right-hand side of Equation (15).

The energy spectrum is related to the spectral density by Equation (7). Accordingly, if we multiply Equation (15) across by $4\pi k^2$, and rearrange terms, the governing equation of the energy spectrum takes the form:

$$\frac{\partial E(k,t)}{\partial t} = W(k) + T(k,t) - 2\nu k^2 E(k,t), \tag{17}$$

where the energy-transfer spectrum is $T(k,t) = 4\pi k^2 Q(k;t,t) \equiv 4\pi k^2 Q(k,t)$. This equation for the energy spectrum is well known, and is nowadays sometimes referred to as the *Lin equation* [27,28]. More detailed discussions of these equations can be found in Chapter 3 of the book [28].

5. The Pioneering Theories of Kraichnan, Edwards, and Herring

These theories are presented in chronological order of their first publication.

5.1. The Direct-Interaction Approximation of Kraichnan

The starting point for Kraichnan was the introduction of the infinitesimal response function $\widehat{R}_{\alpha\beta}(\mathbf{k},t)$, which arises when one assumes that a fluctuation in the stirring forces. thus:

$$f_\alpha(\mathbf{k},t) \to f_\alpha(\mathbf{k},t) + \delta f_\alpha(\mathbf{k},t), \tag{18}$$

leads to a corresponding fluctuation in the velocity field, thus:

$$u_\alpha(\mathbf{k},t) \to u_\alpha(\mathbf{k},t) + \delta u_\alpha(\mathbf{k},t). \tag{19}$$

The infinitesimal response function is then defined by the relationship:

$$\delta u_\alpha(\mathbf{k},t) = \int_{-\infty}^{t} \widehat{R}_{\alpha\beta}(\mathbf{k};t,t')\delta f_\beta(\mathbf{k},t')dt', \tag{20}$$

and its governing equation is obtained by functional differentiation of the Navier–Stokes equation.

Kraichnan obtained his theory by solving this governing equation simultaneously with the NSE, using perturbation theory. It was pointed out by Leslie (see page 56 of [9]) that the DIA was obtained by assuming that $\widehat{R}$ and u are uncorrelated, hence taking the step:

$$\langle \widehat{R}(t,t')u(t)u(t')\rangle = \langle \widehat{R}(t,t')\rangle\langle u(t)u(t')\rangle = R(t,t')\langle u(t)u(t')\rangle, \tag{21}$$

thus, introducing the ensemble-averaged response function $R(t,t') = \langle \widehat{R}(t,t')\rangle$.

This makes the DIA a form of mean-field theory, in that we replace the instantaneous response function by its average value. In fact, Kraichnan's result can be derived directly from conventional renormalized perturbation theory, as given by Wyld [19] (also see page 228 of [14]), thus validating the mean-field assumption. Applying these methods to the equations for the response function and the velocity covariance, leads to the DIA theory.

The governing equation for the response function is found to be,

$$\begin{aligned}
&\left[\frac{\partial}{\partial t} + \nu k^2\right] R(k;t,t') \\
&+ \int d^3j L(\mathbf{k},\mathbf{j}) \int_{t'}^{t} dt'' R(k;t'',t')R(j;t,t'')C(|\mathbf{k}-\mathbf{j}|;t,t'') \\
&= \delta(t-t').
\end{aligned} \tag{22}$$

The second-order equation for the velocity covariance is:

$$\begin{aligned}
&\left[\frac{\partial}{\partial t} + \nu k^2\right] C(k;t,t') \\
&= \int d^3j L(\mathbf{k},\mathbf{j}) \left[\int_0^{t'} ds R(k;t',s)C(j;t,s)C(|\mathbf{k}-\mathbf{j}|;t,s)\right. \\
&- \left.\int_0^{t} ds R(j;t,s)C(k;s,t')C(|\mathbf{k}-\mathbf{j}|;t,s)\right];
\end{aligned} \tag{23}$$

and on the time diagonal:

$$\begin{aligned}&\left(\frac{\partial}{\partial t}+2\nu k^2\right)C(k,t)\\&=2\int d^3jL(\mathbf{k},\mathbf{j})\int_0^t dsR(k;t,s)R(j;t,s)R(|\mathbf{k}-\mathbf{j}|;t,s)\times\\&[C(j,s)C(|\mathbf{k}-\mathbf{j}|,s)-C(k,s)C(|\mathbf{k}-\mathbf{j}|,s)].\end{aligned}\tag{24}$$

The coefficient $L(\mathbf{k},\mathbf{j})$ is defined as:

$$L(\mathbf{k},\mathbf{j})=-2M_{\alpha\beta\gamma}(\mathbf{k})M_{\beta\alpha\delta}(\mathbf{j})P_{\gamma\delta}(\mathbf{k}-\mathbf{j}),\tag{25}$$

and can be evaluated as:

$$L(\mathbf{k},\mathbf{j})=-\frac{[\mu(k^2+j^2)-kj(1+2\mu^2)](1-\mu^2)kj}{k^2+j^2-2kj\mu},\tag{26}$$

where μ is the cosine of the angle between the vectors $\mathbf{k}$ and $\mathbf{j}$. It should be noted that these equations have been rendered in the formalism due to Edwards, rather than that of Kraichnan. For a discussion of this point, see the book [28].

These equations are an exact second-order truncation of the renormalized perturbation theory. Their derivation can rely on either the topology of Feynmann-type diagrams or reversion of power series. They satisfy all the required symmetries and in particular the closure conserves energy, displaying the correct antisymmetric behaviour of the transfer spectrum.

They can predict the free decay of isotropic turbulence, without invoking arbitrary constants, in quite good agreement with experiment and numerical simulation.

5.2. The Self-Consistent-Field (SCF) Theory of Edwards

Edwards used the prescription of the stirring forces to define a turbulence ensemble and made this the basis of a statistical mechanics approach. Taking these forces to have a multivariate normal distribution, and the properties shown in Equation (9), he obtained the mean distribution of the velocities, P, by averaging against the distribution of the forces and in this way obtained a generalization of the Liouville equation to the turbulent case, as:

$$\begin{aligned}\frac{\partial P}{\partial t}=-\sum_{\mathbf{k}}\frac{\partial}{\partial u_\alpha(k)}\Big\{&-\nu k^2u_\alpha(k)+\\&+\sum_{\mathbf{j}}M_{\alpha\beta\gamma}(\mathbf{k})u_\beta(\mathbf{j})u_\gamma)(\mathbf{k}-\mathbf{j})-F(k)\frac{\partial}{\partial u_\alpha(-\mathbf{k})}\Big\}P,\end{aligned}\tag{27}$$

where $F(k)$ is the force spectrum given by Equation (10). Although the non-linearity of the velocity field is incorporated, it is worth bearing in mind the fact that this equation is a linear equation for P and therefore its perturbation treatment is quite a different matter from that of the Navier–Stokes equation as discussed in the preceding section.

Edwards considered the stationary case, so we drop the time derivative in Equation (27) and, for later convenience, rearrange it as:

$$\begin{aligned}&\sum_{\mathbf{k}}\frac{\partial}{\partial u_\alpha(k)}\left\{\nu k^2u_\alpha(k)+F(k)\frac{\partial}{\partial u_\alpha(-\mathbf{k})}\right\}P=\\&\qquad-\sum_{\mathbf{k}}\sum_{\mathbf{j}}M_{\alpha\beta\gamma}(\mathbf{k})u_\beta(\mathbf{j})u_\gamma)(\mathbf{k}-\mathbf{j})\frac{\partial P}{\partial u_\alpha(k)}.\end{aligned}\tag{28}$$

In order to explain the perturbation approach, we introduce a more symbolic notation and write Equation (28) as:

$$LP-VP=0,\tag{29}$$

where the detailed forms of the operators L and V can be deduced from comparison with the above equation. In doing this, we are introducing the more symbolic notation of Herring [5], and this will also be useful when we discuss Herring's actual theory in the next section.

Edwards then sought a solution of Equation (29) in the form of an expansion in a bookkeeping parameter λ, which is put equal to one at the end of the calculation, and is used to order terms in an iterative procedure. For this purpose it is taken to be superficially of the same order as the non-linear terms in the NSE. Thus, expanding the probability distribution P as follows:

$$P = P_0 + \lambda P_1 + \lambda^2 P_2 + \ldots, \tag{30}$$

where the zero-order coefficient satisfies

$$L_0 P_0, \tag{31}$$

while the basis operator L_0 remains to be determined.

This is the key step in the Edwards theory and we will consider it presently. First, we introduce the general procedure, which begins by adding and subtracting $L_0 P$ from Equation (29) to obtain

$$L_0 P - VP - (L_0 - L)P = 0. \tag{32}$$

Further, we can associate the first order in λ with the term VP, as it contains the non linearity of the NSE, second order in λ with $(L_0 - L)P$, as it represents the correction to the perturbation procedure. Hence, we can base the perturbation theory on Equation (32), rewritten as

$$L_0 P - \lambda VP - \lambda^2 (L_0 - L)P = 0, \tag{33}$$

where these assignments can later be shown to be self-consistent.

Now, we substitute the expansion Equation (30) into Equation (33) and equate coefficients of powers of λ. The result at zero order is just Equation (31), while at first order we have

$$L_0 P_1 = VP_0, \tag{34}$$

and at second order

$$L_0 P_2 = VP_1 + (L_0 - L)P_0; \tag{35}$$

and so on, to any order.

Logically, the next step is to decide on a form for the zero-order operator L_0, but in fact it is more intuitive to decide on the zero-order distribution first, and to choose this to be

$$P_0 = N \exp\left\{ -\sum_{\mathbf{k}} \frac{u_\alpha(\mathbf{k}) u_\alpha(-\mathbf{k})}{C(k)} \right\}, \tag{36}$$

where N is the normalization and $C(k)$ is the time-independent covariance of the velocity field. This form of P_0 is chosen to be Gaussian, because we need it to be mathematically tractable. It also means that the zero-order distribution is required to give the *exact* covariance. As formally the exact distribution leads to the exact covariance, we note that P_1 gives zero by symmetry and hence P_2 is required to give zero for self-consistency.

Edwards was guided by the statistical theory of Brownian motion. For instance, he saw the stirring forces as being analogous to the force due to random molecular impacts on the colloidal particles. Accordingly, he assumed that the force spectrum $F(k)$ and the modal decay rate νk^2 could be renormalized by writing:

$$d(k) = F(k) + s(k); \qquad \text{and} \qquad \omega(k) = \nu k^2 + r(k), \tag{37}$$

where the functions $s(k)$ and $r(k)$ remain to be determined; but must satisfy the condition:

$$2\omega(k)C(k) = d(k), \tag{38}$$

for the form Equation (36) for P_0 to hold. Correspondingly, the basis operator must take the Fokker–Planck form:

$$L_0 = \sum_{\mathbf{k}} \frac{\partial}{\partial u_\alpha(k)} \left\{ \omega(k) u_\alpha(k) + d(k) \frac{\partial}{\partial u_\alpha(-\mathbf{k})} \right\} P, \tag{39}$$

where the renormalized quantities $\omega(k)$ and $d(k)$ are the dynamical friction and generalized diffusivity, respectively.

In order to invert the basis operator, Edwards used an eigenfunction expansion, but we will not pursue that here. Details may be found in the original reference [3], and in the books [9,12,14]. The resulting equation for the stationary spectral density function $C(k)$ is

$$F(k) - 2\nu k^2 = -2 \int dj^3 L(\mathbf{k}, \mathbf{j}) \frac{C(\mathbf{k}-\mathbf{j})\{C(j) - C(k)\}}{\omega(k) + \omega(j) + \omega(\mathbf{k}-\mathbf{j})}. \tag{40}$$

The response equation is:

$$\omega(k) = \nu k^2 + \int dj^3 L(\mathbf{k}, \mathbf{j}) \frac{C(\mathbf{k}-\mathbf{j})}{\omega(k) + \omega(j) + \omega(\mathbf{k}-\mathbf{j})}, \tag{41}$$

and these two equations are analogous to Equations (24) and (23) of Kraichnan's DIA, in that order.

In order to complete this presentation of the Edwards theory, we specify the generalized diffusivity $s(k)$ from the constraint (38), thus:

$$s(k) = 2 \int dj^3 L(\mathbf{k}, \mathbf{j}) \frac{C(\mathbf{k}-\mathbf{j})C(j)}{\omega(k) + \omega(j) + \omega(\mathbf{k}-\mathbf{j})}. \tag{42}$$

Equivalence of the Edwards Theory and Kraichnan's DIA

As we have noted, Edwards derived his closure approximation by inverting the Fokker–Planck form of L_0 in terms of an eigenfunction expansion, with the inverse modal lifetimes appearing as eigenvalues. However, instead of doing this, it is a simple matter to recover the Edwards SCF equation for $C(k)$ directly from the DIA Equation (24).

Our starting point is the recognition by Kraichnan in 1964 [20] that the DIA and Edwards covariance equations were equivalent if one assumed the same exponential time dependence for the covariance and the response function. That is,

$$C(k; t - t') = C(k) \exp[\omega(k)|t - t'|], \tag{43}$$

and

$$\begin{aligned} R(k; t - t') &= \exp[\omega(k)(t - t')], \qquad \text{for} \quad t' < t; \\ &= 0, \qquad \text{for} \quad t' > t, \end{aligned} \tag{44}$$

where $\omega(k)$ is the inverse lifetime for correlation and response of mode k. Its prescription is the second part of the Edwards closure approximation. Later, Kraichnan showed that the Edwards theory could be extended to statistically non-stationary states [33], while Herring and Kraichnan [34] demonstrated that the generalized Edwards covariance equation could be obtained from Equation (24), with $C(k,t) \equiv C(k;t,t)$, as:

$$\begin{aligned} \left(\frac{\partial}{\partial t} + 2\nu k^2\right) C(k,t) &= 2 \int d^3 j\, L(\mathbf{k}, \mathbf{j}) D(k, j, |\mathbf{k}-\mathbf{j}|; t) \times \\ &\times [C(j,t)C(|\mathbf{k}-\mathbf{j}|, t) - C(k,t)C(|\mathbf{k}-\mathbf{j}|, t)], \end{aligned} \tag{45}$$

where they introduced a memory function D, given by

$$D(k,j,|\mathbf{k}-\mathbf{j}|;t) = \int_0^t ds\, R(k;t,s)R(j;t,s)R(|\mathbf{k}-\mathbf{j}|;t,s). \quad (46)$$

For the stationary case, this could be written [20]:

$$D(k,j,|\mathbf{k}-\mathbf{j}|) = \frac{1}{\omega(k)+\omega(j)+\omega(|\mathbf{k}-\mathbf{j}|)}. \quad (47)$$

However, when this procedure is applied to the DIA response Equation (22), the result is:

$$\omega(k) = \nu k^2 + \int dj^3 L(\mathbf{k},\mathbf{j})\frac{C(\mathbf{k}-\mathbf{j})}{\omega(j)+\omega(\mathbf{k}-\mathbf{j})}. \quad (48)$$

We see that the DIA form differs from the Edwards form, in that only two lifetimes appear in the denominator. That is, the labelling wavenumber is absent.

5.3. The Self-Consistent-Field (SCF) Theory of Herring

In 1965, Herring introduced a more abstract form of self-consistent-field theory. He took as his starting point the Liouville equation derived by Edwards. In order to discuss this self-consistent-field theory, we shall change our notation to that of Herring and rewrite Equation (29) as

$$L^{(0)}P - VP = 0, \quad (49)$$

where the change from L to $L^{(0)}$ is significant and reflects the fact that we shall be looking at a formal renormalization programme in which $L^{(0)}$ is the bare or viscous operator. In this sense, Herring's SCF will be seen to resemble DIA rather than the Edwards SCF, although there are in fact many procedural similarities between the theories of Herring and Edwards. Herring began by assuming that the basis operator for his expansion method was the sum of single-mode operators. The effect of mode coupling was then taken into account through an expansion in single-mode interactions. In this respect, it can be seen as less phenomenological than the Edwards form, in which the renormalized single-mode (i.e., quasi-particle) form is assumed from the outset.

We begin with some definitions relating to single-mode forms. First, we can generalize the concept of reduced distributions from particle statistical mechanics (e.g., see Section 4.1 of the book [14] or the relevant section of [35]) to single-mode distributions in the turbulent case. Let us write the the distribution of velocities for turbulence in $\mathbf{k}$ -space as

$$P[\mathbf{u}(\mathbf{k})] = \lim_{n\to\infty} P_n[\mathbf{u}(\mathbf{k}_1),\mathbf{u}(\mathbf{k}_2),\ldots,\mathbf{u}(\mathbf{k}_n)], \quad (50)$$

where the limit of infinite space volume (corresponding to $n \to \infty$) will not be taken until the end of the calculation. Here, the notation $\mathbf{k}_n$ is shorthand for the usual definition of a wavevector in a finite system. A particular value of n then corresponds to a particular set, e.g., n_1, n_2, n_3. In order to obtain a single-mode distribution, we integrate out all variables except (say) $\mathbf{u}(\mathbf{k}_n)$; thus

$$P_n[\mathbf{u}(\mathbf{k}_n)] = \int P[\mathbf{u}(\mathbf{k})]d\mathbf{u}(\mathbf{k}_1)d\mathbf{u}(\mathbf{k}_2)\ldots d\mathbf{u}(\mathbf{k}_{n-1})d\mathbf{u}(\mathbf{k}_{n+1})\ldots, \quad (51)$$

which defines the single-mode distribution $P_n[\mathbf{u}(\mathbf{k}_n)]$. Now, define the single-mode projection of an operator A by

$$\langle A\rangle = \sum_n \langle A\rangle_n, \quad (52)$$

where

$$\langle A\rangle_n = \int A \prod_{m\neq n} P[\mathbf{u}(\mathbf{k}_m)]d\mathbf{u}(\mathbf{k}_1)d\mathbf{u}(\mathbf{k}_2)\ldots d\mathbf{u}(\mathbf{k}_{n-1})d\mathbf{u}(\mathbf{k}_{n+1})\ldots. \quad (53)$$

Note that the notation $\langle \ldots \rangle$ can be used here for this specialized purpose, as in this section there should be no confusion with its more usual meaning of ensemble average. Then, a single-mode operator can be defined as one which is invariant under the above averaging process, or one which satisfies the relation

$$\langle A \rangle = A. \tag{54}$$

Now, consider Equation (49). If $V = 0$, then it is clear that there will be no non-linear mixing and the solution P will be a product of single-mode distributions. However, if V is not zero, then non-linear mixing must result in some mode coupling and we should write the general solution as

$$P = \prod_n P_n + R, \tag{55}$$

where P_n is the exact single-mode distributions and R is a remainder term to take account of the mode couplings induced by V. It follows that R must satisfy the condition

$$\langle R \rangle_n = 0, \tag{56}$$

where the operation $\langle \ldots \rangle_n$ is defined by Equation (53) and the general perturbation method then consists of expanding R about $V = 0$. In order to do this, it is convenient to introduce a new operator L, which should satisfy two requirements. First, it should be of single-mode form, or

$$\langle L \rangle = \sum_n L_n = L, \tag{57}$$

and second, in order to conserve probability, it should satisfy

$$\int L_n F[\mathbf{u}(\mathbf{k}_n)] d\mathbf{u}(\mathbf{k}_n) = 0 \tag{58}$$

for any well-behaved test function $F[\mathbf{u}(k_n)]$. We also introduce the associated function P', which is the solution of

$$LP' = 0, \tag{59}$$

such that

$$P' = \prod_n P'_n. \tag{60}$$

In other words, P' is the product of single-mode operators. Lastly, as V tends to zero, consistency requires that $L \to L_0$ and $P' \to P$. We now add the operator L to both sides of Equation (49), thereby leaving it unchanged, to obtain

$$LP = (L - L_0 + V)P. \tag{61}$$

The formal solution of this equation can be written as

$$P = P' + R', \tag{62}$$

and then we must solve Equation (61) perturbatively for R' in terms of $L,\ L_{0,}\ V$, and P_n. Our aim is now to find an operator L which (a) satisfies Equations (57)–(59) and (b) is such that $P'_n = P_n$: this is the self-consistency criterion. Our procedure is as follows. Noting Equation (59), we rewrite Equation (61) as

$$LP = LP' + (\Delta + V)P, \tag{63}$$

where

$$\Delta = L - L_0. \tag{64}$$

Then, inverting the operator on the left-hand side, we obtain

$$P = P' + L^{-1}(\Delta + V)P. \tag{65}$$

At this stage, there are two points worth mentioning. First, there is a technical point here which we are glossing over, in that we should invert the non-singular part of L: see Herring (1965) for a discussion. Second, comparison of Equations (65) and (62) indicates that

$$R' = L^{-1}(\Delta + V)P. \tag{66}$$

Our next step is to solve Equation (65) iteratively. As this is of pedagogic interest, we do this in detail in the next section. The result is:

$$P = P' + L^{-1}TP', \tag{67}$$

where T is given by

$$T = (L - L_0 + V)) + (L - L_0 + V)L^{-1}T. \tag{68}$$

Now, the probability distribution given by Equation (67) is in the following form: 'guessed probability + corrections'. For self-consistency we require that the effect of corrections will vanish, such that $P_n' = P_n$. This implies—along with Equation (55)—that we must have

$$\langle P\rangle_n = \langle\prod_n P_n\rangle + \langle R\rangle_n = \Pi_n P_n \tag{69}$$

where we have used Equations (54) and (56). In addition, from Equations (54) and (60), we note that

$$\langle P'\rangle_n = \prod_n P_n'. \tag{70}$$

Hence, if we operate on both sides of Equation 67 with the single-mode expectation value, we have

$$\prod_n P_n = \prod_n P_n' + \langle L^{-1}TP'\rangle_n \tag{71}$$

and, from the self-consistency condition,

$$\langle L^{-1}TP'\rangle_n = 0 \tag{72}$$

or

$$\langle L^{-1}T\rangle_n P' = 0 \tag{73}$$

as P' is the product of single-mode distributions and therefore is unaffected by the operation $\langle \dots \rangle_n$. This result can also be written as

$$[L_n - L_{0,n} + \langle VL^{-1}T\rangle]P' = 0, \tag{74}$$

where we have used Equation (68) and $L_{0,n}$ is defined by

$$L_0 = \sum_n L_{0,n}. \tag{75}$$

Herring has shown that the self-consistency condition can be satisfied, along with Equations (55) and (58), by choosing L_n such that

$$L_n = L_{0,n} - \langle VL^{-1}T\rangle_n, \tag{76}$$

and Equation (59) is retained as a separate constraint;

$$L_n P_n = 0 \tag{77}$$

where we have put $P'_n = P_n$ in Equation (59), in accordance with our self-consistency requirement. We can obtain the perturbation expansion for $\Delta = L - L_0$ by substituting for T from Equations (68) into (76) and then iterating about $\Delta = 0$. The result is

$$L = L_0 - \langle VL^{-1}V \rangle + O(V^4). \tag{78}$$

Similarly, we can obtain the perturbation expansion for P by substituting Equation (78) into (67) and expanding T:

$$P = P' + L^{-1}VP' + L_{-1}[VL^{-1}V - \langle VL_{-1}V \rangle]P' + \ldots \tag{79}$$

It should be noted that truncation of the perturbation series at any order actually retains infinite powers of V. This is because of the non-linearity of Equation (63). It is at this point that we see the underlying resemblance to the renormalized perturbation theory of the DIA. On the other hand, Herring uses the same eigenfunction expansion as Edwards in order to invert the basis operator, and certainly Equation (65) is very similar (notational changes apart) to the Edwards form. The differences essentially lie in the requirement in Herring's SCF that a particular operator should vanish:

$$\left\{ L_n - L_n^{(0)} + \langle VL_{-1}T \rangle_n \right\} = 0, \tag{80}$$

where the Edwards version would require that the product of the same operator with the P_n should vanish. Furthermore, it involves a weaker assumption, that the basis operator should be the product of single-mode operators, without specifying the form these operators should take. In contrast, Edwards imposes the additional requirement that the basis operator should have the Fokker–Planck form. It should be noted that the basis operator in Herring's SCF is found not to be of this form, and leads to coupled equations for $C(k)$ and $\omega(k)$. The energy equation is identical to Equation (40), which is itself the same as the DIA equation when exponential time dependences are substituted. The equation for $\omega(k)$ is the same as the time-independent DIA response equation, as given by Equation (48).

5.3.1. Solution of Equation (65)

Equation (65) has to be solved by iteration. We write it in the form

$$P = P' + XP,$$

where

$$X = L^{-1}(\Delta + V).$$

Then, we proceed as follows:

- **zero-order approximation**

$$P_0 = P' + XP',$$

- **first-order approximation**

$$P_1 = P' + XP_0 = P' + X(P' + XP') = P' + XP' + X^2P'$$

- **second-order approximation**

$$P_2 = P' + XP' = P' + X(P' + XP' + X^2P') = P' + XP' + X^2P' + X^3P',$$

and so on.

The exact distribution P is given by

$$P = P' + X(1 + X + X^2 + \ldots)P' = P' + X\frac{1}{1-X}P'$$

provided that $1/(1-X)P'$ is non-singular. Then, substituting for X, we find

$$P = P' + L^{-1}TP',$$

which is just Equation (67), and the operator T is given by

$$T = \frac{(\Delta + V)}{1 - L^{-1}(\Delta + V)},$$

which can be shown to reduce to (68).

5.3.2. Time-Dependent SCF

The first step in extending SCF to non-stationary turbulence (Herring 1966) is to restore the time dependence to the Liouville equation. That is, we rewrite Equation (49) as

$$\frac{\partial P}{\partial t} + L^0 P = VP. \tag{81}$$

Then, we generalize Equation (68) to the form

$$\frac{\partial P}{\partial t} + \int_0^\infty L(t,t')P(t')dt' = \int_0^\infty \Omega(t,t')P(t')dt' \tag{82}$$

where

$$\Omega(t,t') = L(t,t') - \delta(t-t')(V - L_0) \tag{83}$$

such that

$$L(t,t') = 0 \quad \text{for} \quad t' > t. \tag{84}$$

It should be noted that this is all fully analogous to the procedure which led to Equation (61) for the stationary case and was based on the introduction of the operator L, as defined by Equation (57). Now, in order to obtain an iterative solution of Equation (82), Herring introduced a new operator, $U(k;t,t'$. The interested reader will find the details of these procedures in the original papers (Herring 1965, 1966); we shall only quote the defining relationship here for completeness, as we shall need it in order to write down the final governing equations. Accordingly, we have

$$\begin{aligned} U(k;t,t') &= 1 - \textstyle\int_{t'}^t ds \int_0^\infty ds' L(s,s')U(k;s',t') \quad \text{for} \quad t > t' \\ &= 0 \quad \text{for} \quad t < t'. \end{aligned} \tag{85}$$

Then, the equation for the single-time correlation $C(k,t)$ is found to be

$$\begin{aligned} \left(\frac{d}{dt} + 2\nu k^2\right)C(k,t) &= 2\int d^3j L(\mathbf{k},\mathbf{j}) \int_0^t g(k;t,s)M(|\mathbf{k}-\mathbf{j}|;t,s)ds; \\ &- \int_0^t g(j;t,s)M(k;t,s)M(|\mathbf{k}-\mathbf{j}|;t,s)ds, \end{aligned} \tag{86}$$

where $g(k;t,t')$ can be interpreted as the response function for mode $\mathbf{k}$, and $M(k;t,t')$ plays the part of a two-time velocity covariance function. The two quantities arise, in fact, as non-trivial functional integrals, and can be written in a rather stylized notation in the form

$$g(k;t,t') = -\int \delta\mathbf{u}(\mathbf{k},t)\mathbf{u}(\mathbf{k},t)U(k;t,t')\frac{\partial P[\mathbf{u}(\mathbf{k},t')]}{\partial \mathbf{u}(\mathbf{k},t)} \tag{87}$$

and

$$M(k;t,t') = \int \delta\mathbf{u}(\mathbf{k},t)\mathbf{u}(\mathbf{k},t)U(k;t,t')\mathbf{u}(\mathbf{k},t)P[\mathbf{u}(\mathbf{k},t)], \tag{88}$$

and satisfy the conditions:

$$g(k;t,t) = 1; \tag{89}$$

$$M(k;t,t) = C(k,t). \tag{90}$$

Herring found governing equations for $g(k;t,t')$ and $M(k;t,t')$ which took the following forms:

$$\left(\frac{d}{dt} + \nu k^2\right) g(k;t,t') = -\int d^3 j L(\mathbf{k},\mathbf{j}) \times \\ \times \int_{t'}^{t} ds\, g(j;t,s) M(|\,\mathbf{k}-\mathbf{j}\,|;t,s) g(k;s,t), \tag{91}$$

$$\left(\frac{d}{dt} + \nu k^2\right) M(k;t,t') = -\int d^3 j L(\mathbf{k},\mathbf{j}) \times \\ \times \int_{t'}^{t} ds\, g(j;t,s) M(|\,\mathbf{k}-\mathbf{j}\,|;t,s) M(k;s,t'), \tag{92}$$

where, as in Equation (86), the coefficient $L(\mathbf{k},\mathbf{j})$ is as defined by Equation (25) and its analytical form is given by Equation (26). It can be deduced by inspection of Equations (91) and (92), that the following relationship holds

$$M(k;t,t') = g(k;t,t')C(k,t'). \tag{93}$$

This allows us to make contact with the DIA equations for $R(k;t,t')$ and $C(k;t,t')$, and $C(k;t,t) = C(k,t)$. It is readily verified that, if we replace the DIA Equation (23) for the two-time correlations $C(k;t,t')$ by Equation (93), then Equations (22) and (24) for $R(k;t,t')$ and $C(k,t)$ become identical with the SCF Equations, Equations (91) and (86) for $g(k;t,t')$ and $C(k,t)$.

5.4. Other Self-Consistent Methods

The Liouville equation was also been taken as a starting point by Balescu and Senatorski [36], who attempted to combine methods applied to two different fields of non-equilibrium statistical mechanics (i.e., statistical electrodynamics and Heisenberg spins on a lattice), arguing that in this way they could take account of the main features of the turbulence problem, i.e., an infinite number of degrees of freedom and strong coupling. The equations of motion are transformed by the introduction of action-angle variables and the basic statistical problem is reformulated in terms of a master equation. This is treated perturbatively and the summation of certain classes of diagrams leads to the recovery of Herring's SCF theory. Self-consistent methods have also been applied directly to the equation of motion (rather than the Liouville equation) by Phythian [37]. Strictly speaking, Phythian worked with the one-dimensional Burgers model equation, but his analysis also goes through for the Navier–Stokes equation. Although the methods used rather resemble those of Edwards and Herring, Phythian did formally introduce an infinitesimal response function which is identical in its basic definition with the corresponding DIA quantity as defined by Equation (20). It is perhaps not altogether surprising, therefore, that Phythian ended up with the DIA equations.

However, the importance of this work turned out to be more than just another way of deriving the DIA theory. Kraichnan found that Phythian's theory leads to a new model representation for the DIA equations and this in turn leads on to the development of 'almost Markovian' theories. It should be noted that the term 'Markovian' here refers to the time dependence, not to energy transfers in wavenumber.

6. Second-Generation Renormalized Perturbation Theories

6.1. The Lagrangian Theories

The Lagrangian theories arose out of Kraichnan's study of the failure of DIA to be compatible with the Kolmogorov $k^{-5/3}$ energy spectrum, which he ultimately interpreted as being due to the lack of invariance under random Galilean transformations. The resulting theories are very complicated and we shall only sketch out some general points here. The key element in Kraichnan's approach was the introduction of a generalized velocity field $\mathbf{u}(\mathbf{x}, t|s)$, with both Eulerian and Lagrangian characteristics. This is defined as the velocity measured at time s of a fluid particle which was at position $\mathbf{x}$ at time t. The two times are known as:

$$t = \text{labelling time}; \qquad \text{and} \qquad s = \text{measuring time}.$$

We may consider two limiting cases:

- Case 1: Set $s = t$ then $\mathbf{u}(\mathbf{x}, t|t) = \mathbf{u}(\mathbf{x}, t)$;
- Case 2: Fix $\mathbf{x} = \mathbf{x}_0$ and $t = t_0$ then $\mathbf{u}(\mathbf{x}_0, t_0|s) = V(\mathbf{x}_0, s)$,

where $V(\mathbf{x}_0, s)$ is the Lagrangian velocity field.

Kraichnan reworked the DIA in terms of this picture and called the result the Lagrangian-history direct-interaction approximation (or LHDI for short) [22]. This turned out to be very complicated and Kraichnan introduced a systematic way of reducing the number of time convolution integrals. This abridged set of equations was referred to, naturally enough, as the abridged LHDI, or ALHDI for short. In 1966, Kraichnan [38] reported the first theoretical value of the Kolmogrov spectral constant, which at $\alpha = 1.77$ lies within the range of experimental values, but nowadays would be regarded as rather large, as the consensus is that it should be about 1.62.

Kraichnan [39] later used the method of reversion (or inversion) of power series and rederived both LHDI and ALHDI. However, from a quantitative point of view, ALHDI did not prove altogether satisfactory, as it tended to overestimate the rate of inertial transfer of energy to higher wavenumbers. That is of interest, as the opposite tendency has been associated with the failure of DIA. This provided a motivation for Kraichnan and Herring [23] to use the same methods with the rate of strain tensor, rather than the velocity field. This led to strain-based ALHDI, or SBALHDI.

In 1981, Kaneda [24] introduced the Lagrangian position function and worked with measuring-time derivatives rather than labelling-time derivatives and derived the Lagrangian renormalized approximation (LRA), which predicted a spectral constant of $\alpha = 1.72$. Later, Kida and Goto [25] applied the sparse direct-interaction pertubation method, as originally used to derive Eulerian DIA, in the Lagrangian framework, to obtain the same result as Kaneda's LRA. More recently, Okumaru [26] has derived yet another Lagrangian theory, which he refers to as a 'Closure model', in a paper which gives a good overview of the theoretical work in this area.

We will return to Lagrangian theories in the section on computations, but their diversity does rather give added emphasis to the 1993 comment of Bazdenkov and Kukharin [40] to the effect that the results of Lagrangian theories depend on whether the velocity, the vorticity, or the rate of strain is taken as the basic field. See also the comments by Frederiksen and Davies [41].

Lastly, as we have diagnosed the problem with the historic Eulerian theories to be that they are Markovian with respect to energy transfer through wavenumber, the obvious question arises: what about the Lagrangian theories? Are they non-Markovian in this respect? In fact it is probably fairly obvious, merely be inspection, that they are in fact non-Markovian. However, perhaps the easiest way to answer this question is to consider the additional question: what steps would be needed to make them Markovian? In fact Kraichnan has answered this question for his Lagrangian-history direct-interaction theory. He has shown that LHDI can be reduced to an eddy-viscosity (i.e., Markovian) form if the dummy wavenumber j is very much larger than the labelling wavenumber k [38]. For further discussion of this point, see Section 10.3.2 of the book [14].

6.2. The Local Energy-Transfer (LET) Theory in Two-Time Form

The LET theory was derived in 1974, in single-time form; and then heuristically in two-time form in 1978. More recently, the two-time LET theory was given a formal derivation in 2017. Here, we will make some brief remarks about this formal derivation [42].

We use the Edwards method, as described in Section 5.2, but we start from the Liouville equation with time-dependence included. The salient point is that we use the simplest possible basis operator that is compatible with a Gaussian distribution. There is still some element of arbitrariness involved in our choice of operator, but we do not follow Edwards and use the Fokker–Planck form as given by (39). It is this choice by Edwards which forces his theory to be Markovian with respect to energy transfers in wavenumber. Our approach is explained in Section 4.3 of [42], where we use Herring's reduced notation for the sake of simplicity. The subsequent perturbation theory is explained in Section 4.4 of [42], and the evaluation of the triple moment in Section 4.5 involves a functional differentiation that leads to the introduction of the response function.

LET Response Function

Kraichnan introduced his infinitesimal response tensor, which related an infinitesimal fluctuation in the velocity field to a corresponding fluctuation in the stirring forces (20). Instead of doing this, the LET introduces an infinitesimal velocity field propagator as determining the system response. In its most general form, this is defined by the relationship between two infinitesimal fluctuations of the velocity field, as:

$$\delta u_\alpha(\mathbf{k},t) = \widehat{R}_{\alpha\beta}(\mathbf{k},\mathbf{k}';t,t')\delta u_\beta(\mathbf{k}',t') \qquad \text{for} \qquad t' \leq t, \tag{94}$$

where the hat symbol indicates that the propagator $\widehat{R}_{\alpha\beta}$ is a random variable. Further, we rewrite Equation (94) in terms of the functional derivative, in order to introduce the ensemble-averaged response function, thus:

$$R_{\alpha\beta}(\mathbf{k};t,t') = \left\langle \frac{\delta u_\alpha(\mathbf{k},t)}{\delta u_\beta(\mathbf{k},t')} \right\rangle = \left\langle \widehat{R}_{\alpha\beta}(\mathbf{k};t,t') \right\rangle = P_{\alpha\beta}(\mathbf{k})R(k;t,t')\,, \qquad t' \leq t\,, \tag{95}$$

where we have in turn invoked homogeneity and then isotropy.

This analysis leads to a general fluctuation-response relation [42], relating the response function to the two-time and single-time covariances, thus:

$$C(k;t,t') = R(k;t,t')C(k;t',t'). \tag{96}$$

The quantity $R(k;t,t')$ is called the response function. We discuss some of the implications of this result in the next section.

6.3. Single-Time Models and Theories

Under this heading we begin with the eddy-damped quasi-normal Markovian (EDQNM) theory. This arose from Orszag's analysis of the failure of the original quasi-normality hypothesis and reflects a later and better understanding of turbulence phenomenology. The EDQNM theory is really a model (compare Kraichnan's similar test-field model [43]), in that it makes a particular assumption which requires the fitting of an adjustable constant. In stationary form it reduces to the Edwards SCF theory. Because it is a single-time theory, it is relatively easy to compute and has been widely applied to practical problems (e.g., [44–47]). For a recent discussion see the book by Sagaut and Cambon [27].

In the interests of completeness we should also mention that McComb and Kiyani [48] derived a Markovianised single-time closure from the LET theory, while Bos and Bertoglio [49] derived a single-time closure from a Lagrangian perspective. Neither of these closures involve the fitting of adjustable constants, and so are classed as theories rather than models.

7. Application of Statistical Closure Theories

The statistical closure theories have been extensively computed for the case of freely decaying turbulence, starting with Kraichnan's pioneering work in 1964 [2]. A concise overview of this activity may be found in [42]. Here, we will just consider their application to passive scalar convection, along with some remarks on the possibility of more general applications.

7.1. Passive Scalar Convection

This is a very big subject and we will restrict our attention to the use of closures in the problem of passive scalar diffusion in isotropic turbulence. We note the pioneering work of Newman and Herring, who applied the test-field model (TFM) to this problem [50]; and later work of Herring et al. [51], which compared the TFM to EDQNM. This work, in particular, underlines the fact that this type of investigation is more about testing the performance of the closures than about practical calculations.

In 1992, as well as introducing the fluctuation–response relationship as the defining step of the LET theory, and providing a new simplified derivation (along with making some minor corrections to the equations), the work was extended to the case of passive scalar convection [52]. Results were obtained for free decay for $5 \leq R_\lambda \leq 1060$ and for Prandtl numbers of 0.1, 0.5, and 1.0. In both velocity and scalar spectra, Kolmogorov-type power laws were found at the higher Reynolds numbers.

For those wishing to pursue this topic further, the recent paper by Kitamura [53] tackles the more advanced problem of passive scalar diffusion in the presence of a mean scalar gradient, and also provides a comprehensive review of spectral methods applied to passive scalar convection.

7.2. Potential for Practical Applications

In view of the great importance of turbulent flow in engineering applications, it is surprising that the two-point statistical closures have been so little exploited for practical applications. To our knowledge, there have only been two exceptions to this. First, there is the activity on climate and atmospheric modelling by Frederiksen and co-workers (see [54] for a recent account), who have used DIA, Herring's SCF, and the LET theory over several decades for this purpose. This work also includes sub-grid modelling.

Secondly, there has been the widespread application of the EDQNM to many situations and over many decades. Its development by Orszag [55] took account of the greater understanding of turbulence phenomenology which had accrued during the 1960s and this is reflected in its success with practical applications. Because of its single-time nature it is easy to compute, and its ability to fit an adjustable constant leads to good quantitative results. As examples we may cite [44–47]; a more general account may be found in the book by Sagaut and Cambon [27].

In contrast, the LET theory, consisting as it does of the covariance equations, (23) for $C(k;t,t')$ and (24) for $C(k;t,t) \equiv C(k,t)$, which are identical to the DIA forms, plus the fluctuation–relaxation relation (96), is a fully two-time theory and has no adjustable parameters. Although rather more complicated than the EDQNM theory, and more demanding to compute, it nevertheless could be applied to non-isotropic and even inhomogeneous problems. Even if presenting too formidable a computational challenge, it could provide a basis for single-point modelling, which might offer a useful extension to existing approaches.

8. Conclusions

8.1. A Point-Wise Summary of the Main Conclusions

As most readers of this article are unlikely to be familiar with quantum field theory, I think that it will be helpful if I try to summarise its application to turbulence in as concise a way as possible. Accordingly, I have set out a series of key points, which I have made as simple as I can without misrepresenting anything. In general my remarks are limited to isotropic turbulence (please note that homogeneity is a necessary condition for isotropy and does not therefore have to be specified). Furthermore, for the most part, I have restricted this summary to Eulerian theories. This is not to denigrate the Lagrangian approach, which is an entirely valid method of seeking a closure approximation, but to recognise that there is still the outstanding problem of converting results into an Eulerian form, which is of course the natural system for experimental work. One thinks of Taylor's theory of turbulent diffusion by continuous movements, which in the past has led to pragmatic attempts to convert it to Eulerian coordinates. In time, perhaps those who espouse the Lagrangian approach will address the problem of applying their results in an Eulerian coordinate system.

1. Every modern two-point closure approximation can be understood as a second-order truncation of a renormalized perturbation theory. The renormalization is achieved in terms of an approximate renormalized response function. All theories lead to the same equation for the energy spectrum, but the form of renormalized response function (and the method of obtaining it) varies from theory to theory.
2. The historic two-time closures (the DIA of Kraichnan and Herring's SCF) perform quite well, both quantitatively and qualitatively, on the test problem posed by freely decaying turbulence from an arbitrary initial state. However, they are not compatible with the Kolmogorov spectrum.
3. There has always been some uncertainty about this failure. Kraichnan argued that it arose because a two-time theory was unable to distinguish between sweeping effects and inertial transfer through wavenumber. In contrast, Edwards interpreted the failure of his single-time theory to be due to an infrared divergence at infinite Reynolds numbers. At the time, there seemed to be some degree of agreement between the two diagnoses, in that they both pointed to a problem at low wavenumbers, but this still left a feeling of uncertainty.
4. A corollary of Kraichnan's argument about the DIA is that no two-time Eulerian theory can be compatible with the Kolmogorov spectrum. However, the LET theory is both a two-time Eulerian theory *and* is compatible with the Kolmogorov spectrum.
5. A corollary of the Edwards diagnosis, in terms of an infrared divergence, is that there may be more to the failure of DIA than Kraichnan's diagnosis would suggest. This is because DIA reduces to a similar form to the Edwards SCF in time-independent form, and hence must also have an infrared divergence.
6. A corollary of the original derivation of the LET theory is that the relation of the DIA's response function to the energy balance is inconsistent with the experimental picture.
7. A corollary of the formal derivation of the LET theory in two-time form [42] is that the unifying explanation of why the historic closures fail to yield the Kolmogorov spectrum is that they are all Markovian with respect to energy transfer through wavenumber, when turbulence is non-Markovian.
8. A conjecture arising from this formal derivation [42] is that the underlying problem with the DIA may be its reliance on the random stirring forces to define the renormalized response function. While it is natural to study a dynamical system by applying a force, in the case of fluid motion it is the turbulent state which is the system. Hence, the applied force first has to create the dynamical system and then maintain it, while also determining its renormalized response. In contrast, the two-time LET determines its response in terms of a random, quasi-entropic force, which is related to the probability distribution of velocities.
9. Although the quantum-style formalisms set out to establish the general nature and existence of a statistical theory of turbulence, they both are dependent (in their different ways) on the use of the stirring forces to formalise their renormalized response. Further, they both reduce to the DIA when truncated at the lowest nontrivial order. Hence, to the extent that the DIA is incorrect, they also are incorrect, at least when applied to fluid turbulence.

10. The LET theory cannot claim to be the theory of turbulence, in that there are some question marks hanging over it. In particular, its prediction of the Kolmogorov spectral constant is almost certainly too large. However, a hybrid theory might be a possibility, with some vertex renormalization included by means of a restriction on wavenumber interaction ranges. In this way, it could, like EDQNM, benefit from the introduction of an adjustable constant.

8.2. Some Personal Reflections

I first met Jack Herring at the NASA-ICASE workshop on turbulence theory at Virginia Beach in October 1984. Taking part were physicists, mathematicians, and engineers, and I, as very much a new boy, was glad to be welcomed into the physicists' group with Jack, Bob Kraichnan, and various others. They were all, particularly Jack, very friendly and I immediately felt at home. In fact, his friendly open manner was the first impression one had of Jack, and I know that others felt the same about him. He was above all a good-humoured, relaxed man and, in view of his great abilities, remarkably unassuming. I shall return to that particular aspect later.

This was a long time ago, and I only have a few memories of the social interactions, but I do recall that there was a 'no-host cocktail hour' when the day's programme finished. I was quite amused by this phrase, which I had not met before, and which simply meant that the organisers were not about to pay for the drinks! I only remember one dinner, where we left the hotel and set off along the sea front to look for somewhere to eat. It turned out that a group of turbulence theorists could not ignore a seafood restaurant called 'Ocean Eddies'. I think we were the only customers and one of us did not like the loud music being played and either switched it off or asked someone to do so. After a few moments of pleasant silence, a small angry-looking man appeared (perhaps Eddy himself?), switched the music back on, glared at us, and then disappeared back behind the scenes. Despite the loud music, it was a pleasant evening and on our way back to the hotel Jack suddenly pounced and scooped something up from the road. It was a small crab that he had spotted scuttling along in front of us. After holding it in his cupped palms for us all to admire, he put it down again to resume its journey to the sea.

My first interaction with Jack was actually the previous year, when I submitted a paper reporting the first computations of my LET theory to the *Journal of Fluid Mechanics*. One morning I received a phone call from George Batchelor to say that Bob Kraichnan and Jack Herring were refereeing my paper and that they would like to discuss it with me. He added, 'You seem to have stirred them up a bit', and he sounded quite approving! In those days, 'discussion' was by means of air-mail letter, and was a relatively slow business. However, it was all very civil and constructive; I managed to satisfy both Jack and Bob, and the paper was duly accepted by JFM.

In 1990, my first book *The Physics of Fluid Turbulence* was published. It was reviewed by Jack for the *Journal of the Atmospheric Sciences*. He gave it a very favourable and quite lengthy review, but I was amused by one of his comments. He pointed out that I had gone into some detail in establishing the foundations of the many-body statistical approach, and it was 'very full, but not too tiresome'. Of course to someone like Jack, this would be very elementary material, but that would not necessarily be the case for most readers. I was glad that it had not been too tiresome!

I encountered Jack on a few other occasions over the years, and found that he was the sort of person with whom you felt that each time you met, you started again from where you last left off, as if it had been the previous day. Discussions were always amicable, even if we did not entirely agree.

While I have no further anecdotes to tell, I do have a couple of points that I would like to make. The first of these arises out of my earlier comment about Jack being unassuming. In his collaboration with Bob Kraichnan from the late 1960s onwards, I felt that Jack to some extent submerged his own interests in favour of Bob's interests. Of course, I have no sense of how the two men worked together on the derivation of the strain-based Lagrangian theory. Nevertheless, the author ordering of the paper reporting the theory [23], and of the paper reporting its subsequent computation [56], rather suggests the minor role for Jack. However, others may disagree, and indeed be better informed than I am. So, I just offer this as an impression that I had at the time.

My second, and final, point is about the three pioneering theories and their authors. I was reflecting on these around the end of the previous millenium, and my conclusion was that these three remarkable theories would always be the foundation of any turbulence theory to come. Each was an outstanding intellectual achievement and they all, in effect, were in agreement. If they had led to a great wave of progress in turbulence, then they would have been in contention for the Nobel prize. However, unfortunately the cultural divide was too great. For instance, intelligent men like Phillip Saffman and John Lumley dismissed them as 'mathematically complicated', and I have seen the anonymous comment 'everyone knows that Kraichnan was wrong and this line of approach is finished.' Well, on the contrary, this line of approach is not finished and in my carefully considered view, Kraichnan, Edwards, and Herring, in their pioneering work on turbulence, were in the Nobel laureate class.

Funding: This research received no external funding.

Institutional Review Board Statement: Not applicable.

Informed Consent Statement: Not applicable.

Data Availability Statement: Not applicable.

Conflicts of Interest: The author declares no conflict of interest.

Appendix A. Quantum-Style Formalisms

We will only touch briefly on this subject here, mainly for the sake of completeness. The formalisms of interest are the diagrammatic method due to Wyld [19], and the functional formalism of Martin et al. [57]. The latter is often referred to as MSR, after the names of its three authors. The aim of a formalism is to establish the existence and nature of a solution to the turbulence problem (this is in the sense of physics, rather than that of pure mathematics), rather than to produce a specific theory. In fact, both of these formalisms reduce to the direct-interaction approximation (DIA) when truncated at the lowest nontrivial order, and, by extension, also can be said to lead to the SCFs of Edwards and Herring.

The paper by Wyld in 1961 was the second publication (after Kraichnan's DIA) to apply the new renormalization methods of physics' turbulence problem, but, unlike the DIA, was presented very much using the terminology of quantum field theory. While this may be off-putting to those who are not themselves theoretical physicists (i.e., most fluid dynamicists), it is worth overcoming this feeling, because the diagrammatic expansion presents quite an intuitive picture of how renormalization can be expected to work. The basic idea is that, although the perturbation series is wildly divergent, some subclasses of terms in the expansion may be summed to all orders. In particular, summing one subclass (with its members determined on topological grounds) leads to an exact expression for the covariance of random stirring forces and the velocity field (also see Section 5.3.1 of the present work, which deals with an exact summation occurring in Herring's SCF). Essentially, the remaining terms lead to a renormalized perturbation series of unknown properties (at least, as regards convergence and possible truncation).

In 1965, Lee pointed out that there was a problem with the double counting of diagrams with Wyld's method and this could be cured by changing the left-hand emergent propagator from each diagram to the bare (i.e., viscous) form [58]. Later, McComb [59], in the course of deriving a two-time version of the LET theory, pointed out that the need for Lee's ad hoc procedure was due to Wyld carrying out more renormalizations than were actually required. A more detailed discussion of Wyld's theory may be found in Section 5.5 of the book [14].

Lastly, Martin et al. [57] argued that even with the improvements introduced by Lee, the Wyld theory was incorrect, due to the way in which it handled vertex corrections. As a result, many people dismissed the Wyld–Lee theory out of hand. This is never a very wise thing to do. After all, the basic perturbation approach is the gold standard for this type of theory. More recently, Berera et al. [60] made a detailed comparison of the two formalisms and concluded that MSR had misinterpreted Wyld's way of handling the vertex corrections. They presented what they called the *Improved Wyld–Lee renormalized perturbation theory* in which there is no conflict with the MSR functional formalism. However, the reader should note the comment about both these formalisms which is made in the Conclusions of this article.

References

1. Kraichnan, R.H. The structure of isotropic turbulence at very high Reynolds numbers. *J. Fluid Mech.* **1959**, *5*, 497–543. [CrossRef]
2. Kraichnan, R.H. Decay of isotropic turbulence in the Direct-Interaction Approximation. *Phys. Fluids* **1964**, *7*, 1030–1048. [CrossRef]
3. Edwards, S.F. The statistical dynamics of homogeneous turbulence. *J. Fluid Mech.* **1964**, *18*, 239. [CrossRef]
4. Edwards, S.F. Turbulence in hydrodynamics and plasma physics. In *Lectures Presented at the Trieste Seminar on Plasma Physics;* IAEA: Vienna, Austria , 1965; p. 595.
5. Herring, J.R. Self-consistent field approach to turbulence theory. *Phys. Fluids* **1965**, *8*, 2219. [CrossRef]
6. Herring, J.R. Self-consistent field approach to nonstationary turbulence. *Phys. Fluids* **1966**, *9*, 2106. [CrossRef]
7. Obukhov, A.M. On the energy distribution in the spectrum of turbulent flow. *Bull. Acad. Sci. USSR Geog. Geophys.* **1941**, *32*, 19.
8. Edwards, S.F.; McComb, W.D. Statistical mechanics far from equilibrium. *J. Phys. A* **1969**, *2*, 157. [CrossRef]
9. Leslie, D.C. *Developments in the Theory of Turbulence;* Clarendon Press: Oxford, UK, 1973.
10. Qian, J. Variational approach to the closure problem of turbulence theory. *Phys. Fluids* **1983**, *26*, 2098. [CrossRef]
11. Shi, J.Z. Qian Jian (1939–2018) and his contribution to small-scale turbulence studies. *Phys. Fluids* **2021**, *33*, 041301. [CrossRef]
12. Beran, M.J. *Statistical Continuum Theories;* Interscience: Hoboken, NJ, USA, 1968.
13. Monin, A.S.; Yaglom, A.M. *Statistical Fluid Mechanics;* MIT Press: Cambridge, MA, USA, 1975.
14. McComb, W.D. *The Physics of Fluid Turbulence*; Oxford University Press: Oxford, UK, 1990.
15. Proudman, I.; Reid, W.H. On the Decay of a Normally Distributed and Homogeneous Turbulent Velocity Field. *Phil. Trans. R. Soc. Lond.* **1954**, *A247*, 163.
16. Tatsumi, T. The theory of decay process of incompressible isotropic turbulence. *Proc. R. Soc. Lond. Ser. A* **1957**, *239*, 16.
17. O'Brien, E.E.; Francis, G.C. A consequence of the zero fourth cumulant approximation. *J. Fluid Mech.* **1962**, *13*, 369. [CrossRef]
18. Ogura, Y. A consequence of the zero-fourth-cumulant approximation in the decay of isotropic turbulence. *J. Fluid Mech.* **1963**, *16*, 33. [CrossRef]
19. Wyld, H.W., Jr. Formulation of the theory of turbulence in an incompressible fluid. *Ann. Phys.* **1961**, *14*, 143. [CrossRef]
20. Kraichnan, R.H. Approximations for steady-state isotropic turbulence. *Phys. Fluids* **1964**, *7*, 1163–1168. [CrossRef]
21. Grant, H.L.; Stewart, R.W.; Moilliet, A. Turbulence spectra from a tidal channel. *J. Fluid Mech.* **1962**, *12*, 241–268. [CrossRef]
22. Kraichnan, R.H. Lagrangian-history closure approximation for turbulence. *Phys. Fluids* **1965**, *8*, 575–598. [CrossRef]
23. Kraichnan, R.H.; Herring, J.R. A strain-based Lagrangian-history turbulence theory. *J. Fluid Mech.* **1978**, *88*, 355. [CrossRef]
24. Kaneda, Y. Renormalized expansions in the theory of turbulence with the use of the Lagrangian position function. *J. Fluid Mech.* **1981**, *107*, 131–145. [CrossRef]
25. Kida, S.; Goto, S. A Lagrangian direct-interaction approximation for homogeneous isotropic turbulence. *J. Fluid Mech.* **1997**, *345*, 307–345. [CrossRef]
26. Okamura, M. Closure model for homogeneous isotropic turbulence in the Lagrangian specification of the flow field. *J. Fluid Mech.* **2018**, *841*, 133. [CrossRef]
27. Sagaut, P.; Cambon, C. *Homogeneous Turbulence Dynamics;* Cambridge University Press: Cambridge, UK, 2008.
28. McComb, W.D. *Homogeneous, Isotropic Turbulence: Phenomenology, Renormalization and Statistical Closures*; Oxford University Press: Oxford, UK, 2014.
29. Verma, M.K. *Energy Transfers in Fluid Flows*; Cambridge University Press: Cambridge, UK, 2019.
30. McComb, W.D. Theory of turbulence. *Rep. Prog. Phys.* **1995**, *58*, 1117–1206. [CrossRef]
31. Kolmogorov, A.N. The local structure of turbulence in incompressible viscous fluid for very large Reynolds numbers. *C. R. Acad. Sci. URSS* **1941**, *30*, 301.
32. Novikov, E.A. Functionals and the random-force method in turbulence theory. *Sov. Phys. JETP* **1965**, *20*, 1290–1294.
33. Kraichnan, R.H. An almost-Markovian Galilean-invariant turbulence model. *J. Fluid Mech.* **1971**, *47*, 513. [CrossRef]
34. Herring, J.R.; Kraichnan, R.H. *Comparison of Some Approximations for Isotropic Turbulence Lecture Notes in Physics;* Chapter Statistical Models and Turbulence; Springer: Berlin, Germany, 1972; Volume 12, p. 148.
35. McComb, W.D. *Study Notes for Statistical Physics: A Concise, Unified Overview of the Subject;* Bookboon: London, UK, 2014.
36. Balescu, R.; Senatorski, A. A new approach to the theory of fully developed turbulence. *Ann. Phys.* **1970**, *58*, 587. [CrossRef]
37. Phythian, R. Self-consistent perturbation series for stationary homogeneous turbulence. *J. Phys. A* **1969**, *2*, 181. [CrossRef]
38. Kraichnan, R.H. Isotropic turbulence and inertial-range structure. *Phys. Fluids* **1966**, *9*, 1728. [CrossRef]
39. Kraichnan, R.H. Eulerian and Lagrangian renormalization in turbulence theory. *J. Fluid Mech.* **1977**, *83*, 349. [CrossRef]

40. Bazdenkov, S.V.; Khukharkin, N.N. On the variational method of closure in the theory of turbulence. *Phys. Fluids A* **1993**, *5*, 2248. [CrossRef]
41. Frederiksen, J.S.; Davies, A.G. The Regularized DIA Closure For Two-Dimensional Turbulence. *Geophys. Astrophys. Fluid Dyn.* **2004**, *98*, 203. [CrossRef]
42. McComb, W.D.; Yoffe, S.R. A formal derivation of the local energy transfer (LET) theory of homogeneous turbulence. *J. Phys. A Math. Theor.* **2017**, *50*, 375501. [CrossRef]
43. Kraichnan, R.H. Test-field model for inhomogeneous turbulence. *J. Fluid Mech.* **1972**, *56*, 287. [CrossRef]
44. Bowman, J.C.; Krommes, J.A.; Ottaviani, M. The realizable Markovian closure. I. General theory, with application to three-wave dynamics. *Phys. Fluids B* **1993**, *5*, 3558–3589. [CrossRef]
45. Bowman, J.C.; Krommes, J.A. The realizable Markovian closure and realizable test-field model 11. Application to anisotropic drift-wave dynamics. *Phys. Plasmas* **1997**, *4*, 3895. [CrossRef]
46. Bos, W.J.T.; Shao, L.; Bertoglio, J.-P. Spectral imbalance and the normalized dissipation rate of turbulence. *Phys. Fluids* **2007**, *19*, 45101. [CrossRef]
47. Bos, W.J.T.; Chevillard, L.; Scott, J.F.; Rubinstein, R. Reynolds number effect on the velocity increment skewness in isotropic turbulence. *Phys. Fluids* **2012**, *24*, 015108. [CrossRef]
48. McComb, W.D.; Kiyani, K. Eulerian spectral closures for isotropic turbulence using a time-ordered fluctuation-dissipation relation. *Phys. Rev. E* **2005**, *72*, 16309–16312. [CrossRef]
49. Bos, W.J.T.; Bertoglio, J.-P. A single-time, two-point closure based on fluid particle displacements. *Phys. Fluids* **2006**, *18*, 031706. [CrossRef]
50. Newman, G.R.; Herring, J.R. A test field model study of a passive scalar in isotropic turbulence. *J. Fluid Mech.* **1979**, *94*, 163. [CrossRef]
51. Herring, J.R.; Schertzer, D.; Lesieur, M.; Newman, G.R.; Chollet, J.P.; Larcheveque, M. A comparative assessment of spectral closures as applied to passive scalar diffusion. *J. Fluid Mech.* **1982**, *124*, 411. [CrossRef]
52. McComb, W.D.; Filipiak, M.J.; Shanmugasundaram, V. Rederivation and further assessment of the LET theory of isotropic turbulence, as applied to passive scalar convection. *J. Fluid Mech.* **1992**, *245*, 279–300. [CrossRef]
53. Kitamurar, T. Spectral theory of passive scalar with mean scalar gradient. *J. Fluid Mech.* **2021**, *923*, A28. [CrossRef]
54. Kitsios, V.; Frederiksen, J.S.; Zidikheri, M.J. Theoretical comparison of subgrid turbulence in atmospheric and oceanic quasi-geostrophic models. *Nonlin. Process. Geophys.* **2016**, *23*, 95–105. [CrossRef]
55. Orszag, S.A. Analytical theories of turbulence. *J. Fluid Mech.* **1970**, *41*, 363. [CrossRef]
56. Herring, J.R.; Kraichnan, R.H. A numerical comparison of velocity-based and strain-based Lagrangian-history turbulence approximations. *J. Fluid Mech.* **1979**, *91*, 581. [CrossRef]
57. Martin, P.C.; Siggia, E.D.; Rose, H.A. Statistical Dynamics of Classical Systems. *Phys. Rev. A* **1973**, *8*, 423–437. [CrossRef]
58. Lee, L.L. A formulation of the theory of isotropic hydromagnetic turbulence in an incompressible fluid. *Ann.Phys* **1965**, *32*, 292. [CrossRef]
59. McComb, W.D. A theory of time dependent, isotropic turbulence. *J. Phys. Math. Gen.* **1978**, *11*, 613. [CrossRef]
60. Berera, A.; Salewski, M.; McComb, W.D. Eulerian Field-Theoretic Closure Formalisms for Fluid Turbulence. *Phys. Rev. E* **2013**, *87*, 013007. [CrossRef]

atmosphere

MDPI

Article

Fog Intermittency and Critical Behavior

Kelly Y. Huang [1,*], Gabriel G. Katul [2], Thomas J. Hintz [1], Jesus Ruiz-Plancarte [3], Qing Wang [3] and Harindra J. S. Fernando [1]

1 Civil and Environmental Engineering and Earth Sciences, University of Notre Dame, Notre Dame, IN 46556, USA
2 Civil and Environmental Engineering, Duke University, Durham, NC 27708, USA
3 Meteorology Department, Naval Postgraduate School, Monterey, CA 93943, USA
* Correspondence: yhuang28@nd.edu

Abstract: The intermittency of fog occurrence (the switching between fog and no-fog) is a key stochastic feature that plays a role in its duration and the amount of moisture available. Here, fog intermittency is studied by using the visibility time series collected during the month of July 2022 on Sable Island, Canada. In addition to the visibility, time series of air relative humidity and turbulent kinetic energy, putative variables akin to the formation and breakup conditions of fog, respectively, are also analyzed in the same framework to establish links between fog intermittency and the underlying atmospheric variables. Intermittency in the time series is quantified with their binary telegraph approximations to isolate clustering behavior from amplitude variations. It is shown that relative humidity and turbulent kinetic energy bound many stochastic features of visibility, including its spectral exponent, clustering exponent, and the growth of its block entropy slope. Although not diagnostic, the visibility time series displays features consistent with Pomeau–Manneville Type-III intermittency in its quiescent phase duration PDF scaling ($-3/2$), power spectrum scaling ($-1/2$), and signal amplitude PDF scaling (-2). The binary fog time series exhibits properties of self-organized criticality in the relation between its power spectrum scaling and quiescent phase duration distribution.

Keywords: dynamical model; fog; intermittency; self-organized criticality; sporadic randomness

Citation: Huang, K.Y.; Katul, G.G.; Hintz, T.J.; Ruiz-Plancarte, J.; Wang, Q.; Fernando, H.J.S. Fog Intermittency and Critical Behavior. *Atmosphere* **2023**, *14*, 875. https://doi.org/10.3390/atmos14050875

Academic Editor: Thierry Bergot

Received: 23 March 2023
Revised: 27 April 2023
Accepted: 15 May 2023
Published: 17 May 2023

1. Introduction

Fog occurrence is associated with saturated water vapor and accompanying water droplets near the earth's surface, causing visibility to drop below 1 km [1]. This reduction in visibility raises numerous logistic and safety issues for aviation, seafaring, and transportation. Ecologically, fog plays an underappreciated role through its transport of water droplets, aerosols, and microorganisms, which impacts the thermodynamic, hydrologic, photosynthetic, and nutrient properties of ecosystems [2,3]. The potential for fog as a potable water source has also been proposed and explored [4,5]. Fog formation continues to receive attention in viticulture, as sustained moist conditions modulate the microclimate and promote mildew disease and fungal infection [3].

A key aspect in fog statistics is its intermittency, the switching between fog and no-fog phases (on–off switching properties), which was shown to be related to the long-term persistence of fog events with a characteristic time scale that appears invariant [6]. Intermittency in precipitation, a process that similarly exhibits on–off switching properties between rain and no-rain events, has received copious attention in literature and has been modeled using stochastic point processes [7,8], multiplicative cascades [9,10], and traditional stochastic models [11]. More recently, Rigby and Porporato [12] found that summer convective rainfall exhibits features of Pomeau–Manneville Type-III intermittency, an example of a sporadically random process (a behavior between periodicity and pure randomness). Further, rainfall has been shown as an example of self-organized criticality (SOC) [13], which started with the original work of Bak and Chen [14]. In contrast, limited attention has been afforded to dynamic models in fog, and this limitation motivated the current work.

A recent study proposed some similarities between rainfall and fog statistics at coarse time scales [15]. Motivated by these results, the work here seeks to explore the extent to which existing dynamical models for precipitation apply to fog occurrences by comparing several scaling relations that describe the fog time series to those that have been noted in the literature. High-resolution time series of visibility Vis taken over a one-month period is used to explore the analytic relations concerning the power laws in the spectral density, the distribution of the off phases, and the distribution of the event sizes for phenomenological comparison to the proposed dynamical models. In addition, the current study also seeks to explore atmospheric variables that could possibly contribute to the intermittency of fog occurrences. To this end, time series of air relative humidity RH and turbulent kinetic energy k, which may provide conditions for the formation and break-up of fog, respectively, are also analyzed herein in the same framework to draw connections between these atmospheric variables and the proposed dynamical models. This analysis will serve as a first step to developing stochastic models for fog, which can be employed in fog simulation studies, to check physical models, and to explore underlying formation processes of fog.

1.1. Types of Dynamical Intermittency

Intermittency is a route to chaos in dynamical systems and is characterized by the irregular alternation between phases of regular, almost periodic behavior ("laminar" phases) and phases of chaotic dynamics ("turbulent" phases). As a control parameter exceeds a critical value, the system eventually becomes fully chaotic without laminar phases. Models of different types of intermittency have been classified and observed in various systems [16,17].

Pomeau and Manneville [18] described three types of intermittency (Types I–III) caused by linear instabilities of periodic trajectories, where Pomeau–Manneville Type I intermittency (PM1I) is associated with an inverse tangent bifurcation, Type II (PM2I) with a Hopf bifurcation, and Type III (PM3I) with a period-doubling bifurcation. PM1I has been observed in numerical studies of ordinary differential equations and discrete dynamical systems [19–21] and in specialized experiments [22–25]; PM2I in numerical and experimental studies of nonlinear oscillators [26,27] and in a hydrodynamical system [28]; and PM3I in biological systems [29,30], electrical applications [31–34], rainfall time series [12], and at the onset of hydrodynamic turbulence [35].

On–off intermittency is another model that describes a time series characterized by switching between near-constant "off" phases punctuated by burst phases of large deviations caused by sudden instabilities [36], and has been observed in many mathematical models [36–39], electrical circuits [40,41], spin wave instabilities [42], gas discharge plasma systems [43], and the electroconvection of nematic liquid crystals [44]. Other models not considered here include Type V [45,46], Type X [47], multi-intermittency [48] and crisis-induced intermittency [49].

For comparison of phenomenological features of these intermittency models with those of fog time series, the analytic properties related to the distribution of the duration of the quiescent or laminar phases, power spectral density, and distribution of event size (or signal amplitudes) can be explored, which are summarized in Table 1. The distribution of the duration of the laminar phases has been used in the literature to differentiate between types of intermittency. PM3I and on–off intermittencies generate probability distribution functions (PDFs) of the quiescent phases $\mathrm{PDF}(l)$ that decay as a power law with an exponent of $-3/2$ (that is, $\mathrm{PDF}(l) \sim l^{-3/2}$) at small l, while PM2I generates ones with an exponent of -2 at small l followed by an exponential tail (or cutoff) at larger l [50]. In contrast, PM1I is expected to show a power law decrease with an exponent of $-1/2$ for small l followed by an increase at large l [50].

Table 1. Phenomenological features of intermittency models under consideration, where PDF refers to the probability density function, l is the duration of the laminar or quiescent phases, f is the frequency or inverse time scale, x is the event size, and E_x is the spectral density function .

	PM Type-I	PM Type-II	PM Type-III	On–off
PDF(l)	$l^{-1/2}$	l^{-2}	$l^{-3/2}$	$l^{-3/2}$
$E_x(f)$	$(f\|\log_{10}f\|^2)^{-1}$	$f^{-1/2}$	$f^{-1/2}$	$f^{-1/2}$
PDF(x)	x^{-1}	x^{-2}	x^{-2}	x^{-1}

Procaccia and Schuster [51] showed that the power spectral density $E_x(f)$ scales as $(f|\log_{10}f|^2)^{-1}$ in a PM1I process and as $f^{-1/2}$ in a PM2I or a PM3I process for $f \to 0$ by considering their 1-D local Poincaré maps [18], where f is the frequency (or inverse time scale). Venkataramani et al. [52] demonstrated that the power spectrum of an on–off intermittent process with no noise exhibits a power law decay with an exponent of $-1/2$ over an intermediate range of frequencies dependent on the system constants. Lastly, the PDF of the signal amplitude (or event size), PDF(x), is considered. On–off intermittency shows a decay in PDF(x) with an exponent of -1 for small x [53], PM1I with an exponent of -1, and PM2I and PM3I with an exponent of -2 [54].

More generally, Pomeau–Manneville intermittency maps and their stochastic counterparts have been shown to be examples of sporadic randomness. Based on the entropic characteristics of trajectories, Gaspard and Wang [54] defined a class of sporadically random dynamical systems that reside between periodicity (processes that generate repeating sequences) and regular randomness (processes whose complexity grows linearly with sequence length, i.e., Markov chains and Brownian motions, which have a convergent entropy rate). Sporadic random processes display a strong memory while retaining a stochastic component, and the complexity grows as a power law with the sequence length.

1.2. Self-Organized Criticality

Generally, phase transitions occur when the control or tuning parameter (an ambient property) is varied across its critical value so that the order parameter (a macroscopic, measurable property) exhibits a jump in its values. The phase transition is of the first order if the order parameter varies discontinuously (i.e., a discontinuous change in entropy at a fixed temperature corresponding to latent heat), and of the second order if it varies continuously but not analytically (i.e., the transition is marked by a sharp corner on the phase diagram).

In second-order continuous phase transitions, the system can reside at criticality, a state between two qualitatively different types of behavior. Since phase transitions typically separate an ordered state from a less ordered one, systems at criticality are often said to be on the edge of chaos, and have been shown to exhibit optimal memory and processing capabilities (for a review, see [55]). Further, in systems where the tuning and order parameters are coupled, the critical point can become an attractor so that the system tunes itself as it evolves toward criticality, in a behavior termed self-organized criticality (SOC) [14]. In SOC, the system is driven toward a second-order phase transition, wherein the dynamics is driven by universal properties. For an SOC system at criticality, a relation is expected between the spectral exponent and the power law decay of the quiescent phases. Controversies notwithstanding, SOC has been shown to be a possible candidate for a wide variety of systems in physical, biological, and social sciences, spanning from climate fluctuations [56] and neural systems [57] to wars [58] (for a complete review, see [59–61]). Similar to the physical analogy laid out by Peters and Neelin [13] for rainfall, it can be argued that a critical value of RH (the tuning parameter) marks the continuous phase transition to a regime of fog events.

2. Methods

2.1. Measurements

Measurements were collected on Sable Island (43.9337° N, 59.9149° W), a small North Atlantic island located approximately 300 km southeast of Halifax, Nova Scotia (see Figure 1), as part of the Fog and Turbulence Interactions in the Marine Atmosphere (FATIMA) field campaign. The site was chosen for its proximity to the Grand Banks region, which is identified to have one of the climatologically highest marine fog occurrences with a peak of 45% during June–July–August [62]. The island's maximum elevation above sea level is 30 m, making it an ideal location for collecting data without significantly disturbing the marine sea fog advecting past it. Data were collected from 1 to 31 July 2022, during which fog (horizontal visibility ≤ 1 km) was observed approximately 25% of the time.

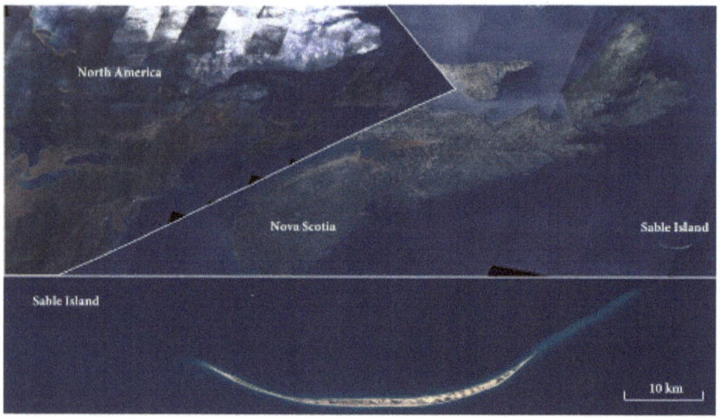

Figure 1. Location of Sable Island in the Grand Banks region of the North Atlantic Ocean (**top**). Sable Island is shown in the inset (**bottom**). Images were taken by the Sentinel-2 multispectral satellite.

Longitudinal (U), transverse (V), and vertical (W) velocities were collected with a triaxial sonic anemometer (Campbell Scientific IRGASON) at a height of 2 m above the ground. Corrections to the sonic anemometer data were applied to account for line path averaging and flow distortion by the sonic transducers [63,64]. A low-pass filter was then applied to avoid high-frequency noise, and subsequent turbulence data were processed using 5-min windows. Gaps in velocity data due to rain were filled using a moving average over a one-hour window, and then again with a one-day window. Turbulence kinetic energy k is then calculated as

$$k = \frac{1}{2}\left(\overline{(u')^2} + \overline{(v')^2} + \overline{(w')^2}\right), \tag{1}$$

where u', v', and w' are the longitudinal, transverse, and vertical fluctuating velocity components, respectively, and the overbar denotes averaging over a 5-min period.

Visibility (Vis) and relative humidity (RH) time series were collected at 15-s intervals using a forward scatter sensor (Vaisala FD70) and the attached humidity probe (Vaisala HMP155), respectively, located approximately 20 m west of the sonic anemometer at a measurement height of 2.5 m (see Figure 2). Here, visibility is used as a proxy for the intensity of fog, with Vis being inversely proportional to the liquid water content [65]. Visibility and relative humidity data are also down-sampled to 5-min averages.

Figure 2. Photos of the IRGASON (**a**) and the FD70 with an attached HMP155 (**b**). The IRGASON is mounted on a measurement tower at a height of 2 m, and the FD70 is located 20 m west of the measurement tower with a measurement height of 2.5 m for *Vis* and 1.7 m for *RH* (**c**). The figure is created by the author.

2.2. Analysis Methods

To analyze intermittency statistics, the telegraph approximation (TA) is used. The TA of a time series $s(t)$ isolates its clustering behavior from its amplitude variations by transforming it into a binary series:

$$\mathrm{TA}(s(t)) = \frac{1}{2}\left(\frac{s(t) - s^*}{|s(t) - s^*|} + 1\right), \tag{2}$$

where $\mathrm{TA}(s(t))$ takes on a value of 1 when $s(t)$ exceeds a threshold s^* and a value of 0 otherwise. As it retains only threshold-crossing (on–off or off–on switching) information, the TA series contains no information about magnitude variation. Here, $k^* = |k| = 0.61\ \mathrm{m}^2/\mathrm{s}^2$, $Vis^* = 1000$ m (to reflect the definition of fog occurrence so that $\mathrm{TA}(Vis) = 1$ when $Vis \leq 1000$ m and $\mathrm{TA}(Vis) = 0$ otherwise), and $RH^* = 100$ % (to reflect saturation so that $\mathrm{TA}(RH) = 1$ when $RH \geq 100\%$ and $\mathrm{TA}(RH) = 0$ otherwise).

Telegraph approximation has been used in turbulence research [66,67] to quantify clustering and intermittency behavior in studies of temperature, velocity, and scalar concentration fluctuations in the atmospheric surface layer, within canopies, and across several surface roughness, density and thermal stratification [68–74]. The same method has also been applied to the study of rainfall intermittency [75]. Sreenivasan and Bershadskii [67] showed empirically from several numerically generated stochastic series that when the power spectra of the original series scales as f^{-n} and that of the TA series as f^{-m},

$$m = \frac{n+1}{2}, \tag{3}$$

where f is, as before, the frequency, n is the spectral exponent of the original series, and m is the spectral exponent of the TA series. This relation appears to hold across a broad range of stochastic processes. Generally, $m < n$ (that the spectra of the TA decays slower than that of the original series) indicates that there is more 'memory' in the TA series such that amplitude variations seem to have "decorrelating" effects on the series. The spectra were calculated using standard fast Fourier transforms from the normalized time series

$$s_n(t) = \frac{s(t) - \overline{s(t)}}{\sigma_s}, \tag{4}$$

where the overbar denotes the mean, and σ_s is the root mean square value of $s(t)$. The window length is set to half of the time series length.

The cluster exponent α characterizes the clustering tendency of the time series and is determined from the scaling of the local standard deviation of the density of threshold-crossings $\langle \delta n_\tau^2 \rangle^{1/2}$ with τ:

$$\langle \delta n_\tau^2 \rangle^{1/2} \sim \tau^{-\alpha}, \tag{5}$$

where $\delta n_\tau = n_\tau - \langle n_\tau \rangle$, n_τ is the average density of zero-crossings within a time interval τ (i.e., the number of threshold crossings normalized by the number of points in that time interval), and angled brackets denote time averaging for window size τ. While a white noise process (that has no clustering tendencies in its zero crossings) has a cluster exponent of $\alpha = 0.5$, Sreenivasan and Bershadskii [67] showed that $\alpha \to 0.1$ (finite clustering persists) as the Taylor microscale Reynolds number $Re_\lambda \to \infty$ in turbulent flows.

The intermittency exponent μ can be calculated from the local average of the variance dissipation rate of a signal $s(t)$ in a time interval τ,

$$\xi_\tau = \frac{1}{\tau} \int_t^{t+\tau} \left| \frac{ds(t)}{dt} \right|^2 dt, \tag{6}$$

as

$$\frac{\langle \xi_\tau^q \rangle}{\langle \xi_\tau \rangle^q} \sim \tau^{-\mu_q}. \tag{7}$$

A uniform random distribution of pulses yields $\mu_q = 0$, while an intermittent series yields non-zero μ_q values. Here, only μ_2 (q = 2) is considered, and the subscript is dropped henceforth. Calculated from the full series, μ_s contains information on both clusterization and amplitude variations, while μ_{TA} calculated from the TA series retains only clusterization information. Thus, their relative magnitudes indicate the role of amplitude variations and clusterization in the observed intermittency. That is, $\mu_{TA} < \mu_s$ indicates that amplitude variations amplify the intermittency; $\mu_{TA} > \mu_s$ indicates that amplitude variations mitigate intermittency; and for $\mu_{TA} \approx \mu_s$, much of the observed intermittency is due to the on–off and off–on properties but not amplitude variability.

The rate of growth of the complexity of a trajectory in time classifies a process as periodic, random, or sporadically random. A conventional method to quantify this growth of complexity for discrete time series is to study the information content of the systems (information theory). Consider a sequence A with symbols a_i, where $i = 0, ..., \lambda$, from a given alphabet. The sequence A is considered periodic if some symbols or sub-sequences occur repeatedly, and disordered (or random) when all of its symbols occur with equal frequency. This can be measured by the Shannon entropy [76]:

$$H = -\sum_i p_i \log p_i, \tag{8}$$

where p_i is the probability that symbol a_i occurs at any position. To reflect the correlations between the symbols, Equation (8) can be generalized to obtain the block entropy

$$H(n) = -\sum_i p_i^{(n)} \log p_i^{(n)}, \tag{9}$$

where $p_i^{(n)}$ are the probabilities of the combinations of n symbols. For periodic processes with period τ, the block entropy reaches a constant and stops growing (entropy rate becomes zero) at $n > \tau$. For purely random processes, the entropy grows linearly with n, reflecting the constant level of uncertainty with each additional step. Thus, for a sporadic random process, where the system is always yielding information and decreasing the uncertainty in the next step but never reaching complete regularity, the entropy grows sub-linearly without reaching a plateau. Therefore, a relation $H(n) \propto n^{\beta}$, with $0 \leq \beta \leq 1$ is expected [54]. Here, to maximize the value of n available without significant finite-size effects, the binary TA signal is used to limit the alphabet to binary (0 and 1) and, subsequently, the number of possible sequences to 2^n. The maximum n is then determined as $n_{max} = \log_2[N/100]$ so that the estimate would be sensitive to sequences of probability on the order of 0.01.

The connection to SOC in the current study arises because the PDF of the duration of the quasi-regular or "laminar" phases, and the PDF of the event sizes both exhibit power law scaling, i.e., $\text{PDF}(l) \sim l^{-\gamma}$ and $PDF(x) \sim x^{-\beta}$. In particular, for SOC systems near critical behavior, Jensen et al. [77] showed that

$$m = 3 - \gamma \tag{10}$$

using the classical sandpile model near critical slopes. To account for intermittency effects, Bershadskii et al. [66] modified this relation to

$$m = 3 - \gamma - \mu/2, \tag{11}$$

where μ is the intermittency correction calculated using squared temporal gradients as in Equation (7).

3. Results

3.1. Overview

Figure 3 presents the time series of RH, k, Vis, and their corresponding TA series over the experimental period. Relative humidity values greater than 100% were observed and could be attributed to the reported sensor inaccuracy of $\pm$1.7% RH. Nonetheless, the study at hand concerns only the switching across saturation, and RH^* values of 100 $\pm$ 1.7% do not significantly alter the conclusions presented below. It can be seen that fog occurrence ($Vis \leq 1000$ m) generally follows high RH and low k. To explore the combined effect of these two criteria, an additional "RH^1k_0" TA series was calculated by assigning a value of 1 only when both $\text{TA}(RH) = 1$ and $\text{TA}(k) = 0$. Analysis of this series will be discussed below.

Binned scatter-plots of Vis against RH and k are presented in Figure 4 with the colormap indicating the density of points. Bin sizes are 0.5% for RH, 0.03 m^2/s^2 for k, and 400 m for Vis. A linear fit between Vis and RH yielded a coefficient of determination $R^2 = 0.63$. In contrast, Vis and k appear uncorrelated ($R^2 = 0.0045$). Nonetheless, the highest density of points for $Vis \leq 1000$ m occurs for low k, indicating that low turbulence levels are well correlated with fog occurrences.

3.2. Spectra

The spectra of both the original and the TA series of RH, k, and Vis are presented in Figure 5. The spectrum for RH shows a peak at 24 h, which could arise from diurnal variation. As a result, the frequency range corresponding to 22–26 h was excluded when calculating the power laws for RH. A mild bump at 24 h can also be seen in the spectrum for Vis, which could be a signature of night-time fog that disappears in the morning.

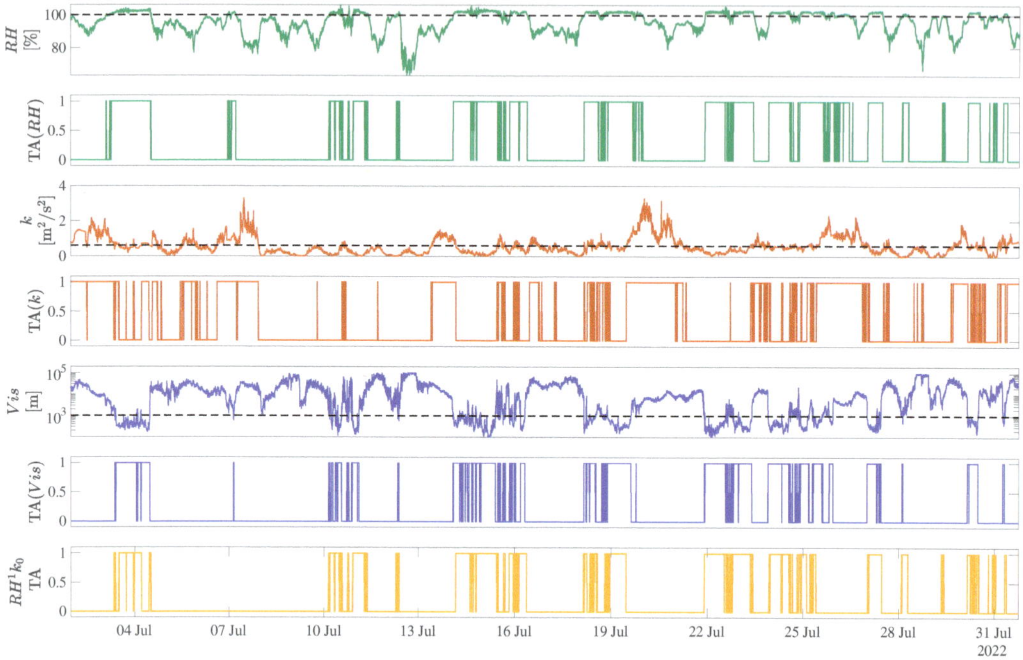

Figure 3. Time series of RH, k, and Vis and their corresponding TA series over the sampling period. Dashed lines in the original series indicate the thresholds ($RH^* = 100$, $k^* = |k| = 0.61$ m^2/s^2, and $Vis^* = 1000$ m) used in Equation (2). The last row presents the combined RH^1k^0 TA series.

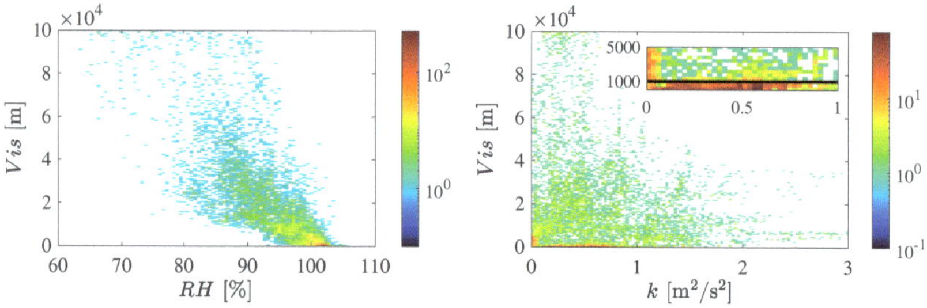

Figure 4. Binned scatter-plots of Vis against RH (**left**) and against k (**right**) with the colormap indicating the density of points.

All spectra appear to exhibit three regimes of power-law scaling, separated at 24 h and at 1 h. Of the three regimes, the fastest decay appears to occur within the frequency range of 1–24 h, while the spectra for time scales >24 h and <1 h decay at a slower rate. Hereafter, we label the frequencies associated with time scales >24 h as slow, <1 h as fine, and bounded between daily and hourly as intermediate. Returning to spectral exponents, the spectral slope n approaches a value of $-1/2$ for the slow regime in Vis but not in RH or k. Generally, for the slow scales, the Vis series exhibits a shallower slope in its spectra than RH or Vis. The spectrum of the combined RH^1k_0 TA series exhibits similar slopes to k in its intermediate and fine scales and to RH in its slow scales.

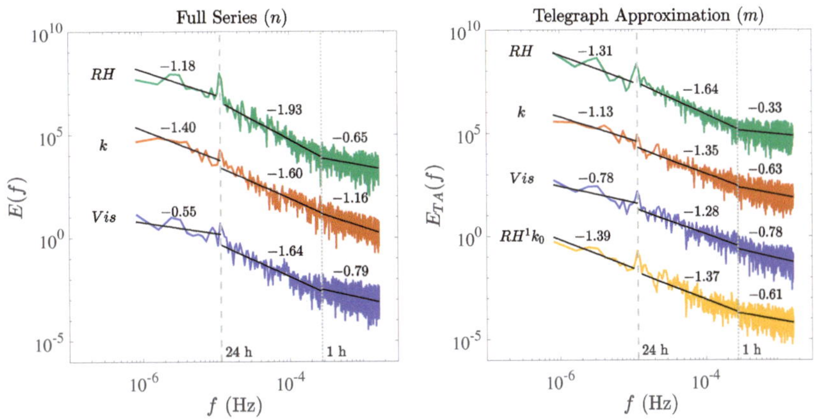

Figure 5. Energy spectra of the full (**left**) and TA (**right**) series of all three variables. The spectra have been offset to permit comparisons. Time scales of 24 h (- - -) and 1 h (···) delineate the scaling regimes.

Overall, the relation between spectral exponents m and n reasonably follow Equation (3) (see Figure 6). For the slow time scales, $m > n$ was observed in both RH and Vis, meaning that there is less "memory" in the TA series and that amplitude variations have correlating effects. For the intermediate and fine time scales (<24 h) in RH and Vis, and for all time scales in k, the spectra of the TA series decay slower than that of its corresponding full series ($m < n$), meaning that amplitude variations seem to have de-correlating effects at those time scales. Thus, much of the power law scaling may be due to the memory in the switching between on–off and off–on instead of the amplitude variations.

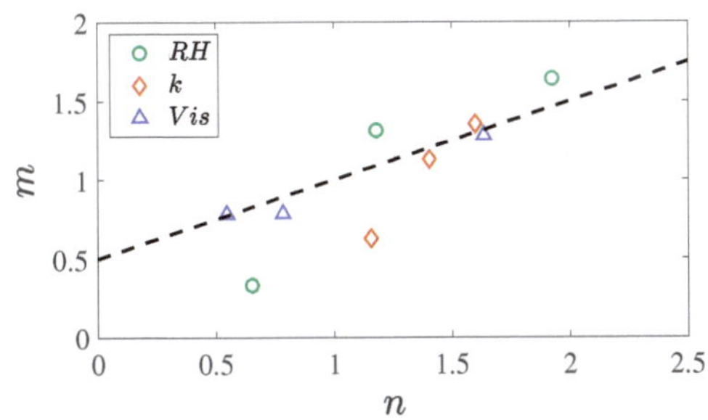

Figure 6. A plot of the spectral exponents of the original series (n) and of the corresponding TA series (m) for all three variables and all three scaling regimes indicated in Figure 5. The dashed line indicates the relation in Equation (3).

As a preliminary indicator of the extent to which variability in RH and k is correlated with Vis, the magnitude-squared coherences C of the full and TA series are presented in Figure 7. In addition, the additive effect of RH and k in explaining Vis is considered in C_{Vis,RH^1k_0} calculated between Vis and a mixed model based on the linear combination of RH and k (with each normalized to span from 0 to 1). Similarly, $C_{TA(Vis),TA(RH^1k_0)}$ considers the coherence between TA(Vis) and the combined RH^1k_0 TA series. The coherences are smoothed through bin averaging to highlight the underlying trends.

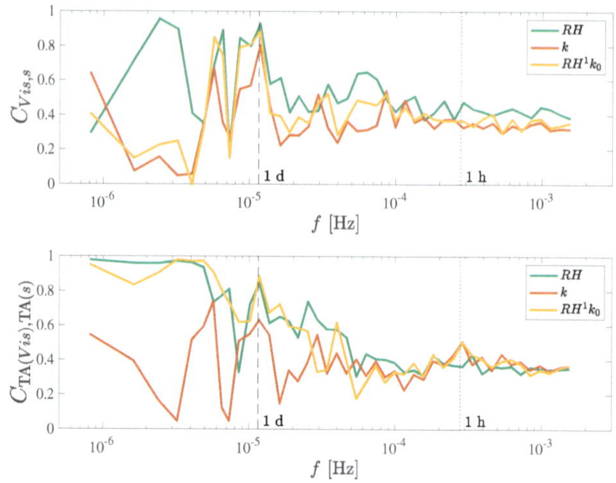

Figure 7. Squared coherence $C_{Vis,s}$ between *Vis* and the two control variables $s = RH$ and $s = k$ (and their combination) as a function of frequency f (**top**). The analysis is repeated for the TA series (**bottom**). The squared coherence measures how correlated *Vis* is to a control variable at a given f. Note the similarity in $C_{Vis,s}$ for the original and TA series at intermediate and fast time scales.

For the slow time scales of the full series, $C_{Vis,RH}$ exhibits peaks (>0.8) at the 5 d, 2 d, and 1 d time scales (showing that the spectra of fog can be reasonably predicted from the spectra of relative humidity at these time scales), while $C_{Vis,k}$ exhibits a minimum (<0.2) so that the spectra of fog are not related to those of k in this region. For the slow time scales of the TA series, it appears that the behavior of *Vis* is captured by that of *RH* ($C_{\mathrm{TA}(Vis),\mathrm{TA}(RH)}$ approaches 1 for time scales >2 d). The combined RH^1k_0 TA series appears to only marginally improve the correlation to *Vis* near the 1 d time scale. For fine and intermediate time scales, both *RH* and k (and their combined effect) exhibit $C \approx 0.4$ in both the full and TA series, indicating that although *RH* and k behaviors are somewhat correlated to that of *Vis*, neither is able to fully explain the variability in *Vis*. These results further prompt the current analysis into its intermittency structure and the possible models beyond the activation and inhibition mechanisms of fog.

3.3. Cluster Exponents

Standard deviations of the zero-crossing density fluctuations δn_τ as a function of τ are featured in Figure 8. All TA series exhibit clear power law scaling with a break in scaling behavior, resulting in two regimes. Generally, there is significant clustering ($\alpha = 0.24 - 0.26$) for larger time scales followed by almost no clustering ($\alpha = 0.5$ consistent with a white noise process). The break between these two regimes occurs at $\tau \approx 12$ h for *RH*, k and RH^1k_0, and at $\tau \approx 24$ h for *Vis*. Surprisingly, the cluster exponent for *Vis* ($\alpha = 0.25$) is consistent with those reported in [15], where $\alpha = 0.26$ and $\alpha = 0.25$ based on deposition data from a fog collector at a coastal and an inland site, respectively, for $\tau < 5$ days.

The intermittency exponents μ_s and μ_{TA} are presented in Figure 9. All TA series show similar levels of intermittency ($\mu_{TA} = 0.42 - 0.44$). For *RH* and *Vis*, $\mu_{TA} > \mu_s$ shows that amplitude variations mitigate intermittency and have 'correlating' effects. In contrast, $\mu_{TA} \approx \mu_s$ for k shows that much of the intermittency in k originates from its on–off and off–on switching properties.

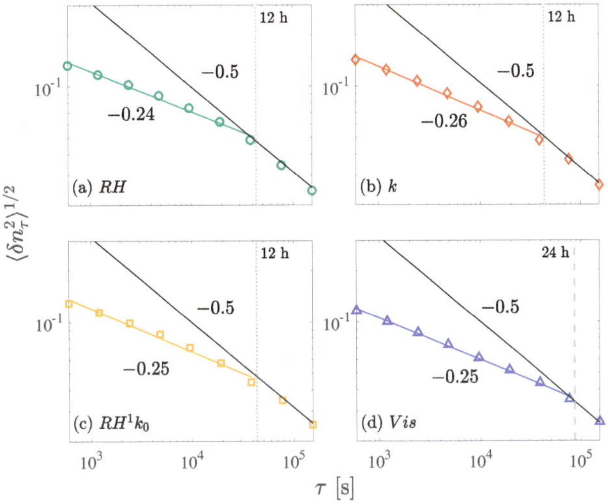

Figure 8. Standard deviations for the running threshold-crossing density fluctuations for *RH* (**a**), *k* (**b**), *Vis* (**c**), and RH^1k_0 (**d**) series as function of window size τ. Best-fits that indicate the power laws and the scaling for a white noise process (α = 0.5) are indicated with straight black lines. The separation of the two scaling regions is also indicated for each variable with a vertical dashed line.

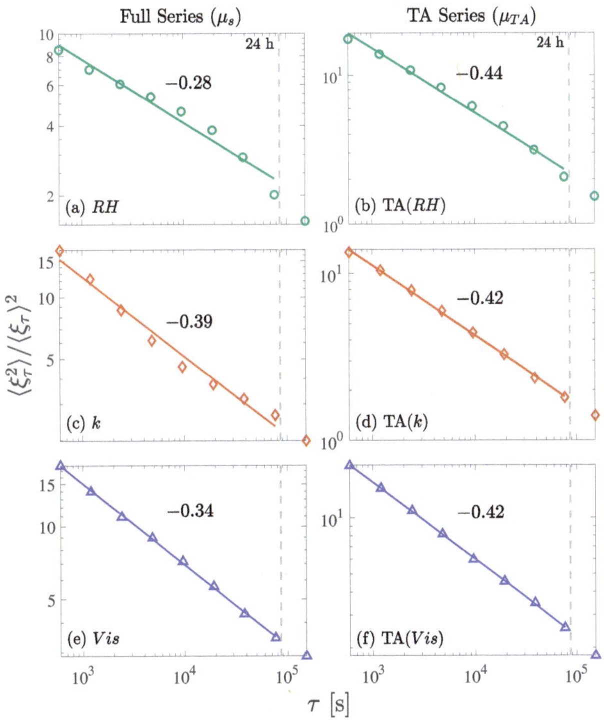

Figure 9. Normalized second moment of the squared temporal gradients for *RH* (**a**), TA(*RH*) (**b**), *k* (**c**), TA(*k*) (**d**), *Vis* (**e**), and TA(*Vis*) (**f**). The straight lines are best-fits that indicate the scaling laws, and the dashed vertical lines indicate a time scale of 24 h.

3.4. Laminar Period Duration and Signal Amplitude PDFs

The PDFs of the duration of quiescent (or laminar) periods for all variables exhibit a power law decay, consistent with systems at criticality (Figure 10). In particular, *Vis* shows a power law scaling exponent γ close to $-3/2$. The scaling exponents for RH, k, and RH^1k_0 are lower at -1.24, -1.25, and -1.26, respectively.

Equation (10) overestimates the measured spectral slope m (as in Figure 5) for all variables by differences of 0.12–0.40. When including intermittency corrections as proposed by [66], μ_s (calculated from the full series) was able to account for the difference between measured and estimated m for RH, and μ_{TA} (calculated from the TA series) was able to account for the difference for Vis. For k, neither μ_s nor μ_{TA} was able to account for the difference, showing that amplitude variations impact the relation, and the SOC model (or its corrected version for intermittency) cannot fully describe the on–off switching properties (Figure 11).

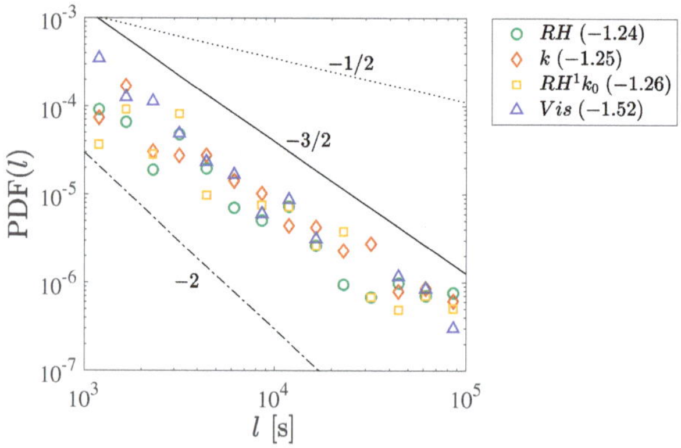

Figure 10. PDFs of the laminar period duration for RH, k, and Vis, with γ values for each variable noted in the legend. Also indicated are the scaling relations expected of a PM1I process ($-1/2$), a PM2I process (-2), and a PM3I or an on–off intermittency process ($-3/2$).

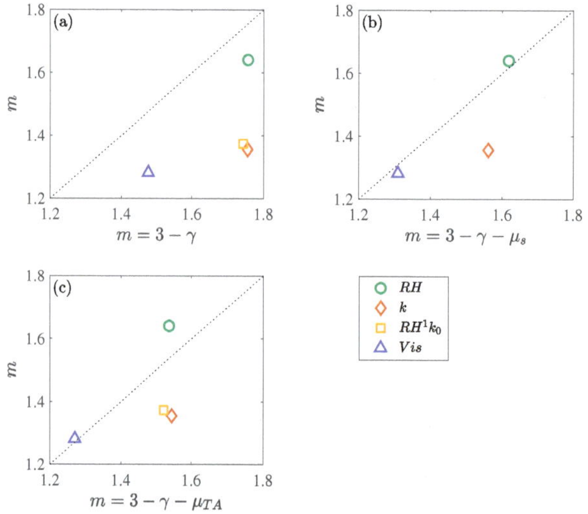

Figure 11. Measured spectral exponents of the TA spectra m (from Figure 5) against expected m of an SOC process near critical point (**a**) with intermittency corrections from the full series (**b**) and from the TA series (**c**). Dotted lines represent a slope of 1.

The PDFs of the signal amplitudes are shown in Figure 12 along with expected −1 and −2 slopes (see Table 1). The *x*-axis for PDF(RH) is reversed to reflect the probability of deviating from the saturation condition. Of the three variables, only PDF(*Vis*) exhibits a clear −2 scaling. The −2 scaling in *Vis* occurs over the approximate range of 300–2000 m, which envelopes the condition for fog occurrence (*Vis* = 1000 m). In PDF(k), there appears to be two regimes separated at approximately 1 m^2/s^2 so that PDF(k) ≈ 1 for $k < 1$ m^2/s^2 followed by a power law decay at $k > 1$ m^2/s^2. A scaling of −2 seems to be present only for a limited range (of approximately $k = 0.8 - 1.2$ m^2/s^2) in the transition between the two scaling regimes. In contrast, PDF(RH) has no such power law scaling.

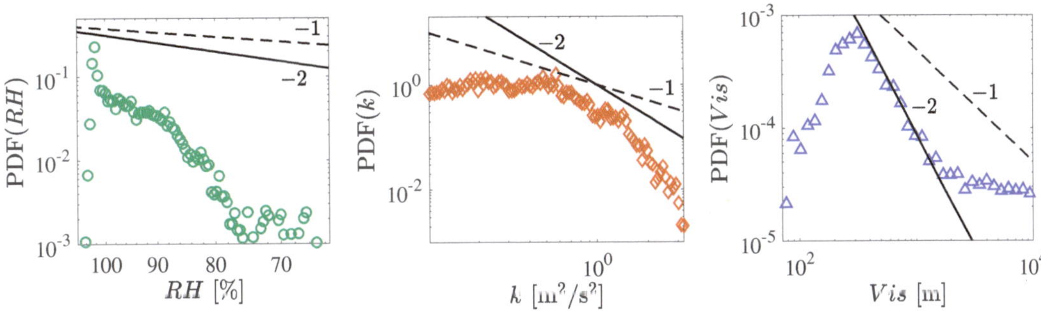

Figure 12. Signal amplitude PDFs for *RH*, *k*, and *Vis*. A power law scaling of −1 consistent with PM1I and on–off intermittency, and of −2 consistent with PM2I and PM3I are indicated for reference.

3.5. Block Entropy

The block entropy $H(n)$ and its growth rate with n (slope $\log\left[\frac{H(n+1)}{H(n)}\right]/\log\left[\frac{n+1}{n}\right]$) for the TA series are presented in Figure 13. For a sporadic process, the slope should converge to a constant value less than 1. All three variables appear to grow nonlinearly, and begin to reach a constant value between 0.5 and 0.6. In the context of block entropy, k appears to be the least ordered (most random), as it seems to asymptotically approach a slope of approximately 0.6. While these results suggest that *RH*, *k*, and *Vis* can be described as sporadically random processes, the limited n_{max} precludes conclusions on their behavior for longer time scales. It can be noted that the slope of $H(n)$ for *Vis* is bounded by that of *RH* and k at all $n > 1$, and that the RH^1k_0 series exhibits similar $H(n)$ values and growth rate to *Vis*.

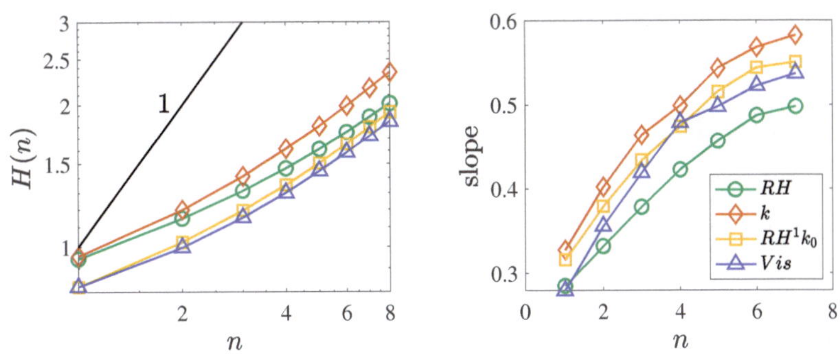

Figure 13. Block entropy $H(n)$ (**left**) and its slope (**right**) as a function of block length n. Solid line in the left panel corresponds to a slope of 1 representative of a purely random process.

4. Study Limitations

Although a longer sampling period than the current one-month dataset is best to make more affirmative conclusions, the agreement between the scaling exponents presented here and in other studies of longer duration lends credence to the results. Räsänen et al. [15] presented a longer-term analysis on fog deposition measured by a tipping bucket between June 2014 and September 2016 at a coastal site in Northern California and between October 2013 and February 2016 at an inland site in Southeastern Kenya. The TA analysis revealed that cluster exponent α, laminar period duration PDF exponent γ, and fog event size scaling exponent agree reasonably well with the exponents presented here (see Table 2), despite the differences in sampling duration. Further, the similarity of the scaling exponents shows that they may also be agnostic to the study site and measurement technique.

Table 2. Comparison of the cluster exponent α, laminar period duration PDF exponent γ, and fog event size scaling exponent from data taken on Sable Island in the current study, an inland site in Southeastern Kenya, and a coastal site in Northern California [15].

	α	γ	PDF(Vis)
Sable Island	0.25	1.52	∼2
Inland	0.25	1.41	1.83
Coastal	0.26	1.33	1.86

Since instrument uncertainties tend to occur at high frequencies and very low frequencies, the scaling relations within the TA framework are relatively insensitive to these uncertainties in the fine scales. In fact, more extensive scaling in the TA spectra than in the full signal spectra have been observed [67]. In the current study, finite clustering persists in the fine scales for all variables considered (Figure 8) so that white noise at the very finest scales does not appear to leave a noticeable signature in the scaling exponents. That is, the fact that clear scaling laws are seen all the way across the fine scales indicates that noise is not propagating much across the time scales.

To further investigate the impact of noise to the presented conclusions, the analysis is repeated for the Vis series with various levels of added noise. As an estimate, the SNR for the measured Vis series is calculated to be 86 by treating frequencies above 1/1000 Hz as white noise (in accordance with the approximate frequency, where the spectra begins to flatten). The signal-to-noise ratio SNR of the measured signal is then progressively deteriorated by adding increasing levels of white Gaussian noise. It can be seen in Figure 14 that all scaling exponents (m, n, α, μ_s, and μ_{TA}) are approximately constant for SNR as low as 40. Similarly, the -2 scaling in PDF(Vis) persists for SNR as low as 40 (Figure 15). The SNR associated with the instrument is much higher than the SNR at which the scaling laws no longer hold, indicating that sensor uncertainties will not impact the conclusions presented here.

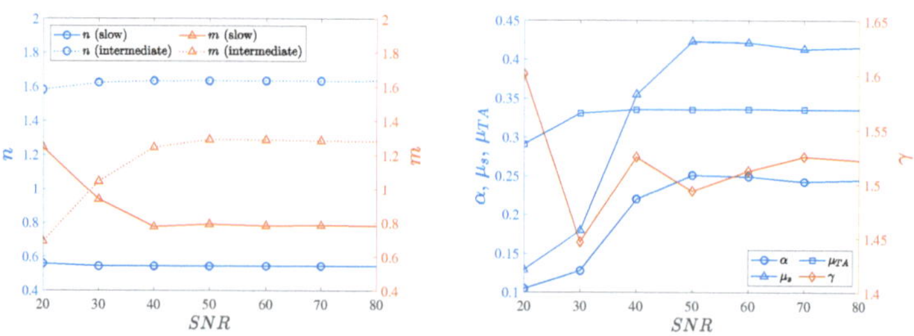

Figure 14. Spectral exponents m and n for the slow and intermediate scaling regimes as functions of SNR (**left**). Cluster exponent α, intermittency exponents μ_s and μ_{TA}, and scaling exponent for the laminar phase PDF γ as functions of SNR (**right**).

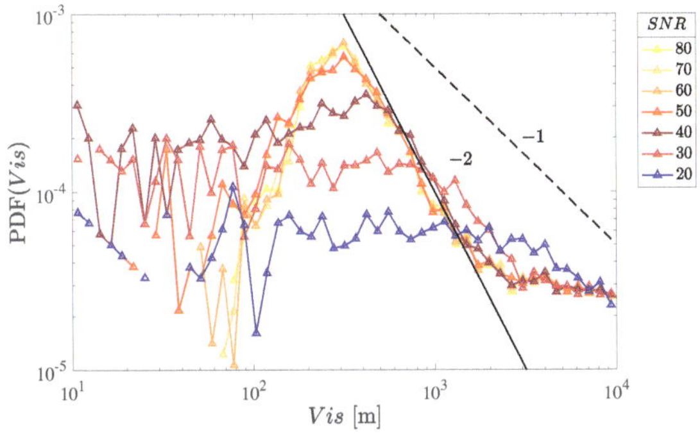

Figure 15. Signal amplitude PDFs for *Vis* with various *SNR*. Clear −2 scaling is present for *SNR* as low as 40.

5. Discussion and Conclusions

A preliminary analysis of the ability of *RH* and *k*, putative variables akin to the formation and breakup conditions of fog, respectively, to explain the variability in *Vis* through the spectra, the squared correlation coefficient, and the magnitude-squared coherence shows that *RH* and *k* play a role in fog occurrence but may not encapsulate its full dynamics. For the slow time scales (>1 d), the spectra for *Vis* exhibit a slower decay rate than either *RH* or *k*, which could indicate the presence of a process perhaps of meso and synoptic scales. When plotted against *Vis*, *RH* appears to be correlated to *Vis* ($R^2 = 0.63$), while *k* presents as a threshold where fog occurrence (low *Vis*) appears to coincide most with low *k*. The coherence analysis indicates that although there are certain scales where *k* and *RH* might be reasonable predictors of *Vis*, the stochastic behavior in *Vis* cannot fully be explained by *RH*, *k*, or a linear combination of the two.

Although not able to fully explain fog occurrence, *RH* and *k* bounds many stochastic features of *Vis*, including its spectral exponent *n*, clustering exponent α, and the growth of its block entropy slope. For intermediate and fine time scales (scales < 1 d), the spectral exponent of *Vis* ($n = 1.64$) is bounded by that of *k* ($n = 1.60$) and *RH* ($n = 1.93$). Similarly, the cluster exponent of *Vis* ($\alpha = 0.25$) is bounded by that of *k* ($\alpha = 0.26$) and *RH* ($\alpha = 0.24$). Lastly, the slope of $H(n)$ with *n* for *Vis* is bounded by *RH* and *k* for $n > 1$. Overall, the combined TA series RH^1k_0 reflected behavior similar to that of *RH* and *k* (i.e., in its spectral exponents, cluster exponent, intermittency exponent, and laminar phase duration distribution scaling) but did not perform significantly better than either *RH* or *k* in reproducing the stochastic features of *Vis*. These observations suggest that the statistics of fog are affected by those of saturation for the formation and those of turbulence for the breakup but that a linear combination of the two is too simplistic to capture the dynamics of fog.

The intermittent nature of fog was then analyzed by hypothesizing that fog is governed by one of the known intermittency models, which would inform future studies on whether and how it can be modeled. Out of the four intermittency models considered (PM1-3I and on–off), the *Vis* series under study presents features of PM3I in its power spectrum scaling (which approaches $-1/2$ as $f \to 0$), signal amplitude distribution scaling (a power law decay with an exponent of −2 over almost a decade), and laminar phase duration distribution scaling (a power law decay with an exponent of $-3/2$ for $l < 24$ h). These features were not observed in the *RH* or *k* series. More generally, PM3I can be viewed as an example of sporadic randomness, which describes systems between purely random and purely deterministic behaviors. Preliminary analysis of the block entropy growth based on the TA series indicates sub-linear behaviors consistent with sporadic randomness. Although limited by the sample length, the slope of $H(n)$ appears to reach a constant as *n* increases for all three variables. Qualitatively, PM3I entails a slow accumulation followed by a random re-injection or relaxation. The intermittency in fog is driven by the dynamics of the atmospheric boundary layer, where the atmospheric water vapor slowly accumulates (the rising limb of the characteristic PM3I processes) under proper moisture and heating conditions (encoded in RH) and is then 're-injected' whenever turbulence levels are high (and presumed to be stochastic). Identifying the physical origin of this intermittency remains an open question, as many variables (e.g., meteorological variables, aerosols, and wave breaking, some of which are interrelated) occurring at various time scales affect the life cycle of fog.

While intermittent behavior emerges as the system heads to chaotic behavior, SOC describes the phase transition between an ordered and a disordered state. Here, the ordered states are the no-fog phases, while the disordered states are fog intensities (represented here by the Vis time series). In finite systems, the critical point between second-order phase transitions is smoothed out over a small interval, so the coexistence between some features of no-fog phases (high k) and the presence of fog could then indicate that the system is at criticality. The integral time scales associated with each variable are calculated and presented in Table 3. For the full series, the integral time scale for k (10.3 h) appears to be slightly longer than that of RH (7.6 h) and Vis (8.0 h). For the binary TA series, the integral time scales for all three variables are commensurate, ranging from 6.9 to 7.6 h such that there seems to be a single relaxation time scale that governs the token formation and breakup process of fog, at least in their clustering tendencies.

Table 3. Integral time scales of the full and TA series in hours.

	RH	k	Vis
Full	7.6	10.3	8.0
TA	7.6	7.1	6.9

Thus, a simplistic physical model that lends itself to such phase transitions and criticality emerges where the activation mechanism for fog arises from thermodynamic interactions while the deactivation mechanism arises from turbulence. For fog events to occur, an air mass must become saturated near the earth's surface, and a sufficient amount of fog condensation nuclei (e.g., sea salts in marine fog) must be present to generate small airborne water droplets (1–30 μm in diameter) that remain suspended. Turbulence (and possibly background winds, which are related to k) then serves as the destructive or deactivating mechanism that dissipates fog. That is, fog is 'deactivated' if the turbulence becomes strong enough, and 'activated' if turbulence is weak enough, provided that air is saturated and condensation nuclei are present. The ensuing dynamics encapsulating these arguments may be expressed by [57]

$$\frac{dA}{dt} = \frac{A}{\tau_o}[\alpha_o(1 - A) - \beta_o], \tag{12}$$

where $A = 0$ is a no-fog state, $A = 1$ is a full fog (or $Vis = 0$), τ_o reflects the fog formation time, α_o is related to the thermodynamic state for fog formation (e.g., $\alpha_o = 0$ for low RH and unity for $RH = 1$), and β_o is related to k ($\beta_o = 0$ for low k and $\beta_o = 1$ for high k). This system exhibits two equilibrium states: $A = 0$ (no fog) and $A = 1 - \beta_o/\alpha_o$ (fog). The activation phase occurs through α_o (or RH) and the deactivation phase is associated with a correlated random β_o (linked to k). In the absence of turbulence ($\beta_o = 0$) but for finite α_o, A will gradually increase until steady-state $dA/dt = 0$ conditions prevail ($A = 1$). These conditions are sporadically 'broken' by either a small α_o (thermodynamic considerations no longer support fog such as advection or entrainment of dry or warm air) or a large β_o (high level of turbulence breaking up the fog). Fog formation necessitates $\beta_o/\alpha_o < 1$. Other mechanisms that dissipate fog beyond turbulence not considered here but worth investigating in future studies include radiation, lack of water vapor sustained in the atmosphere, particle coalescence, or loss of particles near the ground that could serve as condensation nuclei (and all impact α_o to a leading order).

The aim of this paper was to investigate whether SOC describes fog occurrence is based on its universal stochastic patterns in the form of power laws and scale invariance (in time or space or both) in complex systems. The current *Vis* series exhibits power laws in its amplitude distribution, spectra, clustering, and laminar period duration distribution, permitting an evaluation of the relation between the measured exponent of the TA spectra m and the exponent of the laminar period duration scaling γ expected for an SOC system at criticality ($m = 3 - \gamma$). This SOC relation was observed in *Vis* and in RH only after intermittency corrections ($m = 3 - \gamma - \mu$). In RH, this was achieved with the intermittency coefficient calculated from the full series ($\mu = \mu_s$), while incorporating μ_{TA} calculated from the TA series overestimates the measured m. In *Vis*, intermittency correction from the TA series ($\mu = \mu_{TA}$) was needed to achieve the SOC relation. These observations further suggest that fog events, or the underlying formation mechanisms associated with RH, could be SOC processes, albeit the eduction of information needing intermittency corrections.

In celebration of Dr. Herring's career and life work in geophysical fluid dynamics [78,79], stochastic approaches to turbulence [80,81], and the intermittency of high Reynolds number turbulence in particular, this paper presents a statistical analysis of fog and two of its underlying atmospheric variables—turbulent kinetic energy and relative humidity—in relation to its intermittency and self-organized critical behavior. Overall, this study identified the qualitative dynamical features of fog (through its visibility time series) and of the two variables that govern its formation and dissipation. Although not diagnostic, phenomenological properties of PM3I were observed in the *Vis* series, which may be more generally described by a sporadically random process. The results presented also conform with elements of self-organized criticality in its on–off and off–on switching properties. These features and similarities to known dynamical models in literature can be sought in future modeling efforts to reproduce the stochastic properties of fog in a broad range of time scales or as validation. Since many atmospheric variables and physical mechanisms affect the formation, duration, and dissolution of fog, the physical origin and exact analogy to SOC or PM3I, especially under different fog-formation conditions, remain an open question suggested for future inquiries.

Author Contributions: Conceptualization, K.Y.H. and G.G.K.; Data curation, K.Y.H., T.J.H. and J.R.-P.; Formal analysis, K.Y.H.; Funding acquisition, Q.W. and H.J.S.F.; Investigation, K.Y.H., T.J.H., J.R.-P. and Q.W.; Methodology, G.G.K.; Project administration, Q.W. and H.J.S.F.; Software, G.G.K.; Visualization, K.Y.H.; Writing—original draft, K.Y.H.; Writing—review and editing, G.G.K., T.J.H., J.R.-P., Q.W. and H.J.S.F. All authors have read and agreed to the published version of the manuscript.

Funding: K.Y.H., T.J.H., J.R.-P., Q.W. and H.J.S.F. were funded by the Grant N00014-21-1-2296 (Fatima Multidisciplinary University Research Initiative) of the Office of Naval Research, administered by the Marine Meteorology and Space Program. G.G.K. acknowledges partial support from the U.S. National Science Foundation (NSF-AGS-2028633) and the Department of Energy (DE-SC0022072).

Institutional Review Board Statement: Not applicable.

Informed Consent Statement: Not applicable.

Data Availability Statement: Data are available upon request.

Acknowledgments: The authors would like to thank the field research team of Sable Island, Parks Canada—Marvin Willis and Paul LeBlanc—for their invaluable assistance with project operations.

Conflicts of Interest: The authors declare no conflict of interest.

References

1. Myers, J.N. Fog. *Sci. Am.* **1968**, *219*, 74–83. [CrossRef]
2. Shrestha, S.; Moore, G.A.; Peel, M.C. Trends in winter fog events in the Terai region of Nepal. *Agric. For. Meteorol.* **2018**, *259*, 118–130. [CrossRef]
3. Baldocchi, D.; Waller, E. Winter fog is decreasing in the fruit growing region of the Central Valley of California. *Geophys. Res. Lett.* **2014**, *41*, 3251–3256. [CrossRef]

4. Klemm, O.; Schemenauer, R.S.; Lummerich, A.; Cereceda, P.; Marzol, V.; Corell, D.; Van Heerden, J.; Reinhard, D.; Gherezghiher, T.; Olivier, J.; et al. Fog as a fresh-water resource: Overview and perspectives. *Ambio* **2012**, *41*, 221–234. [CrossRef]
5. Montecinos, S.; Carvajal, D.; Cereceda, P.; Concha, M. Collection efficiency of fog events. *Atmos. Res.* **2018**, *209*, 163–169. [CrossRef]
6. Salcedo-Sanz, S.; Piles, M.; Cuadra, L.; Casanova-Mateo, C.; Caamaño, A.; Cerro-Prada, E.; Camps-Valls, G. Long-term persistence, invariant time scales and on-off intermittency of fog events. *Atmos. Res.* **2021**, *252*, 105456. [CrossRef]
7. Foufoula-Georgiou, E.; Krajewski, W. Recent advances in rainfall modeling, estimation, and forecasting. *Rev. Geophys.* **1995**, *33*, 1125–1137. [CrossRef]
8. Georgakakos, K.P.; Kavvas, M.L. Precipitation analysis, modeling, and prediction in hydrology. *Rev. Geophys.* **1987**, *25*, 163–178. [CrossRef]
9. Gupta, V.K.; Waymire, E. Multiscaling properties of spatial rainfall and river flow distributions. *J. Geophys. Res. Atmos.* **1990**, *95*, 1999–2009. [CrossRef]
10. Menabde, M.; Harris, D.; Seed, A.; Austin, G.; Stow, D. Multiscaling properties of rainfall and bounded random cascades. *Water Resour. Res.* **1997**, *33*, 2823–2830. [CrossRef]
11. Olsson, J.; Burlando, P. Reproduction of temporal scaling by a rectangular pulses rainfall model. *Hydrol. Process.* **2002**, *16*, 611–630. [CrossRef]
12. Rigby, J.; Porporato, A. Precipitation, dynamical intermittency, and sporadic randomness. *Adv. Water Resour.* **2010**, *33*, 923–932. [CrossRef]
13. Peters, O.; Neelin, J.D. Critical phenomena in atmospheric precipitation. *Nat. Phys.* **2006**, *2*, 393–396. [CrossRef]
14. Bak, P.; Chen, K. Self-organized criticality. *Sci. Am.* **1991**, *264*, 46–53. [CrossRef]
15. Räsänen, M.; Chung, M.; Katurji, M.; Pellikka, P.; Rinne, J.; Katul, G. Similarity in fog and rainfall intermittency. *Geophys. Res. Lett.* **2018**, *45*, 10–691. [CrossRef]
16. Bergé, P.; Pomeau, Y.; Vidal, C. *Order Within Chaos*; Wiley: Hoboken, NJ, USA, 1987.
17. Lakshmanan, M.; Rajaseekar, S. *Nonlinear Dynamics: Integrability, Chaos and Patterns*; Springer Science & Business Media: Berlin/Heidelberg, Germany, 2012.
18. Pomeau, Y.; Manneville, P. Intermittent transition to turbulence in dissipative dynamical systems. In *Universality in Chaos*; Routledge: Abingdon-on-Thames, UK , 2017; pp. 327–335.
19. Hirsch, J.; Huberman, B.; Scalapino, D. Theory of intermittency. *Phys. Rev. A* **1982**, *25*, 519. [CrossRef]
20. Scholz, H.; Yamada, T.; Brand, H.; Graham, R. Intermittency and chaos in a laser system with modulated inversion. *Phys. Lett. A* **1981**, *82*, 321–323. [CrossRef]
21. Meunier, C.; Bussac, M.N.; Laval, G. Intermittency at the onset of stochasticity in nonlinear resonant coupling processes. *Phys. D Nonlinear Phenom.* **1982**, *4*, 236–243. [CrossRef]
22. Pomeau, Y.; Roux, J.; Rossi, A.; Bachelart, S.; Vidal, C. Intermittent behaviour in the Belousov-Zhabotinsky reaction. *J. De Phys. Lettres* **1981**, *42*, 271–273. [CrossRef]
23. Bergé, P.; Dubois, M.; Mannevillel, P.; Pomeau, Y. Intermittency in Rayleigh-Bénard convection. *J. De Phys. Lettres* **1980**, *41*, 341–345. [CrossRef]
24. Maurer, J.; Libchaber, A. Effect of the Prandtl number on the onset of turbulence in liquid 4He. *J. De Phys. Lettres* **1980**, *41*, 515–518. [CrossRef]
25. Jeffries, C.; Perez, J. Observation of a Pomeau-Manneville intermittent route to chaos in a nonlinear oscillator. *Phys. Rev. A* **1982**, *26*, 2117. [CrossRef]
26. Richetti, P.; Argoul, F.; Arneodo, A. Type-II intermittency in a peroidically driven nonlinear oscillator. *Phys. Rev. A* **1986**, *34*, 726. [CrossRef] [PubMed]
27. Huang, J.Y.; Kim, J.J. Type-II intermittency in a coupled nonlinear oscillator: Experimental observation. *Phys. Rev. A* **1987**, *36*, 1495. [CrossRef] [PubMed]
28. Ringuet, E.; Rozé, C.; Gouesbet, G. Experimental observation of type-II intermittency in a hydrodynamic system. *Phys. Rev. E* **1993**, *47*, 1405. [CrossRef]
29. Velazquez, J.P.; Khosravani, H.; Lozano, A.; Bardakjian, B.L.; Carlen, P.L.; Wennberg, R. Type III intermittency in human partial epilepsy. *Eur. J. Neurosci.* **1999**, *11*, 2571–2576. [CrossRef]
30. Griffith, T.; Parthimos, D.; Crombie, J.; Edwards, D.H. Critical scaling and type-III intermittent chaos in isolated rabbit resistance arteries. *Phys. Rev. E* **1997**, *56*, R6287. [CrossRef]
31. Richter, R.; Peinke, J.; Clauss, W.; Rau, U.; Parisi, J. Evidence of type-III intermittency in the electric breakdown of p-type germanium. *EPL (Europhys. Lett.)* **1991**, *14*, 1. [CrossRef]
32. Kahn, A.; Mar, D.; Westervelt, R. Spatial measurements near the instability threshold in ultrapure Ge. *Phys. Rev. B* **1992**, *45*, 8342. [CrossRef]
33. Ono, Y.; Fukushima, K.; Yazaki, T. Critical behavior for the onset of type-III intermittency observed in an electronic circuit. *Phys. Rev. E* **1995**, *52*, 4520. [CrossRef]
34. Kim, C.M.; Yim, G.S.; Ryu, J.W.; Park, Y.J. Characteristic relations of type-III intermittency in an electronic circuit. *Phys. Rev. Lett.* **1998**, *80*, 5317. [CrossRef]

35. Dubois, M.; Rubio, M.; Berge, P. Experimental evidence of intermittencies associated with a subharmonic bifurcation. *Phys. Rev. Lett.* **1983**, *51*, 1446. [CrossRef]
36. Platt, N.; Spiegel, E.; Tresser, C. On-off intermittency: A mechanism for bursting. *Phys. Rev. Lett.* **1993**, *70*, 279. [CrossRef] [PubMed]
37. Pikovsky, A. On the interaction of strange attractors. *Z. Für Phys. B Condens. Matter* **1984**, *55*, 149–154. [CrossRef]
38. Fujisaka, H.; Yamada, T. A new intermittency in coupled dynamical systems. *Prog. Theor. Phys.* **1985**, *74*, 918–921. [CrossRef]
39. Heagy, J.; Platt, N.; Hammel, S. Characterization of on-off intermittency. *Phys. Rev. E* **1994**, *49*, 1140. [CrossRef]
40. Yu, Y.H.; Kwak, K.; Lim, T.K. On-off intermittency in an experimental synchronization process. *Phys. Lett. A* **1995**, *198*, 34–38. [CrossRef]
41. Hammer, P.W.; Platt, N.; Hammel, S.M.; Heagy, J.F.; Lee, B.D. Experimental observation of on-off intermittency. *Phys. Rev. Lett.* **1994**, *73*, 1095. [CrossRef]
42. Rödelsperger, F.; Čenys, A.; Benner, H. On-off intermittency in spin-wave instabilities. *Phys. Rev. Lett.* **1995**, *75*, 2594. [CrossRef]
43. Feng, D.; Yu, C.; Xie, J.; Ding, W. On-off intermittencies in gas discharge plasma. *Phys. Rev. E* **1998**, *58*, 3678. [CrossRef]
44. John, T.; Stannarius, R.; Behn, U. On-off intermittency in stochastically driven electrohydrodynamic convection in nematics. *Phys. Rev. Lett.* **1999**, *83*, 749. [CrossRef]
45. Bauer, M.; Habip, S.; He, D.R.; Martienssen, W. New type of intermittency in discontinuous maps. *Phys. Rev. Lett.* **1992**, *68*, 1625. [CrossRef] [PubMed]
46. He, D.R.; Wang, D.K.; Shi, K.J.; Yang, C.h.; Chao, L.y.; Zhang, J.Y. Critical behavior of dynamical systems described by the inverse circle map. *Phys. Lett. A* **1989**, *136*, 363–368. [CrossRef]
47. Price, T.; Mullin, T. An experimental observation of a new type of intermittency. *Phys. D Nonlinear Phenom.* **1991**, *48*, 29–52. [CrossRef]
48. San-Martin, J.; Antoranz, J. Type-II intermittency with a double reinjection channel: Multintermittency. *Phys. Lett. A* **1996**, *219*, 69–73. [CrossRef]
49. Grebogi, C.; Ott, E.; Yorke, J.A. Crises, sudden changes in chaotic attractors, and transient chaos. *Phys. D Nonlinear Phenom.* **1983**, *7*, 181–200. [CrossRef]
50. Schuster, H.G.; Just, W. *Deterministic Chaos: An Introduction*; John Wiley & Sons: Hoboken, NJ, USA, 2006.
51. Procaccia, I.; Schuster, H. Functional renormalization-group theory of universal 1/f noise in dynamical systems. *Phys. Rev. A* **1983**, *28*, 1210. [CrossRef]
52. Venkataramani, S.C.; Antonsen ,T.M., Jr.; Ott, E.; Sommerer, J.C. On-off intermittency: Power spectrum and fractal properties of time series. *Phys. D Nonlinear Phenom.* **1996**, *96*, 66–99. [CrossRef]
53. Toniolo, C.; Provenzale, A.; Spiegel, E.A. Signature of on-off intermittency in measured signals. *Phys. Rev. E* **2002**, *66*, 066209. [CrossRef]
54. Gaspard, P.; Wang, X.J. Sporadicity: Between periodic and chaotic dynamical behaviors. *Proc. Natl. Acad. Sci. USA* **1988**, *85*, 4591–4595. [CrossRef]
55. Legenstein, R.; Maass, W. Edge of chaos and prediction of computational performance for neural circuit models. *Neural Netw.* **2007**, *20*, 323–334. [CrossRef] [PubMed]
56. Grieger, B. Quaternary climatic fluctuations as a consequence of self-organized criticality. *Phys. A Stat. Mech. Its Appl.* **1992**, *191*, 51–56. [CrossRef]
57. Hesse, J.; Gross, T. Self-organized criticality as a fundamental property of neural systems. *Front. Syst. Neurosci.* **2014**, *8*, 166. [CrossRef] [PubMed]
58. Roberts, D.C.; Turcotte, D.L. Fractality and self-organized criticality of wars. *Fractals* **1998**, *6*, 351–357. [CrossRef]
59. Turcotte, D.L. Self-organized criticality. *Rep. Prog. Phys.* **1999**, *62*, 1377. [CrossRef]
60. Jensen, H.J. *Self-Organized Criticality: Emergent Complex Behavior in Physical and Biological Systems*; Cambridge University Press: Cambridge, UK, 1998; Volume 10.
61. Pruessner, G. *Self-Organised Criticality: Theory, Models and Characterisation*; Cambridge University Press: Cambridge, UK, 2012.
62. Dorman, C.E.; Mejia, J.; Koračin, D.; McEvoy, D. Worldwide marine fog occurrence and climatology. In *Marine Fog: Challenges and Advancements in Observations, Modeling, and Forecasting*; Springer: Berlin/Heidelberg, Germany, 2017; pp. 7–152.
63. Horst, T.; Oncley, S. Corrections to inertial-range power spectra measured by CSAT3 and Solent sonic anemometers, 1. Path-averaging errors. *Bound. Layer Meteorol.* **2006**, *119*, 375–395. [CrossRef]
64. Horst, T.; Semmer, S.; Maclean, G. Correction of a non-orthogonal, three-component sonic anemometer for flow distortion by transducer shadowing. *Bound. Layer Meteorol.* **2015**, *155*, 371–395. [CrossRef]
65. Gultepe, I.; Milbrandt, J.A.; Zhou, B. Marine fog: A review on microphysics and visibility prediction. In *Marine Fog: Challenges and Advancements in Observations, Modeling, and Forecasting*; Springer: Berlin/Heidelberg, Germany, 2017; pp. 345–394.
66. Bershadskii, A.; Niemela, J.; Praskovsky, A.; Sreenivasan, K. "Clusterization" and intermittency of temperature fluctuations in turbulent convection. *Phys. Rev. E* **2004**, *69*, 056314. [CrossRef]
67. Sreenivasan, K.; Bershadskii, A. Clustering properties in turbulent signals. *J. Stat. Phys.* **2006**, *125*, 1141–1153. [CrossRef]
68. Poggi, D.; Katul, G. Flume experiments on intermittency and zero-crossing properties of canopy turbulence. *Phys. Fluids* **2009**, *21*, 065103. [CrossRef]

69. Cava, D.; Katul, G. The effects of thermal stratification on clustering properties of canopy turbulence. *Bound. Layer Meteorol.* **2009**, *130*, 307–325. [CrossRef]
70. Cava, D.; Katul, G.G.; Molini, A.; Elefante, C. The role of surface characteristics on intermittency and zero-crossing properties of atmospheric turbulence. *J. Geophys. Res. Atmos.* **2012**, *117*, 1–17. [CrossRef]
71. Cava, D.; Mortarini, L.; Giostra, U.; Acevedo, O.; Katul, G. Submeso motions and intermittent turbulence across a nocturnal low-level jet: A self-organized criticality analogy. *Bound. Layer Meteorol.* **2019**, *172*, 17–43. [CrossRef]
72. Heisel, M.; Katul, G.G.; Chamecki, M.; Guala, M. Velocity asymmetry and turbulent transport closure in smooth-and rough-wall boundary layers. *Phys. Rev. Fluids* **2020**, *5*, 104605. [CrossRef]
73. Chowdhuri, S.; Kalmár-Nagy, T.; Banerjee, T. Persistence analysis of velocity and temperature fluctuations in convective surface layer turbulence. *Phys. Fluids* **2020**, *32*, 076601. [CrossRef]
74. Huang, K.Y.; Katul, G.G.; Hultmark, M. Velocity and temperature dissimilarity in the surface layer uncovered by the telegraph approximation. *Bound. Layer Meteorol.* **2021**, *180*, 385–405. [CrossRef]
75. Molini, A.; Katul, G.G.; Porporato, A. Revisiting rainfall clustering and intermittency across different climatic regimes. *Water Resour. Res.* **2009**, *45*. [CrossRef]
76. Shannon, C.E. A mathematical theory of communication. *Bell Syst. Tech. J.* **1948**, *27*, 379–423. [CrossRef]
77. Jensen, H.J.; Christensen, K.; Fogedby, H.C. 1/f noise, distribution of lifetimes, and a pile of sand. *Phys. Rev. B* **1989**, *40*, 7425. [CrossRef]
78. Herring, J.R.; Cermak, J.E.; Neu, W.L.; Richardson, E.V.; Diplas, P.; Rand, R.H.; Yates, G.T.; Spedding, G.R.; de Laurier, J.D.; Hoult, D.P.; et al. Fluid Dynamics in Nature. *Handb. Fluid Dyn. Fluid Mach. Appl. Fluid Dyn.* **1996**, *3*, 1847–1989.
79. Herring, J. A Brief History of the Geophysical Turbulence Program at NCAR. In *Proceedings of the IUTAM Symposium on Developments in Geophysical Turbulence*; Springer: Berlin/Heidelberg, Germany, 2000; pp. 1–4.
80. Herring, J.R. Chapter 4: An Introduction and Overview of Various Theoretical Approaches to Turbulence . In *Applied Mathematical Sciences*; Springer: Berlin/Heidelberg, Germany, 1985; Volume 58, p. 73.
81. Herring, J. The role of statistical models in turbulence. *Stoch. Model. Geosystems* **2012**, *85*, 129.

atmosphere

MDPI

Article

Potential Vorticity Generation in Breaking Gravity Waves

Michael L. Waite * and Nicholas Richardson †

Department of Applied Mathematics, University of Waterloo, Waterloo, ON N2L 3G1, Canada
* Correspondence: mwaite@uwaterloo.ca
† Current address: Department of Mathematics, University of British Columbia, Vancouver, BC V6T 1Z2, Canada.

Abstract: Potential vorticity (PV) is an important quantity in stratified flows because it is conserved following the flow in the absence of forcing and viscous and diffusive effects. However, as shown by previous work for unstratified turbulence, viscosity and diffusion, when present, are not purely dissipative and can create potential vorticity even when none is present initially. In this work, we use direct numerical simulations to investigate the viscous and diffusive generation of potential vorticity and potential enstrophy (integrated square PV) in stratified turbulence. Simulations are initialized with a two-dimensional standing internal gravity wave, which has no potential vorticity apart from some low-level random noise; as a result, all potential vorticity and enstrophy comes from viscous and diffusive effects. Significant potential enstrophy is found when the standing wave breaks, and the maximum potential enstrophy increases with increasing Reynolds number. The mechanism for the initial PV generation is spanwise diffusion of buoyancy perturbations, which grow as the standing wave three-dimensionalizes, into the direction of spanwise vorticity. The viscous and diffusive terms responsible are small-scale and are sensitive to under-resolution, so high resolution is required to obtain robust results.

Keywords: stratified turbulence; potential vorticity; direct numerical simulation

Citation: Waite, M.L.; Richardson, N. Potential Vorticity Generation in Breaking Gravity Waves. *Atmosphere* **2023**, *14*, 881. https://doi.org/10.3390/atmos14050881

Academic Editors: Boris Galperin, Annick Pouquet and Peter Sullivan

Received: 18 April 2023
Revised: 9 May 2023
Accepted: 12 May 2023
Published: 18 May 2023

1. Introduction

The Ertel potential vorticity (PV) is an important quantity in geophysical fluid dynamics. PV characterizes the component of vorticity normal to the isosurfaces of a conserved scalar, and it is materially conserved in the absence of forcing and molecular viscosity and diffusion. In quasi-geostrophic (QG) dynamics, where the PV can be inverted to find the velocity and buoyancy, conservation of PV is everything. Even for the primitive dynamics of the atmosphere and ocean, where large-scale QG turbulence interacts with inertia–gravity waves, weakly rotating stratified turbulence at intermediate scales, and isotropic turbulence at small scales, PV is a useful quantity for analyzing the vortical part of the flow. Indeed, PV is the basis of various decomposition methods for separating slow, quasi-horizontal vortical motion, which has PV, from fast, propagating inertia–gravity waves, which have no PV [1–4].

QG dynamics break down at intermediate length scales in the atmosphere and ocean, where stratification is strong but rotation is weaker than at large scales. Stratified turbulence is often used as an idealization of turbulent dynamics at these scales [5,6]. In stratified turbulence, PV is a useful diagnostic for quantifying the role of vortices and gravity waves. In the Craya–Herring decomposition [7] and the closely related vortical/wave mode decomposition [2], the vortical component of the flow accounts for the linear PV. Lilly [5] proposed that conservation of linear PV might constrain stratified turbulence and lead to an inverse energy cascade, as in two-dimensional turbulence. However, an inverse cascade does not occur in stratified turbulence [6,8], due to a leakage of energy from vortices to gravity waves [9] and the importance of nonlinear terms in the PV [10,11] due to the development of small vertical scales [6,12]. Nevertheless, the question of whether the downscale cascade of energy through such scales in the atmosphere and ocean is dominated by vortices or waves has not yet been fully resolved [13,14].

Since PV is used to diagnose and separate vortices and gravity waves in geophysical flows and stratified turbulence, it is important to know where PV comes from. In QG turbulence, the integrated square PV, called potential enstrophy, cascades conservatively to smaller scales, analogously to the enstrophy cascade in two-dimensional turbulence. As a result, the QG enstrophy cascade from the large-scale circulation is one source of PV at intermediate and small scales in the atmosphere and ocean, where QG breaks down and stratified turbulence may be present.

However, viscosity and diffusion can also create PV. Because PV depends nonlinearly on vorticity and buoyancy, viscous and diffusive effects are not purely dissipative or necessarily restricted to small length scales, as they are for regular vorticity. This effect was explored in unstratified turbulence with a passive scalar by Herring, Kerr, and Rotunno [15]. By analyzing the evolution of PV in decaying turbulence, they found that viscous and diffusive effects on PV may occur at large scales. By considering a Taylor–Green vortex oriented with zero initial PV, they demonstrated that viscosity and diffusion can create large-scale PV even when none is initially present. These findings raised doubts about whether the notion of a potential enstrophy cascade is meaningful in the absence of QG turbulence. Subsequent work suggested that a potential enstrophy cascade may be possible in certain parameter regimes of geophysical turbulence [16], but the role of viscous and diffusive generation of potential vorticity in stratified turbulence has not been explored.

In this paper, inspired by [15], we use direct numerical simulation (DNS) to investigate the role of viscosity and diffusion on the evolution of PV in stratified turbulence. Compared to [15], we explore the effects of stratification and the impact of larger Reynolds numbers. A two-dimensional standing internal gravity wave is used to initialize simulations of decaying stratified turbulence. A standing wave is a linear solution to the equations of motion, but instabilities and nonlinearities inevitably cause the wave to break down into smaller-scale turbulence. Standing waves have been used in laboratory [17] and numerical [18–20] studies of stratified turbulence. More generally, numerical simulation is an important tool for studying breaking gravity waves, which are a key source of turbulence in the atmosphere [21–23]. A breaking standing wave is a useful flow configuration for quantifying the viscous generation of PV because, like the Taylor–Green case in [15], the initial wave has no PV. As a result, any PV in the resulting turbulence must come from viscous generation. We will explore the magnitude and mechanisms of PV generation, the associated length scales, and the dependence on viscosity and stratification.

2. Materials and Methods

2.1. Equations

We use the uniformly stratified, non-rotating Boussinesq equations to describe stratified turbulence. The equations of motion are

$$\frac{\mathrm{D}\boldsymbol{u}}{\mathrm{D}t} = -\nabla p + b\hat{\boldsymbol{z}} + \nu\nabla^2\boldsymbol{u}, \tag{1}$$

$$\nabla \cdot \boldsymbol{u} = 0, \tag{2}$$

$$\frac{\mathrm{D}b}{\mathrm{D}t} + N^2 w = \kappa\nabla^2 b, \tag{3}$$

where $\mathrm{D}/\mathrm{D}t = \partial/\partial t + \boldsymbol{u}\cdot\nabla$ is the material derivative, $\boldsymbol{u} = (u, v, w)$ is the velocity, b is the buoyancy, p is the pressure scaled by a reference density, N is the constant buoyancy frequency, ν is the kinematic viscosity, κ is the buoyancy diffusivity, and $\hat{\boldsymbol{z}}$ is the vertical unit vector. There is no forcing. The buoyancy b is proportional to the negative fluctuation of density from the linear basic state, and N^2 is proportional to the background density gradient; the total buoyancy $b_{\mathrm{tot}} = N^2 z + b$ is proportional to the (negative) density.

The Ertel PV for (1)–(3) is

$$q = \boldsymbol{\omega} \cdot \left(N^2 \hat{\boldsymbol{z}} + \nabla b \right) = N^2 \omega_z + \boldsymbol{\omega} \cdot \nabla b, \tag{4}$$

where $\boldsymbol{\omega} = \nabla \times \boldsymbol{u} = (\omega_x, \omega_y, \omega_z)$ is the vorticity [24]. The PV evolves according to

$$\begin{aligned} \frac{\mathrm{D}q}{\mathrm{D}t} &= \nu \left(N^2 \hat{\boldsymbol{z}} + \nabla b \right) \cdot \nabla^2 \boldsymbol{\omega} + \kappa \boldsymbol{\omega} \cdot \nabla^2 (\nabla b) \\ &= \nu N^2 \nabla^2 \omega_z + \nu \nabla b \cdot \nabla^2 \boldsymbol{\omega} + \kappa \boldsymbol{\omega} \cdot \nabla^2 (\nabla b). \end{aligned} \tag{5}$$

In the absence of viscosity and diffusion, $\mathrm{D}q/\mathrm{D}t = 0$ and the PV is materially conserved. When $\nu, \kappa > 0$, the evolution of q is more complicated. Due to the nonlinear dependence of q on $\boldsymbol{\omega}$ and b, the viscous and diffusive terms in the q Equation (5) are not simply diffusive. Viscous effects can modify q by viscous diffusion of the vorticity into the direction of the buoyancy gradient, either by diffusion of ω_z along the background buoyancy gradient N^2 or by diffusion of $\boldsymbol{\omega}$ in the direction of ∇b. In addition, buoyancy diffusion can modify q by diffusion of the buoyancy gradient into the direction of the vorticity. As a result, as indicated by [15], viscosity and diffusion may create PV even when none is initially present.

The potential enstrophy is the integrated square PV, and is given by

$$V = \frac{1}{2} \langle q^2 \rangle, \tag{6}$$

$$= \frac{1}{2} N^4 \langle \omega_z^2 \rangle + N^2 \langle \omega_z \, \boldsymbol{\omega} \cdot \nabla b \rangle + \frac{1}{2} \langle (\boldsymbol{\omega} \cdot \nabla b)^2 \rangle, \tag{7}$$

where $\langle \cdot \rangle$ denotes domain average. Assuming no flux through the boundary, the potential enstrophy satisfies

$$\begin{aligned} \frac{\mathrm{d}V}{\mathrm{d}t} &= \langle \nu N^2 q \nabla^2 \omega_z + \nu q \nabla b \cdot \nabla^2 \boldsymbol{\omega} + \kappa q \boldsymbol{\omega} \cdot \nabla^2 (\nabla b) \rangle \\ &= D_\nu + D_\kappa, \end{aligned} \tag{8}$$

where D_ν is the term with viscosity ν and D_κ is the term with buoyancy diffusivity κ. The viscous and diffusive terms are not necessarily dissipative, and may be a source or sink of potential enstrophy.

2.2. Scale Analysis

We can estimate the magnitude of the potential enstrophy in stratified turbulence using scale analysis. In stratified turbulence with large buoyancy Reynolds number $Re_b = \epsilon/(\nu N^2)$, where ϵ is the kinetic energy dissipation rate, there should be a Kolmogorov inertial range at small scales between the Ozmidov and Kolmogorov scales

$$L_O = \left(\frac{\epsilon}{N^3} \right)^{1/2}, \quad \eta = \left(\frac{\nu^3}{\epsilon} \right)^{1/4}, \tag{9}$$

respectively [25–27], since $L_O/\eta \sim (Re_b)^{3/4}$ [28,29]. Since the PV is a function of the velocity and buoyancy gradients, it should be dominated by the smallest scales when such an inertial range is present.

Consider the third term in (7), which is likely to be the largest term when the Reynolds number is large since it depends on the squared vorticity and squared buoyancy gradient. Using Kolmogorov scaling gives $\boldsymbol{\omega} \sim v/\eta$, where $v = (\nu\epsilon)^{1/4}$ is the Kolmogorov velocity scale. We assume that the small-scale buoyancy fluctuations are $b \sim vN$. As a result, estimating V using the third term in (7) gives

$$V \sim \frac{N^2 \epsilon^2}{\nu^2} \sim \frac{N^4 U^2}{L^2} Re^2 Fr^2 \sim \frac{N^4 U^2}{L^2} Re_b Re, \tag{10}$$

where the second and third $\sim$ use Taylor's hypothesis $\epsilon \sim U^3/L$ [6], U and L are the large-scale velocity and length scales, $Re = UL/\nu$ is the Reynolds number, and $Fr = U/(NL)$ is the Froude number. Note that N^4U^2/L^2 is the magnitude of the potential enstrophy from the large energy-containing scales. This estimate suggests that stratified turbulence may have significant potential enstrophy at small scales, and that potential enstrophy increases with increasing Re and Re_b.

2.3. Numerical Approach

Simulations are performed in a cubic domain with size 2π and periodic boundary conditions. The initial conditions for a two-dimensional standing wave with unit wavenumber, following [19], are

$$u = a \sin x \cos z, \tag{11}$$
$$w = -a \cos x \sin z, \tag{12}$$
$$b = 0, \tag{13}$$

where $a > 0$ is the amplitude and $v = 0$. In the linear equations, these initial conditions excite a standing wave

$$u = a \sin x \cos z \cos(\omega t), \tag{14}$$
$$w = -a \cos x \sin z \cos(\omega t), \tag{15}$$
$$b = \frac{aN^2}{\omega} \cos x \sin z \sin(\omega t), \tag{16}$$

with frequency $\omega = N/\sqrt{2}$. Since the initial buoyancy and vertical vorticity are zero, the initial standing wave has no PV. Standing waves eventually break and generate small-scale turbulence [19,20]. In particular, the standing wave (14)–(16) is overturning and subject to convective instability when $\sqrt{2}a/N > 1$. In our simulations, the wave amplitude is fixed at $a = 1$. Stratification and the overturning of the standing wave are controlled by varying N.

Equations (1)–(3) are simulated with a Fourier-based spectral transform model. Time-stepping is third-order Adams–Bashforth and the viscous and diffusion terms are handled with a Crank–Nicolson approach [30]. Aliasing is removed using the two-thirds rule, so with n^3 grid points, wave vectors $\boldsymbol{k} = (k_x, k_y, k_z)$ are truncated at $|k_x|, |k_y|, |k_z| = k_T = n/3$, i.e. we keep $-n/3 < k_x, k_y, k_z < n/3$. The resolution n is chosen to resolve the Kolmogorov scale with $\eta k_T \geq 2.5$, where η is calculated using the maximum value of $\epsilon = \nu\langle|\boldsymbol{\omega}|^2\rangle$. As discussed below, high resolution is required to resolve the viscous terms in the q equation properly. While recent studies have shown that stratified turbulence may be sensitive to increasing Prandtl number $Pr = \nu/\kappa$ at Re-values that are attainable in DNS [31,32], in this work we set $Pr = 1$ for computational efficiency, so $\kappa = \nu$. A small amount of random noise is added to the initial velocity, with uniform distribution and magnitude $a/100$, to facilitate three-dimensionalization. This noise introduces a small amount of initial q and V that is quickly removed by viscous and diffusion terms, as shown below.

Since the initial standing wave has unit velocity and unit wavenumber, Equations (1)–(3) can be interpreted as dimensionless with $\nu = 1/Re$, $N = 1/Fr$, and unit turnover time. We use ν, N and Re, Fr interchangeably in what follows to distinguish stratification and viscosity. Several simulations at different Re and Fr are discussed below and summarized in Table 1. The control simulation (Main) has $n = 1024$, $N = 1$ ($Fr = 1$), $\nu = 0.0003$ ($Re = 3333$), and $\sqrt{2}a/N = \sqrt{2}$, so the standing wave overturns and quickly breaks down into stratified turbulence. Sensitivity to Reynolds number is explored in simulations High Visc and Low Visc, in which Re is decreased to 1667 and increased to 4500. Sensitivity to stratification is explored in simulations High Strat and Higher Strat, in which Fr is reduced to $1/\sqrt{2}$ and $1/2$. Finally, sensitivity to resolution is investigated in simulations Low Res and High Res by reducing and increasing the resolution to 512 and 1536 grid points.

Table 1. Summary of simulations.

Run	Fr	$\sqrt{2}a/N$	Re	n	Max ϵ	ηk_T	Max Re_b
Main	1	$\sqrt{2}$	3333	1024	0.0037	3.2	12.2
High Visc	1	$\sqrt{2}$	1667	512	0.0047	2.5	7.9
Low Visc	1	$\sqrt{2}$	4500	1728	0.0060	3.8	26.9
High Strat	$1/\sqrt{2}$	1	3333	1024	0.0052	2.9	8.6
Higher Strat	1/2	$1/\sqrt{2}$	3333	1034	0.0030	3.3	2.5
Low Res	1	$\sqrt{2}$	3333	512	0.0041	1.5	13.8
High Res	1	$\sqrt{2}$	3333	1536	0.0040	4.6	13.2

3. Results

3.1. Main Simulation

We begin by examining the potential enstrophy evolution in the Main simulation, which has $Fr = 1$, $Re = 3333$, and a large, overturning wave amplitude. Time series of kinetic and potential energy, energy dissipation, turbulent Froude number, and potential enstrophy are shown in Figure 1. The wave period for this simulation is $T = 2\pi\sqrt{2} \approx 8.9$. The wave makes two periods with negligible dissipation. By $t = 20$, the energy dissipation begins increasing and the energy begins to decrease more significantly. The potential energy dissipation reaches a maximum at $t = 26$, followed by the maximum kinetic energy dissipation at $t = 30$. Table 1 gives $\eta k_T = 3.2$ and $Re_b = 12.2$, both of which are based on the maximum value of ϵ in the simulation, which is also given in Table 1. Figure 1c shows that the turbulent Froude number $Fr_t = \epsilon/(NE_K) \lesssim 0.1$, where Fr_t is defined using velocity scale $\sqrt{E_K}$, where E_k is the domain-averaged kinetic energy, and length scale from Taylor's hypothesis [29].

The potential enstrophy in this simulation is shown in Figure 1d. The potential enstrophy is negligible until $t \approx 24$. The inset shows V at early times with a log scale; the small potential enstrophy introduced by the random noise decays rapidly until $t \approx 15$, and then begins to increase. A maximum value of $V = 1.9$ is reached at $t = 29$. To put this number in perspective, we can calculate the potential enstrophy that would be associated with the initial velocity field if it was rotated by 90°:

$$V_0 = \frac{1}{2}N^4\langle\left(\omega_y^0\right)^2\rangle = \frac{1}{2}, \tag{17}$$

where ω_y^0 is the vorticity associated with (11)–(13). Therefore, we can see that significant potential enstrophy is generated in this simulation, about four times larger than what one would expect from the initial velocity field and stratification alone.

Vertical (x, z) slices of the total buoyancy b_{tot}, normal component of vorticity ω_y, and potential vorticity q are plotted in Figure 2 for a selection of times from $t = 16$, before the onset of significant dissipation, to $t = 36$, when the turbulence is decaying. First, we discuss the evolution of the buoyancy and normal vorticity (left and centre columns of Figure 2). At $t = 16$, the buoyancy (Figure 2a) shows the structure of the standing wave with some overturning isopycnals, as expected. At $t = 20$, the overturning isopycnals have collapsed (Figure 2b) with significant spanwise vorticity in the overturning regions (Figure 2g). By $t = 24$, the buoyancy field of the standing wave has been significantly altered by breaking (Figure 2c), and smaller spanwise vortices are present (Figure 2h). At $t = 28$ and 36, the wave has largely broken down into turbulence, with the vorticity dominated by small scales (Figure 2d–e,i–j).

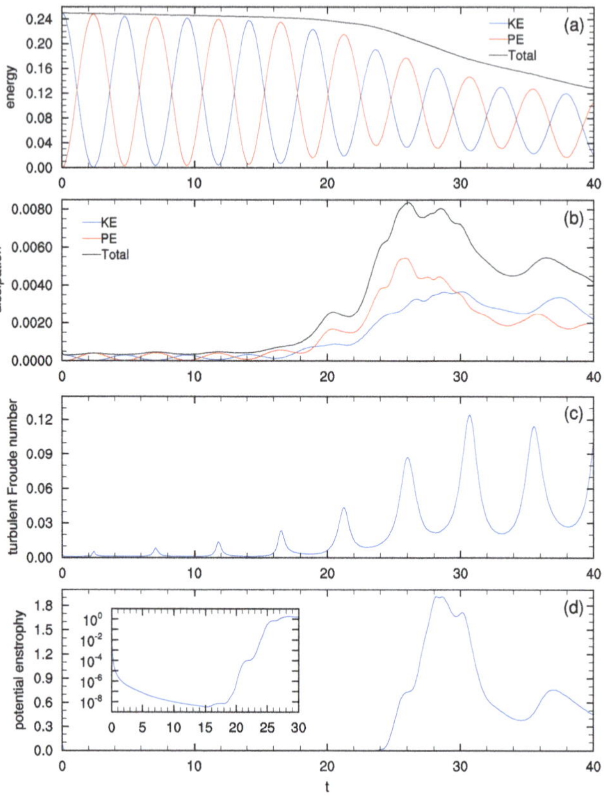

Figure 1. Time series from the Main simulation of (**a**) kinetic and potential energy, (**b**) kinetic and potential energy dissipation, (**c**) turbulent Froude number, and (**d**) potential enstrophy. The inset in (**d**) shows potential enstrophy on a log scale to illustrate the decay of the initial potential enstrophy from random noise.

The potential vorticity is shown in the right column of Figure 2. There is no significant PV visible until $t = 24$. At $t = 24$, thin bands of positive and negative PV appear, mainly in the lower overturning region (similar bands are apparent in the upper overturning region at other y). By $t = 28$, the PV structure has become more rich and complex, with both small-scale bands and vortex structures visible. At this time, the PV is still mainly located in the upper and lower breaking regions, avoiding $z = 0$, π, and 2π. This time corresponds to the maximum V in Figure 1d. The PV occupies more of the domain at $t = 36$, but the amplitude is lower, consistent with the decay of V seen in Figure 1d.

To investigate the increase in PV around $t \approx 24$, time series of the terms in the potential enstrophy Equation (8) are shown in Figure 3. These terms are negligible for $t \lesssim 24$, but begin to deviate from zero for $t \gtrsim 24$. From $24 < t < 26$, the buoyancy diffusivity term D_κ dominates, and is responsible for the initial creation of V as the wave begins to break. At these times, the viscous terms are slightly dissipative. V grows rapidly from $24 < t < 26$, but the growth slows around $t = 26$ (Figure 1d). At $t = 26$, the growth in V increases again, this time, as seen in Figure 3, due to a sharp increase in the viscous term D_ν. After around $t = 28$, the combined viscous and diffusion term is negative and V decreases, apart from a short increase in V around $t = 36$.

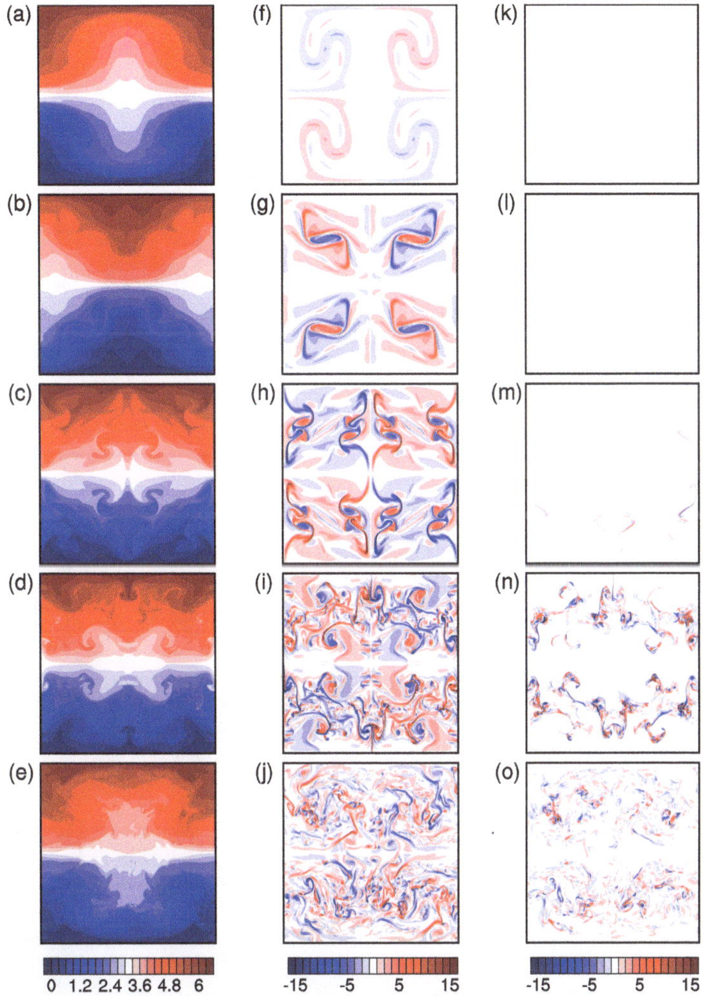

Figure 2. Vertical (x, z) slices of **(a–e)** b_{tot}, **(f–j)** ω_y, and **(k–o)** q at (from top to bottom) $t = 16$, 20, 24, 28, and 36 from Main simulation. All slices are shown at $y = 0$.

Figure 3 indicates that the initial generation of PV, as seen in the rise in V, is a result of the κ term in the V equation, which corresponds to diffusion of the buoyancy gradient into the direction of vorticity. The κ term can be decomposed as

$$D_\kappa = \kappa q \boldsymbol{\omega} \cdot \nabla^2(\nabla b) = \kappa q \omega_x \nabla^2 \frac{\partial b}{\partial x} + \kappa q \omega_y \nabla^2 \frac{\partial b}{\partial y} + \kappa q \omega_z \nabla^2 \frac{\partial b}{\partial z}. \quad (18)$$

For $24 < t < 26$, D_κ is dominated by the second term, which involves ω_y and $\partial b/\partial y$ (not shown). The mechanism of PV generation by this term is illustrated in Figure 4, which shows a vertical slice of q and two horizontal slices of b at $t = 24$. The flow is still mainly two-dimensional at this time, but three-dimensionalization of the buoyancy field is visible along a few lines parallel to the y-axis in Figure 4b,c. The planes $z = z_1$ and $z = z_2$ cut through bands of PV in the lower breaking region. Along z_1, the PV bands occur precisely where the buoyancy field is three-dimensionalizing, as seen in Figure 4b, where the ticks mark the locations of the PV along z_1 in Figure 4a. Similarly, the PV bands along $z = z_2$

occur at the same location as the buoyancy three-dimensionalization at that height. Three-dimensionalization of b promotes growth of $\partial b/\partial y$, which diffuses into the direction of ω_y, which at this time is still the dominant component of the vorticity. Therefore, PV first emerges at the locations where three-dimensionalization of the buoyancy develops, via the second term in (18).

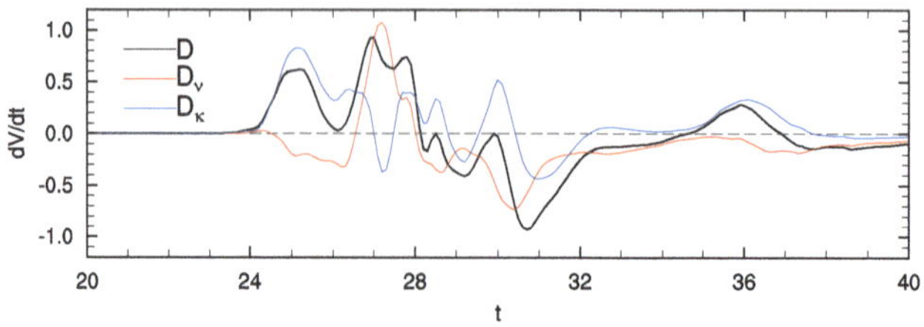

Figure 3. Time series of the terms in the potential enstrophy Equation (8) from the Main simulation.

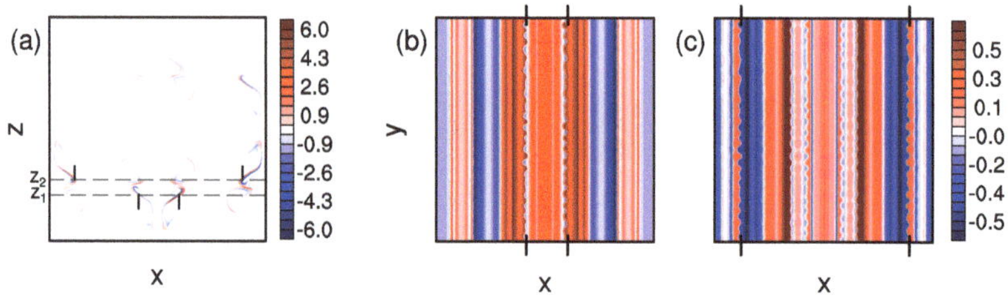

Figure 4. (**a**) Vertical (x, z) slice of q, and horizontal (x, y) slices of b through (**b**) $z = z_1$ and (**c**) $z = z_2$, for the Main simulation at $t = 24$. Note that z_1 and z_2 are shown in (**a**). The vertical line segments in (**a**) mark the locations of significant PV along $z = z_1$ and $z = z_2$; these locations are marked with ticks in (**b**) for z_1 and (**c**) for z_2, respectively.

Kinetic energy spectra $E(k)$ and potential enstrophy spectra $V(k)$, plotted as functions of the total wavenumber $k = |\boldsymbol{k}|$, are shown in Figure 5. Spectra are calculated at $t = 28$, which is the time of maximum potential enstrophy. As expected, the energy spectrum is peaked at the primary standing wave wavenumber $k = 1$, oscillates at the lowest few wavenumbers, and then falls off like a power law until the Kolmogorov wavenumber, which is marked in the figure. By contrast, the potential enstrophy spectrum is peaked at large wavenumbers. It has a positive slope of around 3.3 for $1 \leq k \leq 20$. Despite the fact that V is generated by viscous and diffusive effects, $V(k)$ has a broad peak around $k \approx 20$ to 40, which is just beyond the Ozmidov wavenumber, which is marked in the figure. However, we will see below that this peak is not set by the Ozmidov wavenumber.

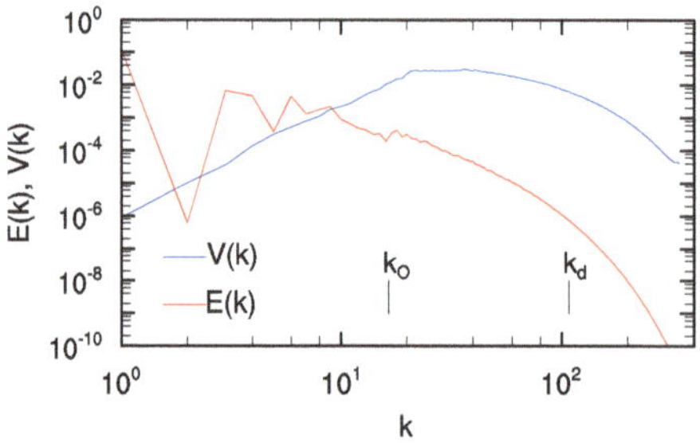

Figure 5. Energy and potential enstrophy spectra from the Main simulation at $t = 28$. The Ozmidov and Kolmogorov wavenumbers $k_O = 1/L_O$ and $k_d = 1/\eta$ are shown.

3.2. Sensitivity to Reynolds and Froude Numbers

We investigate the dependence of viscous and diffusive PV generation on Reynolds number with two additional simulations with lower and higher Reynolds numbers. The Reynolds number is varied by changing ν, and the resolution is adjusted to resolve the Kolmogorov scale. Time series of the potential enstrophy for these simulations, along with the Main simulation, are shown in Figure 6a. With smaller Re, the maximum potential enstrophy is reduced from around 1.9 to 0.4. The time of maximum potential enstrophy is also delayed, from around $t = 28$ in the Main simulation to $t = 37$ in the High Visc simulation. On the other hand, for the Low Visc simulation with larger Re, the maximum potential enstrophy is much larger, reaching a value of around 64 at $t = 27$, which is slightly earlier than the Main simulation. Potential enstrophy spectra from these simulations are shown in Figure 7a. The potential enstrophy spectra for the Main and High Visc experiments are similar at small k but increase with increasing Re at larger k, indicating that the increased potential enstrophy with higher Reynolds number is primarily at small scales going from High Visc to Main. However, the potential enstrophy in the Low Visc case is increased at all scales, not just small scales.

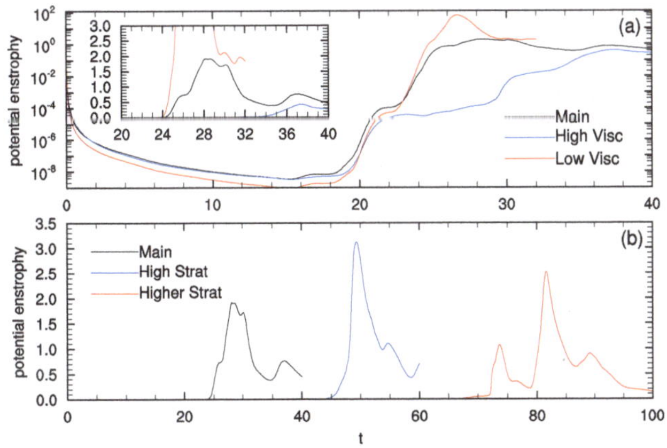

Figure 6. Time series of potential enstrophy from simulations with (**a**) different Re and (**b**) different Fr. A log scale is used in (**a**) with a linear scale in the inset.

Time series of potential enstrophy from simulations with increased stratification are shown in Figure 6b. Increasing the stratification delays the onset of wave breaking and, as a result, the growth in potential enstrophy is also delayed. The High Strat simulation has $\sqrt{2}a/N = 1$ and is therefore on the threshold of convective instability; this simulation reaches a maximum potential enstrophy of 3.1 at $t = 49$, which is larger and later than in the Main simulation. The Higher Strat simulation has $\sqrt{2}a/N = 1/\sqrt{2}$ and reaches a maximum V of 2.5 at $t = 82$. This is a smaller maximum V than in the High Strat simulation, and it also occurs later. Potential enstrophy spectra from these simulations are shown in Figure 7b. The peak in potential enstrophy remains around $k \approx 20$ to 40, even as the Ozmidov wavenumber, which is shown in the figure, increases from 16 to 37 with increasing stratification. The peak in the potential enstrophy spectrum therefore does not scale with the Ozmidov wavenumber. In addition, the large-scale potential enstrophy, with $k < 20$, increases with increasing stratification.

To make sense of the dependence of V on viscosity and stratification, we can compare our simulation results with the prediction (10). The dependence on stratification in particular is difficult to understand, because the maximum V seems to increase, then decrease, with increasing stratification. However, analyzing the dependence of V on Fr or N alone presents an incomplete picture, because the strength of the turbulence, as captured by the maximum ϵ, also changes with stratification (Table 1). Equation (10) suggests that the maximum V should depend on ν, N, and ϵ. Figure 8 shows the maximum V plotted against $N^2\epsilon^2/\nu^2$, where the maximum value of ϵ is used (maximum ϵ generally occurs around the same time as maximum V). While the agreement with (10) is not particularly good, we do find that maximum V increases with increasing $N^2\epsilon^2/\nu^2$, which accounts for the non-monotonic dependence on stratification. In all cases, the maximum potential enstrophy is less than what was predicted by (10), but the Low Visc case, which has the highest Re and Re_b, is closest to (10).

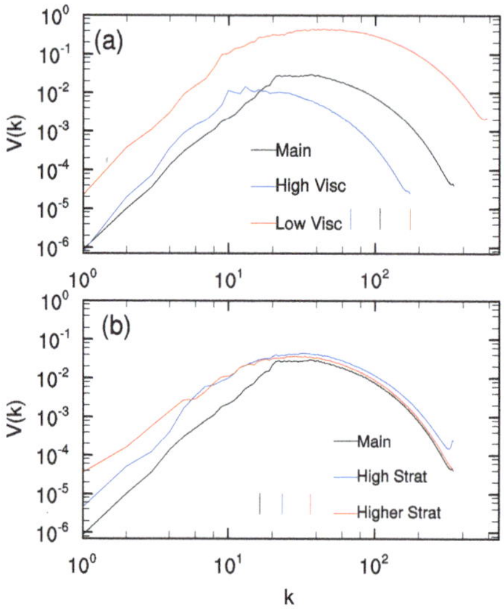

Figure 7. Potential enstrophy spectra from simulations with (**a**) different Re and (**b**) different Fr, at the time of maximum potential enstrophy. The vertical line segments mark the Kolmogorov wavenumbers in (**a**) and the Ozmidov wavenumbers in (**b**).

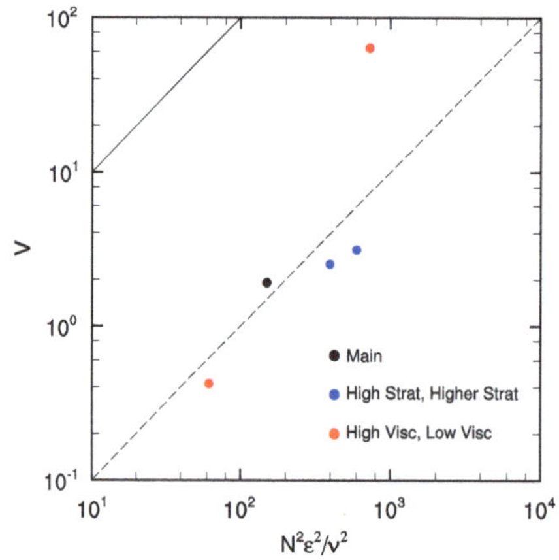

Figure 8. Maximum potential enstrophy V plotted against $N^2\epsilon^2/\nu^2$, where the maximum value of ϵ is used, for the Main simulation (black), simulations with different stratifications (High Strat and Higher Strat, in blue), and simulations with different viscosities (Low Visc and High Visc, in red). The reference lines are $V = N^2\epsilon^2/\nu^2$ (solid) and $V = 0.01N^2\epsilon^2/\nu^2$ (dashed).

3.3. Sensitivity to Numerical Resolution

High spatial resolution is required to accurately resolve the viscous and diffusive terms in the PV equation. Under-resolution can lead to spurious generation of PV. The dependence on resolution is illustrated in Figure 9, which shows time series of V for the Main simulation and simulations with higher and lower resolution. The Low Res simulation, which has $\eta k_T = 1.5$, has a maximum V of around 4.6. This value is more than double the maximum V of 1.9 in the Main simulation, which has $\eta k_T = 3.2$. By contrast, the High Res simulation, which has $\eta k_T = 4.6$, has a maximum V of 1.8, which is just slightly smaller than the value in the Main simulation. The timing of maximum V is slightly accelerated in the Main simulation compared to the High Res simulation, but these simulations are much closer to one another than to the Low Res case. Inadequate resolution results in accelerated and artificially high potential enstrophy growth.

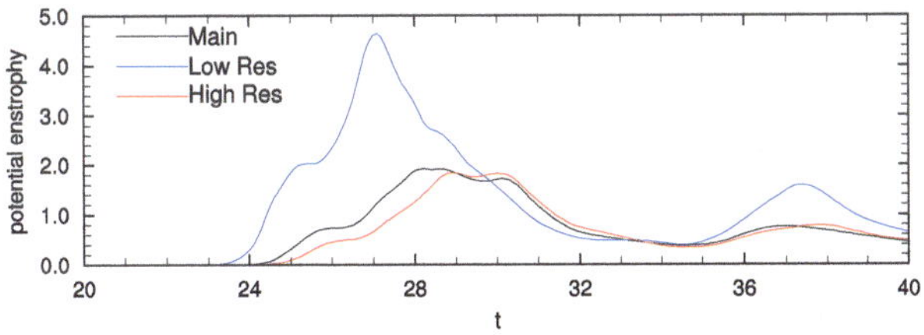

Figure 9. Time series of potential enstrophy from simulations with different resolutions.

4. Discussion

A breaking standing gravity wave is a useful case for studying viscous and diffusive generation of PV in stratified turbulence because such flows have no initial PV. Any PV that develops must come from viscous and diffusive effects. Our simulations indicate that significant PV and potential enstrophy are created as these waves break, much more than would be expected from the velocity and length scales of the breaking wave. Indeed, more potential enstrophy is created at higher Reynolds numbers, because even though the viscosity and diffusion coefficients are smaller, the velocity and buoyancy gradients in the turbulence are correspondingly larger. The Reynolds numbers in this work, while large compared to what was possible in [15], are still small compared to geophysical fluid values. It therefore seems likely that viscous and diffusive generation of PV from breaking gravity waves may be even more significant in the atmosphere and ocean.

The mechanism of the initial PV generation is diffusion of spanwise buoyancy fluctuations, which develop as the waves three-dimensionalize, into the direction of the spanwise vorticity, which is initially dominant. As a result, the initial PV generation occurs at small length scales associated with three-dimensionalization of the flow. Once significant potential enstrophy is present, spectra are peaked at intermediate wavenumbers 20–40, corresponding to scales smaller than the wavelength of the breaking wave, but much larger than the Kolmogorov scale. Interestingly, this intermediate scale does not vary like the Ozmidov scale, and indeed does not change much with Reynolds or Froude number, indicating that it might be related to the scale of the breaking wave. The potential enstrophy at large scales, while relatively small, increases with increasing Reynolds numbers and decreasing Froude numbers.

The mechanism of PV generation in breaking standing waves is somewhat different from the Taylor–Green vortices considered by [15]. The Taylor–Green vortices were oriented so that they had no initial PV, like the standing wave. Potential enstrophy in their case was initially created at large scales, followed by more intense generation at small scales; they confirmed that the small-scale generation was due to viscous and diffusive effects rather than transfer from large scales. The standing wave does not have significant large-scale potential enstrophy at early times. However, for both the Taylor–Green vortex in [15] and our standing wave, viscous and diffusive effects are primarily a source of potential enstrophy; dissipation takes over only at late times. By contrast, for the decaying turbulence with Gaussian initial conditions considered in [15], viscous and diffusive effects were primarily dissipative. More work is required to determine if the differences between our results and the pioneering work of [15] are due to the different Reynolds numbers, initial configuration, or the presence of stratification. In addition, it would be interesting to examine the role of viscous and diffusive effects in stratified turbulence with large-scale PV from random initial conditions or forcing.

Viscous and diffusive generation of potential enstrophy is sensitive to numerical resolution. DNS requires resolution of the Kolmogorov scale. Resolutions with $\eta k_T \approx 1$ are typical for DNS of breaking gravity waves [23] and decaying stratified turbulence [33]. However, our findings show that under-resolution can lead to spurious generation of PV when $\eta k_T = 1.5$. Instead, larger values of $\eta k_T \geq 2.5$ are needed for robust simulations of potential enstrophy generation. Such high resolution is not routinely employed for DNS of stratified turbulence. This finding, while perhaps not surprising given the small-scale mechanisms at play, suggests that some DNS studies of stratified turbulence and gravity wave breaking may have unrealistically high potential enstrophy due to under-resolution of the viscous generation processes. More work is required to understand the implications on the dynamics of turbulence with artificially enhanced potential enstrophy.

Author Contributions: Conceptualization, M.L.W.; methodology, M.L.W.; software, M.L.W. and N.R.; validation, M.L.W. and N.R.; writing—original draft preparation, M.L.W.; writing—review and editing, M.L.W. and N.R.; visualization, M.L.W.; supervision, M.L.W. All authors have read and agreed to the published version of the manuscript.

Funding: This research was funded by the Natural Sciences and Engineering Research Council of Canada grant number RGPIN-07001-2020.

Data Availability Statement: Model data are available upon request to the corresponding author.

Acknowledgments: This research was enabled in part by support provided by Compute Ontario (www.computeontario.ca (accessed on 10 May 2023)) and the Digital Research Alliance of Canada (www.alliancecan.ca (accessed on 10 May 2023)).

Conflicts of Interest: The authors declare no conflict of interest.

References

1. Staquet, C.; Riley, J.J. On the Velocity Field Associated with Potential Vorticity. *Dyn. Atmos. Oceans* **1989**, *14*, 93–123. [CrossRef]
2. Lelong, M.P.; Riley, J.J. Internal Wave-Vortical Mode Interactions in Strongly Stratified Flows. *J. Fluid Mech.* **1991**, *232*, 1–19. [CrossRef]
3. Bartello, P. Geostrophic Adjustment and Inverse Cascades in Rotating Stratified Turbulence. *J. Atmos. Sci.* **1995**, *52*, 4410–4428. [CrossRef]
4. Kafiabad, H.A.; Bartello, P. Balance Dynamics in Rotating Stratified Turbulence. *J. Fluid Mech.* **2016**, *795*, 914–949. [CrossRef]
5. Lilly, D.K. Stratified Turbulence and the Mesoscale Variability of the Atmosphere. *J. Atmos. Sci.* **1983**, *40*, 749–761. [CrossRef]
6. Lindborg, E. The Energy Cascade in a Strongly Stratified Fluid. *J. Fluid Mech.* **2006**, *550*, 207–242. [CrossRef]
7. Herring, J.R. Approach of Axisymmetric Turbulence to Isotropy. *Phys. Fluids* **1974**, *17*, 859–872. [CrossRef]
8. Herring, J.R.; Métais, O. Numerical Experiments in Forced Stably Stratified Turbulence. *J. Fluid Mech.* **1989**, *202*, 97–115. [CrossRef]
9. Waite, M.L.; Bartello, P. Stratified Turbulence Dominated by Vortical Motion. *J. Fluid Mech.* **2004**, *517*, 281–308. [CrossRef]
10. Waite, M.L. Potential Enstrophy in Stratified Turbulence. *J. Fluid Mech.* **2013**, *722*, R4. [CrossRef]
11. Rosenberg, D.; Pouquet, A.; Marino, R. Correlation Between Buoyancy Flux, Dissipation and Potential Vorticity in Rotating Stratified Turbulence. *Atmosphere* **2021**, *12*, 157. [CrossRef]
12. Billant, P.; Chomaz, J.M. Self-Similarity of Strongly Stratified Inviscid Flows. *Phys. Fluids* **2001**, *13*, 1645–1651. [CrossRef]
13. Callies, J.; Bühler, O.; Ferrari, R. The Dynamics of Mesoscale Winds in the Upper Troposphere and Lower Stratosphere. *J. Atmos. Sci.* **2016**, *73*, 4853–4872. [CrossRef]
14. Li, Q.; Lindborg, E. Weakly or Strongly Nonlinear Mesoscale Dynamics Close to the Tropopause. *J. Atmos. Sci.* **2018**, *75*, 1215–1229. [CrossRef]
15. Herring, J.R.; Kerr, R.M.; Rotunno, R. Ertel's Potential Vorticity in Unstratified Turbulence. *J. Atmos. Sci.* **1994**, *51*, 35–47. [CrossRef]

16. Kurien, S.; Smith, L.; Wingate, B. On the two-point correlation of potential vorticity in rotating and stratified turbulence. *J. Fluid Mech.* **2006**, *555*, 131–140. [CrossRef]
17. Benielli, D.; Sommeria, J. Excitation of Internal Waves and Stratified Turbulence by Parametric Instability. *Dyn. Atmos. Oceans* **1996**, *23*, 335–343. [CrossRef]
18. Bouruet-Aubertot, P.; Sommeria, J.; Staquet, C. Breaking of Standing Internal Gravity Waves Through Two-Dimensional Instabilities. *J. Fluid Mech.* **1995**, *285*, 265–301. [CrossRef]
19. Bouruet-Aubertot, P.; Sommeria, J.; Staquet, C. Stratified Turbulence Produced by Internal Wave Breaking: Two-Dimensional Numerical Experiments. *Dyn. Atmos. Oceans* **1996**, *23*, 357–369. [CrossRef]
20. Carnevale, G.F.; Briscolini, M.; Orlandi, P. Buoyancy- to Inertial-Range Transition in Forced Stratified Turbulence. *J. Fluid Mech.* **2001**, *427*, 205–239. [CrossRef]
21. Fritts, D.C.; Isler, J.R.; Andreassen, Ø. Gravity Wave Breaking in Two and Three Dimensions 2. Three-Dimensional Evolution and Instability Structure. *J. Geophys. Res.* **1994**, *99*, 8109–8123. [CrossRef]
22. Dörnbrack, A. Turbulent mixing by breaking gravity waves. *J. Fluid Mech.* **1998**, *375*, 113–141. [CrossRef]
23. Remmler, S.; Fruman, M.D.; Hickel, S. Direct Numerical Simulation of a Breaking Inertia-Gravity Wave. *J. Fluid Mech.* **2013**, *722*, 424–436. [CrossRef]
24. Pedlosky, J. *Geophysical Fluid Dynamics*, 2nd ed.; Springer: Berlin, Germany, 1987.
25. Dougherty, J.P. The Anisotropy of Turbulence at the Meteor Level. *J. Atmos. Terr. Phys.* **1961**, *21*, 210–213. [CrossRef]
26. Ozmidov, R.V. On the Turbulent Exchange in a Stably Stratified Ocean. *Izvestia Akad. Nauk. SSSR Atmospheric and Oceanic Physics Ser.* **1965**, *1*, 853–860.
27. Davidson, P.A. *Turbulence in Rotating, Stratified and Electrically Conducting Fluids*; Cambridge University Press: Cambridge, UK, 2013.
28. Gargett, A.E.; Osborn, T.R.; Nasmyth, P.W. Local Isotropy and the Decay of Turbulence in a Stratified Fluid. *J. Fluid Mech.* **1984**, *144*, 231–280. [CrossRef]
29. Brethouwer, G.; Billant, P.; Lindborg, E.; Chomaz, J.M. Scaling Analysis and Simulation of Strongly Stratified Turbulent Flows. *J. Fluid Mech.* **2007**, *585*, 343–368. [CrossRef]
30. Durran, D.R. *Numerical Methods for Wave Equations in Geophysical Fluid Dynamics*; Springer: Berlin, Germany, 1999.
31. Legaspi, J.D.; Waite, M.L. Prandtl Number Dependence of Stratified Turbulence. *J. Fluid Mech.* **2020**, *903*, A12. [CrossRef]
32. Riley, J.J.; Couchman, M.M.P.; de Bruyn Kops, S.M. The Effect of Prandtl Number on Decaying Stratified Turbulence. *J. Turbul.* **2023**, 1–19. [CrossRef]
33. Maffioli, A.; Davidson, P.A. Dynamics of Stratified Turbulence Decaying From a High Buoyancy Reynolds Number. *J. Fluid Mech.* **2016**, *786*, 210–233. [CrossRef]

atmosphere

Article

Review of Jackson Herring's Early Work on Thermal Convection

Robert M. Kerr

Department of Mathematics, University of Warwick, Coventry CV4 7AL, UK; robert.kerr@warwick.ac.uk

Abstract: Jack Herring had three mid-1960s numerical papers on Rayleigh-Bénard thermal convection that might seem primitive by today's standards, but already encapsulated many of the questions that are still being asked. All of them use severely truncated versions of the incompressible Navier–Stokes–Boussinesq equations with only one, or just a few, horizontal Fourier modes. In the first two papers, 1963 and 1964, the presented results used only one Fourier mode α and three variables. The single mode's variables are its vertical velocity profile $w_\alpha(z,t)$, its temperature profile $\theta_\alpha(z,t)$ and the horizontally uniform vertical profile of the background temperature $\psi(z,t)$. All of the second- and third-order terms are ignored except the convective heat flux $\overline{w\theta}$. The objective was to find asymptotic steady-state solutions. Each paper found evidence for the one-third Nusselt versus Rayleigh scaling of $Nu{\sim}Ra^{1/3}$, originally derived from Malkus' maximum flux principle. The 1963 paper uses free-slip upper and lower boundaries, with magnitudes of Nu that are a factor of three larger than the experiments. In the 1964 paper, by introducing no-slip/rigid boundary conditions, the magnitude of Nu dropped to within 20% of the experimental values. Both $Nu(Ra)$ relations are in good agreement with circa-1990 direct numerical simulations (DNS). This dependence upon the boundary condition at the walls suggests that to obtain physically realistic scaling, no-slip boundary conditions are necessary. The third paper is discussed only in terms of what it might have been aiming to accomplish and its relation to the earlier free-slip results.

Keywords: convection; turbulence; scaling laws

Citation: Kerr, R.M. Review of Jackson Herring's Early Work on Thermal Convection. *Atmosphere* **2023**, *14*, 907. https://doi.org/10.3390/atmos14060907

Academic Editors: Boris Galperin, Annick Pouquet and Peter Sullivan

Received: 6 April 2023
Revised: 19 May 2023
Accepted: 19 May 2023
Published: 23 May 2023

1. Introduction

When Jack Herring's mid-1960s convection papers [1–3] were published, there were already, as there are now, two primary predictions for the dependence of the dimensionless heat flux, the Nusselt number, Nu, upon the Rayleigh number, Ra. From Malkus [4,5] and Kraichnan [6], a scaling law for modest Rayleigh numbers was derived; from Kraichnan, an upper bound on the scaling for asymptotically large Ra was derived, suggesting, respectively,

$$Nu \sim CRa^{1/3} \quad \text{and} \quad Nu \sim CRa^{1/2}/(\log Ra)^{3/2} \quad \text{with} \quad Ra \equiv \frac{\alpha_e g \Delta T d^3}{\kappa\nu}. \tag{1}$$

where the variables are as follows: α_e is the linear coefficient of expansion; g is the gravitational acceleration; $\Delta T = (T_H - T_C)$ is the temperature difference between the hot and cold walls; d is the depth; κ is the thermal conductivity; and ν is the kinematic viscosity. A $Nu{\sim}CRa^{1/2}$ scaling law had also been suggested by Spiegel [7].

To obtain some numerical insight into whether either prediction might hold, Herring used very elementary numerics that incorporated the primary assumptions described by Malkus and Veronis [5]. Using a few horizontal Fourier modes at most, the variables of the reduced dynamics were the vertical profiles of the poloidal vertical velocities and temperatures. Velocities and the vertical vorticity from the toroidal component [8] were neglected. The goal was to find steady-state solutions by iterating in time. For the primary calculations, only one Fourier mode was used and, in each case, the $Nu{\sim}CRa^{1/3}$ scaling prediction was supported, with geometric and Reynolds-number-dependent coefficients,

C, in approximate agreement with 1990s-era direct numerical simulations (DNS) for both free(-slip) [9] and no-slip (rigid) boundaries [10]. No evidence was found for an ultra-high Rayleigh number $Nu{\sim}Ra^{1/2}$ regime.

This review begins by describing the reduced equations and the iteration method and then the results. First the free-slip calculations will be discussed, and then we will move on to those with rigid boundaries. The 1966 paper is not discussed in any detail because the 1963 scaling results are repeated. Then there is a discussion of energy budgets, the relationships to the first DNS direct numerical simulations and the overall message.

2. Governing Equations

The full incompressible Navier–Stokes–Boussinesq equations for the velocity vector $v(r,t)$ and temperature fields $T(r,t)$ for Prandtl number $\sigma = \nu/\kappa$ are:

$$\nabla \cdot v = 0, \tag{2}$$

$$\left(\frac{1}{\sigma}\frac{\partial}{\partial t} - \nabla^2\right)\nabla^2 v = \frac{1}{\sigma}\nabla \times \nabla \times (v \cdot \nabla v) + Ra\nabla \times (\nabla \times kT), \tag{3}$$

$$\left(\frac{\partial}{\partial t} - \nabla^2\right)T = -\nabla \cdot (vT). \tag{4}$$

From these equations, the dimensional variables (denoted by primes) are made non-dimensional using the depth d of the convection cell and κ.

$$v = \frac{d}{\kappa}v', \qquad T = T'/d,$$

$$r = r'/d, \qquad t = \frac{\kappa}{d^2}t'.$$

The double curl has eliminated the pressure from the equations and by considering only the poloidal component, one is left with only one velocity component: the vertical velocity $w(r,t)$. The temperature field $T(r,t)$ is then separated into a fluctuating part θ and a horizontally uniform background field $\psi(z,t)$.

$$T(r,t) = -z + \psi(z,t) + \theta(r,t) \quad \text{with} \quad \beta(z) = -\frac{\partial}{\partial z}\overline{T}(z) = 1 - \frac{\partial\psi}{\partial z}, \tag{5}$$

where $\overline{T}(z)$ is the horizontally averaged temperature. The boundary conditions at $z = 0$ and $z = 1$ on the temperature components are

$$\psi(0,t) = \psi(1,t) = 0 \tag{6}$$

$$\theta(x,y,0,t) = \theta(x,y,1,t) = 0. \tag{7}$$

Upon horizontally averaging the equations, and removing all except one of the horizontal advection and second-order fluctuation terms, the system can be reduced further, with the exception being the vertical heat transport $\overline{w\theta}$ that is forcing the horizontally averaged temperature ψ and its derivative β.

$$\left(\frac{1}{\sigma}\frac{\partial}{\partial t} - \nabla^2\right)\nabla^2 w_\alpha = Ra\nabla_\perp^2\theta_\alpha, \tag{8}$$

$$\left(\frac{1}{\sigma}\frac{\partial}{\partial t} - \nabla^2\right)\theta_\alpha = \beta w, \tag{9}$$

$$\left(\frac{\partial}{\partial t} - \frac{\partial^2}{\partial z^2}\right)\beta = +\frac{\partial^2}{\partial z^2}\overline{w\theta}. \tag{10}$$

where $\nabla_\perp = \partial_x^2 + \partial_y^2$. This is the later 1964 version of the reduced Equations [2].

The horizontally fluctuating terms are then decomposed into horizontal Fourier modes $w(\alpha_j, z, t)$ and $\theta(\alpha_j, z, t)$, where α_j represents a horizontal wavenumber $(k_x, k_y)_j$.

There are two nonlinear terms. βw in (9) and $\overline{w\theta}$ in (10). Today, these terms would be found in physical space, transformed to Fourier space using an FFT, derivatives taken, then transformed back to z-profiles. These calculations were pre-FFT, so the derivatives in the nonlinear terms are included using convolution—that is, summing in Fourier space the nonlinear, multiplied terms.

Once the time derivatives are set, time-iteration using simple forwards Euler is applied to determine if a steady state can be reached. If a steady state is obtained, this question can be posed: Are the steady-state solutions for the two geometries, free-slip and rigid, consistent with our current understanding, numerical or experimental, of convection in those geometries?

3. Herring 1963 Results

In a 1963 paper [1], only one horizontal Fourier mode was used in each simulation and the vertical profiles of all the variables expanded as sums of vertical sine functions. This imposes free-slip boundaries at $z = 0$ and 1 and makes the convolution sums for the nonlinear terms particularly easy.

Choosing values of α and the Rayleigh number that would support convection and not be unstable was not trivial. One restriction due to marginal stability [8] is that the system will not support convection if

$$\frac{(1+\alpha^2)^3}{\alpha^2} \geq \frac{Ra}{\pi^4}. \tag{11}$$

In Section 5 of Ref. [1], the stability is summarized with its Figure 17, including an observation that a small α is enormously unstable. The figures below come from the stable range of α and Ra.

As Ra increases, the variations in the boundary layers become stronger, with profiles that are qualitatively similar to those from free-slip DNS [9]. Furthermore, the fit for $Nu(Ra) = C\,Ra^{1/3}$ (1) gives $C = 0.31$, which with some digging, is exactly what the free-slip DNS finds for $Ra = 5 \times 10^5$ with $Nu = 23.7$ [9].

4. Herring 1964 Results

In the 1964 paper, rigid, no-slip boundary conditions were applied [2].

However, applying the sine function algorithms used for the free-slip calculations directly to the rigid case is impossible because at each wall the vertical velocity has two boundary conditions, not one.

$$w(0,t) = \frac{\partial w}{\partial z}(0,t) = w(1,t) = \frac{\partial w}{\partial z}(1,t) = 0. \tag{12}$$

To handle this, in the appendix of [2], inhomogeneous combinations of sinh and cosh functions [8] of $\pi\alpha z$ are added to terms based upon sines. The inhomogeneous parts allow all four boundary conditions in (12) to be satisfied. Then, Green's functions relating those inhomogeneous plus homogeneous functions to sines are computed.

With those Green functions, a time-advancing iteration procedure analogous to Equations (8)–(10) can be implemented.

Furthermore, to make things simpler, the Prandtl number limit $\sigma \to \infty$, $\sigma^{-1} \to 0$ was taken, hoping that this would include the $\sigma \sim 1$ of the air around us, an assumption that is partially borne out by the DNS of Kerr and Herring [11].

5. Discussion

In the first paper [1], it was shown how a single horizontal Fourier mode with simple parameterizations could generate the proposed $Nu \sim Ra^{1/3}$ scaling [4] for the heat flux. In the second paper [2], by imposing no-slip boundary conditions on the velocity, instead of free-slip, it was shown how two results could be generated and this was consistent with the observations: the coefficient on the $Ra^{1/3}$ scaling and the vertical profiles of w and θ.

However, this unanswered remains question: Does the asymptotic Kraichnan [6] upper bound for high Rayleigh number scaling, $Nu{\sim}Ra^{1/2}$, have any relevance? Today, this hypothetical regime is called the ultimate regime.

The third paper [3] was an attempt to pre-condition and accelerate the iteration towards a steady state in order to determine whether there might be additional scaling laws at much higher Rayleigh numbers, as suggested by that upper bound. The pre-conditioning was only applied to the free-slip geometry and there was no evidence for any new regimes as the earlier free-slip results [1] were reproduced.

Another concept being introduced to numerics was the role of what we now call the energy budget [1]. For Rayleigh-Bénard convection, there are two energy functions,

$$E_K = \tfrac{1}{2}\{w^2\}_v \quad \text{and} \quad E_P = \tfrac{1}{2}\left\{|\theta|^2 + |\psi|^2\right\}_v, \tag{13}$$

the kinetic energy E_K [5] and the potential energy E_P, which is called entropy in the paper. The v-subscripts indicate integration over the volume and the total energy is $E_P + E_K$. The E_P equation is

$$\frac{\partial}{\partial t}E_P + \left\{|\nabla\theta|^2 + |\nabla\psi|^2\right\}_v = \{w\theta\}_v. \tag{14}$$

The energy flow in this reduced system of just w_α, θ_α and β is as follows:

- The inflow begins as the vertical velocity w interacts with the mean temperature gradient β in (9), whose volume integral is $\{\beta w\}_v$.
- E_P then flows to the kinetic energy through the $\nabla_\perp^{-2} Ra \nabla_\perp^2 \theta$ term in (8).
- This is then dissipatively removed by $\nabla_\perp^{-2}$ inverting the $\nabla_\perp^4$ term of (8):

$$\nabla_\perp^{-2}\left\{\nabla_\perp^2 \nabla_\perp^2 w\right\}.$$

- Meanwhile, β is modified by (10).

Similarities and differences. Figures 1 and 2 for no-slip boundaries have the following similarities with, and differences from, the free-slip results in Figures 3 and 4,

- $\overline{T}(z)$ and $\beta(z)$ profiles are similar. Except at the walls where $\partial\beta/\partial z{\sim}0$.
- Similarly for $w(z)$ and $\theta(z)$, except again $\partial w/\partial z = 0$ at the walls.
- **For no-slip (rigid) walls** the coefficient in front of the $Ra^{1/3}$ given in the abstract is $C = 0.1153$ and is within 20% of the experiments. And consistent with $Ra < 10^6$ DNS [10] for which $C \approx 0.09{\sim}0.11$ if strictly $Nu{\sim}Ra^{1/3}$ scaling is assumed.

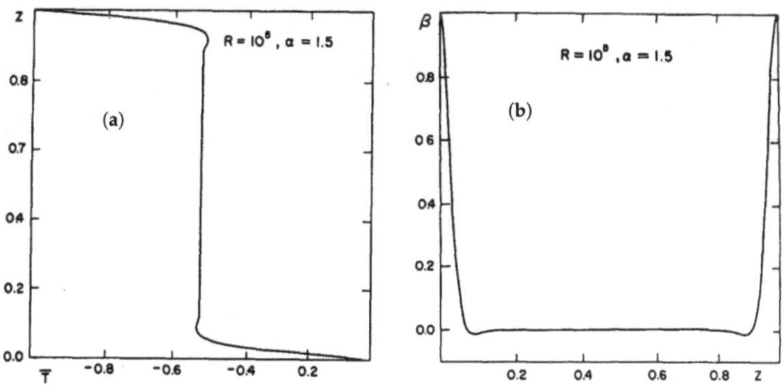

Figure 1. Rigid (no-slip) boundaries with $Ra = 10^5$ and $\pi\alpha = 5$. (**a**) Mean temperature $\overline{T}(z)$. (**b**) Mean gradient $\beta(z)$. $\beta(z)$ has been normalized by the total heat transport $Nu = 6.7$. Figure 6a,b from Ref. [2].

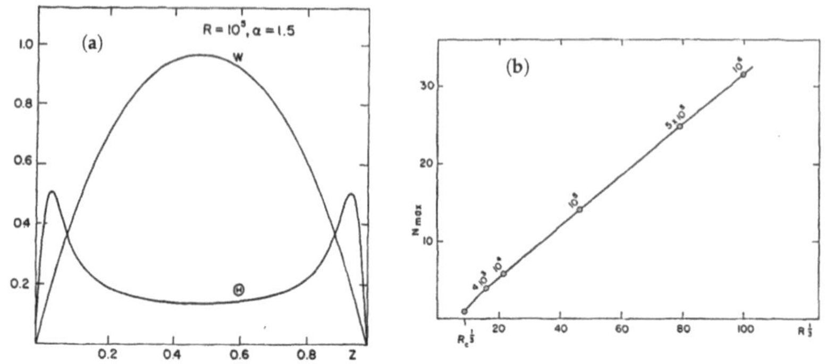

Figure 2. (**a**) $7.85 \times 10^{-3} w$ and $2.23\,\theta$ for $Ra = 10^5$ and $\pi\alpha = 5$. (**b**) The heat flux Nu as a function of $Ra^{1/3}$ from three estimates: $N_{\max}$ is the maximum value of Nu (over different $\pi\alpha$) for each given Ra. N_{RS} is the flux predicted by the relative stability criterion, and N_S is the heat transported by stable single-α solutions. The $N_{\max}$ coefficient given in the abstract is $C = 0.1153$. Figures 5 and 12 from Ref. [2].

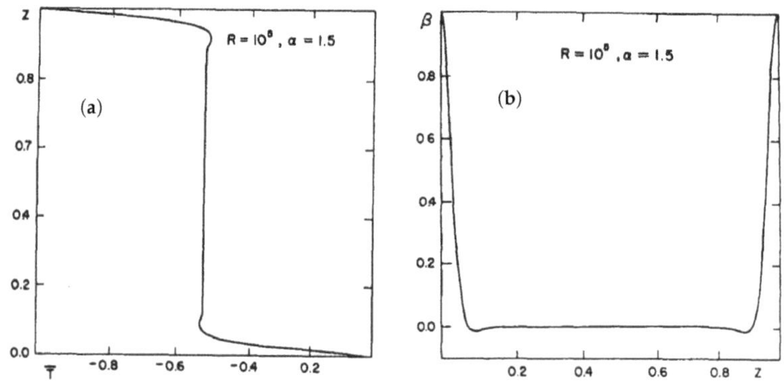

Figure 3. h-slip with $Ra = 10^5$ and $\alpha = 1.5$. (**a**) Mean temperature $\overline{T}(z)$. (**b**) Mean gradient $\beta(z)$. $\beta(z)$ has been normalized by the total heat transport $Nu = 13.82$. Figure 7 from Ref. [1].

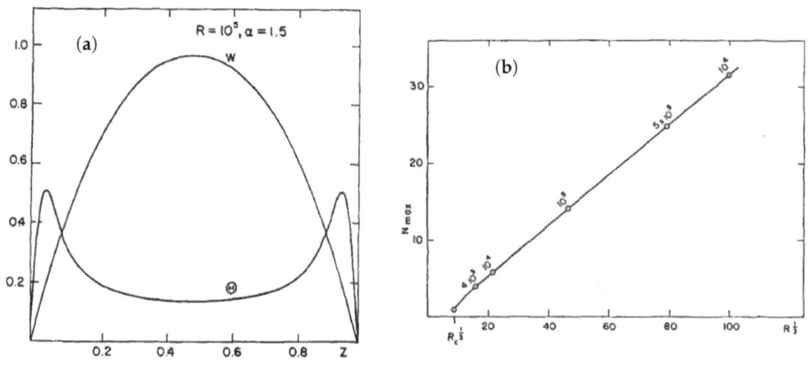

Figure 4. (**a**) $4.33 \times 10^{-3} w$ and $9.42\,\theta$ for $Ra = 10^5$ and $\alpha = 1.5$. (**b**) Maximum total heat flux $Nu_{\max}$ as a function of $Ra^{1/3}$. Figures 6 and 16 from Ref. [1].

The overall message is that, even for these early calculations, to obtain physical convective heat fluxes, rigid (no-slip) boundary conditions were required. This influenced my decision [10] to skip free-slip boundaries and proceed straight to the task of simulating thermal convection with no-slip boundaries.

Funding: This research received no external funding.

Institutional Review Board Statement: Not applicable.

Informed Consent Statement: Not applicable.

Data Availability Statement: Not applicable.

Conflicts of Interest: The authors declare no conflict of interest.

References

1. Herring, J.R. Investigation of problems in thermal convection. *J. Atmos. Sci.* **1963**, *20*, 325–338. [CrossRef]
2. Herring, J.R. Investigation of problems in thermal convection: Rigid boundaries. *J. Atmos. Sci.* **1964**, *21*, 277–290. [CrossRef]
3. Herring, J.R. Some Analytic Results on the Theory of Thermal Convection. *J. Atmos. Sci.* **1966**, *23*, 672–677. [CrossRef]
4. Malkus, W.V.R. The heat transport and spectrum of thermal turbulence. *Proc. R. Soc. A* **1954**, *225*, 196.
5. Malkus, W.V.R.; Veronis, G. Finite amplitude cellular convection. *J. Fluid Mech.* **1958**, *4*, 225. [CrossRef]
6. Kraichnan, R.H. Turbulent thermal convection at arbitrary Prandtl number. *Phys. Fluids* **1962**, *5*, 1374–1389. [CrossRef]
7. Spiegel, E.A. On the Malkus theory of turbulence. In *Mecanique de la Turbulence*; CNRS: Paris, France, 1962; pp. 181–201.
8. Chandrasekar, S. *Hydrodynamic and Hydromagnetic Stability*; Oxford University, Clarendon Press: London, UK, 1961.
9. Balachandar, S.; Maxey, M.R.; Sirovich, L. Numerical simulation of high Rayleigh number convection. *J. Sci. Comp.* **1989**, *4*, 219–236. [CrossRef]
10. Kerr, R.M. Rayleigh Number Scaling in Numerical Convection. *J. Fluid Mech.* **1996**, *310*, 139–179. [CrossRef]
11. Kerr, R.M.; Herring, J. Prandtl number dependence of Nusselt number in DNS. *J. Fluid Mech.* **2000**, *419*, 325–344. [CrossRef]

atmosphere

MDPI

Article

Small-Scale Anisotropy in Stably Stratified Turbulence; Inferences Based on Katabatic Flows

Eliezer Kit [1,*] and Harindra J. S. Fernando [2,3]

1 School of Mechanical Engineering, Tel Aviv University, Tel Aviv 69978, Israel
2 Department of Civil and Environmental Engineering and Earth Sciences, University of Notre Dame, Notre Dame, IN 46530, USA; harindra.j.fernando.10@nd.edu
3 Department of Aerospace and Mechanical Engineering, University of Notre Dame, Notre Dame, IN 46530, USA
* Correspondence: elikit@gmail.com or kit@tauex.tau.ac.il; Tel.: +972-3-6408-929

Abstract: The focus of the current study is on the anisotropy of stably stratified turbulence that is not only limited to large scales and an inertial subrange but also penetrates to small-scale turbulence in the viscous/dissipation subrange on the order of the Kolmogorov scale. The anisotropy of buoyancy forces is well-known, including ensuing effects such as horizontal layering and pancakes structures. Laboratory experiments in the nineties by Van Atta and his students showed that the anisotropy penetrates to very small scales, but their experiments were performed only at a relatively low Re_λ (i.e., at Taylor Reynolds numbers) and, therefore, did not provide convincing evidence of anisotropy penetration into viscous sublayers. Nocturnal katabatic flows having configurations of stratified parallel shear flows and developing on mountain slopes provide high Reynolds number data for testing the notion of anisotropy at viscous scales, but obtaining appropriate time series of the data representing stratified shear flows devoid of unwarranted atmospheric factors is a challenge. This study employed the "in situ" calibration of multiple hot-film-sensors collocated with a sonic anemometer that enabled obtaining a 90 min continuous time series of a "clean" katabatic flow. A detailed analysis of the structure functions was conducted in the inertial and viscous subranges at an Re_λ around 1250. The results of DNS simulations by Kimura and Herring were employed for the interpretation of data.

Keywords: stratified shear flow; structure function; direct numerical simulation (DNS); intermittency; skewness

Citation: Kit, E.; Fernando, H.J.S. Small-Scale Anisotropy in Stably Stratified Turbulence; Inferences Based on Katabatic Flows. *Atmosphere* **2023**, *14*, 918. https://doi.org/10.3390/atmos14060918

Academic Editors: Boris Galperin, Annick Pouquet, Peter Sullivan and Paul D. Williams

Received: 23 April 2023
Revised: 16 May 2023
Accepted: 22 May 2023
Published: 24 May 2023

1. Introduction

Stably stratified turbulence, as already discussed in many previous publications (e.g., [1–3]), is an intriguing phenomenon. The more capabilities there are to delve into smaller scales of stratified turbulence, the more it unravels the inherent complexity [4]. Stably stratified turbulence is ubiquitous in the ocean and atmosphere, and, on larger scales, both stratification and the Earth's rotation become important, leading to geostrophic turbulence. These external effects were discussed by Lumley [5] in their seminal paper, which also considered the effects of an external magnetic field and shear in causing departure from Kolmogorov's regime. It is quite obvious that external anisotropic forces lead to the anisotropy of turbulence, a conspicuous example being the magnetic lines associated with an external magnetic field that lead to directional effects. Columnar vortices with axes parallel to magnetic lines do not naturally interact with magnetic lines, thus curbing the Joule dissipation and increasing the survivability of such vortices. Such behavior was observed in MHD experiments [6], opening up the possibilities for creating two-dimensional turbulence in the laboratory. Attempts were made to employ an analogous approach to strongly (stably) stratified flows. It is unclear, however, why columnar vertical vortices should be more supplant. These vortexes are subject to zig-zag instabilities [7,8], leading to their collapse and the development of thin horizontal layers containing pancake vortexes. These layers, together with the thin layers separating the horizontal layers, form cells of height $L_v = u'/N$, where L_v, u', and N are the characteristic vertical scale, RMS longitudinal turbulence velocity, and the Brunt–Väisälä frequency, respectively. Instead of longitudinal velocity u', the RMS of vertical velocity w' is often used. It is possible that the scaling in this case is strongly dependent on the above characteristic vertical scale.

Similar behavior could be expected when shear acts as an external forcing, although the flow situation is essentially different, since the thin layers, above and below the horizontal layers, are more prone to instability. The emergence of layers in a pure shear flow is, apparently, less plausible. Layering with a long horizontal extent is also not expected in flows with external rotation.

When laboratory grid-generated homogenous isotropic turbulence (HIT) [1] is subjected to (thermally generated) stable stratification, turbulence becomes *anisotropic*, and the relevant controlling parameter of the overall flow is the inverse of the internal/turbulent Froude numbers, $Fr = \varepsilon/Nu'^2$, where ε is the rate of dissipation. The above expression can be modified by assuming that $\varepsilon = u'^3/L_h$ and using $Fr = u'/(NL_h)$, where L_h is the horizontal length scale. The typical inverse Froude numbers assessed in our field experiments, dubbed MATERHORN, were $2 \div 5$ [9], coincidentally in the same range as that observed in [10], though the Re_λ is larger, thus allowing for some comparisons. Additional references are given in the discussion chapter.

2. Methodology

In the nocturnal Atmospheric Boundary Layer (ABL), stable stratification evolves overnight due to radiative cooling, thus leading to interesting small-scale phenomena that was studied during MATERHORN (2011–2016) field experiments [11]. A novel probing system was deployed to capture very small scales, down to Kolmogorov viscous dissipation subrange [12,13]. The finer scales therein were captured by multiple hot-film probes placed in the probe volume of a sonic anemometer, with the latter providing low frequency data to calibrate the former by utilizing a neural network. This sonic- and hot-film anemometer dyad (dubbed the "combo" probe) was placed on a horizontal pole at 6 m height of a 32 m high tower (labeled ES-2), equipped with an array of sonics and thermocouples at various levels; see [12]. ES-2 had seven levels of sonics: 0.4, 4, 10, 16, 20, 25, and 28 m. Careful analysis of data enabled identification of periods that fit the rubric of stratified parallel shear flows, which emerged as katabatic (downslope) flows draining from the nearby Granite Mountain.

After careful processing of sonic records of ES-2, which was a part of a densely instrumented flux tower array (ES-1–ES-5) designed to study katabatic flows at night on the east slope of Granite Mountain (for details see [11,12]), a 90 min period starting from 22:00 MDT (Mountain Daylight Time (local time)) on October 19, 2012 was selected for stratified shear flow studies. Prior to the selected time interval, the wind speed rapidly increased from ~1 m/s to ~4 m/s at the combo probe height. The wind direction changed from its usual oscillations before 22:00 MDT to a nearly constant direction, resembling a stratified parallel flow with relatively low shear. Thereafter, the winds changed quite rapidly between 23:30 MDT and midnight [12]. Although mean quantities were quasi-steady during the selected period, 22:00–23:30 MDT, careful inspection of turbulence statistics showed considerable variability. An analysis of approximately homogeneous subintervals, identified based on stability (Fr), enabled focusing on structure function analysis in the subinterval SI_{devd} (22:50–23:10 MDT). This subinterval was considered as fully developed turbulence, since flow variations were modest, and appearance of small-scale bursting events [12] was very limited (less than 2% of the time).

3. Stratified Turbulence in Field Experiments and DNS

3.1. Puzzling Observations in the MATERHORN Campaign

The detailed results concerning nocturnal turbulence observed during the MATERHORN campaign were published in a series of papers [5,11,12]. In these studies, different data analyses and presentation methods were employed, viz., the spectral approach in Fourier space [12] and the structure functions approach in physical space [4]. Although both approaches are inter-related, each of them better reflects specific features of nocturnal turbulence. In particular, a small-scale bursting phenomenon was uncovered using the spectral approach [12], and the celebrated Kolmogorov $-5/3$ spectrum appropriately described the measured spectral behavior in the inertial subrange for so-called no-bursting intervals, where "bursts" were effectively removed from the time series using proper thresholds based on KE dissipation. At the same time, the structure functions approach enabled the discovery of inertial and viscous (dissipation) subintervals with separate, corresponding power exponents for the bursting and no-bursting intervals. In this paper, we limit ourselves to the time series where the bursting events are removed.

In particular, our recent paper [4], on the stably stratified turbulence occurring in the nocturnal turbulence observed during MATERHORN [11,12], reported an unexpected behavior of the canonical (normalized) third-order longitudinal structure function (Figure 1).

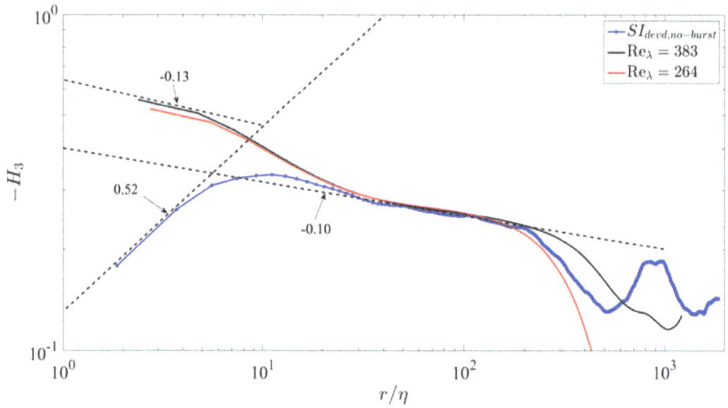

Figure 1. Comparison of the canonical third-order moments obtained in the field campaign [11,12] at Re_λ = 1250 and DNS for the second-highest Re_λ = 264 and 383.

In Figure 1, the canonical third-order structure function measured at a 6 m height from the ground was compared with the same structure functions evaluated for nearly isotropic turbulence obtained by classical box DNS computations [14]. In the field, the Re_λ was 1250, whereas in the DNS computations the Re_λ was 264 and 383. While in the inertial subrange ($r/\eta = 20 \div 200$), both the qualitative and quantitative agreements were very good, in the viscous subrange ($r/\eta \leq 10$) the results were conspicuously different, and the difference increased with the decrease in separation. It is worth noting that at small normalized separations $r/\eta < 10$, the scaling exponents were even of different signs for the field and DNS data. Since, in the viscous subrange, one can expect linear dependence between velocity and separation, e.g., [15], the scaling exponent for the conventional third-order structure function should be about zero to satisfy Kolmogorov's Self-Similar Hypothesis (KSSH) [16].

This difference of scaling exponents in the viscous subrange of field experiments and DNS is puzzling, especially when there is a clear agreement among the scaling exponents in the inertial subrange with each other and with the Kolmogorov inertial subrange scaling. To shed light on this perplexing behavior, the canonical normalized third-order structure function was computed at various separations; it is defined as the third-order structure function L_3 divided by the second-order structure function L_2 to the power 1.5, e.g., $H_3(r) = L_3(r)/(L_2(r))^{3/2}$, which also represents the skewness of the structure function. In Figure 2, we separately consider the third- and second-order structure functions, both normalized according to KSSH [16].

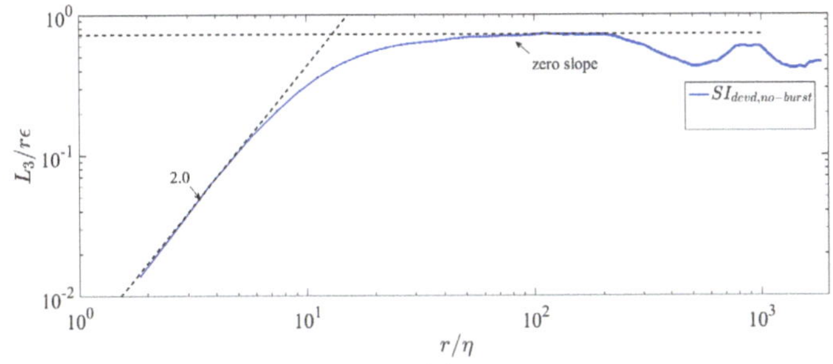

(a)

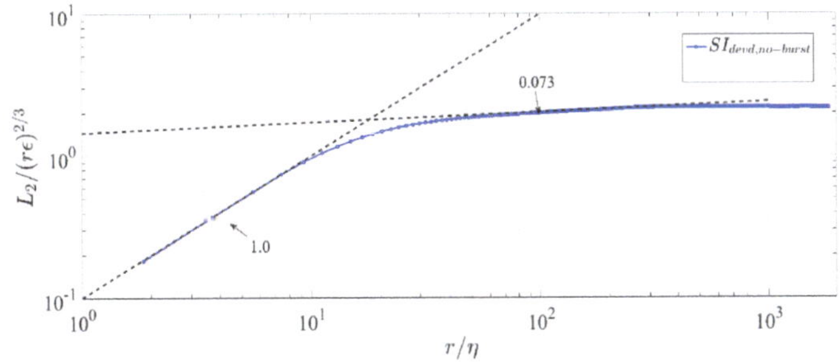

(b)

Figure 2. Normalized according to KSSH third-order (**a**) and second-order (**b**) structure functions obtained from the field data, computed using longitudinal velocity increment $\Delta u(x,r)$.

The third-order structure function in Figure 2a yields, in the viscous subrange, the expected linear dependence of the characteristic velocity increment $\Delta u(x,r)$ on the separation r, e.g., $\Delta u(x,r) \sim r^1$. However, the second-order structure function in Figure 2b corresponds to a different type of dependence, namely, $\Delta u(x,r) \sim r^{5/6}$. This explains the scaling exponent observed for the canonical third-order structure function in Figure 1.

However, as shown in Figure 1, the power exponent of the canonical structure function is relatively small, ~ -0.1 in the DNS computations of the approximately homogeneous isotropic turbulence, compared to $\sim +0.5$ obtained in the field experiment. The determination of the separate scaling exponents from the DNS data, in a manner similar to that of the field data, as shown in Figure 2, yielded the following dependencies, $\Delta u(x,r) \sim r^{0.87}$ for third-order and $\Delta u(x,r) \sim r^{0.91}$, for second-order structure functions. It is worth noting that the expectation of identical behavior by odd and even structure functions is generally unjustified, despite it being widely quoted and often used, in particular, in the Extended Self-Similarity (ESS) approach. In previous research [4], we found that all even (second, fourth, and sixth in this study) structure functions yield $\Delta u(x,r) \sim r^{5/6}$ in the viscous subrange. Then, it was found that all the odd structure functions (first, third, and fifth in this study) constructed using the modulus of velocity increment $|\Delta u(x,r)|$ yield a similar relation, namely, $\Delta u(x,r) \sim r^{5/6}$ in the viscous subrange; here, $\Delta u(x,r)$ is the characteristic velocity increment at selected separation r. The consistent and significant difference between the power exponents of the third- and second-order structure functions is responsible for the substantial power exponent 0.5 at the small scales of the conventional third-order structure function/skewness. In isotropic DNS, this power exponent is about -0.1, which is much closer to zero, though of a different sign. This result is somewhat perplexing, given the expectation that, due to the local isotropy at small scales, the power exponent tends to be zero. Even more puzzling is that, in the inertial subrange, the behavior of the canonical third-order structure function in the DNS and field experiments shows full (qualitative and quantitative) agreement.

This apparently contradicts with the postulate of local isotropy (PLI), which entails better isotropy whence the separation (r/η) is diminishing. Cambon et al. [17], however, predicted such behavior in cases where external anisotropic forces such as rotation (Coriolis) and stable stratification (buoyancy) do not produce turbulent energy. In particular, they state the following: "It is often considered that anisotropy only affects the largest scales of the turbulent flow, so that the cascade of anisotropy only mildly penetrates toward the scales in the inertial range, with eventually no direct impact on the dissipative range, according to the view inherited from Kolmogorov [16]. In this scheme, the large-scale anisotropy induced by body forces and/or mean gradients is just considered as a particular modality of 'forcing' the largest scales. This viewpoint is radically questioned for flows in which the body force responsible for anisotropy is energy conserving." Moreover, they state that "Anisotropy develops from isotropic initial conditions due to nonlinear energy transfer towards the plane of zero frequency in wave-vector space. However, two-dimensionalization does not occur: the fraction of energy at or near the zero-frequency plane remains small at all times. The cascade to small scales is strongly anisotropic, producing angle-dependent spectra which become more and more (not less) anisotropic, the smaller the scale considered".

Obviously, in shear flows, turbulent energy is produced, notwithstanding, the laboratory experiments with homogeneous shear flow by Warhaft and their co-workers [18–20] unambiguously showed that the return to isotropy expected at the small scales does not occur either at a lower Re_λ~O(100) [18] or at a higher Re_λ~O(1000) [19], which was again ascribed to shear penetrating to smaller scales. To quote [19], "The results show that PLI is untenable, both at the dissipation and inertial scales, at least to R_λ~1000, and suggest it is unlikely to be so even at higher Reynolds numbers." See more considerations in the Discussion section.

Our results [4] appear to strongly support the observation that stratified turbulence differs from nearly isotropic turbulence, even at small scales corresponding to the dissipation (viscous) sublayer, and raise a question about the PLI for this important case. It was very desirable to assert the validity of our results by independent measurements at a higher Re_λ~O(1000) or the DNS computations of stably stratified turbulence.

Unfortunately, the laboratory experiments of stably stratified turbulence are limited to a relatively low Re_λ (<100), and, thus, high-quality field experiments with continuous measurements at high sampling rates are called for. In fact, during the entire MATERHORN campaign, only a few records provided "clean" data sets for the nocturnal stably stratified turbulence strongly affected by thermal stratification. It should be stressed that these measurements were possible due to our novel calibration approach that enabled in situ calibration of a multi-hot-film probe based on the simultaneous measurements of low-frequency 3D-velocity data of a collocated sonic anemometer. Employing machine learning (neural network training) enabled the calibration of the hot-film probe in situ, thus avoiding problems with the hot-film's potential deterioration in hostile field environments. The efficacy of this calibration method [21] was tested in a series of papers [22–24] and proved to be very efficient, even in the presence of noise to some degree.

In the current work, the neural network procedure was based on the Multi-Layer Perceptron (MLP) approach. MLP contains one input layer, one or two hidden layers, and one output layer. The number of nodes of the input layer and the number of input signals are the same. We used a fully connected network. In this case, each node of the input layer duplicates and sends its input signal to every neuron of the first hidden layer. The hidden layer consists of a number h of neurons. For the training of MLP, the Back Propagation and Conjugate Gradient Descent methods were sequentially used. In all cases, the number of epochs that was necessary for the generation of the neural network did not exceed 100.

The use of two collocated x-probes (oriented in the horizontal and vertical planes) at a very small separation allowed for the measurement of the full 3D velocity vector. Via the Taylor hypothesis, the time series of the velocity components could be converted into a velocity series in the mean wind direction. As expected, the processed data records were limited in time (about 1 min) to keep the mean transverse velocity components as small as possible. Directly after the end of the record, the probe was aligned with the mean horizontal velocity that was evaluated during the last 5 s of the record. Each record was flagged as successful if the mean wind direction differed by less than 10° from its value at the beginning of the record.

Stratified turbulence is essentially anisotropic (at least at large scales) and is characterized by horizontal pancake-like layering; thus, we posit that the surprising result observed in Figure 1 is related to anisotropy and layering. The DNS of the stably stratified turbulent flow in a box could help in this regard, notwithstanding its idealized nature. DNS simulations were conducted by Jack Herring in collaboration with Yoshi Kimura and were presented in two seminal papers, one in the *Journal of Fluid Mechanics* [2] and the other in *Physica Scripta* [25]. The significance of their work is limited not only to the very detailed and comprehensive presentations of the spectra and structure functions obtained at a relatively high Re_λ but also to the elegant methodology of velocity data analysis based on Craya–Herring decomposition. The latter approach allows for the separation of the entire oscillating flow into 2 types of modes: horizontal and *vortical*; vertical and *wavy* [2,23]. In Section 3.2, the puzzling results (the penetration of the anisotropy caused by stratification into small scales) of our study are discussed in light of [2].

3.2. DNS of Stably Stratified Turbulence by Kimura and Herring 2012 [2]

As shown in [2], the second-order structure function $St_2(d, u, x)$ for the longitudinal velocity component u at separation d (r in our notations) in the x-direction can be presented as the superposition of two terms following Craya–Herring decomposition (expression 4.4 in [2]).

$$St_2(d;u,x) = 2\pi \int_{-\infty}^{\infty} dk_z \int_{0}^{\infty} k_\perp dk_\perp [\Phi_1(k_\perp,k_z)\left(1 - 2\frac{J_1(k_\perp d)}{k_\perp d}\right) + \\ + \frac{3k_z^2}{k_\perp^2+k_z^2}\Phi_2(k_\perp,k_z)\left(1 - \frac{4}{3}J_0(k_\perp d) + \frac{2}{3}\frac{J_1(k_\perp d)}{k_\perp d}\right)]$$

where Φ_1 and Φ_2 are defined as energy densities; the $k_\perp$ and k_z components of vector number $\boldsymbol{k}$ are in the horizontal plane and the vertical direction, respectively; and J_0 and J_1 denote the Bessel function of the corresponding order. While Φ_1 contributes only to velocities in the horizontal plane and is determined by vertical vorticity component ω_z, Φ_2 is contributing to both horizontal and vertical velocity components and can be determined using only the vertical velocity component w, as presented in expressions 2.9 and 2.10, respectively [2]. Since, in a stratified flow, the vertical direction coincides with gravitational acceleration g, the second term of the decomposition may account for the internal waves. Indeed, the behavior of the second-order structure function due to the first term only, evaluated in [2], is practically the same as that of the isotropic turbulence (see Figure 14 in [2]), while adding the second term accounting for buoyancy leads to behavior of the second-order structure function that substantially differs from that of the isotropic turbulence.

The second-order longitudinal structure function in the x-direction (see Figure 15a in [2]) is shown below in Figure 3. Unfortunately, the authors do not use normalization that follows KSSH for data presentation as it is made in our Figure 2b. Notwithstanding, it is clear that, in the inertial subrange, the scaling exponent is greater than Kolmogorov's value of 2/3, which is similar to our result. With the addition of buoyancy effects (the second term), the viscous subrange scaling exponent for the velocity increment's dependence on the separation is close to 5/6 (see the estimated slope 5/3 for the second-order structure at low separations). Unfortunately, [2] does not present the third-order structure function, which could adumbrate Kolmogorov's 4/5 law in the inertial subrange for the horizontal longitudinal third-order structure function of stratified turbulence.

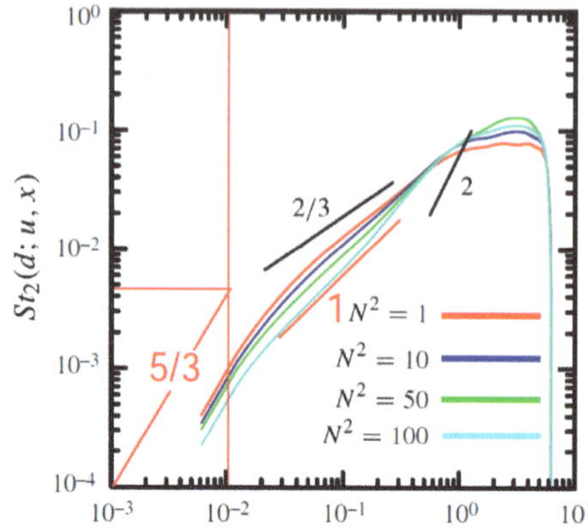

Figure 3. Second-order structure functions for squared normalized buoyancy frequency N^2 = 1, 10, 50, and 100 as a function of separation *d*. The assessed power exponent in viscous subrange is about 5/3. (Reproduced with permission from Kimura and Herring [2], *J. Fluid Mech.* 698, 19. Copyright 2012, Cambridge University Press.)

In recent studies [26,27], the DNS of rotating turbulence in a box was reported as enabling the derivation of the third-order structure functions at different polar angles for examination of their shapes, in particular the penetration of anisotropy to scales in the viscous subrange. It is important to emphasize that rotation, similarly to stratification, leads to modifications, from isotropic to axisymmetric, of the turbulence structure, by generating columnar vortices around the rotation axis. The resemblance is, obviously, not exact and does not include horizontal layering by pancake structures in the horizontal plane, as predicted and observed in stably stratified turbulence. A mechanism of the formation of pancake structures from columnar vortices in stratified turbulence, as mentioned above, is often related to zig-zag instability.

An interesting result observed in the Ph.D. research of Vallefuoco [27] is the development of the clear anisotropy of third-order moments as a function of the polar angle. While DNS without rotation yielded perfect isotropy with moments for polar angles *0* and $\pi/2$ that practically coincided in both inertial and viscous subranges, when rotation is present the moments differ across the scales, including at very small scales on the order of the Kolmogorov scale (see Figure 4.31a,c in [27]).

3.3. A Simplified Model

Inspired by the derivations and results of [2], as presented in Section 3.2., we suggest an (over)simplified model. This model assumes that the longitudinal velocity $u(x)$ measured at a 6 m height consists of two contributions: $u_1(x)$, representing pure HIT (isotropic and asymmetrical), and $u_2(x)$, the strongly stratified turbulence (anisotropic and symmetric, with respect to the vertical), with a low correlation between the two. The PDF of u_1 is essentially asymmetrical for the longitudinal velocity derivative, while the PDF of u_2 may be assumed to be symmetrical. It follows that the third-order structure function for $\Delta u(x, r)$ is determined by u_1 only.

From below expression for $L_3(r)$, following above assumptions,

$$L_3(r) = \langle(\Delta u(x,r))^3\rangle = \langle(\Delta u_1(x,r))^3 + 3(\Delta u_1(x,r))^2 \Delta u_2(x,r) + 3\Delta u_1(x,r)(\Delta u_2(x,r))^2 + (\Delta u_2(x,r))^3\rangle$$

it can be easily assessed that, at the right hand side, the second and third terms are zero, due to the lack of correlation between u_1 and u_2, and the fourth term is zero due to the symmetry of the probability density function of u_2, leaving the first term as the nonzero term. The situation is different for the higher-order odd longitudinal structure functions.

For example, in the fifth-order structure function, the first $\langle(\Delta u_1(x,r))^5\rangle$ and the third terms $10 * \langle(\Delta u_1(x,r))^3\rangle * \langle(\Delta u_2(x,r))^2\rangle$ are nonzero. This simplified model, therefore, may explain the intriguingly different behaviors in the *viscous* subrange of the odd third- and fifth-order structure functions for $\Delta u(x,r)$. That is, the expected linear dependence of velocity increment on separation r for the third-order structure function (Figure 5a in [4]) and an oscillating slope (scaling exponent) for the fifth-order structure functions (Figure 8a in [4]).

The odd first-, third-, and fifth-order structure functions evaluated for $|\Delta u(x,r)|$ in the *viscous* subrange yield the same $r^{5/6}$ dependence (i.e., scaling exponent p*5/6; Figure 9 in [4]) as *all* the even structure functions. It is obvious that all the structure functions for the absolute velocity increments include both contributions (u_1 and u_2). Therefore, the scaling exponent, in general, can be different from that of the homogeneous turbulence. Relatively weak anomalies only start to appear at $p = 6$. However, presently, we are unable to offer a sound explanation for the distinct shape (5/6 scaling exponent) in the above dependence.

Kimura and Herring [2] obtained a surprising result: the ratio of potential energy to kinetic energy for *all N* is about the same, about 0.1. According to [2], "This potential energy is attributable to the three-dimensional turbulence that occupies the space between the quasi-horizontal layers." They note the following: "Why this value asymptotes to ~0.1 of the kinetic energy is a mystery to us." Our belief is that the characteristics of these quasi-horizontal layers are of the utmost importance for the strongly stratified turbulence in the viscous subrange, so the appropriate answer to each one of the above questions might be the key to the mystery. The low Reynolds number laboratory experiment by Fincham et al. [28] illustrates that the interactions between the turbulence in the layers and strong shear between (highly dissipative) layers are characteristics of decaying stratified turbulence, similar to the observations noted by Métais and Herring [29].

The model offered above is simplistic, so a more rigorous attempt to separate isotropic and anisotropic contributions can be made in the framework of *SO*(3) formalism [30], by conducting the decomposition of the appropriately measured structure or correlation functions into spherical harmonics. This is left for future work.

4. Discussion

Our study concerns the anisotropy of stably stratified turbulence that is limited not only to large scales and inertial subrange but also to penetrating the (smaller-scales) domain of the viscous/dissipation subrange on the order of the Kolmogorov scale. The anisotropy of buoyancy forces as well as their propensity for developing horizontal layering and pancakes structures are widely known and may be related to this anisotropy across the scales.

In a recent work [31], direct numerical simulations were performed for the assessment of the local isotropy of dissipative scales for stably stratified flows. It is claimed that the estimation of the dissipation rate of turbulent kinetic energy and density variance can be approximated using the isotropic assumption for the turbulent Froude number $Fr \geq O(1)$. There is a clear departure from isotropy for $Fr < O(1)$. However, in another work [32], it was shown that dissipation-scale isotropy is determined by Re_b, the buoyancy Reynolds number, only. The authors also performed direct numerical simulations to investigate the anisotropy of stratified turbulence and the transition to isotropy at small length scales. Turbulence was generated by forcing large-scale vortical modes, an approach that is broadly consistent with geophysical stratified turbulence. The authors' results suggest that $\mathrm{Re}_b \geq 500$ is required to obtained the same degree of small-scale isotropy seen in the unstratified turbulence at a similar Re.

In relatively old but very comprehensive study [33] in 1981, the authors have already performed direct numerical simulations of decaying homogeneous turbulence in density-stratified fluids. They have examined in particular the energetics, the evolution of characteristic length scales and the importance of nonlinearities in the computed flow fields. They have found that stratification introduces wave-like characteristics into the flow fields. This is exemplified by exchange from (and to) kinetic energy to (and from) potential energy and the development of counter-gradient buoyancy fluxes. Stratification tends to enhance the growth of horizontal scales while inhibiting the growth of vertical scale.

Per a referee, we would also like to point out the very interesting work [34] by Smith and Waleffe, with the title *"Generation of slow large scales in forced rotating stratified turbulence"*, but their focus is on large scales in forced rotating stratified turbulence, so it appears to have a lesser relevance to our work. Indeed, the authors discussed the cases of pure stratified turbulence that was ***randomly forced at small scales***. Their interests are mainly related to the inverse cascade. The same remarks are applicable to the comprehensive study of Delache et al. [35], which dealt with anisotropy in freely decaying rotating turbulence.

Following the comments of reviewers, we carefully evaluated various options for the modeling generation of anisotropic small-scale structures. To do so, we referred to additional studies [36–38] such as Kerstein 1999, Wunsch 2000, and Wunsch and Kerstein 2001 that describe the horizontal layering in stably stratified turbulence, which suggested a one-dimensional turbulence (ODT) model for this analysis. In essence, this model locally applies mixing length theory throughout the simulation domain, defining a wide range of possible mixing lengths and corresponding time scales for each point in space. Turbulent mixing is then randomly applied throughout the system at all length scales, based on the locally appropriate time scales. It is a stochastic model and can be very useful in applications.

Finally, we refer to the most up-to-date monograph [39], which summarized the recent theoretical, computational, and experimental results dealing with homogeneous turbulence dynamics. A large class of flows was covered: flows governed by anisotropic production mechanisms (e.g., shear flows) and flows without production but dominated by waves (e.g., homogeneous rotating or stratified turbulence).

While laboratory experiments in the 1990s showed the penetration of anisotropy to very small scales, such experiments were performed at a relatively low Re_λ and, therefore, did not provide convincing evidence of anisotropy penetration into the viscous subrange. Measurements in the atmosphere during the MATERHORN project could provide such evidence as discussed in this paper, where the viscous subrange was accessed via a specialized (combo) probe, which is an assembly of a high (space-time)-resolution, multi-sensor hot-film probe array collocated with a sonic that measures the full velocity vector at a low-frequency resolution.

By employing machine learning based on a neural network algorithm during post-processing, "in situ" calibration of the combo probe was accomplished. "Clean" stratified turbulence data unblemished by other effects were observed in the Katabatic flows developing near the mountain slopes during nocturnal events, which can provide appropriate data for stratified turbulent studies. As explained in Section 2, 90 min of clean stratified shear flow data could be gleaned, based on which detailed structure function analyses were conducted in the inertial and viscous subranges. Anisotropic behavior in the viscous subrange was revealed at an Re_λ around 1250, which can be considered substantial. The seminal paper of Kimura and Herring [2] on the DNS of stably stratified turbulence was employed for data interpretation.

Our results strongly support that stratified turbulence differs from nearly isotropic turbulence even at a smaller dissipation (viscous) subrange. To the best of our knowledge, this is the first time that such inferences are made at a relatively high Taylor-based Reynolds number of Re_λ ~1250. This result contradicts the hypothesis of the postulate of local isotropy (PLI), which posits enhanced isotropy, whence the separation (r/η) is diminishing. In fact, Cambon et al. [17] predicted kindred behavior when external anisotropic forces, such as rotation (Coriolis) and stable stratification (buoyancy), do not produce turbulent energy.

The Craya–Herring decomposition used in [2] allowed for presenting the second-order structure function for the longitudinal velocity increment at separation r in the x-direction as a superposition of two terms: the turbulence determined by vertical vorticity ω_z and the internal waves determined by vertical velocity component u_z. While the former term is essentially isotropic on the horizontal plane, the latter term is responsible for anisotropy, which we used to qualitatively interpret the observations.

Author Contributions: Both authors equally contributed to the paper. All authors have read and agreed to the published version of the manuscript.

Funding: This work was funded in part by grant number N00014-21-1-2296 (Fatima Multidisciplinary University Research Initiative) of the Office of Naval Research (ONR), administered by the Marine Meteorology and Space Program. The MATERHORN program was funded by ONR award number N00014-11-1-0709. The complex terrain flows research of the authors is also funded by the US National Science Foundation, award number AGS-1921554.

Institutional Review Board Statement: Not applicable.

Informed Consent Statement: Not applicable.

Conflicts of Interest: The authors declare no conflict of interest.

References

1. Thoroddsen, S.T.; Van Atta, C.W. The Influence of Stable Stratification on Small-Scale Anisotropy and Dissipation in Turbulence. *J. Geophys. Res.* **1992**, *97*, 3647–3658. [CrossRef]
2. Kimura, Y.; Herring, J.R. Energy spectra of stably stratified turbulence. *J. Fluid Mech.* **2012**, *698*, 19–50. [CrossRef]
3. Caulfield, C.P. Layering, Instabilities, and Mixing in Turbulent Stratified Flows. *Annu. Rev. Fluid Mech.* **2021**, *53*, 113–145. [CrossRef]
4. Kit, E.; Barami, E.; Fernando, H.J.S. Structure Functions in Nocturnal Atmospheric Boundary Layer Turbulence. *Phys. Rev. Fluids* **2021**, *6*, 084605. [CrossRef]
5. Lumley, J.L. Similarity and the Turbulent Energy Spectrum. *Phys. Fluids* **1967**, *10*, 855. [CrossRef]
6. Kit, L.G.; Tsinober, A.B. On the possibility of realization and investigation of two-dimensional turbulence in a strong magnetic field. *Magnetohydrodynamics* **1971**, *3*, 27–34.
7. Billant, P.; Chomaz, J.M. Theoretical analysis of the zigzag instability of a vertical columnar vortex pair in a strongly stratified fluid. *J. Fluid Mech.* **2000**, *419*, 29–63. [CrossRef]
8. Billant, P.; Chomaz, J.M. Three-dimensional stability of a vertical columnar vortex pair in a stratified fluid. *J. Fluid Mech.* **2000**, *419*, 65–91. [CrossRef]
9. Conry, P.; Kit, E.; Fernando, H.J.S. Measurements of mixing parameters in atmospheric stably stratified parallel shear flow. *Environ. Fluid Mech.* **2020**, *20*, 1177–1197. [CrossRef]
10. Thoroddsen, S.T.; Van Atta, C.W. Experiments on density-gradient anisotropies and scalar dissipation of turbulence in a stably stratified fluid. *J. Fluid Mech.* **1996**, *322*, 383. [CrossRef]
11. Fernando, H.J.S.; Pardyjak, E.R.; Di Sabatino, S.; Chow, F.K.; De Wekker, S.F.J.; Hoch, S.W.; Hacker, J.; Pace, J.C.; Pratt, T.; Pu, Z.; et al. The MATERHORN—unraveling the intricacies of mountain weather. *Bull. Am. Meteorol. Soc.* **2015**, *96*, 1945–1967. [CrossRef]
12. Kit, E.; Hocut, C.; Liberzon, D.; Fernando, H.J.S. Fine-Scale Turbulent Bursts in Stable Atmospheric Boundary Layer in Complex Terrain. *J. Fluid Mech.* **2017**, *833*, 745–772. [CrossRef]
13. Sukoriansky, S.; Kit, E.; Zemach, E.; Midya, S.; Fernando, H.J.S. Inertial range skewness of the longitudinal velocity derivative in locally isotropic turbulence. *Phys. Rev. Fluids* **2018**, *3*, 114605. [CrossRef]
14. Ishihara, T.; Gotoh, T.; Kaneda, Y. Study of high-Reynolds number isotropic turbulence by direct numerical simulation. *Annu. Rev. Fluid Mech.* **2009**, *41*, 165. [CrossRef]
15. Nelkin, M. What do we know about self-similarity in fluid turbulence? *J. Stat. Phys.* **1985**, *4*, 1–15. [CrossRef]
16. Kolmogorov, A.N. The local structure of turbulence in incompressible viscous fluid for very large Reynolds numbers. *Dokl. Akad. Nauk. SSSR.* **1941**, *30*, 301–305.
17. Cambon, C.; Danaila, L.; Godeferd, F.S.; Scott, J.F. Third-order statistics and the dynamics of strongly anisotropic turbulent flows. *J. Turbul.* **2013**, *14*, 121–160. [CrossRef]
18. Garg, S.; Warhaft, Z. On small scale structure of simple shear flow. *Phys. Fluids* **1998**, *10*, 662–673. [CrossRef]
19. Shen, X.; Warhaft, Z. The anisotropy of the small scale structure in high Reynolds number (R_λ~1000) turbulent shear flow. *Phys. Fluids* **2000**, *12*, 2976–2989. [CrossRef]
20. Shen, X.; Warhaft, Z. Longitudinal and transverse structure functions in sheared and unsheared wind-tunnel turbulence. *Phys. Fluids* **2002**, *14*, 370–381. [CrossRef]
21. Kit, E.; Cherkassky, A.; Sant, T.; Fernando, H.J.S. In-situ calibration of hot-film probes using a co-located sonic anemometer: Implementation of a neural network. *J. Atmos. Ocean. Technol.* **2010**, *27*, 23–41. [CrossRef]
22. Kit, E.; Gritz, B. In-situ calibration of hot-film probes using a co-located sonic anemometer: Angular probability distribution properties. *J. Atmos. Ocean. Technol.* **2011**, *28*, 104–110. [CrossRef]

23. Vitkin, L.; Liberzon, D.; Grits, B.; Kit, E. Study of in-situ calibration performance of Hot-Film Probe collocated with a Sonic using "virtual probe" approach. *Meas. Sci. Technol.* **2014**, *25*, 075801. [CrossRef]
24. Kit, E.; Liberzon, D. 3D-calibration of three- and four-sensor hot-film probes based on collocated sonic using neural networks. *Meas. Sci. Technol.* **2016**, *27*, 095901. [CrossRef]
25. Herring, J.R.; Kimura, Y. Some issues and problems of stably stratified turbulence. *Phys. Scr.* **2013**, *T155*, 014031. [CrossRef]
26. Vallefuoco, D.; Naso, A.; Godeferd, F.S. Small-scale anisotropy induced by spectral forcing and by rotation in non-helical and helical turbulence. *J. Turbul.* **2018**, *19*, 107–140. [CrossRef]
27. Vallefuoco, D. Numerical Study of Unconfined and Confined Anisotropic Turbulence. Ph.D. Thesis, Ecole Centrale de Lyon, Lyon, France, November 2017.
28. Fincham, A.M.; Maxworthy, T.; Spedding, G.R. Energy dissipation and vortex structure in freely decaying, stratified grid turbulence. *Dyn. Atmos. Oceans* **1996**, *23*, 155–169. [CrossRef]
29. Métais, O.; Herring, J.R. Numerical simulations of freely evolving turbulence in stably stratified fluids. *J. Fluid Mech.* **1989**, *202*, 117–148. [CrossRef]
30. Biferale, L.; Procaccia, I. Anisotropy in turbulent flows and in turbulent transport. *Phys. Rep.* **2005**, *414*, 43–164. [CrossRef]
31. Garanaik, A.; Venayagamoorthy, S.K. Assessment of small-scale anisotropy in stably stratified turbulent flows using direct numerical simulations. *Phys. Fluids* **2018**, *30*, 126602. [CrossRef]
32. Lang, C.J.; Waite, M.L. Scale-dependent anisotropy in forced stratified turbulence. *Phys. Rev. Fluids* **2019**, *4*, 044801. [CrossRef]
33. Riley, J.J.; Metcalfe, R.W.; Weissman, M.A. Direct numerical simulations of homogeneous turbulence in density-stratified fluids. *AIP Conf. Proc.* **1981**, *76*, 79–112. [CrossRef]
34. Smith, L.M.; Waleffe, F. Generation of slow, large scales in forced rotating, stratified turbulence. *J. Fluid Mech.* **2002**, *451*, 145–168. [CrossRef]
35. Delache, A.; Cambon, C.; Godeferd, F.S. Scale by scale anisotropy in freely decaying rotating turbulence. *Phys. Fluids* **2014**, *26*, 025104. [CrossRef]
36. Kerstein, A.R. One-dimensional turbulence: Model formulation and application to homogeneous turbulence, shear flows, and buoyant stratified flows. *J. Fluid Mech.* **1999**, *392*, 277–334. [CrossRef]
37. Wunsch, S. Scaling laws for layer formation in stably-stratified turbulence. *Phys. Fluids* **2000**, *12*, 672–675. [CrossRef]
38. Wunsch, S.; Kerstein, A. A model for layer formation in stably stratified turbulence. *Phys. Fluids* **2001**, *13*, 702–712. [CrossRef]
39. Sagaut, P.; Cambon, C. *Homogeneous Turbulence Dynamics, Monograph*; Springer: New York, NY, USA, 2018.

atmosphere

MDPI

Article

Turbulence with Magnetic Helicity That Is Absent on Average

Axel Brandenburg [1,2,3,4,*] **and Gustav Larsson** [1,2]

[1] Nordita, KTH Royal Institute of Technology and Stockholm University, Hannes Alfvéns väg 12, 10691 Stockholm, Sweden; gula1839@student.su.se
[2] Oskar Klein Centre, Department of Physics, Stockholm University, AlbaNova, 10691 Stockholm, Sweden
[3] School of Natural Sciences and Medicine, Ilia State University, 0194 Tbilisi, Georgia
[4] McWilliams Center for Cosmology and Department of Physics, Carnegie Mellon University, Pittsburgh, PA 15213, USA
* Correspondence: brandenb@nordita.org

Abstract: Magnetic helicity plays a tremendously important role when it is different from zero on average. Most notably, it leads to the phenomenon of an inverse cascade. Here, we consider decaying magnetohydrodynamic (MHD) turbulence as well as some less common examples of magnetic evolution under the Hall effect and ambipolar diffusion, as well as cases in which the magnetic field evolution is constrained by the presence of an asymmetry in the number density of chiral fermions, whose spin is systematically either aligned or anti-aligned with its momentum. In all those cases, there is a new conserved quantity: the Hosking integral. We present quantitative scaling results for the magnetic integral scale as well as the magnetic energy density and its spectrum. We also compare with cases were a magnetic version of the Saffman integral is initially finite. Rotation in MHD turbulence tends to suppress nonlinearity and thereby also inverse cascading. Finally, the role of the Hosking and magnetic Saffman integrals in shell models of turbulence is examined.

Keywords: decaying turbulence; MHD turbulence; chiral magnetic effect; Hall effect; shell models

Citation: Brandenburg, A.; Larsson, G. Turbulence with Magnetic Helicity That Is Absent on Average. *Atmosphere* **2023**, *14*, 932. https://doi.org/10.3390/atmos14060932

Academic Editors: Boris Galperin, Annick Pouquet and Peter Sullivan

Received: 16 May 2023
Revised: 22 May 2023
Accepted: 23 May 2023
Published: 26 May 2023

1. Introduction

This paper is part of a special issue commemorating the work of Jack Herring. His scientific career started off with papers on the effect of the solar wind on the lunar atmosphere in 1959 [1]. In 1961, he extended this work to exoplanet atmospheres [2]. He also worked on stellar opacities [3]. In all those cases, he was very much ahead of its time. At the time of Parker's prize-winning paper on the discovery of non-static solutions [4], the physical reality and properties of the solar wind were still rather unclear and under-appreciated. Likewise, Herring's work on stellar opacities was well before proper numerical stellar structure and evolution models became available; the Henyey method [5] (solving a matrix equation instead of using an iterative shooting method from both ends) became known only in 1964. Subsequently, Herring turned to hydrodynamic convection and turbulence – topics that then determined much of his future work. During his career, he never really worked on magnetic fields or helicity, but he did interact with people on a daily basis, who were very much involved in these subjects, both early on [6] and also later during his career [7]. It is therefore not surprising that this special issue also extends to topics involving magnetic fields and helicity.

Having helicity in a system usually requires external factors such as stratification and rotation [8–10]. In this sense, the absence of helicity may be regarded as the more generic situation. It may therefore also seem natural that helicity does not play an important role when it is absent on average. This is believed to be the case in hydrodynamic turbulence, but it changes when magnetic fields are involved. Although both kinetic and magnetic helicities are ideal invariants, only the magnetic helicity has a non-ideal dissipation that is slower than that of the magnetic energy. By contrast, the dissipation of kinetic helicity is faster than that of kinetic energy [11,12]. Therefore, in the magnetic case, helicity plays a very important role in a way that is unknown in the hydrodynamic context. But is this still true when the net magnetic helicity is actually zero?

The physical situations of interest here include the decay of primordial magnetic fields in the early Universe during the radiation-dominated era, when the electric conductivity is high and the initially generated magnetic field can only decay. When the plasma is hot enough, the chirality of fermions also plays an important role, leading to an interplay with magnetic helicity. Another situation of interest is when only the Hall effect plays a role, so there are then no fluid motions, but just the flow of electrons. This is relevant in neutron star crusts, which are solid, so the ions are immobile. Another application is to the solar wind on scales below the proton gyroscale, where the ions create a smooth motionless background. The induction equation with just the Hall effect included leads to interesting decay dynamics—remarkably similar to ordinary magnetohydrodynamics (MHD). In all those cases, magnetic helicity can play a role even when it vanishes on average. In those cases, magnetic helicity fluctuations may be responsible for driving an inverse cascade similar to the case of nonvanishing mean magnetic helicity.

Less obvious examples of the dynamics discussed above include the Sun, because here the magnetic helicity is usually nonvanishing on average [13]. Even in the solar wind, where the conditions resemble those of decaying turbulence, the magnetic helicity is observed to be nonvanishing on average and systematically of opposite signs in the northern and southern hemispheres [14]. Near the ecliptic, however, the magnetic helicity fluctuates around zero [11] and may also be in a state of decay, so this may be another example where magnetic helicity fluctuations play an important role.

2. Nonhelical Turbulence and the Hosking Integral

In this section, we discuss the Hosking integral and why it is crucial to understanding nonhelical MHD turbulence with strong magnetic fields. Unlike the case of weak magnetic fields, when the dynamics is still controlled by the presence of hydrodynamic effects, we are dealing here with effects that are specific to the presence of magnetic fields, albeit with zero average. We focus on decaying turbulence.

2.1. Nonhelical Inverse Cascading and Scaling Relations

Already in 2001, it was noted that, even in the nonhelical case of a turbulently decaying magnetic field, there is a small amount of inverse cascading in the sense that for wavenumbers below the peak, the magnetic energy spectrum rises with time uniformly for all lower k [15]. The actual amount of this rise was small and one could have argued that it was just because of numerical inaccuracies. Subsequent simulations [16], however, confirmed such inverse cascading and those authors discussed the potential interplay between the shallower kinetic energy spectrum proportional to k^2 and the steeper magnetic energy spectrum proportional to k^4. The qualitative idea was that the shallower velocity spectrum pushes the magnetic spectrum upward, which then would drive more kinetic energy at small k, and so forth.

The choice of the initial magnetic energy being proportional to k^4 is important here. When such a spectrum was used in the first numerical simulations [15], the authors made reference to the early work in Ref. [17], where causality arguments were put forward. Nowadays, however, Ref. [18] has become the standard reference for the choice of an initial k^4 spectrum. Later, it turned out that with a shallower initial k^2 spectrum, no inverse cascading can be found [19,20]. The reason for this particular aspect will be discussed in more detail in this paper.

In 2014, the idea of an inverse cascade in the nonhelical case with a k^4 spectrum became really very clear [21]. This paper was on the arXiv since April 2014, but the paper was published only in February 2015. The results were reproduced in the relativistic context in Ref. [22]. Their work was on the arXiv since July 2014 and made reference to the 2015 paper. The significance of both findings is that it presents early support for the subsequent discovery of the Hosking integral as a new invariant in MHD turbulence at large magnetic Reynolds numbers.

When the Hosking integral was discovered in Ref. [23], it was originally called the "Saffman helicity invariant". As already pointed out in Ref. [24], H. K. Moffatt informed the community of the fact that this term may be misleading, because the term 'helicity invariant' is reserved for integrals that are chiral in character. He also recalled that Saffman never considered helicity in his papers. The term "magnetic helicity density correlation integral" may be more appropriate, but it is rather clumsy. Following [25], where this quantity was called the Hosking integral, this term continued being used by others [26,27]. It should also be noted that 'integral' instead of 'invariant' is appropriate since applications to turbulence apply always to finite Reynolds and Lundquist numbers. In this connection, it should be emphasized that the Hosking integral tends to decay with time in a power-law fashion and that the exponent decreases with increasing Lundquist number Lu approximately as $\mathrm{Lu}^{-1/4}$ [24].

The energy decay in turbulence is usually characterized by the energy spectrum $E(k,t)$. In the following, we sometimes add the subscripts K and M for kinetic and magnetic energy spectra and other quantities. We focus here on magnetic energy spectra, $E_\mathrm{M}(k,t)$, which are defined such that $\int E_\mathrm{M}(k,t)\,\mathrm{d}k = \langle \boldsymbol{B}^2\rangle/2\mu_0 \equiv \mathcal{E}_\mathrm{M}(t)$ is the magnetic energy, and μ_0 is the vacuum permeability. The decay can then be parameterized by $\mathcal{E}_\mathrm{M}(t)$ and the magnetic integral scale, which is defined in terms of the magnetic energy spectrum as

$$\xi_\mathrm{M}(t) = \int_0^\infty k^{-1} E_\mathrm{M}(k,t)\,dk \Big/ \int_0^\infty E_\mathrm{M}(k,t)\,dk. \tag{1}$$

One can always attempt to describe the relations for $\xi_\mathrm{M}(t)$ and $\mathcal{E}_\mathrm{M}(t)$ through power laws. In addition, the spectrum can evolve underneath an envelope,

$$E_\mathrm{M}(k,t) \le \mathrm{const} \times k^\beta, \tag{2}$$

which is in general different from the initial subinertial range spectrum, $E_\mathrm{M}(k,t_0) = \mathrm{const} \times k^\alpha$, where α is the subinertial range slope. The three relations for $\xi_\mathrm{M}(t)$, $\mathcal{E}_\mathrm{M}(t)$, and $E_\mathrm{M}(k,t)$ can then be constrained through dimensional arguments once we have a good idea about the relevant dimensional quantity that governs the decay.

In 2017, the decay of a nonhelical turbulent magnetic field is found to be described by an exponent β that was determined to be between $\beta = 1$ [28] and $\beta = 2$ [20], but it was unclear why any of those two possibilities, or any other one, would have to be expected. This is what the Hosking integral now explains, namely that $\beta = 3/2$.

Figure 1a shows magnetic energy spectra at four different times for a nonhelical magnetically dominated run corresponding to Run K60D1bc in Ref. [24]. Here, k is normalized by the initial peak wavenumber k_0. We clearly see that the spectrum exhibits inverse cascading in that the spectral magnetic energy *increases* with time at small k, as indicated by the upward arrow on the left. The overall energy does of course decay, as indicated by the decline of the spectral peak and the decrease of spectral energy at large k, as indicated by the downward arrow on the right.

To quantify the temporal changes of $\xi_\mathrm{M}(t)$ and $\mathcal{E}_\mathrm{M}(t)$, it is convenient to compute the instantaneous scaling exponents [28]

$$q(t) = \mathrm{d}\ln\xi_\mathrm{M}/\mathrm{d}\ln t \quad \text{and} \quad p(t) = -\mathrm{d}\ln\mathcal{E}_\mathrm{M}/\mathrm{d}\ln t; \tag{3}$$

see Figure 1b. We see that with time (larger red symbols), the solution evolves toward the point $(q,p) = (4/9,\ 10/9)$, as is also theoretically expected [24]. Although we mainly focus on the case of nonhelical magnetic fields, we also compare in Figure 1b with the expected solution for the fully helical case (orange), and include solutions for hydrodynamic turbulence that are governed either by the Loitsyansky or the Saffman integrals.

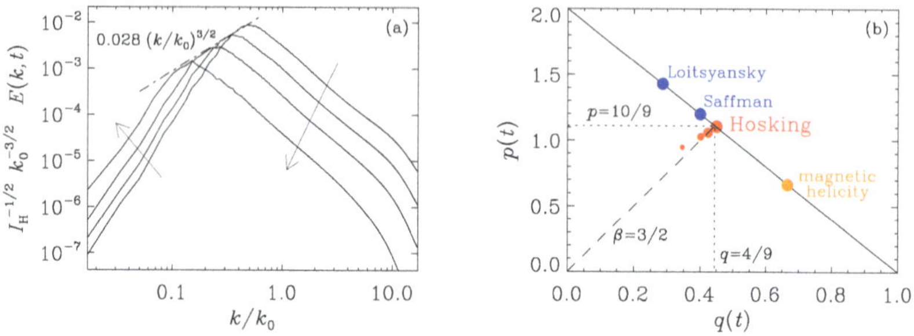

Figure 1. (**a**) magnetic energy spectra, normalized by $I_H^{-1/2} k_0^{-3/2}$; the dashed dotted line shows the envelope $0.028\,(k/k_0)^{3/2}$ under which the spectrum evolves. The times are $ck_1t = 3, 7, 17$, and 58. (**b**) qp diagram showing as red dots the convergence of $p(t)$ versus $q(t)$ toward the Hosking attractor $(q, p) = (4/9,\ 10/9)$. The blue symbols denote the Loitsyansky and Saffman attractors, respectively, and the orange symbol denotes the magnetic helicity attractor.

Before we continue, it is useful to clarify the concept of what we often refer to as a 'governing quantity'. Take, for example, standard hydrodynamic Kolmogorov turbulence. Here, the rate of energy transfer per unit mass ϵ (which is the rate of energy input and also the rate of energy dissipation) is such a quantity and the relevant physical scaling laws can be expressed in terms of powers of ϵ and other relevant variables such as the wavenumber k itself. This then yields for the energy spectrum per unit mass the expression $E(k) = C_K \epsilon^{2/3} k^{-5/3}$, where C_K is a dimensionless coefficient of order unity (the Kolmogorov constant; typically $C_K \approx 1.6$). Other such governing quantities include the mean magnetic helicity density $\langle h \rangle$ and some other quantities that are crucial to the physics. They are usually constant or well conserved.

2.2. The Loitsyansky and Saffman Integrals in Hydrodynamics

In the hydrodynamic case, the decay of turbulence can follow different behaviors depending on the relevant conservation law. (In practice, conserved quantities are usually not perfectly conserved under turbulent conditions, and some are better conserved than others. Which one is the most relevant quantity depends on the relative conservation properties under different circumstances.) One such conserved quantity is the Loitsyansky integral [29,30],

$$I_L = -\int \langle \boldsymbol{u}(\boldsymbol{x}) \cdot \boldsymbol{u}(\boldsymbol{x}+\boldsymbol{r}) \rangle\, r^2\, \mathrm{d}^3\boldsymbol{r}, \tag{4}$$

which is believed to play an important role. This integral reflects the local conservation of angular momentum and has dimensions $[I_L] = \mathrm{m^7\,s^{-2}}$. If this quantity governs the decay of turbulence, the time dependence of the growth of the integral scale can be motivated by dimensional arguments as $\xi(t) \propto I_L^a t^b$, where the exponents a and b must be, on dimensional grounds, $a = 1/7$ and $b \equiv q = 2/7$. The kinetic energy then obeys $\mathcal{E}_K \propto I_L^{2/7} t^{-10/7}$, i.e., $p = 10/7$. The envelope under which the peak of the spectrum evolves obeys $E_K(k,t) \leq C_L I_L k^4$.

Another conserved quantity is the Saffman integral,

$$I_S = \int \langle \boldsymbol{u}(\boldsymbol{x}) \cdot \boldsymbol{u}(\boldsymbol{x}+\boldsymbol{r}) \rangle\, \mathrm{d}^3\boldsymbol{r}, \tag{5}$$

which has dimensions $[I_S] = \mathrm{m^5\,s^{-2}}$. Similarly, if this quantity governs the decay of turbulence, the time dependence of ξ must be $\xi(t) \propto I_S^a t^b$, where $a = 1/5$ and $b \equiv q = 2/5$ on dimensional grounds. The kinetic energy then obeys $\mathcal{E}_K \propto I_L^{2/5} t^{-6/5}$, i.e., $p = 6/5$. The envelope under which the peak of the spectrum evolves obeys in this case $E_K(k,t) \leq C_S I_S k^2$.

Whether I_L or I_S determine the decay depends on the existence of long-range correlations, as can be seen from the Taylor expansion of the kinetic energy spectrum as [23,29]

$$2E_K(k \to 0) \equiv \mathrm{Sp}(\boldsymbol{u})(k \to 0) = \frac{I_S}{2\pi^2}k^2 + \frac{I_L}{12\pi^2}k^4 + \ldots, \tag{6}$$

where an initially non-vanishing Saffman integral automatically implies a k^2 scaling in the subinertial range. Thus, the decay does depend on the infrared part of the initial kinetic energy spectrum. In that case, the slope is the same as that required for the initial spectrum so that the Saffman integral is indeed nonvanishing. Furthermore, as pointed out in Ref. [23], owing to the invariance of I_S and I_L, both an initial k^2 and a k^4 spectrum will remain unchanged. This implies that there can be no inverse cascading in hydrodynamics.

2.3. The Magnetic Saffman Integral: Comparison with the Hosking Integral

As already pointed out in Ref. [23], the formulation of Section 2.2 can also be applied to the magnetic field, except that there is no reason for the magnetic version of the Loitsyansky integral to be conserved. The magnetic Saffman integral (hereafter I_{SM}), on the other hand, might indeed be conserved. Physically, it would reflect the local conservation of magnetic flux. Again, when I_{SM} is non-vanishing initially, we expect a quadratic magnetic energy spectrum, which would also persist at later times. For a steeper k^4 subinertial range magnetic energy spectrum, however, the magnetic Saffman integral must vanish and the Hosking integral is then expected to play a dominant role. It is defined as

$$I_H = \int \langle h(\boldsymbol{x})h(\boldsymbol{x}+\boldsymbol{r})\rangle \, \mathrm{d}^3\boldsymbol{r}, \tag{7}$$

where $h = \boldsymbol{A} \cdot \boldsymbol{B}$ is the magnetic helicity density with dimensions $[h] = [B]^2[x]$. In ordinary MHD, we can express the magnetic field as an Alfvén velocity, i.e., we write the magnetic field in Alfvén units, so $[B] = \mathrm{m\,s^{-1}}$. Therefore, $[h] = [x]^3[t]^{-2}$, and thus $[I_H] = [B]^4[x]^5 = [x]^9[t]^{-4}$. If I_H plays a governing role in the decay, we expect therefore $\xi_M(t) \propto I_H^{1/9}t^{4/9}$, $\mathcal{E}_M \propto I_H^{2/9}t^{-10/9}$, and $E_M(k,t) \le C_H I_H k^{3/2}$.

The Hosking integral is in general expected to be different from zero [23]. This automatically implies a quadratic scaling of the helicity variance spectrum, $\mathrm{Sp}(h)$. Here, $\mathrm{Sp}(h) = \oint_{4\pi} |\tilde{h}|^2 k^2 \mathrm{d}\Omega_k/(2\pi L)^3$ denotes the shell-integrated spectrum, a tilde marks a quantity in Fourier space, and Ω_k is the solid angle in Fourier space, so that $\int \mathrm{Sp}(h)\,\mathrm{d}k = \langle h^2\rangle$. The quadratic scaling for a finite Hosking integral follows from the expansion

$$\mathrm{Sp}(h)(k \to 0) = \frac{I_H}{2\pi^2}k^2 + \ldots \tag{8}$$

In three dimension, a quadratic spectrum corresponds to white noise. We also know that the spectrum of a quadratic quantity cannot be more blue than that of white noise [31], so it seems impossible to have a helicity variance spectrum whose subinertial range is steeper than k^2.

In Figure 2, we show magnetic energy and magnetic helicity variance spectra for initial spectra of the form

$$E_M(k,t_0) = \text{const} \times \frac{k^\alpha}{1+(k/k_0)^{\alpha+5/3}} \propto \begin{cases} k^\alpha & \text{for } k \ll k_0, \\ k^{-5/3} & \text{for } k \gg k_0, \end{cases} \tag{9}$$

for $\alpha = 2$ and $\alpha = 4$. We solve the isothermal compressible MHD equations using the Pencil Code [32] with 1024^3 mesh points. As expected, and as pointed out previously [19,20], there is inverse cascading only for $\alpha = 4$, but not for $\alpha = 2$. Nevertheless, we see that $\mathrm{Sp}(h)$ retains a k^2 spectrum at low wavenumbers in both cases. This suggests that the Hosking integral is indeed always conserved; see Figure 2b,d. It may, however, be less dominant than the magnetic Saffman integral.

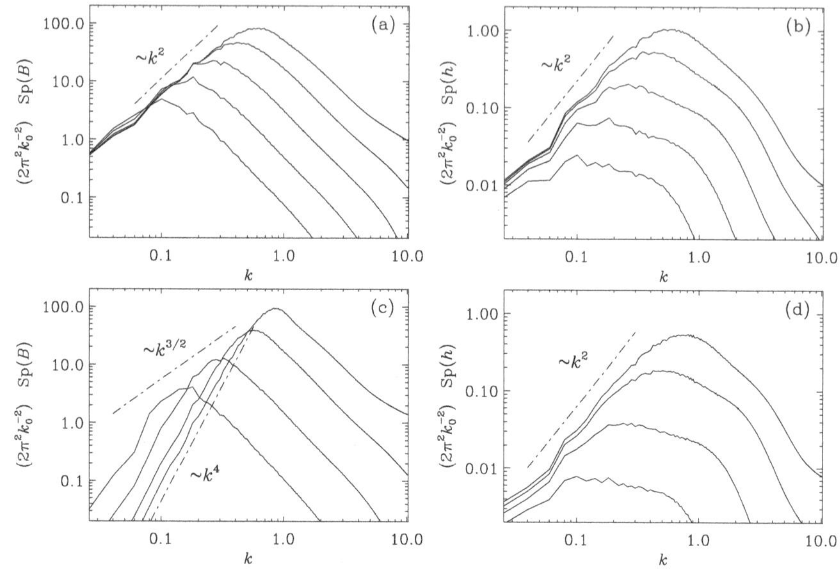

Figure 2. Comparison of $\mathrm{Sp}(\boldsymbol{B})$ (**a**,**c**) and $\mathrm{Sp}(h)$ (**b**,**d**) for $\alpha = 2$ (**a**,**b**) and $\alpha = 4$ (**c**,**d**). The dashed-dotted lines indicate $k^{3/2}$ scaling in (**c**) and k^2 scaling otherwise.

To determine the relevant integrals, I_H and I_SM, it is convenient to plot compensated spectra. Specifically, to determine I_SM and I_H, we scale both $\mathrm{Sp}(\boldsymbol{B})$ and $\mathrm{Sp}(h)$ by $2\pi^2/k^2$. The result is shown in Figure 3. Thus, in summary, we have

$$\xi_\mathrm{M}(t) \approx 0.16\, I_\mathrm{SM}^{1/5} t^{2/5}, \quad \mathcal{E}_\mathrm{M}(t) \approx 4.2\, I_\mathrm{SM}^{2/5} t^{-6/5}, \quad E_\mathrm{M}(k) \approx 0.037\, I_\mathrm{SM} (k/k_0)^2. \tag{10}$$

If the initial spectrum is not $\propto k^2$, but $\propto k^4$, we have

$$\xi_\mathrm{M}(t) \approx 0.15\, I_\mathrm{H}^{1/9} t^{4/9}, \quad \mathcal{E}_\mathrm{M}(t) \approx 3.8\, I_\mathrm{H}^{2/9} t^{-10/9}, \quad E_\mathrm{M}(k) \approx 0.025\, I_\mathrm{H}^{1/2} (k/k_0)^{3/2}. \tag{11}$$

It is remarkable that the prefactors for the Saffman and Hosking scalings are very close to each other; see Table 1 for a summary of the nondimensional prefactors in the relations

$$\xi_\mathrm{M}(t) = C_i^{(\xi)}\, I_i^{\sigma} t^{q}, \quad \mathcal{E}_\mathrm{M}(t) = C_i^{(\mathcal{E})}\, I_i^{2\sigma} t^{-p}, \quad E_\mathrm{M}(k) = C_i^{(E)}\, I_i^{(3+\beta)/\sigma} (k/k_0)^{\beta}, \tag{12}$$

where the index i on the integrals I_i and the coefficients $C_i^{(\xi)}$, $C_i^{(\mathcal{E})}$, and $C_i^{(E)}$ stands for SM or H for magnetic Saffman and Hosking scalings, respectively, and σ is the exponent with which length enters in I_i: $\sigma = 5$ for the magnetic Saffman integral ($i =$ SM) and $\sigma = 9$ for the Hosking integral ($i =$ H). Interestingly, while β and p can uniquely be related to q via $\beta = 2/q - 3$ and $p = 2(1-q)$ [28], the exponent σ is not uniquely linked to q and we have $\sigma q = 2$ for Saffman scaling and $\sigma q = 4$ for Hosking scaling.

The value of $C_\mathrm{SM}^{(E)}$ only makes sense when $\alpha = \beta = 2$, while that of $C_\mathrm{H}^{(E)}$ only makes sense when $\alpha = 4$ and $\beta = 3/2$. For the other cases, the subinertial range spectrum is not parallel to k^β, so α and β are said to be incompatible with each other (see Table 1) and the given values of $C_\mathrm{SM}^{(E)}$ and $C_\mathrm{H}^{(E)}$ only yield crossings in the middle of the subinertial range.

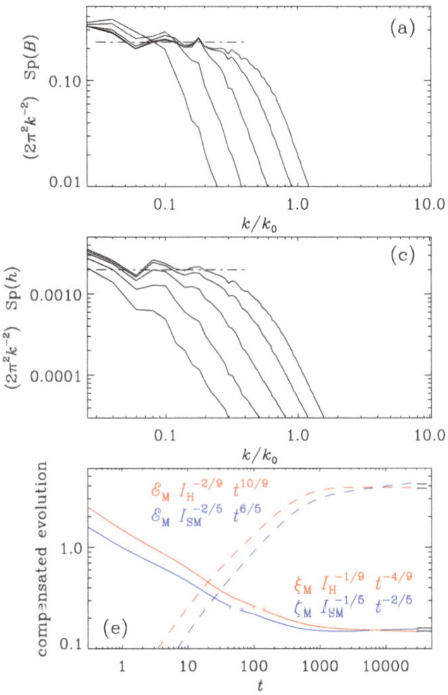

Figure 3. Compensated spectra for $\alpha = 2$ (**a**,**c**,**e**) and $\alpha = 4$ (**b**,**d**,**f**). From (**a**,**b**), the horizontal dashed-dotted lines indicate that $I_{\rm SM} \approx 0.23$ and 0.09, respectively, and from (**c**,**d**) they indicate that $I_{\rm H} \approx 2 \times 10^{-3}$ and 5×10^{-4}, respectively. The asymptotic values estimated from (**e**,**f**) are discussed in the text and in Table 1.

Table 1. Summary of nondimensional prefactors in the relations for $\xi_{\rm M}(t)$, $\mathcal{E}_{\rm M}(t)$, and $E_{\rm M}(k,t)$. The numbers in parentheses indicate that the slope β is incompatible with the value of α.

α	β	$C_{\rm SM}^{(\xi)}$	$C_{\rm H}^{(\xi)}$	$C_{\rm SM}^{(\mathcal{E})}$	$C_{\rm H}^{(\mathcal{E})}$	$C_{\rm SM}^{(E)}$	$C_{\rm H}^{(E)}$
2	2	0.16	0.15	4.2	3.8	0.025	(0.05)
4	3/2	0.15	0.13	4.0	3.5	(0.02)	0.037

We see from Figure 3a that for $\alpha = 2$, the compensated value $(2\pi/k^2)\,{\rm Sp}(\boldsymbol{B}) \to I_{\rm SM} \approx 0.2$. For $\alpha = 4$, on the other hand, we only see a flat envelope, i.e., $(2\pi/k^2)\,{\rm Sp}(\boldsymbol{B}) \leq 0.1$, i.e., $2E_{\rm M}(k,t) \leq 0.1/(2\pi^2/k^2)\,(k/k_0)^2$. From Figure 3c,d, we see that $(2\pi/k^2)\,{\rm Sp}(h) \to I_{\rm H} \approx 0.001$ in both cases, i.e., for $\alpha = 2$ and $\alpha = 4$, respectively.

Given that we now know the values of $I_{\rm SM}$ and $I_{\rm H}$, we can compensate the time evolutions of $\xi_{\rm M}(t) \propto t^q$ with $q = 2/5 = 0.4$ and $q = 4/9 \approx 0.44$, and those of $\mathcal{E}_{\rm M}(t) \propto t^{-p}$ with $p = 6/5 = 1.2$ and $p = 10/9 \approx 1.1$. The results for the corresponding coefficients in Equation (12) are summarized in Table 1.

2.4. The Effect of Rotation

Rotation suppresses hydrodynamics turbulence. This is modelled by including the Coriolis force, $-2\boldsymbol{\Omega} \times \boldsymbol{u}$, on the right-hand side of the momentum equation, which then reads

$$\frac{{\rm D}\boldsymbol{u}}{{\rm D}t} = -c_{\rm s}^2 \boldsymbol{\nabla} \ln \rho - 2\boldsymbol{\Omega} \times \boldsymbol{u} + \frac{1}{\rho}[\boldsymbol{J} \times \boldsymbol{B} + \boldsymbol{\nabla} \cdot (2\rho\nu \mathbf{S})], \quad (13)$$

where ${\rm D}/{\rm D}t = \partial/\partial t + \boldsymbol{u} \cdot \boldsymbol{\nabla}$ is the advective derivative, $c_{\rm s}$ is the isothermal sound speed, $\boldsymbol{\Omega}$ is the angular velocity, $\boldsymbol{J} = \boldsymbol{\nabla} \times \boldsymbol{B}/\mu_0$ is the current density, μ_0 is the permeability, ρ is the density, ν is the viscosity, and $S_{ij} = (\partial_i u_j + \partial_j u_i)/2 - \delta_{ij}\boldsymbol{\nabla}\cdot\boldsymbol{u}/3$ are the components of the rate-of-strain tensor.

In Figure 4, we show $\mathrm{Sp}(\boldsymbol{B})$ and $\mathrm{Sp}(h)$ for $\Omega/c_{\mathrm{s}}k_0 = 10^{-3}$, 0.01, 0.1, and 1 for runs with $\alpha = 4$, which are otherwise the same as that of Figure 2c,d. We see a clear suppression of inverse cascading already for $\Omega/c_{\mathrm{s}}k_0 = 10^{-3}$ and a very strong suppression when this parameter is unity. This is caused by the suppression of the turbulent velocity and thereby of the $\boldsymbol{u}\times\boldsymbol{B}$ nonlinearity in the induction equation.

To express the angular velocity in a physically more meaningful way, we note that in the run with $\Omega/c_{\mathrm{s}}k_0 = 1$, the rms Mach number, $u_{\mathrm{rms}}/c_{\mathrm{s}}$, drops below 10^{-3} by the end of the run, while for $\Omega/c_{\mathrm{s}}k_0 = 10^{-3}$, it still stays well above 10^{-3}. This means that in the latter, the Coriolis number, $\mathrm{Co} \equiv 2\Omega/u_{\mathrm{rms}}k_0$, is around unity when rotational suppression becomes appreciable. At $t = 10^4$, the values of Co are 1.2, 22, 500, and 10^4 for our four runs in Figure 4.

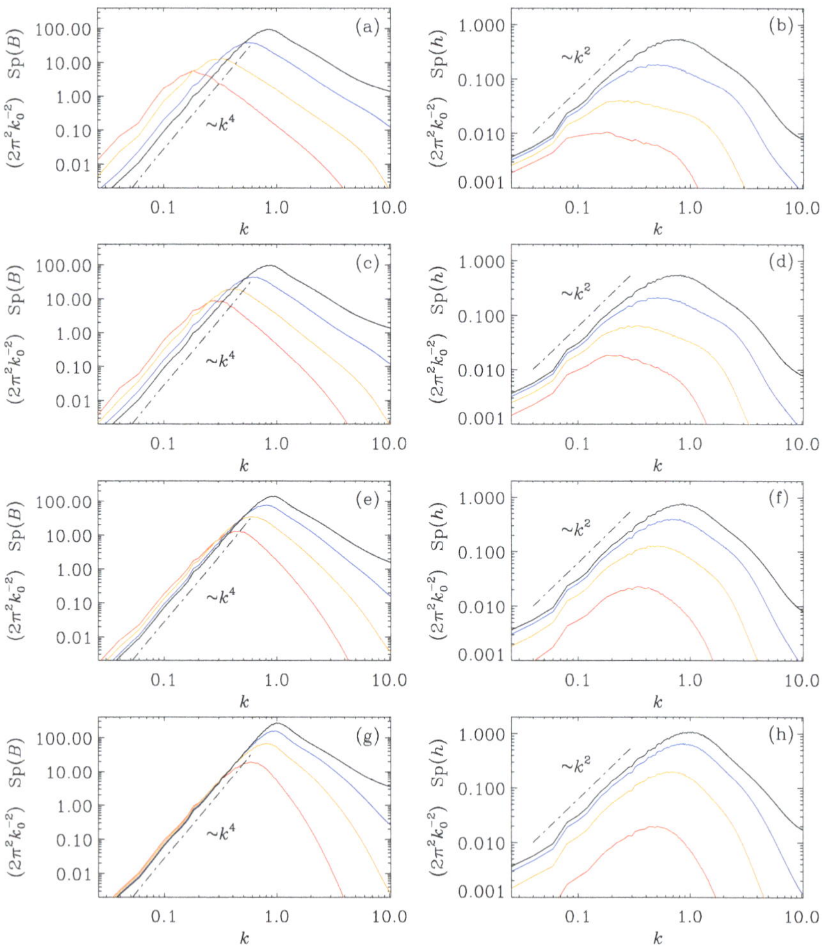

Figure 4. $\mathrm{Sp}(\boldsymbol{B})$ (**a,c,e,g**) and $\mathrm{Sp}(h)$ (**b,d,f,h**) for $\Omega/c_{\mathrm{s}}k_0$ from 10^{-3} (**a,b**: Co = 1.2) to $\Omega/c_{\mathrm{s}}k_0 = 1$ (**g,h**: Co = 10^4). The times are 220 (black), 1000 (blue), 4600 (orange), and 22,000 (red).

3. Extensions of the Hosking Idea

Equation (7) is the Hosking integral in its original form. In the meantime, two further variants of I_{H} have been considered. One is where h has been replaced by $h_{\mathrm{tot}} = \boldsymbol{A}\cdot\boldsymbol{B} + 2\mu_5/\lambda$, where μ_5 is the chiral chemical potential (here in units of an inverse length) and λ is a coefficient that quantifies the coupling between fermions and electromagnetic fields. The case $\langle h_{\rm tot}\rangle = 0$ has been studied recently in Ref. [33]. Another variant of the Hosking integral is that in the case where the magnetic field is controlled by the electromagnetic induction from the Hall effect, which we discuss next.

3.1. Hall Effect

In neutron star crusts, the ions are immobile and the current is only carried by electrons with the velocity $\boldsymbol{u}_{\rm e} = -\boldsymbol{J}/en_{\rm e}$, where e is the electric charge and $n_{\rm e}$ is the electron density. As alluded to in the introduction, a similar situation occurs in the solar wind on scales below the proton gyroscale, where the ions constitute a smooth background [34]. The induction equation with the induction from $\boldsymbol{u}_{\rm e} \times \boldsymbol{B}$ therefore takes the form [35]

$$\frac{\partial \boldsymbol{B}}{\partial t} = \boldsymbol{\nabla} \times \left(-\frac{1}{en_{\rm e}}\boldsymbol{J} \times \boldsymbol{B} - \eta\mu_0 \boldsymbol{J}\right), \tag{14}$$

where η is the magnetic field diffusivity. In the presence of magnetic helicity, one finds an inverse cascade with an overall decay of the magnetic field and a growth of spectral magnetic energy at small wavenumbers below that of the peak of the spectrum [34]. In the present paper, however, the focus is on the nonhelical case, which was already considered in B20, but understood mathematically only later [36].

In this context, it is important to note that the natural dimensions of the magnetic field here are no longer $\mathrm{m\,s^{-1}}$, but $\mathrm{m^2\,s^{-1}}$. This was already emphasized in Ref. [37], who used $e = 1.6 \times 10^{-19}\,\mathrm{A\,s}$, $\mu_0 = 4\pi \times 10^{-7}\,\mathrm{T\,m\,A^{-1}}$, and $n_{\rm e} \approx 2.5 \times 10^{40}\,\mathrm{m^{-3}}$ for neutron star crusts, so we have $en_{\rm e}\mu_0 \approx 5 \times 10^{15}\,\mathrm{T\,s\,m^{-2}}$, and therefore

$$\frac{B}{en_{\rm e}\mu_0} = \frac{B}{5 \times 10^{15}\,\mathrm{T}}\frac{\mathrm{m^2}}{\mathrm{s}}, \tag{15}$$

which is why we say B has dimensions of $\mathrm{m^2\,s^{-1}}$ in the Hall cascade. This modifies all the dimensional arguments related to $\boldsymbol{B}$ correspondingly. In particular, the units of the magnetic helicity are $[h] = \mathrm{m^5\,s^{-2}}$ and those of energy spectra are also $\mathrm{m^5\,s^{-2}}$. Therefore, one has $q = p = 2/5$. This scaling was confirmed in Ref. [37].

In the nonhelical case, the modified Hosking integral has dimensions $m^{13}\,\mathrm{s^{-4}}$, and therefore $q = 4/13$ (instead of $4/9$ in MHD). Furthermore, $p = 10/13$ (instead of $10/9$ in MHD), but still $\beta = 3/2$ (just like in MHD). While such a scaling was already seen in the original simulations of Ref. [37], the work in Ref. [36] showed that the modified Hosking integral is indeed conserved. In Figure 5a,b,d,e, we show that, also for Hall dynamics, the Saffman scaling is obeyed for $\alpha = 2$, while Hosking scaling is obeyed for $\alpha = 4$.

It should be noted that earlier work on the Hall cascade focussed on the concept of Whistler turbulence [38], where the Whistler time $t_{\rm W}$ was identified as the governing timescale [34]. The definition of $t_{\rm W}$ in Ref. [34] involved the electron plasma frequency and the electron gyrofrequency such that the electron mass drops out. Therefore, $t_{\rm W}$ can more easily be written as

$$t_{\rm W} = \frac{L^2}{B_{\rm rms}/en_{\rm e}\mu_0}, \tag{16}$$

which is just the characteristic time based on the magnetic field expressed in units of a diffusivity; see Equation (15). Earlier interpretations in terms of Whistler waves [34] seem therefore artificial and obscured the relevant interpretation of the magnetic field as a quantity with dimensions of $\mathrm{m^2\,s^{-1}}$.

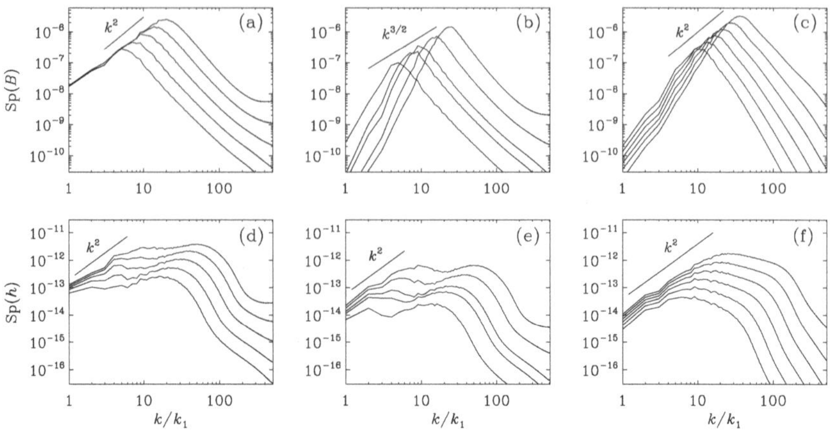

Figure 5. Sp(B) (**a–c**) and Sp(h) (**d–f**) for Hall dynamics with $\alpha = 2$ (**a**,**d**) and $\alpha = 4$ (**b**,**e**), and for ambipolar diffusion with $\alpha = 4$ (**c**,**f**). Note the presence of inverse cascading for $\alpha = 4$ in panels (**b**,**c**), although Sp(h) changes at $k/k_0 \ll 1$ in all cases. The straight lines indicate $k^{3/2}$ scaling in (**b**) and k^2 scaling otherwise.

Observationally, the inverse cascade with helicity may be more easily accessible, and this has already been tried in the context of neutron stars [39]. Measuring inverse cascading in the solar wind is conceptually harder, because the system is statistically steady and more advanced stages of decay would only correspond to larger distances from the Sun. On the other hand, most of the spacecrafts are located near the ecliptic and might therefore allow better insight into nonhelical inverse cascading. Unfortunately, most of the observations focus on the high frequency range of the spectrum [40], while the low frequency range is dominated by noise, making it nearly impossible to say anything about inverse cascading to even larger scales.

3.2. Ambipolar Diffusion

The Hall effect is a two-fluid effect where the two components are the positive and negative charge carriers. Another two-fluid effect is ambipolar diffusion where the charged fluid with positive and negative charge carriers is taken as one component and neutrals are taken as the other component. The governing equation is

$$\frac{\partial \boldsymbol{B}}{\partial t} = \boldsymbol{\nabla} \times \left(-\frac{\boldsymbol{J} \times \boldsymbol{B}}{\rho_{\rm i} \nu_{\rm in}} \times \boldsymbol{B} - \eta \mu_0 \boldsymbol{J} \right), \tag{17}$$

where $\rho_{\rm i}$ is the ion density and $\nu_{\rm in}$ is the ion–neutral collision frequency.

Unlike the Hall effect in neutron star crusts, where the magnetic field is said to have dimensions of $\mathrm{m^2\,s^{-1}}$, we can here write

$$\frac{B}{\sqrt{\rho_{\rm i} \mu_0}} = \frac{B}{1.5 \times 10^{-16}\,\mathrm{T}} \frac{\mathrm{m}}{\mathrm{s}}, \tag{18}$$

where we used $\rho_{\rm i} = 1.7 \times 10^{-26}\,\mathrm{kg\,m^{-3}}$ for the interstellar medium with an ionization fraction of 10^{-5} and a neutral density of one proton per cubic centimeter. This is why we say that with ambipolar diffusion, just like in MHD, B has dimensions of $\mathrm{m\,s^{-1}}$. For this reason, we also see in Figure 5c,f qualitatively the same decay behavior as in ordinary MHD.

3.3. Chiral MHD

For chiral MHD, the induction equation attains an extra term under the curl that leads to a contribution to the electric field proportional to the product of the magnetic field and a pseudoscalar given by the chiral chemical potential, expressed here as a wavenumber [41].

$$\mu_5 = 24\,\alpha_{\rm em}\,(n_{\rm L} - n_{\rm R})\,(\hbar c/k_{\rm B}T)^2, \tag{19}$$

where $\alpha_{\rm em} \approx 1/137$ is the fine structure constant, and $n_{\rm L}$ and $n_{\rm R}$ are the number densities of left- and right-handed fermions, respectively. The uncurled induction equation takes then the form

$$\frac{\partial \boldsymbol{A}}{\partial t} = \eta(\mu_5 \boldsymbol{B} - \mu_0 \boldsymbol{J}) + \boldsymbol{u} \times \boldsymbol{B}, \quad \boldsymbol{J} = \boldsymbol{\nabla} \times \boldsymbol{B}/\mu_0. \tag{20}$$

The term $\eta\mu_5\boldsymbol{B}$ leads to a growth of the magnetic field for wavenumbers $k < \mu_5$, just in the same way as in mean-field dynamo theory [8–10], but here no mean-field theory is invoked. The generated magnetic field is fully helical, but the relevant quantity is now the *total* chirality density

$$h_{\rm tot} = \boldsymbol{A} \cdot \boldsymbol{B} + 2\mu_5/\lambda, \tag{21}$$

and it is its volume average that is conserved, i.e., $\langle h_{\rm tot} \rangle = {\rm const}$, provided the boundary conditions are periodic and/or closed, i.e., perfectly conducting. As the magnetic field grows, μ_5 decreases. The rate of this change is proportional to the parameter λ, which we take here as an adjustable parameter, but in reality is it given by an expression involving the temperature; see Equation (49) of Ref. [41].

It is important to point out that the *physical* chiral chemical potential (which has the units of an energy) is sometimes defined differently. First, the authors of Ref. [33] used Lorentz-Heaviside units, which implies another factor of 4π in the numerator of the conversion factor (or rather the lack of a 4π factor in the denominator), and, second, there is also a factor of 2 in the denominator, so $\hbar c/8\alpha_{\rm em}$ instead of $\hbar c/4\alpha_{\rm em}$ for the conversion factor of [33], because they defined their phsical chiral chemical potential as half the difference between the physical right- and left-handed chiral chemical potentials. In addition, there is a sign difference between Refs. [41] and [33], but this affects only the physical chiral chemical potential and not our equations, where μ_5 has the units of a wavenumber.

It turns out, perhaps not surprisingly, that in this case, when $\langle h_{\rm tot} \rangle = 0$, the turbulence decays again in such a way that $q = 4/9$ and $p = 10/9$ and, again, $\beta = 3/2$. This is just like in ordinary (but nonhelical) MHD. In this case, however, $\langle h \rangle \neq {\rm const}$, but its modulus decays $\propto t^{-r}$ in a way that is compatible with the real-space realizability condition, $|\langle h \rangle| \leq 2\mathcal{E}_{\rm M}\xi_{\rm M}$, i.e., $r = p - q = (10-4)/9 = 2/3$. This was also confirmed in Ref. [33]. This study was then applied to the problem of baryogenesis [42], where one tries to explain the small excess of matter over antimatter in the Universe, which is referred to as baryon asymmetry.

The Hosking scaling was confirmed for $\langle \mu_5 \rangle \xi_{\rm M} \gg 1$ [42], but in the opposite limit of $\langle \mu_5 \rangle \xi_{\rm M} \ll 1$ the Hosking scaling was no longer obeyed and then both $\langle \mu_5 \rangle$ and $\xi_{\rm M}$ are believed to be approximately independently conserved [33]. Trying to understand this more thoroughly must be a goal for future studies, where one may hope to reach much larger scale separation between the different relevant wavenumbers in the system, such as the wavenumber k_0 of the peak of the magnetic energy spectrum and the value of μ_5.

In Figure 6, we plot magnetic energy and magnetic helicity spectra, as well as magnetic helicity variance spectra for a chiral MHD run with balanced chirality and an initial k^4 spectrum for the magnetic field. We see standard inverse cascading with $\beta = 3/2$. Next we compare with the case of an initial k^2 spectrum; see Figure 7. In this case, there is still weak inverse cascading, which is probably a consequence of the strong contribution from mean magnetic helicity conservation over extended spatial patches.

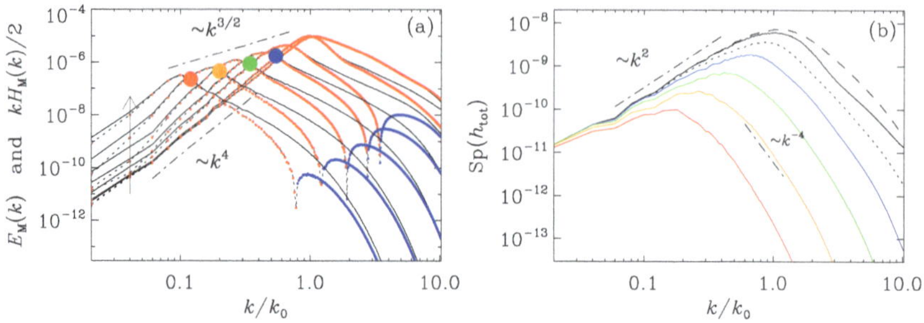

Figure 6. (**a**) Magnetic energy (solid lines) and magnetic helicity spectra (dotted lines), and (**b**) magnetic helicity variance spectra for a chiral MHD run with balanced chirality and an initial k^4 spectrum for the magnetic field. In (**a**), positive (negative) magnetic helicities are indicated by small red (blue) dots. The four large dots denote the positions of ξ_{M}^{-1}. Their colors are the same as those of the solid lines in (**b**) and correspond to the times 1500, 5000, 15,000, and 50,000.

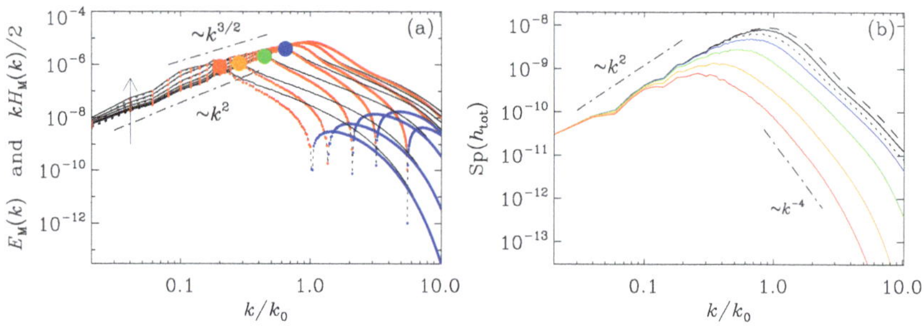

Figure 7. Similar to Figure 6, but with an initial k^2 spectrum. Note the presence of slight inverse cascading in (**a**), although $\mathrm{Sp}(h_{\mathrm{tot}}) = \mathrm{const}$ at $k/k_0 \ll 1$ in (**b**). The different colors refer to the same times as in Figure 6.

Departures from the conservation of the Hosking integral based on h_{tot} have been seen when $\mu_5 < k_0$ [42], but here we have $\mu_5 > k_0$. To understand more thoroughly the regime where $\mu_5 \gg k_0$, we would need to have much larger numerical resolution. A possible alternative is to use shell models [43], as will be discussed next. However, it is unclear whether such models can capture the relevant effects related to the Hosking integral or the chiral magnetic effect.

4. Hosking Integral in Shell Models of Chiral MHD

Shell models describe turbulence through real or complex scalar variables on concentric shells in wavenumber space such that certain conservation laws are obeyed. In MHD, the relevant conservation laws are those of total chirality, total (magnetic plus kinetic) energy, and cross helicity. The Hosking integral describes helicity fluctuations over different scales, and does not have a direct counterpart at the level of shell models. However, the scaling properties resulting from its conservation, could still be manifest in shell models describing the decay of MHD turbulence.

The Hosking integral is particularly important in cases where the mean chirality vanishes. It is also conserved otherwise when the mean total chirality is non-vanishing, but then the conservation of the mean chirality is usually more important. It is also important that the magnetic field is strong, because otherwise the decay properties are dominated by the hydrodynamic turbulent decay. Our goal here is to investigate the decay of magnetic fields with vanishing net chirality in chiral MHD using shell models.

In a shell model, we describe the state of the system in shells of logarithmically spaced wavenumbers $k_n = 2^n$, where $n = 0, 1, 2, \ldots, N$ denotes the shell and N is the truncation level. For $N = 30$, for example, we can span ten orders of magnitude in wavenumber. In MHD, one usually considers complex variables B_n and u_n for the magnetic and velocity fields. The mean magnetic and kinetic energy densities are given by

$$\mathcal{E}_{\rm M} = \tfrac{1}{2} \sum_{n=0}^{N} |B_n|^2 \quad \text{and} \quad \mathcal{E}_{\rm K} = \tfrac{1}{2} \sum_{n=0}^{N} |u_n|^2. \tag{22}$$

In shell models, the fluid density is constant and therefore not indicated in the definition of the kinetic energy. Also the permeability factor in the magnetic energy has been omitted.

Magnetic helicity is a signed quantity, i.e., it can be positive or negative. How to describe this in a standard shell model is a matter of convention. One approach is to associate even and odd shells with the decomposition into positively and negatively polarized modes of the field. This idea was first developed for the kinetic helicity [44]. This then leads to the definition of the magnetic helicity as [45–47]

$$\mathcal{H}_{\rm M} = \sum_{n=0}^{N} (-1)^n |B_n|^2 / k_n, \tag{23}$$

which satisfies the realizability condition

$$k_n |\mathcal{H}_{\rm M}(k_n)| \leq 2E_{\rm M}(k_n). \tag{24}$$

To preserve the preferential growth of positively (negatively), polarized modes on even (odd) shells, we write

$$\left[\eta k(k_n - (-1)^n \mu_5) + \frac{\rm d}{{\rm d}t}\right] B_n = \tfrac{1}{6} {\rm i} k_n \left[M(u, B) - M(B, u)\right], \tag{25}$$

where the μ_5 term leads to a growth of $|b_n|^2$ for even (odd) values of n when μ_5 is positive (negative), and $M(x, y)$ is a nonlinear functional, where x and y stand for the full n-dependent arrays. The essence of shell models is to couple only nearest and next-nearest neighbors. We refer to this model as type I. This prescription then leads to

$$M(x, y) = x_{n+1} y_{n+2} + x_{n-1} y_{n+1} + x_{n-2} y_{n-1} \quad \text{(type I)}. \tag{26}$$

As already emphasized in Section 2.4, the velocity plays a crucial role in producing an inverse cascade. It is governed by the Navier-Stokes equation with the Lorentz force included. There are then two further quadratic nonlinearities for u and B; see Refs. [45–47] for details.

Another approach to treat helicity is to write the equations separately for the positively and negatively polarized modes and thus have evolution equations for $u_n^{\pm}$ and $B_n^{\pm}$. We refer to this model as type II. The helicity density can then be written as [48]

$$\mathcal{H}_{\rm M} = \sum_{n=0}^{N} \left(|B_n^+|^2 - |B_n^-|^2\right) / k_n, \tag{27}$$

and the magnetic energy is $\mathcal{E}_{\rm M} = \sum_{n=0}^{N} (|B_n^+|^2 + |B_n^-|^2)$. The evolution equations for $B_n^{\pm}$ take then the form

$$\left[\eta k(k_n \mp \mu_5) + \frac{\mathrm{d}}{\mathrm{d}t}\right] B_n^{\pm} = \tfrac{1}{6}\mathrm{i}k_n \left[M_{\pm}(u,B) - M_{\pm}(B,u)\right] \quad \text{(type II)}, \tag{28}$$

where [48]

$$M_{\pm}(x,y) = x_{n+1}^{\mp} y_{n+2}^{\pm} + x_{n-1}^{\mp} y_{n+1}^{\mp} + x_{n-2}^{\pm} y_{n-1}^{\mp}. \tag{29}$$

Note that for the intermediate terms, the signs in the superscripts are the same, i.e., $u_{n-1}^{-} B_{n+1}^{-}$ appear in the evolution of B_n^{+} and $u_{n-1}^{+} B_{n+1}^{+}$ in the evolution of B_n^{-}; see also Ref. [49], where such models were proposed independently.

In Figure 8, we present models of types I and II with $N = 30$ shells using $\lambda = 10^{10}$, $k_0 = 2^{14} = 16{,}384 \approx 1.6 \times 10^4$, $\nu = \eta = 5 \times 10^{-11}$, and μ_5 is computed as $\mu_5 = -\mu_{\mathrm{M}}$, where $\mu_{\mathrm{M}} \equiv \mathcal{H}_{\mathrm{M}}\lambda/2 \approx 1.8 \times 10^5$ is the chiral chemical potential equivalent of the magnetic helicity.

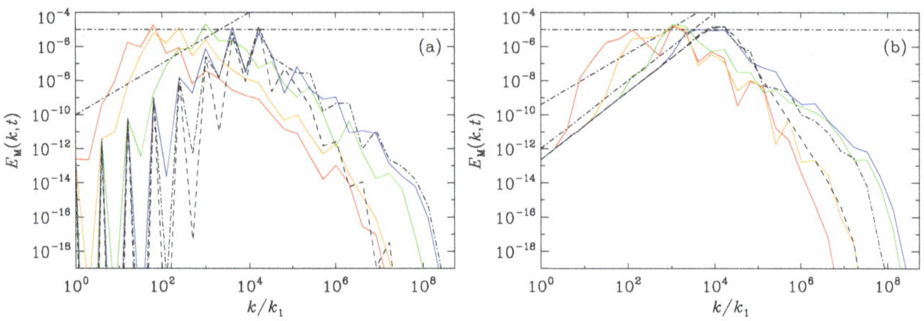

Figure 8. Evolution of $E_{\mathrm{M}}(k,t)$ from shell models of (**a**) type I and (**b**) type II. The times are 10 (red), 1 (orange), 0.1 (green), 0.01 (blue), and earlier times are denoted by black lines of different line types. Note the presence of inverse cascading in both cases.

In all cases, we start with a k^2 spectrum, so we expect to see no inverse cascading. Looking at the results of Figure 8, however, this does not seem to be the case. Our results are still preliminary, but our conclusion so far is that shell models may not capture the same inverse cascade behavior that we have found in the direct numerical simulations. On the other hand, more parameter studies are warranted before one can draw more firm conclusions. One must also remember that departures from the conservation of the Hosking integral have been seen in certain direct numerical simulations [42].

5. Conclusions

In this paper, we have presented a discussion of the Hosking integral in various contexts in which it has been considered so far: ordinary MHD, MHD with chiral fermions, as well as just the induction equation – either with Hall nonlinearity or with ambipolar diffusion nonlinearity. When the total chirality vanishes (non-chiral case with zero magnetic helicity or chiral case with finite magnetic helicity balancing the fermion chirality) it is the correspondingly adapted Hosking integral that governs the decay of $\mathcal{E}_{\mathrm{M}}(t) \propto t^{-p}$ and the increase of $\xi_{\mathrm{M}}(t) \propto t^{q}$ with $p = 10/9$ and $q = 4/9$ for both ordinary MHD and also just the induction equation with ambipolar diffusion. When the nonlinearity is given by the Hall effect, on the other hand, we have $p = 10/13$ and $q = 4/13$. The case with chiral fermions is somewhat special, because now the magnetic field is actually fully helical, but this helicity is balanced by fermion chirality. Again, in that case the Hosking integral determines the decay behavior. However, there is also another decaying quantity: the mean magnetic helicity density, which is now actually finite and balanced by fermion chirality. It is found to decay like $t^{-2/3}$.

In previous work on decaying turbulence, the decay properties of hydrodynamic and MHD turbulence were motivated by the use of self-similarity and invariance of the governing equations under rescaling [28,50]. This is different in the present work where we have just made use of dimensional arguments. Still, the use of invariance under rescaling is necessary to motivate the equilibrium line $p = 2(1 - q)$ in the qp diagram in Figure 1b. It will therefore be interesting to find out whether the existence of this line could also be motivated by other means. It probably can, as implied by the derivation of Equation (12), and this might reveal a more basic relation to the parameter that there was called σ.

An open question is whether the Hosking integral can also play a role in driven MHD turbulence, for example. One possibility could be the production of inverse cascade behavior where magnetic energy grows on wavenumbers below the energy injection wavenumber. This could then leads to a turbulent subinertial range scaling of the form

$$E_{\rm M}(k) \propto I_{\rm H}^{a} k^{b}. \tag{30}$$

Using dimensional arguments, we would find $3 = 9a - b$ and $2 = 4a$ for balancing the dimensions of length and time, respectively. Therefore, $a = 1/2$ and $b = 3/2$. Thus, b is positive and equal to the Kazantsev slope known in kinematic nonhelical small-scale dynamos [51]. Whether or not there is actually a connection with Kazantsev's small-scale dynamo theory remains another open question.

In our work we have also examined whether some aspects of the Hosking integral might be reproducible with shell models. At the moment, this does not seem to be the case, but this could well be a consequence of not having performed sufficiently extensive parameter studies. Thus, more work might be warranted.

Author Contributions: Conceptualization, A.B. and G.L.; methodology, A.B.; software, A.B. and G.L. All authors have read and agreed to the published version of the manuscript.

Funding: This research was funded by Vetenskapsrådet grant number 2019-04234 and NASA ATP award number 80NSSC22K0825.

Data Availability Statement: The source code used for the simulations of this study, the PENCIL CODE [32], is freely available on https://github.com/pencil-code/ (accessed on 1 May 2023). The DOI of the code is https://doi.org/10.5281/zenodo.2315093 (accessed on 1 May 2023). The simulation setups and the corresponding secondary data are available on http://norlx65.nordita.org/~brandenb/projects/Hosking-Shell (accessed on 1 May 2023).

Conflicts of Interest: The authors declare no conflict of interest.

References

1. Herring, J.R.; Licht, A.L. Effect of the solar wind on the lunar atmosphere. *Science* **1959**, *130*, 266. [CrossRef] [PubMed]
2. Herring, J.; Kyle, L. Density in a planetary exosphere. *J. Geophys. Res.* **1961**, *66*, 1980–1982. [CrossRef]
3. Arking, A.; Herring, J. The contribution of absorption lines to the opacity of matter in stellar interiors. *Publ. Astron. Soc. Pac.* **1963**, *75*, 226. [CrossRef]
4. Parker, E.N. Dynamics of the interplanetary gas and magnetic fields. *Astrophys. J.* **1958**, *128*, 664. [CrossRef]
5. Henyey, L.G.; Forbes, J.E.; Gould, N.L. A new method of automatic computation of stellar evolution. *Astrophys. J.* **1964**, *139*, 306. [CrossRef]
6. Pouquet, A.; Frisch, U.; Leorat, J. Strong MHD helical turbulence and the nonlinear dynamo effect. *J. Fluid Mech.* **1976**, *77*, 321–354. [CrossRef]
7. Pouquet, A.; Rosenberg, D.; Stawarz, J.E.; Marino, R. Helicity dynamics, inverse, and bidirectional cascades in fluid and magnetohydrodynamic turbulence: A brief review. *Earth Space Sci.* **2019**, *6*, 351–369. [CrossRef]
8. Moffatt, H.K. *Magnetic Field Generation in Electrically Conducting Fluids*; Cambridge University Press: Cambridge, UK, 1978.
9. Parker, E.N. *Cosmical Magnetic Fields: Their Origin and Their Activity*; Clarendon Press: Oxford, UK, 1979.
10. Krause, F.; Rädler, K.H. *Mean-Field Magnetohydrodynamics and Dynamo Theory*; Pergamon Press: Oxford, UK, 1980.
11. Matthaeus, W.H.; Goldstein, M.L. Measurement of the rugged invariants of magnetohydrodynamic turbulence in the solar wind. *J. Geophys. Res.* **1982**, *87*, 6011–6028. [CrossRef]
12. Brandenburg, A.; Subramanian, K. Astrophysical magnetic fields and nonlinear dynamo theory. *Phys. Rep.* **2005**, *417*, 1–209. [CrossRef]
13. Rust, D.M.; Kumar, A. Evidence for helically kinked magnetic flux ropes in solar eruptions. *Astrophys. J. Lett.* **1996**, *464*, L199. [CrossRef]
14. Brandenburg, A.; Subramanian, K.; Balogh, A.; Goldstein, M.L. Scale dependence of magnetic helicity in the solar wind. *Astrophys. J.* **2011**, *734*, 9. [CrossRef]

15. Christensson, M.; Hindmarsh, M.; Brandenburg, A. Inverse cascade in decaying three-dimensional magnetohydrodynamic turbulence. *Phys. Rev. E* **2001**, *64*, 056405. [CrossRef] [PubMed]
16. Kahniashvili, T.; Tevzadze, A.G.; Brandenburg, A.; Neronov, A. Evolution of primordial magnetic fields from phase transitions. *Phys. Rev. D* **2013**, *87*, 083007. [CrossRef]
17. Durrer, R.; Kahniashvili, T.; Yates, A. Microwave background anisotropies from Alfvén waves. *Phys. Rev. D* **1998**, *58*, 123004. [CrossRef]
18. Durrer, R.; Caprini, C. Primordial magnetic fields and causality. *J. Cosmol. Astropart. Phys.* **2003**, *2003*, 010. [CrossRef]
19. Reppin, J.; Banerjee, R. Nonhelical turbulence and the inverse transfer of energy: A parameter study. *Phys. Rev. E* **2017**, *96*, 053105. [CrossRef]
20. Brandenburg, A.; Kahniashvili, T.; Mandal, S.; Pol, A.R.; Tevzadze, A.G.; Vachaspati, T. Evolution of hydromagnetic turbulence from the electroweak phase transition. *Phys. Rev. D* **2017**, *96*, 123528. [CrossRef]
21. Brandenburg, A.; Kahniashvili, T.; Tevzadze, A.G. Nonhelical inverse transfer of a decaying turbulent magnetic field. *Phys. Rev. Lett.* **2015**, *114*, 075001. [CrossRef]
22. Zrake, J. Inverse cascade of nonhelical magnetic turbulence in a relativistic fluid. *Astrophys. J. Lett.* **2014**, *794*, L26. [CrossRef]
23. Hosking, D.N.; Schekochihin, A.A. Reconnection-controlled decay of magnetohydrodynamic turbulence and the role of invariants. *Phys. Rev. X* **2021**, *11*, 041005. [CrossRef]
24. Zhou, H.; Sharma, R.; Brandenburg, A. Scaling of the Hosking integral in decaying magnetically dominated turbulence. *J. Plasma Phys.* **2022**, *88*, 905880602. [CrossRef]
25. Schekochihin, A.A. MHD turbulence: A biased review. *J. Plasma Phys.* **2022**, *88*, 155880501. [CrossRef]
26. Uchida, F.; Fujiwara, M.; Kamada, K.; Yokoyama, J. New description of the scaling evolution of the cosmological magneto-hydrodynamic system. *arXiv* **2022**, arXiv:2212.14355. [CrossRef]
27. Sharma, R.; Brandenburg, A. Low frequency tail of gravitational wave spectra from hydromagnetic turbulence. *Phys. Rev. D* **2022**, *106*, 103536. [CrossRef]
28. Brandenburg, A.; Kahniashvili, T. Classes of hydrodynamic and magnetohydrodynamic turbulent decay. *Phys. Rev. Lett.* **2017**, *118*, 055102. [CrossRef] [PubMed]
29. Davidson, P.A. Was Loitsyansky correct? A review of the arguments. *J. Turbulence* **2000**, *1*, 6. [CrossRef]
30. Davidson, P.A. The role of angular momentum conservation in homogeneous turbulence. *J. Fluid Mech.* **2009**, *632*, 329. [CrossRef]
31. Brandenburg, A.; Boldyrev, S. The turbulent stress spectrum in the inertial and subinertial ranges. *Astrophys. J.* **2020**, *892*, 80. [CrossRef]
32. Pencil Code Collaboration; Brandenburg, A.; Johansen, A.; Bourdin, P.; Dobler, W.; Lyra, W.; Rheinhardt, M.; Bingert, S.; Haugen, N.; Mee, A.; et al. The Pencil Code, a modular MPI code for partial differential equations and particles: Multipurpose and multiuser-maintained. *J. Open Source Softw.* **2021**, *6*, 2807. [CrossRef]
33. Brandenburg, A.; Kamada, K.; Schober, J. Decay law of magnetic turbulence with helicity balanced by chiral fermions. *Phys. Rev. Res.* **2023**, *5*, L022028. [CrossRef]
34. Cho, J. Magnetic helicity conservation and inverse energy cascade in electron magnetohydrodynamic wave packets. *Phys. Rev. Lett.* **2011**, *106*, 191104. [CrossRef]
35. Goldreich, P.; Reisenegger, A. Magnetic field decay in isolated neutron stars. *Astrophys. J.* **1992**, *395*, 250. [CrossRef]
36. Brandenburg, A. Hosking integral in nonhelical Hall cascade. *J. Plasma Phys.* **2023**, *89*, 175890101. [CrossRef]
37. Brandenburg, A. Hall cascade with fractional magnetic helicity in neutron star crusts. *Astrophys. J.* **2020**, *901*, 18. [CrossRef]
38. Gary, S.P.; Saito, S.; Li, H. Cascade of whistler turbulence: Particle-in-cell simulations. *Geophys. Res. Lett.* **2008**, *35*, L02104. [CrossRef]
39. Sarin, N.; Brandenburg, A.; Haskell, B. Confronting the neutron star population with inverse cascades. *arXiv* **2023**, arXiv:2305.14347. [CrossRef]
40. Smith, C.W.; Hamilton, K.; Vasquez, B.J.; Leamon, R.J. Dependence of the dissipation range spectrum of interplanetary magnetic fluctuations on the rate of energy cascade. *Astrophys. J. Lett.* **2006**, *645*, L85–L88. [CrossRef]
41. Rogachevskii, I.; Ruchayskiy, O.; Boyarsky, A.; Fröhlich, J.; Kleeorin, N.; Brandenburg, A.; Schober, J. Laminar and turbulent dynamos in chiral magnetohydrodynamics. I. Theory. *Astrophys. J.* **2017**, *846*, 153. [CrossRef]
42. Brandenburg, A.; Kamada, K.; Mukaida, K.; Schmitz, K.; Schober, J. Chiral magnetohydrodynamics with zero total chirality. *arXiv* **2023**, arXiv:2304.06612. [CrossRef]
43. Plunian, F.; Stepanov, R.; Frick, P. Shell models of magnetohydrodynamic turbulence. *Phys. Rep.* **2013**, *523*, 1–60. [CrossRef]
44. Kadanoff, L.; Lohse, D.; Wang, J.; Benzi, R. Scaling and dissipation in the GOY shell model. *Phys. Fluids* **1995**, *7*, 617–629. [CrossRef]
45. Brandenburg, A.; Enqvist, K.; Olesen, P. Large-scale magnetic fields from hydromagnetic turbulence in the very early universe. *Phys. Rev. D* **1996**, *54*, 1291–1300. [CrossRef]
46. Frick, P.; Sokoloff, D. Cascade and dynamo action in a shell model of magnetohydrodynamic turbulence. *Phys. Rev. E* **1998**, *57*, 4155–4164. [CrossRef]
47. Basu, A.; Sain, A.; Dhar, S.K.; Pandit, R. Multiscaling in models of magnetohydrodynamic turbulence. *Phys. Rev. Lett.* **1998**, *81*, 2687–2690. [CrossRef]

48. Brandenburg, A.; Enqvist, K.; Olesen, P. The effect of Silk damping on primordial magnetic fields. *Phys. Lett. B* **1997**, *392*, 395–402. [CrossRef]
49. Lessinnes, T.; Plunian, F.; Carati, D. Helical shell models for MHD. *Theor. Comp. Fluid Dyn.* **2009**, *23*, 439–450. [CrossRef]
50. Olesen, P. Inverse cascades and primordial magnetic fields. *Phys. Lett. B* **1997**, *398*, 321–325. [CrossRef]
51. Kazantsev, A.P. Enhancement of a magnetic field by a conducting fluid. *Sov. J. Exp. Theor. Phys.* **1968**, *26*, 1031.

atmosphere

MDPI

Article

Non-Equilibrium Turbulent Transport in Convective Plumes Obtained from Closure Theory

Nobumitsu Yokoi †

Institute of Industrial Science, University of Tokyo, Tokyo 153-8505, Japan; nobyokoi@iis.u-tokyo.ac.jp;
Tel.: +81-3-5452-6113
† Visiting Researcher at Nordic Institute for Theoretical Physics (NORDITA).

Abstract: The non-equilibrium property of turbulence modifies the characteristics of turbulent transport. With the aid of response function formalism, such non-equilibrium effects in turbulent transport can be represented by the temporal variation of the turbulent energy (K) and its dissipation rate (ε) along the mean stream through the advective derivatives of K and ε. Applications of this effect to the turbulent convection with plumes are considered for the first time in this work. The non-equilibrium transport effects associated with plumes are addressed in two aspects. Firstly, the effect associated with a single plume is evaluated using data measured in the recent plume/jet experiments. The second argument is developed for the collective turbulent transport associated with multiple plumes mimicking the stellar convection zone. In this second case, for the purpose of capturing the plume motions into the advective derivatives, use has to be made of the time–space double-averaging procedure, where the turbulent fluctuations are divided into the coherent or dispersion component (which represents plume motions) and the incoherent or random component. With the aid of the transport equations of the coherent velocity stress and the incoherent counterpart, the interaction between the dispersion and random fluctuations are also discussed in the context of convective turbulent flows with plumes. It is shown from these analyses that the non-equilibrium effect associated with plume motions is of a great deal of relevance in the convective turbulence modeling.

Keywords: turbulence; turbulence closure theory; modeling; non-equilibrium effect; convective plumes

Citation: Yokoi, N. Non-Equilibrium Turbulent Transport in Convective Plumes Obtained from Closure Theory. *Atmosphere* **2023**, *14*, 1013. https://doi.org/10.3390/atmos14061013

Academic Editors: Boris Galperin, Annick Pouquet and Peter Sullivan

Received: 26 April 2023
Revised: 25 May 2023
Accepted: 5 June 2023
Published: 12 June 2023

1. Introduction

Turbulent flows encountered in astro/geophysical phenomena as well as laboratory plasma ubiquitously have huge Reynolds numbers. Since the direct numerical simulations (DNSs) of such flows are impossible in the foreseeable future, the turbulence or subgrid-scale (SGS) modeling approach provides a powerful tool for analyzing turbulent flows in these phenomena.

In conventional turbulence modeling, the simple expressions of turbulent fluxes such as the eddy viscosity and eddy diffusivity representations are very useful and have been extensively utilized [1]. For instance, the turbulent momentum flux, defined by the Reynolds stress, is modeled by the turbulent transport coefficients, eddy viscosity, coupled with the mean velocity strain. The eddy viscosity is expressed in terms of the quantities that characterize the dynamics and statistical properties of the turbulence. In the mixing-length formulation, the eddy viscosity ν_T is expressed by the characteristic turbulent velocity v and length scale ℓ as $\nu_T = \nu_T(v, \ell)$. In the standard K–ε model, ν_T is modeled in terms of the turbulent kinetic energy K and its dissipation rate ε as $\nu_T = \nu_T(K, \varepsilon)$. As compared with the mixing-length formulation, the K–ε is more elaborated since the dynamics of the turbulent fields and their interaction with the mean fields are taken into account through the transport equations of K and ε [2,3].

In spite of the notable achievements of simple gradient-diffusion-type models for turbulent fluxes with the eddy viscosity and eddy diffusivity representations, several situations exist wherein such models would completely fail. One such deficiency appears when the eddy viscosity model is applied to a situation with cross-flow configuration. This failure is caused by the lack of vorticity or rotation effects in the eddy viscosity model. An alleviation of this deficiency of the eddy viscosity model was proposed with the aid of an analytical statistical theory of inhomogeneous turbulence applied to a non-reflectional symmetric system with an emphasis on the helicity effects [4,5]. Another kind of deficiency shows up in the case of turbulence with non-equilibrium properties. For instance, in homogeneous-shear turbulence, the evaluation of the eddy viscosity systematically fails because of the overestimates of the turbulent energy K and its dissipation rate ε. There, the energy injection increases with time, and the energy dissipation process cannot catch up with the injection at the energy-containing scale. This is caused by the non-equilibrium property of turbulence induced by the energy injection due to the flow shear. In order to alleviate this flaw in the framework of the K–ε model with the eddy viscosity representation, the non-equilibrium effect has to be incorporated into the eddy viscosity representation [6].

Employing the response function formalism in turbulence closure studies provides a powerful way to construct a self-consistent turbulence theory without resorting to any parameter. There are several classical modern theories of turbulence closure [7–10]. Starting with the Liouville equation derived by Edwards [8], Herring [9,10] constructed a series of self-consistent field (SCF) theories for turbulence closure. In the formulation, the coupled equations of the two-time velocity correlation function and the response function are derived under the requirement that the probability distribution functions satisfy the self-consistency condition. The obtained equations for the correlation and response functions are similar to the those obtained with the direct-interaction approximation (DIA) by Kraichnan [7]. In the combination of the DIA with multiple-scale analysis, we address the convective turbulence with variable density and velocity shear. One important consequence of this multiple-scale DIA analysis lies in the point that it enables us to treat the non-equilibrium effect of turbulence through the Lagrangian derivative of the turbulence fields. If turbulence properties such as the turbulent energy K and its dissipation rate ε are spatially developing in the streamwise direction, the non-equilibrium effects are represented by the advective derivatives of K and ε, $(\mathbf{U} \cdot \nabla)K$ and $(\mathbf{U} \cdot \nabla)\varepsilon$ in the Lagrangian derivatives. In the model for the homogeneous-shear turbulent flow [6], the non-equilibrium properties of turbulence including the variations in the time and length scales of turbulence are taken into account in the expression of the eddy viscosity through the Lagrangian derivative along the mean velocity $\mathbf{U}$ of the turbulent energy and its dissipation rate as $\nu_{\mathrm{T}} = \nu_{\mathrm{T}}(K, \varepsilon, DK/Dt, D\varepsilon/Dt)$ where $D/Dt = \partial/\partial t + \mathbf{U} \cdot \nabla$.

There are several types of buoyant streams in nature. Some buoyant flows are purely driven by thermobuoyancy (plumes) and others are driven by a combination of thermobuoyancy and dynamical forcing (thermobuoyant jets, non-isothermal shear layers). In the buoyancy-dominated turbulent flows, the turbulent convection is well represented by large-scale plumes [11–14]. In order for a turbulence model to treat turbulent convective flows, the turbulence interaction with coherent structures such as a large-scale jet and plume has to be properly represented in the model. Because of the presence of jets and plumes, the turbulent fluxes deviate from the usual gradient diffusivity, viscosity, etc. [14]. The turbulent transport coefficients should properly reflect such coherent structure effects. Another example in astrophysics can be seen in supernova explosion studies. In the core-collapse supernovae (CCSNe) explosion problem, turbulent effects aid the relaunch of blast shock waves due to the neutrino heating [15–21]. In order to identify the turbulence effects most relevant to the explosion condition, Murphy and Meakin [21] extensively examined several types of closure models. By comparing the model simulation results with the 2D simulation counterparts, they found that the standard turbulent flux models based on the gradient-transport approximation failed to reproduce the turbulent transport of the mass, momentum and heat in the stellar interiors. This suggests that the buoyancy-dominated flows of core-collapse convection are characterized by large-scale plumes. The entrainment between the rising and sinking plumes strongly affects the transport properties of neutrino-driven turbulence. In particular, the turbulent dissipation, which is balanced by the buoyancy drive, is dominantly controlled by the entrainment of negatively buoyant plumes.

It has been theoretically and experimentally shown that the dissipation rates in jets and plumes follow the Kolmogorov equilibrium law only if their spreading rates do not vary with the streamwise distance [22,23]. Adopting the Lie symmetry group theory on the statistical turbulent model equations, Sunita and Layek [24] recently argued the case of non-equilibrium turbulent dissipation in buoyant axisymmetric plumes in unstratified stagnant ambient at the infinite-Reynolds-number limit. They considered the relationship between the dissipation rates of the turbulent kinetic energy and turbulent thermal fluctuation, and the entrainment coefficients in the Kolmogorov and non-Kolmogorov regions of turbulent dissipations. They showed that the well-known link between entrainment and dissipation in the Kolmogorov region of turbulence is also established in the non-Kolmogorov region.

As mentioned above, with the aid of the response function formalism, the non-equilibrium properties of plume motions can be expressed by the variations in the turbulent kinetic energy and its dissipation rate along the plume motions. Recent elaborated laboratory experiments with jets and negatively or positively buoyant plumes provide detailed data on the spatial distributions of the turbulent energy and its dissipation rate as well as the jet and plume velocity distributions [25–28]. Utilizing these data, we can evaluate the variation of the turbulent energy and its dissipation rate along the jet/plume motion. Such an evaluation enables us to examine how much non-equilibrium effect may be present in turbulent flows associated with the jet and plume. This is one of the novel points discussed in this work.

Negatively or positively buoyant plumes play a dominant role in mass, momentum and heat mixing in the stellar convection zone. In the stellar convection zone case, we have to simultaneously treat the effects of numerous plumes on the turbulent fluxes, such as the turbulent mass flux, momentum flux and heat or internal-energy flux, in the mean-field equations. The non-equilibrium effect associated with these numerous plumes is investigated with the aid of the numerical simulations of a simplified domain mimicking the stellar convection zone. A turbulence model incorporating the non-equilibrium effect into the expressions of the turbulence fluxes should be validated in comparison with the direct numerical simulations (DNSs) of the turbulent mass, momentum and heat or internal-energy fluxes. In order to properly capture the effects of plume motions, which are part of fluctuating motions in the framework of the mean-field modeling, we introduce some elaborated averaging procedures in this evaluation of the non-equilibrium turbulence model. This is a time–space double-averaging procedure, which has been recently applied to a stellar convection zone [29]. This point will also be discussed in this paper.

The organization of this paper is as follows. In Section 2, we present the mean-field equations with the turbulence correlations, which should be expressed in terms of the mean-field and transport coefficients representing the statistical properties of turbulence. In Section 3, closure models for the turbulent fluxes are promoted with the aid of their theoretical derivations, with special reference to the corrections to the conventional gradient-transport approximations. There, modifications due to the non-equilibrium effects are presented. In Section 4, the roles of the non-equilibrium effects are suggested in the context of recent plume/jet experiments. In Section 5, the role of the plumes in the turbulent transport is discussed in the context of the stellar convection. In addition, the interaction between the coherent and incoherent fluctuations is argued for using the transport equations of each component of fluctuation energy. Concluding remarks are given in Section 6.

2. Equations of Compressible Hydrodynamic Turbulence

The material of this work is plumes in compressible convective turbulence. The turbulent fluxes, such as the turbulent mass flux, the Reynolds stress, the turbulent heat, or internal-energy flux, in the mean-field equations are analyzed with the aid of the two-scale direct-interaction approximation (TSDIA) formulation. In this method, the response or Green's function is used to incorporate the non-equilibrium properties of the turbulence along the mean or grid-scale (GS) flow into the turbulent transport coefficients. By applying the present formulation to the statistical analysis of the turbulent transports in the presence of numerous plumes, a time–space double-averaging procedure is adopted. Since there is no general way of determining the time-domain filtering, we have to make use of the physical considerations on the plume statistics and dynamics including its lifetime for determining the filter in the time domain. This point should be further explored beyond the present formulation described below.

2.1. Mean-Field Equations and Turbulent Correlations

We adopt the conventional Reynolds decomposition procedure

$$f = F + f', \quad F = \langle f \rangle \tag{1}$$

with

$$f = (\rho, \mathbf{u}, p, e) \tag{2}$$

$$F = (\langle \rho \rangle, \mathbf{U}, P, E), \tag{3}$$

$$f = (\rho', \mathbf{u}', p', e'), \tag{4}$$

where ρ is density, $\mathbf{u}$ is the velocity, p is the pressure and e is the internal energy of fluid. With the Reynolds averaging (1), the mean-field equations are obtained in the following form, and the density equation:

$$\frac{\partial \langle \rho \rangle}{\partial t} + \nabla \cdot (\langle \rho \rangle \mathbf{U}) = -\nabla \cdot \langle \rho' \mathbf{u}' \rangle, \tag{5}$$

the mean velocity equation:

$$\begin{aligned} \frac{\partial}{\partial t} \langle \rho \rangle U^\alpha + \frac{\partial}{\partial x^a} \langle \rho \rangle U^a U^\alpha &= -(\gamma - 1) \frac{\partial}{\partial x^\alpha} \langle \rho \rangle E + \frac{\partial}{\partial x^a} \mu S^{a\alpha} \\ &- \frac{\partial}{\partial x^a} \left(\langle \rho \rangle \langle u'^a u'^\alpha \rangle + U^a \langle \rho' u'^\alpha \rangle + U^\alpha \langle \rho' u'^a \rangle \right) + \langle \rho \rangle g^\alpha, \end{aligned} \tag{6}$$

and the mean internal-energy equation:

$$\begin{aligned} \frac{\partial}{\partial t} \langle \rho \rangle E + \nabla \cdot (\langle \rho \rangle \mathbf{U} E) &= \nabla \cdot \left(\frac{\kappa}{C_V} \nabla E \right) - \nabla \cdot \left(\langle \rho \rangle \langle e' \mathbf{u}' \rangle + E \langle \rho' \mathbf{u}' \rangle + \mathbf{U} \langle \rho' e' \rangle \right) \\ &- (\gamma - 1) \left(\langle \rho \rangle E \nabla \cdot \mathbf{U} + \langle \rho \rangle \langle e' \nabla \cdot \mathbf{u}' \rangle + E \langle \rho' \nabla \cdot \mathbf{u}' \rangle \right). \end{aligned} \tag{7}$$

In (6), $\mathcal{S}^{\alpha\beta}$ is the deviatric part of the mean velocity strain defined by

$$\mathcal{S}^{\alpha\beta} = \frac{\partial U^\beta}{\partial x^\alpha} + \frac{\partial U^\alpha}{\partial x^\beta} - \frac{2}{3} \nabla \cdot \mathbf{U} \delta^{\alpha\beta}, \tag{8}$$

and $\mathbf{g} = \{g^\alpha\}$ is the gravitational acceleration.

In the mean-field Equations (5)–(7), we have several second-order turbulence correlations. They are the turbulent mass flux: $\langle \rho' \mathbf{u}' \rangle$, the Reynolds stress: $\langle \mathbf{u}' \mathbf{u}' \rangle$, the turbulent internal-energy flux: $\langle e' \mathbf{u}' \rangle$, the internal-energy dilatation correlation: $\langle e' \nabla \cdot \mathbf{u}' \rangle$, the density dilatation correlation: $\langle \rho' \nabla \cdot \mathbf{u}' \rangle$, and the density–internal-energy correlation: $\langle \rho' e' \rangle$. In the conventional turbulence modeling approach, the expressions of these turbulent correlations are modeled in a heuristic manner. In marked contrast, in a self-consistent turbulence modeling approach based on an analytical theory, the expressions of these correlations are obtained from a closure theory of inhomogeneous turbulence. In the present work, we use the analytical results from a multiple-scale renormalization perturbation expansion method; two-scale direct-interaction approximation (TSDIA) analysis of the fluctuation equations [3,30,31]. The analysis of incompressible and strongly compressible turbulence with a magnetic field using the TSDIA formalism can be seen, for example, in [32–34].

2.2. Simplest Modeling Turbulent Fluxes

In the turbulence modeling approaches, in order to reduce the huge burden of treating complexity of turbulent flows, it is often required that the turbulent fluxes be modeled in the simplest possible manner. A heuristic approach of modeling is to adopt the gradient-diffusion approximations for the turbulent fluxes appearing in (5)–(7). There, the turbulent momentum, mass and energy fluxes are assumed to be written in the form:

$$\langle u'^{\alpha} u'^{\beta} \rangle_{\mathrm{D}} = -\nu_{\mathrm{T}} \mathcal{S}^{\alpha\beta}, \tag{9}$$

$$\langle \rho' \mathbf{u}' \rangle = -\frac{\nu_{\mathrm{T}}}{\sigma_\rho} \nabla \langle \rho \rangle, \tag{10}$$

$$\langle e' \mathbf{u}' \rangle = -\frac{\nu_{\mathrm{T}}}{\sigma_e} \nabla E, \tag{11}$$

where the suffix D denotes the deviatoric or traceless part of a tensor as $\mathcal{A}_{\mathrm{D}}^{\alpha\beta} = \mathcal{A}^{\alpha\beta} - (1/3)\mathcal{A}^{aa}\delta^{\alpha\beta}$. Here, the transport coefficient ν_{T} is the turbulent or eddy viscosity, and σ_ρ and σ_e are the turbulent Prandtl numbers for the density and internal energy, respectively. The turbulent transport coefficients represent the statistical properties of turbulence. For example, the turbulent viscosity is expressed in terms of the intensity of turbulence or turbulent energy (per mass) $K(= \langle \mathbf{u}'^2 \rangle / 2)$.

In the mixing-length theory (MLT) model, the turbulent viscosity is written as

$$\nu_{\mathrm{T}} = v \ell_{\mathrm{C}} \tag{12}$$

apart from the numerical factor, where v is the characteristic intensity of turbulent velocity and ℓ_{C} is the characteristic length in the energy-containing eddies. In practical applications of the model to the geophysical and astrophysical turbulent flows, depending on the problem we consider, the pressure or density-scale height is often adopted as the mixing length. In terms of the turbulent energy K, Equation (12) is expressed as

$$\nu_{\mathrm{T}} = K^{1/2} \ell_{\mathrm{C}} \tag{13}$$

(the Kolmogorov–Prandtl expression).

This model expression for the turbulent viscosity ν_{T} can be obtained from the consideration of the equilibrium energy spectrum of turbulence as follows. The Kolmogorov scaling of turbulence is based on the local equilibrium between the production rate P_K and the dissipation rate ε_K of the turbulent energy as

$$P_K \simeq \varepsilon_K. \tag{14}$$

Hereafter, the dissipation rate of turbulent energy, ε_K, is simply denoted as ε. Assuming the equilibrium scaling of Kolmogorov:

$$E(k) = Ko\, \varepsilon^{2/3} k^{-5/3} \tag{15}$$

(Ko ($\simeq 1.5$): Kolmogorov constant), the turbulent energy is expressed as

$$K = \int dk E(k) = \int_{k=k_{\mathrm{C}}}^{\infty} dk Ko\, \varepsilon^{2/3} k^{-3/5} = C_{K\mathrm{O}} \varepsilon^{2/3} \ell_{\mathrm{C}}^{2/3}, \tag{16}$$

where $\ell_C(= 2\pi/k_C)$ is the largest eddy size. Here, we simply related it to the infrared cut-off wavenumber k_C, and C_{K0} is the model constant linked to the Kolmogorov constant Ko. With (13), the turbulent or eddy viscosity ν_T in (9)–(11) is evaluated as

$$\nu_T = K^{1/2}\ell_C = \varepsilon^{1/3}\ell_C^{4/3} = K^2/\varepsilon \tag{17}$$

apart from the numerical factors. The final equality is the well-known eddy viscosity expression in the standard K–ε model. With this turbulent viscosity expression, Equation (9) constitutes the eddy viscosity representation of the Reynolds stress or turbulent momentum transport.

3. Non-Equilibrium Effect

Non-Equilibrium Effect on Eddy Viscosity

In the standard K–ε model, the turbulent viscosity is expressed in terms of the turbulent energy K and its dissipation rate ε as

$$\nu_T = C_\nu \frac{K^2}{\varepsilon}, \tag{18}$$

where C_ν is model constant usually taken as $C_\nu = 0.09$. However, it has long been known that the standard K–ε model with the turbulent viscosity (18) applied to practical flows has some obvious drawbacks. The representative situations where the standard K–ε model fails include the cross-flow configuration with the circumferential or rotating velocity component as well as the axial one, which impinges flow configuration. Another example is the overestimate of the turbulent energy K and its dissipation rate ε in the homogeneous-shear turbulence. The results of the standard K–ε show a tremendously overestimated turbulent energy K as compared with the counterpart in the direct numerical simulations (DNSs) as schematically depicted in Figure 1. Even in one of the simplest flow geometries such as the homogeneous shear, a simple turbulent viscosity model fundamentally fails to reproduce turbulent flow characteristics.

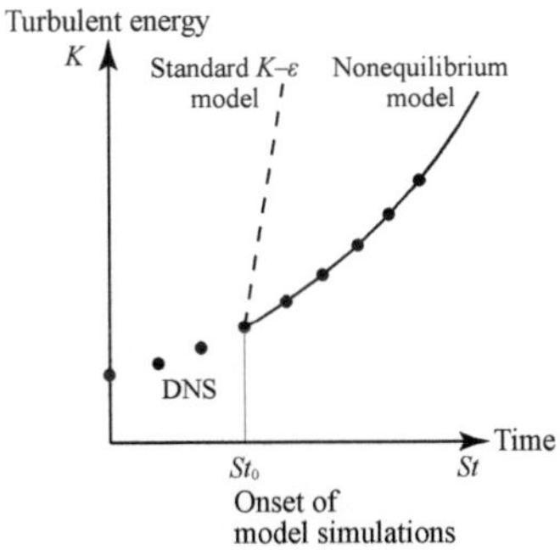

Figure 1. Schematically depicted evolution of the turbulent energy in the simulations of homogeneous-shear flow. •: DNS, – –: standard K–ε model simulation; and ——: non-equilibrium model simulation. Onset of the model simulations is set at a scaled time St_0 using the DNS data (S: shear rate).

In order to rectify such drawbacks of the standard K–ε model, several improvements to the model have been proposed. A possible improvement of the model is the implementation of the non-equilibrium effect on the basis of a statistical theory of inhomogeneous turbulence. The non-equilibrium effect on turbulent viscosity was investigated with the two-scale direct-interaction approximation (TSDIA) [30,35]. In the framework of the TSDIA, the formal solution of the lowest-order turbulent velocity arising from the large-scale inhomogeneities in space and time is given in terms of the homogeneous isotropic basic field $\mathbf{u}'_0(\mathbf{k};\tau_1)$ and the response function $G'(\mathbf{k};\tau,\tau_1)$ as

$$
\begin{aligned}
u_1'^{\alpha}(\mathbf{k};\tau) \;=\; & \frac{\partial U^b}{\partial X^a}\int_{-\infty}^{\tau} d\tau_1 G'^{\alpha b}(\mathbf{k};\tau,\tau_1)u_0'^{a}(\mathbf{k};\tau_1) \\
& -\int_{-\infty}^{\tau} d\tau_1 G'^{\alpha a}(\mathbf{k};\tau,\tau_1)\frac{Du_0'^{a}(\mathbf{k};\tau_1)}{DT} \\
& +2M^{dab}(\mathbf{k})\iint \delta(\mathbf{k}-\mathbf{p}-\mathbf{q})d\mathbf{p}d\mathbf{q}\int_{-\infty}^{\tau} d\tau_1 G'^{\alpha d}(\mathbf{k};\tau,\tau_1)\frac{q^b}{q^2}u_0'^{a}(\mathbf{p};\tau_1)\frac{\partial u'^{c}_0(\mathbf{q};\tau_1)}{\partial X^c} \\
& -M^{abcd}(\mathbf{k})\iint \delta(\mathbf{k}-\mathbf{p}-\mathbf{q})d\mathbf{p}d\mathbf{q}\int_{-\infty}^{\tau} d\tau_1 G'^{\alpha d}(\mathbf{k};\tau,\tau_1)\frac{\partial}{\partial X^c}(u_0'^{a}(\mathbf{p};\tau_1)u_0'^{b}(\mathbf{q};\tau_1)).
\end{aligned}
\tag{19}
$$

Here, the second term on the right-hand side related to the Lagrange derivative represents the non-equilibrium properties of the turbulent field along the flow. The turbulent kinetic energy K with the non-equilibrium effect is written as

$$
\begin{aligned}
K \;=\; & \langle \mathbf{u}'^2\rangle/2 = \frac{1}{2}\int d\mathbf{k}\,\langle u'^{\ell}(\mathbf{k};\tau)u'^{\ell}(\mathbf{k}';\tau)\rangle/\delta(\mathbf{k}+\mathbf{k}') \\
= \; & (\langle u_0'^{\ell}u_0'^{\ell}\rangle + \langle u_0'^{\ell}u_1'^{\ell}\rangle + \langle u_1'^{\ell}u_0'^{\ell}\rangle + \cdots)/2 \\
= \; & C_{K1}\varepsilon^{2/3}\ell_{\mathrm{C}}^{2/3} - C_{K2}\varepsilon^{-3/2}\ell_{\mathrm{C}}^{4/3}\frac{D\varepsilon}{Dt} - C_{K3}\varepsilon^{1/3}\ell_{\mathrm{C}}^{1/3}\frac{D\ell_{\mathrm{C}}}{Dt},
\end{aligned}
\tag{20}
$$

where ℓ_{C} is the length scale of the largest eddy and $D/Dt(=\partial/\partial t + \mathbf{U}\cdot\nabla)$ is the Lagrange or material derivative along the mean-flow velocity. Here, $C_{Kn}(n = 1\text{–}3)$ are the model constants whose magnitudes are related to those of the energy spectrum and the time scale of turbulence (σ_0 in (A5) and ω_0 in (A6)). A brief derivation of (20) is given in Appendix A.

We solve (20) with respect to ℓ_{C} by an iteration method with the first term of (20) as the starting term:

$$K = C_{K1}\varepsilon^{2/3}\ell_{\mathrm{C}}^{2/3}. \tag{21}$$

Then, we have ℓ_{C} as

$$\ell_{\mathrm{C}} = C_{\ell 1}K^{3/2}\varepsilon^{-1}, \tag{22}$$

where $C_{\ell 1}$ is the model constant. With (13), this lowest-order expression of the length scale corresponds to the simplest eddy viscosity model (18).

Substituting (18) into (20), we have the first iteration result as

$$\ell_{\mathrm{C}} = C_{\ell 1}K^{3/2}\varepsilon^{-1} + C_{\ell 2}K^{3/2}\varepsilon^{-2}\frac{DK}{Dt} - C_{\ell 3}K^{5/2}\varepsilon^{-3}\frac{D\varepsilon}{Dt} \tag{23}$$

($C_{\ell 2}$ and $C_{\ell 3}$ are model constants). The second and third terms of (23) represent the non-equilibrium effect. Equation (23) shows that the length scale of turbulence, as well as the time scale and energy of turbulence, is corrected by the D/Dt-related non-equilibrium effect terms. Equation (23) can be approximated as

$$\ell_{\mathrm{C}} = \ell_{\mathrm{E}}\left(1 - C'_{\mathrm{N}}\frac{1}{K}\frac{D}{Dt}\frac{K^2}{\varepsilon}\right), \tag{24}$$

where ℓ_{E} is the equilibrium characteristic length defined by

$$\ell_{\mathrm{E}} = K^{3/2}/\varepsilon, \tag{25}$$

and C'_{N} is the non-equilibrium-effect-related coefficient. Depending on the streamwise development of the turbulent energy K and its dissipation rate ε, the length scale of turbulence changes as compared to the equilibrium cases. At the same time, the time scale of the non-equilibrium turbulence may be expressed as

$$\tau_{\mathrm{NS}} = \frac{\ell_{\mathrm{NS}}}{K^{1/2}} \simeq C_{\mathrm{S}}^{-3/2}\frac{K}{\varepsilon}\left(1 - C'_{\mathrm{N}}\frac{1}{K}\frac{D}{Dt}\frac{K^2}{\varepsilon}\right). \tag{26}$$

This indicates that the effective time scale may change depending on the D/Dt effect.

The non-equilibrium turbulent viscosity ν_{T} can be expressed as

$$\nu_{\mathrm{T}} = C_\nu\frac{K^2}{\varepsilon} - C_{\nu K}\frac{K^2}{\varepsilon^2}\frac{DK}{Dt} + C_{\nu\varepsilon}\frac{K^3}{\varepsilon^3}\frac{D\varepsilon}{Dt}, \tag{27}$$

where C_ν, $C_{\nu K}$ and $C_{\nu\varepsilon}$ are the model constants whose values can be evaluated from theoretical analyses. Equation (27) is rewritten as

$$\nu_{\mathrm{T}} = C_\nu\frac{K^2}{\varepsilon}\left[1 - \left(C_{\mathrm{N}K}\frac{1}{\varepsilon}\frac{DK}{Dt} - C_{\mathrm{N}\varepsilon}\frac{K}{\varepsilon^2}\frac{D\varepsilon}{Dt}\right)\right], \tag{28}$$

or approximately

$$\nu_{\mathrm{T}} = C_\nu\frac{K^2}{\varepsilon}\left(1 - C_{\mathrm{N}}\frac{1}{K}\frac{D}{Dt}\frac{K^2}{\varepsilon}\right) \tag{29}$$

($C_{\mathrm{N}K}$, $C_{\mathrm{N}\varepsilon}$ and C_{N}: model constants). Here, we should note that this non-equilibrium effect does not necessarily require compressibility. This effect shows up even in incompressible cases.

In practical situations wherein turbulent flows are simulated, a large positive D/Dt term may occur locally; in such a case, (29) may lead to a physically undesirable negative turbulent viscosity. In order to alleviate such an unphysical model result to enhance the applicability of the model expression, we use a simple Padé approximation to (29) and have

$$\nu_{\mathrm{T}} = \begin{cases} \nu_{\mathrm{TE}}\left(1 + C_{\mathrm{N}}\frac{1}{K}\frac{D}{Dt}\frac{K^2}{\varepsilon}\right)^{-1} & \text{for} \quad \frac{D}{Dt}\frac{K^2}{\varepsilon} > 0, \\ \nu_{\mathrm{TE}}\left(1 - C_{\mathrm{N}}\frac{1}{K}\frac{D}{Dt}\frac{K^2}{\varepsilon}\right) & \text{for} \quad \frac{D}{Dt}\frac{K^2}{\varepsilon} < 0, \end{cases} \tag{30}$$

where the equilibrium eddy viscosity ν_{TE} is defined in the same form as (18) by

$$\nu_{\mathrm{TE}} = C_\nu\frac{K^2}{\varepsilon}. \tag{31}$$

Equation (30) indicates that, depending on the sign of the Lagrangian or material derivative $(D/Dt)(K^2/\varepsilon)$, the turbulent viscosity is effectively decreased or increased by the non-equilibrium effect. In the case of $(D/Dt)(K^2/\varepsilon) > 0$, ν_{T} is decreased, while in the case of $(D/Dt)(K^2/\varepsilon) < 0$, ν_{T} is increased. This implies that, in this case, the non-equilibrium effect works in the direction to stabilize the deviation from the equilibrium state (return to equilibrium).

The non-equilibrium model expression for the eddy viscosity (30) has been applied to the homogeneous-shear turbulence, where the standard K–ε model with the conventional eddy viscosity (31) leads to an overestimated turbulent energy K and its dissipation rate ε as compared to the DNS results (Figure 1). It was reported that the non-equilibrium model (30) with $C_{\mathrm{N}} = 0.8$ leads to the results consistent with the DNS ones [6]. In this sense, the validity of the non-equilibrium effects in turbulent mixing has been already confirmed in a simple geometrical flow.

4. Non-Equilibrium Effects in Jet and Plume Experiments

As was referred to in Section 1, in the buoyancy-dominated turbulent flows, the turbulent convection is characterized by large-scale plumes. As this consequence, the turbulent fluxes are deviated from the usual gradient-transport approximation. It has been recognized that entrainment dominates the evolution of buoyant plumes [11]. As was shown in a previous study of a supernova convection [21], global turbulence models introducing the entrainment parameters provide the best results for describing the evolution of buoyant plumes. However, such models highly depend on the geometrical configurations of the simulation, and are difficult to apply to a general 3D situation. In this sense, it is preferable to develop a local model without resorting to global parameters, provided that the local model works well enough for reproducing the basic behavior of the convection turbulence with plumes. In order to validate the model, here we consider turbulent flows with much simpler geometrical configurations, namely jets and plumes, utilizing recent experimental results.

The non-equilibrium effect may help develop such a local model. In the presence of a plume, we can expect that the turbulence properties vary in the plume-flow direction. The non-equilibrium effect associated with the plume motion can be incorporated through the variation in the turbulent energy K and its dissipation rate ε along the plume motion in terms of the Lagrangian or advective derivatives, $DK/Dt = [(\partial/\partial t) + \mathbf{U} \cdot \nabla]K$ and $D\varepsilon/Dt = [(\partial/\partial t) + \mathbf{U} \cdot \nabla]\varepsilon$ (Figure 2). The effect of non-equilibrium turbulence can be implemented into the model of turbulent viscosity as (30) with the equilibrium turbulent viscosity given by (31).

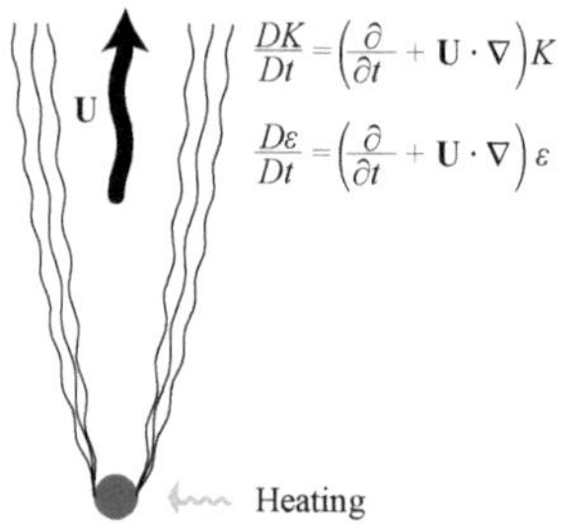

Figure 2. The non-equilibrium effect associated with a plume motion. A plume is thermobuoyantly formed above a heat source. The non-equilibrium effect is taken into account through the Lagrangian or advective derivatives of the turbulent energy K and its dissipation rate ε, $DK/Dt = (\partial/\partial t + \mathbf{U} \cdot \nabla)K$ and $D\varepsilon/Dt = (\partial/\partial t + \mathbf{U} \cdot \nabla)\varepsilon$, where $\mathbf{U}$ is the plume velocity.

Recently, a series of detailed experimental studies of the budgets of turbulent energy, Reynolds stresses and dissipation rate has been performed in single- and multi-phase jets/plumes, including a turbulent round jet discharged into an initially stationary ambient [27], a buoyant multi-phase bubble plume [28], a negatively buoyant multi-phase particle plume [25] and a variable density round jet with co-flow [26].

4.1. Round Jets

Using a set of stereoscopic particle image velocimetry (SPIV) measurements of a turbulent round water jet, the budget of the turbulent energy equation was investigated in Lai and Socolofsky [27]. In this experiment, it was reported that the magnitudes of the normal Reynolds stresses R^{ii} decreased by 30 % over $27 \leq x/D \leq 37$ in the downstream direction (D: exit diameter of the jet nozzle). In particular, the streamwise variation of the axial velocity fluctuation $\langle (u'^x)^2 \rangle / \sqrt{\varepsilon}$ at the jet centerline shows a decreasing tendency at the major region of the axial locations $31 \leq x/D \leq 37$ (Table 1). Although in Lai and Socolofsky [27], these experimental results are argued in the context of the approximate achievement of the self-preserving state in the jet [36]; this decrease tendency may be interpreted as a deviation from the self-preserving state due to the non-equilibrium effect. Since the magnitude of axial velocity fluctuation $\sqrt{\langle (u'^x)^2 \rangle}$ is similar to and just slightly larger than the horizontal counterparts of the velocity fluctuations, $\sqrt{\langle (u'^y)^2 \rangle}$ and $\sqrt{\langle (u'^z)^2 \rangle}$, and since the energy dissipation rate ε is well represented by that of the isotropic turbulence, $\varepsilon_{\mathrm{iso}}$, the evolution of squared axial velocity fluctuation, $\langle (u'^x)^2 \rangle / \sqrt{\varepsilon_{\mathrm{iso}}}$, provides a reasonable surrogate for the basic behavior of K^2/ε:

$$\frac{K^2}{\varepsilon} \propto \left[\frac{\langle (u'^x)^2 \rangle}{\sqrt{\varepsilon_{\mathrm{iso}}}} \right]^2. \tag{32}$$

Table 1. Streamwise variation of normalized turbulent intensities, turbulent energy and turbulent axial velocity fluctuation normalized by the square root of the dissipation rate at the jet centerline. x/D: jet streamwise length, D: jet nozzle exit diameter ($D = 11$ mm), U_c: centerline jet velocity ($U_c = 50$ mm s^{-1}). Data are taken from Figures 7 and 8 of Lai and Socolofsky [27].

x/D	$\sqrt{\langle (u'^x)^2 \rangle}/U_c$	$\sqrt{\langle (u'^y)^2 \rangle}/U_c$	$\sqrt{\langle (u'^z)^2 \rangle}/U_c$	K/U_c^2	$\langle (u'^x)^2 \rangle / \sqrt{\varepsilon_{\mathrm{iso}}}$
31	0.26	0.20	0.19	0.1437	3.25×10^{-2}
37	0.26	0.20	0.19	0.1437	2.65×10^{-2} (−20%)

The decrease in $\langle (u'^x)^2 \rangle / \sqrt{\varepsilon_{\mathrm{iso}}}$ in the downstream direction indicates that the sign of the non-equilibrium effect term is negative as

$$\frac{D}{Dt}\frac{K^2}{\varepsilon} < 0 \quad \text{(in turbulent round jets)}. \tag{33}$$

It follows from (30) that the turbulent viscosity is enhanced by the non-equilibrium effect as

$$\nu_{\mathrm{NE}} = \nu_{\mathrm{E}} \left(1 - C_{\mathrm{N}} \frac{1}{K} \frac{D}{Dt} \frac{K^2}{\varepsilon} \right) > \nu_{\mathrm{E}}. \tag{34}$$

In order to evaluate the non-equilibrium effect, we introduced the non-equilibrium correction factor Λ by

$$\Lambda = C_{\mathrm{N}} \frac{1}{K} \frac{D}{Dt} \frac{K^2}{\varepsilon}, \tag{35}$$

where C_{N} is the model constant estimated as the order of unity ($C_{\mathrm{N}} = 0.8$) through applications to the homogeneous-shear flow turbulence [6]. This factor is approximated in terms of the Lagrangian derivative as

$$\Lambda \simeq C_{\mathrm{N}} \frac{1}{K} (\mathbf{U} \cdot \nabla) \frac{K^2}{\varepsilon}. \tag{36}$$

Using the data of Lai and Socolofsky [27], Λ is estimated as

$$\Lambda \simeq C_{\mathrm{N}} \left(\frac{U_c}{\sqrt{\langle (u'^x)^2 \rangle}} \right)^2 \frac{1}{U_c} \frac{\Delta\{[\langle (u'^x)^2 \rangle / \sqrt{\langle \varepsilon_{\mathrm{iso}} \rangle}]^2\}}{D\Delta(x/D)} \simeq 1.7 \tag{37}$$

with $C_{\mathrm{N}} = 1$ (Table 1). Here, $\Delta\{[\langle (u^x)^2 \rangle / \sqrt{\langle \varepsilon_{\mathrm{iso}} \rangle}]^2\}$ in the nominator and $\Delta(x/D)$ in the denominator are the increments in $[\langle (u^x)^2 \rangle / \sqrt{\langle \varepsilon_{\mathrm{iso}} \rangle}]^2$ and x/D along the jet streamwise direction.

In Lai and Socolofsky [27], the budgets of the turbulent energy and dissipation rate equations were also investigated. Using the balance in the budgets, the coefficients of each term of the K–ε model were examined. They suggested the befitting directions for the change in model coefficients. The eddy viscosity constant C_ν was increased from the standard value of $C_\nu = 0.09$ to 0.09–0.27, whereas a reduction of C_ν to 0.07 is suggested from the matching for the jet spreading rate. This implies that it is inadequate to use a constant model coefficient C_ν for the turbulent viscosity. The enhancement of eddy viscosity due to the non-equilibrium effect is basically in the correct direction to the evaluation of C_ν, and the estimated value of Λ (37) is consistent with the suggested change of C_ν. The degree of enhancement/suppression should be determined by the non-equilibrium property of the turbulence, which depends on the spatial location.

4.2. Buoyant Bubble Plumes

In Lai and Socolofsky [28], the turbulent energy budgets in bubble plume were experimentally investigated. There, the turbulent energy evolution was measured both in the adjustment phase of the plume dependent on the source conditions (source size and geometry) and in the asymptotic phase independent of the source condition.

These experimental results show that, in the asymptotic phase (Case A), the turbulent energy K, its dissipation rate ε and K^2/ε in the plume core region ($r \leq b_g$) decrease in the axial or downstream direction along the plume velocity, while in the adjustment phase (Case B) K, ε and K^2/ε increase (Table 2). However, the relative variation in the turbulent kinetic energy at the downstream to the upstream is much higher in the adjustment phase (Case B) than in the asymptotic phase (Case A). This means that the turbulent energy inhomogeneity along the plume flow, which may lead to the non-equilibrium effect is more prominent in the adjustment phase than in the asymptotic phase. This is natural since the non-equilibrium property is expected to be much higher in the adjustment phase than in the asymptotic phase. In the adjustment phase, the Lagrangian derivative of K^2/ε is positive as

$$\frac{D}{Dt}\frac{K^2}{\varepsilon} > 0 \quad \text{(in the adjustment phase of buoyant bubble plume).} \tag{38}$$

In this case, from (30)

$$\nu_{\rm NE} = \nu_{\rm E}(1+\Lambda)^{-1} < \nu_{\rm E} \quad \text{(in the adjustment phase of buoyant bubble plume).} \tag{39}$$

In the experiment of Lai and Socolofsky [28], Λ is estimated as

$$\Lambda \simeq C_{\rm N}\frac{1}{K}W_{\rm c}\frac{\Delta(K^2/\varepsilon)}{D\Delta(z/D)} \simeq 0.20 \tag{40}$$

with $C_{\rm N} = 1$ (Table 2). Here, $\Delta(K^2/\varepsilon)$ and $\Delta(z/D)$ are the increments of K^2/ε and z/D along the vertical streamwise height, respectively. This suggests that due to the non-equilibrium effect, the turbulent mixing in the buoyant bubble plumes is expected to be suppressed by a few 10% as compared to the standard equilibrium turbulent transport.

Table 2. Evolutions of turbulent energy and its dissipation rate along the plume flow. Case A is the asymptotic phase and Case B is the adjustment phase. z/D: vertical streamwise height; D: dynamic length scale (D = 6.8 cm for Case A and D = 20.4 cm for Case B); W_c: plume centerline velocity; and b_g: Gaussian plume radius. Data are taken from Figures 20 and 25 of Lai and Socolofsky [28].

Case A (Asymptotic Phase)							
z/D	W_c (cm s^{-1})	b_g	K/W_c^2	$(\varepsilon/W_c^3)b_g \times 10^3$	K (cm^2 s^{-2})	ε (cm^2 s^{-3})	K^2/ε (cm^2 s^{-1})
6.6	20.64	5.46	0.135	0.04	57.51	64.42	51.34
11.0	17.90	8.00	0.165	0.086	52.87	61.65	45.34
					(−8.1%)	(−4.3%)	(−11.7%)
Case B (Adjustment Phase)							
z/D	W_c (cm s^{-1})	b_g	K/W_c^2	$(\varepsilon/W_c^3)b_g \times 10^3$	K (cm^2 s^{-2})	ε (cm^2 s^{-3})	K^2/ε (cm^2 s^{-1})
2.19	28.50	5.27	0.15	0.035	121.84	153.74	96.56
3.66	25.14	8.53	0.225	0.0896	142.20	165.78	121.97
					(+16.7 %)	(+7.8%)	(+26.3 %)

4.3. Variable Density Jets

There are several studies on the buoyancy-driven variable density turbulence with jets and plumes [26,37]. In recent experiments by Charonko and Prestridge [26], variable density effects in turbulent round jets with co-flow at high- and low-density ratios are experimentally investigated. There, the axial and radial mixing mechanisms of jet are compared between the cases with low and high variable density contrast. It was confirmed that the Reynolds stress is suppressed by the turbulent mass flux in the increased variable density case. It was also reported that the radial transport of momentum and the energy cascade down to small scales are significantly suppressed in the variable density case. These suppressions mostly take place in locations where the density fluctuation is large.

At the upstream region ($x_1/d_0 < 16.3$, x_1: axial location; d_0: inner diameter), the turbulent energy—regardless of both whether averaging is mass-weighted or the Reynolds and whether turbulent energy is normalized by excess momentum or the initial kinetic energy of the mean flow—develops with a downstream distance along the centerline as

$$\frac{DK}{Dt} > 0 \quad \text{(at upstream region of variable density jets)}. \tag{41}$$

This increase in turbulent energy along the mean flow takes place in both the small- and large-density variation cases. Since the evolution of the dissipation rate ε with the axial distance is not explicitly presented in Charonko and Prestridge [26], that of K^2/ε cannot be accurately evaluated. However, the analysis of several flows later in Section 5.1 implies that the trend of the K^2/ε evolution along the mean flow or the sign of K^2/ε can be surmised from the evolution of K itself. From the non-equilibrium effect $D(K^2/\varepsilon)/Dt[\sim (K/\varepsilon)DK/Dt]$ in (30), we see that the turbulent transport is suppressed at the upstream region.

At the same time, Charonko and Prestridge [26] showed that the variable density effects decrease both the peak and asymptotic values of turbulent energy. By considering the budget in the evolution equation of turbulent energy, they suggested that in the high variable density case, the density fluctuation contributes to the suppression of energy transfer into small-scale fluctuations and preserving the mean-flow structures further downstream. These features can be well understood from the viewpoint of a density-variance effect; the presence of density variance $K_\rho(= \langle \rho'^2 \rangle)$ leads to the suppression of turbulent transport as in (30). Charonko and Prestridge [26] concluded that the variable density effect must be modeled to accurately capture dynamical mixing in jets and other variable density flow phenomena. The combination of the non-equilibrium and density-variance effects proposed in the present work is a candidate for modeling the variable density effect in these flows.

5. Plumes in Stellar Convection Zone

In the previous section, we argued the non-equilibrium effect associated with a single plume and jet. In modeling such a situation, the velocity of the single plume/jet is treated as the local mean velocity, and the non-equilibrium effect associated with the mean velocity can be captured by the Lagrangian or advective derivative $D/Dt = \partial/\partial t + \mathbf{U} \cdot \nabla$ [$\mathbf{U}(\equiv \langle \mathbf{u} \rangle)$: mean velocity]. However, in the practical application of the turbulence model to geo/astrophysical flow phenomena, we often encounter situations where the effect of numerous plumes and jets on the global or averaged flow dynamics must be properly evaluated.

One example is the turbulent transport in the stellar convection zone. In some flow configurations, such as in the convective motion in a closed domain, the mean velocity formulated with a simple ensemble average or averaging in the statistically homogeneous plane is very small because of the statistical cancellations due to symmetry. Furthermore, such an averaged quantity cannot capture localized flow structures such as plumes. Because of the small mean flow ($\langle \mathbf{u} \rangle \equiv \mathbf{U} \simeq 0$), the non-equilibrium effect represented by the mean flow advection, $(\mathbf{U} \cdot \nabla) K^2/\varepsilon$ in DK/Dt, is not suitable for capturing the plume effects. With a simple ensemble or space-averaging procedure, the direct impact due to the non-equilibrium effect might be negligible. However, the presence of grid-scale (GS) or large-eddy velocity component suggests that the counterpart of the non-equilibrium effect at GS may arise as a result of the alternation of the subgrid-scale (SGS) turbulent transport.

5.1. Non-Equilibrium Effect in the Stellar Convection

Recently, in the context of the stellar convection, the non-equilibrium effects associated with the convective plume motion were investigated in the framework of double averaging [29]. For the purpose of exploring the non-equilibrium effect associated with the plumes, fluid dynamics in the stellar convection zone was investigated with the aid of direct numerical simulations (DNSs). In the setup of the simulations, the diving plumes are driven by the cooling layer at the upper surface of the stellar convection zone. Because of the cooling, the surface layer is convectively unstable.

In the non-locally driven convection case, there exist a lot of downward plumes in the shallow region of the convection zone as in the lower plot of Figure 3b.

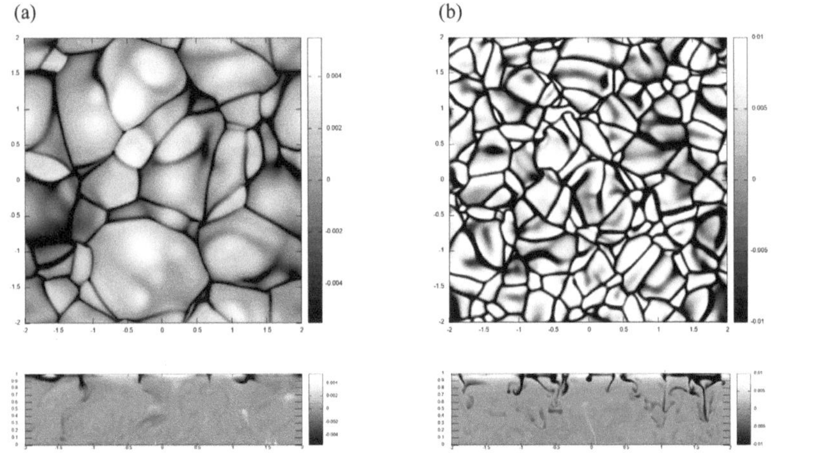

Figure 3. Entropy distributions in the direct numerical simulations (DNSs) for the locally driven case (**a**) and the non-locally driven case (**b**). The horizontal cross-sections of the entropy fluctuation $s'(= s - \langle s \rangle)$ at the top surface (**Top**). In the non-locally or cooling-driven case, the horizontal extension of the cell structures is much more limited than the counterpart in the locally driven case. The vertical cross-sections of the entropy fluctuation s' from the horizontal mean (**bottom**). In the non-locally driven case, the low-entropy down flow or plume structures produced at the surface are prominent in the upper region.

In order to capture the dynamics of plume motion, the time–space double-averaging procedure was adopted there. A field quantity f is divided into

$$f = \overbrace{\langle \overline{f} \rangle + \underbrace{\tilde{f}}_{\substack{\overline{f} - \langle \overline{f} \rangle \\ \text{coherent} \\ \text{fluctuation}}}}^{\overline{f}} + \underbrace{f''}_{\substack{f - \overline{f} \\ \text{incoherent} \\ \text{fluctuation}}}, \tag{42}$$

where $\overline{f}$ denotes the time average, $\langle f \rangle$ denotes the space or ensemble average and $\tilde{f}(= \overline{f} - \langle \overline{f} \rangle)$ denotes the residual of the temporal average subtracted by the ensemble average part. This part may be called the dispersion or coherent fluctuation of f, $f''(= f - \overline{f})$, the fluctuation around the time average $\overline{f}$. As for the time averaging, the resolved part of a quantity f, $\overline{f}$, is defined by the relation

$$\overline{f}(\mathbf{x};t) = \int f(\mathbf{x};s)G(t-s)ds \tag{43}$$

with the simplest top-hat time filter function $G(t-s)$ defined by

$$G(t-s) = \begin{cases} 1/T & (|t-s| \leq T/2), \\ 0 & (\text{otherwise}). \end{cases} \tag{44}$$

Here, T is the averaging time window, which should be put in the range:

$$\tau \ll T \ll \Xi, \tag{45}$$

where τ is the eddy turnover time of turbulence and Ξ is the time scale of the mean-field evolution.

In this formalism, the fluctuation around the space or ensemble average, f', is divided into the coherent and incoherent parts as

$$f' = \tilde{f} + f''. \tag{46}$$

With this double-averaging procedure, the time-averaged velocity $\overline{\mathbf{u}}$ is divided into the space-averaged velocity $\langle\overline{\mathbf{u}}\rangle$ and the dispersion velocity $\tilde{\mathbf{u}}$ as

$$\overline{\mathbf{u}} = \langle\overline{\mathbf{u}}\rangle + \tilde{\mathbf{u}}. \tag{47}$$

The non-equilibrium effect along the plume motion is represented by the factor

$$\widetilde{\Lambda}_{\mathrm{D}} = \left\langle (\tilde{\mathbf{u}} \cdot \nabla)\overline{\mathbf{u}'^2} \right\rangle. \tag{48}$$

This means that, if the turbulent energy is inhomogeneous along the plume motion $\tilde{\mathbf{u}}$, the turbulent fluxes are enhanced or suppressed depending on the sign of $\widetilde{\Lambda}_{\mathrm{D}}$.

Realizations of the non-equilibrium factor $\widetilde{\Lambda}_{\mathrm{D}} = \langle (\tilde{\mathbf{u}} \cdot \nabla)\overline{\mathbf{u}'^2} \rangle$ in the stellar convection simulations are plotted in Figure 4. In this case, statistically, the plume velocity is in the downward direction ($\tilde{u}^z < 0$) while the turbulent energy $\overline{\mathbf{u}'^2}$ decreases in the downward direction ($\partial\overline{\mathbf{u}'^2}/\partial z > 0$). As a result, the non-equilibrium factor $\widetilde{\Lambda}_{\mathrm{D}}$ is statistically negative as

$$\widetilde{\Lambda}_{\mathrm{D}} = \left\langle (\tilde{\mathbf{u}} \cdot \nabla)\overline{\mathbf{u}'^2} \right\rangle = \left\langle \tilde{u}^z \frac{\partial \overline{\mathbf{u}'^2}}{\partial z} \right\rangle < 0. \tag{49}$$

The spatial distribution of $\widetilde{\Lambda}_{\mathrm{D}}$ is concentrated in the region near the surface where the diving plume structures are prominently observed. The magnitude of $\widetilde{\Lambda}_{\mathrm{D}}$ shows a much larger value in the case with the smaller averaging time window T defined in (44). This clearly shows that the non-equilibrium effect associated with the plume motion can be captured by the Lagrangian derivative of the turbulent energy along the plume motion.

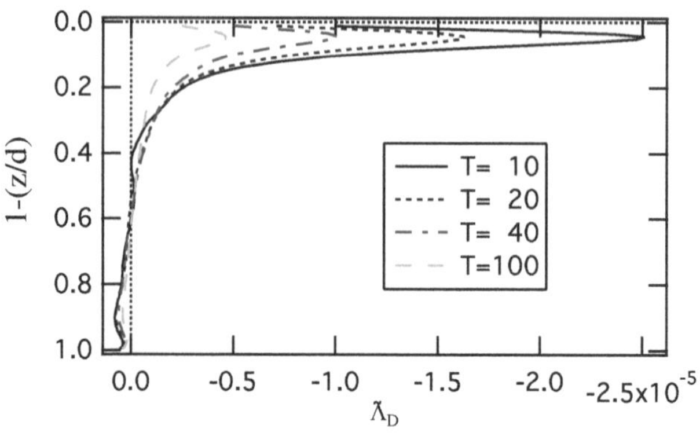

Figure 4. Spatial distributions of the non-equilibrium effect factor $\widetilde{\Lambda}_{\mathrm{D}} = \langle (\widetilde{\mathbf{u}} \cdot \nabla) \overline{\mathbf{u}'^2} \rangle$. The left axis $1 - (z/d)$ is the depth from the surface ($z = d$). The magnitude of the non-equilibrium effect factor $|\widetilde{\Lambda}_{\mathrm{D}}|$ depends on the time averaging window T.

In the framework of the time–space-averaging procedure, the non-equilibrium effect is incorporated into the turbulent internal-energy flux. On the basis of the eddy viscosity expression (30) in the usual ensemble averaging procedure, the turbulent internal-energy diffusivity κ_{T} with the non-equilibrium effect in the time–space-averaging procedure may be formulated as

$$\kappa_{\mathrm{T}} = \kappa_{\mathrm{TE}} \left(1 - \widetilde{C} \langle \overline{\rho} \rangle^{-1/3} \widetilde{\Lambda}_{\mathrm{D}} \right), \tag{50}$$

where κ_{TE} is the equilibrium turbulent internal-energy diffusivity, $\langle \overline{\rho} \rangle$ is the horizontally averaged density and $\widetilde{C}$ is the model constant. The mean density dependence $\langle \overline{\rho} \rangle^{-1/3}$ in (50) is obtained from the argument that the coherent dissipation rate $\widetilde{\varepsilon}$ is proportional to the cubic root of the buoyancy flux associated with a plume. See Section 6.3 and Appendix C of Yokoi, Masada and Takiwaki [29] for the detailed arguments on this dependence.

The turbulent internal-energy flux $\langle e' u'^z \rangle$ obtained from the DNSs in the locally and non-locally driven convection cases, which, respectively, correspond to (a) and (b) of Figure 3, are plotted in Figure 5. We see from the figure that the spatial profiles of these two cases are fairly different. First, the magnitude of $\langle e' u'^z \rangle$ is much larger in the non-locally driven convection case. At the same time, turbulent transport is significantly localized near the surface region in the non-locally driven case. These prominent properties of turbulent transport in the non-locally driven convection cannot be properly reproduced by the usual gradient-diffusion model with the mixing-length theory.

The spatial distribution of $\langle e' u'^z \rangle$ calculated by the model with the non-equilibrium effect (50):

$$\langle e' u'^z \rangle = -\kappa_{\mathrm{T}} \nabla E \tag{51}$$

which is also plotted in Figure 5 (E: mean internal energy). The comparison with the DNS results shows that the spatial distribution of $\langle e' u'^z \rangle$ in the non-locally driven convection is well reproduced by the present model with the non-equilibrium effect.

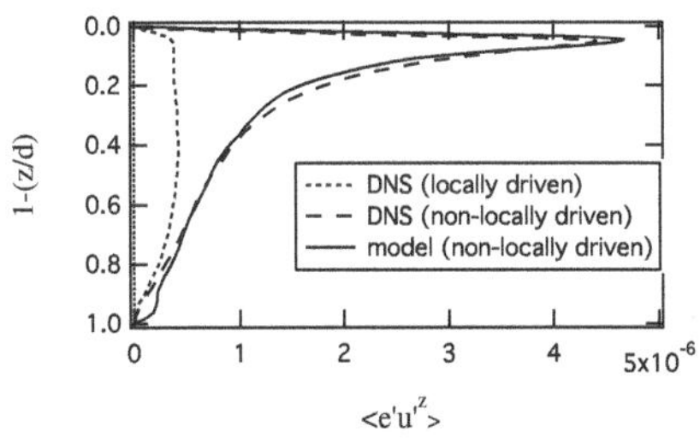

Figure 5. Spatial distributions of the turbulent internal-energy flux $\langle e' u'^z \rangle$ obtained by the direct numerical simulations (DNSs) and the model with the non-equilibrium effect. The left axis $1 - (z/d)$ is the depth from the surface ($z = d$). The DNS results for the locally driven convection case ($\cdots$), for the non-locally driven convection case ($--$) and the result from the model with the non-equilibrium effect for the non-locally driven convection case (—).

5.2. Interaction between Coherent and Incoherent Fluctuations

In this double-averaging formulation, the fluctuation around the space or ensemble average, f', is divided into the coherent and incoherent components as $f' = \widetilde{f} + f''$. These two components are not independent but interact with each other. For instance, let us consider the Reynolds stress in the time–space double-averaging procedure. The evolutions of the Reynolds stress of the coherent velocity fluctuation, $\langle \widetilde{\mathbf{u}}\widetilde{\mathbf{u}} \rangle$, are subject to

$$\left(\frac{\partial}{\partial t} + \langle \overline{\mathbf{u}} \rangle \cdot \nabla \right) \langle \overline{\widetilde{u}^i \widetilde{u}^j} \rangle = \widetilde{P}^{ij} + \widetilde{\Pi}^{ij} - \widetilde{\varepsilon}^{ij} + \frac{\partial \widetilde{T}^{ij\ell}}{\partial x^\ell} + \widetilde{\mathcal{P}}^{ij}, \tag{52}$$

where $\widetilde{\mathbf{P}}(= \{\widetilde{P}^{ij}\})$, $\widetilde{\mathbf{\Pi}}(= \{\widetilde{\Pi}^{ij}\})$, $\widetilde{\varepsilon}(= \{\widetilde{\varepsilon}^{ij}\})$ and $\nabla \cdot \widetilde{\mathbf{T}}(= \{\partial \widetilde{T}^{ij\ell}/\partial x^\ell\})$ are the production, re-distribution, dissipation and the transport rates of the coherent components, respectively. They are defined as

$$\widetilde{P}^{ij} = -\langle \widetilde{u}^j \widetilde{u}^\ell \rangle \frac{\partial \langle u \rangle^i}{\partial x^\ell} - \langle \widetilde{u}^i \widetilde{u}^\ell \rangle \frac{\partial \langle u \rangle^j}{\partial x^\ell}, \tag{53}$$

$$\widetilde{\Pi}^{ij} = -\frac{1}{\rho_0} \left\langle \widetilde{p} \left(\frac{\partial \widetilde{u}^i}{\partial x^j} + \frac{\partial \widetilde{u}^j}{\partial x^i} \right) \right\rangle, \tag{54}$$

$$\widetilde{\varepsilon}^{ij} - 2\nu \left\langle \frac{\partial \widetilde{u}^i}{\partial x^\ell} \frac{\partial \widetilde{u}^j}{\partial x^\ell} \right\rangle, \tag{55}$$

$$\frac{\partial \widetilde{T}^{ij\ell}}{\partial x^\ell} = +\frac{\partial}{\partial x^\ell} \left(-\left\langle \widetilde{u}^\ell \widetilde{u}^i \widetilde{u}^j \right\rangle + \langle \widetilde{p} \widetilde{u}^i \rangle \delta^{\ell j} + \langle \widetilde{p} \widetilde{u}^j \rangle \delta^{\ell i} + \nu \frac{\partial}{\partial x^\ell} \langle \widetilde{u}^i \widetilde{u}^j \rangle \right). \tag{56}$$

On the other hand, the evolution of the Reynolds stress of the incoherent velocity fluctuation, $\langle \mathbf{u}'' \mathbf{u}'' \rangle$, is subject to

$$\left(\frac{\partial}{\partial t} + \langle \overline{\mathbf{u}} \rangle \cdot \nabla \right) \langle \overline{u''^i u''^j} \rangle = P''^{ij} + \Pi''^{ij} - \varepsilon''^{ij} + \frac{\partial T''^{ij\ell}}{\partial x^\ell} + \mathcal{P}''^{ij}, \tag{57}$$

where $\mathbf{P}''(= \{P''^{ij}\})$, $\mathbf{\Pi}''(= \{\Pi''^{ij}\})$, $\varepsilon''(= \{\varepsilon''^{ij}\})$ and $\nabla \cdot \mathbf{T}''(= \{\partial T''^{ij\ell}/\partial x^\ell\})$ are defined as

$$P''^{ij} = -\langle u''^j u''^\ell \rangle \frac{\partial \langle u \rangle^i}{\partial x^\ell} - \langle u''^i u''^\ell \rangle \frac{\partial \langle u \rangle^j}{\partial x^\ell}, \tag{58}$$

$$\Pi''^{ij} = -\frac{1}{\rho_0} \left\langle p'' \left(\frac{\partial u''^i}{\partial x^j} + \frac{\partial u''^j}{\partial x^i} \right) \right\rangle - 2\nu \left\langle \frac{\partial u''^i}{\partial x^\ell} \frac{\partial u''^j}{\partial x^\ell} \right\rangle, \tag{59}$$

$$\varepsilon''^{ij} = -2\nu\left\langle \frac{\partial u''^i}{\partial x^\ell}\frac{\partial u''^j}{\partial x^\ell}\right\rangle, \tag{60}$$

$$\frac{\partial T''^{ij\ell}}{\partial x^\ell} = +\frac{\partial}{\partial x^\ell}\left(-\left\langle u''^\ell u''^i u''^j\right\rangle + \langle p'' u''^i\rangle\delta^{\ell j} + \langle p'' u''^j\rangle\delta^{\ell i} + \nu\frac{\partial}{\partial x^\ell}\langle u''^i u''^j\rangle\right). \tag{61}$$

The terms (53)–(56) and (58)–(61) are similar to the counterparts of the Reynolds stress equation without using the double-averaging procedure. Actually, for example, the addition of $\widetilde{P}^{ij}$ and P''^{ij} just gives the usual production term of $\langle u'^i u'^j\rangle$. On the other hand, the final terms in (52) and (57) originate from the double-averaging procedure. These terms, $\widetilde{\mathcal{P}}^{ij}$ and $\mathcal{P}''^{ij}$, are defined by

$$\mathcal{P}''^{ij} = -\left\langle \widetilde{u''^\ell u''^j}\frac{\partial \widetilde{u}^i}{\partial x^\ell}\right\rangle - \left\langle \widetilde{u''^\ell u''^i}\frac{\partial \widetilde{u}^j}{\partial x^\ell}\right\rangle = -\widetilde{\mathcal{P}}^{ij}, \tag{62}$$

and represent the transfer rates between the coherent and incoherent components. Here, the dispersion part of the incoherent Reynolds stress is defined by

$$\widetilde{u''^i u''^j} = \overline{u''^i u''^j} - \left\langle \overline{u''^i u''^j}\right\rangle. \tag{63}$$

If we add (52) and (57), these two terms cancel each other out, and will not contribute to the budget of the total Reynolds stress $\langle \mathbf{u}'\mathbf{u}'\rangle$ but to the transfer between the coherent and incoherent components. The dispersion part of the incoherent Reynolds stress $\widetilde{u''^\ell u''^j}$ (63) coupled with the spatial structure of the coherent motion, $\partial\widetilde{u}^j/\partial x^\ell$, the turbulent stress are transferred between the coherent and incoherent components.

Taking the contraction of i and j in (52) and (57), we obtain the evolution equations of the coherent and incoherent energies as

$$\left(\frac{\partial}{\partial t} + \langle\overline{\mathbf{u}}\rangle\cdot\nabla\right)\left\langle\frac{1}{2}\widetilde{\mathbf{u}}^2\right\rangle = \widetilde{P} - \widetilde{\varepsilon} + \nabla\cdot\widetilde{\mathbf{T}} + \left\langle \widetilde{u''^\ell u''^m}\frac{\partial\widetilde{u}^m}{\partial x^\ell}\right\rangle, \tag{64}$$

$$\left(\frac{\partial}{\partial t} + \langle\overline{\mathbf{u}}\rangle\cdot\nabla\right)\left\langle\frac{1}{2}\mathbf{u}''^2\right\rangle = P'' - \varepsilon'' + \nabla\cdot\mathbf{T}'' - \left\langle \widetilde{u''^\ell u''^m}\frac{\partial\widetilde{u}^m}{\partial x^\ell}\right\rangle, \tag{65}$$

where $\widetilde{P}$ and P'' are the production rates, $\widetilde{\varepsilon}$ and ε'' are the dissipation rates, $\nabla\cdot\widetilde{\mathbf{T}}$ and $\nabla\cdot\mathbf{T}''$ is the transport rate of the coherent and incoherent turbulent energies, respectively. The final terms in (64) and (65) represent the energy transfer rates between the coherent and incoherent turbulent energies. They are denoted by

$$P_{K''} = -\left\langle \widetilde{u''^\ell u''^m}\frac{\partial\widetilde{u}^m}{\partial x^\ell}\right\rangle = -P_{\widetilde{K}}. \tag{66}$$

If we approximate the dispersion part of the Reynolds stress by the eddy viscosity-type model by

$$\widetilde{u''^\ell u''^m} \simeq -\widetilde{\nu}\frac{\partial\widetilde{u}^m}{\partial x^\ell}, \tag{67}$$

the energy transfer rate from the coherent to incoherent fluctuations, $P_{K''}$, is expressed as

$$P_{K''} = -\left\langle \widetilde{u''^\ell u''^m}\frac{\partial\widetilde{u}^m}{\partial x^\ell}\right\rangle \simeq +\widetilde{\nu}\left(\frac{\partial\widetilde{u}^m}{\partial x^\ell}\right)^2 > 0. \tag{68}$$

As (66) shows, the sink of the coherent fluctuation energy, $P_{\widetilde{K}} < 0$, corresponds to the production of the incoherent fluctuation energy, $P_{K''} > 0$. In this case, the kinetic energy of the plume or coherent fluctuation motion driven by the surface cooling is transferred towards the kinetic energy of the random or incoherent fluctuation motions as schematically depicted in Figure 6.

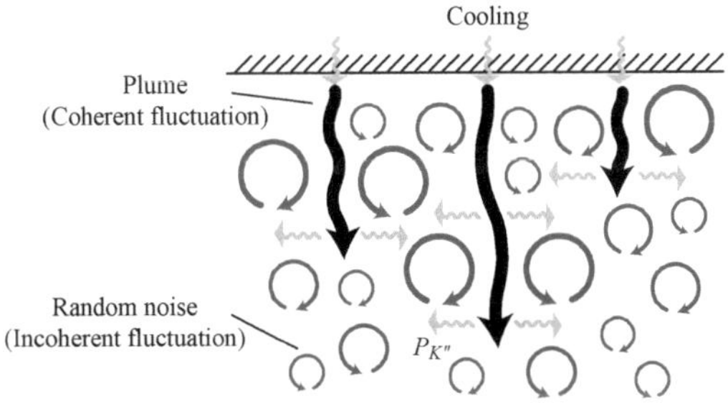

Figure 6. Schematic picture of the interaction between the coherent and incoherent fluctuations. Plumes (coherent fluctuations, depicted by thick black curved lines) are driven by the surface cooling. The energy of the plume motions are transferred to the energy of the random noise (incoherent fluctuation, depicted by gray circle eddies) by the interaction $P_{K''}$ if $P_{K''} > 0$.

In this picture, the diving plumes driven by the surface cooling enhances the turbulent transport through the non-equilibrium effect along the plume motions. During this process, the energy of plumes (coherent fluctuations) is transferred to the energy of random noises (incoherent fluctuations) through the inhomogeneous motions of plumes $\nabla \mathbf{u}$ coupled with the incoherent fluctuations $\widetilde{\mathbf{u}''\mathbf{u}''}$. Of course, there are several other mechanisms such as the production P'', dissipation ε'', etc. in (65), that contribute to the evolution of the incoherent fluctuation energy $\langle \mathbf{u}''^2 \rangle$. In this sense, the present picture is a fairly simplified one.

With the non-equilibrium effect, the energy transfer rate to the incoherent component of the fluctuation is written as

$$P_{K''} = -\left\langle \widetilde{u''^{\ell} u''^{m}} \frac{\partial \tilde{u}^m}{\partial x^\ell} \right\rangle \simeq +\tilde{\nu}\left(1 - \tilde{\Lambda}_{\mathrm{D}}\right)\left(\frac{\partial \tilde{u}^m}{\partial x^\ell}\right)^2. \tag{69}$$

As the DNSs of the surface cooling case show (Figure 4), the non-equilibrium-effect coefficient $\tilde{\Lambda}_{\mathrm{D}}$ is negative:

$$\tilde{\Lambda}_{\mathrm{D}} < 0. \tag{70}$$

In this case, the energy transfer to the incoherent or random fluctuation is enhanced by the non-equilibrium turbulence effect along the plume motion. This matches the tendency of the surface cooling-driven convection case. The presence of the diving plumes strongly enhances the turbulence mixing through the non-equilibrium effect. This picture is based on several simplifications, but still suggests the importance of the non-equilibrium effect associated with the plume motions in the turbulent transport in convection.

6. Conclusions

Linear gradient-transport-type turbulence models for the turbulent fluxes with the eddy viscosity and diffusivity representation work poorly in some turbulent flows. For example, the K–ε turbulence model based on the eddy viscosity $\nu_{\mathrm{T}} = C_\nu K^2/\varepsilon$ cannot be applied to the homogeneous-shear turbulence; it leads to the overestimation of K and ε. Convective turbulence with plumes/jets and supersonic turbulence accompanying shocks constitute other examples. Simulations with a simple linear model provide substantially over- or underestimated turbulent transports (turbulent viscosity, mass flux, heat flux, etc.). To rectify this deficiency of linear gradient-transport models, in this work, the non-equilibrium effect mediated by the streamwise variation in turbulence are incorporated into the turbulence model. A turbulence model with these effects implemented in combination was proposed. In the model, in addition to the usual eddy viscosity and diffusivity representations for the turbulent transports, deviations from the simple representations due to the non-equilibrium effect are taken into account.

Relevance of the non-equilibrium effects was argued with utilizing recent elaborated experiments in turbulent jets and plumes. It was pointed out that the expected results of the non-equilibrium effect on turbulent transports in the turbulent jets and plumes are transport enhancement or suppression due to the non-equilibrium effect (depending on the turbulence variation along the jets/plumes). These evaluations of turbulent transports seem to be in preferable directions to correct the discrepant results obtained from simple conventional gradient-transport-type models.

Funding: This research was funded by the Japan Society for the Promotion of Science (JSPS) Grants-in-Aid for Scientific Research: JP18H01212 and JP23H01199. The Isaac Newton Institute (INI) for Mathematical Sciences program: "Frontier in dynamo theory: From the Earth to the stars" (DYT2) was supported by EPSRC grant no EP/R014604/1. The Nordic Institute for Theoretical Physics (NORDITA) program: "Toward a comprehensive model for the galactic magnetic field" was partly supported by NordForsk and the Royal Astronomical Society.

Institutional Review Board Statement: Not applicable.

Informed Consent Statement: Not applicable.

Data Availability Statement: Data from the numerical simulations and analysis presented in this paper are available from the author upon reasonable request.

Acknowledgments: The author would like to thank Youhei Masada and Tomoya Takiwaki for providing data from numerical simulations and valuable discussions on density variance and non-equilibrium effects in stellar convection simulations. Part of this work was started during the author's stay at the INI, Cambridge, during the DYT2 program from September to December 2022. The author also would like to thank the NORDITA for support and hospitality during the program: Towards a comprehensive model for the galactic magnetic field, in April 2023. The author would like to cordially dedicate this work to the memory of Jack Herring, who had kept encouraging the author in working turbulence closure theory and convection modeling.

Conflicts of Interest: The author declares no conflict of interest.

Appendix A. Non-Equilibrium Turbulent Energy from the Two-Scale Direct-Interaction Approximation Analysis

From the two-scale direct-interaction approximation (TSDIA) analysis, the turbulent energy $K(\equiv \langle \mathbf{u}'^2 \rangle /2)$ is known to be expressed in terms of as [30]

$$K = \int d\mathbf{k}\; Q(k;\tau,\tau) - \int d\mathbf{k} \int_{-\infty}^{\tau} d\tau_1\; G(k,\mathbf{X};\tau,\tau_1) \frac{DQ(k,\mathbf{X};\tau,\tau_1)}{DT}, \tag{A1}$$

where $Q(k,\mathbf{X};\tau,\tau',T)$ and $G(k,\mathbf{X};\tau,\tau',T)$ are the energy spectral function and the response function defined by

$$Q(k,\mathbf{X};\tau,\tau',T) = \sigma(k,\mathbf{X};T) \exp\left[-\omega(k,\mathbf{X};T)|\tau-\tau'|\right], \tag{A2}$$

$$G(k,\mathbf{X};\tau,\tau',T) = H(k,\tau-\tau') \exp\left[-\omega(k,\mathbf{X};T)(\tau-\tau')\right], \tag{A3}$$

respectively. Here the Heaviside step function defined by

$$H(x) = \begin{cases} 1 & \text{for } x \geq 0, \\ 0 & \text{for } x < 0, \end{cases} \tag{A4}$$

and

$$\sigma(k,\mathbf{X};T) = \sigma_0 \varepsilon(\mathbf{X},T)^{2/3} k^{-11/3}, \tag{A5}$$

$$\omega(k,\mathbf{X};T) = \omega_0 \varepsilon(\mathbf{X},T)^{1/3} k^{2/3}. \tag{A6}$$

As we see below, Equation (A5) corresponds to the Kolmogorov scaling of homogeneous and isotropic equilibrium turbulence.

Substituting (A2)–(A6) into Equation (A1), we have

$$\begin{aligned} K &= \int d\mathbf{k}\,\sigma(\mathbf{k},\mathbf{X};T)\exp[-\omega(k,\mathbf{X};T)|\tau-\tau|] \\ &- \int d\mathbf{k}\int_{-\infty}^{\tau} d\tau_1 H(\tau-\tau_1)\exp[-\omega(k,\mathbf{X};T)(\tau-\tau_1)] \\ &\qquad \times\frac{D}{DT}\sigma(k,\mathbf{X};T)\exp[-\omega(k,\mathbf{X};T)|\tau-\tau_1|] \end{aligned} \tag{A7}$$

The first term of (A7) denoted by K_E is evaluated as

$$\begin{aligned} K_E &= \int d\mathbf{k}\,\sigma(\mathbf{k},\mathbf{X};T)\exp[-\omega(k,\mathbf{X};T)|\tau-\tau|] \\ &= 4\pi\int_{k_C}^{\infty} dk\,k^2\sigma_0\varepsilon^{2/3}k^{-11/3} = 4\pi\sigma_0\int_{k_C}^{\infty} dk\,\varepsilon^{2/3}k^{-5/3}, \end{aligned} \tag{A8}$$

where

$$k_C = 2\pi/\ell_C \tag{A9}$$

is the infrared cut-off wave number representing the largest eddy size of turbulence, ℓ_C. We further introduce the scale transformation

$$s = k/k_C \tag{A10}$$

to express the relative scale based on the infrared cut-off wave number. With this length scale, the equilibrium energy part can be expressed as

$$\begin{aligned} K_E &= 4\pi\sigma_0\varepsilon^{2/3}k_C\int_{s\geq 1} ds\,s^{-5/3} = 6\pi\sigma_0\varepsilon^{2/3}k_C^{-2/3} \\ &= 3(2\pi)^{1/3}\sigma_0\varepsilon^{2/3}\ell_C^{2/3}. \end{aligned} \tag{A11}$$

The second term in (A7) represents the non-equilibrium effect on the turbulent energy. This part is evaluated as

$$\begin{aligned} K_N &= -\int d\mathbf{k}\int_{-\infty}^{\tau} d\tau_1 H(\tau-\tau_1)\exp[-\omega(k,\mathbf{X};T)(\tau-\tau_1)] \\ &\qquad \times\frac{D}{DT}\sigma(k,\mathbf{X};T)\exp[-\omega(k,\mathbf{X};T)|\tau-\tau_1|] \\ &= -\int d\mathbf{k}\int_{-\infty}^{\tau} d\tau_1\frac{D\sigma(k,\mathbf{X};T)}{DT}\exp[-2\omega(k,\mathbf{X};T)(\tau-\tau_1)] \\ &\quad +\int d\mathbf{k}\int_{-\infty}^{\tau} d\tau_1(\tau-\tau_1)\sigma(k,\mathbf{X};T)\frac{D\omega}{DT}\exp[-2\omega(k,\mathbf{X};T)(\tau-\tau_1)] \\ &= -\left[4\pi\int k^2dk\frac{1}{2\omega}\frac{D\sigma}{DT}\exp[-2\omega(k,\mathbf{X},T)(\tau-\tau_1)]\right]_{\tau_1=-\infty}^{\tau} \\ &\quad +\left[4\pi\int dk\,k^2\frac{\sigma}{2\omega^2}\frac{D\omega}{DT}\exp[-2\omega(k,\mathbf{X};T)(\tau-\tau_1)]\right]_{\tau_1=-\infty}^{\tau} \\ &= -2\pi\int dk\,k^2\left(\frac{1}{\omega}\frac{D\sigma}{DT}-\frac{\sigma}{2\omega^2}\frac{D\omega}{DT}\right) \end{aligned} \tag{A12}$$

Here, in integration with respect to τ_1, we integrated by parts, and used the fact that neither σ nor ω explicitly depend on τ_1. With Equations (A5) and (A6), (A12) is calculated as

$$K_N = -\pi\frac{\sigma_0}{\omega_0}k_C^{-4/3}\varepsilon^{-2/3}\frac{D\varepsilon}{DT} + \frac{\pi}{6}\frac{\sigma_0}{\omega_0}\varepsilon^{1/3}k_C^{-7/3}\frac{Dk_C}{DT}. \tag{A13}$$

It follows from Equation (A9) that the non-equilibrium part of the energy is expressed in terms of ε and ℓ_C as

$$K_N = -\frac{1}{2}(2\pi)^{-1/3}\frac{\sigma_0}{\omega_0}\varepsilon^{-2/3}\ell_C^{4/3}\frac{D\varepsilon}{DT} - \frac{1}{12}(2\pi)^{-4/3}\frac{\sigma_0}{\omega_0}\varepsilon^{1/3}\ell_C^{1/3}\frac{D\ell_C}{DT}. \tag{A14}$$

With the addition of (A11), the turbulent energy is expressed as

$$K = 3(2\pi)^{1/3}\sigma_0\varepsilon^{2/3}\ell_C^{2/3} - \frac{1}{2}(2\pi)^{-1/3}\frac{\sigma_0}{\omega_0}\varepsilon^{-2/3}\ell_C^{4/3}\frac{D\varepsilon}{DT} - \frac{1}{12}(2\pi)^{-4/3}\frac{\sigma_0}{\omega_0}\varepsilon^{1/3}\ell_C^{1/3}\frac{D\ell_C}{DT}, \tag{A15}$$

or with model constants C_{K1}, C_{K2} and C_{K3}, the expression for the turbulent kinetic energy is written as

$$K = C_{K1}\varepsilon^{2/3}\ell_C^{2/3} - C_{K2}\varepsilon^{-3/2}\ell_C^{4/3}\frac{D\varepsilon}{Dt} - C_{K3}\varepsilon^{1/3}\ell_C^{1/3}\frac{D\ell_C}{Dt}, \tag{A16}$$

where the model constants are defined as

$$C_{K1} = 3(2\pi)^{1/3}\sigma_0, \ C_{K2} = \frac{1}{2}(2\pi)^{-1/3}\frac{\sigma_0}{\omega_0}, \ C_{K3} = \frac{1}{12}(2\pi)^{-4/3}\frac{\sigma_0}{\omega_0}. \tag{A17}$$

Equation (A16) is Equation (20) in Section 3.

References

1. Launder, B.E.; Spalding, D.B. *Lectures in Mathematical Models of Turbulence*; Academic Press: London, UK, 1972.
2. Rodi, W. *Turbulence Models and Their Application in Hydraulics: A State-of-the-Art Review*, 3rd ed.; Taylor and Francis: London, UK, 2000.
3. Yokoi, N. Turbulence, transport and reconnection. In *Topics in Magnetohydrodynamic Topology, Reconnection and Stability Theory: CISM International Centre for Mechanical Sciences, 591*; MacTaggart, D., Hillier, A., Eds.; Springer: Berlin/Heidelberg, Germany, 2020; pp. 177–265.
4. Yokoi, N.; Yoshizawa, A. Statistical analysis of the effects of helicity in inhomogeneous turbulence. *Phys. Fluids A* **1993**, *5*, 464–477. [CrossRef]
5. Yokoi, N.; Brandenburg, A. Large-scale flow generation by inhomogeneous helicity. *Phys. Rev. E* **2016**, *93*, 033125. [CrossRef] [PubMed]
6. Yoshizawa, A.; Nisizima, S. A non-equilibrium representation of the turbulent viscosity based on the two-scale turbulence theory. *Phys. Fluids A* **1993**, *5*, 3302–3304. [CrossRef]
7. Kraichnan, R.H. The structure of isotropic turbulence at very high Reynolds numbers. *J. Fluid Mech.* **1959**, *5*, 497–543. [CrossRef]
8. Edwards, S.F. The statistical dynamics of homogeneous turbulence. *J. Fluid Mech.* **1964**, *18*, 239–273. [CrossRef]
9. Herring, J.R. Self-consistent-field approach to turbulence theory. *Phys. Fluids* **1965**, *8*, 2219–2225. [CrossRef]
10. Herring, J.R. Self-consistent-field approach to nonstationary turbulence. *Phys. Fluids* **1966**, *9*, 2106–2110. [CrossRef]
11. Spruit, H.C. Convection in stellar envelopes: A changing paradigm. *Mem. Soc. Astron. Ital.* **1997**, *68*, 397–413.
12. Rast, M.P. Compressible plume dynamics and stability. *J. Fluid Mech.* **1998**, *369*, 125–149. [CrossRef]
13. Cossette, J.-F.; Rast, M.P. Supergranulation as the largest buoyant driven convective scale of the Sun. *Astrophys. J. Lett.* **2016**, *829*, L17. [CrossRef]
14. Brandenburg, A. Stellar mixing length theory. *Astrophys. J.* **2016**, *832*, 6. [CrossRef]
15. Bethe, H.A.; Wilson, J.R. Revival of stalled supernova shock by neutrino heating. *Astrophys. J.* **1985**, *295*, 14–23. [CrossRef]
16. Couch, S.M.; Warren, M.L.; O'Connor, E.P. Simulating turbulence-aided neutrino-driven core-collapse supernova explosions in one dimension. *Astrophys. J.* **2020**, *890*, 127. [CrossRef]
17. Janka, H.-T.; Müller, E. Neutrino heating, convection, and mechanism of Type-II supernova explosions. *Astron. Astrophys.* **1996**, *306*, 167–198.
18. Mabanta, J.W.; Murphy, Q.A. How turbulence enables core-collapse supernova explosion. *Astrophys. J.* **2018**, *856*, 22. [CrossRef]
19. Marek, A.; Janka, H.-T. Delayed neutrino-driven supernova explosions aided by the standing accretion-shock instability. *Astrophys. J.* **2009**, *694*, 664–696. [CrossRef]
20. Müller, B. A critical assessment of turbulence models for 1D core-collapse supernova simulations. *Mon. Not. R. Astron. Soc.* **2019**, *487*, 5304–5323. [CrossRef]
21. Murphy, Q.A.; Meakin, C. A global turbulence model for neutrino-driven convection in core-collapse supernovae. *Astrophys. J.* **2011**, *742*, 74. [CrossRef]
22. Layek, G.C.; Sunita, V. Non-Kolmogorov scaling and dissipation laws in planar turbulent plume. *Phys. Fluids* **2018**, *30*, 115105. [CrossRef]

23. Cafiero, G.; Vassilicos, J.C. Nonequilibrium turbulence scalings and self-similarity in turbulent planar jets. *Proc. R. Soc. Lond. A* **2019**, *475*, 20190038.
24. Sunita, V.; Layek, G.C. Nonequilibrium turbulent dissipation in buoyant axisymmetric plume. *Phys. Rev. Fluids* **2021**, *6*, 104602. [CrossRef]
25. Bordoloi, A.D.; Lai, C.C.K.; Clark, L.; Veliz, G.; Variano, E. Turbulence statistics in a negatively buoyant multiphase plume. *J. Fluid Mech.* **2020**, *896*, A19. [CrossRef]
26. Charonko, J.J.; Prestridge, K. Variable-density mixing in turbulent jets with coflow. *J. Fluid Mech.* **2017**, *825*, 887–921. [CrossRef]
27. Lai, C.C.K.; Socolofsky, S.A. Budgets of turbulent kinetic energy, Reynolds stresses, and dissipation in a turbulent round jet discharged into a stagnant ambient. *Environ. Fluid Mech.* **2019**, *19*, 349–377. [CrossRef]
28. Lai, C.C.K.; Socolofsky, S.A. The turbulent kinetic energy budget in a bubble plume. *J. Fluid Mech.* **2019**, *865*, 993–1041. [CrossRef]
29. Yokoi, N.; Masada, Y.; Takiwaki, T. Modelling stellar convective transport with plumes—I. Non-equilibrium turbulence effect in double-averaging formulation. *Mon. Not. R. Astron. Soc.* **2022**, *516*, 2718–2735. [CrossRef]
30. Yoshizawa, A. Statistical analysis of the deviation of the Reynolds stress from its eddy-viscosity representation. *Phys. Fluids* **1984**, *27*, 1377–1387. [CrossRef]
31. Yoshizawa, A. *Hydrodynamic and Magnetohydrodynamic Turbulent Flows: Modeling and Statistical Theory*; Kluwer: Dordrecht, The Netherlands, 1998; pp. 265–303.
32. Yokoi, N. Cross helicity and related dynamo. *Geophys. Astrophys. Fluid Dyn.* **2013**, *107*, 114–184. [CrossRef]
33. Yokoi, N. Electromotive force in strongly compressible magnetohydrodynamic turbulence. *J. Plasma Phys.* **2018**, *84*, 735840501. [CrossRef]
34. Yokoi, N. Mass and internal-energy transports in strongly compressible magnetohydrodynamic turbulence. *J. Plasma Phys.* **2018**, *84*, 735840603. [CrossRef]
35. Okamoto, M. Theoretical investigation of an eddy-viscosity-type representation of the Reynolds stress. *J. Phys. Soc. Jpn.* **1994**, *63*, 2101–2122. [CrossRef]
36. Tennekes, H.; Lumley, J.L. *A First Course in Turbulence*; The MIT Press: Cambridge, MA, USA, 1972.
37. Livescu, D.; Ristorcelli, J. R. Variable-density mixing in buoyancy-driven turbulence. *J. Fluid Mech.* **2008**, *605*, 145–180. [CrossRef]

atmosphere

MDPI

Review

The Decay of Energy and Scalar Variance in Axisymmetric Turbulence

Peter A. Davidson

Department of Engineering, University of Cambridge, Cambridge CB2 1PZ, UK; pad3@cam.ac.uk

Abstract: We review the recent progress in our understanding of the large scales in homogeneous (but anisotropic) turbulence. We focus on turbulence which emerges from Saffman-like initial conditions, in which the vortices possess a finite linear impulse. Such turbulence supports long-range velocity correlations of the form $\langle u_i u'_j \rangle = O(r^{-3})$, where $\mathbf{u}$ and $\mathbf{u}'$ are separated by a distance r, and these long-range interactions dominate the dynamics of large eddies. We show that, for axisymmetric turbulence, the energy and integral scales evolve as $u_\perp^2 \sim u_{//}^2 \sim t^{-6/5}$ and $\ell_\perp \sim \ell_{//} \sim t^{2/5}$, where $\perp$ and $//$ indicate directions that are perpendicular and parallel to the symmetry axis, respectively. These predictions are consistent with the evidence of direct numerical simulations. Similar results are obtained for the passive scalar variance, where we find that $\langle \theta^2 \rangle \sim t^{-6/5}$. The primary point of novelty in our discussion of passive scalar decay is that it is based in real (rather than spectral) space, making use of an integral invariant which is a generalization of the isotropic Corrsin integral.

Keywords: scalar variance; Saffman turbulence; axisymmetric turbulence

Citation: Davidson, P.A. The Decay of Energy and Scalar Variance in Axisymmetric Turbulence. *Atmosphere* **2023**, *14*, 1019. https://doi.org/10.3390/atmos14061019

Academic Editors: Boris Galperin, Annick Pouquet, Peter Sullivan and Doris Folini

Received: 16 March 2023
Revised: 31 May 2023
Accepted: 5 June 2023
Published: 13 June 2023

1. Introduction

1.1. The Summer of 2002

In the summer of 2002, the author visited NCAR to discuss the behavior of the large scales in freely decaying, homogeneous turbulence with Jack Herring. It was to be the first of many memorable visits to NCAR. The primary focus at that time was on the behavior of Loitsyansky's integral in isotropic turbulence, but it left the author with a longstanding interest in the dynamics of large scales in all types of homogeneous turbulence. Initially, that interest centered on large scales in so-called Batchelor turbulence, whose energy spectrum satisfies $E(k \to 0) \sim k^4$, and where Loitsyansky's integral was shown to be conserved in fully developed turbulence (see Jackson et al. [1] and Ishida, Davidson, and Kaneda [2]). However, interest soon spread to include Saffman ($E(k \to 0) \sim k^2$) turbulence (Davidson, Okamoto, and Kaneda [3]; Krogstad and Davidson [4]), as well as the evolution of the large scales in rotating, stratified, and MHD turbulence (Davidson [5,6]). Finally, as part of an international network, the author began a study on the large scales in passive scalar mixing (Yoshimatsu and Kaneda [7]). Many of these studies have their roots in those discussions at NCAR in the summer of 2002.

This paper reviews some of the more recent developments that have been made in the large-scale behavior of freely decaying Saffman turbulence, with particular emphasis on energy decay and the kinematics of passive scalars. It particularly builds on the work of Davidson [5,6] and Davidson, Okamoto, and Kaneda [3], and on the fruitful discussions the author had with Yukio Kaneda and Katsunori Yoshimatsu in 2016 and 2017 as part of an international network.

1.2. Some Important Questions about the Decay of Passive Scalar Fluctuations

Let us start with some background information on the decay of passive scalar fluctuations in homogeneous turbulence, particularly the decay of scalar variance. Perhaps the first point to make is that, although there are points of contact with Richardson's celebrated theory of relative dispersion (Richardson [8]), passive scalar decay and Richardson dispersion are distinct phenomena, with the former focusing on the integral scales and the latter focusing on the inertial sub-range (Monin and Yoglam [9]).

Perhaps the most important paper in passive scalar decay is that written by Corrsin [10]. Crucially, Corrsin noted that, under certain conditions, a passive scalar, θ, evolving in freely decaying *isotropic* turbulence, possesses an integral invariant of the two-point correlation function $\langle \theta\theta' \rangle = \langle \theta(\mathbf{x})\theta(\mathbf{x}') \rangle$. Specifically, he noted that

$$L_\theta = \int_{V_\infty} \langle \theta\theta' \rangle d\mathbf{r}, \ \mathbf{r} = \mathbf{x}' - \mathbf{x}, \tag{1}$$

is conserved (when it exists), provided that the triple correlation $\langle \mathbf{u}\theta\theta' \rangle$ falls off faster than $O(r^{-2})$ at large separations, where $r = |\mathbf{r}| = |\mathbf{x}' - \mathbf{x}|$. The possible existence of such an invariant immediately raises a host of questions, including the following.

(i) Under what conditions is this integral convergent? In other words, when does L_θ exist?
(ii) If the integral exists, under what conditions would we expect to find $\langle \mathbf{u}\theta\theta' \rangle \leq O(r^{-3})$, so that L_θ is an invariant?
(iii) If the invariant exists in isotropic turbulence, how does this generalize to anisotropic turbulence?
(iv) Can we use the existence of the invariant to predict the decay rate of passive scalar variance?

We shall address all of these questions in this review. In particular, we generalize Corrsin's integral to anisotropic and axisymmetric turbulence and use it to predict the decay rate of passive scalar variance.

Of course, the statistics of θ depend on those of the velocity field, and so the answers to the questions above must depend on the nature of the turbulence in which the scalar finds itself. The strongest long-range correlations are to be found in anisotropic Saffman turbulence, where the two-point velocity correlations fall off slowly as $O(r^{-3})$, as discussed in Saffman [11,12] and Davidson, Okamoto, and Kaneda [3]. Consequently, the most stringent test for the existence of Corrsin's integral invariant is in anisotropic Saffman turbulence, and so that is the case that we consider here. To focus our thoughts, we follow Davidson, Okamoto, and Kaneda [3] and restrict the discussion to statistically axisymmetric (but anisotropic) turbulence.

We shall see that, in axisymmetric Saffman turbulence, L_θ is indeed convergent, though only conditionally so. It is also conserved. Moreover, the fact that L_θ is conserved in Saffman turbulence strongly suggests that it is an invariant in all other forms of homogeneous turbulence. Finally, we combine the conservation of L_θ with the self-similarity of large scales to show that $\langle \theta^2 \rangle \ell^\theta_{//} (\ell^\theta_\perp)^2$ is an invariant, where $\ell^\theta_{//}$ and $\ell^\theta_\perp$ are integral scales parallel and perpendicular to the axis of symmetry, respectively. This in turn demands that the passive scalar variance decays as $\langle \theta^2 \rangle \sim t^{-6/5}$, at least in Saffman turbulence.

1.3. The Structure of the Paper

The structure of the paper is as follows. We start, in Section 2, by reminding readers about the properties of certain integral invariants (Saffman invariants) associated with the two-point velocity correlations. Such invariants exist in so-called Saffman turbulence (turbulence in which the eddies possess a statistically significant amount of linear impulse) and are a statistical manifestation of linear momentum conservation. These invariants are important as their conservation constrains the way in which the integral scales can evolve, and this in turn has consequences for the rate of energy decay and the rate of growth of integral length scales.

Next, in Sections 3 and 4, we generalize Corrsin's passive scalar integral to anisotropic axisymmetric turbulence and note that its invariance reflects the conservation of the total amount of passive scalar within a large control volume embedded in the turbulence. We also note the analogy between our generalized Corrsin integral and the Saffman invariants.

The results of Sections 2–4 are then combined in Section 5, where we use these various integral invariants to determine the behavior of integral scales in axisymmetric turbulence. In particular, we deduce the rate of energy decay and passive scalar variance, as well as the evolution of various integral scales. We close, in Sections 6 and 7, with the discussion and conclusions, respectively. As noted earlier, the primary point of novelty in the paper lies in the use of our generalized Corrsin integral to deduce the integral properties of a passive scalar.

2. Invariants of Freely Decaying, Axisymmetric Saffman Turbulence

As a prelude to our discussion of invariants in passive scalar mixing, it seems appropriate to review the large-scale dynamics of freely decaying Saffman turbulence. The review is based on the results of Saffman [11] for the most general anisotropic case, and on the findings of Davidson [5] and Davidson, Okamoto, and Kaneda [3], hereafter denoted as DOK12, for the case where the fields are statistically axisymmetric.

2.1. Anisotropic Saffman Turbulence

Saffman [11] considered homogeneous turbulence which emerges from an initial condition in which there are no long-range algebraic correlations in the vorticity field. Thus, he assumed that $\langle \omega_i \omega'_j \rangle_{r\to\infty}$ and $\langle \omega_i \omega_j \omega'_k \rangle_{r\to\infty}$, as well as all higher-order long-range vorticity cumulants, are exponentially small at $t = 0$. (From now on, we shall abbreviate expressions such as $\langle \omega_i(\mathbf{x})\omega_j(\mathbf{x}') \rangle_{r\to\infty}$ and $\langle \omega_i \omega'_j \rangle_\infty$). This leads to a low-wavenumber energy spectrum of the form $E(k \to 0) \sim k^2$ and so provides an alternative to the more traditional analysis of Batchelor and Proudman [13], who found $E(k \to 0) \sim k^4$. Interestingly, recent measurements taken by Krogstad and Davidson [4] support the idea that, at least for certain types of grid turbulence, the dynamics of freely decaying turbulence are indeed consistent with $E(k \to 0) \sim k^2$, rather than the more traditional assumption of $E(k \to 0) \sim k^4$.

The spectral tensor,

$$\Phi_{ij}(\mathbf{k}) = \frac{1}{(2\pi)^3} \int \langle u_i u'_j \rangle(\mathbf{r}) e^{-j\mathbf{k}\cdot\mathbf{r}} d\mathbf{r},$$

in Saffman turbulence takes the low-k form

$$\Phi_{ij}(\mathbf{k} \to 0) = \left\{ \delta_{i\alpha} - \frac{k_i k_\alpha}{k^2} \right\} \left\{ \delta_{j\beta} - \frac{k_j k_\beta}{k^2} \right\} M_{\alpha\beta} + O(k), \tag{2}$$

where $M_{\alpha\beta}$ is symmetric and independent of $\mathbf{k}$. Moreover, the coefficients $M_{\alpha\beta}$ is related to the second moment of the two-point vorticity correlation, $\langle \omega_i \omega'_j \rangle$, by

$$(2\pi)^3 M_{ij} = \frac{1}{2}\Omega_{ij} - \frac{1}{4}\delta_{ij}\Omega_{kk},\ \Omega_{ij} = \int r^2 \langle \omega_i \omega'_j \rangle d\mathbf{r}, \tag{3}$$

(Saffman [11]). Evidently, $\Phi_{ij}(\mathbf{k})$ is non-analytic at $\mathbf{k} = 0$ as the value depends on the way in which we approach the origin. Expansion (2) yields

$$\Phi_{ii}(\mathbf{k}) = \left\{ \delta_{\alpha\beta} - \frac{k_\alpha k_\beta}{k^2} \right\} M_{\alpha\beta} + O(k),\ E(k \to 0) = \frac{4}{3}\pi M_{\alpha\alpha} k^2, \tag{4}$$

and the corresponding asymptotic form of $\langle u_i u'_j \rangle_\infty$ is

$$\langle u_i u'_j \rangle_\infty = -M_{\alpha\beta}\pi^2 \left\{ \delta_{i\alpha}\nabla^2 - \frac{\partial^2}{\partial r_i \partial r_\alpha} \right\} \left\{ \delta_{j\beta}\nabla^2 - \frac{\partial^2}{\partial r_j \partial r_\beta} \right\} r. \tag{5}$$

Note that $\langle u_i u'_j \rangle_\infty = O(r^{-3})$.

The origin of the r^{-3} tails in $\langle u_i u'_j \rangle_\infty$ reverts back to Biot–Savart law, in which an eddy with a finite linear impulse induces irrotational far-field velocity fluctuations of the form

$$\mathbf{u}_\infty(\mathbf{x}) = \frac{1}{4\pi}\left(\mathbf{L}_{\text{eddy}} \cdot \nabla\right)\nabla(1/r) + O(r^{-4}), \tag{6}$$

where $\mathbf{L}_{\text{eddy}}$ is the linear impulse of the eddy, $\mathbf{L}_{\text{eddy}} = \frac{1}{2}\int_{V_{\text{eddy}}} \mathbf{x} \times \boldsymbol{\omega} dV$. Thus, individual eddies cast a long shadow which falls off slowly as $|\mathbf{u}_\infty| = O(r^{-3})$, leading to the long-range algebraic correlations $\langle u_i u'_j \rangle_\infty = O(r^{-3})$ and $\langle u_i u_j u'_k \rangle_\infty = O(r^{-3})$. Since these far-field fluctuations are irrotational, they do not contribute to the leading-order terms in $\langle \omega_i \omega'_j \rangle_\infty$, which are of order $\langle \omega_i \omega'_j \rangle_\infty \leq O(r^{-6})$, rather than $\langle \omega_i \omega'_j \rangle_\infty \leq O(r^{-5})$, for $t > 0$ (Saffman [11]).

Since $\langle \omega_i \omega'_j \rangle_\infty \leq O(r^{-6})$, the integral Ω_{ij} defined by (3) is absolutely convergent, and so M_{ij} is well defined. On the other hand, the integrals

$$L_{ij} = \lim_{V\to\infty} \int_V \langle u_i u'_j \rangle d\mathbf{r} \tag{7}$$

are only conditionally convergent. That is to say, they are convergent in the sense that they are independent of the size of V as $V \to \infty$. They are also uniquely determined by M_{ij}, since L_{ij} may be written as a surface integral whose integrand is the far-field Expansion (5):

$$L_{ij} = \lim_{V\to\infty} \int_V \langle u_i u'_j \rangle d\mathbf{r} = \lim_{S\to\infty} \oint_S \langle u_i u'_k \rangle_\infty r_j dS_k. \tag{8}$$

However, the numerical value of L_{ij} depends on the shape of the surface S as it recedes to infinity. This conditional convergence is a direct consequence of the non-analytic form of $\Phi_{ij}(\mathbf{k})$ at $\mathbf{k} = 0$. For the particular case of a large spherical volume of radius R and volume V_R, Saffman [11] established that

$$L_{ij} = \lim_{V_R\to\infty} \int_{V_R} \langle u_i u'_j \rangle d\mathbf{r} = (2\pi)^3 \left[\frac{7}{15} M_{ij} + \frac{1}{15}\delta_{ij} M_{kk}\right]. \tag{9}$$

The integral L_{ij} is particularly important because, as Saffman [11] suggested, L_{ij}, and hence M_{ij}, are invariants:

$$L_{ij} = \text{constant}, \; M_{ij} = \text{constant}. \tag{10}$$

This idea is as follows. Homogeneity tells us that

$$L_{ij} = \lim_{V_R\to\infty} \left\langle \left\{ \frac{1}{V_R^{1/2}} \int_{V_R} u_i d\mathbf{x} \right\} \left\{ \frac{1}{V_R^{1/2}} \int_{V_R} u'_j d\mathbf{x}' \right\} \right\rangle \tag{11}$$

(see, for example, Davidson [6]), where V_R is a large spherical control volume with surface S_R and radius R. So, the dynamical behavior of L_{ij} depends on that of the linear momentum $\mathbf{P} = \int_{V_R} \mathbf{u} dV$. However, $\mathbf{P}$ can only change as a result of pressure forces acting on the surface S_R, or because momentum is convected across S_R. Moreover, these are both random events spread over a large surface, and so the central limit theorem tentatively suggests that their cumulative effects are scaled as $O\left(S_R^{1/2}\right) \sim R$, i.e., the momentum change in any finite time is $O\left(V_R^{1/3} t\right)$. This in turn suggests that the pressure forces and momentum fluxes across S_R are too weak to change $\mathbf{P}/V_R^{1/2}$ in the limit of $V_R \to \infty$, and that consequently L_{ij} is an invariant in accordance with (11). This is essentially the argument of Saffman [11].

Unfortunately, it is rare that one satisfies the strict requirements for the central limit theorem to hold, especially in turbulence where the long-range interactions mean that remote events are rarely statistically independent. Consequently, we must regard the argument above as merely suggestive. However, formal proof of the invariance of L_{ij} is given in the study conducted by Davidson [5,6].

2.2. Statistically Axisymmetric Saffman Turbulence

We now turn to the case of statistically axisymmetric Saffman turbulence, with or without reflectional symmetry. In particular, we need to summarize the findings of Davidson [5,6] and DOK12, as these directly lead to invariants of passive scalar mixing in axisymmetric turbulence. As usual, we use subscripts $//$ and $\perp$ to indicate quantities that are parallel and perpendicular to the symmetry axis, respectively, which we take to be the z axis. Here, it may be shown that $M_{ij} = 0$ if $i \neq j$, and so the only non-zero components of M_{ij} are $M_{//}$ and $M_{xx} = M_{yy} = \frac{1}{2}M_{\perp}$. This supports the idea that $M_{//}$ and $M_{\perp}$ are the only invariants of axisymmetric Saffman turbulence. Similarly, (3) tells us that Ω_{ij} is diagonal, and that

$$(2\pi)^3 M_{//} = \frac{1}{4}\Omega_{//} - \frac{1}{4}\Omega_{\perp},\ (2\pi)^3 M_{\perp} = -\frac{1}{2}\Omega_{//}. \tag{12}$$

Therefore, the most general form of $\Phi_{ij}(\mathbf{k} \to 0)$ for Saffman turbulence is:

$$\Phi_{//} = \Phi_{zz} = M_{//}\frac{k_{\perp}^4}{k^4} + \frac{1}{2}M_{\perp}\frac{k_z^2 k_{\perp}^2}{k^4} + O(k), \tag{13}$$

$$\Phi_{\perp} = \Phi_{xx} + \Phi_{yy} = M_{//}\frac{k_z^2 k_{\perp}^2}{k^4} + \frac{1}{2}M_{\perp}\left[1 + \frac{k_z^4}{k^4}\right] + O(k), \tag{14}$$

(Davidson [5]), where $k_{\perp}^2 = k^2 - k_z^2$, and we omit the off-diagonal terms. Thus, with regards to the spectral tensor, our two invariants can be written as follows:

$$M_{//} = \Phi_{//}(k_z = 0, k_{\perp} \to 0) = \frac{1}{(2\pi)^3}\underset{k_{\perp}\to 0}{\mathrm{Lim}}\int e^{-j\mathbf{k}_{\perp}\cdot\mathbf{r}_{\perp}}\langle u_{//}u'_{//}\rangle d\mathbf{r} = \text{constant}, \tag{15}$$

$$\frac{1}{2}M_{\perp} = \Phi_{\perp}(k_z = 0, k_{\perp} \to 0) = \frac{1}{(2\pi)^3}\underset{k_{\perp}\to 0}{\mathrm{Lim}}\int e^{-j\mathbf{k}_{\perp}\cdot\mathbf{r}_{\perp}}\left\langle \mathbf{u}_{\perp}\cdot\mathbf{u}'_{\perp}\right\rangle d\mathbf{r} = \text{constant}, \tag{16}$$

$$M_{\perp} = \Phi_{\perp}(k_{\perp} = 0, k_z \to 0) = \frac{1}{(2\pi)^3}\underset{k_z\to 0}{\mathrm{Lim}}\int e^{-jk_z r_z}\left\langle \mathbf{u}_{\perp}\cdot\mathbf{u}'_{\perp}\right\rangle d\mathbf{r} = \text{constant}, \tag{17}$$

while

$$\Phi_{//}(k_{\perp} = 0, k_z \to 0) = 0.$$

Similarly, the only non-zero components of L_{ij} are $L_{//}$ and $L_{\perp}$. As for the general anisotropic case, $L_{//}$ and $L_{\perp}$ are uniquely determined by M_{ij}, although the relationship between $L_{//}$, $L_{\perp}$, $M_{//}$, and $M_{\perp}$ depends on the shape of V in (7). For example, in the particular case of a large cylindrical control volume of radius R and height $2H$, the following can be deduced:

$$L_{//} = (2\pi)^3 M_{//}\frac{1 + \frac{1}{2}(R/H)^2}{\left[1 + (R/H)^2\right]^{3/2}} + (2\pi)^3 M_{\perp}\frac{\frac{1}{4}(R/H)^2}{\left[1 + (R/H)^2\right]^{3/2}}, \tag{18}$$

$$L_{\perp} = (2\pi)^3 M_{\perp}\left\{1 - \frac{\frac{1}{2} + \frac{3}{4}(R/H)^2}{\left[1 + (R/H)^2\right]^{3/2}}\right\} + (2\pi)^3 M_{//}\frac{\frac{1}{2}(R/H)^2}{\left[1 + (R/H)^2\right]^{3/2}}, \tag{19}$$

(see DOK12), from which

$$L_{//}(H/R \to \infty) = (2\pi)^3 M_{//} = \text{constant, (long cylinder)}, \tag{20}$$

$$L_{\perp}(H/R \to \infty) = (2\pi)^3 \frac{1}{2} M_{\perp} = \text{constant, (long cylinder)}, \tag{21}$$

$$L_{//}(R/H \to \infty) = 0, \text{(thin disc)}, \tag{22}$$

$$L_{\perp}(R/H \to \infty) = (2\pi)^3 M_{\perp} = \text{constant, (thin disc)}. \tag{23}$$

These are equivalent to the spectral results (15)→(17), where $H/R \to \infty$ corresponds to $k_z = 0$ and $k_\perp \to 0$, while $R/H \to \infty$ is equivalent to $k_\perp = 0$ and $k_z \to 0$. The constraint that L_{ij} = constant is now reduced to

$$L_{//} = \int \langle u_{//} u'_{//} \rangle d\mathbf{r} = \text{constant}, \tag{24}$$

$$L_{\perp} = \int \left\langle \mathbf{u}_{\perp} \cdot \mathbf{u}'_{\perp} \right\rangle d\mathbf{r} = \text{constant}, \tag{25}$$

and this is true, irrespective of the shape of V. The invariance of $L_{//}$ and $L_\perp$, or equivalently $M_{//}$ and $M_\perp$, is confirmed in the direct numerical simulations of DOK12.

3. Passive Scalar Statistics in Axisymmetric Saffman Turbulence

Let us now turn to passive scalar mixing in Saffman turbulence. Let θ be the passive scalar which has a mean of zero, $\langle\theta\rangle = 0$, and is governed by

$$\frac{D\theta}{Dt} = \alpha \nabla^2 \theta. \tag{26}$$

We shall assume that, like the velocity field, its initial energy spectrum takes the form $E_\theta(k \to 0) \sim k^2$. (As we shall see, such an initial condition ensures that the spectrum $E_\theta(k \to 0) \sim k^2$ then persists for all time.) In such a case, the spectral function

$$\Phi_\theta(\mathbf{k}) = \frac{1}{(2\pi)^3} \int \langle \theta\theta' \rangle(\mathbf{r}) e^{-j\mathbf{k}\cdot\mathbf{r}} d\mathbf{r}, \tag{27}$$

takes the low-k form

$$\Phi_\theta(\mathbf{k} \to 0) = \left\{ \delta_{\alpha\beta} - \frac{k_\alpha k_\beta}{k^2} \right\} N_{\alpha\beta} + O(k), \tag{28}$$

and

$$E_\theta(k \to 0) = \frac{4}{3}\pi N_{\alpha\alpha} k^2,$$

where $N_{\alpha\beta}$ is symmetric and independent of $\mathbf{k}$. This has the same structural form as $\Phi_{ii}(\mathbf{k} \to 0)$ in (4), provided that we substitute $N_{\alpha\beta}$ for $M_{\alpha\beta}$, and so we conclude that $\langle\theta\theta'\rangle_\infty = O(r^{-3})$.

In statistically axisymmetric turbulence, $N_{ij} = 0$ if $i \neq j$, and so the only non-zero components of N_{ij} are $N_{//}$ and $N_{xx} = N_{yy} = \frac{1}{2}N_\perp$. The low-$k$ form of the spectral function $\Phi_\theta(\mathbf{k})$ is then simplified to

$$\Phi_\theta(\mathbf{k} \to 0) = \left[N_{//} + \frac{1}{2} N_\perp \right] - \left[N_{//} - \frac{1}{2} N_\perp \right] \frac{k_{//}^2}{k^2} + O(k), \tag{29}$$

which is analytic at $\mathbf{k} = 0$ only for the isotropic case of $N_{//} = \frac{1}{2}N_{\perp}$. Note that $N_{//}$ and $N_{\perp}$ can be written in terms of the spectral function as

$$N_{//} + \frac{1}{2}N_{\perp} = \Phi_\theta(k_{//} = 0, k_{\perp} \to 0) = \frac{1}{(2\pi)^3} \operatorname*{Lim}_{k_{\perp} \to 0} \int e^{-j\mathbf{k}_{\perp} \cdot \mathbf{r}_{\perp}} \langle \theta\theta' \rangle d\mathbf{r}, \tag{30}$$

and

$$N_{\perp} = \Phi_\theta(k_{\perp} = 0, k_{//} \to 0) = \frac{1}{(2\pi)^3} \operatorname*{Lim}_{k_{//} \to 0} \int e^{-jk_{//}r_{//}} \langle \theta\theta' \rangle d\mathbf{r}. \tag{31}$$

Thus, we see that $N_{//}$ and $N_{\perp}$ are closely related to the integral $\operatorname*{Lim}_{V \to \infty} \int_V \langle \theta\theta' \rangle d\mathbf{r}$, which we shall shortly see is an invariant of scalar mixing.

Let us now turn to the Corrsin integral

$$L_\theta = \operatorname*{Lim}_{V \to \infty} \int_V \langle \theta\theta' \rangle d\mathbf{r}, \tag{32}$$

which is central to scalar mixing analysis. Since $\langle \theta\theta' \rangle_\infty = O(r^{-3})$, L_θ is, in general, only conditionally convergent, and so it is important to determine the precise form of $\langle \theta\theta' \rangle$ in the far-field. Here, the analogy to $\Phi_{ii}(\mathbf{k})$ is again helpful. Based on (5), we have

$$\left\langle \mathbf{u} \cdot \mathbf{u}' \right\rangle_\infty = -\pi^2 [2M_{//} - M_{\perp}] \frac{r^2 - 3r_{//}^2}{r^5}, \tag{33}$$

and so the structural similarity between (4) and (28) tells us that

$$\langle \theta\theta' \rangle_\infty = -\pi^2 [2N_{//} - N_{\perp}] \frac{r^2 - 3r_{//}^2}{r^5}. \tag{34}$$

Note that the long-range correlation, $\langle \theta\theta' \rangle_\infty = O(r^{-3})$, vanishes only in the isotropic case.

The key point about (34) is that its right-hand side may be written as a divergence of a vector of magnitude r^{-2},

$$\langle \theta\theta' \rangle_\infty = \pi^2 [2N_{//} - N_{\perp}] \, \nabla \cdot \left[\left(\frac{\partial}{\partial r_{//}} \frac{1}{r} \right) \hat{\mathbf{e}}_z \right]. \tag{35}$$

Thus, if we integrate $\langle \theta\theta' \rangle_\infty(\mathbf{r})$ over the volume bounded by two large concentric surfaces of the same shape but different sizes, the net result is zero. Moreover, it remains zero as the outer surface recedes to infinity. This is important because it means that L_θ is convergent, if only conditionally convergent (in the sense defined above). For example, in the case of a large cylindrical control volume of radius R and height $2H$, (30) and (31) suggest the following:

$$\int \langle \theta\theta' \rangle d\mathbf{r} = (2\pi)^3 \left[N_{//} + \frac{1}{2}N_{\perp} \right], \qquad (R/H \to 0, \text{ long cylinder}), \tag{36}$$

$$\int \langle \theta\theta' \rangle d\mathbf{r} = (2\pi)^3 N_{\perp}, \qquad (R/H \to \infty, \text{ thin disc}). \tag{37}$$

In summary, in the isotropic case of $N_{//} = \frac{1}{2}N_{\perp}$, there are no long-range correlations of the form $\langle \theta\theta' \rangle_\infty = O(r^{-3})$, and so L_θ is absolutely convergent and given by either (30) or (31):

$$L_\theta = \int \langle \theta\theta' \rangle d\mathbf{r} = (2\pi)^3 \left[N_{//} + \frac{1}{2}N_{\perp} \right] = (2\pi)^3 N_{\perp}. \tag{38}$$

However, in the anisotropic case of $N_{//} \neq \frac{1}{2}N_{\perp}$, we have $\langle \theta\theta' \rangle_\infty = O(r^{-3})$ and L_θ is only conditionally convergent, and its value depends on the shape of V.

4. An Invariant for Passive Scalar Mixing in Axisymmetric Saffman Turbulence

Crucially, integral L_θ is an invariant of freely decaying Saffman turbulence. This may be shown as follows. From (26), we have the two-point evolution equation:

$$\frac{\partial}{\partial t}\langle\theta\theta'\rangle = -\frac{\partial}{\partial r_i}\langle(\delta u_i)\theta\theta'\rangle + 2\alpha\nabla^2\langle\theta\theta'\rangle, \tag{39}$$

where $\delta\mathbf{u} = \mathbf{u}' - \mathbf{u}$ is a velocity increment. By integrating over a volume V in $\mathbf{r}$-space, and noting that $\langle\theta\theta'\rangle_\infty \leq O(r^{-3})$, we obtain

$$\frac{d}{dt}L_\theta = -\oint_{S_\infty}\langle\delta\mathbf{u}\theta\theta'\rangle\cdot d\mathbf{S}, \tag{40}$$

where S_∞ is the surface of V as that surface recedes to infinity. However, in Saffman turbulence, the long-range correlations are no stronger than order r^{-3}, and so $\langle\delta\mathbf{u}\theta\theta'\rangle_\infty \leq O(r^{-3})$. The surface integral in (40) now vanishes, yielding

$$L_\theta = \lim_{V\to\infty}\int_V\langle\theta\theta'\rangle d\mathbf{r} = \text{constant}. \tag{41}$$

The physical interpretation of (40) and (41) is clear. Homogeneity allows us to rewrite L_θ as

$$L_\theta = \lim_{V\to\infty}\int_V\langle\theta\theta'\rangle d\mathbf{r} = \lim_{V\to\infty}\frac{1}{V}\left\langle\left[\int_V\theta dV\right]^2\right\rangle, \tag{42}$$

and so L_θ is conserved if the scalar flux across the bounding surface of V, say S, is less than order $V^{1/2}$ (or equivalently $S^{3/4}$) as $V\to\infty$. However, the central limit theorem suggests that this flux should be no larger that $S^{1/2}$, consistent with L_θ being an invariant. In fact, (40) is nothing more than a concise form of the scalar budget equation

$$\frac{d}{dt}L_\theta = \frac{d}{dt}\frac{1}{V}\left\langle\left[\int_V\theta dV\right]^2\right\rangle = -\frac{2}{V}\left\langle\int_V\theta' dV'\oint_S\theta\mathbf{u}\cdot d\mathbf{S}\right\rangle = -\frac{2}{V}\int_V\left\langle\oint_S\theta'\theta\mathbf{u}\cdot d\mathbf{S}\right\rangle dV', \tag{43}$$

and the vanishing of the surface integral in (40) amounts to the statement that there is negligible convection of the passive scalar in or out of V in the limit of $V\to\infty$.

Perhaps some comments are in order at this point. First, since $N_{//}$ and $N_\perp$ are related to L_θ through (30) and (31), we would expect the invariance of L_θ to translate to the statement that $N_{//}$ and $N_\perp$ are both invariants. This is indeed the case (Yoshimatsu and Kaneda, 2018 [14]). Second, if $\ell^\theta_{//}$ and $\ell^\theta_\perp$ are integral scales for the passive scalar, say

$$\ell^\theta_{//} = \frac{1}{\langle\theta^2\rangle}\int\left\langle\theta(\mathbf{x})\cdot\theta(\mathbf{x}+r\hat{\mathbf{e}}_z)\right\rangle dr, \tag{44}$$

$$\ell^\theta_\perp = \frac{1}{\langle\theta^2\rangle}\int\left\langle\theta(\mathbf{x})\cdot\theta(\mathbf{x}+r\hat{\mathbf{e}}_x)\right\rangle dr, \tag{45}$$

then, since L_θ, $N_{//}$, and $N_\perp$ are all dominated by large scales, we have

$$L_\theta \sim N_{//} \sim N_\perp \sim \left\langle\theta^2\right\rangle\ell^\theta_{//}\left(\ell^\theta_\perp\right)^2. \tag{46}$$

Moreover, when the *large scales* in the passive scalar field evolve in a self-similar way, i.e., when $\langle\theta\theta'\rangle/\langle\theta^2\rangle$ is a function of $r_{//}/\ell^\theta_{//}$ and $\mathbf{r}_\perp/\ell^\theta_\perp$ only, then the invariance of $N_{//}$ and $N_\perp$ demands

$$\left\langle\theta^2\right\rangle\ell^\theta_{//}\left(\ell^\theta_\perp\right)^2 = \text{constant}, \tag{47}$$

which places a strong constraint on the evolution of the scalar variance $\langle\theta^2\rangle$.

Third, the constraint (47) is similar to that imposed on the velocity field as a result of (15)→(17). That is to say, $M_{//}$ and $M_\perp$ are dominated by large scales and so

$$M_{//} \sim \left\langle \mathbf{u}_{//}^2 \right\rangle \ell_{//}^u (\ell_\perp^u)^2, \; M_\perp \sim \left\langle \mathbf{u}_\perp^2 \right\rangle \ell_{//}^u (\ell_\perp^u)^2, \tag{48}$$

where $\ell_{//}^u$ and $\ell_\perp^u$ are suitable integral scales for the velocity field (see Section 5). Moreover, when the large-scale velocity correlations are self-similar, as is typical of fully developed turbulence, (15) →(17) demand

$$\left\langle \mathbf{u}_{//}^2 \right\rangle \ell_{//}^u (\ell_\perp^u)^2 = \text{constant}, \; \left\langle \mathbf{u}_\perp^2 \right\rangle \ell_{//}^u (\ell_\perp^u)^2 = \text{constant}. \tag{49}$$

The existence of invariants $M_{//}$ and $M_\perp$ (or, equivalently, $L_{//}$ and $L_\perp$), as well as the associated conservation relationships (49), is confirmed in DOK12.

Finally, we combine $\langle \delta \mathbf{u} \theta \theta' \rangle_\infty \leq O(r^{-3})$ with (39). This tells us that, if $\langle \theta\theta' \rangle_\infty = O(r^{-3})$ at $t = 0$, then we have $\langle \theta\theta' \rangle_\infty = O(r^{-3})$ for all $t > 0$. Thus, the initial Saffman-like spectrum $E_\theta(k \to 0) \sim k^2$ is preserved for all time, as noted above. Indeed, this has to be the case since

$$E_\theta(k \to 0) = \frac{4}{3}\pi N_{\alpha\alpha} k^2, \tag{50}$$

and $N_{ii} = N_{//} + N_\perp$ is an invariant. Conversely, if the initial spectrum takes an alternative form, say $E_\theta(k \to 0) \sim k^4$, then the Saffman spectrum $E_\theta(k \to 0) \sim k^2$ remains kinematically inaccessible to the passive scalar.

5. The Predicted Decay Rate of Energy and Scalar Variance

We shall now use the invariants $M_{//}$, $M_\perp$, $N_{//}$, and $N_\perp$, or equivalently $L_{//}$, $L_\perp$, and L_θ, combined with the assumption of self-similarity of large scales, to predict the evolution of kinetic energy and scalar variance in freely evolving turbulence. The approach is the same as that used by Saffman [12] and DOK12, where the invariance of $M_{//}$ and $M_\perp$ is used to predict the decay rate of $\left\langle \mathbf{u}_{//}^2 \right\rangle$ and $\langle \mathbf{u}_\perp^2 \rangle$.

5.1. The Rate of Energy Decay

Before discussing the decay of scalar variance, we should note how the kinetic energy and integral length scales $\ell_{//}^u$ and $\ell_\perp^u$ evolve, as the passive scalar is slave to the velocity field.

Let us start by introducing some notation. We define the velocity integral scales as follows:

$$\ell_{//}^u = \frac{1}{u_{//}^2} \int \left\langle u_{//}(\mathbf{x}) u_{//}(\mathbf{x} + r\hat{\mathbf{e}}_z) \right\rangle dr, \; u_{//} = \left\langle \mathbf{u}_{//}^2 \right\rangle^{1/2}, \tag{51}$$

$$\ell_\perp^u = \frac{1}{2u_\perp^2} \left\{ \int \left\langle u_x(\mathbf{x}) u_x(\mathbf{x} + r\hat{\mathbf{e}}_x) \right\rangle dr + \int \left\langle u_y(\mathbf{x}) u_y(\mathbf{x} + r\hat{\mathbf{e}}_y) \right\rangle dr \right\}, \; u_\perp = \left\langle \frac{1}{2}\mathbf{u}_\perp^2 \right\rangle^{1/2}. \tag{52}$$

Next, we assume that, once the turbulence is fully developed, the large scales in the velocity field (scales of order $\ell_{//}^u$ and $\ell_\perp^u$) are self-similar, i.e., $\langle u_{//} u'_{//} \rangle / u_{//}^2$ and $\langle \mathbf{u}_\perp \cdot \mathbf{u}'_\perp \rangle / u_\perp^2$ are functions of $r_{//}/\ell_{//}^u$ and $\mathbf{r}_\perp/\ell_\perp^u$ only. This is certainly observed in the numerical experiments of DOK12. We shall refer to this as *partial self-similarity*, as it does not extend down to Kolmogorov scales, nor perhaps to the far-field correlations $\langle u_i u'_j \rangle_\infty$ given by (5). Now, $M_{//}$ and $M_\perp$ are absolutely convergent and so are independent of the far-field contribution $\langle u_i u'_j \rangle_\infty$. (The long-wavelength oscillation associated with $e^{-j\mathbf{k}_\perp \cdot \mathbf{r}_\perp}$ and $e^{-j\mathbf{k}_{//} \cdot \mathbf{r}_{//}}$ kills off the far-field contributions to these integrals, leaving (15)→(17) well behaved.) It follows that partial self-similarity may be applied to (15)→(17), which then demands that, in fully developed turbulence,

$$M_\perp = c_\perp u_\perp^2 (\ell_\perp^u)^2 \ell_{//}^u = \text{constant}, \; M_{//} = c_{//} u_{//}^2 (\ell_\perp^u)^2 \ell_{//}^u = \text{constant}, \tag{53}$$

where $c_\perp$ and $c_{//}$ are dimensionless constants. In such a situation, we have

$$u_\perp^2/u_{//}^2 = \text{constant}. \tag{54}$$

This suggests that, rather surprisingly, any velocity anisotropy which is present at the beginning of the fully developed stage will be locked into the turbulence thereafter. The persistence of anisotropy in the form of (54) is confirmed in the numerical simulations of DOK12.

We now estimate the decay rate of energy under the assumption that partial self-similarity holds and that, as usual, the flux of energy to small scales is controlled by large scales. Since $u_{//}^2/u_\perp^2 = \text{constant}$, we only need to consider one component of energy, say $u_\perp^2$, and dimensional analysis then gives us

$$\frac{du_\perp^2}{dt} = -A\frac{u_\perp^3}{\ell_\perp^u}, \qquad A = A\left(\ell_{//}^u/\ell_\perp^u\right), \quad \text{[fully developed turbulence only]} \tag{55}$$

where A is a dimensionless function of $\ell_{//}^u/\ell_\perp^u$. Now, certain types of axisymmetric turbulence satisfy $u_\perp/\ell_\perp^u \sim u_{//}/\ell_{//}^u$ as a result of continuity. In such cases, the constraint $u_{//}^2/u_\perp^2 = \text{constant}$ demands that $\ell_{//}^u/\ell_\perp^u = \text{constant}$, and this is certainly observed in the fully developed state in the numerical experiments of DOK12. Equations (53) and (55) then simplify to

$$u_\perp^2(\ell_\perp^u)^3 = \text{constant}, \quad \frac{du_\perp^2}{dt} = -A\frac{u_\perp^3}{\ell_\perp^u}, \tag{56}$$

which results in the following decay laws:

$$u_\perp^2 \sim u_{//}^2 \sim t^{-6/5}, \ \ell_\perp^u \sim \ell_{//}^u \sim t^{2/5}, \quad \text{[fully developed turbulence only]} \tag{57}$$

as suggested by Saffman [12], and as observed in the direct numerical simulations of DOK12.

5.2. The Decay of Scalar Variance

Given (57), we can now predict the decay rate of scalar variance in fully developed turbulence. The analysis differs somewhat depending on whether $\ell_{//}^u$ and $\ell_\perp^u$ are greater or smaller than the scalar integral scales $\ell_{//}^\theta$ and $\ell_\perp^\theta$. Consider first the case where the initial conditions have $\ell_{\min}^\theta >> \ell_{\max}^u$. (Here, the subscripts indicate whether the minimum or maximum of the two possible length scales is taken.) Then, mixing occurs on the scale of $\ell_{\max}^u$, and dimensional analysis, combined with $u_{//}^2/u_\perp^2 = \text{constant}$ and $\ell_{//}^u/\ell_\perp^u = \text{constant}$, gives us

$$\frac{d\langle\theta^2\rangle}{dt} = -C\frac{u_\perp}{\ell_\perp^u}\langle\theta^2\rangle, \tag{58}$$

where C is a constant. Given (57), and the fact that mixing occurs on a scale much smaller than $\ell_{\min}^\theta$, we expect $\langle\theta^2\rangle$ to decay as a power law, while $\ell_{//}^\theta$ and $\ell_\perp^\theta$ will be more or less constant, and this will continue for as long as $\ell_{\min}^\theta >> \ell_{\max}^u$. Note that the scalar field is not self-similar during this stage, so that (47) does not apply. Eventually, $\ell_{\max}^u$ will grow to be of the same order as $\ell_{\min}^\theta$, and thereafter $\ell_{//}^\theta$ and $\ell_\perp^\theta$ will grow along with $\ell_{//}^u$ and $\ell_\perp^u$, as $t^{2/5}$. If the scalar spectrum exhibits partial self-similarity during this growth, then invariant (47) combined with (57) yields

$$\langle\theta^2\rangle \sim t^{-6/5}, \ \ell_\perp^\theta \sim \ell_{//}^\theta \sim t^{2/5}. \quad \text{[fully developed turbulence, } \ell^\theta \sim \ell^u\text{]}. \tag{59}$$

Now consider the initial conditions in which $\ell^\theta_{\max} << \ell^u_{\min}$. In this case, mixing is no longer accomplished by the integral-scale eddies, but rather by inertial-range eddies of scale $\ell^\theta_{\max}$. In such a situation, we have

$$\frac{d\langle\theta^2\rangle}{dt} \sim -C\frac{\langle\theta^2\rangle}{\tau_\theta}, \quad C = C\left(\ell^\theta_{//}/\ell^\theta_\perp\right), \tag{60}$$

where C is a dimensionless function of $\ell^\theta_{//}/\ell^\theta_\perp$ and τ_θ is the eddy turnover time of scales of order $\ell^\theta_{\max}$. Now, τ_θ can be estimated from Kolmogorov's 2/3 law as $\left(\ell^\theta_{\max}\right)^{2/3}/\varepsilon^{1/3}$. (Here, ε is the rate of dissipation of kinetic energy.) Since the anisotropy of the velocity integral scales is fixed in fully developed turbulence, i.e., $u^2_{//}/u^2_\perp$ = constant and $\ell^u_{//}/\ell^u_\perp$ = constant, we would expect the velocity anisotropy at scale $\ell^\theta_{\max}$ to also be constant. This in turn suggests that $\ell^\theta_{//}/\ell^\theta_\perp$ will be constant, or equivalently $\ell^\theta_{//} \sim \ell^\theta_\perp \sim \ell^\theta$, provided that the velocity anisotropy is not too large at scale $\ell^\theta_{\max}$. In such cases, (47) and (60) simplify to

$$\frac{d\langle\theta^2\rangle}{dt} \sim -C\frac{\varepsilon^{1/3}\langle\theta^2\rangle}{\left(\ell^\theta\right)^{2/3}}, \quad \left\langle\theta^2\right\rangle\left(\ell^\theta\right)^3 = \text{constant}, \tag{61}$$

from which

$$\frac{d\langle\theta^2\rangle}{dt} \sim -\varepsilon^{1/3}\left\langle\theta^2\right\rangle^{11/9}, \tag{62}$$

where (57) has $\varepsilon \sim t^{-11/5}$. Once again, we conclude that

$$\left\langle\theta^2\right\rangle \sim t^{-6/5},\ \ell^\theta_\perp \sim \ell^\theta_{//} \sim t^{2/5}. \quad \text{[fully developed turbulence, } \ell^\theta << \ell^u\text{]} \tag{63}$$

In summary, we would expect the scalar variance to decay as $\left\langle\theta^2\right\rangle \sim t^{-6/5}$, and the scalar integral scales to grow as $\ell^\theta_\perp \sim \ell^\theta_{//} \sim t^{2/5}$, whenever the turbulence is fully developed and $\ell^\theta \leq O(\ell^u)$. These various predictions are confirmed in the direct numerical simulations of Yoshimatsu and Kaneda [7].

6. Discussion

Our understanding of the large-scale dynamics of homogeneous turbulence has come a long way since those early discussions at NCAR in 2002, and this author has greatly enjoyed the journey, particularly those collaborations with Jack Herring, Yukio Kaneda, and Per-Age Krogstad. We now have a reasonably complete understanding of large-scale dynamics, both in terms of kinetic energy and passive scalar mixing, and this is true of both Batchelor $E(k \to 0) \sim k^4$ and Saffman $E(k \to 0) \sim k^2$ turbulence.

As noted by Kolmogorov, in Batchelor turbulence, the kinetic energy should decay as $u^2 \sim t^{-10/7}$, provided that Loitsyansky's integral is conserved. This decay law was confirmed in the direct numerical simulations of Ishida, Davidson, and Kaneda [2], but only for fully developed turbulence. (Loitsyansky's integral is not conserved during an initial transient, before the turbulence becomes fully developed.) Similarly, for isotropic Saffman turbulence, the prediction of $u^2 \sim t^{-6/5}$ is observed in the wind tunnel experiments, as stipulated by Krogstad and Davidson [4]. Finally, in anisotropic (but axisymmetric) Saffman turbulence, the theoretical predictions of

$$u^2_\perp \sim u^2_{//} \sim t^{-6/5},\ \ell^u_\perp \sim \ell^u_{//} \sim t^{2/5},$$

are well supported by the direct numerical simulations of DOK12, while the estimates

$$\left\langle\theta^2\right\rangle \sim t^{-6/5},\ \ell^\theta_\perp \sim \ell^\theta_{//} \sim t^{2/5},$$

are consistent with the direct numerical simulations of Yoshimatsu and Kaneda [7], as discussed above.

Ironically, the one question which has proved to be most resistant to theoretical attack turns out to be the very subject of those early discussions with Jack Herring back in 2002, i.e., the observation that Loitsyansky's integral is effectively constant in fully developed Batchelor turbulence. The point is this: Batchelor and Proudman [13] noted that the long-range velocity correlations in $E(k \to 0) \sim k^4$ turbulence fall off sufficiently slowly with r so as to invalidate the usual proof that Loitsyansky's integral is an invariant. However, the direct numerical simulations show that, once the turbulence is fully developed, but not before, Loitsyansky's integral is indeed constant (Herring et al. [1]; Ishida, Davidson, and Kaneda, [2]). It seems that there is something particular about the structure of fully developed turbulence that suppresses the long-range interactions of Batchelor and Proudman [13], and we still do not understand what that might be. Evidently, there is still much to accomplish, and so this is a story without an ending.

7. Conclusions

We reviewed the integral invariant theory of Saffman turbulence, showing that the various statistical invariants, both the Saffman and Corrsin integrals, have a simple physical interpretation. We have also shown that, under a wide range of conditions, the integral scales of a passive scalar evolve as

$$\left\langle \theta^2 \right\rangle \sim t^{\ 6/5},\ \ell_\perp^\theta \sim \ell_{//}^\theta \sim t^{2/5}.$$

The primary point of novelty of our approach lies in its use of various integral invariants to deduce the evolution of integral scales.

Funding: This research received no external funding.

Institutional Review Board Statement: Not applicable.

Informed Consent Statement: Not applicable.

Acknowledgments: The author would like to thank the Leverhulme Trust for its generous support through the grant RPG-2015-195/RG77943, as well as Yukio Kaneda for their fruitful discussions on passive scalar mixing.

Conflicts of Interest: The author declares no conflict of interest.

References

1. Herring, J.R.; Kimura, Y.; James, R.; Davidson, P.A. Statistical and dynamical aspects of isotropic and stably stratified turbulence. In *Mathematical and Physical Theory of Turbulence*; Cannnon, J., Shivamoggi, B.K., Eds.; Chapman & Hall: London, UK, 2006.
2. Ishida, T.; Davidson, P.A.; Kaneda, Y. On the decay of isotropic turbulence. *J. Fluid Mech.* **2006**, *564*, 455–475. [CrossRef]
3. Davidson, P.A.; Okamoto, N.; Kaneda, Y. On freely-decaying, anisotropic, axisymmetric, Saffman turbulence. *J. Fluid Mech.* **2012**, *706*, 150–172. [CrossRef]
4. Krogstad, P.-A.; Davidson, P.A. Is grid turbulence Saffman turbulence? *J. Fluid Mech.* **2010**, *642*, 373–394. [CrossRef]
5. Davidson, P.A. On the Decay of Saffman Turbulence Subject to Rotation, Stratification or an Imposed Magnetic Field. *J. Fluid Mech.* **2010**, *633*, 268–292. [CrossRef]
6. Davidson, P.A. *Turbulence in Rotating, Stratified and Electrically Conducting Fluids*; Cambridge University Press: Cambridge, UK, 2013.
7. Yoshimatsu, K.; Kaneda, Y. Anisotropy freezing of passive scalar fields in anisotropy growing homogeneous turbulence. *Phys. Rev. Fluids* **2020**, *5*, 014604. [CrossRef]
8. Richardson, L.F. Atmospheric diffusion shown on a distance-neighbour graph. *Proc. R. Soc. A* **1926**, *110*, 709–737.
9. Monin, A.S.; Yaglom, A.M. *Statistical Fluid Mechanics II*; MIT Press: Cambridge, MA, USA, 1975.
10. Corrsin, S. The decay of isotropic temperature fluctuations in an isotropic turbulence. *J. Aeronaut. Sci.* **1951**, *18*, 417–423. [CrossRef]
11. Saffman, P.G. The large-scale structure of homogeneous turbulence. *J. Fluid Mech.* **1967**, *27*, 581–593. [CrossRef]
12. Saffman, P.G. Note on decay of homogenous turbulence. *Phys. Fluids* **1967**, *10*, 1349. [CrossRef]

13. Batchelor, G.K.; Proudman, I. The large-scale structure of homogeneous turbulence. *Philos. Trans. R. Soc. Lond. A* **1956**, *248*, 369–405.
14. Yoshimatsu, K.; Kaneda, Y. Large-scale structure of velocity and passive scalar fields in freely decaying homogeneous anisotropic turbulence. *Phys. Rev. Fluids* **2018**, *3*, 104601. [CrossRef]

atmosphere

Article

Statistics of the Inertial Energy Transfer Range in d-Dimensional Turbulence ($2 \le d \le 3$) in a Lagrangian Renormalized Approximation

Toshiyuki Gotoh [1,2,*,†] **and Yukio Kaneda** [3,†]

1 Department of Engineering Physics, Graduate School of Engineering, Nagoya Institute of Technology, Nagoya 466-8555, Japan
2 Research and Education Center for Natural Sciences, Keio University, Yokohama 223-8521, Japan
3 Graduate School of Mathematics, Nagoya University, Nagoya 464-8602, Japan; kaneda@math.nagoya-u.ac.jp
* Correspondence: gotoh.toshiyuki@nitech.ac.jp
† These authors contributed equally to this work.

Abstract: Statistics in the inertial energy transfer range (IETR) of d-dimensional turbulence ($2 \le d \le 3$) are studied using a Lagrangian renormalized approximation (LRA). The LRA suggests that the energy spectrum in the IETR is given by $K_d|\bar{\epsilon}|^{2/3}k^{-5/3}$, where K_d is a constant and $\bar{\epsilon}$ is the energy flux across wave-number k; the energy transfer is forward for $d_c < d \le 3$ but inverse for $2 \le d < d_c$, where $d_c \approx 2.065$; at $d = d_c$, K_d diverges and the skewness of the longitudinal velocity difference vanishes; and the d-dependence of the two-time Lagrangian velocity correlation spectra under appropriate normalization is weak in the IETR.

Keywords: d-dimensional turbulence; inertial energy transfer range; Lagrangian renormalized approximation

PACS: 47.27.Ak; 47.27.Jv; 47.27.Gs; 05.20.Jj

Citation: Gotoh, T.; Kaneda, Y. Statistics of the Inertial Energy Transfer Range in d-Dimensional Turbulence ($2 \le d \le 3$) in a Lagrangian Renormalized Approximation. *Atmosphere* **2023**, *14*, 1053. https://doi.org/10.3390/atmos14061053

Academic Editors: Boris Galperin, Annick Pouquet and Peter Sullivan

Received: 10 May 2023
Revised: 15 June 2023
Accepted: 16 June 2023
Published: 20 June 2023

1. Introduction

It was at an IUTAM Symposium held at Kyoto on 1983 when the second author (Y.K.) first met Jack Herring, while it was in 1990 at Boulder when the first author (T.G.) first met him. Since then, both authors have had chances to enjoy discussions with Jack, including during the workshops of the Geophysical Turbulence Program at the National Center for Atmospheric Research and his visits to Nagoya. The discussions covered various topics, but particularly statistical theories of turbulence. Our common interests included the spectral closure theories, in particular two-point two-time closure approximations. We were inspired by his works, especially those on spectral closure approaches. The discussions with him have been friendly and stimulating.

In this paper, we consider the application of a spectral closure approach to turbulence in the d space dimension with $2 \le d \le 3$. In practice, we use a spectral closure theory based on a view point of Lagrangian renormalization. The closure theory is called Lagrangian renormalized approximation (LRA) [1] here. The LRA is a two-point two-time Lagrangian closure approximation free from any adjustment parameter, and similar in this sense to the abridged Lagrangian-history direct-interaction approximation (ALHDIA) of Kraichnan [2] and strain-based ALHDIA of Kraichnan and Herring [3] and Herring and Kraichnan [4]. The equations of the LRA are simpler than those of the ALHDIA and strain-based ALHDIA.

In general, the dynamics of a fluid may be fundamentally different in different space dimension. For example, vortex stretching plays important dynamic roles in three, but not two, dimensions. It is not surprising that, given such observations, the statistics of turbulence may differ fundamentally in different space dimensions. Indeed, studies so far suggest that energy in the so-called inertial energy cascade range is transferred in opposite directions in two and three dimensions, i.e., from large to small eddies (scales) (forward) in three dimensions, but from small to large scales (backward or inverse) in two dimensions, if the turbulence has a sufficiently large Reynolds number Re. One may ask: What happens in spaces of noninteger dimension d between two and three dimensions?

It is difficult to realize turbulence in such noninteger dimension by experiments or numerical simulation, so that one might think that such a question is purely academic. However, the history of the statistical mechanics of critical phenomena suggests that understanding the physics of noninteger dimension may improve our understanding of physics in three (and two) dimensions, as noted by Fournier and Frisch [5] in a pioneering study of turbulence in noninteger dimension. The study of turbulence in noninteger dimension via a closure-theoretical approach may extend the field of the applications of spectral closures in which both Jack Herring and the authors were/are interested.

In this paper, we begin by considering the homogeneous and isotropic turbulences (HIT) of an incompressible fluid that obeys the Navier–Stokes (NS) equation in integer dimensions d with $d = 2, 3, 4 \cdots$. After reviewing certain exact relations in physical and wave-vector spaces of d-dimension in Section 2, we briefly review, in Section 3, the LRA equations for HIT in d-dimensions, following Gotoh et al. [6]. In the LRA equations, the parameter d representing the space dimension plays a key role. The dimension d is limited to integers in Sections 2 and 3. In Section 4, we take a bold step; we relax d to allow noninteger values. Assuming that the LRA equations are applicable to noninteger values d, we analyze (in Section 4) the statistics in the wavenumber range wherein the energy flux across wavenumber k is independent of k. Here, we refer to this range as the inertial energy transfer range (IETR). A particular attention is paid to statistics associated with Lagrangian two-point two-time velocity correlation and the skewness factor of the longitudinal velocity increment in the ITER; two-time statistics are, in general, not captured in single time closures such as the eddy-damped quasinormal Markovian (EDQNM) approximation [7] used in [5,8].

2. Exact Relations

2.1. Navier–Stokes Equation in d-Dimension with $d = 2, 3, 4 \cdots$

We begin with the NS equation for an incompressible fluid

$$\frac{\partial u_i}{\partial t} + u_j \frac{\partial u_i}{\partial x_j} = -\frac{\partial p}{\partial x_i} + \nu \Delta u_i + f_i^{ext}, \tag{1}$$

and the incompressibility condition

$$\frac{\partial u_i}{\partial x_i} = 0, \tag{2}$$

in d-dimensional space ($d = 2, 3, 4 \cdots$) where u_i, x_i and f_i^{ext} are the components of the velocity $\boldsymbol{u}$, the position $\boldsymbol{x}$, and the external force in the i-th Cartesian direction ($i = 1, 2, 3, \cdots, d$), respectively, t is the time, p is the pressure divided by the constant fluid density, and ν is the kinematic viscosity. Throughout, we employ the summation convention for repeated indices.

Multiplying Equation (1) by u_i and taking the summation over i gives

$$\begin{aligned} &\frac{\partial}{\partial t}\frac{1}{2}u_i^2 + \frac{\partial}{\partial x_j}\left[\left(\frac{1}{2}u_i^2 + p\right)u_j\right] \\ &= \nu\frac{\partial}{\partial x_j}\left(u_i\frac{\partial u_i}{\partial x_j}\right) - \nu\frac{\partial u_j}{\partial x_i}\frac{\partial u_j}{\partial x_i} + u_i f_i^{ext}, \end{aligned} \tag{3}$$

under the incompressibility condition Equation (2). Integrating Equation (1) over the fluid domain V, one obtains

$$\frac{\partial}{\partial t}\int_V \frac{1}{2}u_i^2(x)dx = -\int_V [\varepsilon(x) - \varepsilon_f(x)]dx, \tag{4}$$

where the energy dissipation rate $\varepsilon(x)$ and energy injection rate $\varepsilon_f(x)$ at position x are defined by

$$\varepsilon(x) = \nu\frac{\partial u_j}{\partial x_i}\frac{\partial u_j}{\partial x_i}, \tag{5}$$

$$\varepsilon_f(x) = u_i(x)f_i^{ext}(x), \tag{6}$$

and the integration over the boundary surface of the domain V is assumed to be negligible. This assumption can be justified under the assumption of periodic boundary conditions, for example.

2.2. Spectral Relations

In homogeneous and isotropic turbulence (HIT), the single-time two-point velocity correlation $\langle u_i(x,t)u_j(x',t)\rangle$depends on x and x' only via $r \equiv x - x'$, and its Fourier transform $Q_{ij}(k,t)$ with respect to r may be written in the form

$$Q_{ij}(k,t) = \frac{1}{d-1}P_{ij}(k)Q(k,t), \tag{7}$$

where $Q(k,t)$ depends on k only via $k \equiv |k|$ and $P_{ij}(k) \equiv \delta_{ij} - k_ik_j/k^2$. The Fourier transform $\hat{u}(k)$ of $u(r)$ with respect to r is related to $u(r)$ by

$$u(r) = \int \hat{u}(k)e^{ik\cdot r}d^dk, \tag{8}$$

in which the integration is over the entire k-space. The total energy $E(t)$ is given by

$$E(t) = \frac{\langle u^2(t)\rangle}{2} = \int_0^\infty E(k,t)\,dk, \tag{9}$$

where the energy spectral density $E(k,t)$ is defined as

$$E(k,t) = \frac{1}{2}S_dk^{d-1}Q(k,t), \quad S_d = \frac{2\pi^{d/2}}{\Gamma(d/2)}, \tag{10}$$

and S_d is the surface area of a d-dimensional sphere of unit radius. The brackets $\langle\ \rangle$ indicate an appropriate average such as the ensemble or space average.

According to the NS equation, Equation (1), the energy spectrum $E(k,t)$ obeys

$$\left(\frac{\partial}{\partial t} + 2\nu k^2\right)E(k,t) = T(k,t) + W(k,t), \tag{11}$$

where $W(k,t)$ is the energy injected by the external force f_i^{ext}. $T(k,t)$ is the so-called energy transfer function that represents the energy transfer attributable to nonlinear triad interactions in the wave-vector space. The energy conservation by the nonlinear term in the NS equation implies that

$$\int_0^\infty T(k,t)dk = 0. \tag{12}$$

The so-called energy flux across k is defined by

$$\Pi(k,t) = \int_k^\infty T(k',t)dk' = -\int_0^k T(k',t)dk'. \tag{13}$$

The average energy dissipation rate per unit mass is given by

$$\langle \varepsilon \rangle = 2\nu \int_0^\infty k^2 E(k) dk, \tag{14}$$

and the average energy injection rate per unit mass is given by

$$\left\langle \varepsilon_f \right\rangle = \int_0^\infty W(k) dk. \tag{15}$$

Here and hereinafter, we omit the time-arguments at will.

2.3. Statistics in the Inertial Energy Transfer Ranges of Wave-Vector and Physical Spaces

Equations (11) and (13) imply that

$$\Pi(k,t) = \int_k^\infty \left[\frac{\partial}{\partial t} E(k',t) + 2\nu k^2 E(k',t) - W(k',t) \right] dk'. \tag{16}$$

Suppose that Re is very large, so that $T(k',t)$, i.e., the sum of the three terms in the integrand of Equation (16) is negligibly small (in an appropriate sense) over a certain range of k', say $k_{\mathcal{L}} \ll k' \ll k_\ell$. Equation (16) then implies that the energy flux $\Pi(k)$ is almost constant independent of k, and we may therefore approximate it as

$$\Pi(k) = \bar{\varepsilon}, \tag{17}$$

over the range wherein we have set the constant to $\bar{\varepsilon}$. We assume here that the time dependence of $\bar{\varepsilon}$ is negligible.

As shown in Appendix A, if such a range k is sufficiently wide, we obtain the approximation

$$D_{LLL}(r) = -\frac{12\bar{\varepsilon}}{d(d+2)} r, \tag{18}$$

over the corresponding range of r, i.e., $\ell \equiv (1/k_\ell) \ll r \ll \mathcal{L}\ (\equiv 1/k_{\mathcal{L}})$, where $D_{LLL} = \langle u_r^3 \rangle$ is the third-order structure function of the longitudinal velocity increment $u_r = (\boldsymbol{r}/r) \cdot (\boldsymbol{u}(\boldsymbol{x}+\boldsymbol{r}) - \boldsymbol{u}(\boldsymbol{x}))$. We call this range inertial energy transfer range (IETR), and the wave-number range $k_{\mathcal{L}} \ll k \ll k_\ell$ is called the IET wave-number range (simply, IETR). The relation (18) with $\bar{\varepsilon} = \langle \varepsilon \rangle$ can be derived from a generalization of the Kármán–Howarth–Kolmogorov (KHK) equation for $d = 3$ to any arbitrary integer $d > 0$ for freely decaying turbulence with $f_i^{ext} = 0$ [9–11].

For freely decaying turbulence in three dimensions ($d = 3$), Equation (18) with $\bar{\varepsilon} = \langle \varepsilon \rangle$ gives Kolmogorov's 4/5-th law [12],

$$D_{LLL}(r) = -\frac{4}{5} \bar{\varepsilon} r. \tag{19}$$

Kolmogorov noted that under the assumption that the skewness S of u_r defined by

$$S \equiv \frac{D_{LLL}(r)}{[D_{LL}(r)]^{3/2}}, \tag{20}$$

is constant, Equation (19) gives

$$D_{LL}(r) = C_3 (\bar{\varepsilon} r)^{2/3}, \tag{21}$$

where $D_{LL} \equiv \langle u_r^2 \rangle$, $C_3 = [-4/(5S)]^{2/3}$ is a nondimensional constant, and we have used $\bar{\varepsilon} > 0$. A similar statement holds true for $d \neq 3$, i.e., if the skewness factor is constant, then Equation (18) gives

$$D_{LL}(r) = C_d (|\bar{\varepsilon}| r)^{2/3}, \tag{22}$$

where C_d is a nondimensional constant, but may depend on d. Here, we use $|\bar{\varepsilon}|$ instead of $\bar{\varepsilon}$ to allow the possibility that $\bar{\varepsilon}$ may be negative.

In spectral space, Equation (21) gives

$$E(k) = K_d |\bar{\varepsilon}|^{2/3} k^{-5/3}, \tag{23}$$

provided that the IETR is sufficiently wide, where K_d is a nondimensional constant that may depend on d. Straightforward algebra shows that C_d is related to K_d by

$$C_d = 2.56\Gamma(d/2)K_d/\Gamma(d/2+4/3), \tag{24}$$

where Γ is the gamma function [6]. Equations (18), (20), (22) and (24) then give

$$S^{(d)} = -\frac{12}{d(d+2)}\frac{\bar{\varepsilon}}{|\bar{\varepsilon}|}\left(\frac{\Gamma(d/2+4/3)}{2.56\Gamma(d/2)}\right)^{3/2} K_d^{-3/2}, \tag{25}$$

where we have replaced S in Equation (20) by $S^{(d)}$.

3. LRA Equations in Integer Dimension *d*

3.1. LRA Equations

One of the main interests in the study of turbulence is how to theoretically predict the transfer $T(k,t)$. Extensive studies have been made on this problem. Readers may refer to, for example, the review by Zhou [13] for a recent and comprehensive review of the studies.

The LRA is one of the closure theories that give approximations for the transfer $T(k,t)$ [1]. Let $Q_{ij}(\boldsymbol{x},\boldsymbol{x}',t,s)$ be the two-time two-point Lagrangian velocity correlation defined by

$$Q_{ij}(\boldsymbol{x},\boldsymbol{x}',t,s) \equiv \langle v_i(\boldsymbol{x},s|t)v_j(\boldsymbol{x}',s|s)\rangle, \quad (t \geq s) \tag{26}$$

and $G_{ij}(\boldsymbol{x},\boldsymbol{x}',t,s)$ is the Lagrangian response function defined by

$$G_{ij}(\boldsymbol{x},\boldsymbol{x}',t,s) \equiv \langle \delta v_i(\boldsymbol{x},s|t)/\delta f_j(\boldsymbol{x}',s)\rangle, \quad (t \geq s) \tag{27}$$

where $\boldsymbol{v}(\boldsymbol{x},s|t)$ is the generalized velocity defined as the fluid velocity at time t of the fluid particle that was at position $\boldsymbol{x}$ at time s, $\delta f_i(\boldsymbol{x},t)$ is an infinitesimal force term added to the right-hand side of Equation (1), and δ/δ denotes functional differentiation. Since $\boldsymbol{v}(\boldsymbol{x},t|t)$ is just the Eulerian velocity $\boldsymbol{u}(\boldsymbol{x},t)$, we have $Q_{ij}(\boldsymbol{x},\boldsymbol{x}',t,t) = \langle u_i(\boldsymbol{x},t)u_j(\boldsymbol{x}',t)\rangle$.

In statistically homogeneous turbulence, $Q_{ij}(\boldsymbol{x},\boldsymbol{x}',t,s)$ and $G_{ij}(\boldsymbol{x},\boldsymbol{x}',t,s)$ depend on $\boldsymbol{x}$ and $\boldsymbol{x}'$ only via $\boldsymbol{r} \equiv \boldsymbol{x}-\boldsymbol{x}'$. Let the Fourier transforms with respect to $\boldsymbol{r}$ of $Q_{ij}(\boldsymbol{x},\boldsymbol{x}',t,s)$ and $G_{ij}(\boldsymbol{x},\boldsymbol{x}',t,s)$ after an appropriate normalization be, respectively, $Q_{ij}(\boldsymbol{k},t,s)$ and $G_{ij}(\boldsymbol{k},t,s)$. The LRA gives a closed set of equations for $P_{im}(\boldsymbol{k})Q_{mj}(\boldsymbol{k},t,s)$ and $P_{im}(\boldsymbol{k})G_{mj}(\boldsymbol{k},t,s)$. (Since the pressure in Equation (1) kills the compressible part of δf_i, it is shown that $P_{im}(k)G_{mj}(k,t,s) = P_{im}(k)G_{mn}(k,t,s)P_{nj}(k)$. The latter expression instead of the former was used in Kaneda (1981)). In HIT, they may be written in the form

$$P_{im}(\boldsymbol{k})Q_{mj}(\boldsymbol{k},t,s) = \frac{1}{d-1}P_{ij}(\boldsymbol{k})Q(k,t,s), \tag{28}$$

$$P_{im}(\boldsymbol{k})G_{mj}(\boldsymbol{k},t,s) = P_{ij}(\boldsymbol{k})G(k,t,s), \tag{29}$$

where $Q(k,t,s)$ and $G(k,t,s)$ depend on $\boldsymbol{k}$ only via k. Since $\boldsymbol{v}(\boldsymbol{x},t|t) = \boldsymbol{u}(\boldsymbol{x},t)$, as noted above, and the fluid is incompressible, we have $P_{im}(\boldsymbol{k})Q_{mj}(\boldsymbol{k},t,t) = Q_{ij}(\boldsymbol{k},t,t) = Q_{ij}(\boldsymbol{k},t)$, where $Q_{ij}(\boldsymbol{k},t)$ is the Fourier transform of $\langle u_i(\boldsymbol{x},t)u_j(\boldsymbol{x}',t)\rangle$ with respect to $\boldsymbol{r} \equiv \boldsymbol{x}-\boldsymbol{x}'$. For $t = s$, Equation (28) with $Q(k,t,t) = Q(k,t)$ therefore reduces to Equation (7). The same closure equations as those in the LRA were derived by Kida and Goto [14] for $d = 3$.

We quote here the LRA equations for $T(k,t)$, $Q(k,t,s)$ and $G(k,t,s)$ in integer dimension $d \geq 2$ from Gotoh et al. [6] as follows. Readers may refer to their paper for the details of the derivation, and to Kaneda [15] for the basic ideas and properties of the LRA.

The single time correlation $Q(k,t) \equiv Q(k,t,t)$ obeys

$$\left(\frac{\partial}{\partial t} + 2\nu k^2\right) Q(k,t) = T_Q(k,t) + W_Q(k,t), \tag{30}$$

where

$$T_Q(k,t) = 2N_d k^{4-d} \iint_\Delta dpdq\ (pq)^{d-2}(1-x^2)^{\frac{d-3}{2}} b_{kpq}^{(d)} \\ \times \int_{-\infty}^{t} G_d(k,t,s)G_d(p,t,s)G_d(q,t,s)Q(q,s)[Q(p,s)-Q(k,s)]\, ds, \tag{31}$$

$$W_Q(k,t) = \frac{1}{S_d k^{d-1}} W(k.t), \tag{32}$$

and

$$b_{kpq}^{(d)} = d\left(\frac{p}{k}\right)^2(1-x^2) + \frac{p}{k}(2z^3 - 3z - xy), \tag{33}$$

$$N_d = \frac{S_{d-1}}{(d-1)^2}, \tag{34}$$

$\iint_\Delta dpdq$ denotes the integral over the domain of p, q such that (k, p, q) can be the length of the sides of a triangle, and (x, y, z) denotes the cosines of the interior angles opposite the triangle sides (k, p, q), respectively. Here, we assume that the time t is large enough, so that we may let the lower bound of the time integral in Equation (31) be $-\infty$.

The Lagrangian two-time velocity correlation $Q(k,t,s)$ satisfies the fluctuation–dissipation relation

$$Q(k,t,s) = G_d(k,t,s)Q(k,s,s), \qquad t \geq s \tag{35}$$

and the Lagrangian response function $G_d(k,t,s)$ satisfies $G_d(k,s,s) = 1$ and

$$\left(\frac{\partial}{\partial t} + \nu k^2 + \mu_d(k,t,s)\right) G_d(k,t,s) = 0, \qquad t \geq s \tag{36}$$

where

$$\mu_d(k,t,s) = \int_0^\infty dp\ kp\ J_d\left(\frac{p}{k}\right) \int_s^t G_d(p,t,s')E(p,s')ds', \tag{37}$$

$$J_d(x) = M_d \hat{J}_d(x), \tag{38}$$

$$\hat{J}_d(x) = \hat{J}_d(1/x) = \begin{cases} xF(1,-d/2,d/2+2,x^2), & x \leq 1 \\ \frac{1}{x}F(1,-d/2,d/2+2,1/x^2), & x > 1 \end{cases} \tag{39}$$

$$M_d = \frac{4S_{d-1}}{(d-1)^2 S_d} B\left(\frac{d+3}{2}, \frac{1}{2}\right), \tag{40}$$

F is Gauss' hypergeometric function and B is the beta function. Here, we omit terms that disappear when $f_i^{ext} = 0$.

According to Equation (31), the energy flux is given by

$$\Pi_d(k,t) = S_d N_d \int_k^\infty dk' \iint_\Delta dpdq\ k'^3 (pq)^{d-2}(1-x^2)^{\frac{d-3}{2}} b_{k'pq}^{(d)} \\ \times \int_{-\infty}^{t} G_d(k',t,s)G_d(p,t,s)G_d(q,t,s)Q(q,s)[Q(p,s)-Q(k,s)]\, ds, \tag{41}$$

in the LRA. It can be confirmed that, as in three dimensions, the detailed balance and the energy conservation by the nonlinear term in the NS equation hold in the LRA in any integer dimensions ($\geq$2).

3.2. LRA Equations in the IETR

A simple dimensional analysis using the idea of Kolmogorov's similarity theory [16] suggests that $E(k)$ and $G_d(k,t,s)$ in the IETR, where Equation (17) holds, are given by the following similarity forms:

$$\begin{aligned} &E(k) = K_d|\bar{\epsilon}|^{2/3}k^{-5/3}, \quad G_d(k,t,s) = G_d(k,\tau) = g_d(\sigma), \\ &\tau = t - s, \quad \sigma = \gamma|\bar{\epsilon}|^{1/3}k^{2/3}\tau, \quad \gamma = \sqrt{M_d K_d}, \end{aligned} \tag{42}$$

where K_d is the Kolmogorov constant in d-dimension. Here and hereafter we assume the statistical stationarity in the IETR.

The LRA equations are confirmed to be consistent with the similarity forms in Equation (42). Upon substitution of the similarity forms (42) of Equation (10) into Equation (37), the integral converges appropriately at small and large wave-number limits. Via substitution, one can reduce Equation (36) to

$$\frac{dg_d(\sigma)}{d\sigma} + \left(\int_0^\infty dx\, x^{-2/3}\hat{f}_d(x) \int_0^\sigma d\xi\; g_d(x^{2/3}\xi)\right) g_d(\sigma) = 0. \tag{43}$$

Similarly, by substituting the similarity forms, one can reduce the expression Equation (41) for the energy flux in the IETR to

$$\Pi_d(k) = |\bar{\epsilon}| K_d^{3/2} \frac{4d\, S_{d-1}}{(d-1)^2 S_d \sqrt{M_d}} A_d, \tag{44}$$

where

$$A_d = \int_0^1 du \log(1/u) \int_{max(u,|1-u|)}^{u+1} dv\, \Theta_d(1,u,v)\bar{D}_d(1,u,v), \tag{45}$$

$$\bar{D}_d(1,u,v) = (1-x^2)^{\frac{d-3}{2}} u^{-8/3}\left[(\bar{b}^{(d)}_{1uv} + \bar{b}^{(d)}_{1vu})v^{-8/3} - (\bar{b}^{(d)}_{1uv}u^{d+2/3}v^{-8/3} + \bar{b}^{(d)}_{1vu}v^{d-2})\right], \tag{46}$$

$$b^{(d)}_{1uv} = d\,\bar{b}^{(d)}_{1uv} = d\left(u^2(1-x^2) + \frac{u}{d}(2z^3 - 3z - xy)\right), \tag{47}$$

and Θ_d is the triple relaxation time defined by

$$\Theta_d(1,u,v) = \int_0^\infty g_d(s) g_d(u^{2/3}s) g_d(v^{2/3}s)\; ds. \tag{48}$$

4. Statistics in the IETR for $2 \leq d \leq 3$

Up to this point, we have assumed that the space dimension d is an integer, and we derived the relations in Sections 2 and 3. We now take the bold step of assuming that we can continue to use Equation (25) and the LRA equations in a noninteger dimension d with $2 < d < 3$. If so, we can solve Equation (43) to obtain g_d numerically for any noninteger dimension d, in $2 < d < 3$, in the same way as in Refs. [17–19]. Note that Expression (9) in Ref. [18] for a damping factor η is too small by a factor of 2; thus, some factors must be corrected. For example, C (K_2 in the present notation) is too small by a factor of $2^{1/3}$, and, thus, should be $C = 7.41$.

Figure 1a shows the response function $G_d(k, t-s)$ plotted against $|\bar{\epsilon}|^{1/3}k^{2/3}\tau$ for $d = 2.0, 2.1, \cdots, 3$ and $d_{c\pm} = d_c \pm 0.0001$, where $d_c \approx 2.065$. The function G_d decays more rapidly as d decreases from 3 and rapidly moves to zero just above and below d_c. However, it decays more slowly as d changes from d_c to 2. Figure 1b shows that the values of g_d as a function of $\sigma = \sqrt{M_d K_d}|\bar{\epsilon}|^{1/3}k^{2/3}\tau$ are almost independent of d. Given the fluctuation–dissipation relation (35), this implies that the normalized two-point two-time Lagrangian-correlation spectrum $Q(k,t,s)/Q(k,s,s)$ is also almost independent of d under an appropriate normalization of time.

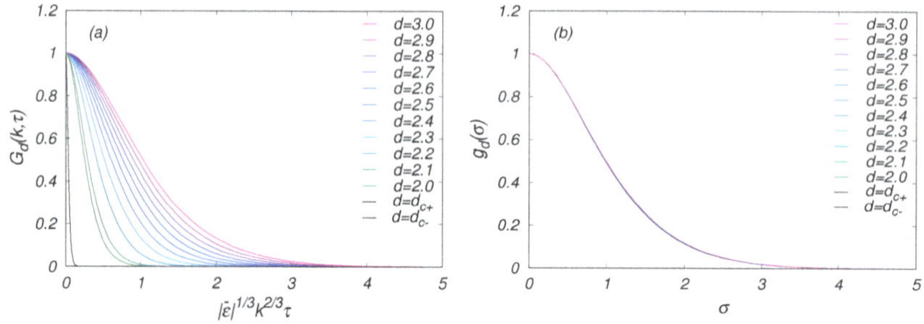

Figure 1. Response function for $d = 2, 2.1, 2.2, \cdots, 3$ and $d_{c\pm} = d_c \pm 0.0001$ and $d_c = 2.065$. (**a**) $G_d(k, \tau)$ plotted against $|\bar{\varepsilon}|^{1/3}k^{2/3}\tau$. (**b**) Normalized response function $g_d(\sigma)$.

Substituting the values of g_d thus obtained and integrating Equation (41) numerically, we derive estimates of K_d for the given d, as in previous studies. The numerical code used in this study is based on that of Ref. [6]. The values of K_d thus obtained are plotted in Figure 2a,b. It is seen that K_d as a function of d tends to $K_2 = 7.43$ for $d = 2$ and $K_3 = 1.73$ for $d = 3$, respectively, and diverges at a point d_c. These numerical estimates, 7.43 and 1.73, are in agreement with the estimates 7.41 and 1.72 [6,14,17–19] within a difference of less than 0.6%.

It is numerically clear that A_d is positive, so Π_d given by Equation (44) is, thus, also positive; the energy transfer is forward at $d_c < d \leq 3$ but negative (backward) at $2 \leq d < d_c$. If one assumes $A_d \propto (d - d_c)$ at small $|d - d_c|$, we obtain $K_d \propto |d - d_c|^\sigma$ with $\sigma = -2/3$, which is in good agreement with the numerical result of Figure 2b and agrees also with that of Fouriner and Frisch [5], who argued that K_d diverges at $d = d_{FF}$ and the exponent is $-2/3$, on the basis of the EDQNM approximation, where $d_{FF} \approx 2.05$, which is very close to $d_c = 2.065$ noted above. In the EDQNM approximation in Fouriner and Frisch [5], $\Pi(k)$ in the IETR is given by Equation (44) but with $G_d(k, \tau)$ in Equation (42) replaced by $G_{EDQNM}(k, \tau) = \exp(-\mu_k \tau)$, and $1/\mu_k$ represents the local eddy turnover time and is given by

$$\mu_k = \lambda_d \left(\int_0^k r^2 E(r) dr \right)^{1/2}, \tag{49}$$

and λ_d is an adjusting parameter. The critical dimension d_{FF} is seen to be independent of the choice of λ_d.

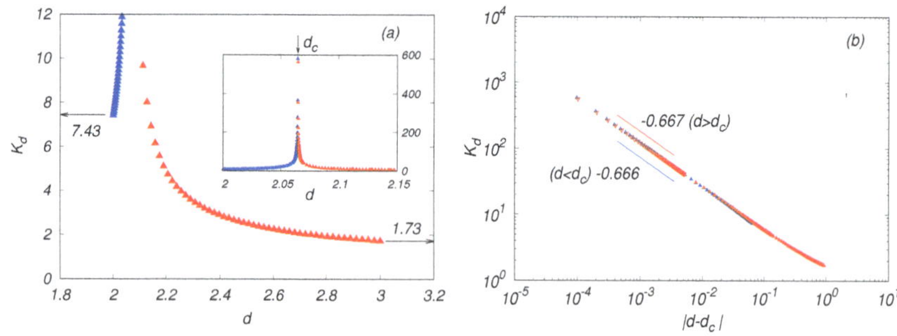

Figure 2. Variation of K_d by d. (**a**) K_d vs. d for $2 \leq d \leq 3$. Values at $d > d_c$ ($d < d_c$) are shown in red (blue). The numbers are $K_2 = 7.43$ and $K_3 = 1.73$. The inset figure shows a close-up of K_d near $d_c \approx 2.065$. (**b**) K_d vs. $|d - d_c|$ in logarithms. The numbers are the slopes (red: $d > d_c$, blue: $d < d_c$).

Substitution of the values of K_d thus obtained from Equation (25) gives the values of the skewness factor $S^{(d)}$, which are plotted in Figure 3. It is seen that $S^{(d)}$ decreases as d decreases from $d = 3$, and changes its sign at $d = d_c$ where K_d diverges.

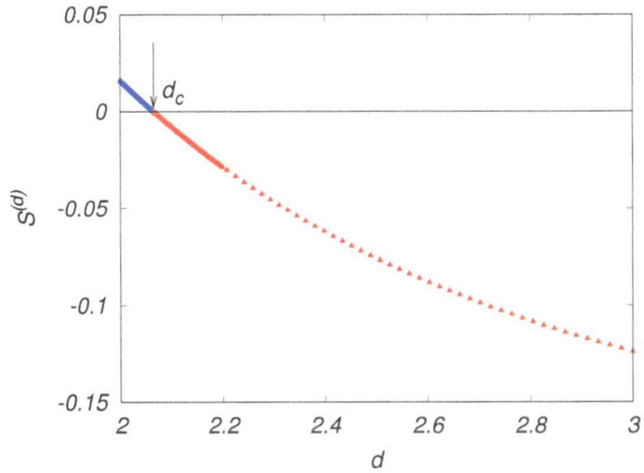

Figure 3. Variation of the skewness factor $S^{(d)}$ with d. Values at $d > d_c$ ($d < d_c$) are shown in red (blue). The arrow indicates $d_c \approx 2.065$.

5. Discussion

The overlap of the curves of g_d for different d in Figure 1b is impressive. This implies that the time dependence of the normalized Lagrangian two-point two-time correlation $Q_d(k,t,s)/Q_d(k,s,s)$ and that of the Lagrangian response function $G_d(k,t,s)$ under an appropriate normalization of time are insensitive to the difference in the space dimension. It would be interesting to explore the physics behind the overlap and implications thereof.

The skewness $S^{(d)}$ of the velocity increment must be 0 if the statistics of the velocity field are Gaussian. Hence, $S^{(d)}$ is a measure of the degree of the non-Gaussianity of the turbulence statistics. The results in Section 4 show that $S^{(d)}$ monotonically approaches the Gaussian value, i.e., 0, with the departure of the dimension d from 3, in the range $d_c < d < 3$, and increases from 0 in the range $2 < d < d_c$.

Recently studies have been made on turbulence in noninteger space dimensions $d < 3$ using a fractal Fourier decimation model; see, for example, [20–25], and references cited therein. Studies so far made suggest that the non-Gaussianity of small-scale statistics depends on the space dimension d. In this respect, the results in Section 4 noted above are consistent with this, although the model used here is based on continuation of the LRA at integer dimensions to noninteger dimension $2 < d < 3$; this differs from the decimation model. It would be interesting to apply the LRA to the decimation model to see if the LRA can well capture the influence of d on the statistics related to the skewness $S^{(d)}$ in the model.

Author Contributions: Both authors contributed equally to this work. All authors have read and agreed to the published version of the manuscript.

Funding: This work was supported by JSPS KAKENHI Grant Numbers JP20H00225 and JP23K03245.

Institutional Review Board Statement: Not applicable.

Informed Consent Statement: Not applicable.

Data Availability Statement: No data were used.

Acknowledgments: This paper is dedicated to the memory of Jack Herring. Y.K. is grateful to S.S. Ray for informing Y.K. of references [20–25], and for interesting discussions on the fractal Fourier decimation model, which stimulated Y.K.'s interest in turbulence in noninteger dimension. Y.K. would also like to thank the Isaac Newton Institute for Mathematical Sciences, Cambridge, for support and hospitality during the program "Mathematical aspects of turbulence: Where do we stand?" when work on this paper was undertaken. This work was supported by EPSRC grant number EP/R014604/1.

Conflicts of Interest: The authors declare no conflict of interest.

Appendix A. Relation between $\Pi_d(k)$ and $D_{LLL}(r)$

In this Appendix, we consider the relation between the energy flux in wavenumber space and the third-order moment $D_{LLL}(r)$ of the longitudinal velocity increment. Let us consider the following quantity [26],

$$\varepsilon_d(\boldsymbol{r}) \equiv -\frac{1}{4}\nabla_{\boldsymbol{r}} \cdot \left\langle |\delta \boldsymbol{u}|^2 \delta \boldsymbol{u} \right\rangle, \qquad \delta \boldsymbol{u}(\boldsymbol{r}) = \boldsymbol{u}(\boldsymbol{x}+\boldsymbol{r}) - \boldsymbol{u}(\boldsymbol{x}). \tag{A1}$$

In terms of $\varepsilon_d(\boldsymbol{r})$, the energy transfer flux $\Pi_d(K)$ in the wavenumber space is given by

$$\Pi_d(K) = \frac{1}{(2\pi)^d}\int_0^K \mathrm{d}k k^{d-1} \int \mathrm{d}\Omega_d(\boldsymbol{k}) \int_0^\infty \mathrm{d}r r^{d-1} \int \mathrm{d}\Omega_d(\boldsymbol{r})\, \varepsilon_d(\boldsymbol{r}) e^{-i\boldsymbol{k}\cdot\boldsymbol{r}}. \tag{A2}$$

where $\Omega_d(\boldsymbol{r})$ ($\Omega_d(\boldsymbol{k})$) denotes the surface of the unit sphere centered at $\boldsymbol{r}=0$ ($\boldsymbol{k}=0$) in the d-dimensional physical (wave-vector) space. If statistical isotropy is assumed, we can write $\varepsilon_d(\boldsymbol{r}) = \varepsilon_d(r)$ and perform the surface integrals. The result is

$$\Pi_d(K) = Z_d K \int_0^\infty \mathrm{d}r (Kr)^{\frac{d}{2}-1} J_{\frac{d}{2}}(Kr) \varepsilon_d(r), \qquad Z_d = \frac{S_d^2}{(2\pi)^d} 2^{\frac{d}{2}-1}\Gamma\left(\frac{d}{2}\right), \tag{A3}$$

where $J_{\frac{d}{2}}(x)$ is the $d/2$-th-order Bessel function. The function $\varepsilon_d(r)$ for the isotropic turbulence is expressed in terms of the cubic moment of the longitudinal velocity increment $D_{LLL}(r)$ as

$$\varepsilon_d(r) = -\frac{1}{12}\left(d + r\frac{d}{dr}\right)\left(d+2+r\frac{d}{dr}\right)\frac{D_{LLL}(r)}{r}. \tag{A4}$$

Substituting Equation (A4) into Equation (A3) and performing integration by parts, assuming that the integrands rapidly move to zero for small and large r, we obtain

$$\Pi_d(K) = -\frac{1}{12} Z_d K^{\frac{d}{2}+2} \int_0^\infty \mathrm{d}r\, r^{\frac{d}{2}+1} J_{\frac{d}{2}+2}(Kr) \frac{D_{LLL}(r)}{r}. \tag{A5}$$

Conversely, $\varepsilon_d(\boldsymbol{r})$ can also be expressed in terms of $\Pi_d(k)$ as

$$\varepsilon_d(\boldsymbol{r}) = 2^{\frac{d}{2}-1}\Gamma\left(\frac{d}{2}\right) r \int_0^\infty \mathrm{d}k\, (kr)^{1-\frac{d}{2}} J_{\frac{d}{2}}(kr) \Pi_d(k). \tag{A6}$$

Substituting Equation (A4) into the above equation, and integrating in r over the domain $[0, r]$ under the conditions $D_{LLL}(r) \propto r^3$ for $r \to 0$, we obtain

$$D_{LLL}(r) = -12\Gamma\left(\frac{d}{2}\right)\left(\frac{r}{2}\right)^{1-\frac{d}{2}} \int_0^\infty \mathrm{d}k\, k^{-1-\frac{d}{2}} J_{\frac{d}{2}+2}(kr) \Pi_d(k). \tag{A7}$$

Expressionsin terms of the transfer function $T_d(k)$ were obtained for $d = 2$ by [27] and for integer dimensions $d \geq 2$ by [8]. Equation (A7) indicates that the skewness $S^{(d)}(r)$ changes sign depending on $\Pi_d(K) = \bar{\varepsilon}$, being negative in the forward energy cascade range, but negative in the inverse energy cascade range.

Suppose that $\Pi_d(k) = \bar{\varepsilon}$ in the IETR. Then, the integrals of Equations (A6) and (A7) converge, and yield

$$\varepsilon_d(\boldsymbol{r}) = -\frac{1}{4}\nabla_{\boldsymbol{r}} \cdot \left\langle |\delta \boldsymbol{u}|^2 \delta \boldsymbol{u} \right\rangle = \bar{\varepsilon}, \tag{A8}$$

and

$$D_{LLL}(r) = -\frac{12}{d(d+2)}\bar{\varepsilon} r, \tag{A9}$$

respectively, which are consistent with Equation (A4), as expected.

References

1. Kaneda, Y. Renormalized expansions in the theory of turbulence with the use of the Lagrangian position function. *J. Fluid Mech.* **1981**, *107*, 131–145. [CrossRef]
2. Kraichnan, R.H. Lagrangian-history closure approximation for turbulence. *Phys. Fluids* **1965** *8*, 575–598. Erratum in *Phys. Fluids* **1966**, *9*, 1884. [CrossRef]
3. Kraichnan, R.H.; Herring, J.R. A strain-based Lagrangian-history turbulence theory. *J. Fluid Mech.* **1978**, *88*, 355–367. [CrossRef]
4. Herring, J.R.; Kraichnan, R.H. Numerical comparison of velocity-based and strain-based Lagrangian-history turbulence approximations. *J. Fluid Mech.* **1979**, *91*, 581–597. [CrossRef]
5. Fournier, J.D.; Frisch, U. *d*-dimensional turbulence. *Phys. Rev. A* **1978**, *17*, 747–762. [CrossRef]
6. Gotoh, T.; Watanabe, Y.; Shiga, Y.; Nakano, T.; Suzuki, E. Statistical properties of four-dimensional turbulence. *Phys. Rev. E* **2007**, *75*, 016310. [CrossRef]
7. Orszag, S.A. Analytical theories of turbulence. *J. Fluid Mech.* **1970**, *41*, 363–386. [CrossRef]
8. Clark, D.; Ho, R.D.; Berera, A. Effect of spatial dimension on a model of fluid turbulence. *J. Fluid Mech.* **2021**, *912*, A40. [CrossRef]
9. Fukayama, D.; Oyamada, T.; Nakano, T.; Gotoh, T.; Yamamoto, K. Longitudinal structure functions in decaying and forced turbulence. *J. Phys. Soc. Jpn.* **2000**, *69*, 701–715. [CrossRef]
10. Hill, R.J. Equations relating structure functions of all orders. *J. Fluid Mech.* **2001**, *434*, 379–388. [CrossRef]
11. Yakhot, V. Mean-field approximation and small parameter in turbulence theory. *Phys. Rev. E* **2001**, *63*, 026307. [CrossRef]
12. Kolmogorov, A.N. Dissipation of energy in locally isotropic turbulence. *C. R. Acad. Sci. USSR* **1941**, *32*, 16–18.
13. Zhou, Y. Turbulence theories and statistical closure approaches. *Phy. Rep.* **2021**, *935*, 1–117.
14. Kida, S.; Goto, S. A Lagrangian direct-interaction approximation for homogeneous isotropic turbulence. *J. Fluid Mech.* **1997**, *345*, 307–345. [CrossRef]
15. Kaneda, Y. Lagrangian renormalized approximation of turbulence. *Fluid Dyn. Res.* **2007**, *39*, 526–551. [CrossRef]
16. Kolmogorov, A.N. The local structure of turbulence in incompressible viscous fluid for very Large Reynolds numbers. *C. R. Acad. Sci. USSR* **1941**, *30*, 301–305.
17. Kaneda, Y. Inertial range structure of turbulent velocity and scalar fields in a Lagrangian renormalized approximation. *Phys. Fluids* **1986**, *29*, 701–708. [CrossRef]
18. Kaneda, Y. Inertial range of two-dimensional turbulence in a Lagrangian renormalized approximation. *Phys. Fluids* **1987**, *30*, 2672–2675. [CrossRef]
19. Gotoh, T. Passive scalar diffusion in two dimensional turbulence in the Lagrangian renormalized approximation. *J. Phys. Soc. Jpn.* **1989**, *58*, 2365–2379. [CrossRef]
20. Frisch, U.; Pomyalov, A.; Procaccia, I.; Ray, S.S. Turbulence in noninteger dimensions by fractal Fourier Decimation. *Phys. Rev. Lett.* **2012**, *108*, 074501. [CrossRef] [PubMed]
21. Lanott, A.S.; Benzi, R.; Malapaka, S.K.; Toschi, F.; Biferale, L. Turbulence on a fractal Fourier set. *Phys. Rev. Lett.* **2015**, *115*, 264502. [CrossRef] [PubMed]
22. Buzzicotti, M.; Biferale, L.; Frisch, U.; Ray, S.S. Intermittency in fractal Fourier hydrodynamics: Lessons from the Burgers equation. *Phys. Rev. E* **2016**, *93*, 033109. [CrossRef] [PubMed]
23. Buzzicotti, M.; Bhatnagar, A.; Biferale, L.; Lanotte, A.S.; Ray, S.S. Lagrangian statistics for Navier–Stokes turbulence under Fourier mode reduction: Fractal and homogeneous decimation. *New J. Phys.* **2016**, *18*, 113047. [CrossRef]
24. Ray, S.S. Non-intermittent turbulence: Lagrangian chaos and irreversibility. *Phys. Rev. Fluids* **2018**, *3*, 072601. [CrossRef]
25. Picardo, J.R.; Bhatnagar, A.; Ray, S.S. Lagrangian irreversibility and Eulerian dissipation in fully developed turbulence. *Phys. Rev. Fluids* **2020**, *5*, 042601. [CrossRef]
26. Frisch, U. *Turbulence*; Cambridge University Press: Cambridge, UK, 1995.
27. Cerbus, R.T.; Chakraborty, P. The third-order structure function in two dimensions: The Rashomon effect. *Phys. Fluids* **2017**, *29*, 111110. [CrossRef]

atmosphere

Article

Realizable Eddy Damped Markovian Anisotropic Closure for Turbulence and Rossby Wave Interactions

Jorgen S. Frederiksen [1,*] and Terence J. O'Kane [2]

[1] CSIRO Environment, Aspendale, Melbourne 3195, Australia
[2] CSIRO Environment, Hobart 7004, Australia
* Correspondence: jorgen.frederiksen@csiro.au

Abstract: A realizable Eddy Damped Markovian Anisotropic Closure (EDMAC) is presented for the interaction of two-dimensional turbulence and transient waves such as Rossby waves. The structure of the EDMAC ensures that it is as computationally efficient as the eddy damped quasi normal Markovian (EDQNM) closure but, unlike the EDQNM, is guaranteed to be realizable in the presence of transient waves. Jack Herring's important contributions to laying the foundations of statistical dynamical closure theories of fluid turbulence are briefly reviewed. The topics covered include equilibrium statistical mechanics, Eulerian and quasi-Lagrangian statistical dynamical closure theories, and the statistical dynamics of interactions of turbulence with topography. The impact of Herring's work is described and placed in the context of related developments. Some of the further works that have built upon Herring's foundations are discussed. The relationships between theoretical approaches employed in statistical classical and quantum field theories, and their overlap, are outlined. The seminal advances made by the pioneers in strong interaction fluid turbulence theory are put in perspective by comparing related developments in strong interaction quantum field theory.

Keywords: Eulerian closures; Lagrangian closures; Markovian closures; turbulence; Rossby waves; statistical mechanics; homogeneous flows; inhomogeneous flows; topography; field theory

Citation: Frederiksen, J.S.; O'Kane, T.J. Realizable Eddy Damped Markovian Anisotropic Closure for Turbulence and Rossby Wave Interactions. *Atmosphere* **2023**, *14*, 1098. https://doi.org/10.3390/atmos14071098

Academic Editors: Boris Galperin, Annick Pouquet and Peter Sullivan

Received: 6 June 2023
Accepted: 29 June 2023
Published: 30 June 2023

1. Introduction

Jack Herring (1975) [1] developed a theory of two-dimensional (2D) anisotropic turbulence by generalizing the eddy-damped quasi-normal Markovian (EDQNM) closure that had been derived by Orszag (1970) [2] for three-dimensional (3D) homogeneous isotropic turbulence (HIT). The EDQNM was numerically implemented and studied for 2D HIT by Leith (1971) [3]. Herring's interest was in examining the relaxation of 2D homogeneous anisotropic turbulence (HAT) back to isotropy in comparison with the return of 3D HAT to HIT. The study was performed without the presence of transient Rossby waves. Indeed, incorporating transient waves in the EDQNM closure, while guaranteeing realizability, has been a long-standing problem that we will discuss in detail.

Our aim in this article is to present a variant of the EDQNM closure for HAT, which is realizable in the presence of transient waves. The study focuses on Rossby waves, although the same approach can be used for other waves. We call the model the realizable eddy damped Markovian anisotropic closure (EDMAC). The EDMAC model is as computationally efficient as the EDQNM and so expands the applicability Markovian closures to cater for transient waves at very little computational cost.

A second aim is to summarize some of Jack Herring's major achievements in laying the foundations of the statistical dynamical theory of turbulence. It is also to place his work in the context of related developments, to note the impacts of his work and how it has allowed further advances in this complex and difficult field.

The article is structured as follows. In Section 2, we review some of Jack Herring's major pioneering contributions to formulating the foundations of the statistical dynamical closure theory for fluid turbulence. His impact on the field is also discussed and his work placed in the context of related works and some of the further advances that have built on these foundations are discussed. A major aim of this article is also to present a generalization of the EDQNM closure that is guaranteed to be realizable for 2D HAT in the presence of transient waves like Rossby waves. This work thus extends the closures of Orszag [2] and Herring [1] and provides a resolution of a long-standing problem.

In Section 3, the dynamical equations for 2D HAT interacting with Rossby waves on a β–plane are summarized. The corresponding direct numerical simulation equations are displayed in the spectral space of Fourier coefficients on the doubly periodic domain in Section 4. In Section 5, non-Markovian closures for 2D HAT are presented and three variants of corresponding Markovian closures with auxiliary evolution equations for the triad relaxation functions are derived in Section 6. The realizable EDMAC closure for 2D HAT interacting with transient Rossby waves is presented in Section 7. The EDMAC model is constructed to be as numerically efficient as the EDQNM closure since it has an analytical expression for the triad relaxation function like the EDQNM. Section 8 contains a few of Jorgen's personal reflections on Jack Herring and perspectives on strong interaction statistical field theories and a comparison of progress in strong interaction fluid turbulence with that in hadron physics in quantum field theory. Our conclusions are summarized in Section 9. Appendix A establishes the conditions under which the real part of the EDMAC triad relaxation function is positive semi-definite and Appendix B presents the Langevin equation that underpins the EDMAC model and ensures that it is realizable for 2D HAT in the presence of transient Rossby waves.

2. Herring's Statistical Dynamical Theories and Their Impacts and Extensions

Jack Herring made giant pioneering steps in laying the foundations of the statistical dynamical theories of fluid turbulence. In this Section, we briefly summarize some of his major achievements, the impacts of his works and some of the related and further developments that have occurred.

2.1. *Equilibrium Statistical Mechanics*

Equilibrium statistical mechanics is perhaps the simplest theoretical framework that gives some insight into the more complex phenomena of turbulence in the presence of forcing and dissipation. The appeal of equilibrium statistical mechanics is that the inviscid unforced equations of motion, with conservation laws like energy and enstrophy, have exact analytical solutions.

2.1.1. Complete Statistical Mechanics Theories

Herring (1977) [4] developed the statistical mechanics theory of two-dimensional flows over random topography with ensemble averages taken over both the flow fields and the topography. He formulated the canonical equilibrium solutions based on planar geometry spectral representations of the flow fields and topography. The aim was to provide guidance and understanding of the more complicated statistical dynamical closure theories that he formulated and solved numerically for forced dissipative turbulence over random topography. Herring's thermal equilibrium solutions for ensembles of random topography led on from the earlier statistical mechanics of point vortices by Onsager (1949) [5], Kraichnan's (1967) [6] and (1975) [7] planar geometry spectral solutions without topography, and the planar geometry statistical mechanics for quasi geostrophic flows over single realization topography by Salmon et al., (1976) [8].

These seminal works were extended to formulate canonical equilibrium theory in spherical geometry both with and without single realization topography [9,10]. Indeed, the relationships between equilibrium statistical mechanics solutions in planar geometry and on the sphere have been further elucidated and reviewed in this special issue dedicated to Jack Herring by Salmon and Pizzo [11].

Bretherton and Haidvogel (1976) [12] developed a different model for understanding turbulence over topography in which the dynamical solutions were minimum enstrophy states that are nonlinearly stable [13]. Frederiksen and Carnevale (1986) [14] established the relationships between canonical equilibrium states and minimum enstrophy states for barotropic flows over topography on the sphere and this was subsequently generalized to baroclinic flows [15,16]. The corresponding equivalence was formulated for barotropic flows over topography in planar geometry by Carnevale and Frederiksen (1987) [17]. They also considered the thermodynamic limit of infinite resolution and showed that then the canonical equilibrium state is statistically sharp and identical to the nonlinearly stable minimum enstrophy state. In their study, it was also noted that for the continuum dynamics of fluids, more general nonlinear stable states than the minimum enstrophy states are possible since an infinity of invariants exists in the inviscid case. Moreover, they pointed out that statistical mechanics theory can also be generalized to account for these invariants to be consistent with the many-invariant nonlinearly stable flows. The in-principle and practical difficulty is developing a realizable numerical model with an infinity of invariants on which theoretical formulations and solutions can be based.

2.1.2. Empirical Statistical Mechanics Theories

Miller and collaborators [18,19] and Robert and collaborators [20,21] made attempts at formulating complete statistical mechanics theories with an infinity of invariants. However, the shortcomings of these attempts have been noted in several studies. Chorin [22] and Turkington [23] pointed out that the lattice models of Miller et al. [19] always have a shortest scale, while this is not the case for continuum fluids. Majda and Wang [24] regard the many-invariant approaches as empirical rather than complete statistical theories like the energy-enstrophy theories with underpinning realizable spectral models. More extensive discussions of statistical mechanics methods including recent works and applications are provided in the reviews in Refs. [24–27].

2.2. *Eulerian Statistical Dynamical Closure Theories*

The late 1950s to mid-1960s was a time of extraordinary advances in the theory of strong turbulence, one of the most difficult problems in classical physics. In the vanguard was the Eulerian direct interaction approximation (DIA) closure theory of Kraichnan [28,29] for homogeneous isotropic turbulence (HIT). The DIA was based on formal, rather heuristic, renormalized perturbation theory and, in the language of modern physics, is a bare vertex approximation [30–32]. This was followed by the nonequilibrium steady state theory of Edwards [33] and the self-consistent field theory (SCFT) of Herring [34,35] in steady state and time-dependent forms. Both the Edwards and Herring statistical closures for HIT are based on Liouville or Fokker–Planck formalisms and are original in their approaches. Carnevale and Frederiksen [36] and McComb [37] in this special issue in commemoration of Jack Herring compare these different formalisms at nonequilibrium steady states. Another Eulerian non-Markovian closure theory for HIT that was developed in the 1970s by McComb [38–40] is the local energy transfer theory (LET).

The three time-dependent Eulerian non-Markovian closure theories developed by Kraichnan, Herring and McComb are based on quite different physical reasoning and theoretical approaches. However, the final closure equations are in fact simply related. The theories are all bare vertex approximation theories. The theories all have the same single-time cumulant equation and differ just in how the two-time cumulant and response functions are treated. Kraichnan's DIA has separate equations for the response functions and two-time cumulants while both Herring's SCFT and McComb's LET effectively assume a fluctuation dissipation theorem (FDT) between the response and two-time cumulant functions [41]. In principle the FDT should only be strictly valid in statistical mechanical equilibrium as described in Section 2.1. The SCFT and LET closures can be derived from the DIA by employing the *prior-time FDT* [36] (Equation (3.5)) defined by

$$C_{\mathbf{k}}(t,t') \equiv R_{\mathbf{k}}(t,t')C_{\mathbf{k}}(t',t') \tag{1}$$

for $t \geq t'$. Here, $C_{\mathbf{k}}(t,t')$ is the two-time spectral cumulant at wavenumber $\mathbf{k}$, $R_{\mathbf{k}}(t,t')$ is the response function and $C_{\mathbf{k}}(t',t')$ the prior time single-time cumulant. The SCFT and DIA have the same response function equation, but the SCFT then calculates the two-time cumulant from Equation (1). On the other hand, the LET and the DIA have the same two-time cumulant equation with the LET determining the response function from Equation (1).

The DIA, SCFT and LET closures are quite skillful in capturing the large energy containing scales, which is probably the most important property for geophysical fluids and plasmas. However, the focus of much of the statistical closure theory for HIT has been on the power law behaviour of the closures as well as their realizability. The realizability of the DIA closure was a major triumph but the power laws of the DIA, and the SCFT, were found to differ a little from the inertial range power laws of $k^{-\frac{5}{3}}$ for 3D turbulence and from the k^{-3} enstrophy cascading inertial range for two-dimensional turbulence (Herring et al., 1974 [42]). Again, the power law fall-off of the LET closure and general performance are very similar to the DIA and SCFT closures for two-dimensional turbulence (Frederiksen and Davies 2000 [43]). The power law deficiencies of the DIA closure were ascribed by Kraichnan [29] to spurious sweeping effects of the small eddies by the large eddies.

The study of Herring et al. [42] established the properties of the DIA closure for two-dimensional HIT in comparison with direct numerical simulations (DNS) on the doubly periodic domain. The DNS used the discrete Fourier transform but for the sake of efficiency the DIA closure, as with most closure codes, was formulated for the continuum problem and used logarithmic discretization to reach high wavenumbers. Frederiksen and Davies [43] made a series of similar comparisons of the DIA, SCFT and LET closures for two-dimensional HIT with DNS and for a range of large-scale Reynolds numbers between $R_L \approx 50$ and $R_L \approx 4000$ that included experimental setups very similar to those of Herring et al. [32]. They used closures formulated on the same discrete spectral space as the DNS which involved a much larger computational task. It meant that all interactions were included in both closures and DNS and a direct comparison was possible. It was found that the discrete closures were in much better agreement with DNS at the low to moderate Reynolds numbers used by Herring et al. [42]. As well, the DIA, SCFT and LET closures had very similar performance and similar deficits in kinetic energy and palinstrophy at the smallest scales compared with DNS.

Reviews of the subsequent development of closures for HIT are given in Refs. [27,37–40,44–47].

2.3. *Quasi-Lagrangian Statistical Dynamical Closure Theories*

The recognition of the inertial range discrepancies of the Eulerian DIA compared with the observed power laws of 2D and 3D turbulence led to the development of quasi-Lagrangian closures for HIT by Kraichnan (1965) [48] and (1977) [49], Herring and Kraichnan (1979) [50], Kaneda (1981) [51] and Gotoh et al. (1988) [52]. This was a truly epic effort of theoretical and numerical model development. Unfortunately, the outcome was not wholly successful. Unlike the Eulerian DIA, which is independent of field variable formulation, or norm, the quasi-Lagrangian closures depend on whether the derivations use labelling time derivatives [48] or measuring time derivatives [51]. As well, the quasi-Lagrangian closures depend on the choice of field variable, or representative. This is effectively equivalent to specifying one two-state parameter, the labelling or measuring time, and one continuous parameter that determines the field variable or linear combinations of field variables.

The quasi-Lagrangian closures are still second order in perturbation theory, and the aims have been to make transformations that avoid the spurious convection effects identified by Kraichnan [29]. They do not, however, provide a fundamental solution to the vertex renormalization problem "which is the whole problem of strong turbulence" (Martin et al., 1973 [53]).

Kraichnan (1964) [54] had earlier recognized that the power law discrepancies of the Eulerian DIA could be overcome by cutting off the wavenumber interactions between the larger and smaller scales in the response functions and two-time cumulants. This results in a one parameter regularized DIA closure with empirical vertex renormalization depending on the cut-off parameter α (see Section 5.3). However, it turns out that α is only weakly dependent on whether the turbulence is 2D or 3D and on whether the turbulence is homogeneous or inhomogeneous. For 3D HIT, the regularized DIA closure captures the DNS statistics closely, including the $k^{-\frac{5}{3}}$ power law, with α between 3 and 3.5 [54,55]. For 2D turbulence, the regularized DIA (Frederiksen and Davies 2004 [56]) and regularized QDIA (O'Kane and Frederiksen 2004 [57]) have employed values of α between 4 and 6. Closures for 2D HIT have used $\alpha = 6$ to give good agreement with DNS including the k^{-3} power law, and for inhomogeneous closures, $\alpha = 4$ has been used. However, the closures in the HIT simulations were used with repeated cumulant update restarts [56] after relatively short times to optimize computational resources, and that allowed a larger value of α to be used. We expect that, with sufficiently long-time integrals in the non-Markovian closures, taking $\alpha \approx 4$ may be a reasonable universal value for both 2D and 3D turbulence.

The quasi-Lagrangian closures do of course improve on the small-scale behaviour of the Eulerian closures, but there are still significant differences from the statistics of DNS. These differences depend on both the formulations and on whether the turbulence is 2D or 3D. The differences also generally become more apparent at higher Reynolds numbers. For the case of 2D HIT, Frederiksen and Davies [56] compared their regularized DIA calculations with quasi-Lagrangian closure results based on both labelling time derivatives and measuring time derivatives and for several choices of field, or representative, formulations. Compared with the statistics of DNS, the regularized DIA performed better than the abridged Lagrangian-history direct interaction (ALHDI) approximation, the strain-based abridged Lagrangian-history direct interaction (SALHDI) approximation of Herring and Kraichnan [50], and the Lagrangian renormalized approximation (LRA) of Gotoh et al. [52].

2.4. Homogeneous Closures for Turbulent Flows over Topography

Herring (1977) [4] generalized the Eulerian DIA and the test field model (TFM) (Kraichnan 1971 [58]) to study the statistics of turbulent 2D flows over ensembles of random topography. He considered rotating flows on an f-plane, so there was no differential rotation with latitude, or β-effect, or propagating Rossby waves. The aim was to shed further light into the findings of Bretherton and Haidvogel [12] that slowly decaying turbulence over topography tends to progress through a sequence of minimum enstrophy states for a fixed energy (in the absence of the β-effect).

Herring [4] was able to establish broad parameter ranges for which there was general agreement between his forced dissipative closure results and the exact static solutions for inviscid flows. The locking of the flow to the topography was determined for the DIA and TFM closures with very similar behaviour at the larger scales.

2.5. Inhomogeneous Closures for Turbulent Flows over Topography

Frederiksen (1999) [59] continued the work on understanding the effects of topography on 2D turbulence. He developed a generalization of the Eulerian DIA closure theory that applies to inhomogeneous turbulence over single realization topography. This quasi-diagonal direct interaction approximation (QDIA) has a statistical dynamical equation for the mean flow unlike homogeneous closure equations, for which the mean flow is zero. The mean flow also couples to the single-time cumulant, the two-time cumulant and the response function that describe the turbulence. The QDIA expresses the off-diagonal elements of the cumulants and response functions, in Fourier space, in terms of the diagonal elements and the mean flow and topography. Consequently, the QDIA is only a few times more computationally demanding than the corresponding DIA for homogeneous turbulence.

The inhomogeneous QDIA closure theory has been generalized to multi-field classical systems including quasi-geostrophic and 3D turbulence [60,61] and to quantum field theories [32]. O'Kane and Frederiksen (2004) [57] implemented the closure for numerical studies in comparison with ensembles of DNS on an f-plane. Frederiksen and O'Kane

(2005) [62] generalized the QDIA closure for the interaction of inhomogeneous turbulent flows with Rossby waves and topography on a β-plane and noted the remarkable agreement between the closure and large ensembles of DNS for low Reynolds number flows. The QDIA closure has been extensively applied in bare vertex form at lower Reynolds numbers and in regularized form with $\alpha = 4$ at higher Reynolds numbers. It has been used to study the dynamics of turbulence, Rossby wave and topography interactions, the predictability of blocking regime transitions, data assimilation, and has been extensively applied for developing subgrid scale parameterizations. Reviews of the literature are given in refs. [44,63,64].

2.6. Markovian Statistical Closure Theories without Waves

The non-Markovian closures, the DIA, SCFT and LET for homogeneous turbulence, and the QDIA for inhomogeneous turbulence, present a large computational task that scales as $O(T^3)$ where T is the length of the time integration. This can be improved to scaling like $O(T^2)$ by restarting the integrations periodically and using the updated three-point cumulant in the restart [43,61,65]. However, Markovian closures such as Orszag's (1970) [2] EDQNM closure are still very much more computationally efficient since they scale like $O(T)$. The EDQNM is a one parameter theory with the eddy damping specified analytically to satisfy inertial range power laws. It was first implemented numerically by Leith (1971) [3] for 2D HIT with a form equivalent to that given in our Section 7.1. Herring (1975) [1] generalized the EDQNM closure to develop a theory of 2D anisotropic turbulence without the presence of waves. The EDQNM has an underpinning Langevin equation representation, as noted by Leith [3] and Herring and Kraichnan [66], that guarantees realizability for HIT and HAT without transient waves.

The EDQNM in its most basic form is represented by just the single-time cumulant spectral equation. It can be arrived at from the DIA closure by invoking the *current-time FDT*

$$C_{\mathbf{k}}(t,t') \equiv R_{\mathbf{k}}(t,t')C_{\mathbf{k}}(t,t) \tag{2}$$

for $t \geq t'$. As well an analytical form for the response function is assumed and the time-history integrals of the DIA are then performed analytically to determine a triad relaxation function that enters the cumulant equation (see Section 7.1).

In the presence of transient waves, it is possible for the EDQNM closure to have unphysical behaviour and blow up (Bowman et al., 1993 [67]). To avoid this, the steady state form of the triad relaxation function has often then been used [46,68–70] or modified quasi-normal Markovian closures employed [71,72].

2.7. Markovian Statistical Closure Theories with Waves

Bowman et al. [67] made detailed analyses of the EDQNM and the problems with realizability that can occur in the presence of transient waves. In particular, in the standard formulation, when waves, such as drift-waves or Rossby waves, are included in the response function it is not possible to guarantee that the real part of the triad relaxation function will be non-negative. This is also the case if the prior time FDT in Equation (1) is used instead of the current time FDT in Equation (2). Bowman et al. [67] found that they could derive a realizable Markovian closure (RMC) by using a fluctuation dissipation theorem that involves both current and prior time cumulants and that we have termed the *correlation FDT*

$$C_{\mathbf{k}}(t,t') \equiv [C_{\mathbf{k}}(t,t)]^{\frac{1}{2}} R_{\mathbf{k}}(t,t')[C_{\mathbf{k}}(t',t')]^{\frac{1}{2}} \tag{3}$$

for $t \geq t'$. Note though that while the EDQNM triad relaxation function has an analytical form for the RMC, a time-dependent equation must be solved to determine it.

The works of Bowman et al. [67], Hu et al. [73] and Bowman and Krommes [74] describe the further development of realizable Markovian closures for homogeneous turbulence with transient waves, in which the triad relaxation functions are determined by auxiliary differential equations.

Markovian inhomogeneous closures (MICs) were developed and tested against large ensembles of DNS by Frederiksen and O'Kane [44,75]. They started their formulations with the inhomogeneous QDIA closure for turbulent 2D flow interacting with Rossby waves and topography and employed the three versions of the FDT that they combined as:

$$C_{\mathbf{k}}(t,t') \equiv [C_{\mathbf{k}}(t,t)]^{1-X} R_{\mathbf{k}}(t,t')[C_{\mathbf{k}}(t',t')]^{X} \quad (4)$$

for $t \geq t'$ and $C_{\mathbf{k}}(t,t') = C_{-\mathbf{k}}(t',t)$ for $t' > t$. Here, $X = 0$ for the current-time FDT used for the EDQNM of Orszag {1970}, $X = \frac{1}{2}$ for the correlation FDT used for the RMC of Bowman et al. [67], $X = 1$ for the prior-time FDT used for the SCFT of Herring [35] and the LET of McComb [38,39].

All the MICs developed by Frederiksen and O'Kane [44,75] performed remarkably well compared with the DNS ensembles in low Reynolds number numerical experiments. However, it is desirable to be sure that the Markovian closures employed will be realizable under all circumstances as is the case for the formulations using the correlation FDT ($X = \frac{1}{2}$). Determining the relaxation functions through time integration of differential equations is nevertheless a considerable computational overhead. Frederiksen and O'Kane [75], therefore, developed the eddy damped Markovian inhomogeneous closure (EDMIC) that generalizes the EDQNM to inhomogeneous flows. The EDMIC has analytical forms for the relaxation functions and is realizable under the same conditions as the EDQNM.

A major aim of this article is to formulate a realizable eddy damped Markovian anisotropic closure (EDMAC) with analytical triad relaxation function in the presence of transient Rossby waves that thus generalizes the EDQNM.

2.8. Classical and Quantum Statistical Field Theory Formalisms

Herring's [34,35] theoretical approach to deriving the statistical dynamical closure equations based on Liouville or Fokker–Planck formalisms was original. It was different from the approaches that had become standard in quantum field theory since the remarkable success of renormalized perturbation theory in quantum electrodynamics by Feynman, Schwinger and Tomonaga in the mid-20th century [32]. This is also the case for the works of Edwards [33], McComb [38,39], and the published work of Kraichnan [28,29]. However, Martin et al. [53] (Footnote 11) note "It seems that Kraichnan's rules for calculating the renormalized vertices to a given order generate the quantities which are given exact non-perturbative definitions here. We are grateful to Dr. Kraichnan for providing us with old unpublished notes ... ".

Wyld [30] and Lee [31] reconstructed Kraichnan's Eulerian DIA closure through renormalized perturbation theory with a diagrammatic representation similar to Feynman diagrams. Lee considered magneto-hydrodynamics to sixth order and pointed out that Wyld had mistakenly replaced the bare propagator by the renormalized propagator in some terms in his fourth order representation. The DIA in these works appears as a bare vertex approximation in second order renormalized perturbation theory.

Martin et al. [53] (hereafter MSR) generalized the Schwinger–Dyson functional operator approach to classical statistical dynamics by introducing an adjoint operator that generates the response function. They also introduced an associated non-Hermitian Hamiltonian, in terms of the field variable and the adjoint operator, and a generating functional from which the mean field, two-point cumulants, response functions, self-energies and vertex functions can be derived through functional differentiation.

The MSR formalism was expanded by Rose [76] to include random forcing including non-Gaussian noise and non-Gaussian initial conditions. The full power of renormalized perturbation theory was achieved by the reformulation of the MSR approach through the Feynman path integral formalism [77] by Phythian [78] and most generally by Jensen [79].

Berera et al. [80] made an important reconciliation between the Wyld diagrammatic approach at fourth order and the MSR formalism. As further discussed by McComb [37] they corrected some errors in both the Wyld and MSR formalisms to show the consistency between the diagrammatic approach and the functional method, as would be expected.

The Schwinger–Dyson functional formalism was designed for developing a statistical theory of scattering from asymptotic 'in' states to 'out' states. This is analogous to the steady state theories of Edwards [33] and Herring [34] with the fluctuation dissipation theorem (FDT) imposed. The time-dependent nonequilibrium statistical dynamical theories are more complex and do not satisfy the FDT exactly; that is why MSR needed to introduce the adjoint operator to generate the separate response function. The adjoint operator incidentally arises naturally in the path integral generalization [78,79]. In fact, the MSR (1973) [53] formalism and its path integral generalizations for classical systems are much more similar to the closed time path (CTP) formalism of Schwinger (1961) [81] and Keldysh (1965) [82] for time-dependent nonequilibrium quantum field theories. The earlier Schwinger–Keldysh CTP theory contains all the aspects of adjoint operators, resulting in operator doubling, Pauli matrices, two-point matrix Greens functions consisting of cumulants and response functions and matrix self-energies that appear in the MSR formalism. Indeed, it is surprising that the Schwinger–Keldysh formalism was not referenced by MSR.

Analyses of the relationships between the Schwinger–Keldysh CTP quantum field theory formalism and the MSR classical field theory approach are described in the works of Cooper et al. [83] and Blagoev et al. [84]. Frederiksen [32] presents a parallel development of classical and quantum statistical field theories based on these formalisms, and the path integral generalization, to include non-Gaussian noise, non-Gaussian initial conditions, and quantum effects. It is shown that the approaches are equivalent and that the classical approach can be applied to generate the second order statistical equations for quantum systems provided a suitable non-Gaussian noise term is included. For the Klein–Gordon equation with interaction Lagrangian of $g\phi^3$, the quantum effects correspond to a pure skewness non-Gaussian noise, and for $\lambda\phi^4$, it corresponds to pure kurtosis. Thus, the differences between classical and quantum field theories manifest themselves through an additional quantum self-energy term proportional to Planck's constant squared, and in the initial conditions.

Frederiksen [32] also developed the inhomogeneous QDIA closure for quantum field theories. For both classical and quantum field theories, the QDIA closure is computationally tractable. This is because the off-diagonal elements of the two-point cumulants and response functions—the propagators—in Fourier space are represented in terms of the diagonal cumulants and response functions and the mean fields (and possibly topography).

3. Two-Dimensional Barotropic Flows on a β-Plane

In this study we focus on 2D turbulent flows in planar geometry and with differential rotation in the latitudinal direction that generates the β-effect. The equations of motion are most conveniently described by a single equation for the vorticity, ζ. The vorticity is of course the Laplacian of the stream function ψ whose gradient, in turn, is related to the zonal and meridional wind fields. Throughout this article, we develop our results on the doubly periodic domain $0 \leq x \leq 2\pi$, $0 \leq y \leq 2\pi$ with $\mathbf{x} = (x,y)$.

The barotropic vorticity equation is given by

$$\frac{\partial \zeta}{\partial t} = -J(\psi, \zeta + \beta y) + \hat{\nu}_0 \nabla^2 \zeta + f^0. \tag{5}$$

Here, the Jacobian is

$$J(\psi, \zeta) = \frac{\partial \psi}{\partial x}\frac{\partial \zeta}{\partial y} - \frac{\partial \psi}{\partial y}\frac{\partial \zeta}{\partial x} \tag{6}$$

and the vorticity ζ is the Laplacian of the stream function

$$\zeta = \nabla^2 \psi \equiv \left(\frac{\partial^2}{\partial x^2} + \frac{\partial^2}{\partial y^2}\right)\psi. \tag{7}$$

In these equations, β denotes the beta-effect, f^0 specifies any external forcing, and $\hat{\nu}_0$ is the viscosity. The dissipation in Equation (5) is represented by the Laplacian, but we shall, in fact, also consider higher order dissipation operators so that in spectral space, $\hat{\nu}_0$ depends on the wavenumber.

Rossby waves, and superpositions of Rossby waves proportional to $\exp \mathrm{i}(\mathbf{k}.\mathbf{x} - \omega_{\mathbf{k}} t)$, are solutions to Equation (5) (in the absence of forcing, topography, and viscosity). The Rossby wave frequency satisfies the dispersion relationship

$$\omega_{\mathbf{k}} = \omega_{\mathbf{k}}^{\beta} = -\frac{\beta k_x}{k^2}. \tag{8}$$

4. Dynamical Equations in Fourier Space

Our analysis and theoretical developments will be performed in Fourier space with each of the fields having a spectral representation similar to that of the vorticity:

$$\zeta(\mathbf{x}, t) = \sum_{\mathbf{k} \in \mathbf{R}} \zeta_{\mathbf{k}}(t) \exp(i\mathbf{k}.\mathbf{x}), \tag{9}$$

where

$$\zeta_{\mathbf{k}}(t) = \frac{1}{(2\pi)^2} \int_0^{2\pi} d^2\mathbf{x} \zeta(\mathbf{x}, t) \exp(-i\mathbf{k}.\mathbf{x}). \tag{10}$$

Here, $\mathbf{x} = (x, y)$, $\mathbf{k} = (k_x, k_y)$, $k = \left(k_x^2 + k_y^2\right)^{1/2}$. The reality of the physical space fields implies that in spectral space $\zeta_{-\mathbf{k}} = \zeta_{\mathbf{k}}^*$. The summations in Equation (9) are over the domain $\mathbf{R}$ which is a circular wavenumber domain excluding the origin $\mathbf{0}$. In spectral space, the resulting dynamical equation is:

$$\left(\frac{\partial}{\partial t} + \nu_0(\mathbf{k}) k^2\right) \zeta_{\mathbf{k}}(t) = \sum_{\mathbf{p} \in \mathbf{R}} \sum_{\mathbf{q} \in \mathbf{R}} \delta(\mathbf{k}, \mathbf{p}, \mathbf{q}) K(\mathbf{k}, \mathbf{p}, \mathbf{q}) \zeta_{-\mathbf{p}}(t) \zeta_{-\mathbf{q}}(t) + f_{\mathbf{k}}^0(t). \tag{11}$$

The generalized delta function is defined by $\delta(\mathbf{k},\mathbf{p},\mathbf{q}) = 1$ if $\mathbf{k} + \mathbf{p} + \mathbf{q} = 0$ and 0 if $\mathbf{k} + \mathbf{p} + \mathbf{q} \neq 0$. It is also convenient to define the complex $\nu_0(\mathbf{k})k^2$ that represents both the viscosity and the Rossby wave frequency $\omega_{\mathbf{k}}$:

$$\nu_0(\mathbf{k}) k^2 = \hat{\nu}_0(k) k^2 + \mathrm{i}\omega_{\mathbf{k}}. \tag{12}$$

The Rossby wave frequency is defined in Equation (8). We have also generalized the form of the viscosity $\hat{\nu}_0 \rightarrow \hat{\nu}_0(k)$ to allow for more general dissipation operators in Equation (5). The interaction coefficient $K(\mathbf{k},\mathbf{p},\mathbf{q})$ is given by

$$K(\mathbf{k}, \mathbf{p}, \mathbf{q}) = \frac{1}{2}[p_x q_y - p_y q_x](p^2 - q^2)/p^2 q^2. \tag{13}$$

5. Eulerian Non-Markovian Statistical Dynamical Closures

Statistical dynamical closures encapsulate the dynamics of infinite ensembles of characteristic flows. Our focus here is on 2D flows described by the DNS in Equation (5) (in physical space) and Equation (11) (in Fourier space). A given member making up the ensemble, the particular flow field, here given by the vorticity spectral coefficient $\zeta_{\mathbf{k}}(t)$ at wavenumber $\mathbf{k}$, is conveniently represented by its mean $< \zeta_{\mathbf{k}}(t) > \equiv \overline{\zeta}_{\mathbf{k}}(t)$ and deviation from the ensemble mean $\widetilde{\zeta}_{\mathbf{k}}(t)$. For homogeneous turbulence, with zero mean flow, we have

$$< \zeta_{\mathbf{k}}(t) > \equiv \overline{\zeta}_{\mathbf{k}}(t) = 0\,; \zeta_{\mathbf{k}}(t) = \widetilde{\zeta}_{\mathbf{k}}(t). \tag{14}$$

The spectral equation for $\widetilde{\zeta}_{\mathbf{k}}$ is thus given by Equation (11) with $\zeta_{\mathbf{k}}(t) \rightarrow \widetilde{\zeta}_{\mathbf{k}}(t)$. Given that the mean field is zero, this also means that the mean forcing is zero and that

$$< f_{\mathbf{k}}^0(t) > \equiv \overline{f}_{\mathbf{k}}^0(t) = 0\,; f_{\mathbf{k}}^0(t) = \widetilde{f}_{\mathbf{k}}^0(t)\,. \tag{15}$$

5.1. The DIA Closure for Homogeneous Turbulence

A simple formal derivation of the DIA closure equations for homogeneous 2D turbulence is given by Frederiksen [85]. The DIA closure consists of coupled equations for the two-time two-point cumulant

$$C_{\mathbf{k}}(t,t') = < \widetilde{\zeta}_{\mathbf{k}}(t)\widetilde{\zeta}_{-\mathbf{k}}(t') > \tag{16}$$

and the ensemble average response function

$$R_{\mathbf{k}}(t,t') = \left\langle \widetilde{R}_{\mathbf{k}}(t,t') \right\rangle. \tag{17}$$

Here, the response function for an individual disturbance is

$$\widetilde{R}_{\mathbf{k}}(t,t') = \frac{\delta\widetilde{\zeta}_{\mathbf{k}}(t)}{\delta\widetilde{f}^{0}_{\mathbf{k}}(t')}, \tag{18}$$

where δ denotes the functional derivative. The response function measures the change in the individual field $\widetilde{\zeta}_{\mathbf{k}}(t)$ at time t due to an infinitesimal change in the forcing $\widetilde{f}^{0}_{\mathbf{k}}(t')$ at the earlier time t'.

We can obtain the equation for the two-time cumulant $C_{\mathbf{k}}$ from Equation (11), by multiplying each term by $\widetilde{\zeta}_{-\mathbf{k}}(t')$ and averaging. This gives the cumulant equation

$$\begin{aligned}&\left(\tfrac{\partial}{\partial t} + \nu_0(\mathbf{k})k^2\right)C_{\mathbf{k}}(t,t') \\ &= \sum_{\mathbf{p}}\sum_{\mathbf{q}} \delta(\mathbf{k},\mathbf{p},\mathbf{q})K(\mathbf{k},\mathbf{p},\mathbf{q}) < \widetilde{\zeta}_{-\mathbf{p}}(t)\widetilde{\zeta}_{-\mathbf{q}}(t)\widetilde{\zeta}_{-\mathbf{k}}(t') > + < \widetilde{f}^{0}_{\mathbf{k}}(t)\widetilde{\zeta}_{-\mathbf{k}}(t') >\end{aligned} \tag{19}$$

where $t > t'$ and $C_{\mathbf{k}}(t,t') = C_{-\mathbf{k}}(t',t)$ for $t' > t$. It can also be shown that

$$< \widetilde{f}^{0}_{\mathbf{k}}(t)\widetilde{\zeta}_{-\mathbf{k}}(t') > = \int_{t_0}^{t'} ds F^{0}_{\mathbf{k}}(t,s)R_{-\mathbf{k}}(t',s) \tag{20}$$

where t_0 is the initial time and

$$F^{0}_{\mathbf{k}}(t,s) = < \widetilde{f}^{0}_{\mathbf{k}}(t)\widetilde{f}^{0*}_{\mathbf{k}}(s) > . \tag{21}$$

Thus, the two-point cumulant equation is coupled to the response function through the random forcing term, but it also turns out through the closure for the three-point cumulant in Equation (19). When this closure is performed [85], the two-time cumulant equation becomes

$$\begin{aligned}&\left(\tfrac{\partial}{\partial t} + \nu_0(\mathbf{k})k^2\right)C_{\mathbf{k}}(t,t') + \int_{t_0}^{t} ds\eta_{\mathbf{k}}(t,s)C_{-\mathbf{k}}(t',s) \\ &= \int_{t_0}^{t'} ds(S_{\mathbf{k}}(t,s) + F^{0}_{\mathbf{k}}(t,s))R_{-\mathbf{k}}(t',s)\end{aligned} \tag{22}$$

where $t > t'$ with $C_{\mathbf{k}}(t,t') = C_{-\mathbf{k}}(t',t)$ for $t' > t$. In a similar way, the response function equation can be derived [85] as

$$\left(\frac{\partial}{\partial t} + \nu_0(\mathbf{k})k^2\right)R_{\mathbf{k}}(t,t') + \int_{t'}^{t} ds\eta_{\mathbf{k}}(t,s)R_{\mathbf{k}}(s,t') = \delta(t-t') \tag{23}$$

for $t \geq t'$ and $R_{\mathbf{k}}(t,t) = 1$.

In Equations (22) and (23)

$$\eta_{\mathbf{k}}(t,s) = -4\sum_{\mathbf{p}}\sum_{\mathbf{q}}\delta(\mathbf{k},\mathbf{p},\mathbf{q})K(\mathbf{k},\mathbf{p},\mathbf{q})K(-\mathbf{p},-\mathbf{q},-\mathbf{k})R_{-\mathbf{p}}(t,s)C_{-\mathbf{q}}(t,s), \tag{24}$$

and

$$S_{\mathbf{k}}(t,s) = 2\sum_{\mathbf{p}}\sum_{\mathbf{q}}\delta(\mathbf{k},\mathbf{p},\mathbf{q})K(\mathbf{k},\mathbf{p},\mathbf{q})K(-\mathbf{k},-\mathbf{p},-\mathbf{q})C_{-\mathbf{p}}(t,s)C_{-\mathbf{q}}(t,s). \tag{25}$$

The above two terms are known in field theory as self-energies. They modify or renormalize the damping or forcing in the two-point cumulant and response function equations. The term $\eta_{\mathbf{k}}(t,s)$ is the nonlinear damping that appears in the two-point cumulant and response function equations. The nonlinear noise term, $S_{\mathbf{k}}(t,s)$, renormalizes the bare noise spectrum $F^0_{\mathbf{k}}(t,s)$ in the two-point cumulant equation. The noise terms are positive semi-definite.

The system of statistical dynamical equations is finally closed by the equation for the single-time two-point cumulant:

$$\begin{aligned}&\left(\frac{\partial}{\partial t} + 2\mathrm{Re}\nu_0(\mathbf{k})k^2\right)C_{\mathbf{k}}(t,t) + 2\mathrm{Re}\int_{t_0}^{t} ds\eta_{\mathbf{k}}(t,s)C_{-\mathbf{k}}(t,s)\\ &= 2\mathrm{Re}\int_{t_0}^{t} ds(S_{\mathbf{k}}(t,s) + F^0_{\mathbf{k}}(t,s))R_{-\mathbf{k}}(t,s)\end{aligned} \tag{26}$$

where the initial conditions $C_{\mathbf{k}}(t_0,t_0)$ are to be specified.

5.2. The SCFT and LET Closures for Homogeneous Turbulence

McComb [37] has discussed in detail the historical developments of the SCFT closure of Herring and the LET closure of McComb. However, as noted in the introduction, and in Ref. [43], the SCFT and LET closures can be obtained from the DIA by imposing the prior-time FDT. The three closures have in common the single-time cumulant prognostic in Equation (26). The SCFT also has the same expression as the DIA for the evolution of the response function in Equation (23), while the LET and DIA two-time cumulants are both determined by Equation (22). For the SCFT, the two-time cumulant is then determined by the prior-time FDT in Equation (1), and for the LET, the response function is instead obtained by this FDT.

5.3. Regularized non-Markovian Closures for Homogeneous Turbulence

Kraichnan [29] attributed the incorrect inertial ranges of the Eulerian DIA to spurious convection (advection) effects of the large eddies on the small-scale eddies. He showed that this could be overcome by restricting the ranges of interactions in the response function and two-time cumulant equations. He argued that this approach was analogous to using a quasi-Lagrangian formulation. We see the procedure as a regularization that corresponds to an empirical vertex renormalization. Specifically, we define

$$\breve{K}(\mathbf{k},\mathbf{p},\mathbf{q}) = \theta(p - k/\alpha)\theta(q - k/\alpha)K(\mathbf{k},\mathbf{p},\mathbf{q}), \tag{27}$$

where α is a wavenumber cut-off parameter and θ is the Heaviside step function. As noted in the introduction, α only depends weakly on whether the turbulence is 2D or 3D or whether it is homogeneous or inhomogeneous. The regularized non-Markovian closures are obtained by replacing the interaction coefficient $K(\mathbf{k},\mathbf{p},\mathbf{q})$ by $\breve{K}(\mathbf{k},\mathbf{p},\mathbf{q})$ in Equation (23) for the response function and in Equation (22) for the two-time cumulant, but not in Equation (26) for the single-time cumulant.

6. Statistical Dynamical Equations for Markovian Anisotropic Closures

Next, we develop the theory of Markovian anisotropic closures (MACs) that consist of the single-time cumulant equation and an auxiliary differential equation for the evolution of the triad relaxation function. The MACs are designed to describe the statistical dynamics of anisotropic turbulence, and one of the variants is guaranteed realizability, even in the presence of transient Rossby waves. Based on the non-Markovian DIA in Section 5, we formulate three variants of the MACs that we refer to as MAC^X. Here, the superscript X relates to that used in the combined FDT in Equation (4), with $X = 0$ being the current-time FDT, $X = \frac{1}{2}$ the correlation FDT, and $X = 1$ the prior-time FDT.

The single-time two-point cumulant equation for the DIA, SCFT and LET closures is the same and can be written in the following form:

$$\frac{\partial}{\partial t} C_{\mathbf{k}}(t,t) + 2\mathrm{Re}[\mathcal{N}^{\eta}_{\mathbf{k}}(t) + \mathcal{N}^{0}_{\mathbf{k}}] = 2\mathrm{Re}[\mathcal{F}^{S}_{\mathbf{k}}(t) + \mathcal{F}^{0}_{\mathbf{k}}(t)] \tag{28}$$

where $C_{\mathbf{k}}(t,t)$ is real. The $\mathcal{F}_{\mathbf{k}}(t)$ and $\mathcal{N}_{\mathbf{k}}(t)$ functions have the following expressions:

$$\mathcal{F}^{S}_{\mathbf{k}}(t) = 2\sum_{\mathbf{p}}\sum_{\mathbf{q}} \delta(\mathbf{k},\mathbf{p},\mathbf{q}) K(\mathbf{k},\mathbf{p},\mathbf{q}) K(-\mathbf{k},-\mathbf{p},-\mathbf{q}) \boldsymbol{\Delta}(-\mathbf{k},-\mathbf{p},-\mathbf{q})(t), \tag{29}$$

with

$$\boldsymbol{\Delta}(-\mathbf{k},-\mathbf{p},-\mathbf{q})(t) = \int_{t_0}^{t} ds R_{-\mathbf{k}}(t,s) C_{-\mathbf{p}}(t,s) C_{-\mathbf{q}}(t,s), \tag{30}$$

and

$$\mathcal{F}^{0}_{\mathbf{k}}(t) = \int_{t_0}^{t} ds F^{0}_{\mathbf{k}}(t,s) R_{-\mathbf{k}}(t,s). \tag{31}$$

Also

$$\mathcal{N}^{\eta}_{\mathbf{k}}(t) = -4\sum_{\mathbf{p}}\sum_{\mathbf{q}} \delta(\mathbf{k},\mathbf{p},\mathbf{q}) K(\mathbf{k},\mathbf{p},\mathbf{q}) K(-\mathbf{p},-\mathbf{q},-\mathbf{k}) \Delta(-\mathbf{p},-\mathbf{q},-\mathbf{k})(t), \tag{32}$$

and

$$\mathcal{N}^{0}_{\mathbf{k}}(t) = \nu_0(\mathbf{k}) k^2 C_{\mathbf{k}}(t,t) = \mathcal{D}^{0}_{\mathbf{k}} C_{\mathbf{k}}(t,t), \tag{33}$$

where

$$\mathcal{D}^{0}_{\mathbf{k}} = \nu_0(\mathbf{k}) k^2. \tag{34}$$

We now apply the FDTs in Equation (4) to simplify the nonlinear noise and damping terms in Equations (29) and (32). The time history integrals can then be expressed by relaxation functions $\boldsymbol{\Theta}^X$. The expressions for the relaxation functions can, in turn, be expressed through time dependent differential equations. This then effects the Markovianization with the single-time cumulant equation augmented by the differential equations for $\boldsymbol{\Theta}^X$. Thus,

$$\begin{aligned} \mathcal{F}^{S}_{\mathbf{k}}(t) = {} & 2\sum_{\mathbf{p}}\sum_{\mathbf{q}} \delta(\mathbf{k},\mathbf{p},\mathbf{q}) K(\mathbf{k},\mathbf{p},\mathbf{q}) K(-\mathbf{k},-\mathbf{p},-\mathbf{q}) \\ & \times C^{1-X}_{-\mathbf{p}}(t,t) C^{1-X}_{-\mathbf{q}}(t,t) \boldsymbol{\Theta}^{X}(-\mathbf{k},-\mathbf{p},-\mathbf{q})(t), \end{aligned} \tag{35}$$

with

$$\boldsymbol{\Theta}^{X}(-\mathbf{k},-\mathbf{p},-\mathbf{q})(t) = \int_{t_0}^{t} ds R_{-\mathbf{k}}(t,s) R_{-\mathbf{p}}(t,s) R_{-\mathbf{q}}(t,s) C^{X}_{-\mathbf{p}}(s,s) C^{X}_{-\mathbf{q}}(s,s). \tag{36}$$

Also

$$\mathcal{N}^{\eta}_{\mathbf{k}}(t) = \mathcal{D}^{\eta}_{\mathbf{k}}(t) C_{\mathbf{k}}(t,t), \tag{37}$$

with

$$\begin{aligned} \mathcal{D}^{\eta}_{\mathbf{k}}(t) = {} & -4\sum_{\mathbf{p}}\sum_{\mathbf{q}} \delta(\mathbf{k},\mathbf{p},\mathbf{q}) K(\mathbf{k},\mathbf{p},\mathbf{q}) K(-\mathbf{p},-\mathbf{q},-\mathbf{k}) \\ & \times C^{1-X}_{-\mathbf{q}}(t,t) C^{-X}_{-\mathbf{k}}(t,t) \boldsymbol{\Theta}^{X}(-\mathbf{p},-\mathbf{q},-\mathbf{k})(t). \end{aligned} \tag{38}$$

The single-time cumulant equation then simplifies to

$$\left(\frac{\partial}{\partial t} + 2\mathrm{Re}(\mathcal{D}^r_{\mathbf{k}}(t))\right) C_{\mathbf{k}}(t,t) = 2\mathrm{Re}(\mathcal{F}^r_{\mathbf{k}}(t)) \tag{39}$$

and the response function equation becomes

$$\frac{\partial}{\partial t} R_{\mathbf{k}}(t,t') + \mathcal{D}^r_{\mathbf{k}}(t) R_{\mathbf{k}}(t,t') = \delta(t-t'). \tag{40}$$

Here,

$$\mathcal{D}^r_{\mathbf{k}}(t) = \mathcal{D}^0_{\mathbf{k}} + \mathcal{D}^\eta_{\mathbf{k}}(t)\,;\, \mathcal{F}^r_{\mathbf{k}}(t) = \mathcal{F}^0_{\mathbf{k}}(t) + \mathcal{F}^S_{\mathbf{k}}(t) \tag{41}$$

define the renormalized dissipation operator $\mathcal{D}^r_{\mathbf{k}}(t)$ and the renormalized stochastic force $\mathcal{F}^r_{\mathbf{k}}(t)$.

The integral form for the relaxation functions $\mathbf{\Theta}^X$ in Equation (36) can be replaced by differential equations since the response functions are simplified in Equation (40). The ordinary differential equation for $\mathbf{\Theta}^X$ is:

$$\begin{aligned} &\tfrac{\partial}{\partial t}\, \mathbf{\Theta}^X(\mathbf{k},\mathbf{p},\mathbf{q})(t) + (\mathcal{D}^r_{\mathbf{k}}(t) + \mathcal{D}^r_{\mathbf{p}}(t) + \mathcal{D}^r_{\mathbf{q}}(t))\mathbf{\Theta}^X(\mathbf{k},\mathbf{p},\mathbf{q})(t) \\ &= C^X_{\mathbf{p}}(t,t) C^X_{\mathbf{q}}(t,t) \end{aligned} \tag{42}$$

with $\mathbf{\Theta}^X(\mathbf{k},\mathbf{p},\mathbf{q})(0) = 0$ and $\mathcal{D}^r_{\mathbf{k}}$ given in Equation (41).

The three MACs with $X = 0,\ \frac{1}{2},\ 1$ are specified by Equation (39) for the single-time cumulant $C_{\mathbf{k}}(t,t)$, together with Equation (42) for the relaxation function $\mathbf{\Theta}^X(\mathbf{k},\mathbf{p},\mathbf{q})(t)$. The MACs with $X = 0,\ \frac{1}{2},\ 1$ are all realizable in the absence of transient waves like Rossby waves and the variant with $X = \frac{1}{2}$ can also be shown to be realizable even in the presence of such waves [67].

We show, in the next section, that it is possible to derive a Markovian closure with analytical representation of the triad relaxation function, which is, therefore, even more computationally efficient, from the MAC with $X = 0$. Note that when $X = 0$, Equation (39) for the single-time cumulant becomes:

$$\begin{aligned} &\left(\tfrac{\partial}{\partial t} + 2\hat{\nu}_0(k)k^2\right) C_{\mathbf{k}}(t,t) \\ &= 8\sum_{\mathbf{p}}\sum_{\mathbf{q}} \delta(\mathbf{k},\mathbf{p},\mathbf{q}) K(\mathbf{k},\mathbf{p},\mathbf{q}) K(\mathbf{p},\mathbf{q},\mathbf{k}) \mathrm{Re}\mathbf{\Theta}^0(\mathbf{k},\mathbf{p},\mathbf{q})(t) \\ &\times C_{\mathbf{q}}(t,t)\left[C_{\mathbf{k}}(t,t) - C_{\mathbf{p}}(t,t)\right]. \end{aligned} \tag{43}$$

Here, we have used the properties of the interaction coefficients that $K(\mathbf{k},\mathbf{p},\mathbf{q}) = K(\mathbf{k},\mathbf{q},\mathbf{p})$, and $K(\mathbf{k},\mathbf{p},\mathbf{q}) + K(\mathbf{p},\mathbf{q},\mathbf{k}) + K(\mathbf{q},\mathbf{k},\mathbf{p}) = 0$, the fact that the single-time cumulants are real, and the symmetry properties of the triad relaxation functions $\mathbf{\Theta}^0$.

7. Realizable Eddy-Damped Markovian Anisotropic Closure

The MAC closures with $X = 0$ and in Equations (36) and (43) can be simplified in a similar way to the derivation of the EDQNM closure [2,3,67–69]. In this section, we establish the EDMAC model which is a suitable realizable generalization of the EDQNM for homogeneous turbulent flows interacting with transient Rossby waves. We seek to replace the differential equation in Equation (42) with an analytical parameterized expression that generalizes that used for the EDQNM model.

Firstly, we note that the solution to the response function differential equation in Equation (40) is

$$R_{\mathbf{k}}(t,t') = \exp\left(-\int_{t'}^{t} ds \mathcal{D}^r_{\mathbf{k}}(s)\right) \tag{44}$$

where $\mathcal{D}^r_{\mathbf{k}}$ is defined in Equation (41). Thus, from Equation (36), the triad relaxation function with $X = 0$ simplifies to

$$\mathbf{\Theta}^0(\mathbf{k},\mathbf{p},\mathbf{q})(t) = \int_{t_0}^{t} dt' \exp\left(-\int_{t'}^{t} ds\left[\mathcal{D}^r_{\mathbf{k}}(s) + \mathcal{D}^r_{\mathbf{p}}(s) + \mathcal{D}^r_{\mathbf{q}}(s)\right]\right). \tag{45}$$

Next, we make the Markov approximation for the $\mathcal{D}^r_{\mathbf{k}}$ such $\mathcal{D}^r_{\mathbf{k}}(s) \to \mathcal{D}^r_{\mathbf{k}}(t)$. The generalized dissipation terms can therefore be taken outside the integrals in Equation (44). Thus,

$$R_{\mathbf{k}}(t,t') = \exp(-\mathcal{D}^r_{\mathbf{k}}(t)(t-t')), \tag{46}$$

and

$$\begin{aligned}\Theta^0(\mathbf{k},\mathbf{p},\mathbf{q})(t) &= \int_{t_0}^{t} dt' \exp\Big(-\Big[\mathcal{D}^r_{\mathbf{k}}(t)+\mathcal{D}^r_{\mathbf{p}}(t)+\mathcal{D}^r_{\mathbf{q}}(t)\Big](t-t')\Big) \\ &= \frac{1-\exp\Big(-\Big[\mathcal{D}^r_{\mathbf{k}}(t)+\mathcal{D}^r_{\mathbf{p}}(t)+\mathcal{D}^r_{\mathbf{q}}(t)\Big](t-t_0)\Big)}{\Big[\mathcal{D}^r_{\mathbf{k}}(t)+\mathcal{D}^r_{\mathbf{p}}(t)+\mathcal{D}^r_{\mathbf{q}}(t)\Big]}.\end{aligned} \tag{47}$$

have simpler analytical forms.

7.1. Analytical Triad Relaxation Function for EDQNM

In the EDQNM the only prognostic equation is Equation (43) for the second order cumulant $C_{\mathbf{k}}(t,t)$. Moreover, the eddy damping $\mathcal{D}^\eta_{\mathbf{k}}$ that appears through $\mathcal{D}^r_{\mathbf{k}} = \mathcal{D}^\eta_{\mathbf{k}} + \mathcal{D}^0_{\mathbf{k}}$ in the triad relaxation function in Equation (47) is generally specified by an analytical form that is consistent with the k^{-3} enstrophy cascading inertial range:

$$\mathcal{D}^\eta_{\mathbf{k}}(t) \to \mu^{eddy}_{\mathbf{k}}(t) = \gamma\Big[k^2 C_{\mathbf{k}}(t,t)\Big]^{\frac{1}{2}}. \tag{48}$$

Integral forms over wavenumbers have also been used for the eddy damping [86], and our arguments here and in Section 7.2 apply equally for those forms. We note that $C_{\mathbf{k}}(t,t)$ is real and positive and γ is a positive empirically determined dimensionless coefficient. Thus, the EDQNM for homogeneous turbulence has the considerable simplification and computational efficiency of having an analytical expression for the triad relaxation time $\Theta^{EDQNM}(\mathbf{k},\mathbf{p},\mathbf{q})(t)$, given in Equation (47), with the superscript $0 \to EDQNM$.

For homogeneous turbulence on an f–plane, without Rossby waves, including for HIT, $\mathcal{D}^0_{\mathbf{k}} = \hat{\nu}_0(k)k^2 > 0$ is real as is $\mathcal{D}^\eta_{\mathbf{k}}(t) \to \mu^{eddy}_{\mathbf{k}}(t) > 0$ in Equation (47). Thus,

$$\Theta^{EDQNM}(\mathbf{k},\mathbf{p},\mathbf{q})(t) = \frac{1-\exp(-[\mu_{\mathbf{k}}(t)+\mu_{\mathbf{p}}(t)+\mu_{\mathbf{q}}(t)](t-t_0))}{\mu_{\mathbf{k}}(t)+\mu_{\mathbf{p}}(t)+\mu_{\mathbf{q}}(t)} \tag{49}$$

where

$$\mu_{\mathbf{k}}(t) = \hat{\nu}_0(k)k^2 + \mu^{eddy}_{\mathbf{k}}(t) = \hat{\nu}_0(k)k^2 + \gamma\Big[k^2 C_{\mathbf{k}}(t,t)\Big]^{\frac{1}{2}}. \tag{50}$$

Now, $\Theta^{EDQNM}(\mathbf{k},\mathbf{p},\mathbf{q})(t) = \mathrm{Re}\Theta^{EDQNM}(\mathbf{k},\mathbf{p},\mathbf{q})(t) \geq 0$ is both real and non-negative, and this ensures that the cumulant $C_{\mathbf{k}}(t,t)$ is also real and non-negative, and thus realizable, as also shown in Appendix B with superscript $EDMAC \to EDQNM$.

For homogeneous anisotropic turbulence interacting with transient Rossby waves on a β-plane $\mathcal{D}^0_{\mathbf{k}} = \hat{\nu}_0(k)k^2 + i\omega_{\mathbf{k}}$ and taking $\mathcal{D}^\eta_{\mathbf{k}}(t) \to \mu^{eddy}_{\mathbf{k}}(t)$ again we have

$$R_{\mathbf{k}}(t,t') \doteq R^{EDQNM}_{\mathbf{k}}(t,t') = \exp(-[\mu_{\mathbf{k}}(t)+i\omega_{\mathbf{k}}](t-t')), \tag{51}$$

and

$$\begin{aligned}&\Theta^{EDQNM}(\mathbf{k},\mathbf{p},\mathbf{q})(t) \\ &= \frac{1-\exp\big(-\big[\mu_{\mathbf{k}}(t)+\mu_{\mathbf{p}}(t)+\mu_{\mathbf{q}}(t)+i(\omega_{\mathbf{k}}+\omega_{\mathbf{p}}+\omega_{\mathbf{q}})\big](t-t_0)\big)}{\mu_{\mathbf{k}}(t)+\mu_{\mathbf{p}}(t)+\mu_{\mathbf{q}}(t)+i(\omega_{\mathbf{k}}+\omega_{\mathbf{p}}+\omega_{\mathbf{q}})}.\end{aligned} \tag{52}$$

Unfortunately, the wave terms mean it is not possible to guarantee that $\mathrm{Re}\Theta^{EDQNM}(\mathbf{k},\mathbf{p},\mathbf{q})(t) \geq 0$, as noted by Bowman et al. [67], and so there may be situations where $C_{\mathbf{k}}(t,t)$ is no longer realizable.

7.2. Analytical Relaxation Function for EDMAC

The EDMAC generalizes the EDQNM model by using an analytical form for the triad relaxation function that is realizable in the presence of transient Rossby waves (or drift-waves or, indeed, other waves). It thus solves the problem that was the focus of Bowman et al. [67] but without the introduction of an auxiliary Markovian evolution equation for the triad relaxation function. As we have noted above, although the auxiliary equation makes the whole system Markovian, it is still a large computational overhead since the triad relaxation function depends on time and six spectral space dimensions that reduce to four when the Kronecker delta functions in Equations (35) and (38) are implemented. This is for 2D turbulence with corresponding larger computational effort required for 3D turbulence.

In the EDMAC model for homogeneous anisotropic turbulence interacting with transient Rossby waves on a β-plane we make the replacement

$$\mathcal{D}^r_{\mathbf{k}}(t) \to \rho_{\mathbf{k}}(t) + i\omega_{\mathbf{k}} \tag{53}$$

where

$$\rho_{\mathbf{k}}(t) = \mu_{\mathbf{k}}(t) + c\frac{\omega_{\mathbf{k}}^2}{\mu_{\mathbf{k}}(t)} \tag{54}$$

with $\mu_{\mathbf{k}}(t)$ given in Equation (50) and a typical value of $c = \frac{1}{2}$. Thus, the response function equation becomes

$$R_{\mathbf{k}}(t,t') \doteq R^{EDMAC}_{\mathbf{k}}(t,t') = \exp(-[\rho_{\mathbf{k}}(t) + i\omega_{\mathbf{k}}](t-t')), \tag{55}$$

and

$$\begin{aligned} &\Theta^{EDMAC}(\mathbf{k},\mathbf{p},\mathbf{q})(t) \\ &= \frac{1-\exp\big(-\big[\rho_{\mathbf{k}}(t)+\rho_{\mathbf{p}}(t)+\rho_{\mathbf{q}}(t)+i(\omega_{\mathbf{k}}+\omega_{\mathbf{p}}+\omega_{\mathbf{q}})\big](t-t_0)\big)}{\rho_{\mathbf{k}}(t)+\rho_{\mathbf{p}}(t)+\rho_{\mathbf{q}}(t)+i(\omega_{\mathbf{k}}+\omega_{\mathbf{p}}+\omega_{\mathbf{q}})}. \end{aligned} \tag{56}$$

As shown in Appendices A and B, the EDMAC model is realizable for all $c \geq \frac{1}{4}$. The eddy damping in Equation (50) was of course chosen by Orszag [2] on empirical grounds and from a practical point of view it would probably not matter if the frequency renormalized eddy damping was arrived at in the same way. However, in a sequel where we study the performance of the EDMAC model with transient Rossby waves, we also aim to arrive at the form above through renormalized perturbation theory. It is clear that for small $\omega_{\mathbf{k}}$ the frequency squared term in Equation (54) is negligible. As well, for large wave numbers k, such as in typical geophysical enstrophy cascading inertial ranges, the frequency squared term also becomes negligible. Thus, we expect that the EDMAC model will be as computationally efficient as the EDQNM but will in addition be realizable in the presence of transient Rossby waves (and other waves) since $\mathrm{Re}\Theta^{EDMAC}(\mathbf{k},\mathbf{p},\mathbf{q})(t) \geq 0$ for all $c \geq \frac{1}{4}$. As well, we expect that at least for small amplitude waves the efficient performance and veracity of the EDMAC will be like that of the EDQNM but without the possibility of blow up.

7.3. EDMAC for Three-Dimensional Turbulent Flows

There are various possible extensions of the frequency renormalized eddy damping parameterization for the realizable EDMAC that we have presented for interaction of 2D turbulence with Rossby waves. Of course, the EDQNM was initially formulated for 3D HIT [2] and has been applied to a variety of problems in 2D and 3D HIT and HAT, including with rotation [46,47]. We expect similar generalizations should be possible for the realizable EDMAC. Indeed, our results are easily extended to the model of 3D quasigeostrophic flows in Appendix B of Frederiksen [60], suitably generalized for flow on a β-plane with Rossby waves.

8. Reflections and Perspectives

8.1. Jorgen's Personal Reflections

Jack Herring first loomed large as a giant intellect and pioneer of statistical fluid dynamics during my nine-month-long visit to NCAR in 1980 as part of a year's sabbatical from CSIRO Division of Atmospheric Physics. I was an early career research scientist at

CSIRO, having switched from quantum field theory to classical geophysical fluid dynamics five years earlier. My Ph.D. and post doc had involved determining the analytical properties of Feynman diagrams, particularly the class of loop diagrams, which form the basis of dispersion relations. These dispersion relations included the quadratic statistical closure equations that describe meson-meson scattering—the core and iconic problem of strong interaction quantum field theories. I was, therefore, very keen to learn as much as possible about statistical turbulence closure theory and geophysical fluid dynamics as possible. I suggested that the 'Downunder' tradition of afternoon tea and coffee be instituted at the Climate and Global Dynamics Group and at it, and at lunch, learnt an amazing amount of science from the great scientists at NCAR (a list too long to produce) and from more junior staff and post docs. During this time, my friendship with Jack developed and continued over the years.

As documented in Section 2, Jack has been an enormous influence on my career with much of my work following in his footsteps. When the time came to launch into the complexities of the numerical formulation of turbulence closure codes, Jack very generously showed and explained all the tricks needed to efficiently implement the non-Markovian integro-differential equations. Jack was always very friendly, cheerful, and unassuming, always interested, and posed perceptive questions, including at seminars. Always there during my short and longer visits to NCAR. Jack took a lot of interest, particularly in our work on the inhomogeneous QDIA closures, from being thesis examiner of Terry's Ph.D. on the numerical implementation and further development of the closure to our most recent publication in Fluids 2022 [44].

8.2. Perspectives on Strong Interaction Theories

The non-Markovian closure theories of Kraichnan, Herring, Edwards and McComb for HIT tackle the core and iconic problem of strong turbulence without the complexities of inhomogeneity. They are accurate at the large energy containing scales but their power law behaviour in inertial ranges have, in principle or in practice, deficiencies since they do not fundamentally address the vertex renormalization problem. However, restricting the wavenumber ranges in the two-time cumulant and response function equations in one parameter regularized versions of some of these closures gives excellent results. This is also the case for the QDIA closure for inhomogeneous turbulence.

To put into perspective the state of strong interaction turbulence closure theories, it is perhaps useful to look at the corresponding situation in strong interaction quantum field theory. As noted in the previous subsection, pion-pion scattering is the core and iconic problem of strong interaction hadron physics. Mandelstam (1958) [87] developed closure equations for hadron scattering which are quadratic in the partial wave spectral scattering amplitudes. A question of major interest was whether the scattering equations have unique solutions that can be calculated through convergent iteration. To answer this question, it was necessary to reformulate the equations, and the proof was established for pion-pion scattering (Frederiksen 1975 [88]; Atkinson et al., 1976 [89]). This was done using functional analysis contraction mapping theorems in Banach space of doubly Hölder continuous scattering amplitudes. However, it has taken until 2023 for the iteration procedure to be numerically implemented to generate the nonperturbative scattering amplitudes in a remarkable effort by Tourkine and Zhiboedov (2023) [90].

While scattering between asymptotic 'in' and 'out' states has been the focus of much of strong interaction hadron physics, often with introduction of some empirical data, time-dependent non-equilibrium quantum field theories have become of increasing interest in the last two decades. This is particularly so in studies of Bose–Einstein condensation far from equilibrium [91,92], in studies of cosmology and inflation [93–95], and quark-gluon plasma [96]. In these problems, the generalization from homogeneous to inhomogeneous statistical quantum field theory is required to describe the time evolution of quantum fluctuations in spatially inhomogeneous dynamical fields. The Schwinger–Keldysh CTP approach has generally been the basis for these studies. The Schwinger–Keldysh [81,82] equations, like Kraichnan's [97] inhomogeneous DIA (IDIA), and the MSR [53] and path integral extensions [78,79], require the full covariance matrix to close the mean field equation. This is a severe restriction on the size of the problems that can be solved since if the field has N degrees of freedom the full covariance matrix has N^2 (the QDIA gets around this problem as described in Section 2). Consequently, the full inhomogeneous problem has generally only been tackled in one space dimension [98–100]. Interestingly in studies of the evolution of a Bose gas, Cooper et al. [99] found that the bare vertex approximation performs better than other closure schemes like the two-particle irreducible expansion. The bare vertex approximation is the basis of the DIA, SCFT, LET and Edward's closures and guarantees the realizability of the DIA, the IDIA, the QDIA and the bare vertex quantum closure equations.

The statistical dynamics of strongly interacting fields is an enormously difficult problem. The achievements of the pioneers in fluid turbulence closure theory have been truly outstanding by any comparison.

9. Conclusions

Jack Herring made remarkable pioneering steps and contributions to laying the foundations of the statistical dynamical closure theory of fluid turbulence. Some of his important works have been briefly reviewed, their impacts and related developments discussed, and some of the further developments and extensions summarized.

A particular focus of this article has also been to present the eddy damped Markovian anisotropic closure (EDMAC), a generalization of the eddy damped quasi normal Markovian (EDQNM) that is realizable for anisotropic turbulence interacting with transient waves such as Rossby waves. By construction, the EDMAC is as computationally efficient as the EDQNM but overcomes a long-standing problem of realizability with transient waves present. This builds on Herring's (1975) [1] generalization of the EDQNM for anisotropic 2D turbulence and his studies of the relaxation to isotropy.

Author Contributions: Conceptualization, J.S.F. and T.J.O.; methodology, J.S.F.; formal analysis, J.S.F.; investigation, J.S.F. and T.J.O.; writing—original draft preparation, J.S.F.; writing—review and editing, J.S.F. and T.J.O.; supervision, J.S.F.; project administration, J.S.F.; funding acquisition, T.J.O. All authors have read and agreed to the published version of the manuscript.

Funding: T.J.O. was funded by CSIRO Environment.

Institutional Review Board Statement: Not applicable.

Informed Consent Statement: Not applicable.

Data Availability Statement: Data sharing is not applicable since no new data were created or analysed in this article.

Conflicts of Interest: The authors declare no conflict of interest.

Appendix A. Positive Semi-Definite Triad Relaxation Time with Rossby Waves

As noted in Section 7, lack of realizability of the EDQNM model in the presence of transient waves can be overcome by using a frequency renormalized eddy damping. The response function is then:

$$R_{\mathbf{k}}(t,t') = \exp -[\rho_{\mathbf{k}}(t) + i\omega_{\mathbf{k}}](t-t') \tag{A1}$$

where $\rho_{\mathbf{k}}(t) = \mu_{\mathbf{k}}(t) + c\omega_{\mathbf{k}}^2/\mu_{\mathbf{k}}(t) > 0$ is the frequency renormalized damping, $\mu_{\mathbf{k}}(t) > 0$ is the damping, $\omega_{\mathbf{k}}$ is the Rossby wave frequency and a typical value of $c = \frac{1}{2}$. The triad relaxation time is then given by

$$\Theta^{EDMAC}(\mathbf{k}_1, \mathbf{k}_2, \mathbf{k}_3)(t) = \frac{1 - \exp -[\rho + i\omega](t - t_0)}{\rho + i\omega} \tag{A2}$$

where, without loss of generality, we take $t_0 = 0$. Here,

$$\begin{aligned} \mu &= \mu_{\mathbf{k}_1}(t) + \mu_{\mathbf{k}_2}(t) + \mu_{\mathbf{k}_3}(t), \\ \omega &= \omega_{\mathbf{k}_1} + \omega_{\mathbf{k}_2} + \omega_{\mathbf{k}_3} \\ \rho_j &= \mu_{\mathbf{k}_j} + c\frac{\omega^2_{\mathbf{k}_j}}{\mu_{\mathbf{k}_j}},\ j = 1,2,3 \\ \rho &= \rho_1 + \rho_2 + \rho_3. \end{aligned} \tag{A3}$$

Then

$$\begin{aligned} &\mathrm{Re}\boldsymbol{\Theta}^{EDMAC}(\mathbf{k}_1, \mathbf{k}_2, \mathbf{k}_3)(t) \\ &= \tfrac{1}{\rho^2+\omega^2}[\rho\{1 - [\exp -\rho t]\cos\omega t\} + \omega[\exp -\rho t]\sin\omega t] \end{aligned} \tag{A4}$$

where

$$\frac{\partial \mathrm{Re}\boldsymbol{\Theta}^{EDMAC}(\mathbf{k}_1, \mathbf{k}_2, \mathbf{k}_3)(t)}{\partial t} = [\exp -\rho t]\cos\omega t. \tag{A5}$$

Next, we examine the conditions under which $\mathrm{Re}\boldsymbol{\Theta}^{EDMAC}(\mathbf{k}_1, \mathbf{k}_2, \mathbf{k}_3)(t) \geq 0$ so that the EDMAC model of Section 7 (Equations (39) and (43)) is realizable as outlined in Appendix B. Firstly, we note that for $t \geq 0$, $\cos\omega t$ is then an even function of ω and both ω and $\sin\omega t$ are odd functions with their product $\omega\sin\omega t$ being even. Thus, for $0 < |\omega| t \leq \pi$

$$1 - [\exp -\rho t]\cos|\omega|t > 0;\ |\omega|\ [\exp -\rho t]\sin|\omega|t \geq 0. \tag{A6}$$

Further, for $|\omega|t > \pi$,

$$\exp -\rho t < \exp -(\rho\frac{\pi}{|\omega|}) < \exp -(\pi) < 0.05 \tag{A7}$$

provided

$$\rho \geq |\omega|. \tag{A8}$$

Of course, in general

$$-1 \leq \cos|\omega|t \leq 1; -1 \leq \sin|\omega|t \leq 1 \tag{A9}$$

and thus for $|\omega|t > \pi$,

$$\left[\rho^2 + \omega^2\right]\mathrm{Re}\boldsymbol{\Theta}^{EDMAC}(\mathbf{k}_1, \mathbf{k}_2, \mathbf{k}_3)(t) > 0.9\rho > 0. \tag{A10}$$

The relationship in Equation (A8) holds if

$$\left(\mu_{\mathbf{k}_j}(t) + c\frac{\omega^2_{\mathbf{k}_j}}{\mu_{\mathbf{k}_j}(t)}\right) \geq |\omega_{\mathbf{k}_j}|, j = 1,2,3 \tag{A11}$$

and the inequalities in Equation (A11) are valid provided $c \geq \frac{1}{4}$ which is established by solving simple quadratic equations.

Appendix B. Langevin Equation for EDMAC Model

The EDMAC model in Equations (39) and (43) can be shown to be realizable since it is underpinned by a stochastic model as is the EDQNM as discussed by Leith [3] and Herring and Kraichnan [66]. The Langevin equation, which allows precise construction of the EDMAC model, is given by:

$$\left(\frac{\partial}{\partial t} + \mathcal{D}^0_{\mathbf{k}} + \mathcal{D}^\eta_{\mathbf{k}}(t)\right)\tilde{\zeta}_{\mathbf{k}}(t) = f^0_{\mathbf{k}}(t) + f^S_{\mathbf{k}}(t) \tag{A12}$$

where $\mathcal{D}^0_{\mathbf{k}}$ is given in Equation (34) and $\mathcal{D}^{\eta}_{\mathbf{k}}(t)$ is given in Equation (38) with $X = 0$ and $\mathbf{\Theta}^0 \to \mathbf{\Theta}^{EDMAC}$. As well,

$$\not{f}^{0}_{\mathbf{k}}(t) = \tilde{f}^{0}_{\mathbf{k}}, \tag{A13}$$

and

$$\begin{aligned} \not{f}^{S}_{\mathbf{k}}(t) = \sqrt{2}\sum_{\mathbf{p}}\sum_{\mathbf{q}}\delta(\mathbf{k}+\mathbf{p}+\mathbf{q})K(\mathbf{k},\mathbf{p},\mathbf{q})\left[\mathrm{Re}\mathbf{\Theta}^{EDMAC}(\mathbf{k},\mathbf{p},\mathbf{q})(t)\right]^{\frac{1}{2}} \\ \times w(t)\rho^{(1)}_{-\mathbf{p}}(t)\rho^{(2)}_{-\mathbf{q}}(t). \end{aligned} \tag{A14}$$

The variables $\rho^{(i)}_{\mathbf{k}}(t)$, where $i = 1, 2$ or 3, and $w(t)$ are independent random variables that satisfy the following relationships:

$$< \rho^{(i)}_{\mathbf{k}}(t)\rho^{(j)}_{-\mathbf{l}}(t') >= \delta_{ij}\delta_{\mathbf{kl}}C_{\mathbf{k}}(t,t'), \tag{A15}$$

with

$$< \tilde{\zeta}_{\mathbf{k}}(t)\tilde{\zeta}_{-\mathbf{k}}(t') >= C_{\mathbf{k}}(t,t'), \tag{A16}$$

and

$$< w(t)w(t') >= \delta(t-t'). \tag{A17}$$

Here, δ is the Kronecker delta function in Equation (A15), and in Equation (A17) it is the Dirac delta function.

The realizability of the cumulants $C_{\mathbf{k}}(t,t)$, in the EDMAC model is established by the Langevin equation provided $\mathrm{Re}\mathbf{\Theta}^{EDMAC}(\mathbf{k},\mathbf{p},\mathbf{q})(t) \geq 0$. This is in turn is shown in Appendix A to be the case provided $c \geq \frac{1}{4}$. The EDMAC equations also preserve conservation of kinetic energy and potential enstrophy (in the absence of forcing and dissipation).

References

1. Herring, J.R. Theory of two-dimensional anisotropic turbulence. *J. Atmos. Sci.* **1975**, *32*, 2252–2271. [CrossRef]
2. Orszag, S.A. Analytical theories of turbulence. *J. Fluid Mech.* **1970**, *41*, 363–386. [CrossRef]
3. Leith, C.E. Atmospheric predictability and two-dimensional turbulence. *J. Atmos. Sci.* **1971**, *28*, 145–161. [CrossRef]
4. Herring, J.R. On the statistical theory of two-dimensional topographic turbulence. *J. Atmos. Sci.* **1977**, *34*, 1731–1750. [CrossRef]
5. Onsager, L. Statistical hydrodynamics. *Nouvo Cimento* **1949**, *6*, 279–287. [CrossRef]
6. Kraichnan, R.H. Inertial ranges in two-dimensional turbulence. *Phys. Fluids* **1967**, *10*, 1417–1423. [CrossRef]
7. Kraichnan, R. Statistical dynamics of two-dimensional flow. *J. Fluid Mech.* **1975**, *67*, 155–175. [CrossRef]
8. Salmon, R.; Holloway, G.; Hendershott, M. The equilibrium statistical mechanics of simple quasigeostrophic models. *J. Fluid Mech.* **1976**, *75*, 691–703. [CrossRef]
9. Frederiksen, J.S.; Sawford, B.L. Statistical dynamics of two-dimensional inviscid flow on a sphere. *J. Atmos Sci.* **1980**, *37*, 717–732. [CrossRef]
10. Frederiksen, J.S.; Sawford, B.L. Topographic waves in nonlinear and linear spherical barotropic models. *J. Atmos. Sci.* **1981**, *38*, 69–86. [CrossRef]
11. Salmon, R.; Pizzo, N. Two-dimensional flow on the sphere. *Atmosphere* **2023**, *14*, 747. [CrossRef]
12. Bretherton, F.; Haidvogel, D. Two-dimensional turbulence above topography. *J. Fluid Mech.* **1976**, *78*, 129–154. [CrossRef]
13. Arnold, V. On conditions for nonlinear stability of plane stationary curvilinear flows of an ideal fluid. *Sov. Math. Dokl.* **1965**, *162*, 773–777.
14. Frederiksen, J.S.; Carnevale, G.F. Stability properties of exact nonzonal solutions for flow over topography. *Geophys. Astrophys. Fluid Dyn.* **1986**, *35*, 173–207. [CrossRef]
15. Frederiksen, J.S. Nonlinear stability of baroclinic flows over topography. *Geophys. Astrophys. Fluid Mech.* **1991**, *57*, 85–97. [CrossRef]
16. Frederiksen, J.S. Nonlinear studies on the effect of topography on baroclinic zonal flows. *Geophys. Astrophys. Fluid Dyn.* **1991**, *59*, 57–82. [CrossRef]
17. Carnevale, G.F.; Frederiksen, J.S. Nonlinear stability and statistical mechanics of flow over topography. *J. Fluid Mech.* **1987**, *175*, 157–181. [CrossRef]
18. Miller, J. Statistical mechanics of Euler equations in two dimensions. *Phys. Rev. Lett.* **1990**, *65*, 2137–2140. [CrossRef]
19. Miller, J.; Weichman, P.; Cross, M. Statistical mechanical mechanics, Euler equation, and Jupiter's red spot. *Phys. Rev. A* **1992**, *45*, 2328–2359. [CrossRef]

20. Robert, R. A maximum entropy principle for two-dimensional Euler equations. *J. Stat. Phys.* **1991**, *65*, 531–553. [CrossRef]
21. Sommeria, J.; Nore, C.; Dumont, T.; Robert, R. Theorie statistique de la tache rouge de Jupiter. *C. R. Acad. Sci. Paris* **1991**, *312*, 999–1005.
22. Chorin, A.J. Partition functions and equilibrium measures in two-dimensional and quasi-three-dimensional turbulence. *Phys. Fluids* **1996**, *8*, 2656–2660. [CrossRef]
23. Turkington, B. Statistical equilibrium measures and coherent states in two-dimensional turbulence. *Commun. Pure Appl. Math.* **1999**, *52*, 781–809. [CrossRef]
24. Majda, A.J.; Wang, X. *Nonlinear Dynamics and Statistical Theories for Basic Geophysical Flows*; Cambridge University Press: Cambridge, UK, 2006; ISBN -9780511616778. [CrossRef]
25. Frederiksen, J.S.; O'Kane, T.J. Entropy, closures and subgrid modeling. *Entropy* **2008**, *10*, 635–683. [CrossRef]
26. Bouchet, F.; Venaille, A. Statistical mechanics of two-dimensional and geophysical flows. *Phys. Rep.* **2012**, *515*, 227–295. [CrossRef]
27. Zhou, Y. Turbulence theories and statistical closure approaches. *Phys. Rep.* **2021**, *935*, 1–117. [CrossRef]
28. Kraichnan, R.H. The structure of isotropic turbulence at very high Reynolds numbers. *J. Fluid Mech.* **1959**, *5*, 497–543. [CrossRef]
29. Kraichnan, R.H. Decay of isotropic turbulence in the direct-interaction approximation. *Phys. Fluids* **1964**, *7*, 1030–1048. [CrossRef]
30. Wyld, H.W., Jr. Formulation of the theory of turbulence in an incompressible fluid. *Ann. Phys.* **1961**, *14*, 143–165. [CrossRef]
31. Lee, L.L. A formulation of the theory of isotropic hydromagnetic turbulence in an incompressible fluid. *Ann. Phys.* **1965**, *32*, 292–321. [CrossRef]
32. Frederiksen, J.S. Quasi-diagonal inhomogeneous closure for classical and quantum statistical dynamics. *J. Math. Phys.* **2017**, *58*, 103303. [CrossRef]
33. Edwards, S.F. The statistical dynamics of homogeneous turbulence. *J. Fluid Mech.* **1964**, *18*, 239–273. [CrossRef]
34. Herring, J.R. Self-consistent-field approach to turbulence theory. *Phys. Fluids* **1965**, *8*, 2219–2225. [CrossRef]
35. Herring, J.R. Self-consistent-field approach to nonstationary turbulence. *Phys. Fluids* **1966**, *9*, 2106–2110. [CrossRef]
36. Carnevale, G.F.; Frederiksen, J.S. Viscosity renormalization based on direct-interaction closure. *J. Fluid Mech.* **1983**, *131*, 289–304. [CrossRef]
37. McComb, W.D. Jackson R. Herring and the Statistical Closure Problem of Turbulence: A Review of Renormalized Perturbation Theories. *Atmosphere* **2023**, *14*, 827. [CrossRef]
38. McComb, W.D. A local energy-transfer theory of isotropic turbulence. *J. Phys. A* **1974**, *7*, 632–649. [CrossRef]
39. McComb, W.D. *The Physics of Fluid Turbulence*; Oxford University Press: Oxford, UK, 1990; ISBN 9780198562566.
40. McComb, W.D. *Homogeneous, Isotropic Turbulence: Phenomenology, Renormalization and Statistical Closures*; Oxford University Press: Oxford, UK, 2014; ISBN 9780199689385. [CrossRef]
41. Frederiksen, J.S.; Davies, A.G.; Bell, R.C. Closure theories with non-Gaussian restarts for truncated two-dimensional turbulence. *Phys. Fluids* **1994**, *6*, 3153–3163. [CrossRef]
42. Herring, J.R.; Orszag, S.A.; Kraichnan, R.H.; Fox, D.G. Decay of two-dimensional homogeneous turbulence. *J. Fluid Mech.* **1974**, *66*, 417–444. [CrossRef]
43. Frederiksen, J.S.; Davies, A.G. Dynamics and spectra of cumulant update closures for two-dimensional turbulence. *Geophys. Astrophys. Fluid Dyn.* **2000**, *92*, 197–231. [CrossRef]
44. Frederiksen, J.S.; O'Kane, T.J. Statistical Dynamics of Mean Flows Interacting with Rossby Waves, Turbulence, and Topography. *Fluids* **2022**, *7*, 200. [CrossRef]
45. Lesieur, M. *Turbulence in Fluids*; Springer: Dordrecht, The Netherlands, 2008.
46. Cambon, C.; Mons, V.; Gréa, B.-J.; Rubinstein, R. Anisotropic triadic closures for shear-driven and buoyancy-driven turbulent flows. *Computers Fluids* **2017**, *151*, 73–84. [CrossRef]
47. Sagaut, P.; Cambon, C. *Homogeneous Turbulence Dynamics*; Springer Nature: Cham, Switzerland, 2018.
48. Kraichnan, R.H. Lagrangian-history approximation for turbulence. *Phys. Fluids* **1965**, *8*, 575–598. [CrossRef]
49. Kraichnan, R.H. Eulerian and Lagrangian renormalization in turbulence theory. *J. Fluid Mech.* **1977**, *83*, 349–374. [CrossRef]
50. Herring, J.R.; Kraichnan, R.H. A numerical comparison of velocity-based and strain-based Lagrangian-history turbulence approximations. *J. Fluid Mech.* **1979**, *91*, 581–597. [CrossRef]
51. Kaneda, Y. Renormalised expansions in the theory of turbulence with the use of the Lagrangian position function. *J. Fluid Mech.* **1981**, *107*, 131–145. [CrossRef]
52. Gotoh, T.; Kaneda, Y.; Bekki, N. Numerical integration of the Lagrangian renormalized approximation. *J. Phys. Soc. JPN.* **1988**, *57*, 866–880. [CrossRef]
53. Martin, P.C.; Siggia, E.D.; Rose, H.A. Statistical dynamics of classical systems. *Phys. Rev. A* **1973**, *8*, 423–437. [CrossRef]
54. Kraichnan, R.H. Kolmogorov's hypothesis and Eulerian turbulence theory. *Phys. Fluids* **1964**, *7*, 1723–1733. [CrossRef]
55. Sudan, R.N.; Pfirsch, D. On the relation between "mixing length" and "direct interaction approximation" theories of turbulence. *Phys. Fluids* **1985**, *28*, 1702–1718. [CrossRef]
56. Frederiksen, J.S.; Davies, A.G. The regularized DIA closure for two-dimensional turbulence. *Geophys. Astrophys. Fluid Dyn.* **2004**, *98*, 203–223. [CrossRef]
57. O'Kane, T.J.; Frederiksen, J.S. The QDIA and regularized QDIA closures for inhomogeneous turbulence over topography. *J. Fluid Mech.* **2004**, *65*, 133–165. [CrossRef]
58. Kraichnan, R.H. An almost-Markovian Galilean-invariant turbulence model. *J. Fluid Mech.* **1971**, *47*, 513–524. [CrossRef]

59. Frederiksen, J.S. Subgrid-scale parameterizations of eddy-topographic force, eddy viscosity and stochastic backscatter for flow over topography. *J. Atmos. Sci.* **1999**, *56*, 1481–1494. [CrossRef]
60. Frederiksen, J.S. Statistical dynamical closures and subgrid modeling for QG and 3D inhomogeneous turbulence. *Entropy* **2012**, *14*, 32–57. [CrossRef]
61. Frederiksen, J.S. Self-Energy Closure for Inhomogeneous Turbulent Flows and Subgrid Modeling. *Entropy* **2012**, *14*, 769–799. [CrossRef]
62. Frederiksen, J.S.; O'Kane, T.J. Inhomogeneous closure and statistical mechanics for Rossby wave turbulence over topography. *J. Fluid Mech.* **2005**, *539*, 137–165. [CrossRef]
63. Frederiksen, J.S.; Kitsios, V.; O'Kane, T.J.; Zidikheri, M.J. Stochastic Subgrid Modelling for Geophysical and Three-Dimensional Turbulence. In *Nonlinear and Stochastic Climate Dynamics*; Franzke, C.J.E., O'Kane, T.J., Eds.; Cambridge University Press: Cambridge, UK, 2017; pp. 241–275. [CrossRef]
64. Kitsios, V.; Frederiksen, J.S.; O'Kane, T.J. Subgrid parameterization of eddy, meanfield and topographic interactions in simulations of an idealized Antarctic Circumpolar Current. *J. Adv. Model. Earth Syst.* **2023**, *15*, e2022MS003412. [CrossRef]
65. Rose, H.A. An efficient non-Markovian theory of non-equilibrium dynamics. *Physica D* **1985**, *14*, 216–226. [CrossRef]
66. Herring, J.R.; Kraichnan, R.H. Statistical Models and Turbulence. In *Lecture Notes in Physics: Proceedings of a Symposium Held at the University of California*; Ehlers, J.K., Hepp, K.H.A., Weidenmuller, H.A., Eds.; Springer: Berlin/Heidelberg, Germany, 1972; pp. 148–194.
67. Bowman, J.C.; Krommes, J.A.; Ottaviani, M. The realizable Markovian closure. I. General theory, with application to three-wave dynamics. *Phys. Fluids* **1993**, *5*, 3558–3589. [CrossRef]
68. Carnevale, G.F.; Martin, P.C. Field theoretic techniques in statistical fluid dynamics: With application to nonlinear wave dynamics. *Geophys. Astrophys. Fluid Dyn.* **1982**, *20*, 131–164. [CrossRef]
69. Carnevale, G.F.; Frederiksen, J.S. A statistical dynamical theory of strongly nonlinear internal gravity waves. *Geophys. Astrophys. Fluid Dyn.* **1983**, *23*, 175–207. [CrossRef]
70. Zhou, Y.; Matthaeus, W.H.; Dmitruk, P. Magnetohydrodynamic turbulence and time scales in astrophysical and space plasmas. *Rev. Mod. Phys.* **2004**, *76*, 1015–1034. [CrossRef]
71. Sukoriansky, S.; Galperin, B. QNSE theory of turbulence anisotropization and onset of the inverse energy cascade by solid body rotation. *J. Fluid Mech.* **2016**, *805*, 384–421. [CrossRef]
72. Galperin, B.; Sukoriansky, S.; Qiu, B. Seasonal oceanic variability on meso- and sub-mesoscales: A turbulence perspective. *Ocean Dynam.* **2021**, *71*, 475–489. [CrossRef]
73. Hu, G.; Krommes, J.A.; Bowman, J.C. Statistical theory of resistive drift-wave turbulence and transport. *Phys. Plasmas* **1997**, *4*, 2116–2133. [CrossRef]
74. Bowman, J.C.; Krommes, J.A. The realizable Markovian closure and realizable test-field model. II. Application to anisotropic drift-wave dynamics. *Phys. Plasmas* **1997**, *4*, 3895–3909. [CrossRef]
75. Frederiksen, J.S.; O'Kane, T.J. Markovian inhomogeneous closures for Rossby waves and turbulence over topography. *J. Fluid Mech.* **2019**, *858*, 45–70. [CrossRef]
76. Rose, H.A. Aspects of Statistical Dynamics of Classical Systems. Ph.D. Thesis, Harvard University, Cambridge, MA, USA, 1974.
77. Feynman, R.P.; Hibbs, A.R. *Quantum Mechanics and Path Integrals*; Dover Publications: Mineola, NY, USA, 1965.
78. Phythian, R. The functional formalism of classical statistical dynamics. *J. Phys. A Math. Gen.* **1977**, *10*, 777–789. [CrossRef]
79. Jensen, R.V. Functional integral approach to classical statistical dynamics. *J. Stat. Phys.* **1981**, *25*, 183–201. [CrossRef]
80. Berera, A.; Salewski, M.; McComb, W.D. Eulerian field-theoretic closure formalisms for fluid turbulence. *Phys. Rev. E* **2013**, *87*, 013007. [CrossRef] [PubMed]
81. Schwinger, J. Brownian motion of a quantum oscillator. *J. Math. Phys.* **1961**, *2*, 407–432. [CrossRef]
82. Keldysh, L.V. Diagram technique for nonequilibrium processes. *Sov. Phys. JETP* **1965**, *20*, 1018–1026.
83. Cooper, F.; Khare, A.; Rose, H. Classical limit of time-dependent quantum field theory—A Schwinger-Dyson approach. *Phys. Lett. B* **2001**, *515*, 463–469. [CrossRef]
84. Blagoev, K.B.; Cooper, F.; Dawson, J.F.; Mihaila, B. Schwinger-Dyson approach to nonequilibrium classical field theory. *Phys. Rev. D* **2001**, *64*, 125033. [CrossRef]
85. Frederiksen, J.S. Renormalized closure theory and subgrid-scale parameterizations for two-dimensional turbulence. In *Nonlinear Dynamics: From Lasers to Butterflies, World Scientific Lecture Notes in Complex Systems*; Ball, R., Akhmediev, N., Eds.; World Scientific: Singapore, 2003; Volume 1, pp. 225–256. ISBN 978-981-4486-36-1. [CrossRef]
86. Pouquet, A.; Lesieur, M.; Andre, J.C.; Basdevant, C. Evolution of high Reynolds number two-dimensional turbulence. *J. Fluid Mech.* **1975**, *72*, 305–319. [CrossRef]
87. Mandelstam, S. Determination of the pion-pion scattering amplitude from dispersion relations and unitarity. General theory. *Phys. Rev.* **1958**, *112*, 1344–1360. [CrossRef]
88. Frederiksen, J.S. Sommerfeld-Watson representation for double spectral functions. II. Crossing symmetric pion-pion scattering amplitude without Regge poles. *Commun. Math. Phys.* **1975**, *43*, 1–16. [CrossRef]
89. Atkinson, D.; Kaekebeke, M.; Frederiksen, J.S.; Johnson, P.W. Sommerfeld-Watson representation for double-spectral functions. III. Crossing symmetric pion-pion scattering amplitude with Regge poles. *Commun. Math. Phys.* **1976**, *51*, 67–84. [CrossRef]
90. Tourkine, P.; Zhiboedov, A. Scattering amplitudes from dispersive iterations of unitarity. *arXiv* **2023**, arXiv:2303.088339v1.

91. Gasenzer, T. Ultracold gases far from equilibrium. *Eur. Phys. J. Spec. Top.* **2009**, *168*, 89–148. [CrossRef]
92. Berges, J.; Sexty, D. Bose-Einstein condensation in relativistic field theories far from equilibrium. *Phys. Rev. Lett.* **2012**, *108*, 161601. [CrossRef] [PubMed]
93. Kofman, L.; Linde, A.; Starbinsky, A.A. Towards the theory of reheating after inflation. *Phys. Rev. D* **1997**, *56*, 3258–3295. [CrossRef]
94. Micha, R.; Tkachev, I.I. Relativistic turbulence: A long way from preheating to equilibrium. *Phys. Rev. Lett.* **2003**, *90*, 121301. [CrossRef]
95. Kofman, L. Preheating after Inflation. In *Inflationary Cosmology*; Lemoine, M., Martin, J., Peter, P., Eds.; Lecture Notes Phys; Springer: Berlin/Heidelberg, Germany, 2007; Volume 738, pp. 55–79. ISBN 978-3-540-74352-1.
96. Arnold, P. Quark-gluon plasmas and thermalization. *Int. J. Mod. Phys. E* **2007**, *16*, 2555–2594. [CrossRef]
97. Kraichnan, R.H. Test-field model for inhomogeneous turbulence. *J. Fluid Mech.* **1972**, *56*, 287–304. [CrossRef]
98. Aarts, G.; Berges, J. Classical aspects of quantum fields far from equilibrium. *Phys. Rev. Lett.* **2002**, *88*, 041603. [CrossRef]
99. Cooper, F.; Dawson, J.F.; Mihaila, B. Quantum dynamics of phase transitions in broken symmetry $\lambda\varphi^4$ field theory. *Phys. Rev. D* **2003**, *67*, 056003. [CrossRef]
100. Boyanovsky, D.; Destri, C.; Vega, H.J. The approach to thermalization in the classical φ^4 theory in 1+1 dimensions: Energy cascades and universal scaling. *Phys. Rev. D* **2004**, *69*, 045003. [CrossRef]

atmosphere

MDPI

Article

Stable Boundary Layers and Subfilter-Scale Motions

James C. McWilliams [1], Charles Meneveau [2], Edward G. Patton [3] and Peter P. Sullivan [3,*]

[1] Department of Atmospheric and Oceanic Sciences, University of California, Los Angeles, CA 90095, USA; jcm@atmos.ucla.edu
[2] Department of Mechanical Engineering, Johns Hopkins University, Baltimore, MD 21218, USA; meneveau@jhu.edu
[3] National Center for Atmospheric Research, Boulder, CO 80301, USA; patton@ucar.edu
* Correspondence: pps@ucar.edu

Abstract: Recent high-resolution large-eddy simulations (LES) of a stable atmospheric boundary layer (SBL) with mesh sizes $N = (512^3, 1024^3, 2048^3)$ or mesh spacings $\triangle = (0.78, 0.39, 0.2)$ m are analyzed. The LES solutions are judged to be converged based on the good collapse of vertical profiles of mean winds, temperature, and low-order turbulence moments, i.e., fluxes and variances, with increasing N. The largest discrepancy is in the stably stratified region above the low-level jet. Subfilter-scale (SFS) motions are extracted from the LES with $N = 2048^3$ and are compared to sonic anemometer fields from the horizontal array turbulence study (HATS) and its sequel over the ocean (OHATS). The results from the simulation and observations are compared using the dimensionless resolution ratio $\Lambda_w / \triangle_f$ where $\triangle_f$ is the filter width and Λ_w is a characteristic scale of the energy-containing eddies in vertical velocity. The SFS motions from the observations and LES span the ranges $0.1 < \Lambda_w / \triangle_f < 20$ and are in good agreement. The small, medium, and large range of $\Lambda_w / \triangle_f$ correspond to Reynolds-averaged Navier–Stokes (RANS), the gray zone (a.k.a. "Terra Incognita"), and fine-resolution LES. The gray zone cuts across the peak in the energy spectrum and then flux parameterizations need to be adaptive and account for partially resolved flux but also "stochastic" flux fluctuations that represent the turbulent correlation between the fluctuating rate of strain and SFS flux tensors. LES data with mesh 2048^3 will be made available to the research community through the web and tools provided by the Johns Hopkins University Turbulence Database.

Keywords: stratified turbulence; large-eddy simulation; subfilter-scale motions; surface layer observations; gray zone

Citation: McWilliams, J.C.; Meneveau, C.; Patton, E.G.; Sullivan, P.P. Stable Boundary Layers and Subfilter-Scale Motions. *Atmosphere* **2023**, *14*, 1107. https://doi.org/10.3390/atmos14071107

Academic Editor: Stephan De Wekker

Received: 20 April 2023
Revised: 15 May 2023
Accepted: 19 May 2023
Published: 4 July 2023

1. Preface

Along with others in this collection, this paper is dedicated to our colleague and friend, Jack(son) Herring. It is also an encomium for the long history of turbulence research centered around the National Center for Atmospheric Research (NCAR), where Jack worked from 1972. One of the better purposes of NCAR was as a gathering place for researchers from around the world, especially in the summer outside of teaching terms. Another was as a national supercomputing center beginning even before such a term was invented (i.e., punch cards in the basement of the Mesa Laboratory), and from the outset turbulence simulations were one of the premier scientific applications [1]. The summer gatherings were often informal and barely scripted or sometimes formally funded and convened as workshops. They did a great deal to fertilize the development of turbulence theories, simulations, and experiments among the physics, engineering, and geophysical fluid dynamics communities and to entrain younger scientists into the subject. A conspicuous event was a summer-long school and symposium in 1987 [2]. It was partly managed under the auspices of NCAR's Advanced Study Program, where Jack was first hired, and at some point in the 1980s it was more formally designated as the Geophysical Turbulence Program (GTP) that continues today. Its scientific scope was broad and its human reach was wide, involving hundreds of scientists whose roster probably no archaeology could recover. Among its luminaries from outside NCAR were R. Kraichnan, U. Frisch, E. Lorenz, D. Montgomery, H. Tennekes, M. Lesieur, and J. Riley, and from within, C. Leith, D. Lilly, J. Deardorff, J. Wyngaard, D. Lenschow, and A. Pouquet; there were many others in and outside NCAR, ourselves, and of course Jack.

2. Introduction

Stratified turbulence impacts many applications and the topic attracts researchers in the physics, geophysical, and engineering communities; see reviews [3–6]. The present contribution is narrower in scope, focusing on stably stratified turbulent boundary layers (SBLs). The SBL is a key component of earth-system modeling, viz., large-scale weather, climate, and ocean models [4,5,7–12]. Stratified boundary-layer turbulence is also important in electromagnetic wave propagation in the lower atmosphere, wind energy, and the dispersion of pollutants [13–15].

Observations, along with direct numerical simulation (DNS) and large-eddy simulation (LES), are important tools for studying SBLs, but each face impediments; simulation results are often at low Reynolds number [16] and require long computational run times, and observations are impacted by inherent atmospheric variability and surface heterogeneity; the latter can result in global impacts on the SBL [17]. Observations and simulations must cope with spatial and temporal intermittency in SBLs with strong stratification [5] and coherent structures [18].

This paper is a report of recent, very-high-resolution simulations of stable atmospheric boundary layers (SBL) underneath a steady geostrophic wind. It is one of the canonical idealized problems in geophysical fluid dynamics, closely related to Jack's interest in homogeneous stratified turbulence, but with the complication of an adjacent rough, flat boundary. The narrow purpose here is to report on the resolution dependency in LES and make detailed statistical comparisons with near-surface measurements from instrument arrays. A larger goal is to establish the validity of SBL LES as a test-bed for evaluating turbulent closure schemes near the boundary, although we have not yet gone very far in that direction.

The outline of the rest of the paper is the following: Section 3 presents the LES equations and introduces recent LES of SBLs with fine-mesh resolution; Section 4 briefly describes field observations collected in the atmospheric surface layer specifically designed to measure subfilter-scale (SFS) fields over a range of atmospheric stability; Section 5 analyzes the LES solutions and their sensitivity to the mesh resolution and compares SFS motions extracted from the LES and field observations; Section 6 summarizes the findings; and Section 7 is a brief commentary on the continuing science of turbulence.

3. LES

3.1. Governing Equations

The LES model equations for an SBL under the Boussinesq approximation with system rotation and stable stratification with a flat bottom boundary are well documented [19,20]. The equations are briefly introduced here:

$$\frac{\partial \overline{\mathbf{u}}}{\partial t} + \overline{\mathbf{u}} \cdot \nabla \overline{\mathbf{u}} = -\mathbf{f} \times (\overline{\mathbf{u}} - \mathbf{U}_g) - \nabla \overline{\pi} + \hat{\mathbf{z}}\beta(\overline{\theta} - \theta_{ref}) - \nabla \cdot \mathbf{T} \tag{1a}$$

$$\frac{\partial \overline{\theta}}{\partial t} + \overline{\mathbf{u}} \cdot \nabla \overline{\theta} = -\nabla \cdot \mathbf{B} \tag{1b}$$

$$\frac{\partial e}{\partial t} + \overline{\mathbf{u}} \cdot \nabla e = \mathcal{P} + \mathcal{B} + \mathcal{D} - \mathcal{E} \tag{1c}$$

$$\nabla \cdot \overline{\mathbf{u}} = 0\,. \tag{1d}$$

The equation set includes transport equations: for momentum $\rho\overline{\mathbf{u}}$ (1a); for virtual potential temperature $\overline{\theta}$ (1b); and for SFS turbulent kinetic energy e (1c). The divergence-free (incompressible) condition (1d) determines the elliptic pressure variable $\overline{\pi}$. The variables that appear in (1) are: velocity components $\overline{\mathbf{u}} \equiv \overline{u}_i = (\overline{u}, \overline{v}, \overline{w})$, geostrophic winds $\mathbf{U}_g = (U_g, V_g)$, rotation vector $\mathbf{f} = (0, 0, f)$ with Coriolis parameter f, unit vector $\hat{\mathbf{z}}$ in the vertical direction, and buoyancy parameter $\beta = g/\theta_{ref}$, where g is gravity and θ_{ref} is a reference temperature. Pressure $\overline{p}$ and air density ρ do not appear explicitly in (1). The terms on the right-hand side of (1c) for e are shear production $\mathcal{P}$, buoyancy production–destruction $\mathcal{B}$, diffusion $\mathcal{D}$, and dissipation $\mathcal{E}$. The modeling of these terms is described in [19]. The overbar notation $\overline{(\;)}$ denotes a spatially filtered quantity.

The LES equations are formally derived by applying a spatial filter term-by-term to the governing equations of motion, although physics-inspired alternatives are possible [21]. This operation introduces the unknown SFS kinematic momentum and temperature fluxes

$$\mathbf{T} \equiv \tau_{ij} = \overline{u_i u_j} - \overline{u}_i \overline{u}_j \quad ; \quad \mathbf{B} \equiv \tau_{i\theta} = \overline{u_i \theta} - \overline{u}_i \overline{\theta} \tag{2}$$

For the SBL we adopt the two-part SFS model proposed by Sullivan et al. [22], which utilizes the transport Equation (1c) and an eddy viscosity approach in the parameterization of the SFS fluxes given by (2). This parameterization is specifically tailored to a high-Reynolds-number LES that uses rough-wall surface boundary conditions based on Monin–Obukhov (MO) similarity theory. In the eddy viscosity prescription $\nu_t \sim \gamma \sqrt{e}\, \ell$ the isotropy factor $\gamma = S'/(\langle S \rangle + S')$ where $(\langle S \rangle, S')$ are the resolved strain rate average and fluctuation, respectively; the strain rate average is over an x–y plane. The isotropy factor varies with the grid resolution and essentially reduces the length scale ℓ as the wall is approached depending on the magnitude of the resolved turbulence fluctuations. The boundary conditions, solution algorithm, and further details are provided in Sullivan et al. [20]. To streamline the notation and text in the following discussion, we now drop the overbar symbol on all spatially filtered (i.e., resolved) variables and simply refer to the virtual potential temperature θ as "temperature".

3.2. LES of Stable Boundary Layers

The present work targets the LES of the so-called GABLS1 intercomparison case described in detail by Holtslag [23] and Beare et al. [24]. The problem design provides an excellent test-bed to study stratified turbulence in a high-Reynolds-number boundary-layer flow using LES. GABLS1 is a canonical high-latitude SBL driven by constant geostrophic winds $U_g = 8$ m s^{-1} with Coriolis parameter $f = 1.39 \times 10^{-4}$ s^{-1} above a horizontally homogeneous rough surface $z_o = 0.1$ m. The simulations are initiated from a neutral state with an overlying stable inversion $\partial_z \theta = 0.01$ K m^{-1} imposed at an initial height $z = 100$ m. The computational domain size is $(400 \times 400 \times 400)$ m. A constant rate of surface cooling $C_r = 0.25$ K h^{-1} is applied starting at $t = 0$ to generate stably stratified turbulence. The specification of surface temperature fully couples the momentum and temperature relations in the Monin–Obukhov bottom boundary conditions, which is a more demanding test for LES compared to a specified temperature flux: further description of the simulation design is given in Sullivan et al. [20]. Previous LESs find a quasi-steady SBL featuring a low-level jet with wind profile veering in the SBL after 8 h of simulation; the SBL depth $h \approx 200$ m, the surface temperature flux $Q_* \approx -9$ K m s^{-1}, and friction velocity $u_* \approx 0.25$ m s^{-1}, which vary with the mesh resolution [20]. The GABLS1 SBL is weakly stable with continuous turbulence and the boundary-layer stability metric is $h/L_{mo} \sim 1.7$, where $L_{mo} \approx 118$ m is the Monin–Obukhov length; see Table 1 and definitions of h, L_{mo} in Section 5.1.

The first GABLS1 intercomparison used LES models with coarse and fine meshes of $N = 128^3$ and 200^3 gridpoints [24,25], suitable for the computational capabilities at that time. Although there was qualitative agreement amongst the LES models, there are hints that an increased resolution resulted in shallower SBLs inducing a change in the surface friction velocity u_* and surface cooling flux Q_*. This motivated further study and Sullivan et al. [20] explored the LES solution sensitivity using meshes of $N = (200^3, 512^3, 1024^3)$ gridpoints along with four different cooling rates $C_r = (0.25, 0.375, 0.5, 1.0)$ K h^{-1}; the 1024^3 simulation required approximately 2 million core hours to complete a 9 h simulation.

Table 1. Bulk simulation properties, with entries: LES case, mesh points N, cooling rate C_r, mesh spacing $\triangle$, friction velocity u_*, temperature flux Q_*, boundary-layer height h, Monin–Obukhov stability length L_{mo}, boundary-layer stability h/L_{mo}.

Case	N	C_r (K h^{-1})	$\triangle$ (m)	u_* (m s^{-1})	$Q_* \times 10^3$ (K m s^{-1})	h (m)	L_{mo} (m)	h/L_{mo}
B	512^3	0.25	0.78	0.265	−10.5	205.3	119.3	1.72
C	1024^3	0.25	0.39	0.255	−9.63	197.5	116.4	1.70
C2	2048^3	0.25	0.20	0.249	−8.83	187.2	118.6	1.56
F2	2048^3	1.0	0.20	0.177	−14.82	143.7	24.9	5.77

The present work expands on Sullivan et al. [20] using an unprecedented resolution of $\triangle = 0.2$ m in all three directions with a mesh of $N = 2048^3$. Because of the computational cost a two-step recipe is used to perform the LES. First, a seed simulation with $N = (512 \times 512 \times 2048)$ is run from scratch for nearly 8 h. The grid in this simulation is anisotropic but has the advantage that the wind and temperature profiles vary smoothly in the vertical direction. Next, the last volume from the seed simulation is archived and the field variables are interpolated in the horizontal x–y directions using zero-padded Fourier transforms to generate a restart volume with $N = 2048^3$ points. The fine-mesh solution is then started and run for an additional 0.5 h. This simulation strategy results in smooth restarts with the high-wavenumber part of the spectrum rapidly filled with small-scale turbulence. Two simulations with weak and strong cooling rates $C_r = (0.25, 1)$ K h^{-1} are carried out.

Selected snapshots from the two simulations as well as time sequences with the 2048^3 mesh are being included as part of datasets available to the community at the Johns Hopkins Turbulence Databases (JHTDB). The system enables scientists' access to world-class simulations using the approach of "Immersive Analysis", which allows the user to insert immersive "virtual sensors" into the simulated flow. This approach facilitates access to large datasets without having to download large amounts of data, as well as performing analysis close to the data [26,27]. The ability to query the data in a spatially localized fashion is particularly relevant to the SBL, whose fields are characterized by strong spatially and temporally intermittent events. At the time of this writing, the data ingestion and data publishing process has not yet been finished, but completion is expected soon.

4. Observations of Subfilter-Scale Motions

An alternative to using low-Reynolds-number DNS datasets for a priori tests [28] of SFS models is to use laboratory measurements of higher-Reynolds-number turbulence collected at multiple spatial points. In principle, the measurement data can then be spatially filtered, mimicking the spatial-filtering operation in LES [29]. At still-higher Reynolds numbers a similar approach can be undertaken using measurements in the atmospheric surface layer. The latter was frequently championed in seminars and discussions with John Wyngaard when visiting NCAR. The horizontal array turbulence study (HATS) was a field campaign [30] specifically designed to construct SFS motions from turbulence data collected at multiple locations in the atmospheric surface layer over a range of stability. HATS, carried out in the central California valley, was a collaborative community effort between NCAR, GTP, Johns Hopkins University, Pennsylvania State University, and Clemson University. The field campaign built on work by several groups, including Meneveau and Lund [31], Tong et al. [32,33], Tao et al. [34]; see the many cited references in these papers and the review [35].

HATS used two horizontal crosswind arrays of sonic anemometers mounted at different vertical locations, as shown in Figure 1; four array configurations varying the sonic anemometer horizontal separation and vertical location were used in HATS. The anemometer arrays measured time series of the turbulent velocity components (u, v, w) (streamwise, crosswind, and vertical, respectively) and virtual potential temperature θ. The multi-point measurements in the crosswind direction y are combined with Taylor's frozen-field assumption in the streamwise direction x [36] to create x–y planes of stratified turbulence $(\mathbf{u},\theta)(x,y,z,t)$. The wind speed and diurnal variations combined with spatial variations in the instrument positioning created a rich dataset for studying spatially filtered turbulence in the atmospheric surface layer. For example, Sullivan et al. [37], Kleissl et al. [38,39], Wyngaard [40], Chen and Tong [41], Hatlee and Wyngaard [42], Chamecki et al. [43] and Ramachandran and Wyngaard [44] studied SFS dynamics and evaluated eddy viscosity models, dynamic Smagorinsky models, subfilter-scale budgets, and rate-equation models using HATS data; the studies employed single and double filtering available from the HATS arrays. Triple filtering was explored in Higgins et al. [45] using a 4×4 array of sonic anemometers that also enabled vertical (albeit more sparsely sampled) filtering, confirming the accuracy of double filtering. Kumar et al. [46] reproduced HATS conditions during a daily cycle using low-resolution LES. The success of HATS motivated follow-on campaigns focused on atmospheric turbulence in the surface layer; over the ocean OHATS [47], over a glacier SnoHATS [48], in an orchard canopy CHATS [49], and over land accounting for horizontal advection AHATS [50].

Figure 1. Twin horizontal arrays of sonic anemometers used to measure subfilter-scale variables in the atmospheric surface in the HATS field campaign. The upper array is located at $z_d = 8.66$ m with anemometer horizontal spacing $\delta y_d = 2.17$ m. The vertical location and horizontal spacing of anemometers in the lower array are $(z_s, \delta y_s) = (4.33, 1.08)$ m. Photo courtesy of Thomas Horst (NCAR).

5. Results

5.1. SBL Low-Order Moments

The advent of high-performance computing allows mesh convergence to be studied using 3D simulations. Solution convergence in LES is complicated by the mesh dependence in the SFS model, e.g., holding the domain size fixed, the resolved flux in a 100^3 simulation is of course different than the resolved flux in a 2048^3 simulation. Judging convergence in LES is also challenging as the subgrid-scale model and numerical discretization errors are intertwined since both depend explicitly on the mesh spacing [51]. Here we use physically based metrics based on the vertical profiles of low-order statistical moments to judge convergence, as discussed in Sullivan and Patton [52] for convectively driven boundary layers; also see Geurts and Fröhlich [53]. Table 1 provides a summary of the bulk parameters for SBL simulations with $C_r = 0.25\ \mathrm{K\,h^{-1}}$, grid meshes $N = (512^3, 1024^3, 2048^3)$,

$C_r = 1$ K h^{-1}, $N = 2048^3$. The variables in Table 1 are the LES case, grid mesh points N, mesh spacing $\triangle$, cooling rate C_r, friction velocity u_*, surface temperature flux Q_*, boundary-layer depth h, Monin–Obukhov stability length $L_{mo} = -u_*^3/\kappa\beta Q_*$ with von Kármán constant $\kappa = 0.4$, and boundary-layer stability h/L_{mo}. The SBL depth h is defined as the vertical location where the vertical gradient of temperature $\partial_z\langle\theta\rangle$ reaches a maximum (see [20,54]). There is variation in (u_*, Q_*) with the grid resolution because of variability in the SBL depth h. Dai et al. [55] speculates that this is a consequence of the stability length-scale correction used in the LES. Statistics, denoted by angle brackets $\langle\,\rangle$, are formed by averaging in the x–y planes and over the time period $8 < t < 9$ h, except for simulation C2, which used a shorter 30 min time average. A turbulent fluctuation from a horizontal mean is denoted by a superscript prime $(\,)'$.

Figure 2 compares the profiles of winds and temperature for three different resolutions $(512^3, 1024^3, 2048^3)$ with fixed surface cooling $C_r = 0.25$ K h^{-1}. To eliminate the slight variability with the boundary-layer depth we introduce the dimensionless vertical coordinate z/h, as shown in the right panel of Figure 2. The flux and variance profiles are normalized by the surface values (u_*^2, Q_*) as appropriate. Under this normalization the wind profiles collapse well (the right panel of Figure 2). The low-level jet is positioned at $z/h = 0.8$ and its magnitude is 1.2 U_g.

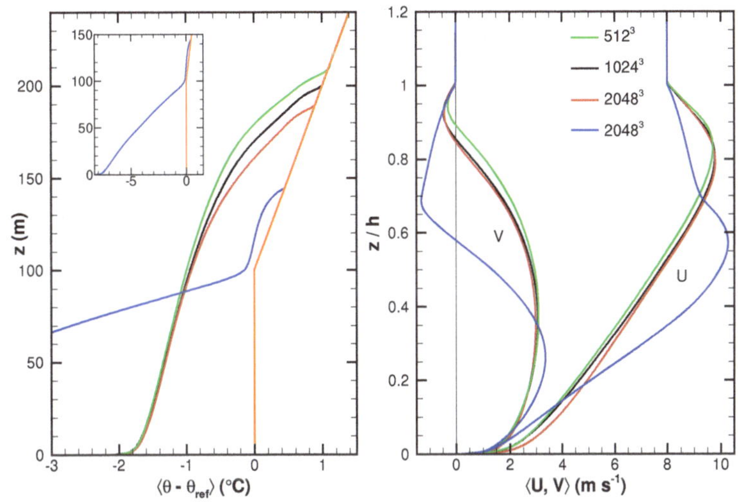

Figure 2. Vertical profiles of average temperature $\langle\theta - \theta_{ref}\rangle$ (**left panel**). Simulations (B, C, C2) with cooling rate $C_r = 0.25$ K h^{-1} and $N = (512^3, 1024^3, 2048^3)$ are represented by line colors (green, black, red), respectively. Simulation F2 with $C_r = 1$ K h^{-1} and $N = 2048^3$ is presented in blue. The inset figure shows the variation of temperature in F2 in the lower SBL. The initial temperature field at $t = 0$ is shown as the orange line and $\theta_{ref} = 265$ K. Vertical profiles of average winds $\langle U, V\rangle$ are shown in the **right panel**. Note that the vertical coordinate for the wind profiles is normalized by the SBL height h from each simulation.

The vertical profiles of the Richardson number Ri and the squared shear and buoyancy frequency (S^2, N^2) are displayed in Figure 3 for the three resolutions considered. Here

$$Ri(z) = \frac{N^2}{S^2} \;;\; N^2 = \frac{g}{\theta_{ref}}\frac{\partial\langle\theta\rangle}{\partial z} \;;\; S^2 = \left(\frac{\partial\langle\mathbf{u}_\perp\rangle}{\partial z}\right)^2. \tag{3}$$

where the horizontal wind aligned with the mean wind direction is given by $u_h = \mathbf{u}_\perp(\mathbf{x}, t)\cdot\langle\mathbf{u}_\perp\rangle/|\langle\mathbf{u}_\perp\rangle|$. The profiles of (Ri, S^2, N^2) are very well converged for the three different mesh resolutions below $z/h < 0.5$. In the upper SBL, above the low-level jet the profiles for simulations C2 and C are converged. These profiles show that the simulation is in the weakly stable regime as in Sullivan et al. [20]. In particular, the profile $Ri(z) \approx 0.2$ shows the approximate validity of the very simple RANS parameterization of a constant Richardson number above the Monin–Obukhov surface layer.

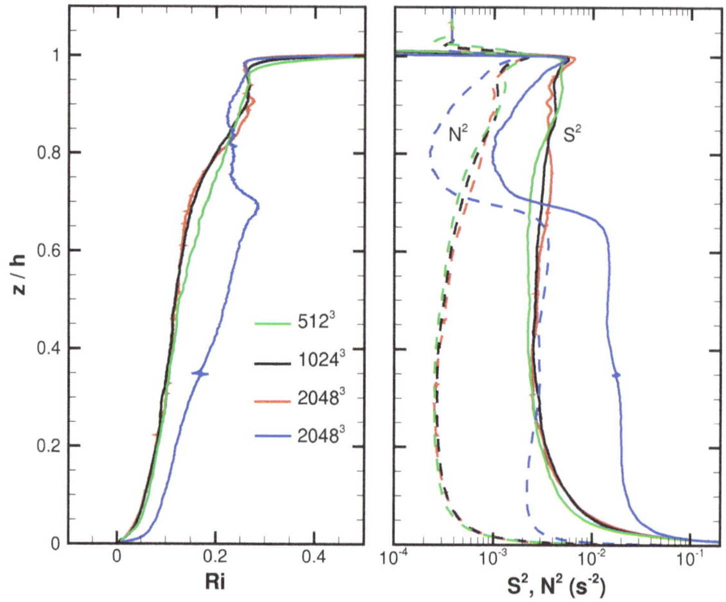

Figure 3. Vertical profiles of average Richardson number (**left panel**). Simulations with cooling rate $C_r = 0.25\ \mathrm{K\,h^{-1}}$ (B, C, C2) with $N = (512^3, 1024^3, 2048^3)$ are denoted by line colors (green, black, red), respectively. Simulation F2 with $C_r = 1\ \mathrm{K\,h^{-1}}$ and $N = 2048^3$ is presented in blue. Vertical profiles of average shear and buoyancy frequency squared (S^2, N^2), indicated by (solid, dashed) lines, respectively, are shown in the **right panel**.

Figures 2 and 3 also illustrate a strong dependence on bulk stratification in the SBL. The bulk stability measure h/L_{mo} increases from 1.56 to 5.77 as the cooling rate varies from $C_r = 0.25$ to $1\ \mathrm{K\,h^{-1}}$. Increasing stratification leads to a decrease in the SBL turbulence level and as a result the height of the low-level jet (LLJ) descends from $z/h = 0.8$ to 0.58, i.e., from 105 m to 83 m. Then the wind veering in F2 is sharper and compressed in the lower SBL compared to C2. In F2, notice that the SBL is nearly equally split between the vertical layers below and above the LLJ. The mean wind, temperature, shear, and buoyancy profiles all change markedly above the LLJ compared to their counterparts in the lower SBL. Compensating for changes in the shear and buoyancy profiles leads to a relatively uniform $Ri \sim 0.2$ profile above the LLJ, which supports weak stratified turbulence. A further discussion of the impacts of increasing stratification on SBL statistics is given in [20].

The velocity variances from simulations (B, C, C2), which include the SFS contribution $2\langle e \rangle/3$, collapse reasonably well for the three mesh resolutions considered, as shown in Figure 4. The momentum and temperature fluxes $\langle u'w', v'w', w'\theta' \rangle$ that include both the resolved and SFS contributions are in close agreement as the mesh spacing varies from $\triangle = (0.78, 0.39, 0.20)$; see Figure 5. The profile of SFS energy $\langle e \rangle$, shown in Figure 6, shows a systematic decrease with resolution over the bulk of the SBL. For example, at $z/h = 0.5$ the scaling relationship $e \propto \triangle^{2/3}$ holds [36] (p. 589). The results in Figure 6 find a 40% decrease in e for a mesh-size reduction from 0.78 m to 0.2 m. Based on these results, we judge the LES results to be well converged.

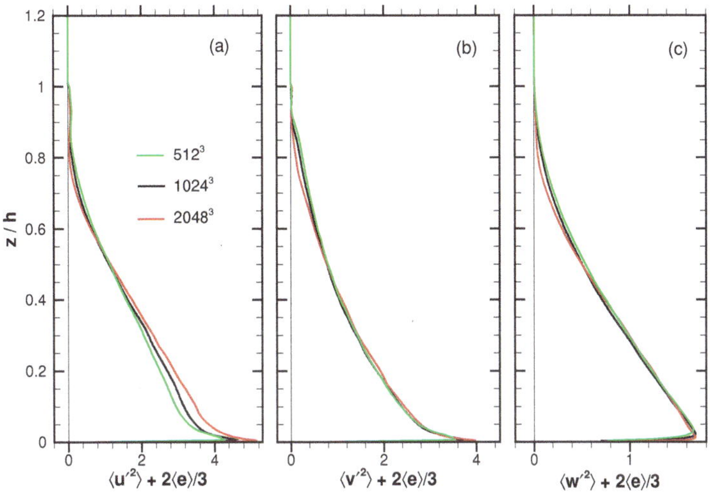

Figure 4. Vertical profiles of velocity variances $\langle u'^2, v'^2, w'^2 \rangle / u_*^2$ (panels (**a**)–(**c**), respectively). The variances include an estimate of the SFS contribution $2\langle e \rangle / 3$. Results are from simulations (B, C, C2) with $N = (512^3, 1024^3, 2048^3)$ denoted by line colors (green, black, red), respectively. The vertical coordinate is normalized by the SBL height h from each simulation.

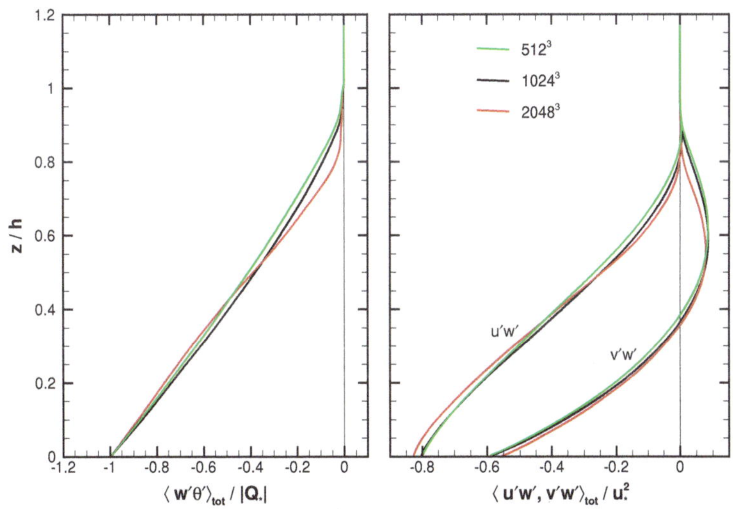

Figure 5. Vertical profiles of average temperature flux (**left panel**) and momentum fluxes (**right panel**). The fluxes are normalized by Q_* and u_*^2 as appropriate. The fluxes include the SFS contributions. Results are from simulations (B, C, C2) with $N = (512^3, 1024^3, 2048^3)$, denoted by line colors (green, black, red), respectively. The vertical coordinate is normalized by the SBL height h from each simulation.

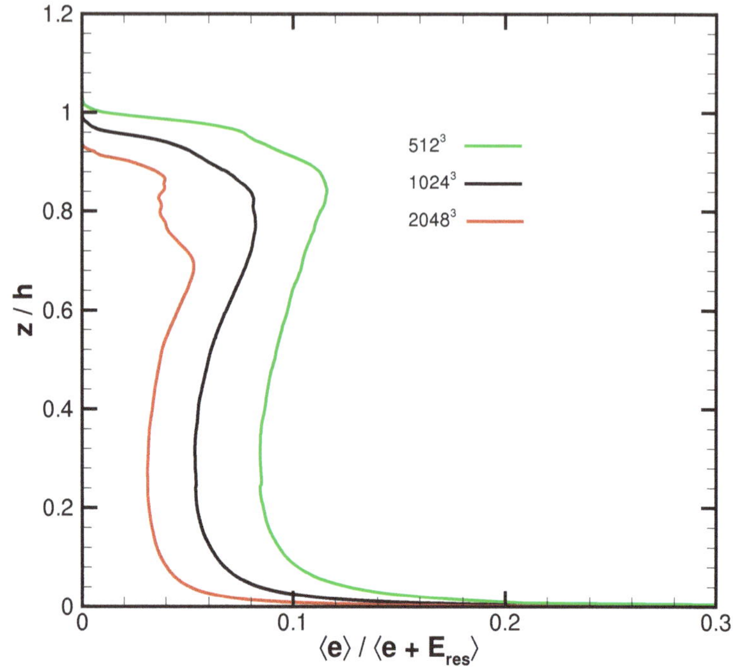

Figure 6. Vertical profile of SFS energy in the LES as a fraction of the total energy $\langle e\rangle/\langle e + E_{res}\rangle$ for different meshes $N = (512^3, 1024^3, 2048^3)$ denoted by (green, black, red) lines, respectively. The resolved kinetic energy $E_{res} = (u'^2 + v'^2 + w'^2)/2$.

5.2. Structures in Stable Boundary Layers

Flow visualization is extensively utilized to identify turbulent structures in the SBL; an example is provided in Figure 7. This image shows the instantaneous temperature isolines in an x–z plane from simulations C2 and F2 with weak and strong stratification, respectively. Inspection of the images shows an abundance of tightly compressed contour lines sprinkled throughout the SBL; the sharp gradients in θ are signatures of warm–cold temperature fronts passing through the domain. Notice that the temperature fronts are very sharp and tilted in the downstream direction, primarily a consequence of the sheared streamwise velocity $\langle u\rangle(z)$. Animations show the spatial and temporal evolution of the fronts (not shown). Frequently, the fronts are observed to extend over the full depth of the SBL and nearly the full horizontal extent of the domain, as shown in the upper panel of Figure 7. Turbulent mixing between fronts at different z levels leaves voids with nearly constant temperatures. Thus, at a fixed x–y location a vertical profile of θ displays a staircase pattern as different front families are crossed. As the stratification increases, the fronts tilt farther downstream and the separation between fronts shrinks considerably, as shown in the lower panel of Figure 7. A horizontal slice through the domain at $z/L_{mo} = 0.164$ m from simulation C2 provides a sense of the front coherence in the spanwise direction as well as the spatial randomness of the fronts. Figure 8 shows multiple fronts at various positions in the horizontal domain; the spanwise extent of a front is $\sim$50 m or less. At this instance in the simulation the warm upstream side of the front is modestly stronger than the cool downstream side of the front.

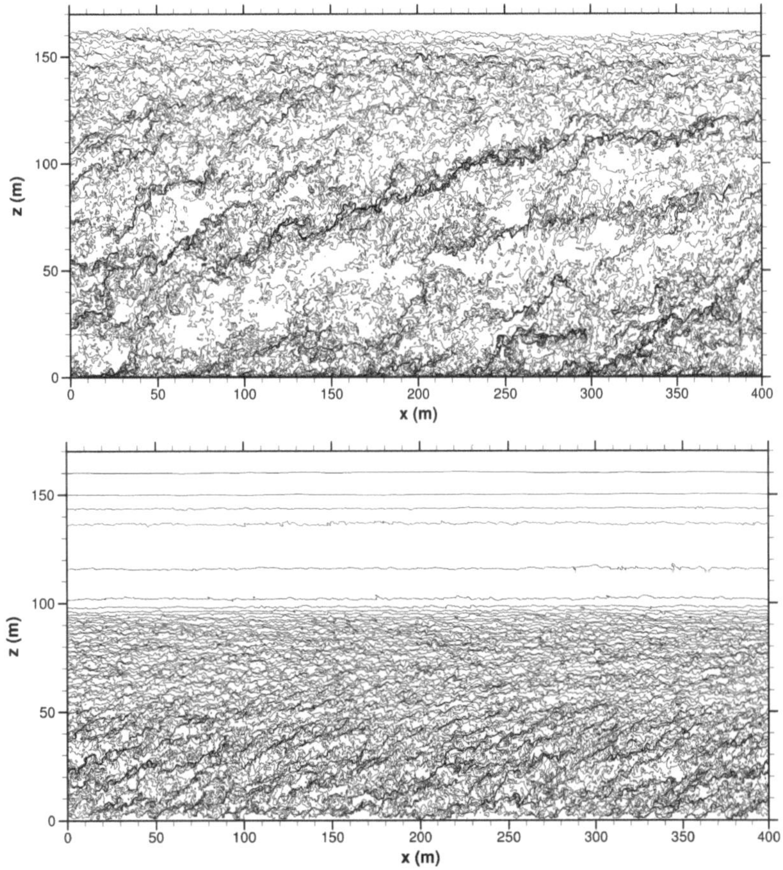

Figure 7. Temperature isolines in an x–z plane from simulations with weak surface cooling (**C2 upper panel**) and strong surface cooling (**F2 lower panel**) at $t \sim (8.556, 8.352)$ h, respectively. In the upper panel there are 51 contour levels between $(-2 < \theta - \theta_{ref} < 0)$ °C. In the lower panel there are 71 contour levels between $(-7.5 < \theta - \theta_{ref} < 0)$ °C. In addition, in the lower panel, contour levels $(0.2, 0.4, 0.5, 0.6)$ °C are shown for $z > 135$ m. The reference temperature $\theta_{ref} = 265$ K.

Linear stochastic estimation (LSE) pioneered by Adrian [56] is used to compute the conditional averages of the turbulent fields in the SBL. Our application of LSE [20] uses an event trigger based on a positive–negative temperature jump separated by a finite distance in a horizontal plane. This event choice is guided by the instantaneous flow visualization of θ' shown in Figure 8, which depicts numerous warm–cool temperature fronts. As an example, our LSE temperature event with $E = \pm 0.1$ K corresponds to a scalar flux about five times the surface value $|Q_*|$. The conditional fields $(\widehat{\mathbf{u}}, \widehat{p}, \widehat{\theta})$, velocity gradient tensor $\partial_j \widehat{u}_i$, and vorticity $\nabla \times \widehat{\mathbf{u}}$ are estimated for a range of vertical locations, spatial separation, and event amplitudes [20]. Our conditional sampling finds pairs of counter-rotating vortices in the SBL, as shown in Figure 9. The vortices are aligned with the mean wind direction and are tilted forward in the downstream direction. The upstream vortices act in concert to pump warm fluid forward while the downstream vortices pump cool fluid backwards, resulting in a near-stagnation point in the region between the vortices. The front boundary is very sharp but the vortices creating the fronts are well-resolved coherent structures. Sullivan et al. [20] show that the scale of the vortices lies near the peak in the spectrum for θ and w. The coherent structures in the interior of a sheared SBL are considerably different than the pancake vortices in large-scale stably stratified turbulence away from a boundary [57].

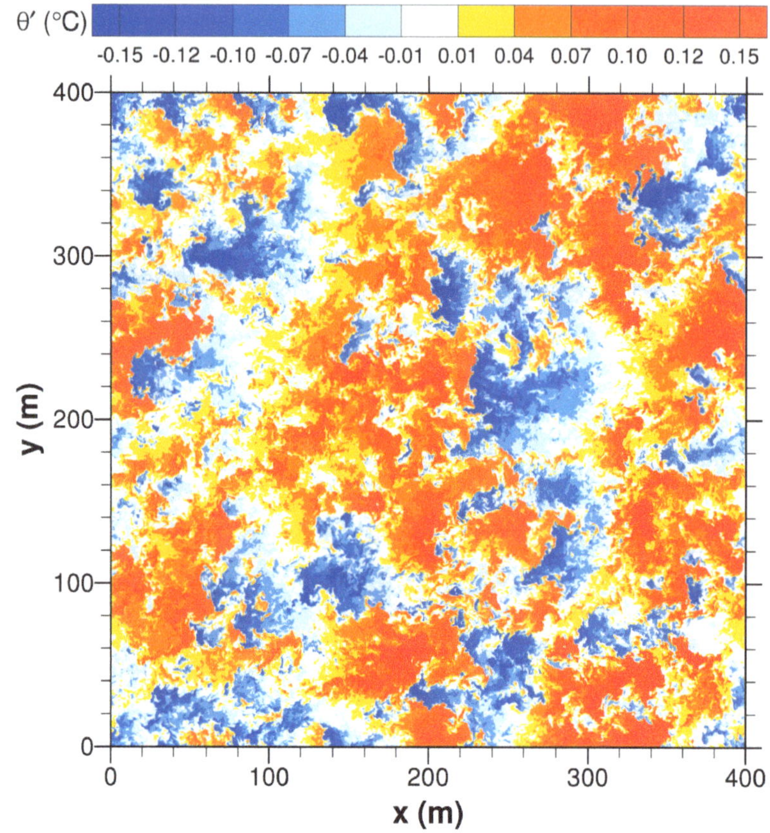

Figure 8. Fluctuating temperature field θ' at $z/L_{mo} = 0.164$ from simulation C2 at $t = 8.556$ h. Examples of sharp warm–cold temperature fronts are located at (x, y) = (20, 230), (200, 270), (230, 50), (230, 220), (330, 340) m. The color bar for $\theta' = \theta - \langle\theta\rangle$ is in units of °C.

5.3. SFS Motions in Observations and LES

The HATS dataset contains a mix of variations in wind speed, stratification, and perturbations in the array configuration, i.e., the vertical location z_d and horizontal separation δy_d of the anemometers shown in Figure 1. To account for these variations in computing the SFS motions Sullivan et al. [37] introduced a resolution ratio $\Lambda_w/\triangle_f$. The filter scale $\triangle_f \propto \delta y_d$ and is isotropic in x–y; only 2D filtering is employed in HATS. The vertical velocity is used to define Λ_w for the following reasons: w is the least-resolved field in LES, w statistics closely follow Monin–Obukhov similarity relationships in the surface layer, and w impacts the vertical fluxes, which are key ingredients in LES and also large-scale models. The length scale $\Lambda_w = 2\pi\langle U\rangle\tau_p$ where $\langle U\rangle$ is the mean wind speed and τ_p is the Eulerian integral time scale for vertical velocity. The autocorrelation function for w is fit to the form $R_w(t) = \exp(-t/\tau_p)$. This equates to finding a spectral peak in the vertical velocity spectrum assuming a high wavenumber spectrum k^{-2}; see Kaimal and Finnigan [58] (p. 63), Sullivan et al. [37], and Horst et al. [30]. This definition of Λ_w is robust and accounts for the distance z above the surface and stratification. In HATS the resolution ratio spans the range $0.1 < \Lambda_w/\triangle_f < 10$.

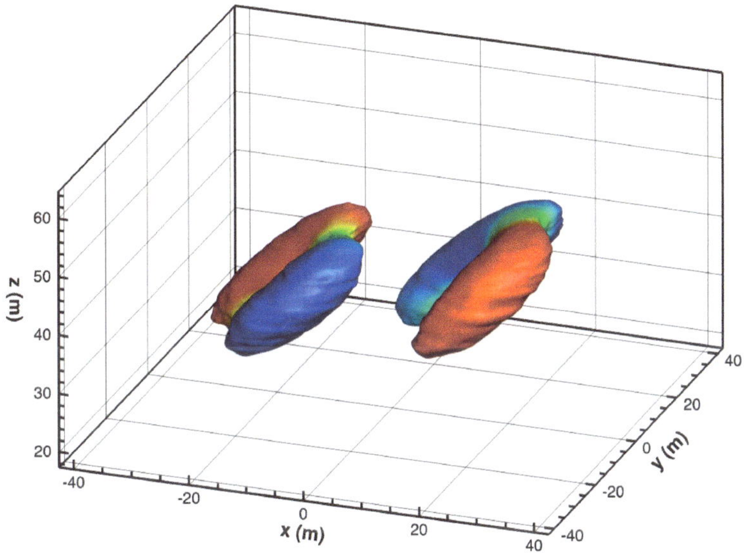

Figure 9. Oblique view of the typical 3D vortical structures in the SBL at a height $z/z_i = 0.2$ from simulation C with grid mesh of 1024^3 points [20] obtained using linear stochastic estimation-based conditional averaging. To indicate the sign of vortical rotation, the surface is colored by the vertical component of the vorticity vector $\zeta\,\hat{\mathbf{z}}$ with deep red (blue) colors corresponding to positive–upward (negative–downward) rotation, respectively. The mean horizontal winds are from left to right and at this height turn 30.6 degrees towards the $+y$ direction.

Physically, $\Lambda_w/\triangle_f$ is a measure of the scale separation between the large-scale energy-containing eddies and eddies near the filter scale, as sketched in Figure 10. For $\Lambda_w/\triangle_f \gg 1$ the separation is wide and the turbulence is well-resolved with the filter scale in the inertial range. Meanwhile, for $\Lambda_w/\triangle_f \leq 1$, the filter scale is near or left of the energy-containing eddies and the energy-containing turbulence is thus under-resolved. The ratio $\Lambda_w/\triangle_f$ is a measure of the simulation resolving power. When $\Lambda_w/\triangle_f \gg 1$ the simulation is LES and when $\Lambda_w/\triangle_f \ll 1$ the simulation is akin to unsteady Reynolds-Averaged Navier–Stokes (RANS). Wyngaard [40] calls the intermediate regime $\Lambda_w/\triangle_f \sim 1$ "Terra Incognita", which has similarity with the so-called gray zone [59,60]; it is unknown if the SFS closures used in LES or RANS are applicable in the gray zone. As $\Lambda_w/\triangle_f$ varies between the LES and RANS limits the length scale ℓ in an eddy-viscosity closure is predicted to vary linearly with the grid spacing $\ell \propto \min(\triangle, \ell_t)$ until $\triangle$ exceeds the scale of the energy-containing eddies ℓ_t, then ℓ becomes constant [61]. The crossover from LES to RANS is also recognized in the engineering community: Perot and Gadebusch [62] describe a two-equation self-adapting closure that spans the regime from DNS to LES to RANS for a mixing layer. Our estimate of a scale-aware length scale for dissipation in the SBL is discussed below.

To make a quantitative comparison between HATS and LES, we define the LES length scale Λ_w in terms of the spatial autocorrelation of vertical velocity $R_w(x,y)$, but aligned with the mean wind direction (see Figure 11). For LES, $\Lambda_w = 2\pi L_w$, where L_w is the integral scale obtained by integration of the autocorrelation $R_w(x', y' = 0)$ from $x' = 0$ to its first-zero crossing; Taylor's hypothesis is not used. The resolved fields $(\mathbf{u}, \theta)$, used to compute SFS motions, are also rotated into the mean-wind coordinate frame (x', y').

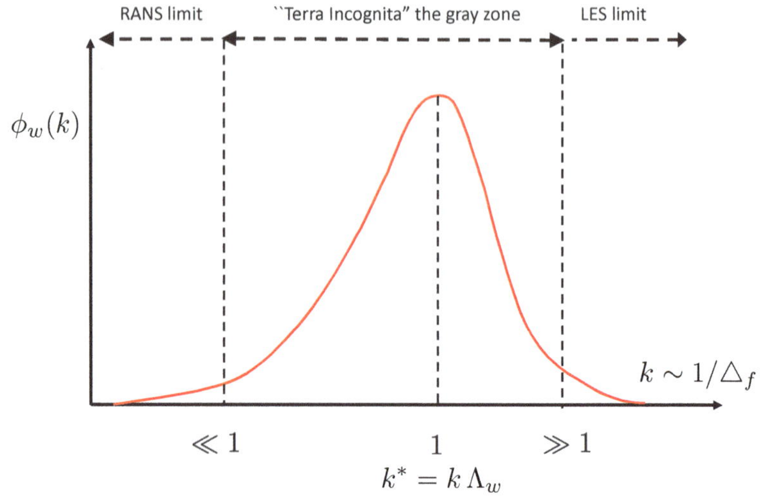

Figure 10. A sketch of the vertical velocity spectrum $\phi_w(k)$ in a horizontal plane as function of the horizontal wavenumber magnitude k. Its peak is at the non-dimensional wavenumber $k^* = k\,\Lambda_w = \Lambda_w/\triangle_f = 1$. The limits $k^* = \Lambda_w/\triangle_f \gg 1$ and $k^* = \Lambda_w/\triangle_f \ll 1$ are the LES and RANS regimes, respectively. In between these two limits is "Terra Incognita" or the gray zone. Figure is adapted from Wyngaard [40].

Analysis of the LES fields uses a filter scale $\triangle_f$ referenced to the scale $\triangle_c$ used in its subgrid model. For simulation C2 with the 2048^3 mesh and uniform spacing in all three directions the length scale in the subgrid model is $\triangle_c = (9/4)^{1/3}\triangle = 0.25$ m. Filter scales in the range $5 < \triangle_f/\triangle_c < 1000$ are used in the present analysis. The lower limit is chosen to help minimize contamination from the subgrid model while the upper limit is constrained by the horizontal dimensions of the LES box, 400 m in each direction. The turbulent fields at vertical levels $z/L_{mo} = (0.083, 0.164)$ inside and near the top of the surface layer are analyzed; at these two levels $\Lambda_w = (21.1, 26.7)$ m. Thus, the resolution ratio possible with the LES data spans $0.1 < \Lambda_w/\triangle_f < 20$, which nicely overlaps with the HATS data. The spatial filtering is a simple top-hat filter applied sequentially in the x–y directions, and the results from ten volumes are averaged to construct the SFS statistics. The SFS momentum and temperature fluxes, in both observations and LES, are constructed from:

$$\tau_{ij} = \widehat{u_i u_j} - \widehat{u}_i \widehat{u}_j \quad \text{and} \quad \tau_{i\theta} = \widehat{u_i \theta} - \widehat{u}_i \widehat{\theta}\,, \tag{4}$$

where the overhat notation $\widehat{(\;)}$ denotes two-dimensional x–y spatial filtering at $\triangle_f$.

The variation in the SFS velocity variances $\langle \tau_{11}, \tau_{22}, \tau_{33} \rangle$, momentum fluxes $\langle \tau_{12}, \tau_{13} \rangle$, and temperature fluxes $\langle \tau_{1\theta}, \tau_{3\theta} \rangle$ with varying resolution ratio $\Lambda_w/\triangle_f$ from HATS, OHATS, and LES are shown in Figures 12–14, respectively. Because of physical constraints, OHATS employed only a single array with horizontal spacing $\delta y_d = 0.58$ m. As a result, the variation in $\Lambda_w/\triangle_f$ from OHATS is a consequence of weak diurnal changes and possible wave effects over the ocean. The small value of $\triangle_f$ pushes all the OHATS results towards large $\Lambda_w/\triangle_f \sim 6$ to 10. HATS employed four different array configurations and, coupled with a vigorous diurnal cycle, results in a wide range of the resolution ratio. Inspection of the HATS data shows stratification is important, as the spectral peak in vertical velocity shifts from large to small scales as the stratification transitions from unstable to stable [58]. For example, the results from HATS array-1 at $z_d = 4.18$ m, blue bullets in Figure 12, span nearly a decade in $\Lambda_w/\triangle_f$ as $-0.3 < z_d/L_{mo} < 1.6$ (see Sullivan et al. [37]).

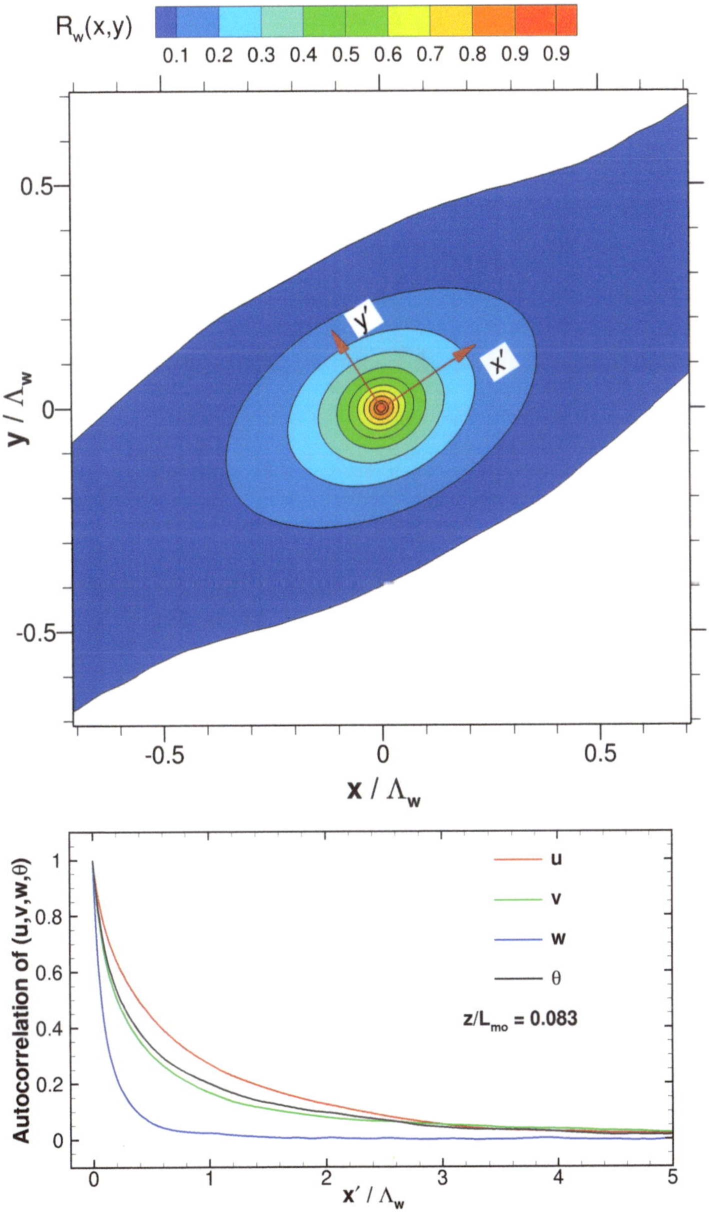

Figure 11. Two-dimensional autocorrelation for vertical velocity $R_{ww}(x, y)$ at $z/L_{mo} = 0.083$ from simulation C2 at $t = 8.556$ h (**upper panel**). The 1d correlation, extracted from the 2d correlation, for (u, v, w, θ) aligned with the mean wind direction x' at $y' = 0$ (**lower panel**). The peak wavelength for the w correlation is $\Lambda_w = 21.0$ m.

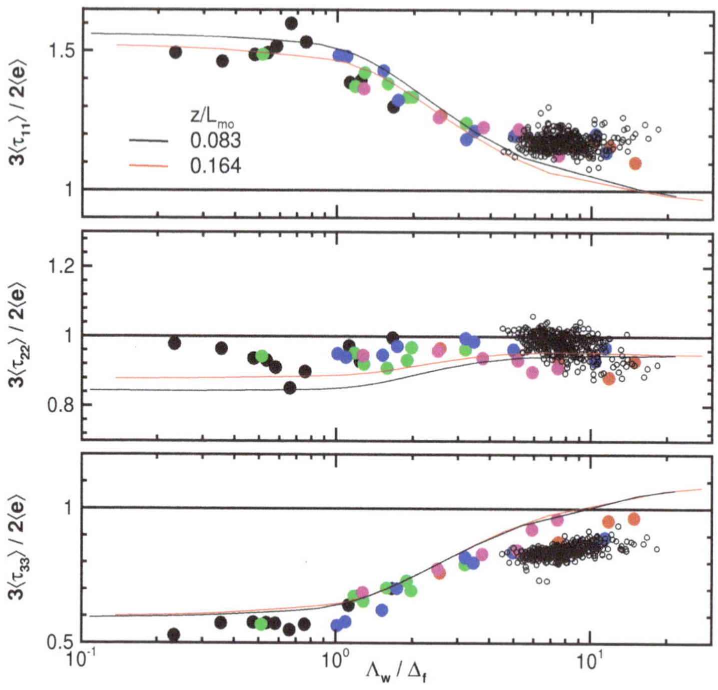

Figure 12. Variation of subfilter-scale velocity variances with resolution ratio $\Lambda_w/\triangle_f$. Measurements are collected in the atmospheric surface layer over land (colored bullets are from different sonic arrays in HATS) and over the ocean (open circles from OHATS) [37,47]. The (red, black) lines are filtered results from simulation C2 with mesh 2048^3. The (red, black) lines are in the surface layer $z/L_{mo} = (0.164, 0.083)$, respectively. The measurements and simulation results highlight the deviation from isotropy when the resolution ratio is $\Lambda_w/\triangle_f \sim \mathcal{O}(1)$.

Overall, the qualitative agreement between the simulation results and observations for variances and fluxes is good, and in some instances the comparison is quantitatively good. The normal variances τ_{ii} in Figure 12 are essentially a bulk metric of SFS isotropy. The observations and simulations both show that the variances $\langle\tau_{ii}\rangle$ tend to unity only for large $\Lambda_w/\triangle_f \geq 10$; isotropy at small scales is an implicit assumption in most Smagorinsky closures for LES, and thus fine computational grids are apparently needed to satisfy this metric. For $\Lambda_w/\triangle_f \leq 1$ the ratio $\langle\tau_{11}\rangle/\langle\tau_{33}\rangle > 2.3$ indicates high anisotropy in the peak energy-containing eddies in the streamwise and vertical components. The momentum fluxes $\langle\tau_{12}, \tau_{13}\rangle$ computed from LES fields are in good agreement with the observations. Notice the vertical momentum flux shows an expected steady approach towards $-u_*^2$ at small $\Lambda_w/\triangle_f$ at the level $z/L_{mo} = 0.083$. The results also suggest that the $z/L_{mo} = 0.164$ level is slightly outside the surface layer in the SBL. At small values of $\Lambda_w/\triangle_f \leq 0.4$, the LES results show a departure of the horizontal variance $\langle\tau_{22}\rangle$ from isotropy, qualitatively consistent with the measurements but perhaps to a somewhat greater degree. The LES also shows non-zero values of horizontal momentum flux $\langle\tau_{12}\rangle$, while the measurement scatter is too great to confidently assess this.

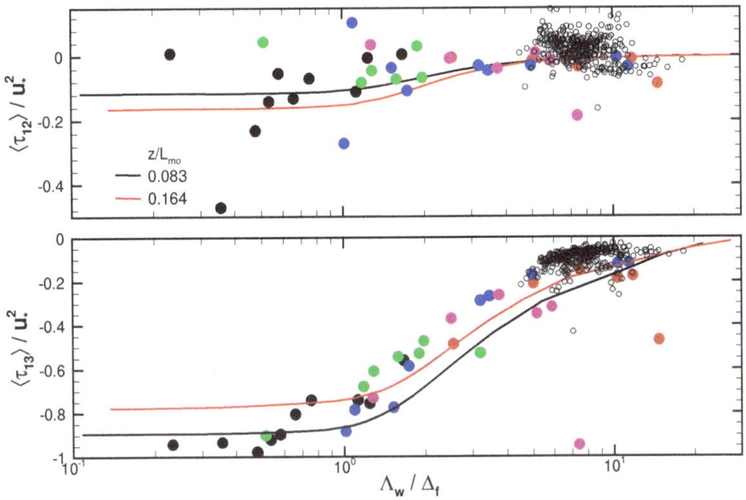

Figure 13. Variation of subfilter-scale momentum fluxes with varying resolution ratio $\Lambda_w/\triangle_f$. Horizontal momentum flux $\langle u'v'\rangle_{sfs}/u_*^2$ (**upper panel**) and vertical momentum flux $\langle u'w'\rangle_{sfs}/u_*^2$ (**lower panel**). Measurements are collected in the atmospheric surface layer over land (colored bullets) and ocean (open circles) [37,47]. The (red, black) lines show filtered results from simulation C2 with mesh 2048^3 and spacing $\triangle = 0.2$ m. The (red, black) lines are in the surface layer $z/L_{mo} = (0.164, 0.083)$, respectively. Note the different vertical scale in the upper and lower panels.

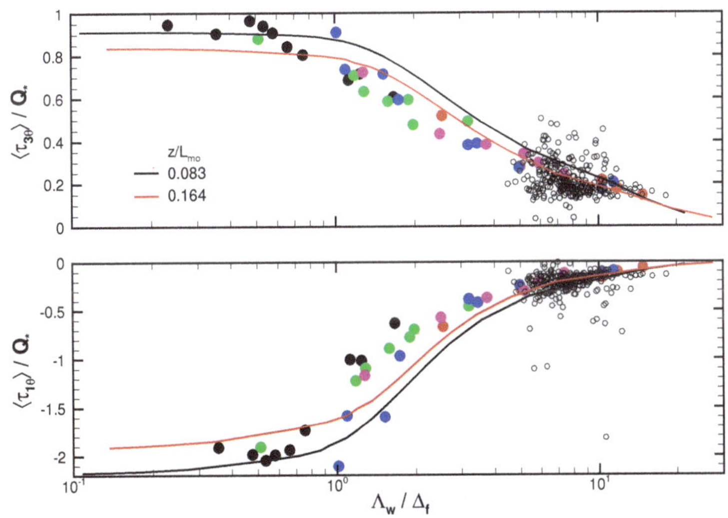

Figure 14. Variation of subfilter-scale temperature fluxes with varying resolution ratio $\Lambda_w/\triangle_f$. Horizontal flux $\langle u'\theta'\rangle_{sfs}/Q_*$ (**upper panel**) and vertical flux $\langle w'\theta'\rangle_{sfs}/Q_*$ (**lower panel**). Measurements are collected in the atmospheric surface layer over land (colored bullets) and over the ocean (open circles) [37,47]. The (red, black) lines show the filtered results from simulation C2 with mesh 2048^3. The (red, black) lines are in the surface layer $z/L_{mo} = (0.164, 0.083)$, respectively.

The horizontal and vertical scalar fluxes $\langle\tau_{1\theta}, \tau_{3\theta}\rangle$ are interesting. These SFS fluxes from LES are in good agreement with the observations. Notice as $\Lambda_w/\triangle_f \to 0$ the vertical scalar flux tends to the surface flux Q_* as expected while the horizontal scalar flux tends to −2.5, which agrees very well with the HATS observations and also with the classic Kansas results, where Wyngaard et al. [63] find the ensemble average of the total scalar flux $-\langle u'\theta' \rangle / \langle w'\theta' \rangle \approx 2.5$ at large positive values of the stability parameter $z/L_{mo} \sim 0.6$. Based on the equations for scalar flux, Wyngaard [61] points out that $\langle \tau_{1\theta} \rangle$ is produced by tilting of the vertical scalar flux by vertical shear $\partial_z \langle u \rangle$, not by a horizontal gradient $\partial_x \langle \theta \rangle$, i.e., an eddy viscosity model is inadequate for horizontal scalar flux when $\Lambda_w / \triangle_f$ is small. Dynamically, the tilted vortices in Figure 9 are the agent producing horizontal scalar flux, as discussed in [20].

The average SFS variances and fluxes from LES agree well with the HATS and OHATS observations over a wide range of $\Lambda_w / \triangle_f$, and thus we expect the instantaneous spatially varying SFS motions from LES are also representative of atmospheric surface-layer flows. The visualizations in Figures 15–17 show the vertical momentum flux and vertical and horizontal temperature flux in the horizontal planes at level $z/L_{mo} = 0.164$ m for resolution ratios $\Lambda_f / \triangle_f = (0.15, 0.91, 2.09, 10.4)$, i.e., spanning either side of the gray zone. Inspection of the figures shows a smooth transition with varying $\Lambda_w / \triangle_f$. A parameterization needs to model all of the flux when $\Lambda_w / \triangle_f$ is small and stochastic fluctuations at large $\Lambda_w / \triangle_f$. In the intermediate gray zone the SFS motions contain a fraction of the total flux (see Figure 13), but stochastic fluctuations are also clearly present at the same time, e.g., see the visualization in Figure 15.

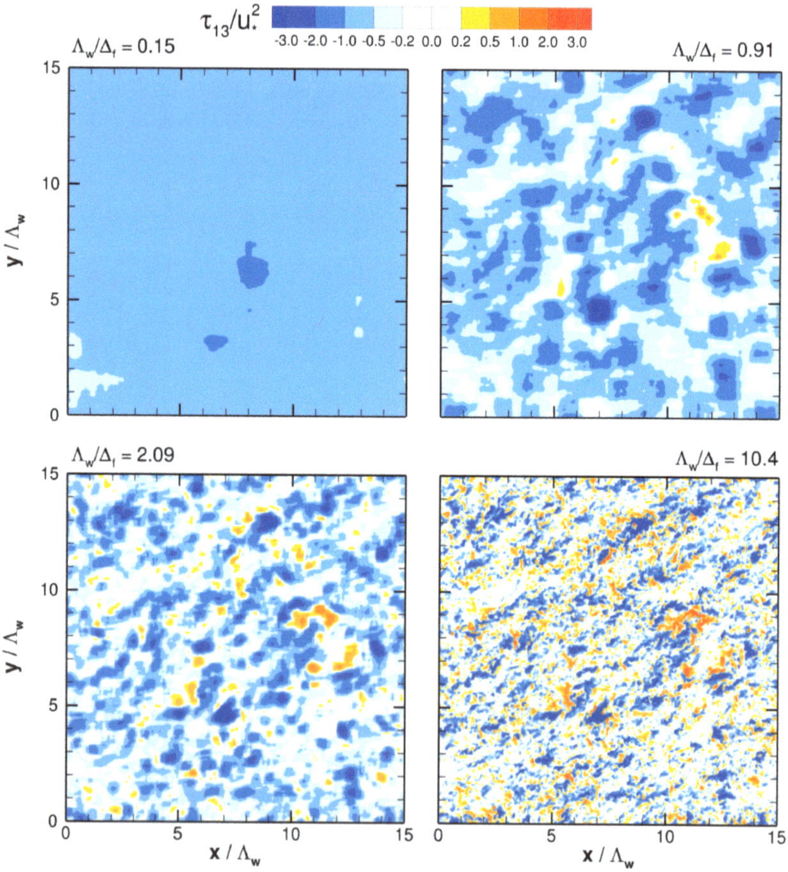

Figure 15. Subfilter-scale momentum flux τ_{13} at $z/L_{mo} = 0.164$ m with varying resolution ratio $\Lambda_w / \triangle_f = (0.52, 0.91, 2.09, 10.4)$ from simulation C2. The normalization is by u_*^2.

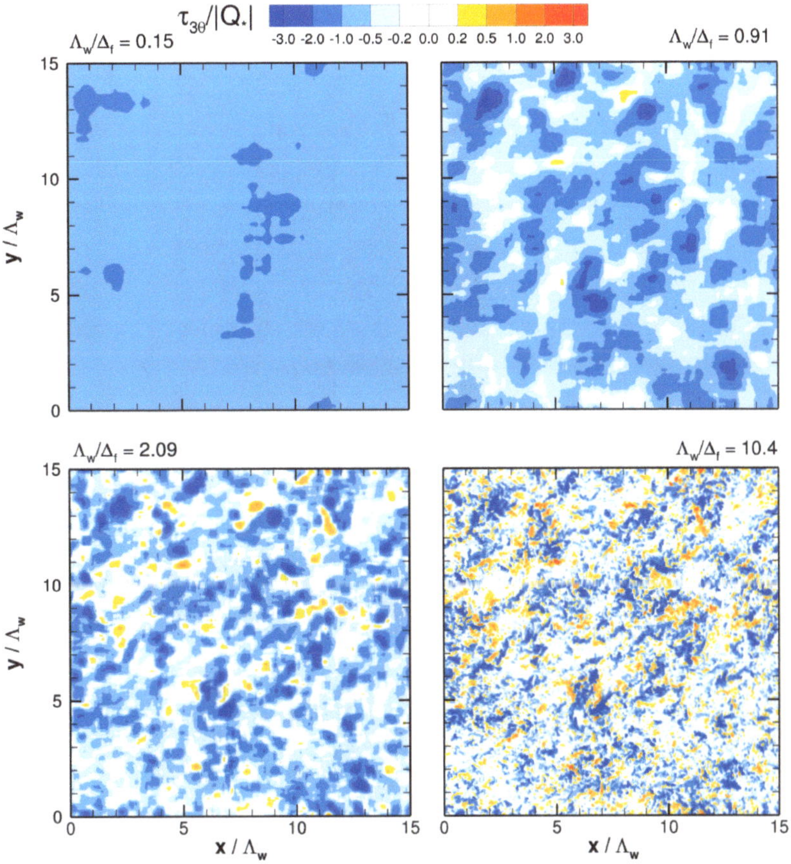

Figure 16. Subfilter-scale vertical temperature flux $\tau_{3\theta}/|Q_*|$ at $z/L_{mo} = 0.164$ with varying resolution ratio $\Lambda_w/\triangle_f = (0.15, 0.91, 2.09, 10.4)$ from simulation C2. To preserve the sign of the flux the normalization is by $|Q_*|$.

In the gray zone, to ensure the proper energy transfer between the resolved and subgrid fields, a SFS flux parameterization needs to account for a fraction of the net flux plus fluctuations correlated with the fluctuating strain rate. To illustrate the idea, consider the energy equation for the resolved motions, which contains the transfer term $\tau_{ij}S_{ij}$, where S_{ij} is the resolved rate of strain tensor; recall this term also appears in the SFS energy equation with opposite sign [64]. Decomposing the SFS flux and resolved rate of strain tensors into a mean and fluctuation yields the expression

$$\langle \tau_{ij} S_{ij} \rangle = \langle \tau_{ij} \rangle \langle S_{ij} \rangle + \langle \tau'_{ij} S'_{ij} \rangle ; \tag{5}$$

notice that $\tau'_{ij}S'_{ij}$ allows both forward and backward (backscatter) motion of energy, which was observed in HATS [37]. Inspection of (5) shows the difficulty: energy transfer at small and large $\Lambda_w/\triangle_f$ is $\langle \tau_{ij} \rangle \langle S_{ij} \rangle$ or the turbulent correlation $\langle \tau'_{ij} S'_{ij} \rangle$, respectively. However, Figure 18 shows that in the intermediate gray zone both terms on the right-hand side of (5) contribute to the energy transfer. Thus, a SFS flux parameterization needs to account for a fraction of the net flux but also important stochastic fluctuations because of the energy transfer between the resolved and subgrid fields. The SFS momentum fluctuations are not random noise but are clearly correlated with fluctuations in the strain rate. The transfer of scalar variance faces a similar dilemma in the gray zone.

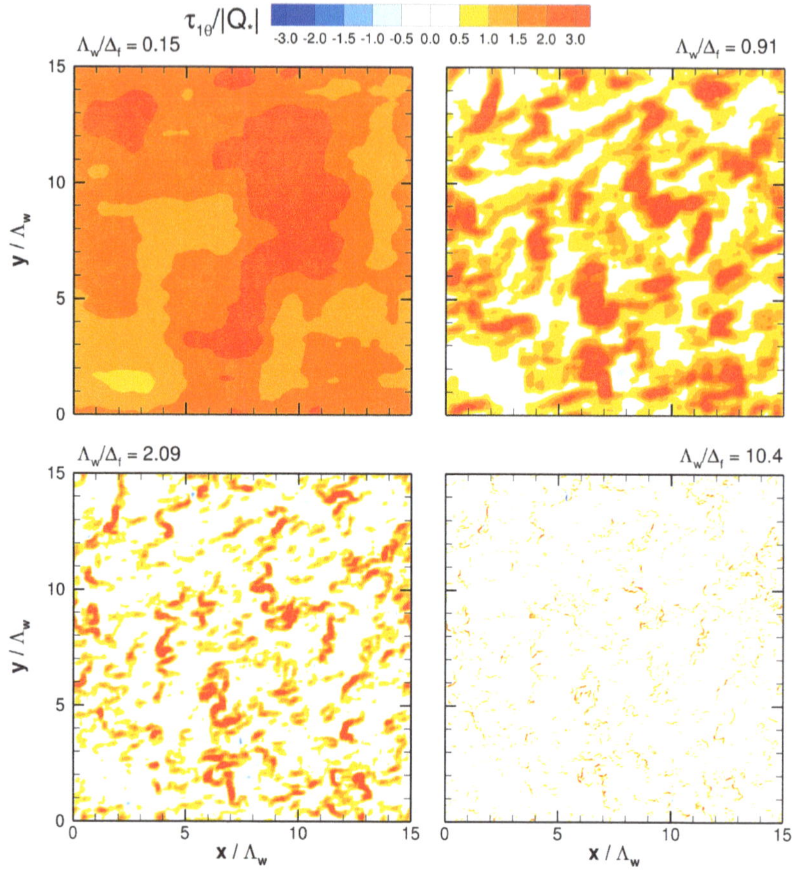

Figure 17. Subfilter-scale horizontal temperature flux $\tau_{1\theta}/|Q_*|$ at $z/L_{mo} = 0.164$ with varying resolution ratio $\Lambda_w/\triangle_f = (0.52, 0.91, 2.09, 10.4)$ from simulation C2. To preserve the sign of the flux the normalization is by $|Q_*|$.

Parameterizations in the gray zone are further confronted by the possible "double counting" of the momentum and temperature fluxes. This can result when a SFS paramterization is used outside its design range in the space $\Lambda_w/\triangle_f$. For example, this can occur in the gray zone $\Lambda_w/\triangle_f \sim 1$ when a single-column model, designed for RANS vertical flux, is used at the same time the model grid resolution is sufficient to support the resolved turbulence. Under these conditions the estimate of flux can be double counted [60]. Flux double counting spoils the energy and scalar transfer between the resolved and SFS fields.

A cursory inspection of the patterns in τ_{13} (Figure 15) and $\tau_{3\theta}$ (Figure 16) suggests a possible amplitude correlation, i.e., $\tau_{3\theta}$ is approximately related to τ_{13} by a constant [65]. The assumption that the scalar diffusivity and momentum eddy viscosity are related is routinely adopted by LES models. The visualization of the streamwise scalar flux shows a near-zero mean and small fluctuations at large $\Lambda_w/\triangle_f$; $\tau_{1\theta}$ is routinely neglected in SFS modeling [66]. At small $\Lambda_w/\triangle_f$ the horizontal scalar flux is clearly non-zero, but its impact disappears under the usual assumption of a periodic horizontally homogeneous flow $\partial_x\langle u'\theta'\rangle = 0$.

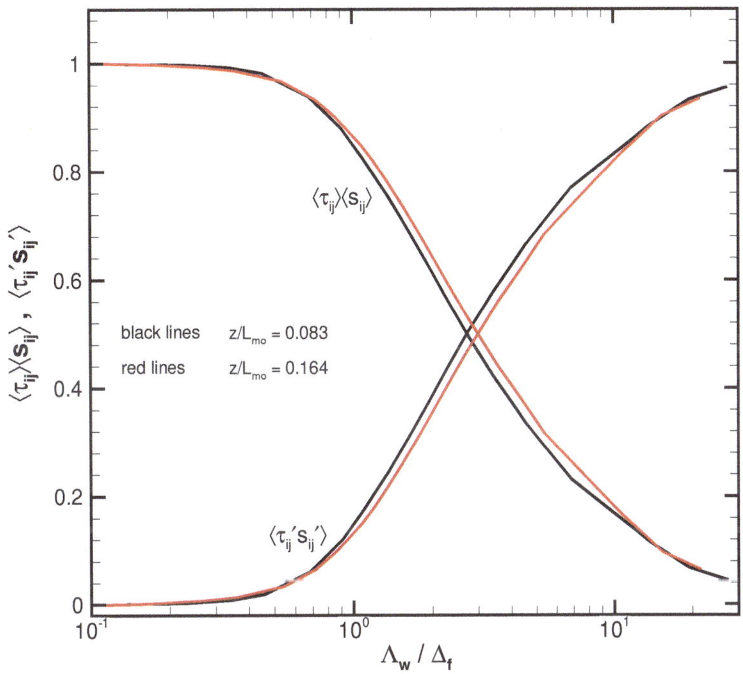

Figure 18. Variation of the production terms in Equation (5) for varying filter width $\triangle_f$ at two heights near the surface $z/L_{mo} = (0.083, 0.164)$ denoted by (black, red) lines, respectively, from simulation C2. The individual production terms are normalized by the total production $\langle \tau_{ij} S_{ij} \rangle$ at each filter width. Total production $\langle \tau_{ij} S_{ij} \rangle$ is nearly constant with varying $\triangle_f$, but slopes downward at small filter widths because of the actual subgrid model used in the LES.

The fine-mesh LES results can also be used to craft parameterizations that span the gray zone. For example, consider a $q^2 - \ell$ second-order closure that uses a prognostic equation for the unresolved SFS turbulence kinetic energy $q^2 = (u'^2 + v'^2 + w'^2)/2$ with ℓ a prescribed length scale. This closure uses a TKE equation with a model for viscous dissipation ϵ [65], typically of the form

$$\epsilon = \frac{q^3}{L_\epsilon} \, . \tag{6}$$

The dissipation model requires specification of the length scale L_ϵ. Figure 19 shows the variation of L_ϵ with inverse resolution ratio $\triangle_f / \Lambda_w$ computed from the LES results; this result is obtained by assuming a constant dissipation across $\triangle_f$. For small filter widths in the LES regime, q^2 is a small fraction of the total energy and $L_\epsilon \propto \triangle_f$. At large filter widths approaching the RANS regime, q^2 is the total kinetic energy and the length scale saturates, i.e., $L_\epsilon / \Lambda_w \sim 1.4$ in the surface layer. A smooth variation of L_ϵ is found at intermediate filter widths between the RANS and LES limits. In the middle of the gray zone $L_\epsilon \sim \Lambda_w$ at $\triangle_f = \Lambda_w$. Thus, in general one needs an adaptive parameterization $L_\epsilon = L_\epsilon(z, \triangle_f / \Lambda_w)$ as the filter width varies across the RANS to LES regimes.

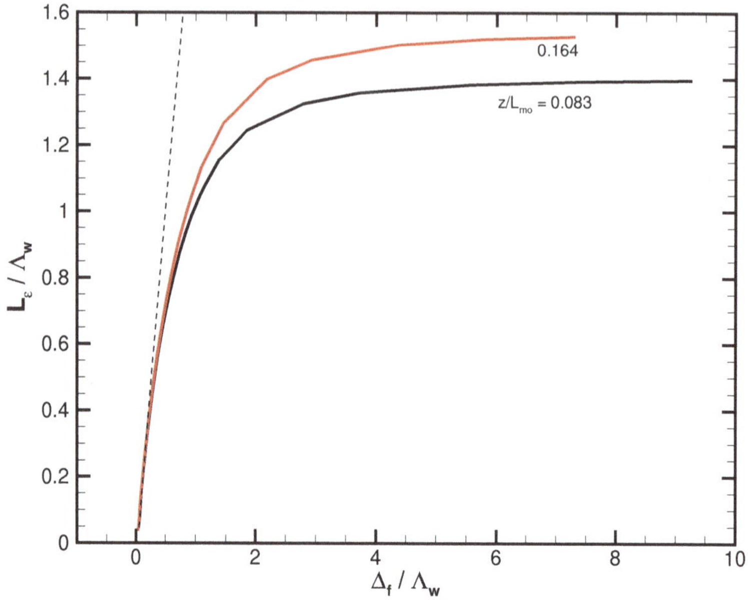

Figure 19. Variation of the dissipation length scale L_ϵ for varying filter width $\triangle_f$ at two heights near the surface $z/L_{mo} = (0.083, 0.164)$ denoted by (black, red) lines, respectively, from simulation C2. The black dashed line is a linear fit to L_ϵ for small filter widths similar to the predictions by Wyngaard [61].

6. Summary and Discussion

Recent high-resolution large-eddy simulations (LES) of a stable atmospheric boundary layer (SBL) with mesh sizes $N = (512^3, 1024^3, 2048^3)$, or mesh spacings $\triangle = (0.78, 0.39, 0.2)$ m, are described and analyzed. The SBL posing is the GABLS1 intercomparison case for LES driven by a constant surface cooling rate $C_r = 0.25$ K h^{-1} described by Beare et al. [24] and Holtslag [23]. An SBL with a nearly four-times-stronger stratification on a mesh with $N = 2048^3$ is also analyzed. These recent high-resolution datasets with 2048^3 mesh points are being added to the Johns Hopkins University Turbulence Database (JHTDB) for further analysis by the research community.

The LES solutions on meshes $N = (512^3, 1024^3, 2048^3)$ are judged converged based on the good collapse of vertical profiles of mean winds, temperature, and low-order turbulence moments, i.e., fluxes and variances. The largest discrepancy is in the stably stratified region above the low-level jet (LLJ) where the turbulence is small-scale and weak. The boundary-layer height between the simulations varies by less than 10%, which induces small changes in the friction velocity and surface temperature flux. These comparisons with different N indicate that further convergence of low-order moments is likely with even larger values. With increasing stratification, $C_r = 1$ K h^{-1}, the LLJ descends and the SBL is nearly equally split into two vertical layers. Above the LLJ the Richardson number is uniform $Ri \sim 0.2$, which supports only weak turbulence.

The flow visualization shows the SBL is filled with ubiquitous temperature fronts tilted in the streamwise direction. The spatial distance separating the warm–cold fronts is thin, i.e., the fronts are very sharp. Mesh refinement decreases the spatial separation between the fronts, and increasing stratification causes the fronts to tilt farther forward in a shallower SBL. At a fixed horizontal location x–y a temperature profile exhibits a staircase pattern in the vertical direction where sharp jumps are separated by a nearly uniform (constant) temperature. The flow visualization highlights the SBL dynamics across scales. Pairs of counter-rotating, well-resolved "horseshoe" vortices tilted in the downstream–upward direction are the agents inducing the sharp temperature fronts; the vortices induce both vertical and horizontal temperature and momentum fluxes. The scale of the vortices coincides with the scale of the peak in the θ and w spectra, as shown by Sullivan et al. [20].

The subfilter-scale (SFS) motions deduced from the LES solutions are compared with observations from the horizontal array turbulence study (HATS) and the ocean horizontal turbulence study (OHATS); these multi-point measurements were collected from cross-wind arrays of sonic anemometers in the atmospheric surface layer over a range of wind speeds, stratification, and sensor spacings. The SFS momentum and temperature fluxes, $\tau_{ij} = \widehat{u_i u_j} - \widehat{u}_i \widehat{u}_j$ and $\tau_{i\theta} = \widehat{u_i \theta} - \widehat{u}_i \widehat{\theta}$, are constructed from the observations and LES using similar 2D low-pass filters with varying filter width $\triangle_f$. Their comparisons are carried out using a dimensionless resolution ratio $\Lambda_w / \triangle_f$, where Λ_w is the characteristic length scale of the energy-containing eddies for vertical velocity w. Over the range $0.1 < \Lambda_w / \triangle_f < 10$, the SFS fluxes and variances from the fine-mesh LES and observations are in good agreement. At first blush this seems surprising, but notice that when $\triangle_f \gg \triangle$ very-well-resolved turbulent motions are used to compute the SFS variances and fluxes.

The limits $\Lambda_w / \triangle_f > 10$ and $\Lambda_w / \triangle_f < 0.1$ correspond to the LES and unsteady Reynolds-averaged Navier–Stokes (RANS) regimes, respectively. The intermediate regime is encountered near walls and in simulations with coarse LES grids or fine mesoscale grids. The regime with $\Lambda_w / \triangle_f \sim 1$ is termed "Terra Incognita" or the gray scale by Wyngaard [40]. The SFS fluxes from HATS and LES both vary smoothly across the gray scale. The SFS parameterization in the gray scale with partially resolved flux needs to account for both a "RANS" flux in addition to fluctuations. Fluctuations in the SFS momentum fluxes are not random noise but are well correlated with the strain rate. A scale-aware parameterization needs to set the proper energy or scalar transfer between the resolved and subgrid fields.

The good agreement between the LES results and observations is encouraging and allows fine-mesh LES fields to be used for guidance in formulating SFS models and for a priori tests for the SBL and other high-Reynolds-number flows. These would be more relevant than similar tests with DNS that are limited to geophysically unrealistically low values of the Reynolds number. Given the large number of SFS parameterization models that have been proposed and the several canonical geophysical flow regimes, it would be a substantial enterprise to make systematic tests, but well worthwhile. Moreover, in the spirit of our age, one could ask whether artificial intelligence algorithms should supervise the tests or even search beyond the present SFS proposals.

7. Whither Turbulence?

This section title is shamelessly borrowed from Lumley [67], which is a collection of essays about the state and direction of turbulence research at that time. (See also the biographical discussion of Lumley's life and views in Leibovich and Warhaft [68]).

Turbulent flows are ubiquitous in natural fluids, and their science is centuries old. It is an unusual science in several ways. Its governing partial differential equation, Navier–Stokes, is known and seems to be valid after many experimental tests. Yet its mathematical nature is still incompletely characterized (n.b., the unclaimed Clay Mathematics Institute Millennium Prize for proof of existence and smoothness of its solutions). It is highly accessible to measurements in laboratories and Earth's atmosphere and ocean. Yet these measurements always seem to be insufficiently complete. It has been and continues to be a premier computational problem with various discrete approximations to the known continuum equations. Yet solution convergence remains elusive in pursuit of the very large Reynolds numbers in natural flows, and turbulence seems to have an endless variety of outcomes in different physical regimes. We remember Horace Lamb's quip about whether we would even understand the nature of turbulence if it were explained to us in heaven [69].

Thus, we just keep plugging away. Turbulence is far too important to abandon, and measurements and simulations can always answer whatever a well-conceived next question is. We can continue to make models of how turbulence behaves in the subgrid-scale regime at high Reynolds number; they are at least useful, if not provably correct, and they are falsifiable with a next set of measurements and simulations. The solutions in this paper could play such a role for many extant models. The goal is to identify and solve problems that enable us to infer generic behaviors, recognizing that many different paradigms are needed for geophysical turbulence.

The present paper is a happy confluence of multi-point atmospheric measurements, apparently adequate subgrid parameterizations, and large computational grids that takes yet another step up the tall ladder toward heaven, where maybe we will meet Jack again.

Author Contributions: All authors contributed to the work. J.C.M. wrote critical sections of the manuscript, suggested analyses, and edited the manuscript. C.M. participated in HATS, edited the manuscript, and suggested analyses. E.G.P. ran simulations, edited the manuscript, and suggested analyses. P.P.S. participated in HATS and OHATS, ran simulations, analyzed the observational and simulation data, and wrote large sections of the manuscript. All authors have read and agreed to the published version of the manuscript.

Funding: CM, EGP, and PPS were supported by NSF award number 2103874 with the JHU to NCAR, subaward 2005117865. PPS and JCM were supported by the Office of Naval Research through the Physical Oceanography Program, award numbers N00014-17-1-2334 and N00014-18-1-2599, and by the National Oceanic and Atmospheric Administration, award number NA19OAR4310378. PPS and EGP acknowledge support from the National Science Foundation and the Geophysical Turbulence Program at the National Center for Atmospheric Research (NCAR). We also acknowledge high-performance computing support from Cheyenne (https://doi.org/10.5065/D6RX99HX, accessed on 1 May 2023) provided by NCAR's Computational and Information Systems Laboratory, sponsored by the National Science Foundation.

Institutional Review Board Statement: Not applicable.

Informed Consent Statement: Not applicable.

Data Availability Statement: The numerical simulation data used in this manuscript will be accessible via the Johns Hopkins Turbulence Database: https://turbulence.pha.jhu.edu (accessed on 1 May 2023). The HATS and OHATS data can be accessed at: https://doi.org/10.5065/D6X34VWV (accessed on 1 May 2023) and https://doi.org/10.5065/D6959G0N (accessed on 1 May 2023), respectively.

Acknowledgments: The authors thank James Edson (Woods Hole Oceanographic Institute), John Wyngaard (The Pennsylvania State University), and Thomas Horst (NCAR) for their participation in the OHATS field campaign. The five reviewers provided useful comments that improved the manuscript.

Conflicts of Interest: The authors declare no conflict of interest. The funders had no role in the design of the study; in the collection, analyses, or interpretation of data; in the writing of the manuscript; or in the decision to publish the results.

References

1. Patterson, G.; Orszag, S. Spectral calculations of isotropic turbulence: Efficient removal of aliasing interactions. *Phys. Fluids* **1971**, *14*, 2538. [CrossRef]
2. Herring, J.R.; McWilliams, J.C. Lecture Notes on Turbulence. In *Lecture Notes from the NCAR-GTP Summer School*; Herring, J.R., McWilliams, J.C., Eds.; World Scientific: Singapore, 1989; p. 371. [CrossRef]
3. Fernando, H.J.S. Turbulent mixing in stratified fluids. *Annu. Rev. Fluid Mech.* **1991**, *23*, 455–493. [CrossRef]
4. Fernando, H.J.S.; Weil, J.C. Whither the stable boundary layer? A shift in the research agenda. *Bull. Am. Meteorol. Soc.* **2010**, *91*, 1475–1481. [CrossRef]
5. Mahrt, L. Stably stratified atmospheric boundary layers. *Annu. Rev. Fluid Mech.* **2014**, *46*, 23–45. [CrossRef]
6. Caulfield, C.P. Layering, instabilities, and mixing in turbulent stratified flows. *Annu. Rev. Fluid Mech.* **2021**, *53*, 113–145. [CrossRef]
7. Large, W.G.; McWilliams, J.C.; Doney, S.C. Oceanic vertical mixing: A review and a model with a nonlocal boundary layer parameterization. *Rev. Geophys.* **1994**, *32*, 363–403. [CrossRef]
8. McWilliams, J.C. Phenomenological hunts in two-dimensional and stably stratified turbulence. In *Atmospheric Turbulence and Mesoscale Meteorology*; Federovich, E., Rotunno, R., Stevens, B., Eds.; Cambridge University Press: Cambridge, UK, 2004; pp. 35–49. [CrossRef]

9. Cuxart, J.; Holtslag, A.A.M.; Beare, R.J.; Bazile, E.; Beljaars, A.; Cheng, A.; Conangla, L.; Ek, M.; Freedman, F.; Hamdl, R.; et al. Single-column model intercomparison for a stably stratified atmospheric boundary layer. *Bound.-Layer Meteorol.* **2006**, *118*, 273–303. [CrossRef]
10. Svensson, G.; Holtslag, A.A.M. Analysis of model results for the turning of the wind and related momentum fluxes in the stable boundary layer. *Bound.-Layer Meteorol.* **2009**, *132*, 261–277. [CrossRef]
11. Holtslag, A.A.M.; Svensson, G.; Baas, P.; Basu, S.; Beare, B.; Beljaars, A.C.M.; Bosveld, F.C.; Cuxart, J.; Lindvall, J.; Steeneveld, G.J.; et al. Stable atmospheric boundary layers and diurnal cycles: Challenges for weather and climate models. *Bull. Am. Meteorol. Soc.* **2013**, *94*, 1691–1706. [CrossRef]
12. Heisel, M.; Sullivan, P.P.; Katul, G.G.; Chamecki, M. Turbulence organization and mean profile shapes in the stably stratified boundary layer: Zones of uniform momentum and air temperature. *Bound.-Layer Meteorol.* **2023**, *186*, 533–565. [CrossRef]
13. Wyngaard, J.C.; Seaman, N.; Kimmel, S.J.; Otte, M.; Di, X.; Gilbert, K.E. Concepts, observations, and simulation of refractive index turbulence in the lower atmosphere. *Radio Sci.* **2001**, *36*, 643–669. [CrossRef]
14. Kelley, N.; Shirazi, M.; Buhl, M.; Jager, D.; Wilde, S.; Adams, J.; Bianchi, J.; Sullivan, P.; Patton, E. *Lamar Low-Level Jet Project Interim Report*; Technical Report NREL/TP-500-34593; National Renewable Energy Laboratory: Golden, CO, USA, 2003. [CrossRef]
15. Weil, J.C. Stable boundary layer modeling for air quality applications. In *Air Pollution Modeling and Its Application XXI*; Castelli, S.T., Steyn, D., Eds.; NATO Science for Peace and Security Series C: Environmental Security; Springer: Dordrecht, The Netherlands, 2012; pp. 57–61. [CrossRef]
16. Nieuwstadt, F.T.M. Direct numerical simulation of stable channel flow at large stability. *Bound.-Layer Meteorol.* **2005**, *116*, 277–299. [CrossRef]
17. van de Wiel, B.J.H.; Moene, A.F.; Hartogensis, O.K.; de Bruin, H.A.R.; Holtslag, A.A.M. Intermittent turbulence in the stable boundary layer over land. Part III. A classification for observations during CASES-99. *J. Atmos. Sci.* **2003**, *60*, 2509–2522. [CrossRef]
18. van Haren, H.; Gostiaux, L. A deep-ocean Kelvin-Helmholtz billow train. *Geophys. Res. Lett.* **2010**, *37*, L03605. [CrossRef]
19. Moeng, C.H.; Sullivan, P.P. Large-eddy simulation. In *Encyclopedia of Atmospheric Sciences*, 2nd ed.; North, G.R., Zhang, F., Pyle, J., Eds.; Academic Press: Cambridge, MA, USA, 2015; Volume 4, pp. 232–240. [CrossRef]
20. Sullivan, P.P.; Weil, J.C.; Patton, E.G.; Jonker, H.J.J.; Mironov, D.V. Turbulent winds and temperature fronts in large-eddy simulations of the stable atmospheric boundary layer. *J. Atmos. Sci.* **2016**, *73*, 1815–1840. [CrossRef]
21. Johnson, P.L. A physics-inspired alternative to spatial filtering for large-eddy simulations of turbulent flows. *J. Fluid Mech.* **2022**, *934*, A30. [CrossRef]
22. Sullivan, P.P.; McWilliams, J.C.; Moeng, C.H. A subgrid-scale model for large-eddy simulation of planetary boundary-layer flows. *Bound.-Layer Meteorol.* **1994**, *71*, 247–276. [CrossRef]
23. Holtslag, A.A.M. GEWEX atmospheric boundary-layer study GABLS on stable boundary layers. *Bound.-Layer Meteorol.* **2006**, *118*, 243–246. [CrossRef]
24. Beare, R.J.; Macvean, M.K.; Holtslag, A.A.M.; Cuxart, J.; Esau, I.; Golaz, J.C.; Jimenez, M.A.; Khairoutdinov, M.; Kosovic, B.; Lewellen, D.; et al. An intercomparison of large-eddy simulations of the stable boundary layer. *Bound.-Layer Meteorol.* **2006**, *118*, 242–272. [CrossRef]
25. Huang, J.; Bou-Zeid, E. Turbulence and vertical fluxes in the stable atmospheric boundary layer. Part I: A large-eddy simulation study. *J. Atmos. Sci.* **2013**, *70*, 1513–1527. [CrossRef]
26. Li, Y.; Perlman, E.; Wan, M.; Yang, Y.; Meneveau, C.; Burns, R.; Chen, S.; Szalay, A.; Eyink, G. A public turbulence database cluster and applications to study Lagrangian evolution of velocity increments in turbulence. *J. Turbul.* **2008**, *9*, N31. [CrossRef]
27. Graham, J.; Kanov, K.; Yang, X.; Lee, M.; Malaya, N.; Lalescu, C.; Burns, R.; Eyink, G.; Szalay, A.; Moser, R.; et al. A web services accessible database of turbulent channel flow and its use for testing a new integral wall model for LES. *J. Turbul.* **2016**, *17*, 181–215. [CrossRef]
28. Piomelli, U.; Moin, P.; Ferziger, J.H. Model consistency in large eddy simulation of turbulent channel flows. *Phys. Fluids* **1988**, *31*, 1884–1891. [CrossRef]
29. Liu, S.; Meneveau, C.; Katz, J. On the properties of similarity subgrid-scale models as deduced from measurements in a turbulent jet. *J. Fluid Mech.* **1994**, *275*, 83–119. [CrossRef]
30. Horst, T.W.; Kleissl, J.; Lenschow, D.H.; Meneveau, C.; Moeng, C.H.; Parlange, M.B.; Sullivan, P.P.; Weil, J.C. Field observations to obtain spatially-filtered turbulence fields from transverse arrays of sonic anemometers in the atmospheric surface layer. *J. Atmos. Sci.* **2004**, *61*, 1566–1581. [CrossRef]
31. Meneveau, C.; Lund, T.S. The dynamic Smagorinsky model and scale-dependent coefficients in the viscous range of turbulence. *Phys. Fluids* **1997**, *9*, 3932–3934. [CrossRef]
32. Tong, C.; Wyngaard, J.C.; Khanna, S.; Brasseur, J.G. Resolvable- and subgrid-scale measurement in the atmospheric surface layer: Technique and issues. *J. Atmos. Sci.* **1998**, *55*, 3114–3126. [CrossRef]
33. Tong, C.; Wyngaard, J.C.; Brasseur, J.G. Experimental study of the subgrid-scale stresses in the atmospheric surface layer. *J. Atmos. Sci.* **1999**, *56*, 2277–2292. [CrossRef]
34. Tao, B.; Katz, J.; Meneveau, C. Geometry and scale relationships in high Reynolds number turbulence determined from three-dimensional holographic velocimetry. *Phys. Fluids* **2000**, *12*, 941–944. [CrossRef]

35. Meneveau, C.; Katz, J. Scale-invariance and turbulence models for large-eddy simulations. *Annu. Rev. Fluid Mech.* **2000**, *32*, 1–32. [CrossRef]
36. Pope, S.B. *Turbulent Flows*; Cambridge University Press: Cambridge, UK, 2000; p. 771. [CrossRef]
37. Sullivan, P.P.; Horst, T.W.; Lenschow, D.H.; Moeng, C.H.; Weil, J.C. Structure of subfilter-scale fluxes in the atmospheric surface layer with application to large-eddy simulation modeling. *J. Fluid Mech.* **2003**, *482*, 101–139. [CrossRef]
38. Kleissl, J.; Meneveau, C.; Parlange, M.B. On the magnitude and variability of subgrid-scale eddy-diffusion coefficients in the atmospheric surface layer. *J. Atmos. Sci.* **2003**, *60*, 2372–2388. [CrossRef]
39. Kleissl, J.; Parlange, M.; Meneveau, C. Field experimental study of dynamic Smagorinsky models in the atmospheric surface layer. *J. Atmos. Sci.* **2004**, *61*, 2296–2307. [CrossRef]
40. Wyngaard, J.C. Toward numerical modeling in the Terra Incognita. *J. Atmos. Sci.* **2004**, *61*, 1816–1826. [CrossRef]
41. Chen, Q.; Tong, C. Investigation of the subgrid-scale stress and its production rate in a convective atmospheric boundary layer using measurement data. *J. Fluid Mech.* **2006**, *547*, 65–104. [CrossRef]
42. Hatlee, S.C.; Wyngaard, J.C. Improved subfilter-scale models from the HATS field data. *J. Atmos. Sci.* **2007**, *64*, 1694–1705. [CrossRef]
43. Chamecki, M.; van-Hout, R.; Meneveau, C.; Parlange, M.B. Concentration profiles of particles settling in the neutral and stratified atmospheric surface layer. *Bound.-Layer Meteorol.* **2007**, *125*, 25–38. [CrossRef]
44. Ramachandran, S.; Wyngaard, J.C. Subfilter-scale modelling using transport equations: Large-eddy simulation of the moderately convective atmospheric boundary layer. *Bound.-Layer Meteorol.* **2011**, *139*, 1–35. [CrossRef]
45. Higgins, C.W.; Meneveau, C.; Parlange, M.B. The effect of filter dimension on the subgrid-scale stress, heat flux, and tensor alignments in the atmospheric surface layer. *J. Atmos. Ocean. Technol.* **2007**, *24*, 360–375. [CrossRef]
46. Kumar, V.; Kleissl, J.; Meneveau, C.; Parlange, M.B. Large-eddy simulation of a diurnal cycle of the atmospheric boundary layer: Atmospheric stability and scaling issues. *Water Resour. Res.* **2006**, *42*, W06D09. [CrossRef]
47. Sullivan, P.P.; Edson, J.B.; Horst, T.W.; Wyngaard, J.C.; Kelly, M. Subfilter scale fluxes in the marine surface layer: Results from the Ocean Horizontal Array Turbulence Study (OHATS). In Proceedings of the 17th Symposium on Boundary Layer and Turbulence, American Meteorological Society, San Diego, CA, USA, 23 May 2006; Paper 4.1, pp. 1–8. Available online: http://ams.confex.com/ams/BLTAgFBioA/techprogram/paper_110884.htm (accessed on 1 May 2023).
48. Bou-Zeid, E.; Higgins, C.; Huwald, H.; Meneveau, C.; Parlange, M.B. Field study of the dynamics and modelling of subgrid-scale turbulence in a stable atmospheric surface layer over a glacier. *J. Fluid Mech.* **2010**, *665*, 480–515. [CrossRef]
49. Patton, E.G.; Horst, T.W.; Sullivan, P.P.; Lenschow, D.H.; Oncley, S.P.; Brown, W.O.J.; Burns, S.P.; Guenther, A.B.; Held, A.; Karl, T.; et al. The canopy horizontal array turbulence study. *Bull. Am. Meteorol. Soc.* **2011**, *92*, 593–611. [CrossRef]
50. Nguyen, K.X.; Horst, T.W.; Oncley, S.P.; Tong, C. Measurements of the budgets of the subgrid-scale stress and temperature flux in the convective atmospheric surface layer. *J. Fluid Mech.* **2013**, *729*, 388–422. [CrossRef]
51. Chow, F.K.; Moin, P. A further study of numerical errors in large-eddy simulations. *J. Comp. Phys.* **2003**, *184*, 366–380. [CrossRef]
52. Sullivan, P.P.; Patton, E.G. The effect of mesh resolution on convective boundary-layer statistics and structures generated by large-eddy simulation. *J. Atmos. Sci.* **2011**, *68*, 2395–2415. [CrossRef]
53. Geurts, B.J.; Fröhlich, J. A framework for predicting accuracy limitations in large-eddy simulation. *Phys. Fluids* **2002**, *14*, L41–L44. [CrossRef]
54. Sullivan, P.P.; Moeng, C.H.; Stevens, B.; Lenschow, D.H.; Mayor, S.D. Structure of the entrainment zone capping the convective atmospheric boundary layer. *J. Atmos. Sci.* **1998**, *55*, 3042–3064. [CrossRef]
55. Dai, Y.; Basu, S.; Maronga, B.; de Roode, S.R. Addressing the grid-size sensitivity issue in large-eddy simulations of stable boundary layers. *Bound.-Layer Meteorol.* **2021**, *178*, 63–89. [CrossRef]
56. Adrian, R.J. Stochastic estimation of the structure of turbulent fields. In *Eddy Structure Identification*; Bonnet, J.P., Ed.; Springer: Vienna, Austria, 1996; pp. 145–196. [CrossRef]
57. Kimura, Y.; Herring, J.R. Diffusion in stably stratified turbulence. *J. Fluid Mech.* **1996**, *328*, 253–269. [CrossRef]
58. Kaimal, J.C.; Finnigan, J.J. *Atmospheric Boundary Layer Flows*; Oxford University Press: New York, NY, USA, 1994. [CrossRef]
59. Bryan, G.H.; Morrison, H. Sensitivity of a simulated squall line to horizontal resolution and parameterization of microphysics. *Mon. Weather Rev.* **2012**, *140*, 202–225. [CrossRef]
60. Doubrawa, P.; Muñoz-Esparza, D. Simulating real atmospheric boundary layers at gray-zone resolutions: How do currently available turbulence parameterizations perform? *Atmosphere* **2020**, *11*, 345. [CrossRef]
61. Wyngaard, J.C. Changing the face of small-scale meteorology. In *Atmospheric Turbulence and Mesoscale Meteorology*; Federovich, E., Rotunno, R., Stevens, B., Eds.; Cambridge University Press: Cambridge, UK, 2004; pp. 17–34. [CrossRef]
62. Perot, J.B.; Gadebusch, J. A self adapting turbulence model for flow simulation at any mesh resolution. *Phys. Fluids* **2007**, *19*, 115105. [CrossRef]
63. Wyngaard, J.C.; Cóte, O.R.; Izumi, Y. Local free convection, similarity, and the budgets of shear stress and heat flux. *J. Atmos. Sci.* **1971**, *28*, 1171–1182. [CrossRef]
64. Moeng, C.H. A large-eddy simulation model for the study of planetary boundary-layer turbulence. *J. Atmos. Sci.* **1984**, *41*, 2052–2062. [CrossRef]
65. Moeng, C.H.; Wyngaard, J.C. Evaluation of turbulent transport and dissipation closures in second-order modeling. *J. Atmos. Sci.* **1989**, *46*, 2311–2330. [CrossRef]

66. Deardorff, J.W. Three-dimensional numerical modeling of the planetary boundary layer. In *Workshop on Micrometeorology*; Haugen, D.A., Ed.; American Meteorological Society: Boston, MA, USA, 1972; pp. 271–311.
67. Lumley, J.L. (Ed.) *Whither Turbulence? Turbulence at the Crossroads*; Springer: Berlin/Heidelberg, Germany, 1990; Volume 357. [CrossRef]
68. Leibovich, S.; Warhaft, Z. John Leask Lumley: Whither Turbulence? *Annu. Rev. Fluid Mech.* **2018**, *50*, 1–23. [CrossRef]
69. Goldstein, S. Fluid mechanics in the first half of this century. *Annu. Rev. Fluid Mech.* **1969**, *1*, 1–29. [CrossRef]

atmosphere

MDPI

Article

Average Turbulence Dynamics from a One-Parameter Kinetic Theory

Hudong Chen [1], Ilya Staroselsky [1], Katepalli R. Sreenivasan [2,3,*] and Victor Yakhot [2,4]

1 Dassault Systemes Simulia Corp, 175 Wyman Street, Waltham, MA 02451, USA
2 Department of Mechanical and Aerospace Engineering, New York University, New York, NY 11201, USA
3 Department of Physics, Courant Institute of Mathematical Sciences, New York University, New York, NY 11201, USA
4 Department of Mechanical Engineering, Boston University, Boston, MA 02215, USA
* Correspondence: krs3@nyu.edu

Abstract: We show theoretically that the mean turbulent dynamics can be described by a kinetic theory representation with a single free relaxation time that depends on space and time. A proper kinetic equation is constructed from the Klimontovich-type kinetic equation for fluid elements, which satisfies the Navier–Stokes hydrodynamics exactly. In a suitably averaged form, the turbulent kinetic energy plays the role of temperature in standard molecular thermodynamics. We show that the dynamics of turbulent fluctuations resembles a collision process that asymptotically drives the mean distribution towards a Gaussian (Maxwell–Boltzmann) equilibrium form. Non-Gaussianity arises directly from non-equilibrium shear effects. The present framework overcomes the bane of most conventional turbulence models and theoretical frameworks arising from the lack of scale separation between the mean and fluctuating scales of the Navier-Stokes equation with an eddy viscous term. An averaged turbulent flow in the present framework behaves more like a flow of finite Knudsen number with finite relaxation time, and is thus more suitably described in a kinetic theory representation.

Keywords: turbulence modeling; kinetic theory; finite Knudsen flows

Citation: Chen, H.; Staroselsky, I.; Sreenivasan, K.R.; Yakhot, V. Average Turbulence Dynamics from a One-Parameter Kinetic Theory. *Atmosphere* **2023**, *14*, 1109. https://doi.org/10.3390/atmos14071109

Academic Editors: Boris Galperin, Annick Pouquet and Peter Sullivan

Received: 19 May 2023
Revised: 28 June 2023
Accepted: 28 June 2023
Published: 4 July 2023

1. Introduction

Turbulent flows differ from laminar flows fundamentally via the appearance of large fluctuations on scales much larger than the microscale thermal fluctuations [1]. The effect of turbulent fluctuations on the mean flow has been modeled as an eddy viscosity since Boussinesq [2] and Reynolds [3], based on the analogy to thermal fluctuations of molecules in a gas; see also Prandtl [4]. These eddy-viscosity ideas have been a pillar of theoretical and engineering models of turbulence to this day. It has, however, been clear for some time that eddy viscosity is largely a metaphor, and that more faithful physics is required for proper turbulence modeling. One hope had been that the limitations of eddy viscosity could be circumvented by starting with the Boltzmann equation [5], but that approach has been unsuccessful for reasons to be described shortly [6]. Instead, by starting with the Klimontovich-type kinetic equation [7], whose average is the exact Navier–Stokes equation, we obtain a BGK-type relaxation equation for one-point probability density function (pdf) of turbulent fluctuation. This is the major contribution of this paper. Eddy viscosity as well as all higher order corrections for the Reynolds stress are determined by this only unknown parameter in the present framework.

As noted already, essentially all conventional turbulence models describing the mean turbulent flow dynamics have been based on the Navier–Stokes equation with an eddy viscosity together with the regular molecular viscosity. Specifically, one writes

$$\partial_t \mathbf{U} + \mathbf{U} \cdot \nabla \mathbf{U} = -\nabla \bar{p} + \nu \nabla^2 \mathbf{U} + \nabla \cdot (\nu_{eddy} \bar{\mathbf{S}}), \tag{1}$$

where ν and ν_{eddy} are the molecular and the eddy viscosities, respectively. The latter is a consequence of modeling the (deviatoric part of) the Reynolds stress term of turbulent fluctuations by an eddy-viscosity term according to the Boussinesq approximation

$$\langle \delta \mathbf{u} \delta \mathbf{u} \rangle \approx -\nu_{eddy} \bar{\mathbf{S}}, \tag{2}$$

where "$\langle\ \rangle$" denotes an ensemble average. In Equations (1) and (2), $\mathbf{U}$ denotes the average flow velocity, $\delta \mathbf{u}$ the fluctuation, and $\bar{\mathbf{S}}$ the rate-of-strain tensor of the mean flow.

It is well recognized that underlying the concept of eddy viscosity is the requirement of a separation of scales between the mean and fluctuating fields. But simple estimates [8] reveal that the effective mean free path of eddies (e.g., Prandtl's mixing length [4]) is comparable to the characteristic scale of the mean flow. Nevertheless, because of its simplicity, eddy viscosity remains the foundation of most models of the Reynolds averaged Navier–Stokes (RANS) type, as well as large eddy simulations (LES) [9].

There have been extensions beyond the framework of eddy viscosity to nonlinear models [10–13], but the level of empiricism increases substantially. Alternatively, it may appear at first sight that the lack of separation of scales in turbulence can be accounted for more naturally by seeking a description in terms of the kinetic theory. Indeed, the mean dynamics generated by averaging the Boltzmann equation has flow features with finite Knudsen number (a measure of the ratio of the mean free path of molecules to the characteristic flow length scale). Using an expansion in small Knudsen number [14] one can typically obtain non-Newtonian terms, whose transport coefficients agree with the above nonlinear turbulence models. In fact, one can formally generate equations for the turbulent velocity as well as turbulent kinetic energy [6,15,16], and match the conditions to some known turbulent eddy-viscosity model. But this approach has a difficulty of principle. If we agree that turbulence is a property of the Navier–Stokes equations, the averaged equations arising from the Boltzmann equation are "imposed" instead of being a description of turbulence from first principles.

Another way of deriving continuum equations from the Boltzmann equation was first attempted over two decades ago by successively averaging the regular Boltzmann equation to remove small scales [17]. Unfortunately, this attempt was unsuccessful: among other problems, the main issue is that the resulting averaged equilibrium distribution contains the temperature in the same place as the turbulent kinetic energy. Hence, the thermal molecular effects cannot be removed via some successive averaging procedure. The fundamental reason for this bottleneck is that molecules move approximately with the speed of sound ($\sim\sqrt{T}$), so that no naive averaging procedure can pick out the tiny hydrodynamic velocity from a microscopic velocity distribution of molecules, which is essentially isotropic with a standard deviation of $\sqrt{T}$. As a consequence, the resulting eddy viscosity is erroneously dependent on (thermal) temperature as opposed to turbulent hydrodynamic properties alone. The existence of dimensional quantities other than molecular viscosity, outside of the intrinsic turbulent flow properties, is also inconsistent with turbulent scaling properties in the limit of infinite Reynolds number (vanishing molecular viscosity). Therefore, the regular Boltzmann equation is at least an inconvenient starting point, if not an inappropriate one; or, one could say that there is a need for a fundamentally different averaging process.

In the current study, we take an alternative starting point. Instead of averaging the Boltzmann equation for molecules, we start from the framework of a micro-kinetic equation known as the Klimontovich equation [7]. For a turbulent flow that comprises an infinite number of fluid elements of infinitesimal sizes, the Klimontovich equation describes the evolution of the probability density function of such fluid elements. This equation is exactly the (unaveraged) Navier–Stokes equation. As we will show below, averaging the Klimontovich equation leads to a desired kinetic equation, without suffering from the mixing of dynamic scales with the micro thermal temperature. Equally important, the resulting kinetic equation provides a new framework for understanding the nature of turbulent fluctuations and turbulent eddies. This new formulation also puts the long-standing and heuristic analogy of turbulent eddies with molecules on better theoretical grounds and offers deeper insights into the nature of turbulent fluctuations and their effects on the averaged turbulent hydrodynamics.

The formulation described in Sections 2–5 is followed by a discussion of its implications in Section 6.

2. Basic Formulation

For a particle that moves exactly according to the Navier–Stokes equation, in the description of Klimontovich, the pdf of particles, $f = f(\mathbf{x}, \mathbf{v}, t)$, obeys the equation of motion given by

$$\partial_t f + \mathbf{v} \cdot \nabla f + \mathbf{a} \cdot \nabla_{\mathbf{v}} f = 0. \tag{3}$$

If the fluid flow is self-consistently generated by the motion of such particles, we have

$$\int d\mathbf{v} f(\mathbf{x}, \mathbf{v}, t) = 1, \tag{4}$$

corresponding to the normalization of the pdf for an incompressible flow. We also have

$$\int d\mathbf{v}\mathbf{v} f(\mathbf{x}, \mathbf{v}, t) = \mathbf{u}(\mathbf{x}, t), \tag{5}$$

where $\mathbf{u}(\mathbf{x}, t)$ is the fluid velocity. Note here that the "particles" are *not* the actual molecules that makes up the fluid, but an ensemble of infinitesimal fluid elements. Integrating Equation (3) and using Equation (5), we obtain

$$\partial_t \mathbf{u} + \nabla \cdot (\mathbf{u}\mathbf{u}) - \mathbf{a} = 0. \tag{6}$$

In the above, we have used the relation

$$\mathbf{u}\mathbf{u} = \int d\mathbf{v}\mathbf{v}\mathbf{v} f, \tag{7}$$

reflecting the fact that a particle is moving exactly according to the fluid flow field. Equivalently, it also indicates that the pdf $f(\mathbf{x}, \mathbf{v}, t)$ of fluid elements has zero temperature (i.e., root-mean-square (rms) deviation of local fluid velocity), in contrast to the pdf of molecules corresponding to thermal temperature.

Equation (6) recovers the exact Navier–Stokes equation if the body-force term $\mathbf{a}$ is defined by

$$\mathbf{a} = -\nabla p + \nu \nabla^2 \mathbf{u}, \tag{8}$$

where ν is the molecular viscosity, and the pressure $p(\mathbf{x}, t)$ is determined by the incompressibility constraint.

Now let us consider an ensemble averaged (or a coarse-grained) pdf, $F(\mathbf{x}, \mathbf{v}, t) \equiv \langle f(\mathbf{x}, \mathbf{v}, t) \rangle$. Obviously, the normalization condition remains valid as

$$\int d\mathbf{v} F(\mathbf{x}, \mathbf{v}, t) = 1. \tag{9}$$

The averaged pdf gives the averaged velocity field, i.e., $\mathbf{U}(\mathbf{x}, t) \equiv \langle \mathbf{u}(\mathbf{x}, t) \rangle$, so we have

$$\int d\mathbf{v}\mathbf{v} F(\mathbf{x}, \mathbf{v}, t) = \mathbf{U}(\mathbf{x}, t). \tag{10}$$

Taking the average over Equation (3), we obtain

$$\partial_t F + \mathbf{v} \cdot \nabla F + \bar{\mathbf{a}} \cdot \nabla_{\mathbf{v}} F = -\nabla_{\mathbf{v}} \cdot \langle \delta\mathbf{a} \delta f \rangle. \tag{11}$$

In the above, $\mathbf{a} = \langle \mathbf{a} \rangle + \delta\mathbf{a}$ and $f = F + \delta f$, $F = \langle f \rangle$. Due to linearity and incompressibility, we have

$$\langle \mathbf{a} \rangle = -\nabla \bar{p} + \nu \nabla^2 \mathbf{U} \tag{12}$$

Because of the existence of the unknown term on the right-hand side, Equation (11) for the averaged pdf is not closed.

Taking the average over the Navier–Stokes equation (i.e., Equation (6)), we obtain the known Reynolds-averaged equation

$$\partial_t \mathbf{U} + \mathbf{U} \cdot \nabla \mathbf{U} + \nabla \cdot \sigma - \langle \mathbf{a} \rangle = 0, \tag{13}$$

where $\langle \mathbf{a} \rangle$ is given by (12). The additional quantity σ is the so-called Reynolds stress tensor defined as

$$\sigma \equiv \langle \delta \mathbf{u} \delta \mathbf{u} \rangle. \tag{14}$$

From the averaged pdf, we have

$$\begin{aligned} \int d\mathbf{v} \mathbf{v} \mathbf{v} F &= \langle \int d\mathbf{v} \mathbf{v} \mathbf{v} f \rangle \\ &= \langle \mathbf{u} \mathbf{u} \rangle = \mathbf{U} \mathbf{U} + \sigma. \end{aligned} \tag{15}$$

In the above, the basic definition $\mathbf{u} = \mathbf{U} + \delta \mathbf{u}$ and $\langle \delta \mathbf{u} \rangle = 0$. Consequently, taking the trace of Equation (15), we have

$$\int d\mathbf{v} \frac{1}{2} \mathbf{v}^2 F = \frac{1}{2} \langle \mathbf{u}^2 \rangle = \frac{1}{2} \mathbf{U}^2 + K, \tag{16}$$

where $K = \frac{1}{2} \langle (\delta \mathbf{u})^2 \rangle$ is the turbulent kinetic energy.

From Equations (9), (10), and (15) as well as Equation (16), one can recognize that σ and K have the following definitions based on the kinetic theory:

$$\int d\mathbf{v} (\mathbf{v} - \mathbf{U})(\mathbf{v} - \mathbf{U}) F = \sigma, \tag{17}$$

and

$$\int d\mathbf{v} \frac{1}{2} (\mathbf{v} - \mathbf{U})^2 F = K. \tag{18}$$

It is important to note from Equation (17) that the approach we developed here shows that the Reynolds stress tensor is fully determined by the averaged pdf, F. We emphasize the following non-trivial point. We have made no assumption so far that goes beyond those usually attributed to the ensemble of turbulent fluctuations $\delta \mathbf{u}$; the entire information about this ensemble that is necessary for the full knowledge of turbulent Reynolds stress components, and thus for the full macroscopic description of turbulent flow, is contained in a single function $F(\mathbf{x}, \mathbf{v}, t)$ that is determined by the kinetic theory. Further, Equation (18) implies that the turbulent kinetic energy plays a role similar to the thermal energy in a regular kinetic theory for molecules.

Integrating Equation (11) by $\int d\mathbf{v}\mathbf{v}$, we have

$$\partial_t \mathbf{U} + \mathbf{U} \cdot \nabla \mathbf{U} + \nabla \cdot \sigma - \langle \mathbf{a} \rangle = - \int d\mathbf{v} \mathbf{v} \nabla_{\mathbf{v}} \cdot \langle \delta \mathbf{a} \delta f \rangle. \tag{19}$$

Comparing Equations (13) and (19), the term on the right hand side of (19) should vanish. That is,

$$- \int d\mathbf{v} \mathbf{v} \nabla_{\mathbf{v}} \cdot \langle \delta \mathbf{a} \delta f \rangle = 0.$$

This is also seen as a direct consequence of the fluctuation term on the right hand side of Equation (19). After integration by parts, we obtain

$$- \int d\mathbf{v} \mathbf{v} \nabla_{\mathbf{v}} \langle \delta \mathbf{a} \delta f \rangle = \int d\mathbf{v} \langle \delta \mathbf{a} \delta f \rangle.$$

Since $\mathbf{a}$ is not a function of $\mathbf{v}$, we have

$$\int d\mathbf{v}\langle\delta\mathbf{a}\delta f\rangle = \langle\delta\mathbf{a}\int d\mathbf{v}\delta f\rangle$$
$$= \langle\delta\mathbf{a}\int d\mathbf{v}(f - F)\rangle = 0. \tag{20}$$

The result in the last step vanishes because $\int d\mathbf{v}(f - F) = 0$ from Equations (4) and (9). Therefore, the additional term on the right-hand side of Equation (11) is formally like a "collision term" in the Boltzmann equation that conserves both mass and momentum:

$$\int d\mathbf{v}\nabla_{\mathbf{v}}\cdot\langle\delta\mathbf{a}\delta f\rangle = 0$$
$$\int d\mathbf{v}\mathbf{v}\nabla_{\mathbf{v}}\cdot\langle\delta\mathbf{a}\delta f\rangle = 0. \tag{21}$$

Consequently, we may rename this collision term in Equation (11) as

$$C \equiv -\nabla_{\mathbf{v}}\cdot\langle\delta\mathbf{a}\delta f\rangle, \tag{22}$$

but also point out that we do not know its explicit form. The central task in this turbulence representation is to find an appropriate closure for the collision term C.

3. The Energy Equation

Before investigating a possible form of C, we review the averaged energy equation here. Taking a dot product of $\mathbf{u}$ with Equations (6) and (8), we have

$$\partial_t(\frac{\mathbf{u}^2}{2}) + \nabla\cdot(\frac{\mathbf{u}^2}{2}\mathbf{u}) = -\mathbf{u}\cdot\nabla p + \nu\mathbf{u}\cdot\nabla^2\mathbf{u}. \tag{23}$$

Likewise, taking a dot product of $\mathbf{U}$ with Equation (13), we have,

$$\partial_t(\frac{\mathbf{U}^2}{2}) + \nabla\cdot(\frac{\mathbf{U}^2}{2}\mathbf{U}) + \mathbf{U}\cdot(\nabla\cdot\sigma) = -\mathbf{U}\cdot\nabla\bar{p} + \nu\mathbf{U}\cdot\nabla^2\mathbf{U}. \tag{24}$$

Substituting $\mathbf{u} = \mathbf{U} + \delta\mathbf{u}$ into Equation (23), taking the average, and subtracting Equation (24), after some algebra, we obtain

$$\partial_t K + \mathbf{U}\cdot\nabla K + \sigma : \nabla\mathbf{U} + \nabla\cdot(\mathbf{Q} + \mathbf{W}) = \nu\nabla^2 K - \epsilon. \tag{25}$$

Here the dissipation term ϵ is given by

$$\epsilon \equiv \nu\langle(\nabla\delta\mathbf{u}) : (\nabla\delta\mathbf{u})\rangle, \tag{26}$$

and the effective flux of turbulent kinetic energy due to turbulent fluctuations is given by

$$\mathbf{Q} \equiv \langle\delta\mathbf{u}\frac{(\delta\mathbf{u})^2}{2}\rangle. \tag{27}$$

There is also a "work"-related flux term $\nabla\cdot\mathbf{W}$,

$$\nabla\cdot\mathbf{W} \equiv \langle\delta\mathbf{u}\cdot\nabla\delta p\rangle, \tag{28}$$

defined using the incompressibility condition $\nabla\cdot\delta\mathbf{u} = 0$.

Equation (25) governing the turbulent kinetic energy can also be derived straightforwardly from Equation (11). It can be shown that the contribution from the collision term C contains ϵ. Therefore, unlike the case of a regular molecular system, the turbulent kinetic energy is not conserved in its collision process among turbulent fluctuations, as expected from energy transfer from fluid fluctuations to thermal energy. Multiplying Equation (11) by $\frac{1}{2}\mathbf{v}^2$ and performing $\int d\mathbf{v}$, we obtain

$$\partial_t \int d\mathbf{v}\frac{1}{2}\mathbf{v}^2 F + \nabla \cdot \int d\mathbf{v}\mathbf{v}\frac{1}{2}\mathbf{v}^2 F + \int d\mathbf{v}\frac{1}{2}\mathbf{v}^2 \bar{\mathbf{a}} \cdot \nabla_{\mathbf{v}} F = \int d\mathbf{v}\frac{1}{2}\mathbf{v}^2 C, \tag{29}$$

where C is given by Equation (22). According to Equation (16),

$$\int d\mathbf{v}\frac{1}{2}\mathbf{v}^2 F = \frac{1}{2}\mathbf{U}^2 + K,$$

where K is defined in Equation (18). Integrating by parts, it can be shown that

$$\int d\mathbf{v}\frac{1}{2}\mathbf{v}^2 \langle \mathbf{a} \rangle \cdot \nabla_{\mathbf{v}} F = \mathbf{U} \cdot \langle \mathbf{a} \rangle.$$

With some straightforward algebra, one can show that

$$\nabla \cdot \int d\mathbf{v}\mathbf{v}\frac{1}{2}\mathbf{v}^2 F = \nabla \cdot \{\mathbf{U}[\frac{1}{2}\mathbf{U}^2 + K] + \mathbf{U} \cdot \sigma + \mathbf{Q}\},$$

where σ is defined in Equation (17); the flux $\mathbf{Q}$, defined by (27), can be rewritten as

$$\mathbf{Q} \equiv \langle \delta\mathbf{u}\frac{(\delta\mathbf{u})^2}{2} \rangle = \int d\mathbf{v}\frac{1}{2}(\mathbf{v} - \mathbf{U})(\mathbf{v} - \mathbf{U})^2 F. \tag{30}$$

Taking all these into Equation (29), we obtain

$$\partial_t[\frac{1}{2}\mathbf{U}^2 + K] + \nabla \cdot \{\mathbf{U}[\frac{1}{2}\mathbf{U}^2 + K] + \mathbf{U} \cdot \sigma + \mathbf{Q}\} + \mathbf{U} \cdot \bar{\mathbf{a}} = \int d\mathbf{v}\frac{1}{2}\mathbf{v}^2 C. \tag{31}$$

Subtracting Equation (24) from (31) and using the definition of average force (12), we obtain the following equation for the turbulent kinetic energy evolution from this kinetic theory formulation:

$$\partial_t K + \mathbf{U} \cdot \nabla K + \sigma : \nabla\mathbf{U} + \nabla \cdot \mathbf{Q} = \int d\mathbf{v}\frac{1}{2}\mathbf{v}^2 C. \tag{32}$$

By recalling the definition (22), the right side of Equation (32) can be recast as

$$\int d\mathbf{v}\frac{1}{2}\mathbf{v}^2 C = -\nabla \cdot \mathbf{W} + \nu\nabla^2 K - \epsilon, \tag{33}$$

to recover Equation (25) for incompressible flows.

4. The Collision Term

It is difficult to derive an explicit form for the collision term C in Equation (11). In our previous work [16], we used an experimental observation that the single-particle pdf F^{eq} in a homogeneous turbulent flow exhibits a local Gaussian form

$$F^{eq} = W\, exp[-3(\mathbf{v} - \mathbf{U})^2/2K], \tag{34}$$

where W is the normalization factor for $\int d\mathbf{v}F^{eq} = 1$. See also [18,19]. Let us now present some further semi-theoretical arguments for this form of the equilibrium distribution function, and argue that it is the asymptotic limit of the collision process C. Suppose we start evolving the distribution function F according to Equation (11) towards its local equilibrium F^{eq}. Based on the expression for the collision integral in Equation (22), the ensemble average could be reinterpreted as

$$C = -\langle \delta \mathbf{a} \cdot \nabla_{\mathbf{v}} \delta f \rangle = -\frac{1}{N} \sum_{n}^{N} \delta \mathbf{a}^{(n)} \cdot \nabla_{\mathbf{v}} \delta f^{(n)}. \tag{35}$$

Here the ensemble is explicitly written as the average of N realizations of deviations from the mean, for $n = 1, \ldots, N \to \infty$. Therefore Equation (35) defines C as the summation of N small increments to the distribution function as it evolves towards equilibrium, as defined by Equation (11). By an informal appeal to the central limit theorem, the argument can be made that F should converge, in the most probable sense, to a Gaussian form, since it is the average sum of infinitely many uncorrelated processes. We realize that, while realizations are uncorrelated, their deviations from the mean are not; we are thus fully self-critical with respect to this argument. If, however, one accepts that these processes are uncorrelated to a leading order, then the only Gaussian that can be a candidate for the equilibrium is that given by Equation (34), since it satisfies the conservation of mass and momentum, as well as the energy constraint based on (9), (10) and (16):

$$\begin{aligned}
&\int d\mathbf{v} F^{eq} = \int d\mathbf{v} F = 1 \\
&\int d\mathbf{v}\mathbf{v} F^{eq} = \int d\mathbf{v}\mathbf{v} F = \mathbf{U} \\
&\int d\mathbf{v} \frac{1}{2} \mathbf{v}^2 F^{eq} = \frac{1}{2} \mathbf{U}^2 + K'.
\end{aligned} \tag{36}$$

These constraints together with a Gaussian distribution completely determine F^{eq} given by Equation (34). Here K' is the value of turbulent kinetic energy in equilibrium. Therefore, the collision process C drives the distribution towards an equilibrium distribution which is exactly defined by Equation (34). We note that this semi-theoretical argument does not quantify the rate of approach to equilibrium; that is, we do not know the time it takes for the collision process C to drive F to F^{eq}.

Based on the argument above, we could assume that the collision term can be described as a relaxation process towards the local Gaussian equilibrium with an unspecified relaxation time τ, namely a BGK form [20]

$$C = -\frac{F - F^{eq}}{\tau}. \tag{37}$$

Equation (11) together with (37) gives a closed kinetic Boltzmann–BGK representation for the averaged fluid motion with an unspecified relaxation time τ. Obviously, $\tau = \tau(\mathbf{x}, t)$, a function of local and nonlocal properties and structures of turbulence. Understanding and the determination of this fundamental time scale is work for the future, but a few comments are in order.

First, τ must have similar characteristics as K/ϵ but it must be kept in mind that the deviation of F from the equilibrium value, F^{eq}, occurs when there exists a shear in the mean flow, as dictated by Equation (11). This is in agreement with the experimental observation of the Gaussian single-point pdf F^{eq} (Equation (34)) in homogeneous turbulence. Second, one can in principle write down a dynamic equation for τ as done in a different context in Ref. [21], where a wake behind a twin-airfoil, driven away from equilibrium by an impulsive pressure gradient, was allowed to relax back to equilibrium.

As discussed above, the collision operator C obeys the conservation of mass and momentum in (21) as

$$\begin{aligned}
&\int d\mathbf{v} [F - F^{eq}] = 0 \\
&\int d\mathbf{v}\mathbf{v} [F - F^{eq}] = 0.
\end{aligned} \tag{38}$$

In contrast, C does not conserve the turbulent kinetic energy K. As shown above, the turbulent kinetic energy should change, in the sense of a global space average, according to

$$\int d\mathbf{v}\frac{(\mathbf{v}-\mathbf{U})^2}{2}C = -\epsilon. \tag{39}$$

Consequently, the K value appearing in F^{eq} (Equation (34)) should be slightly smaller than the K value derived using F according to Equation (16) or Equation (18). If we assume the BGK form for the collision term, then the difference between the two K values is simply $\Delta K = \tau\epsilon$.

5. Kinetic Representation of the Average Turbulence Dynamics

Equations (9), (10), (18), (11), (12), (22), (34), (36), (37) and (39) fully define a kinetic theoretic representation of the averaged turbulent flow dynamics in the form of Boltzmann–BGK with one unspecified collision relaxation time, τ. Solving the kinetic system, one obtains the averaged pdf F, and the latter fully determines fundamental quantities such as the Reynolds stress tensor σ according to Equation (17). Note that for homogeneous flow the derivatives in the kinetic Equations (11) and (33) disappear so that the solution corresponding to the BGK collision integral (39) is the Gaussian F^{eq} (34), which is consistent with experimental observations for single point pdf. Therefore, from this perspective, the non-Gaussian statistics are due to spatial inhomogeneities of the averaged flow, generating non-equilibrium properties.

It is straightforward to derive that Equations (13) and (25) are the results of the appropriate moments of the Boltzmann-BGK model. The resulting Reynolds stress term has an explicitly expanded form in terms of turbulent hydrodynamic quantities via the Chapman–Enskog expansion of a finite mean-free path [6]; its leading order (i.e., Navier–Stokes) expression (in the small mean-free path limit) is the usual eddy-viscosity form

$$\sigma_{ij} = \frac{2}{3}K\delta_{ij} - \nu_{eddy}\bar{S}_{ij}, \tag{40}$$

where $\nu_{eddy} = 2\tau K/3$. Here δ_{ij} is the Kronecker delta function, and $\bar{S}_{ij}$ is the averaged strain rate tensor

$$\bar{S}_{ij} = \frac{1}{2}(\frac{\partial U_i}{\partial x_j} + \frac{\partial U_j}{\partial x_i}).$$

The trace of $\bar{S}_{ij}$ vanishes due to incompressibility. The next order contributions to σ include both time and space derivatives and nonlinear functions of $\bar{S}_{ij}$ as well as the anti-symmetric tensor $\bar{\Omega}_{ij}$ [6], as given below:

$$\begin{aligned} \sigma^{(2)}_{ij} = &- 2\nu_{eddy}\frac{D}{Dt}[\tau\bar{S}_{ij}] \\ &- 6\frac{\nu^2_{eddy}}{K}[\bar{S}_{ik}\bar{S}_{kj} - \frac{1}{3}\bar{S}_{kl}\bar{S}_{kl}] \\ &+ 3\frac{\nu^2_{eddy}}{K}[\bar{S}_{ik}\bar{\Omega}_{kj} + \bar{S}_{jk}\bar{\Omega}_{ki}]. \end{aligned} \tag{41}$$

Here, $D/Dt \equiv \partial_t + \mathbf{U}\cdot\nabla$, and

$$\bar{\Omega}_{ij} = \frac{1}{2}(\frac{\partial U_i}{\partial x_j} - \frac{\partial U_j}{\partial x_i})$$

is the vorticity tensor. The summation convention for repeated indices is adopted in Equation (41). It is interesting to mention that by associating τ to a known turbulent eddy-viscosity model, the coefficients in the next order (i.e., (40) and (41)) are in quantitative agreement with some representative nonlinear turbulence models that were constructed or proposed from various other means [6]. Using the same analysis, one can also expect, to the leading order,

$$\mathbf{Q} = -\eta_{eddy}\nabla K, \tag{42}$$

with $\eta_{eddy}{\sim}\nu_{eddy}$.

In short, we have theoretically constructed a kinetic theory representation for the averaged turbulent flow dynamics. However, the equation is not closed due to the unspecified time τ. The latter is central for formulating any specific kinetic-theory-based turbulent models. This requires an understanding of the structures and interactions (collisions) of turbulent fluctuations in terms of $\langle \delta\mathbf{a} \cdot \nabla_{\mathbf{v}}\delta f \rangle$.

6. Discussion

6.1. Past Closure Attempts

Closure models for the Reynolds stress tensor with an eddy-viscous form as in (40) (and (42)) have been a long-standing foundation for mainstream turbulence models. The underlying concept has been the heuristic analogy between turbulent eddy fluctuations and molecular motions. For the mean motion of turbulent flows, this analogy results in the Navier–Stokes equation with an added eddy viscosity. However, this procedure is questionable because of the lack of scale separation between the mean flow and the so-called turbulent eddies. (Separate studies [22,23] have also demonstrated weak correlation between the Reynolds stress tensor σ and the mean turbulent rate-of-strain $\bar{\mathbf{S}}$ tensor). Extensions beyond the eddy-viscosity term, which include higher order non-Newtonian effects with the framework of a modified Navier–Stokes equation [10–13], have offered useful insights but limited success.

In the past, kinetic theory representation based on the Boltzmann equation has also been attempted [6,15,16]. However, these attempts are imposed externally with the intent to generate the a priori known averaged turbulent hydrodynamic equations, although the former contains more information beyond the viscous order [6]. There is a question as to why such a kinetic equation should be justified other than that it produces the already expected hydrodynamic equations. There is also the question of how such a kinetic representation related to the first principles of fluid turbulence, if the latter is a property of the Navier–Stokes equations. On the other hand, one main issue related to averaging the regular Boltzmann equation is that the resulting averaged equilibrium distribution contains the temperature in the same place as the turbulent kinetic energy. Hence, the thermal molecular effects cannot be removed by successive averaging procedure that can pick out the tiny hydrodynamic velocity from a microscopic velocity of molecules. Thus, the most straightforward approach of performing averages of the regular Boltzmann equation over molecular motions has been without commensurate gain [17]). For an interesting effort in the direction of using the Boltzmann equation for obtaining the dynamics of filtered fluid equations, see [24].

6.2. Present Work

We have shown here that the starting point for deriving a desired kinetic equation for averaged dynamics of turbulence can be a Klimontovich-type self-consistent body-force-based equation. This equation describes the motion of particles (i.e., fluid elements of infinitesimal size) corresponding exactly to the (unaveraged) Navier–Stokes hydrodynamics. Averaging the Klimontovich equation results in a kinetic equation with a collision term representing the effect of turbulent fluctuations on mean dynamics. Our present kinetic-theoretic representation does not suffer from the problem of scale separation. Besides including the eddy viscosity as well as the next order effects, effects from turbulent fluctuations of all orders of a scale expansion are also naturally included.

An attractive feature of this kinetic-based approach is that a single distribution function $F(\mathbf{x}, \mathbf{v}, t) \equiv \langle f(\mathbf{x}, \mathbf{v}, t) \rangle$, governed by (11), contains the entire information needed for the (turbulent ensemble) averaging of all properties, as long as they are based on single point pdfs. Indeed, in addition to second order moments (17), (18), the knowledge of F provides full information about the turbulent energy diffusion flux (30) which is a third order quantity—and, indeed, about any arbitrary N-th order moment of the single point pdf of turbulent fluctuations. In particular, the non-Gaussian properties of the pdf such as flatness and skewness are all contained in the kinetic-based distribution function F. Note also that the all these conclusions are obtained under some of the weakest assumptions possible. Indeed nothing stronger than the mere existence of ensemble average (such that $\langle \delta \mathbf{u} \rangle = \mathbf{0}$) is required to exactly derive expressions such as (17), (18), and (30); in particular no assumptions about the specific form of the equilibrium distribution function and/or collision integral are invoked. This self-consistent approach can be applied for the kinetic-theory reformulation of equations other than Navier–Stokes (e.g., the Burgers equation) by an appropriate choice of self consistently generated body force $\mathbf{a}(\mathbf{x}, t)$ (Equation (8)).

In this formulation, the turbulent kinetic energy K, which measures the square of the mean turbulent fluctuation, plays a role like the thermal energy or temperature in the regular Boltzmann equation for motion of molecules. It is interesting to note one important difference between a turbulent kinetic-energy-based temperature T_{turb} (=2K/3) and a conventionally known thermal temperature T: Unlike a regular low Mach number flow with $\sqrt{T} >> |\mathbf{u}|$, the turbulent kinetic-energy-based temperature is often less than the mean flow velocity, $\sqrt{T_{turb}} < |\mathbf{U}|$. Therefore, an averaged turbulent flow bears some character of a supersonic flow, even though it is still incompressible because of the pressure term from the incompressibility constraint.

We have shown that the averaged kinetic equation generates the same form of the governing equations for the mean turbulent velocity $\mathbf{U}$ and the turbulent kinetic energy K, as that from direct averaging of the Navier–Stokes equation. Furthermore, interactions of turbulent fluctuations play a role like a molecular collision process that drives the mean pdf towards a Maxwell–Boltzmann (Gaussian) equilibrium distribution. The single relaxation time scale determines not only the eddy viscosity, but also the turbulent transport coefficients of all orders in the expansion of turbulent flow scales. The collision process obeys the conservation of mass and momentum. On the other hand, it generates a loss of flow energy that directly corresponds to the turbulent dissipation ϵ. The present analysis justifies a relaxation process to equilibrium. The full kinetic equation for describing the dynamics of a mean turbulent flow is closed when the relaxation time τ, the dissipation ϵ, as well as the flux $\mathbf{W}$ (cf. Equations (28) and (33)), are specified. However, whether all turbulent moments relax with the same relaxation time τ (i.e., BGK model) is a model assumption, although there is some supporting evidence for this view [25]. The ultimate determination of τ should be obtained from an analysis of the full expression of the collision term in (22). One may measure its value experimentally or numerically from, for instance, the time evolution of turbulent kinetic energy in a rapid distortion problem (cf. [13]).

6.3. The Collision Term and the Relaxation Effects

An estimate, based on a typical conventional turbulence model, indicates that the effective mean-free path, L_{mfp}, or the collision relaxation time τ are, respectively, about one tenth of the length and time scale of the energy containing eddies, namely $L_{mfp} {\sim} C_\mu K^{3/2}/\epsilon$ and $\tau {\sim} C_\mu K/\epsilon$. $C_\mu {\sim} 0.1$. Therefore, a mean turbulent flow is more resembling of a flow with finite Knudsen number, $Kn{\sim}0.1$. This suggests that the representation of the mean turbulent flow dynamics by the Navier–Stokes equation with an eddy viscosity is not sufficient, in general. Instead, contributions from the higher order terms (in the small Knudsen number expansion) become non-negligible, and are responsible for non-Newtonian fluid dynamic behavior beyond the Navier–Stokes representation. Indeed, averaged turbulence exhibits the known phenomena of finite Knudsen number and non-Newtonian character, such as the slip velocity near wall [16], the secondary flow structures in a square duct and the known retarded rapid distortion behavior (cf. [10,12,13]). A kinetic-theory-based representation for turbulence naturally includes finite Knudsen effects of all orders, and thus all the non-Newtonian phenomena mentioned above are automatically captured.

Lastly, we wish to make the following observation. Microscopic motion of molecules is described by a Boltzmann-type kinetic equation. The averaged motion of the molecules is described by the Navier–Stokes hydrodynamic equation. Yet, when further averaged, a suitable kinetic formulation comes back as a desirable description—completing a virtuous circle.

Author Contributions: Conceptualization, H.C., I.S., K.R.S. and V.Y.; formal analysis, H.C., I.S., K.R.S. and V.Y.; writing—original draft preparation, H.C., I.S. and K.R.S.; writing—review and editing, H.C., I.S. and K.R.S. All authors have read and agreed to the published version of the manuscript.

Funding: This research received no external funding.

Institutional Review Board Statement: Not applicable.

Informed Consent Statement: Not applicable.

Data Availability Statement: Not applicable.

Acknowledgments: We are appreciative of the opportunity to contribute to this volume memorializing Jack Herring. In particular, H.C. has fond memories of interacting and co-authoring with him (together with Robert Kraichnan and Robert Kerr). K.R.S. values the unassuming and collegial ways of Jack on scientific matters both central and peripheral to his own interests. We acknowledge the influence on this work of extended interactions with our late friend, Steve Orszag.

Conflicts of Interest: The authors declare no conflict of interest.

References

1. Saint-Venant, A.J.C. Formules et tables nouvelles pour les eaux courantes. *Ann. Mines* **1851**, *20*, 49.
2. Boussinesq, L. Essai theorique sur les lois trouvees experimentalement par M. Bazin pour l'ecoulement uniforme de l'eau dans les canaux decouverts. *CR Acad. Sci. Paris* **1870**, *71*, 389–393.
3. Reynolds, O. On the dynamical theory of turbulent incompressible viscous fluids and the determination of the criterion. *Phil. Trans. R. Soc. Lond.* **1894**, *A186*, 123–161.
4. Prandtl, L. Bericht uber Untersuchungen zur ausgebildeten. *Z. Angew. Math. Mech.* **1925**, *5*, 136–139. [CrossRef]
5. Boltzmann, L. Weiere Studien uber das Warmegleichgewicht unter Gasmolekulen. *Sitzungberichte Kais. Akad. Wiss. Wien Math.-Naturwiss. Classe* **1872**, *66*, 275–370.
6. Chen, H.; Orszag, S.; Staroselsky, I.; Succi, S. Expanded analogy between Boltzmann kinetic theory of fluids and turbulence. *J. Fluid Mech.* **2004**, *519*, 301. [CrossRef]
7. Klimontovich, Y. *The Statistical Theory of Non-Equilibrium Processes in a Plasma*; M.I.T. Press: Cambridge, MA, USA, 1967.
8. Wilcox, D. *Turbulence Modeling for CFD*; DCW Industries: La Canada, CA, USA, 1993.
9. Lesieur, M. *Turbulence in Fluids*, 3rd ed.; Kluwer Academic Publishers: Alphen am Rhein, The Netherlands, 1997.
10. Speziale, C. On the nonlinear K-l and K-ϵ models of turbulence. *J. Fluid Mech.* **1987**, *178*, 459–467. [CrossRef]
11. Yoshizawa, A. Statistical modeling of a transport-equation for the kinetic energy dissipation rate. *Phys. Fluids* **1987**, *30*, 628–631. [CrossRef]
12. Rubinstein, R.; Barton, J.M. Nonlinear Reynolds stress models and the renormalization group. *Phys. Fluids* **1990**, *A2*, 1472–1476. [CrossRef]
13. Yakhot, V.; Orszag, S.; Thangam, S.; Gatski, T.; Speziale, C. Development of turbulence models for shear flows by a double expansion technique. *Phys. Fluids A* **1992**, *4*, 1510–1520. [CrossRef]
14. Chapman, S.; Cowling, T. *The Mathematical Theory of Non-Uniform Gases*, 3rd ed.; Cambridge University Press: Cambridge, UK, 1990.
15. Degond, P.; Lemou, M. Turbulence models for incompressible fluids derived from kinetic theory. *J. Math. Fluid Mech.* **2002**, *4*, 257–284. [CrossRef]
16. Chen, H.; Staroselsky, I.; Yakhot, V. On non-perturbative formulation of hydrodynamics using kinetic theory. *Phys. Scr.* **2013**, *T155*, 014040. [CrossRef]
17. Chen, H.; Succi, S.; Orszag, S. Analysis of subgrid scale turbulence using the Boltzmann Bhatnagar-Gross-Krook kinetic equation. *Phys. Rev. E* **1999**, *59*, R2527–R2530. [CrossRef]
18. Sreenivasan, K.R.; Yakhot, V. Dynamics of three-dimensional turbulence from Navier-Stokes equations. *Phys. Rev. Fluids* **2021**, *6*, 104604. [CrossRef]
19. Sreenivasan, K.R.; Yakhot, V. Saturation of exponents and the asymptotic fourth state of turbulence. *arXiv* **2022**, arXiv:208.09561.
20. Bhatnagar, P.L.; Gross, E.P.; Krook, M. A model for collision process in gases. I. Small amplitude processes in charged and neutral one-component systems. *Phys. Rev.* **1954**, *94*, 511–525. [CrossRef]
21. Prabhu, A.; Narasimha, R. Turbulent non-equilibrium wakes. *J. Fluid Mech.* **1972**, *54*, 19–38. [CrossRef]

22. Borue, V.; Orszag, S.A. Local energy flux and subgrid-scale statistics in three-dimensional turbulence. *J. Fluid Mech.* **1998**, *366*, 1–31. [CrossRef]
23. Schmitt, F.G. About Boussinesq's turbulent viscosity hypothesis: Historical remarks and a direct evaluation of its validity. *Comptes Rendus Mec.* **2007**, *335*, 617–627. [CrossRef]
24. Girimaji, S. Boltzmann kinetic equation for filtered fluid turbulence. *Phys. Rev. Lett.* **2007**, *99*, 034501. [CrossRef]
25. Yakhot, V.; Orszag, S. Renormalization group analysis of turbulence, I. Basic theory. *J. Sci. Comput.* **1986**, *1*, 3–52. [CrossRef]

atmosphere

MDPI

Article

Segregation of Fast-Reactive Species in Atmospheric Turbulent Flow

Guy P. Brasseur [1,2,3,*], Mary Barth [1], Jan Kazil [4,5], Edward G. Patton [1] and Yuting Wang [3]

[1] National Center for Atmospheric Research, Boulder, CO 80307, USA; barthm@ucar.edu (M.B.); patton@ucar.edu (E.G.P.)
[2] Max Planck Institute for Meteorology, 20146 Hamburg, Germany
[3] Department of Civil and Environmental Engineering, The Hong Kong Polytechnic University, Hung Hom, Kowloon, Hong Kong 999077, China; yuting.wang@polyu.edu.hk
[4] Cooperative Institute for Research in Environmental Sciences, University of Colorado, Boulder, CO 80309, USA; jan.kazil@noaa.gov
[5] National Oceanic and Atmospheric Administration, Chemical Sciences Laboratory, Boulder, CO 80305, USA
* Correspondence: brasseur@ucar.edu or guy.brasseur@mpimet.mpg.de

Abstract: Atmospheric turbulence, which produces chaotic motions in the planetary boundary layer, can inhibit mixing between fast-reacting species produced or released at different locations. This segregation process modifies the effective rate at which reactions occur between these species and is not appropriately accounted for in coarse-resolution models, since these models assume complete mixing of tracers within each grid box. Here, we present a few examples of large-eddy simulations (LES) applied to chemically reactive species in a forested area with high emissions of biogenic hydrocarbons, an urban area rich in anthropogenic emissions, and a maritime area with high emissions of reduced sulfur species.

Keywords: turbulence; chemistry; segregation; LES

Citation: Brasseur, G.P.; Barth, M.; Kazil, J.; Patton, E.G.; Wang, Y. Segregation of Fast-Reactive Species in Atmospheric Turbulent Flow. *Atmosphere* **2023**, *14*, 1136. https://doi.org/10.3390/atmos14071136

Academic Editors: Boris Galperin, Annick Pouquet and Peter Sullivan

Received: 10 May 2023
Revised: 6 July 2023
Accepted: 8 July 2023
Published: 11 July 2023

1. Introduction

The simulation of chemically reacting systems in turbulent flows has received considerable attention in recent years. The first applications focused on combustion processes and have allowed for progress, for example, in the design of a large number of industrial systems. At the same time, models with an explicit representation of turbulence in the atmosphere have been used to investigate the dispersion and fate of air pollutants. For example, Herring and Wyngaard [1] examined the behavior of a tracer undergoing a first-order decay in a convective flow and, using a detailed simulation, showed that the eddy diffusion applied to species concentrations must be represented by a nonlocal diffusion operator. Advances in computational fluid dynamics (CFD) techniques and the development of detailed chemical mechanisms have allowed for accurate air pollution simulations for the atmospheric boundary layer, especially in regions with inhomogeneous surface emissions. Furthermore, flux measurements obtained from surface or airborne instrumentation using eddy covariance techniques have provided experimental information on surface emissions and on convective exchanges in the lowest layers of the atmosphere.

In many applications, the spatial distribution and temporal evolution of chemical species in the atmosphere are derived by relatively coarse mathematical models that take into account emissions of primary constituents, in situ transformations, multiscale transport processes, and surface deposition. Models that describe these processes at a global or regional scale are constrained by the spatial resolution adopted to solve the model equations and by the resolution at which driving quantities such as surface emissions are provided. In such models, the reactive species are assumed to be entirely mixed within each grid box, and the likely subgrid segregation between these species is therefore ignored.

Such segregation, which is most prominent for constituents that chemically transform at similar time scales as the air motions, results from the small-scale inhomogeneities in the emissions and the complex nature of the turbulent flow. Figure 1 demonstrates how boundary-layer turbulence organizes itself differently under shear vs. buoyancy influences [2].

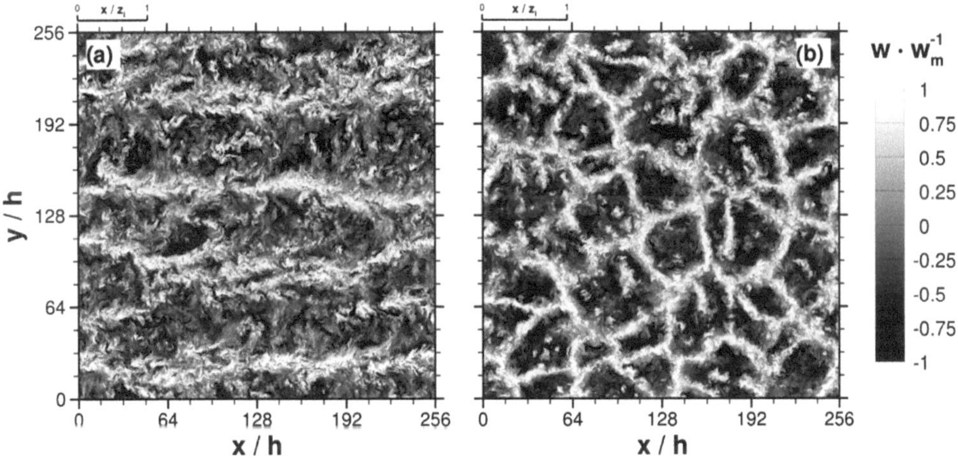

Figure 1. Instantaneous horizontal slices of vertical velocity (normalized by a characteristic velocity scale (w_m)) at a height of 120 m ($z/h = 6$) above a 20 m tall forest (h) from two turbulence-resolving atmospheric boundary layer simulations of differing stabilities. The left panel (**a**) is from an atmospheric boundary layer simulation with near-neutral stability, where the vertical shear of horizontal wind speed dominates the turbulence ($z_i/L \sim -1$; z_i is the depth of the atmospheric boundary layer, and L is the Monin–Obukhov length), and the right panel (**b**) is from a similar simulation but where buoyancy dominates the turbulence ($z_i/L \sim -\infty$) (see Patton et al. [2] for additional details). Of importance for turbulence–chemistry interactions is (1) that the character of organization in the turbulence transitions from elongated roll-like structures in shear-dominated near-neutral conditions to a cellular hexagon-like structure under buoyantly dominated conditions and (2) that surface-emitted species are most likely found within the narrow upwelling motions depicted in white, while species entrained from above the boundary layer are found primarily in the more broad, darker-colored regions. Figure adapted from Patton et al. [2] with the permission of © American Meteorological Society.

The mathematical concept of segregation can be introduced by first expressing the concentration [A] of a chemical species (A) as the sum of its averaged value and the deviation relative to this average. Following the Reynolds decomposition formulation, we write $[\mathrm{A}] = \overline{[\mathrm{A}]} + [\mathrm{A}']$, where the overbar refers to the average values, and the prime sign refers to the deviation from this average. If two species (A and B) react with each other, the corresponding reaction rate is expressed as

$$\frac{\mathrm{d}[\mathrm{A}]}{\mathrm{dt}} = \frac{\mathrm{d}[\mathrm{B}]}{\mathrm{dt}} = -\mathrm{k}[\mathrm{A}][\mathrm{B}],$$

where k represents the rate constant of the reaction (assumed here to be a constant). The time evolution of the mean concentrations of A and B becomes

$$\frac{\mathrm{d}\overline{[\mathrm{A}]}}{\mathrm{dt}} = \frac{\mathrm{d}\overline{[\mathrm{B}]}}{\mathrm{dt}} = -\mathrm{k}\left\{\left(\overline{[\mathrm{A}]}\,.\,\overline{[\mathrm{B}]}\right) + \overline{[\mathrm{A}'][\mathrm{B}']}\right\} = -\mathrm{k}\left(\overline{[\mathrm{A}]}\,.\,\overline{[\mathrm{B}]}\right)(1 + \mathrm{I_{AB}}),$$

where

$$I_{AB} = \frac{\left\{\overline{[A'][B']}\right\}}{\left(\overline{[A]}.\overline{[B]}\right)}$$

represents the segregation intensity [3], which is proportional to the covariance of the two reactants' concentrations. If chemical species A and B are entirely mixed, the segregation intensity is equal to zero, and the Damköhler number (D_a) [4], defined as the ratio between the turbulence time scale τ_{turb}) and the chemical timescale (τ_{chem}),

$$D_a = \frac{\tau_{turb}}{\tau_{chem}}$$

is smaller than 1. If, however, the chemical lifetime of the two species is small, with a Damköhler number larger than one, the segregation intensity is negative and varies between 0 and −1. In this case, for species whose emissions are not colocated, the reaction rate is considerably lower than if the two species were perfectly mixed. A value of −1 refers to fully segregated species with no reactions occurring between them. Reactants can also have positive covariance, thereby having an increased effective reaction rate compared to the well-mixed condition and a Damköhler number larger than 1. Mousavi et al. [5] showed that, in coarse models, this number can reach values close to 50, especially in the vicinity of pollution point sources, and recommended that irregular or adaptive grids be used to improve the simulation in the vicinity of these sources.

Chemical segregation in the lower layers of the atmosphere is affected by the structure and the strength of the flow, specifically by the boundary layer stability; the length scales of turbulence; the size of shear-driven and thermal structures; and, more generally, by the organization of turbulence in the dry and cloudy boundary layers. In numerical models, efficient mixing and tracer dispersion, which tends to reduced by segregation, are therefore favored by high Reynolds numbers, high turbulence intensity, and high subscale diffusivity. Segregation is also influenced by the flow velocity and by buoyancy effects. The spatial and temporal distributions of surface emissions of reactive species impact how these species disperse and segregate.

Atmospheric chemistry is controlled by oxidation reactions between constituents that are emitted near the surface and oxidants that are not directly emitted from the surface but rather produced by photochemistry. It is more likely that segregation occurs between atmospheric reactants near strong source regions and an oxidant with higher mixing ratios above the boundary layer. An example of this scenario is the reaction between nitric oxide (NO) and ozone (O_3), where NO is emitted from combustion sources (anthropogenic and natural) and naturally from soils. Because ozone is produced through photochemistry in the atmosphere and deposited on the Earth's surface, its mixing ratios often increase with height above the surface. Other examples include reactions between volatile organic compounds (VOCs) and the hydroxyl radical (OH). The turbulence timescale varies diurnally and seasonally and can be different over land compared to oceanic regions. Typically, in the summer, the mid-day convective boundary layer over land (τ_{turb}) is 15 min, which means that chemical constituents with a lifetime of about 15 min may become segregated from their oxidants.

The effect of segregation on the mean atmospheric production or destruction of chemical species can be expressed in a coarse model by correcting the value of the original rate constants (k_{AB}). Using the Krol et al. [6] chemistry mechanism, Vinuesa and Vila-Guerau de Arellano [7] introduced the concept of an effective reaction rate (k^{eff}_{AB}) to parameterize the effect of segregation on chemical reactivity in coarse models. They wrote:

$$k^{eff}_{AB} = k_{AB}(1 + I_{AB})$$

In practice, such a correction is difficult to implement, since the value of the segregation intensity is constantly changing with the state of the turbulent flow. In fact, their approach has not been tested with complex chemical mechanisms or in regional or global-scale chemistry transport models.

The covariance between the concentration of different species and, hence, the segregation coefficient can be derived from conventional Reynolds-averaged Navier–Stokes (RANS)-based simulations with an appropriate closure of the higher-order equations that involve the interactions between various turbulent fluctuations. Such an approach may be inaccurate if the averaged chemical reaction rates affected by the strong nonlinearity associated with the chemical mechanism are calculated on the basis of averaged parameters only. Although more computationally expensive than a RANS models, large-eddy simulation (LES) models spatially filter the equations of motion at sufficiently small scales compared to RANS that the largest energy-containing scales of the turbulence, which extracts energy from the mean flow, are resolved. In this case, only the smallest and nearly isotropic eddies that act primarily to dissipate energy are parameterized through a subgrid scale model [8–11]. In other words, RANS models completely parameterize the influence of turbulence, while LES models resolve the dominant turbulent motions.

Investigations in the atmospheric boundary layer using LES to study turbulence–chemistry interactions began with generic species A and B, where one species was emitted at the surface and the other entrained into the boundary layer from the free troposphere [12,13]. This simple configuration was necessitated by computational constraints in the 1990s. As computational capacity increased, studies emerged examining chemical systems of reactions, albeit using a simple representation [6].

In this paper written in honor of Jack Herring's lifelong scientific accomplishments, we present a few examples of LES-based simulations applied to chemically reactive species in the atmospheric boundary layer under different environments: (1) in a forested area with high emissions of biogenic hydrocarbons, (2) in an urban area rich in anthropogenic emissions, and (3) over a maritime area with high emissions of reduced sulfur species.

2. Boundary Layer over a Forest

Forested land covers ~30% of Earth's land surface [14]. Trees are strong emitters of volatile organic compounds (VOCs), especially isoprene emitted from broadleaf deciduous trees and terpenes from needleleaf trees. The emissions of these biogenic VOCs comprise 80% of the total global VOC emissions [15]. Isoprene and monoterpenes participate in atmospheric photochemistry, contributing to the production of ozone [16,17] and secondary organic aerosol (SOA; e.g., [18–21]). Thus, the emissions and chemistry of isoprene and monoterpenes are important sources and sinks to represent in chemistry transport models. For example, during daytime, isoprene emissions increase, depending on the temperature and amount of photosynthetic active radiation (PAR) reaching the vegetation. Isoprene readily reacts with hydroxyl radicals (OHs) to form peroxy radicals. The rate of the reaction between isoprene and OH is fast, so isoprene has a lifetime of 15–30 min, which is comparable to the turbulence turnover time of the atmospheric boundary layer (ABL).

Several studies have investigated the interactions between turbulence and chemistry over forested regions without considering the effects of the forest canopy on turbulent motions. Krol et al. [6] were the first to use a simple atmospheric chemistry mechanism (7 predicted trace gases using 10 photochemical reactions) to address segregation effects between reactants. Their simulations revealed that a highly reactive VOC with an ~18 min chemical lifetime (like that of isoprene) had an ABL average intensity of segregation of 20% for a domain with homogeneous emissions. Krol et al. [6] also examined the impact of heterogeneous emissions and found segregation to be even stronger for the VOC + OH reaction compared to the case with homogeneous emissions.

The segregation of isoprene and OH was suggested as a possible cause for large OH observations that could not be explained with known sources/sinks in the tropics [22,23]. To determine if this was a viable explanation, Ouwersloot et al. [24] performed LES simulations with 18 reacting trace gases and 19 chemical reactions and found that the segregation effect alone was not sufficient to reconcile modeled and observed OH concentrations. They also showed that heterogeneous emissions enhance segregation of reactants (Figure 2) in accordance with Krol et al. [6] and Kaser et al. [25]. By obtaining high-temporal-resolution measurements of isoprene and OH just above a mixed deciduous forest near Julich, Germany, analyses of the 2003 Emission and CHemical transformation of biogenic volatile Organic compounds (ECHO 2003, [26]) case study produced estimates from measurements of the intensity of segregation between isoprene and OH for the 25 July 2003 case study [27,28]. Dlugi et al. [27] calculated 9–15% segregation between isoprene and OH during mid-day (1200–1300 CET) and <10% segregation for other analysis calculations between 1000 and 1400 CET. Dlugi et al. [28] extended the analysis of isoprene and OH segregation with the ECHO 2003 data, showing that both chemical transformations and dynamical processes (turbulent and convective mixing, as well as advection) are responsible for observed segregation. Kaser et al. [25] calculated a measurement-based budget of isoprene in the atmospheric boundary layer using aircraft data from the 2013 Southern Oxidant and Aerosol Study (SOAS). Combining measurement analysis, detailed chemistry box modeling, and large-eddy simulations to separate the roles of chemistry and heterogeneous emissions, Kaser et al. [25] found that the surface heterogeneity of isoprene emissions segregated isoprene and OH by up to 30%.

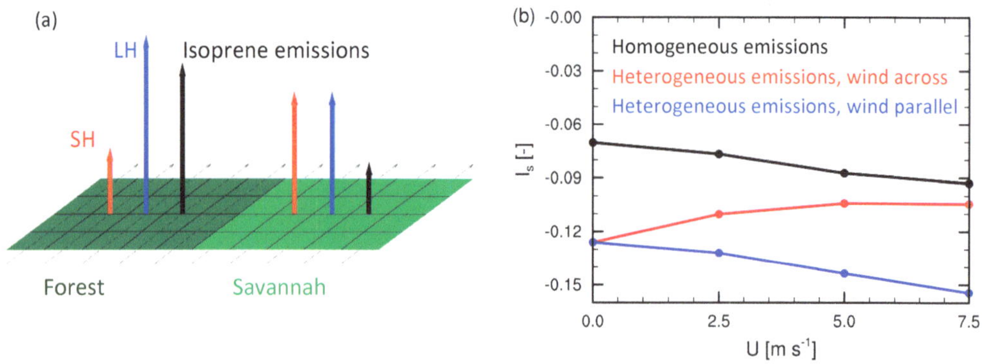

Figure 2. (**a**) Schematic of imposing heterogeneous emissions of sensible heat (red), latent heat (blue), and isoprene (black) surface emissions used by Ouwersloot et al. [24]. (**b**) Intensity of segregation for isoprene and OH in the ABL as a function of the horizontal wind for LES studies of homogeneous and heterogeneous emissions. In (**b**), the blue line corresponds to wind blowing parallel between the forest and savannah patches, while the red line corresponds to wind blowing across from forest to savannah. Figure adapted from Ouwersloot et al. [24], © Author(s) 2011, licensed under a Creative Commons Attribution (CC BY) 3.0 License.

Many of these early LES applications of ABL chemistry used simple chemical mechanisms, leaving questions as to whether more complex chemistry reduces the effect of segregation through its role in producing OH from other reactions. As computational capabilities increased, additional studies were pursued utilizing LES models that included atmospheric chemistry reaction schemes of varying complexity. These studies investigated the impact of fair-weather cumulus clouds [29–31], the role of varying nitrogen oxide ($NO_x = NO + NO_2$) scenarios (from very low NO_x found in remote regions to very high NO_x found in urban forest centers [32]), and the effect of varying weather scenarios [33]. These studies found that the presence of fair-weather cumulus clouds causes venting of the ABL, increasing the segregation between the surface-sourced isoprene and in situ-formed OH reactants in the cloud layer [29]. Kim et al. [30] applied LES coupled with a more complex chemistry scheme (52 reacting trace gases and 142 reactions) to summertime, fair-weather cumulus convective ABL. They found that the combined effect of modified photodissociation rates and isoprene emissions via cloud shading due to cloud scattering of solar radiation did not impact isoprene and OH radical mixing ratios substantially [30], as the two effects tended to offset each other. However, the dissolution of soluble trace gases and subsequent aqueous-phase chemistry within the cloud droplets enhanced the segregation between reactants [31] within the cloud layer (Figure 3). Kim et al. [32] showed that the covariance of isoprene and OH in the mixed layer (z = 500 m) depended on NO_x levels, with the largest segregation of 18% occurring for NO_x = 8–10 ppbv, the smallest segregation of 5% for NO_x = 1–2 ppbv, and ~6% segregation for NO_x 0.1–0.3 ppbv. By conducting LES with chemistry runs for three case studies of the DISCOVER-AQ field campaign, Li et al. [33] found that segregation between isoprene and OH was larger (10% segregation) for a hot, humid day than for a clear-sky, cool summer day scenario and a moderately warm day with fair-weather cumulus clouds. Under hot, humid, and convective conditions, the isoprene and oxygenated VOC lifetimes lengthened due to higher isoprene emissions, elevated initial chemical concentrations, and competition among VOCs for reaction with OH.

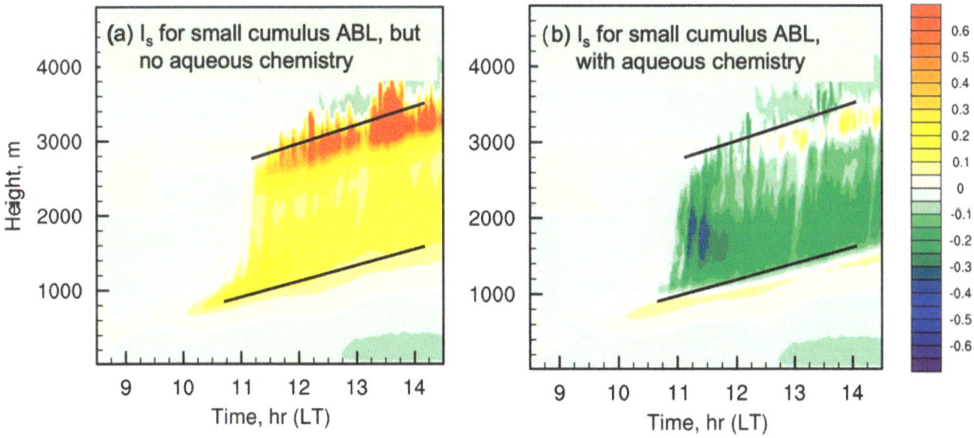

Figure 3. Temporal evolution of the horizontally averaged intensity of segregation for isoprene and OH from the NCAR LES using (**a**) relatively complex gas-phase chemistry only and (**b**) gas-phase and aqueous-phase chemistry. The black lines in panels (**a**,**b**) roughly mark the cloud base and cloud top of the cloudy layer. The cloud top and base are defined as the maximum and minimum heights where liquid water is present in the model domain, respectively. The figure is based on that from Li et al. [31], with the permission of © WILEY.

By absorbing momentum throughout its distributed depth, tall trees alter ABL turbulent motions by creating a hydrodynamically unstable inflection point in the mean vertical wind profile. Large ABL-scale turbulent motions trigger instability when bringing high-momentum fluid down to the canopy top, thereby enhancing the vertical shear of the horizontal wind, which produces pairs of head-up and head-down vortices whose scales are thought to be set by the vorticity thickness at the canopy top [34,35]. These canopy-induced turbulent motions are thought to perform 60–80% of the exchange between the canopy layers and aloft and can be regularly identified in tower-based field data as sweep (downwelling) and ejection (upwelling) motions separated by a scalar microfront [36,37]. In weak-wind, buoyantly dominated conditions, buoyant plumes from the ground and canopy drive the exchange between the canopy layers and aloft [2,38,39].

Organized turbulent motions therefore transport trace gases from the canopy into the ABL, which affects the above-canopy tropospheric chemistry. Trace gases are also transported into the forest canopy, where within-canopy photochemistry and dry deposition can remove important trace gases. Within the canopy, organized turbulent structures lead to spatial variability of reactants, resulting in segregations of reactants. Clifton et al. [40] used an LES coupled to a multilayer canopy model with a simple chemical mechanism (19 trace gases with 41 reactions) to examine the interactions between the canopy, turbulence, and atmospheric chemistry. Their results for specific cases showed that segregation within the canopy altered reaction rates substantially (from 48% segregation to 23% positive covariance). They illustrated that high soil NO emissions promote more segregation among reactants and that high horizontal spatial variability of reactants within the forest canopy plays a role in separating reactants, as previously shown by Krol et al. [6], Dlugi et al. [28], Kaser et al. [25], and Patton et al. [2].

These previous LES studies either employed simple atmospheric chemistry schemes with highly detailed turbulence representation or reduced the resolution of the LES to include more complex chemistry. There is a strong need to represent both turbulence and chemistry in detail (i.e., high-resolution LES with complex chemistry mechanisms) and represent in-canopy and above-canopy gas-phase and aerosol chemistry. The role of aerosols in affecting heating rates, surface-sensible and latent heat fluxes, and ABL structure has been examined for idealized convective boundary-layer regimes [41,42]. The impact of aerosols on the ABL needs to be extended to learn about the subsequent impact on chemical reactivity. Aerosol impacts on ABL composition and chemical reactivity should be investigated further through combined field experiments and modeling analysis, as aerosols are complex in terms of their spatial variability, optical properties, and composition. To evaluate and gain further insight into the processes affecting chemical composition in the ABL, field observations with strategically placed towers and aircraft measurements must be pursued.

3. Urban Boundary Layer

Segregation effects are significant when (1) the reaction rate is fast [43], (2) the emission sources of the reacting species are spatially inhomogeneous [24,44], or (3) the emissions of one chemical compound are intense [32,45]. In urban areas, the emissions produced by human activities are usually characterized by heterogeneous distributions. An example is provided by the traffic emissions taking place on different roads. Furthermore, in polluted cities, anthropogenic emissions are intense, especially road traffic emissions, while industrial facilities and power plants produce additional pollution. Therefore, the segregation effect is an important factor for understanding air pollution in large cities.

Atmospheric chemistry in urban areas involves OH oxidation of anthropogenic VOCs and biogenic VOCs (similar to the processes discussed for the forested boundary layer) and their chemistry with NO and NO_2, which produces O_3. The segregation effect of fast-reacting chemical reactants in the turbulent urban environment with spatially heterogeneous emissions needs to be addressed. In some previous studies, a simple NO–NO_2–O_3 mechanism (with three reactions) was coupled with microscale models because of the potential for NO, which is emitted from anthropogenic surface sources, and O_3 to be segregated. For instance, Baker et al. [46] introduced O_3–NO_X chemistry into a CFD-LES model and applied it in an idealized street canyon. They showed that the variations in the chemical species were largely affected by the turbulent structures at the different locations of the canyon. Auger and Legras [47] adopted a more comprehensive chemical mechanism with 44 species in an LES model for urban application and pointed out that the segregation effect was strong if the emissions were restricted to a limited area, especially during the morning, when mixing is incomplete. Bright et al. [48] compared their CFD-LES model with coupled chemistry to a zero-dimensional box model to investigate the impact of segregation in street canyons. Their LES simulation resulted in lower NO_X (−3%), OH (−11%), and HO_2 (−8%) concentrations but higher O_3 (6%) values relative to the results of a box model for their designed scenario. Zhong et al. [49,50] compared the segregation effect between deep and regular urban street canyons, which is dependent on the aspect ratio, and revealed that this effect was greater in deeper street canyons due to the vertically aligned vortex structures present in the poorly mixed environment.

Li et al. [44] conducted a more systematic analysis of the segregation effect under urban-like conditions through a series of DNS experiments. A simple second-order $A + B \rightarrow C$

reaction was adopted, and the effects on the segregation intensity resulting from the reaction rate, the strength of the emission fluxes, and the heterogeneity in the emissions were calculated. Their study showed that the segregation intensity increased considerably when the surface emissions of one chemical compound was enhanced from rural to urban values because the availability of the other tracer became limited. Spatial heterogeneity also has an important impact on the segregation intensity.

The above studies were performed either by CFD models in an idealized street block or by LES/DNS models in a flat domain. Wang et al. [51] developed an LES experiment based on a realistic case in the vicinity of Hong Kong Island with complex topography, land use, and emission distributions (Figure 4a,b). A simplified O_3 photochemical mechanism was adopted with 15 species and 18 reactions. As Hong Kong Island is characterized by a special landscape with large forest areas in the mountainous region in the island's center and a dense built-up urban canopy along the coast, both anthropogenic and biogenic emission sources are important, although geographically separated. This work discussed the factors that affect segregation in a polluted situation. The results showed that heterogeneity in emissions is the dominant factor that causes large segregation in the urban area. Topography has an influence on the turbulent structure and can therefore affect segregation locally. The segregation intensity for the reaction of $NO + O_3 \rightarrow NO_2$ can reach about -40% in the urban area, especially on the leeward slope of the mountains (Figure 4c,d).

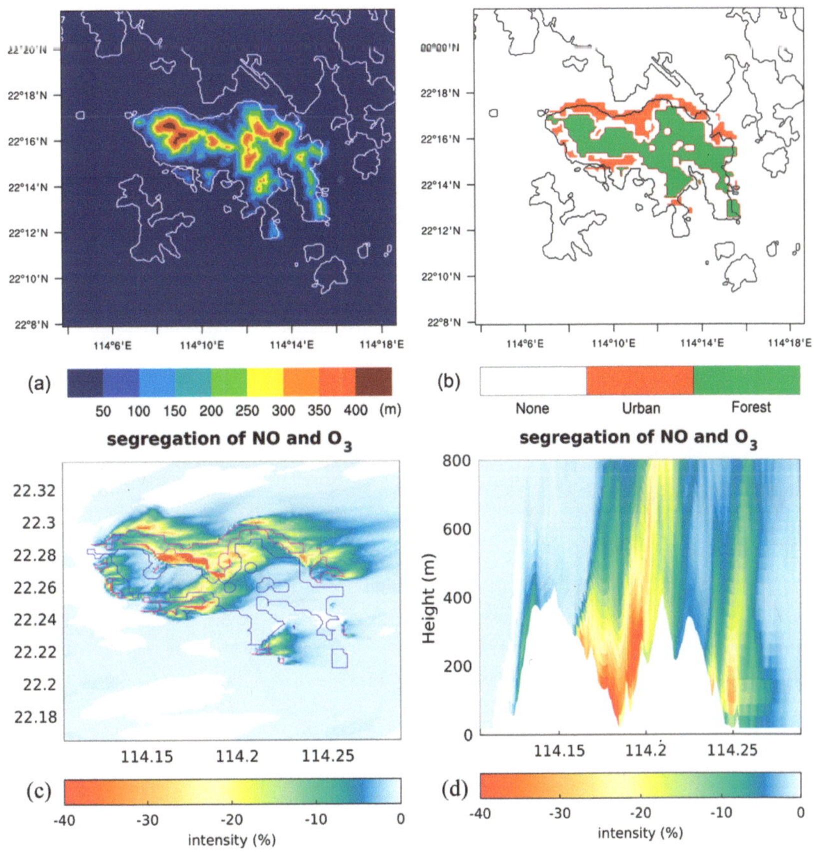

Figure 4. (**a**) Terrain height of Hong Kong Island; (**b**) emissions map adopted based on the land use (red: anthropogenic emissions from urban areas; green: biogenic emissions from forested areas); (**c**) segregation intensity near the surface (the magenta line shows the urban area, and the blue line shows

the forested area); (**d**) vertical cross section along latitude 22.275° N. Figure adapted from Wang et al. [51] as an illustration of the idealized LES modeling for a convective PBL over a polluted urban environment. This study adopted two nested domains with the outer domain (Δx = 300 m) to provide the initial and boundary conditions for the inner domain (Δx = 100 m), as shown in the figure. The experiment was initialized from the output of the mesoscale WRF (Δx = 1 km), and the prevailing wind is westerly.

In a subsequent study, Wang et al. [52] set up several control runs with different strengths of emission fluxes to account for clean and less polluted conditions (the anthropogenic emission rates were reduced by two orders of magnitude for NO, CO, and VOCs). Their study shows that the pollution levels have an important impact on the distributions of O_3 and OH oxidants due to their reactions with the primary pollutants. The segregation intensities between the anthropogenic VOCs and OH radicals are also largely influenced by the pollution level, while the biogenic VOCs are less affected.

Some recent studies reported realistic simulations of chemical species using coupled meso-to-microscale models. For instance, Wang et al. [53] performed multiscale simulations with coupled WRF-LES-Chem for Hong Kong with the highest spatial resolution of 33 m. Figure 5a,b show the near-surface noontime distribution of NO and O_3 in the innermost domain covering the Kowloon peninsula and the northern coast of Hong Kong Island. Figure 5c highlights the calculated segregation between NO and O_3. High segregation intensity is found near the location of road traffic emissions and on the leeward slope of the mountain, which is consistent with the idealized experiment discussed above.

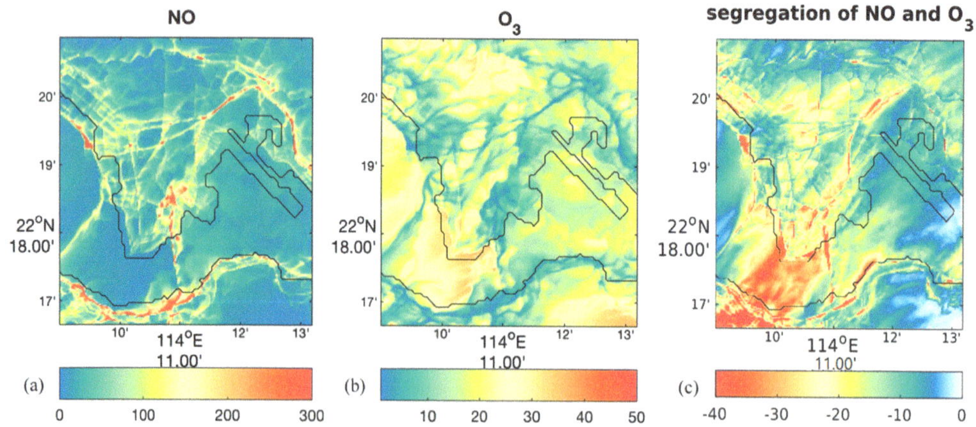

Figure 5. The distribution of NO (**a**) and O_3 (**b**) near the surface at noontime adapted from the 33 m WRF-LES-Chem simulation by Wang et al. [53], © Author(s) 2023, licensed under a Creative Commons Attribution (CC BY) 4.0 License. and the segregation intensities between NO and O_3 (**c**) calculated from the concentrations of NO and O_3 shown in this figure.

4. Clouds, Turbulence, and Biogenic Sulfur in the Marine Boundary Layer

The marine boundary layer (MBL) extends from the ocean surface to a capping inversion that separates it from the free troposphere. Ocean-emitted chemical species are transformed in the MBL by radiation, as well as gas- and aqueous-phase chemistry, and by aerosol and cloud processes. Gas-phase conversion is commonly started by reactions with short-lived, photochemically produced reactants, such as the hydroxyl radical (OH), which forms from ozone that is entrained into the MBL from the free troposphere. Thus, biogenic sulfur compounds emitted from the oceans, primarily dimethyl sulfide (CH_3SCH_3, DMS), hydrogen sulfide (H_2S), carbonyl sulfide (OCS), and carbon disulfide (CS_2), are oxidized to sulfur dioxide (SO_2), sulfuric acid (H_2SO_4), methanesulfonic acid (CH_3SO_3H), hydroperoxymethyl thioformate ($HOOCH_2SCHO$, HPMTF [54]), and ultimately sulfate ($SO_4{}^{2-}$). This alters the aerosol population by nucleation of new particles in the gas phase and by growth through gas-phase condensation and aqueous-phase mass accumulation. Condensation of water vapor may turn aerosol particles into cloud droplets. This connects biogenic sulfur emissions from the oceans to MBL clouds, which are sensitive to aerosol and exert leverage over the Earth's radiation budget [55].

Chemical segregation is determined by MBL dynamics, which are driven by surface-sensible and latent heat fluxes, by radiative heating and cooling, and by wind shear. Decoupling, a common [56] phenomenon that stratifies the MBL into a lower layer above the ocean and an upper, cloud-free or cloudy layer below the inversion, contributes to segregation. Due to the strong radiative cooling at their tops and the process of latent heating and cooling, clouds reinforce MBL circulation, increase mixing, and counteract segregation. Thus, clouds play an important role in the segregation of chemical species in the MBL.

We explore how trade cumulus cloud organization and the associated mesoscale dynamics shape the distribution of an ocean-emitted species and modulate its segregation relative to a free-tropospheric species. Figure 6 shows snapshots taken on 2 February 2020 at 17 h 00 m 00 s UTC from an LES of the sugar-to-flower transition following Narenpitak et al. [57]. The simulation starts on 2 February 2020 at 00 h 00 m 00 s UTC and tracks two tracers with a lifetime of 24 h. One tracer is initialized with 0.6 ppt in the boundary layer and 0 ppt in the free troposphere and is emitted at 8.6 μmol m^{-2} d^{-1} from the ocean. The other tracer is initialized with 0 ppt in the boundary layer and 3 ppt in the free troposphere and has no surface source or sink.

The cloud field contains several "Flower" clouds [58] with stratiform outflows near their tops and cold pools at the surface surrounded by small cumuli (Figure 6a). Figure 6b shows the segregation intensity of the two tracers in the MBL. Segregation intensity is reduced at the locations of the Flower clouds relative to the surrounding areas with isolated trade cumulus clouds. The reduced segregation is caused by mesoscale circulation and vertical transport in the Flowers clouds. Figure 6c,d show the distribution of the surface-emitted tracer, and Figure 6e,f show the distribution of cloud water along two transects through the cloud field (Figure 6a). At the location of cumulus cluster "A", an early stage of Flower clouds, circulation on the mesoscale that is associated with the formation of moist patches in which Flowers form [57] creates a tracer patch in the cumulus layer (Figure 6b). The tracer is lofted from the mixed layer by individual cumulus updrafts, some of which rise as high as 2600 m (Figure 6e). At the location of mature Flower cloud "B", the aggregated cumulus updrafts forming the Flower loft large amounts of the tracer from the mixed layer upward (Figure 6d). The aggregated nature of the Flower is evident from the distribution of cloud water (Figure 6f). Additionally, smaller structures in the tracer distribution in the cloud layer shaped by the mesoscale circulation of the Flower cloud state appear along both transects (Figure 6c,d).

The tracer distribution in the Flower cloud state (Figure 6) is an idealized approximation of the distribution of a surface-emitted species such as DMS. In the case of DMS, a part is scavenged inside the cumulus clouds. The remainder undergoes rapid oxidation by OH near cloud tops due to enhanced actinic fluxes, possibly with a lifetime shorter than 24 h, to SO_2, H_2SO_4, and other sulfur species and potentially initiates aerosol nucleation. Despite the idealization, the example illustrates that trade cumulus clouds act as a valve for the transport of ocean-emitted species from the mixed layer into the boundary layer above. Trade cumulus cloud organization shapes the resulting distribution, with potential consequences for chemical conversion and new particle formation.

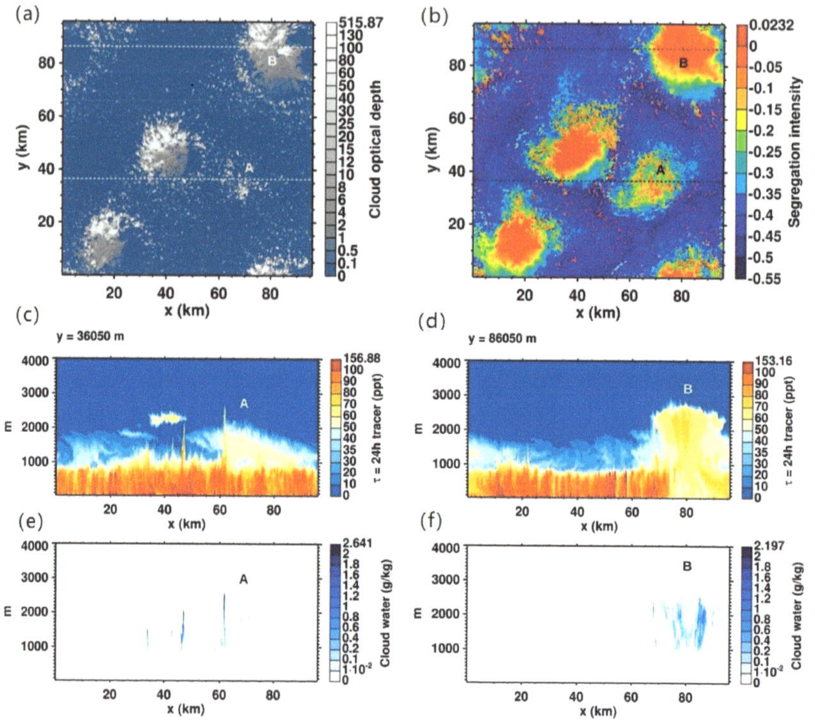

Figure 6. "Flower"-type trade cumulus organization in the Caribbean east of Barbados on 2 February 2020 at 13 h 00 m 00 s local solar time simulated by Narenpitak et al. [57]. (**a**) Cloud optical depth. (**b**) Segregation intensity of a surface-emitted tracer and a free-tropospheric tracer with lifetimes of τ = 24 h. The mixing ratio of the surface-emitted tracer (**c**,**d**) and cloud water content (**e**,**f**) are shown along two transects. "A" denotes the location of a cumulus cluster, an early stage of "Flower" development, and "B" denotes the location of a mature "Flower" aggregate.

5. Outlook

Spatial segregation of reactants by organization in turbulence clearly participates the evolution of those reactants in the ABL. The importance of segregation in modulating reaction rates in the ABL varies with the scale of the separation distance of sources/sinks and with the chemical or atmospheric regime.

Due to the need to simultaneously sample multiple species at sufficiently fast time scales to interrogate the role of the organized structures comprising the turbulence (with time scales ranging from milliseconds to hours) in controlling reaction rates, species–species covariances (and, hence, segregation) have proven difficult to measure. As such, much of what the community currently knows about species–species segregation comes from idealized modeling.

To advance predictive skill, the community needs to surmount critical hurdles in observing and simulating turbulence–chemistry interactions. While instruments like PTR-ToF-MS (proton-transfer reaction time-of-flight mass spectrometry) enable rapid sampling across a range of chemical masses, they remain prohibitively expensive and are not conducive to sampling in regimes with condensed cloud water, which limits their utility due to the ubiquitous spatial heterogeneity of reactant sources/sinks in the ABL. When attempting to interrogate the spatial variation of species–species covariances, sampling with a single instrument requires substantial compromises. Sampling strategies need to be designed to measure both the turbulence and the chemistry. While profiles on a single tower provide information on locally occurring near-surface processes controlling species–species segregation, reactions occurring near the surface can be strongly influenced by the diurnal evolution of the boundary layer as it grows and decays each day in response to solar forcing (through both dilution due to the larger volume into which species are mixed and entrainment of reacting species from aloft, e.g., [59,60]). While flying an instrument like a PTR-ToF-MS through the boundary layer on an aircraft can provide vertical profile information, flying sufficiently long horizontal legs to sample the full range of scales within the ABL turbulence ensures that only a limited number of heights can be sampled within a timeframe that one might be able to consider the turbulence and chemistry reasonably stationary. Therefore, the community would truly benefit from the development of remote (e.g., lidar) and in situ sensing capabilities with sufficient power to continuously sample highly range- and time-resolved reactant profiles in both cloudy and cloudless conditions toward understanding of the spatially and diurnally varying evolution of species–species segregation across varying climates and chemical regimes.

Although LES resolves the largest energy-containing scales of turbulence performing reactant transport in the ABL (and the role of those large-scale structures in modulating reactivity), the influence of scales smaller than the grid resolution must still be parameterized. At those small scales, current LES practice generally assumes that reactants are treated as if they are passive, nonreacting scalars (e.g., [24,61]). Hence, segregation and chemical modification of flux–gradient relationships (e.g., [62,63]) are ignored at these scales. This assumption is made primarily for simplicity's sake, since the community does not yet sufficiently understand the spectral decomposition of turbulence–chemistry interactions necessary for the development of a parameterization that only applies at a certain scale and smaller. While numerous chemically aware subgrid models for LES have been proposed in the combustion engineering literature (e.g., [64,65]), these subgrid models typically rely on a priori knowledge of the joint probability density distribution of the reactants of interest; in the atmospheric boundary layer, such joint PDFs vary with atmospheric stability, height, and the chemical regime. For near-surface applications, a key opportunity to advance chemically aware subgrid models for LES could lie in adapting what have become known as horizontal array turbulence study strategies (HATS, e.g., [66–70]) to reactants. The HATS strategy uses time-synchronized spatial arrays of fast instrumentation to explicitly filter planes of field data into large and small scales. If such arrays were complemented with an equal number of PTR-ToF-MS instruments, one could interrogate spatially filtered species–species covariances (i.e., the joint PDF of the small scales) toward the development of a chemically aware subgrid-scale model and its variation with Damköhler number and atmospheric stability. Similar strategies could be undertaken by filtering DNS simulation data, but the role of the largest scales of motion would be lost.

At coarser resolutions, where turbulence is not resolved (i.e., in regional air quality and climate models), techniques to parameterize the influence of segregation on reactivity have been proposed (e.g., [7,61]); however, these techniques have only been evaluated for highly reduced chemistry. Attempts to incorporate the SOMCRUS framework [61] into today's weather and climate models have proven difficult due to the need to solve so many additional equations for the species–species covariance. Pathways to reduce the computational requirements remain elusive because of the continually evolving variation of the processes that dominate turbulence–chemistry interactions through diurnal and seasonal time scales and a lack of turbulence-chemistry-focused observations spanning the full ABL depth.

Nevertheless, there are abundant opportunities to advance the understanding and representation of the turbulence–chemistry interactions that control reactivity in the atmospheric boundary layer and in turbulent cloudy environments. Success requires research spanning traditional disciplinary boundaries; Jack Herring was one of the unique individuals who recognized this need.

Author Contributions: All authors were directly involved in the conceptualization, methodology, and writing of this paper. All authors have read and agreed to the published version of the manuscript.

Funding: The National Center for Atmospheric Research is sponsored by the US National Science Foundation. This research was supported in part by NOAA cooperative agreement NA22OAR4320151. YW contribution to this work is supported by the Hong Kong Research Grants Council (grant no. T24-504/17-N).

Data Availability Statement: This paper reviews studies published in the scientific literature and is does not make use of new research data.

Conflicts of Interest: The authors declare no conflict of interest.

References

1. Herring, J.R.; Wyngaard, J.C. Convection with a first-order chemically reactive passive scalar. In Proceedings of the Fifth International Symposium on Turbulent Shear Flows, Cornell University, Ithaca, NY, USA, 7–9 August 1987; pp. 324–336.
2. Patton, E.G.; Sullivan, P.P.; Shaw, R.H.; Finnigan, J.J.; Weil, J.C. Atmospheric stability influences on coupled boundary layer and canopy turbulence. *J. Atmos. Sci.* **2016**, *73*, 1621–1647. [CrossRef]
3. Danckwerts, P.V. The definition and measurement of some characteristics of mixtures. *Appl. Sci. Res.* **1952**, *3*, 279–296. [CrossRef]
4. Damköhler, G. Influence of turbulence on the velocity flames in gas mixtures. *Z. Elektrochem. Angew. Phys. Chem.* **1940**, *46*, 601–626. [CrossRef]
5. Mousavi, M.; Soltanieh, M.; Badakhshan, A. Influence of turbulence and atmospheric chemistry on grid size with respect to location in modeling and simulation of photochemical smog formation and transport. *Environ. Model. Softw.* **1999**, *14*, 657–663. [CrossRef]
6. Krol, M.C.; Molemaker, M.J.; de Arellano, J.V.G. Effects of turbulence and heterogeneous emissions on photochemically active species in the convective boundary layer. *J. Geophys. Res. Atmos.* **2000**, *105*, 6871–6884. [CrossRef]
7. Vinuesa, J.-F.; Arellano, J.V.D. Fluxes and (co-)variances of reacting scalars in the convective boundary layer. *Tellus B Chem. Phys. Meteorol.* **2003**, *55*, 935–949. [CrossRef]
8. Smagorinsky, J. General circulation experiments with the primitive equations: I. The basic experiment. *Mon. Weather Rev.* **1963**, *91*, 99–164. [CrossRef]
9. Lilly, D.K. On the application of eddy viscosity concept in the inertial sub-range of turbulence. *NCAR Tech. Note* **1966**, *123*, 19. [CrossRef]
10. Lilly, D.K. The representation of small-scale turbulence in numerical simulation experiments. *Proc. IBM Sci. Comput. Symp. Environ. Sci.* **1967**, *23*, 195–210. [CrossRef]
11. Deardorff, J.W. A numerical study of three-dimensional turbulent channel flow at large Reynolds numbers. *J. Fluid Mech.* **1970**, *41*, 453–480. [CrossRef]
12. Schumann, U. Large-eddy simulation of turbulent diffusion with chemical reactions in the convective boundary layer. *Atmos. Environ.* **1989**, *23*, 1713–1727. [CrossRef]
13. Sykes, R.I.; Parker, S.F.; Henn, D.S.; Lewellen, W.S. Turbulent mixing with chemical reaction in the planetary boundary layer. *J. Appl. Meteorol. Climatol.* **1994**, *33*, 825–834. [CrossRef]
14. FAO. *Global Forest Resources Assessment 2020: Main Report*; 978-92-5-132974-0; Food and Agriculture Organization of the United Nations: Rome, Italy, 2020; p. 184.
15. Olivier, J.G.J.; Berdowski, J.J.M. Global Emissions Sources and Sinks. In *The Climate System*; Berdowski, J., Guicherit, R., Heij, B.J., Eds.; AA Balkema Publishers: Lisse, The Netherlands, 2001; pp. 33–77.
16. Chameides, W.L.; Lindsay, R.W.; Richardson, J.; Kiang, C.S. The role of biogenic hydrocarbons in urban photochemical smog: Atlanta as a case study. *Science* **1988**, *241*, 1473–1475. [CrossRef] [PubMed]
17. Trainer, M.; Williams, E.J.; Parrish, D.D.; Buhr, M.P.; Allwine, E.J.; Westberg, H.H.; Fehsenfeld, F.C.; Liu, S.C. Models and observations of the impact of natural hydrocarbons on rural ozone. *Nature* **1987**, *329*, 705–707. [CrossRef]
18. Carlton, A.G.; Wiedinmyer, C.; Kroll, J.H. A review of Secondary Organic Aerosol (SOA) formation from isoprene. *Atmos. Chem. Phys.* **2009**, *9*, 4987–5005. [CrossRef]
19. Griffin, R.J.; Cocker III, D.R.; Flagan, R.C.; Seinfeld, J.H. Organic aerosol formation from the oxidation of biogenic hydrocarbons. *J. Geophys. Res. Atmos.* **1999**, *104*, 3555–3567. [CrossRef]
20. Kanakidou, M.; Seinfeld, J.H.; Pandis, S.N.; Barnes, I.; Dentener, F.J.; Facchini, M.C.; Van Dingenen, R.; Ervens, B.; Nenes, A.; Nielsen, C.J.; et al. Organic aerosol and global climate modelling: A review. *Atmos. Chem. Phys.* **2005**, *5*, 1053–1123. [CrossRef]
21. Pandis, S.N.; Paulson, S.E.; Seinfeld, J.H.; Flagan, R.C. Aerosol formation in the photooxidation of isoprene and β-pinene. *Atmos. Environ.* **1991**, *25*, 997–1008. [CrossRef]
22. Butler, T.M.; Taraborrelli, D.; Brühl, C.; Fischer, H.; Harder, H.; Martinez, M.; Williams, J.; Lawrence, M.G.; Lelieveld, J. Improved simulation of isoprene oxidation chemistry with the ECHAM5/MESSy chemistry-climate model: Lessons from the GABRIEL airborne field campaign. *Atmos. Chem. Phys.* **2008**, *8*, 4529–4546. [CrossRef]

23. Karl, T.G.; Christian, T.J.; Yokelson, R.J.; Artaxo, P.; Hao, W.M.; Guenther, A. The Tropical Forest and Fire Emissions Experiment: Method evaluation of volatile organic compound emissions measured by PTR-MS, FTIR, and GC from tropical biomass burning. *Atmos. Chem. Phys.* **2007**, *7*, 5883–5897. [CrossRef]
24. Ouwersloot, H.G.; Vilà-Guerau de Arellano, J.; Van Heerwaarden, C.C.; Ganzeveld, L.N.; Krol, M.C.; Lelieveld, J. On the segregation of chemical species in a clear boundary layer over heterogeneous land surfaces. *Atmos. Chem. Phys.* **2011**, *11*, 10681–10704. [CrossRef]
25. Kaser, L.; Karl, T.; Yuan, B.; Mauldin III, R.; Cantrell, C.; Guenther, A.B.; Patton, E.; Weinheimer, A.J.; Knote, C.; Orlando, J.; et al. Chemistry-turbulence interactions and mesoscale variability influence the cleansing efficiency of the atmosphere. *Geophys. Res. Lett.* **2015**, *42*, 10894–10903. [CrossRef]
26. Koppmann, R.; Kesselmeier, J.; Meixner, F.; Warnke, J.; Bandur, R.; Hoffmann, T.; Aubrun, S.; Leitl, B.; Schatzmann, M.; Dlugi, R.; et al. Emission and chemical transformation of biogenic volatile organic compounds investigations in and above a mixed forest stand (ECHO): An overview. In Proceedings of the Integrated Land Ecosystem–Atmosphere Processes Study (ILEAPS) International Open Science Conference 2003, Helsinki, Finland, 29 September–3 October 2003; pp. 198–201.
27. Dlugi, R.; Berger, M.; Zelger, M.; Hofzumahaus, A.; Siese, M.; Holland, F.; Wisthaler, A.; Grabmer, W.; Hansel, A.; Koppmann, R.; et al. Turbulent exchange and segregation of HO_x radicals and volatile organic compounds above a deciduous forest. *Atmos. Chem. Phys.* **2010**, *10*, 6215–6235. [CrossRef]
28. Dlugi, R.; Berger, M.; Zelger, M.; Hofzumahaus, A.; Rohrer, F.; Holland, F.; Lu, K.; Kramm, G. The balances of mixing ratios and segregation intensity: A case study from the field (ECHO 2003). *Atmos. Chem. Phys.* **2014**, *14*, 10333–10362. [CrossRef]
29. de Arellano, J.V.-G.; Kim, S.-W.; Barth, M.C.; Patton, E.G. Transport and chemical transformations influenced by shallow cumulus over land. *Atmos. Chem. Phys.* **2005**, *5*, 3219–3231. [CrossRef]
30. Kim, S.W.; Barth, M.C.; Trainer, M. Influence of fair-weather cumulus clouds on isoprene chemistry. *J. Geophys. Res. Atmos.* **2012**, *117*, D10302. [CrossRef]
31. Li, Y.; Barth, M.C.; Patton, E.G.; Steiner, A.L. Impact of in-cloud aqueous processes on the chemistry and transport of biogenic volatile organic compounds. *J. Geophys. Res. Atmos.* **2017**, *122*, 11131–11153. [CrossRef]
32. Kim, S.W.; Barth, M.C.; Trainer, M. Impact of turbulent mixing on isoprene chemistry. *Geophys. Res. Lett.* **2016**, *43*, 7701–7708. [CrossRef]
33. Li, Y.; Barth, M.C.; Chen, G.; Patton, E.G.; Kim, S.W.; Wisthaler, A.; Mikoviny, T.; Fried, A.; Clark, R.; Steiner, A.L. Large-eddy simulation of biogenic VOC chemistry during the DISCOVER-AQ 2011 campaign. *J. Geophys. Res. Atmos.* **2016**, *121*, 8083–8105. [CrossRef]
34. Finnigan, J.J.; Shaw, R.H.; Patton, E.G. Turbulence structure above a vegetation canopy. *J. Fluid Mech.* **2009**, *637*, 387–424. [CrossRef]
35. Raupach, M.R.; Finnigan, J.J.; Brunet, Y. Coherent eddies and turbulence in vegetation canopies: The mixing-layer analogy. *Bound. Layer Meteorol.* **1996**, *78*, 351–382. [CrossRef]
36. Gao, W.; Shaw, R.H.; Paw, U.K.T. Observation of organized structure in turbulent flow within and above a forest canopy. *Bound. Layer Meteorol.* **1989**, *47*, 349–377. [CrossRef]
37. Gao, W.; Shaw, R.H.; Paw, U.K.T. Conditional analysis of temperature and humidity microfronts and ejection/sweep motions within and above a deciduous forest. *Bound. Layer Meteorol.* **1992**, *59*, 35–57. [CrossRef]
38. Brunet, Y. Turbulent flow in plant canopies: Historical perspective and overview. *Bound. Layer Meteorol.* **2020**, *177*, 315–364. [CrossRef]
39. Dupont, S.; Patton, E.G. Influence of stability and seasonal canopy changes on micrometeorology within and above an orchard canopy: The CHATS experiment. *Agric. For. Meteorol.* **2012**, *157*, 11–29. [CrossRef]
40. Clifton, O.E.; Patton, E.G.; Wang, S.; Barth, M.; Orlando, J.; Schwantes, R.H. Large eddy simulation for investigating coupled forest canopy and turbulence influences on atmospheric chemistry. *J. Adv. Model. Earth Syst.* **2022**, *14*, e2022MS003078. [CrossRef]
41. Barbaro, E.; de Arellano, J.V.-G.; Krol, M.C.; Holtslag, A.A.M. Impacts of aerosol shortwave radiation absorption on the dynamics of an idealized convective atmospheric boundary layer. *Bound. Layer Meteorol.* **2013**, *148*, 31–49. [CrossRef]
42. Barbaro, E.; de Arellano, J.V.G.; Ouwersloot, H.G.; Schröter, J.S.; Donovan, D.P.; Krol, M.C. Aerosols in the convective boundary layer: Shortwave radiation effects on the coupled land-atmosphere system. *J. Geophys. Res. Atmos.* **2014**, *119*, 5845–5863. [CrossRef]
43. Patton, E.G.; Davis, K.J.; Barth, M.C.; Sullivan, P.P. Decaying scalars emitted by a forest canopy: A numerical study. *Bound. Layer Meteorol.* **2001**, *100*, 91–129. [CrossRef]
44. Li, C.W.Y.; Brasseur, G.P.; Schmidt, H.; Mellado, J.P. Error induced by neglecting subgrid chemical segregation due to inefficient turbulent mixing in regional chemical-transport models in urban environments. *Atmos. Chem. Phys.* **2021**, *21*, 483–503. [CrossRef]
45. Molemaker, M.J.; de Arellano, J.V.-G. Control of chemical reactions by convective turbulence in the boundary layer. *J. Atmos. Sci.* **1998**, *55*, 568–579. [CrossRef]
46. Baker, J.; Walker, H.L.; Cai, X. A study of the dispersion and transport of reactive pollutants in and above street canyons—A large eddy simulation. *Atmos. Environ.* **2004**, *38*, 6883–6892. [CrossRef]
47. Auger, L.; Legras, B. Chemical segregation by heterogeneous emissions. *Atmos. Environ.* **2007**, *41*, 2303–2318. [CrossRef]
48. Bright, V.B.; Bloss, W.J.; Cai, X. Urban street canyons: Coupling dynamics, chemistry and within-canyon chemical processing of emissions. *Atmos. Environ.* **2013**, *68*, 127–142. [CrossRef]

49. Zhong, J.; Cai, X.-M.; Bloss, W.J. Modelling the dispersion and transport of reactive pollutants in a deep urban street canyon: Using large-eddy simulation. *Environ. Pollut.* **2015**, *200*, 42–52. [CrossRef] [PubMed]
50. Zhong, J.; Cai, X.-M.; Bloss, W.J. Large eddy simulation of reactive pollutants in a deep urban street canyon: Coupling dynamics with O_3-NO_x-VOC chemistry. *Environ. Pollut.* **2017**, *224*, 171–184. [CrossRef]
51. Wang, Y.; Ma, Y.-F.; Muñoz-Esparza, D.; Li, C.W.Y.; Barth, M.; Wang, T.; Brasseur, G.P. The impact of inhomogeneous emissions and topography on ozone photochemistry in the vicinity of Hong Kong Island. *Atmos. Chem. Phys.* **2021**, *21*, 3531–3553. [CrossRef]
52. Wang, Y.; Brasseur, G.P.; Wang, T. Segregation of atmospheric oxidants in turbulent urban environments. *Atmosphere* **2022**, *13*, 315. [CrossRef]
53. Wang, Y.; Ma, Y.F.; Muñoz-Esparza, D.; Dai, J.; Li, C.W.Y.; Lichtig, P.; Tsang, R.C.W.; Liu, C.H.; Wang, T.; Brasseur, G.P. Coupled mesoscale–microscale modeling of air quality in a polluted city using WRF-LES-Chem. *Atmos. Chem. Phys.* **2023**, *23*, 5905–5927. [CrossRef]
54. Veres, P.R.; Neuman, J.A.; Bertram, T.H.; Assaf, E.; Wolfe, G.M.; Williamson, C.J.; Weinzierl, B.; Tilmes, S.; Thompson, C.R.; Thames, A.B.; et al. Global airborne sampling reveals a previously unobserved dimethyl sulfide oxidation mechanism in the marine atmosphere. *Proc. Natl. Acad. Sci. USA* **2020**, *117*, 4505–4510. [CrossRef]
55. Boucher, O. Clouds and Aerosols. In *Climate Change 2013: The Physical Science Basis Contribution of Working Group I to the Fifth Assessment Report of the Intergovernmental Panel on Climate Change*; Cambridge Univ Press: Cambridge, UK, 2013; pp. 571–657.
56. Luo, T.; Wang, Z.; Zhang, D.; Chen, B. Marine boundary layer structure as observed by A-train satellites. *Atmos. Chem. Phys.* **2016**, *16*, 5891–5903. [CrossRef]
57. Narenpitak, P.; Kazil, J.; Yamaguchi, T.; Quinn, P.; Feingold, G. From sugar to flowers: A transition of shallow cumulus organization during ATOMIC. *J. Adv. Model. Earth Syst.* **2021**, *13*, e2021MS002619. [CrossRef]
58. Stevens, B.; Bony, S.; Brogniez, H.; Hentgen, L.; Hohenegger, C.; Kiemle, C.; L'Ecuyer, T.S.; Naumann, A.K.; Schulz, H.; Siebesma, P.A.; et al. Sugar, gravel, fish and flowers: Mesoscale cloud patterns in the trade winds. *Q. J. R. Meteorol. Soc.* **2020**, *146*, 141–152. [CrossRef]
59. Vilà-Guerau de Arellano, J.; Van den Dries, K.; Pino, D. On inferring isoprene emission surface flux from atmospheric boundary layer concentration measurements. *Atmos. Chem. Phys.* **2009**, *9*, 3629–3640. [CrossRef]
60. Kaser, L.; Patton, E.G.; Pfister, G.G.; Weinheimer, A.J.; Montzka, D.D.; Flocke, F.; Thompson, A.M.; Stauffer, R.M.; Halliday, H.S. The effect of entrainment through atmospheric boundary layer growth on observed and modeled surface ozone in the Colorado Front Range. *J. Geophys. Res. Atmos.* **2017**, *122*, 6075–6093. [CrossRef]
61. Lenschow, D.H.; Gurarie, D.; Patton, E.G. Modeling the diurnal cycle of conserved and reactive species in the convective boundary layer using SOMCRUS. *Geosci. Model Dev.* **2016**, *9*, 979–996. [CrossRef]
62. Vilà-Guerau De Arellano, J.; Duynkerke, P.G. Influence of chemistry on the flux-gradient relationships for the NO-O_3-NO_2 system. *Bound. Layer Meteorol.* **1992**, *61*, 375–387. [CrossRef]
63. Hamba, F. A modified *K* model for chemically reactive species in the planetary boundary layer. *J. Geophys. Res. Atmos.* **1993**, *98*, 5173–5182. [CrossRef]
64. Cook, A.W.; Riley, J.J. A subgrid model for equilibrium chemistry in turbulent flows. *Phys. Fluids* **1994**, *6*, 2868–2870. [CrossRef]
65. Cook, A.W.; Riley, J.J. Subgrid-scale modeling for turbulent reacting flows. *Combust. Flame* **1998**, *112*, 593–606. [CrossRef]
66. Tong, C.; Wyngaard, J.C.; Khanna, S.; Brasseur, J.G. Resolvable-and subgrid-scale measurement in the atmospheric surface layer: Technique and issues. *J. Atmos. Sci.* **1998**, *55*, 3114–3126. [CrossRef]
67. Tong, C.; Wyngaard, J.C.; Brasseur, J.G. Experimental study of the subgrid-scale stresses in the atmospheric surface layer. *J. Atmos. Sci.* **1999**, *56*, 2277–2292. [CrossRef]
68. Horst, T.W.; Kleissl, J.; Lenschow, D.H.; Meneveau, C.; Moeng, C.-H.; Parlange, M.B.; Sullivan, P.P.; Weil, J.C. HATS: Field observations to obtain spatially filtered turbulence fields from crosswind arrays of sonic anemometers in the atmospheric surface layer. *J. Atmos. Sci.* **2004**, *61*, 1566–1581. [CrossRef]
69. Sullivan, P.P.; Horst, T.W.; Lenschow, D.H.; Moeng, C.-H.; Weil, J.C. Structure of subfilter-scale fluxes in the atmospheric surface layer with application to large-eddy simulation modelling. *J. Fluid Mech.* **2003**, *482*, 101–139. [CrossRef]
70. Patton, E.G.; Horst, T.W.; Sullivan, P.P.; Lenschow, D.H.; Oncley, S.P.; Brown, W.O.; Burns, S.P.; Guenther, A.B.; Held, A.; Karl, T.; et al. The canopy horizontal array turbulence study. *Bull. Amer. Meteor. Soc.* **2011**, *92*, 593–611. [CrossRef]

atmosphere

MDPI

Article

Velocity Fluctuations Spectra in Experimental Data on Rayleigh–Taylor Mixing

Kurt C. Williams and Snezhana I. Abarzhi *

Department of Mathematics and Statistics, The University of Western Australia, Crawley, Perth, WA 6009, Australia; kurt.williams1997@gmail.com
* Correspondence: snezhana.abarzhi@gmail.com

Abstract: Rayleigh–Taylor (RT) interfacial mixing plays an important role in nature and technology, including atmospheric flows. In this work, we identify the physics properties of Rayleigh–Taylor mixing through the analysis of unprocessed experimental data. We consider the fluctuations spectra of the specific kinetic energy of each of the velocity components, and identify their spectral shapes, by employing the group theory guided foundations and the rigorous statistical method. We find the spectral shape parameters, including their mean values and relative errors, and apply the Anderson–Darling test to inspect the residuals and the goodness-of-fit. We scrupulously study the effect of the fitting window and identify, for each velocity component, the best fit interval, where the relative errors are small and the goodness of fit is excellent. We reveal that the fluctuations spectra in RT mixing experiments can be identified by a compound function, being a product of a power-law and an exponential. The data analysis results unambiguously discovered the dynamic anisotropy and the dynamic bias of RT mixing and displayed the necessity to improve the design of experiments on RT mixing.

Keywords: Rayleigh–Taylor instability; interfacial mixing; self-similarity; fluctuations spectra; anomalous scaling; goodness of fit

PACS: 47.20.Ma; 47.20.-k; 52.35.-g; 52.35.Py

Citation: Williams, K.C.; Abarzhi, S.I. Velocity Fluctuations Spectra in Experimental Data on Rayleigh–Taylor Mixing. *Atmosphere* **2023**, *14*, 1178. https://doi.org/10.3390/atmos14071178

Academic Editors: Boris Galperin, Annick Pouquet and Peter Sullivan

Received: 8 June 2023
Revised: 12 July 2023
Accepted: 18 July 2023
Published: 21 July 2023

1. Introduction

This paper was invited to contribute to *Turbulence from Earth to Planets, Stars and Galaxies, the Commemorative Issue Dedicated to the Memory of Jackson Rae Herring*, of the *Atmosphere*. Dr. Jackson Herring provided very important contributions to our understanding and modeling of realistic turbulent processes. This may include his earlier works on the growth of uncertainty in decaying isotopic turbulence [1], the decay of two-dimensional homogeneous turbulence [2], and the strain-based Lagrangian-history turbulence theory [3]. This may also include his later investigations of the Lagrangian velocity correlations in homogeneous isotropic turbulence [4] and the far-dissipation range of turbulence studies of [5]. The research interests of Dr. Jackson Herring ranged broadly, from the use and testing of closures in fluid turbulence, such as the Prandtl number dependence of Nusselt number in direct numerical simulations [6], to the issues and problems of stably stratified turbulence [7].

In as much as turbulence is considered to be the last unsolved problem of classical physics, Rayleigh–Taylor interfacial mixing are its more complex counterparts [8]. To better understand self-similar Rayleigh–Taylor mixing, bias free interpretation of data is necessary [8–11]. Our work yields physics properties and anomalous scaling of Rayleigh–Taylor mixing and identifies directions for advancements in experimental design based on analysis of experimental data [12]. We are inspired by approaches developed by colleagues [3,7,13–17].

Rayleigh–Taylor instability (RTI) is a hydrodynamic instability that develops at the interface between fluids of differing densities that are accelerated against their density gradients [18,19]. First investigated by Lord Rayleigh in 1883, RTI is an ubiquitous element of many natural and engineered processes. Of particular interest to modern science is the nature of late-stage interfacial Rayleigh–Taylor (RT) mixing. The myriad of potential factors and expected dynamics means that a bias-free, statistically sound method of data analysis is needed to appreciate the full dynamical picture of RT mixing.

In the natural world, RT mixing is a critical factor for phenomena at a huge disparity of scales, from microscopic to stellar scales. It is responsible for the fingering of interstellar media in the matter pulled along the edges of black holes, and the accelerated ejection matter in both core-collapse supernovae and solar flares. On the Earth, RT mixing is observed in polar deep ocean convection currents, undersea volcano formation, the dispersion of pollutants in the atmosphere, and in coastal boundary up-welling systems in the ocean. In the industrial world the processes of, nano-fabrication, fossil fuel extraction, plasma fusion and premixed combustion all exhibit interfacial RT mixing [8,9,11,17,20–23].

The late-stage of RT dynamics, interfacial RT mixing, is a continuing subject of active research in contemporary science, mathematics, and engineering. RT mixing is uniquely challenging to understand, as it exhibits incredibly complex behaviors and coupling of scales. It is purported that interfacial RT mixing is self-similar, with the length-scale in the acceleration direction growing quadratic with time. However, RT mixing remains a phenomenon of incredibly complex, rich dynamics, evading accurate capture by simplistic models [9,24].

In this work, we employ a rigorous data analysis method, guided by group theory, to analyze the spectral behavior of an RT mixing experiment [12]. We consider the fluctuations of the specific kinetic energy of each of three velocity components. We find that the experimental data revealed properties of RT mixing in good agreement with group theory results. The dynamics observed in experiments exhibit dynamic anisotropy and the dynamic bias of self-similar RT mixing [12], and reveal the necessity of improving the design of experiments on RT mixing [8,10].

The need for statistically rigorous data analysis is borne out of the extreme challenge that RTI, and, in particular, RT mixing, presents to theory, simulations and experiments. Experiments must contend with the vast array of scales that are significant to the dynamics, and must tightly control the implementation of the flow, since even microscopic deviations from the idealized case may rapidly grow into structures and deformations at later times. Furthermore, the need for both highly precise and non-invasive diagnostics further restricts the efficacy of experiments, upon which, for example, numerical simulations are highly dependent for low-level subgrid assumptions [8,9,11,24].

To combat these difficulties, and advance understanding of RT mixing, rigorous data analysis is required [25,25]. Substantial success has recently been achieved, with the data analysis method employed in this paper characterizing the fluctuations spectra of specific kinetic energy, density and mass flux in the direction of the acceleration for an RTI experiment, which has evaded theory analysis for some time due to the inhomogeneity in the fluid densities [26–28]. Careful data analysis found excellent agreement with the group theory predictions [28]. It was also consistent with the group theory discovery that RT mixing can keep order and, in fact, retain memory of deterministic conditions, even at late times [9,24,28,29].

For isotropic homogeneous turbulence, accurate quantification of fluctuations spectra in experiments and observations led to the discovery of anomalous scaling. Anomalous scaling is the generic name for departures of the fluctuations spectra in realistic turbulent processes from the canonical Kolmogorov's scaling laws [13,17,30,31]. Such anomalies are also observed in turbulent boundary layers, passive scalar mixing, buoyancy-driven turbulence, turbulent convection, and compressible turbulence. The classical works find that, in realistic turbulent processes, the spectral shapes of the fluctuations can be more complex than Kolmogorov's spectra [13,15,16,32,33]. As a function of the mode number, the spectral shape can have a distinct power-law dependence, can be described by a product of a power-law and an exponential, and can also be expressed by a stretched exponential function. Models, including anomalous scaling, accurately capture the properties of fluctuations in experiments and observations. They assert that Kolmogorov turbulence, as an ideal stochastic process, can be an extreme challenge to implement in realistic environments [13,17,25,34,35].

In this work, we investigate data on fluctuations in self-similar RT mixing in the experiments [12,25]. The data are obtained through hot-three-wire anemometry probing. We analyze the fluctuations of the specific kinetic energy of the three velocity components. We employ and further develop the method [25] to identify, with statistical confidence, the spectral shapes of the fluctuations of each of the following three quantities: the mean values of the spectral shape's parameters, their relative errors, and the goodness of fit. To inspect the residuals, which should be χ_2^2 distributed, the Anderson–Darling test is employed [36]. We find the domains of the mode numbers, where the relative errors of the spectral shape parameters are small and the goodness of fit is excellent. The data analysis unambiguously identifies the dynamic anisotropy, and the dynamic bias, of RT mixing in the experiments [12]. Group theory is employed to explain these properties quantitatively and qualitatively, and to link them to other experiments on RT mixing and realistic turbulent processes [9,16,17,24,32,33,37–39]. Our results demonstrate the need for improvements in experimental design for studies in RT mixing, and may help chart directions for future research [8,10,11,24,40,41].

2. Methodology and Foundations

2.1. Theory

2.1.1. Group Theory Methodology

Rayleigh–Taylor dynamics is governed by equations in the bulk, a boundary value problem and an initial value problem. The equations in the bulk of each fluid are the laws of conservation of mass, momentum and energy. They are nonlinear partial differential equations in three-dimensional space and time, and are the Euler equations in the case of ideal fluids. The equation of state closes the system of equations. The singular boundary value problem consists of the boundary conditions at the nonlinear freely evolving unstable fluid interface. These conditions are represented by a sub-set of non-linear partial differential equations. For ideal fluids, the mass flux, the normal component of velocity and the pressure are continuous at the interface, whereas the tangential component of velocity and the specific enthalpy may have a discontinuity. The boundary value problem is also augmented with the boundary conditions at the outside boundaries of the domain. The ill-posed initial value problem is represented by initial perturbations of the interfaces and the flow fields [9,26,42].

The theoretical problem of Rayleigh–Taylor dynamics is intellectually challenging. Success in solving the problem is achieved with the group theory approach, including accurate analytical solutions for the scale-dependent early-time and late-time nonlinear dynamics, as well as for the self-similar Rayleigh–Taylor mixing in a broad range of conditions. Particularly, the dynamics of Rayleigh–Taylor mixing can be captured within the framework of the momentum model that has the same symmetries and scaling transformations as, and is directly linked to, the governing equation [9,24,26,29,43–46].

Particularly, RT dynamics is driven by a specific, per unit mass, balance of the rate of momentum gain $\tilde{\mu}$ and the rate of momentum loss μ. The gain and loss of specific momentum are associated with the gain $\tilde{\epsilon}$ and dissipation ϵ of the specific energy as $\tilde{\mu} = \tilde{\epsilon}/v$ and $\mu = \epsilon/v$, where v is the velocity of the parcel of fluid in the acceleration direction. For constant acceleration with the magnitude g, the rate of momentum loss is $\mu = Cv^2/L$ and the rate of energy dissipation is $\epsilon = Cv^3/L$, where C is the drag value and L is the length scale. The rates of gain of momentum and energy are $\tilde{\mu} \sim g$ and $\tilde{\epsilon} \sim gv$. In the regime of self-similar mixing, the length scale is $L \sim |h|$, where h is the amplitude in the acceleration direction increasing quadratic with time $|h| \sim gt^2$. The rates of loss and gain of momentum relate as $\mu \sim \tilde{\mu} \sim g$, whereas the rates of energy dissipation and gain are time-dependent with $\epsilon \sim \tilde{\epsilon} \sim g^2 t$ [9,24,26,29,43–46].

2.1.2. Scaling Laws and Sensitivity to Deterministic Conditions

In RT mixing, the invariant form of the scaling transformation is the rate of loss of specific momentum $\mu \sim v^2/L \sim v_l^2/l$, where $v(v_l)$ is the velocity scale at large (small) length scale $L(l)$. This leads to the velocity scaling $v_l/v \sim (l/L)^{1/2}$, the Reynolds number $Re \sim g^2 t^3/\nu$, the viscous scale $l_\nu \sim (\nu^2/\mu)^{1/3}$ and the span of scales $L/l_\nu \sim Re^{2/3}$, where $L \sim gt^2$ and ν is the kinematic viscosity.

These properties depart from those of Kolmogorov turbulence [9,29,34,35,42,43]. Furthermore, in contrast to stochastic canonical turbulence, Rayleigh–Taylor mixing is sensitive to deterministic (initial and flow) conditions [9,24,26,29,42].

To evaluate the sensitivity of fluctuations induced by the RT mixing process to those set by the deterministic conditions, we consider two parcels of fluid involved in the flow with a time delay $\tilde{\tau}$ one after another. The invariance of the rate of momentum loss $\mu \sim v^2/L$ leads to ballistic dynamics, $v \sim \mu t$. The process-induced velocity fluctuations, $\sim\mu\tilde{\tau}$, are comparable to the fluctuations set deterministically by the parcels' relative velocities, $\sim(\tilde{\mu} - \mu)\tilde{\tau}$. The ratio of the velocity fluctuations and the mean velocity decays with time, $\sim(\tilde{\tau}/t)$. This suggests that RT mixing can re-laminarize, akin to accelerated turbulent flows [15,24,29,37].

2.1.3. Fluctuations Spectra

In Rayleigh–Taylor mixing, [26,27,29] the invariance of the rate of momentum loss leads the fluctuations of the specific kinetic energy v^2 and the spectral density $E(k)$

$$v^2 \sim \int E(k)dk \sim \int \mu k^{-2} dk, \quad E(k) \sim \mu k^{-2}. \tag{1}$$

Since RT dynamics is anisotropic, the dynamics in the direction of the acceleration can differ from those in the normal plane. Particularly, while in the mixing regime, the vertical length scale can change self-similarly, $h \sim gt^2$, and the horizontal scale can remain nearly invariable, $\lambda \sim const$. In addition to the invariant quantity μ, the fluctuations of the specific kinetic energy corresponding to the velocity components $v_\perp^2$ in the direction normal to the acceleration can be driven by the velocity scale $\sqrt{g\lambda}$. This leads to

$$v_\perp^2 \sim \int E(k)dk \sim \int (g\lambda) k^{-1} dk, \quad E(k) \sim g\lambda k^{-1}. \tag{2}$$

In the experiments, [12], due to sensitivity of RT dynamics to deterministic conditions, the fluctuations of the specific kinetic energy corresponding to the velocity components $v_\perp^2$ in the directions normal to the acceleration can also be driven by the velocity scale $\hat{U}$. Here, $\hat{U} \sim o(U)$, with U being the velocity of the co-flowing streams. The scale $\hat{U}$ is set by the accuracy of the velocity U, being the deterministic noise; which is within a few percent in the experiments [12]. This provides:

$$v_\perp^2 \sim \int E(k)dk \sim \int \hat{U}^2 k^{-1} dk, \quad E(k) \sim \hat{U}^2 k^{-1}. \tag{3}$$

Finally, in the experiments [12], the fluctuations of the specific kinetic energy corresponding to the velocity components $v_\perp^2$ can also be driven by the interplay of the velocity scale $\hat{U}$ and the rate of momentum loss $\mu \sim g$, as

$$v_\perp^2 \sim \int E(k)dk \sim \int \hat{U}\sqrt{\mu} k^{-3/2} dk, \quad E(k) \sim \hat{U}\sqrt{\mu} k^{-3/2}. \tag{4}$$

The spectral density in Equation (4) is the geometric mean of the spectral densities of the kinetic energy velocity components, along with, and normal to, the acceleration in Equations (2) and (3), with $\tilde{U}\sqrt{\tilde{\mu}}k^{-3/2} = \sqrt{(\mu k^{-2})(\hat{U}^2 k^{-1})}$.

2.1.4. Spectral Shapes in Experiments

Quantification of power-laws is a key objective of experimental diagnostics of canonical turbulence and RT mixing. Reliable identification of an exponent and a prefactor of a power-law requires a substantial span of resolved scales; it is a challenge to achieve this in experiments and simulations [10,15,26,41,42,45,46].

In this work, in order to analyze unprocessed raw data from the hot-three-wire anemometry measurements of Rayleigh–Taylor mixing experiments [12,25], and to identify the data-based properties of RT mixing, we apply a compound function represented by a product of a power-law and an exponential

$$E(k) \sim k^{\alpha} \exp(\beta k) \tag{5}$$

The compound function $E(k) \sim k^{\alpha} \exp(\beta k)$ behaves as a power-law for scales $k \ll k_{\nu}$ and as an exponential for scales $\sim k_{\nu}$, where k_{ν} is the wave-vector corresponding to the viscous scale, with $k_{\nu} \sim 1/l_{\nu}$. The signs of spectral shape parameters are $\alpha, \beta < 0$, because, in RT mixing with constant acceleration, larger velocities correspond to larger length scales. The use of the compound function in RT spectra is discussed in detail in our earlier works [25,25,28]. The compound function $E(k) \sim k^{\alpha} \exp(\beta k)$ is applied in turbulence to describe realistic spectra in experiments and observations [9,13,14,16,17,25,30–32].

2.2. Outline of Experiments

In this work, we analyze data in the experiments [12] to further understand the properties of RT mixing. Our earlier works studied the fluctuations of the cross-stream velocity component, the density and the mass flux [25,28]. Here we analyze the fluctuations of each of three velocity components. We consider data from the pure Rayleigh–Taylor setup A1S0 and we focus on data taken at very late times to ensure that the RT flow reaches the self-similar mixing [12]. The Reynolds number of the mixing flow is estimated to be $\sim 3.4 \times 10^4$ [12].

The details of the experiments, the experimental setups, the diagnostics and the data can be found in the works [12,25,25,28].

2.2.1. Experimental Setup

For the purpose of completeness and the reader's convenience, Figure 1 illustrates the schematics of the experiments [12]. The flow is three-dimensional, with the (x, y, z) spatial directions and with the (u, v, w) corresponding velocity components. In the pure RT setup A1S0, the velocities of the co-flowing streams equal one another, $U = U_1 = U_2$, within a few percent accuracy [12].

In Figure 1, as in the experiments [12], the x-direction is the direction of the co-flowing streams. In the x-direction, the velocity component is u. We call it the stream-wise velocity. The y-direction is the direction of the acceleration, which is the Earth's gravity, and is directed from the heavy (top) to the light (bottom) fluid. In the y-direction, the velocity component is v. We call it cross-stream velocity. The z-direction is normal to the directions of the acceleration and the co-flowing streams. In the z-direction, the velocity component is w. We call it cross-tank velocity [28].

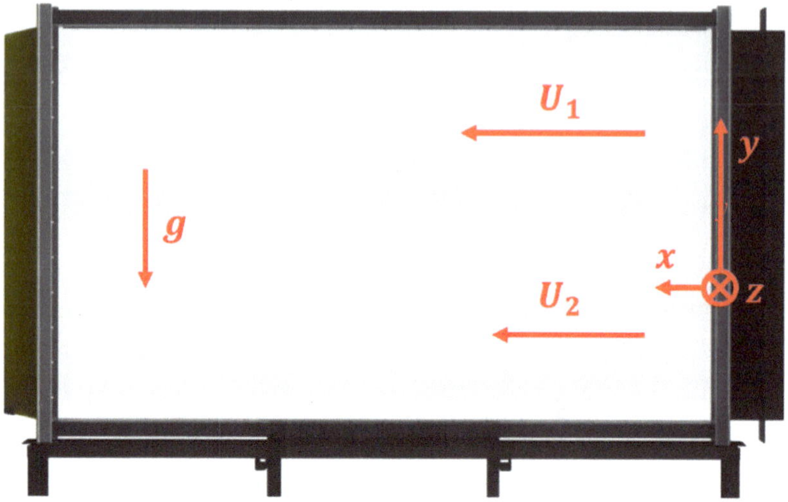

Figure 1. Configuration of the experimental tank. The gas inflow is stream-wise (in the x-direction), with the gravity driving the flow against the cross-stream direction (y-direction) and with both the inflow and gravity being incident to the cross-tank (z) direction.

2.2.2. Characteristic Scales in the Experiments

Based on the experimental parameters in the setup A1S0 [12], we evaluated the wave-vector $k_\nu = (g/\nu^2)^{1/3}$ as $k_\nu = (1.13 - 1.08) \times 10^3$ $[\mathrm{m}^{-1}]$, and the viscous length scale $l_\nu = 2\pi/k_\nu$ as $l_\nu = (5.58 - 5.84) \times 10^{-3}$ [m]. The scales k_ν and l_ν were comparable to those of the mode of fastest growth in RTI.

We evaluated the scales which were set by the deterministic noise; particularly, by the 1% accuracy in the magnitude of the velocity of the co-flowing streams $\hat{U} = 0.01$ $U = 6.30 \times 10^{-3}$ [m/s], where $U = U_1 = U_2 = 6.3 \times 10^{-1}$ [m/s]. This yielded the wave-vector $\hat{k}_\nu = \hat{U}/\nu = (4.06 - 3.79) \times 10^2$ $[\mathrm{m}^{-1}]$ and the corresponding length scale $\hat{\lambda}_\nu = 2\pi/\hat{k}_\nu = (1.55 - 1.65) \times 10^{-2}$ [m].

We evaluated the scales which were set by the velocity of the co-flowing streams with magnitude $U = U_1 = U_2 = 6.3 \times 10^{-1}$ [m/s]. This included the wave-vector $\check{k}_\nu = U/\nu = (4.06 - 3.79) \times 10^4$ $[\mathrm{m}^{-1}]$, the corresponding length scale $\check{\lambda}_\nu = 2\pi/\check{k}_\nu = (1.55 - 1.66) \times 10^{-4}$ [m], and the corresponding time scale $\check{t}_\nu = \nu/U^2 = (3.91 - 4.18) \times 10^{-5}$ [s].

The largest horizontal length scale and vertical length scale corresponded to $L = 31.5$ [m] and $H = 1.2$ [m], respectively. In this data set, the total sampling time was $T = 5 \times 10^1$ [s]. The horizontal length-scale L and the total sampling time T related as $L = UT$. The scale H was the tunnel width and it was used to scale the lengths [12].

For further details, including the resolution and the level of instrumental noise, the reader is referred to the works [25,25,28].

2.3. Method of Data Analysis

The data analysis method we employed was successfully applied in [28], predicated on an earlier method in [25,25].

The method is derived from the statistically sound Whittle Maximum Likelihood Estimation (MLE) technique but includes an Anderson–Darling test applying a newly calculated test statistics. The method finds estimations of the model parameters, verifies the goodness of fit and probes the effect of the range of wave-vectors on the estimation values and statistical uncertainty in the fit parameters. By interrogating the effect of the wave-vector range, it is possible to identify the most-likely parameters, the uncertainty in the estimation, and the wave-vector range the model is valid for.

2.3.1. Spectrum Fitting Method

The data analysis methodology is explained in detail in [28], as well as in [25,25]. For the convenience of the reader, herein we provide a brief overview of the spectrum fitting method.

The analyzed data constitutes a time-series of measurements from the anemometer. To analyze the spectral properties of the fluctuations, we constructed a Fast Fourier transform (FFT) of the fluctuations. Assuming the real and imaginary components of the FFT represent a pair of identical, independent, normally distributed random values, the residuals should be chi-square distributed. MLE was employed to the fit parameters for a given fitting window [47]. To verify the statistical validity of the fit, a goodness-of-fit score p_{AD} was computed [25,36,48–50]. The confidence intervals were computed by a Monte-Carlo method from a previous work [28].

The process allows for rapid computation of the fit parameters yielding the lowest uncertainties and their associated goodness-of-fit score.

2.3.2. Effect of the Fitting Interval

We approximated the data with a compound function $S(k) \sim k^{\alpha} \exp(\beta k)$. The exponential component of this function $\exp(\beta k)$ is scale-dependent, and defines a natural length scale $\sim |\beta|^{-1}$. The power-law component k^{α} is scale-invariant, and a substantial span of k values is required for confident parameter fitting and accurate fit evaluation. In the realistic experimental data considered in our work, the range of k values is relatively short, and the choice of the fit interval may influence the compound function parameters α and β. In order to identify the best fitting interval, to ensure the accuracy of our results and to preempt misinterpretations, we scrupulously investigated the dependence of the compound function parameters α and β on the the left cut-off k_l and the right cut-off k_r of the interval $k \in [k_l, k_r]$ over which the fit was performed [25,28]. Such effects are known to exist in realistic turbulent processes [14,16,17,32].

3. Data Analysis Results

In this section, we employed the spectrum fitting method to scrupulously analyze the RT mixing data for fluctuations of the three velocity components in the experiments [12]. We investigated the properties of the fluctuations of the specific kinetic energy for the stream-wise u^2 and cross-tank w^2 velocity components, and we outlined those for the cross-stream velocity component v^2 studied in detail earlier [28]. We aimed to understand and answer the following questions: Can RT mixing exhibit dynamic anisotropy at very late times?; Can RT mixing be similar to isotropic and homogeneous canonical turbulence?; What are the properties of RT mixing in the environment of the experiments [12]? These three questions motivated and framed our study.

3.1. *Stream-Wise Velocity*

3.1.1. Spectral Properties of the Data

We first analyzed fluctuations of the kinetic energy of the stream-wise velocity component u^2 in the direction of the co-flowing streams of the heavy and the light fluids. In this case, in the interval $k \in [110, 3700]$, an MLE for compound spectra $S(k) \sim \exp(\alpha \ln(k) + \beta k)$ provided fit parameters $\alpha = -1.06 \pm 0.06, \beta = -(2.14 \pm 0.05) \times 10^{-3}$. These results are illustrated by Figure 2. The fitted periodogram found excellent agreement in $k \in [110, 3700]$ (shaded in magenta), whereas, on the right, a noise level of $S_{noise} = 10^{-7}$ accurately captured the large wave-number ($k > 3000$) data.

For the stream-wise velocity component u^2 fluctuations spectra, Figure 3 presents the dependence of the parameters α and β of the compound function $S(k) \sim k^{\alpha} \exp(\beta k)$ on the left and right cut-offs $k_{l(r)}$, and on the fitting window, $k \in [k_l, k_r]$. The Figure illustrates how α and β varied for all the data (Figure 3a), and for the data in the fitting window with the strictest values of relative errors and with high goodness-of-fit (Figure 3b).

Figure 4 further illustrates the measures of the uncertainty in the estimations of the compound function parameters and the goodness-of-fit, including the dependence of the relative errors σ_α/α (Figure 4a) and σ_β/β (Figure 4c) and the Anderson–Darling test score p_{AD} (Figure 4b) on the left and the right cut-offs $k_{l(r)}$ for all the data considered.

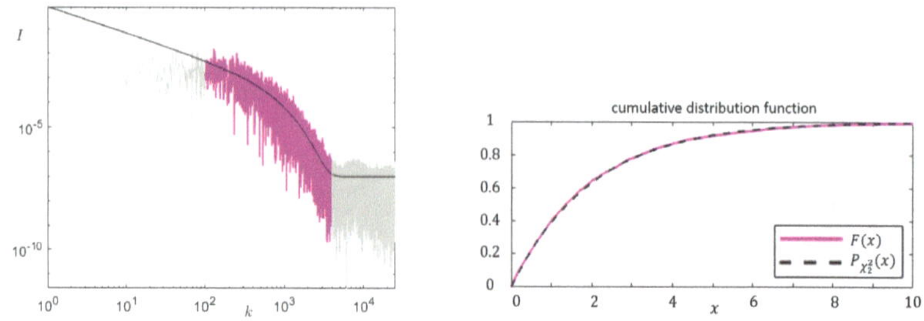

(**a**) Periodogram (**b**) Cumulative Distribution Function

Figure 2. (**a**) The estimated fluctuation spectra $S(k)$ (black line) found tight agreement with the periodogram of the streamwise (u^2) velocity fluctuation spectra in the region considered (magenta). (**b**) The Cumulative Distribution Function of the residuals closely followed that of the χ^2_2 distribution.

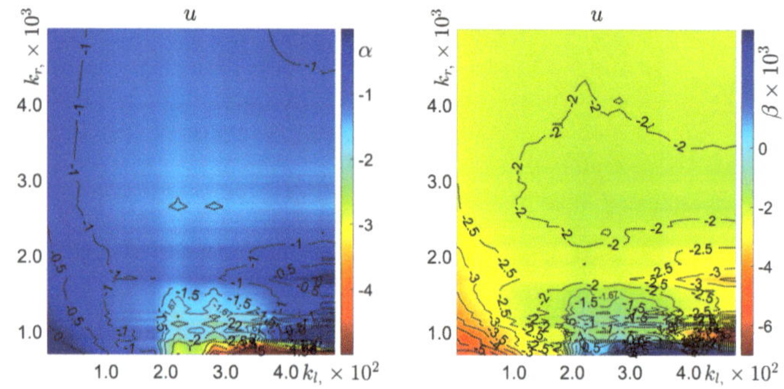

(**a**) No restrictions

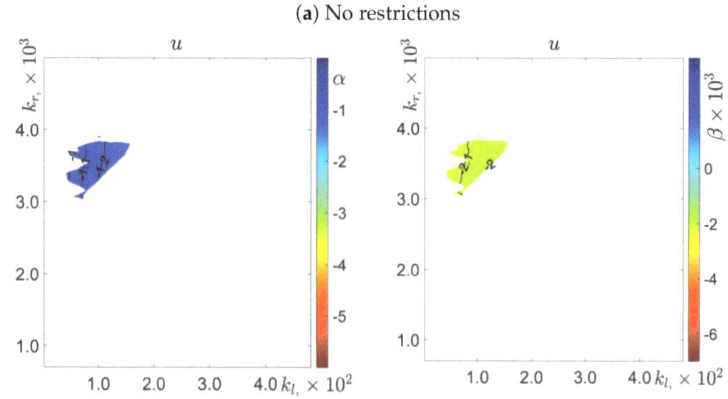

(**b**) $\sigma_\alpha/\alpha < 7\%, \sigma_\beta/\beta < 3\%, p_{AD} > 0.5$

Figure 3. Variations in the estimation of α (**left**) and β (**right**) as a function of the fit window for the stream-wise velocity u^2 fluctuations.

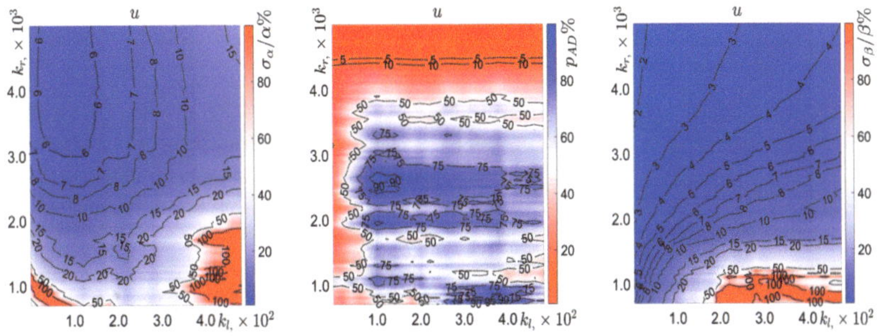

(**a**) $|\sigma_\alpha/\alpha\%|$ (**b**) $p_{AD}\%$ (**c**) $|\sigma_\beta/\beta\%|$

Figure 4. Measures of uncertainty in obtaining α (**a**), the uncertainty in the estimation of β (**c**) and the goodness of fit score p_{AD} (**b**) for stream-wise velocity u^2 fluctuations.

3.1.2. Analysis of Residuals and Goodness of Fit

Applying the Anderson–Darling test between the empirical Cumulative Distribution Function (CDF) of the observed residuals Y_k and a χ_2^2 distribution yielded a goodness of fit value of $p_{AD} = 0.5073 \gg 0.05$, suggesting that the observed residuals were χ_2^2 distributed, and, thus, the fit was statistically valid. The uncertainties in the fit parameters were small with $\sigma_\alpha/\alpha = 5.52\%$ and $\sigma_\beta/\beta = 2.43\%$, and the spectral fit was well contained by the data. The certainty in the fit was further justified in the near-perfect overlap between the empirical CDF of the residuals Y_k and the χ_2^2 distribution. This result is illustrated in Figure 2b.

3.1.3. Effect of the Left and Right Cut-Off

The α values for the stream-wise velocity estimations ranged from $\alpha = -0.5$ to $\alpha = -2.5$ over the majority of the fit window, with very small outlier regions, in which $\alpha > -0.5$ and $\alpha < -5$. For the vast majority of the fitting window, $-1.5 < \alpha < -1$, with divergences for the left cut-off, $k_l < 100$ (for which $\alpha > -1$), right cut-off, $k_r < 200$ (with α varying from as high as $\alpha > 0$ to as low as $\alpha < -5$) and very large fit windows $k_l > 400, k_r > 4500$ (with $\alpha > -1$).

The estimations of β were somewhat more varied, ranging from $\beta > -2 \times 10^{-3}$ down to $\beta < -5 \times 10^{-3}$. The estimated values were relatively stable for right cut-offs $k_r > 2000$, with the estimated β values being $-2.5 \times 10^{-3} < \beta < -2 \times 10^{-3}$ in this region, except for a central window region with $k_l > 100$ and $2500 < k_r < 3500$), in which $-2 \times 10^{-3} < \beta \ll -1.75 \times 10^{-3}$. For right cut-offs below the region with $k_r < 2000$ estimates, the values of β varied wildly, being highly affected by the variations over the source and integral parts of the spectrum, and with estimations ranging from $\beta < -6 \times 10^{-3}$ up to $\beta > 4 \times 10^{-3}$.

The uncertainty in estimating α was inverse to the width of the fit window, with the lowest uncertainty $\sigma_\alpha/\alpha < 6\%$ being achieved for $20 < k_l < 150$ and $k_r > 3000$. The uncertainty was relatively symmetric about $k_l = 80$, with the uncertainty increasing as the left cut-off deviated from $k_l = 80$ and as the right cut-off k_r decreased. There were two small regions for which the estimation for α should be discounted: the corner defined by $k_r < 1000, k_l < 200$ and the corner defined by $k_r < 2000, k_l > 350$.

The uncertainty of β was similarly inverse to the fit window, with the minimum uncertainty $\sigma_\beta/\beta < 2\%$ being achieved in the region $k_l < 20, k_r > 3500$. The dependence on the left cut-off k_l was stronger than on the right cut-off k_r for $k_r < 2000$ and for $k_l < 200$. There was a small region, $k_l > 200, k_r < 1250$, for which constructing a fit was impossible, with uncertainties starting at $\sigma_\beta/\beta > 50\%$ and ranging to $\sigma_\beta/\beta \sim 100\%$.

Overall, the goodness of fit value in the estimating the fit parameters was above the threshold for rejection. The resultant range is presented in Figure 3b.

3.2. *Cross-Tank Velocities*

3.2.1. Spectral Properties of the Data

We now consider the fluctuations spectra of the kinetic energy of the cross-tank velocity component w^2 in the direction normal to the acceleration and to the direction of the co-flowing streams. In this case, for a broad interval $k \in [150, 3700]$, the MLE fit of the data with a compound function $S(k) \sim \exp(\alpha \ln(k) + \beta k)$ with $\alpha = -1.49 \pm 0.07$ and $\beta = (1.30 \pm 0.05) \times 10^{-3}$ and with $p_{AD} = 0.6382 \gg 0.05$, which implied excellent goodness of fit and confidence in the fit parameters. The estimated noise $S_{noise} \sim 10^{-7}$ captured the behavior of the right of the periodogram ($k > 4000$). These results are illustrated in Figure 5.

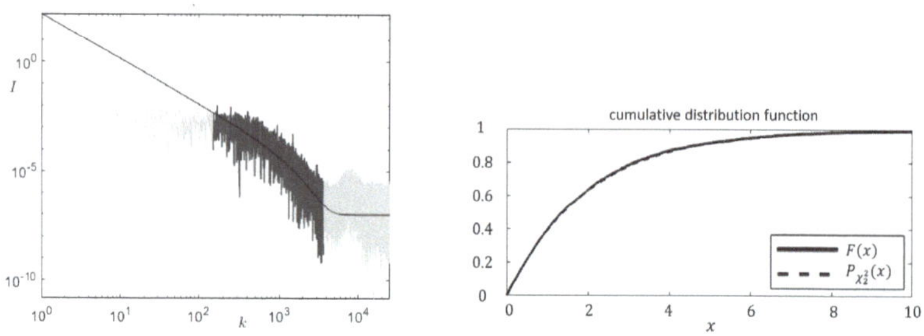

(**a**) Periodogram (**b**) Cumulative Distribution Function

Figure 5. (**a**) The estimated fluctuation spectra $S(k)$ (black line) found tight agreement with the periodogram of the cross-tank velocity (w^2) fluctuation spectra in the region considered (gray). (**b**) The Cumulative Distribution Function of the residuals closely followed that of the χ_2^2 distribution.

For the cross-tank velocity w^2 fluctuations spectra, Figure 6 presents the dependence of the parameters α and β of the compound function $S(k) \sim k^{\alpha} \exp(\beta k)$ on the left and right cut-offs $k_{l(r)}$ and on the fitting window, $k \in [k_l, k_r]$. The Figure illustrates how α and β varied for all the data (Figure 6a); and for the data in the fitting window with the strictest values of relative errors and with high goodness-of-fit (Figure 6b).

Figure 7 further illustrates the measures of the uncertainty in the estimations of the compound function parameters and the goodness-of-fit, including the dependence of the relative errors σ_α / α (Figure 7a) and σ_β / β (Figure 7c) and the Anderson–Darling test score p_{AD} (Figure 7b) on the left and the right cut-offs $k_{l(r)}$ for all the data considered.

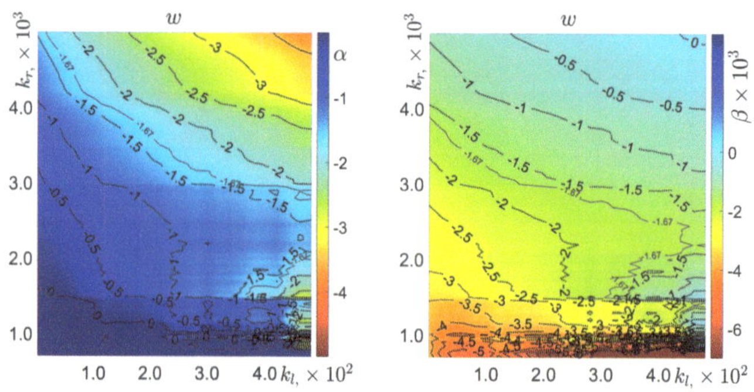

(**a**) No restrictions

Figure 6. *Cont.*

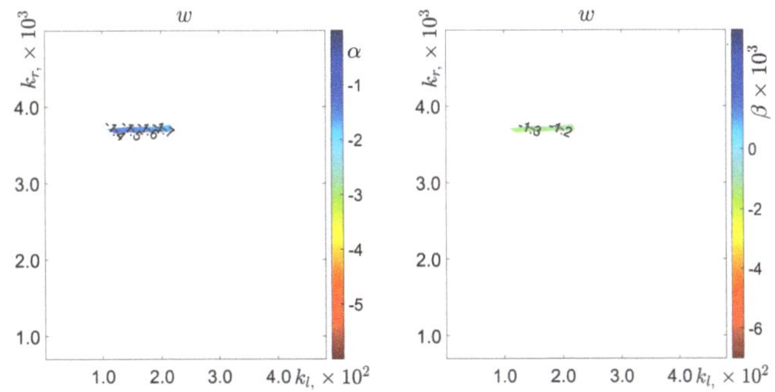

(**b**) $\sigma_\alpha/\alpha < 7\%, \sigma_\beta/\beta < 3\%, p_{AD} > 0.5$

Figure 6. Variations in the estimation of α (**left**) and β (**right**) as a function of the fit window for the cross-tank velocity w^2 fluctuations.

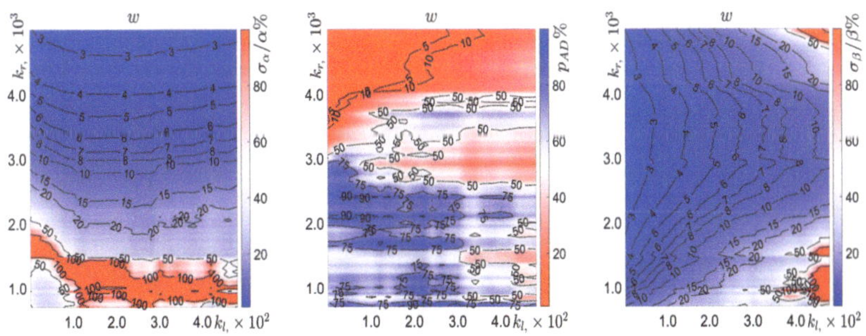

(**a**) $|\sigma_\alpha/\alpha\%|$ (**b**) $p_{AD}\%$ (**c**) $|\sigma_\beta/\beta\%|$

Figure 7. Measures of uncertainty in obtaining α (**a**), the uncertainty in the estimation of β (**c**) and the goodness of fit score p_{AD} (**b**) for the cross-tank velocity w^2 fluctuations.

3.2.2. Analysis of Residuals and Goodness of Fit

We applied the resultant fitted function to determine the observed residuals Y_k, and tested the null hypothesis that the residuals were χ_2^2 distributed. Testing the hypothesis between the empirical CDF of the residuals and the CDF of the χ_2^2 distribution yielded a goodness of fit value of $p_{AD} = 0.5382 \gg 0.05$. Hence, the null hypothesis could not be rejected. This confidence in the fit was corroborated by the tight visual agreement between the empirical Cumulative Distribution Function (CDF) of the residuals and the CDF of the χ_2^2 distribution. This is illustrated by Figure 5b. In addition, the fitted function precisely captured the behavior of the periodogram and the uncertainty in the fit parameters was small, with $\sigma_\alpha/\alpha = 2.96\%$ and $\sigma_\beta/\beta = 3.98\%$. This is illustrated by Figure 5a. The resultant fitted function was, thus, an accurate prescription of RT mixing dynamics.

3.2.3. Effect of the Left and Right Cut-Offs

The distribution of estimated α values varied from $\alpha < -3$ to $\alpha > 0$. The estimation of α decreased as the right cut-off k_r increased, and increased as the left cut-off k_l increased, with the dependence on the left cut-off k_l being more significant than the dependence on the right cut-off k_r for $k_l < 350, k_r > 1500$ and for $k_r > 4000$. For the small region $k_r < 1500$ and for $k_r > 4000$. For the small region $k_r < 1500, k_l > 450$, the estimations for α varied greatly, due to the high sensitivity of the dynamics to the integral and inertial ranges. For small left and right cut-offs $k_l < 400, k_r < 1200$, corresponding to only the integral and inertial ranges, the estimation provided $\alpha \sim 0$, over which range the velocity fluctuations appeared almost entirely scale-independent.

The distribution of estimated β values ranged from $\beta \sim 0$ to $\beta < -6 \times 10^{-3}$, but only varied from $\beta \sim 0$ to $\beta \sim -3 \times 10^{-3}$ for $k_r > 1500$. For $k_l > 200$ (and $k_r > 1500$), the dependence on the left cut-off was significantly greater than on the right cut-off. In a similar manner to the estimations for α, the fit yielded greatly fluctuating parameter estimations in the region $k_r < 1500, k_l > 450$, even more so than in the rest of the $k_r < 1500$ range. In the regions wherein α was estimated as $\alpha \sim 0$, the MLE yielded $\beta \sim -3 \times 10^{-3}$, in agreement with existing literature.

The uncertainty in the estimation of α was nearly independent of the left cut-off k_l, except in the regions $k_l < 100$ ($k_l > 40$), wherein the uncertainty in α decreased (increased) with increasing k_l. The uncertainty in α generally decreased with increasing k_r, in correspondence with the larger contribution of the near-dissipation and dissipative regimes. For $k_r < 1500$, $k_l > 100$ and for $1500 < k_r < 2000$ and $k_l < 100$, the uncertainty in estimating α was $\sigma_\alpha/\alpha \sim 100\%$, and was excluded from consideration.

The uncertainty in estimating β had a more complex dependence on the left and right cut-offs than the uncertainty in α. For the fixed left cut-off, k_l, the uncertainty in estimating β was minimized in the region $3000 < k_r < 3500$. For the right cut-offs k_r greater than (less than) the bounds of this region, the uncertainty was increasing (decreasing) in k_r. For the fixed right cut-off k_r, the uncertainty in estimating β was increasing for increasing left cut-off. Overall, the uncertainty gave the appearance of a minimum uncertainty at $k_l = 0, k_r = 3250$, outward from which the uncertainty increased with distance from this point, with a stronger dependence on the left cut-off k_l than the right cutoff k_r. There existed regions, with $k_l > 400$ and $k_r > 4750$, and with $k_l > 400$ and $k_r > 1500$ m in which the fit became untenable, with uncertainties of $\sigma_\beta/\beta \sim 100\%$.

The goodness of fit score, as evaluated by the Anderson–Darling test, was much less regular than for the other parameters and varied a great deal in nearly all regions. For the region bounded by the line segment between $k_l = 0, k_r = 3500$ and $k_l = 280, k_r = 5000$, the goodness of fit value was $p_{AD} < 0.05 < 0.10$ and necessarily rejected. For $k_l > 100, k_r > 3600$ and $k_l < 100, k_r > 3000$, the goodness of fit $p_{AD} < 0.5$. There was a large region of $p_{AD} < 0.5$ for $k_l > 170, 1600 < k_r < 3400$ and several small patches of uncertain fits ($p_{AD} < 0.5$) in the domain $k_l > 300, k_r < 2000$. For the rest of the domain, the goodness of fit varied from $0.5 < p_{AD} < 0.95$, with various, seemingly uncorrelated, regions with $p_{AD} > 0.75$; commenting on these is, statistically, meaningless.

3.3. Cross-Stream Velocity

The properties of fluctuations of the kinetic energy v^2 of the cross-stream velocity component v in the direction of the acceleration g were scrupulously analyzed in our earlier work [28]. We briefly outline the obtained results. In this case, for the compound function $S(k) \sim k^\alpha \exp(\beta k)$, the MLE yielded the fit parameters $\alpha = -2.04 \pm 0.06$ and $\beta = (1.02 \pm 0.05) \times 10^{-3}$ in the fitting interval $k \in [100, 4000]$. The fit had a goodness of fit score $p_{AD} = 0.5699 \gg 0.05$, confirming the validity of the fit. The presumed noise level $S_{noise} = 10^{-7}$ excellently captured the behavior of the data for $k > 4000$. The fit was in excellent agreement with the periodogram of the data for the wave numbers considered $k \in [100, 4000]$.

4. RT Mixing Characteristics in the Experiments

In this section, we identify the physics features of RT mixing in the experiments [12], from the results of our scrupulous data analysis guided by the group theory approach.

4.1. Flow Characteristics

To obtain the quantitative characteristics of RT mixing, we summarize our data analysis results in Table 1. Table 1 presents the characteristic quantities for each of the fluctuation spectra diagnosed in the experiments; namely, these are the estimated values of the parameters α and β of the compound function $S(k) \sim k^{\alpha} \exp(\beta k)$, the left k_l and the right k_r cut-offs of the fitting interval $k \in [k_l, k_r]$, and the span of scales $lg(k_r/k_l)$, with $lg = log_{10}$.

We further evaluated the characteristic scales set by the exponential decay factor $1/|\beta|$ of the compound function $S(k) \sim k^{\alpha} \exp(\beta k)$. Table 2 presents the value $1/|\beta|$, the corresponding wave-vector $k^* = 1/|\beta|H$ and the length scale $l^* = 2\pi/k^*$ of the fluctuations spectra of each of the velocity components.

We also provide, here, for the reader's convenience, the characteristic scales in the setup A1S0 in the RT mixing experiments Table 3 presents the scales set by the acceleration g, including the values of the wave-vector k_ν and the length scale l_ν. It also provides the scales set by the deterministic noise $\hat{U}$, including the values of the wave-vector $\hat{k}_\nu$ and the length scale $\hat{l}_\nu$. The deterministic noise was due to the inaccuracy of the magnitude of the velocity of the co-flowing streams, $\hat{U} = 0.01\ U$, in the experiments [12].

Table 1. Characteristics of the fluctuation spectra for each of the velocity components, including the parameters α and β of the compound function $S(k) \sim k^{\alpha} \exp(\beta k)$, the left k_l and the right boundaries k_r of the fitting interval, and the span of scales $lg(k_r/k_l)$.

Quantity	α	β	$k_l H$	$k_r H$	$lg(k_r/k_l)$
$\langle u^2 \rangle$	-1.06 ± 0.06	$(-2.14 \pm 0.05) \times 10^{-3}$	110	3700	1.53
$\langle w^2 \rangle$	-1.49 ± 0.07	$(-1.30 \pm 0.05) \times 10^{-3}$	150	3700	1.39
$\langle v^2 \rangle$	-2.04 ± 0.06	$(-1.02 \pm 0.05) \times 10^{-3}$	100	4000	1.60

Table 2. Characteristic scales set by the exponential decay factor β, including the value $1/|\beta|$, the corresponding wave-vector $k^* = 1/|\beta|H$ and the length scale $l^* = 2\pi/k^*$ in the fluctuations spectra for each of the velocity components.

Quantity	$1/\|\beta\|$	k^*, [m^{-1}]	l^*, [m]
$\langle u^2 \rangle$	4.67×10^2	3.89×10^2	1.61×10^{-2}
$\langle w^2 \rangle$	7.69×10^2	6.41×10^2	9.80×10^{-3}
$\langle v^2 \rangle$	9.80×10^2	8.17×10^2	7.69×10^{-3}

Table 3. Characteristic scales in RT mixing experiments in the setup A1S0 set by the acceleration g (k_ν and l_ν) and by the deterministic noise $\hat{U}$ ($\hat{k}_\nu$ and $\hat{l}_\nu$).

k_ν, [m^{-1}]	l_ν, [m]	$\hat{k}_\nu$, [m^{-1}]	$\hat{l}_\nu$, [m]
1.11×10^3	5.71×10^{-3}	3.93×10^2	1.50×10^{-2}

4.2. Anomalous Scaling

According to our data analysis results, the fluctuations spectra in RT mixing in realistic fluids can be represented by a product of a power-law and an exponential, $S(k) \sim k^{\alpha} \exp(\beta k)$. In the compound function $S(k)$, the power-law k^{α} captured the self-similar dynamics at relatively large scales, whereas the exponential $\exp(\beta k)$ described the scale-dependent dynamics at small scales. These results were consistent with our theoretical foundations, and with the classical works, Equations (1)–(4) [9,13,14,17,26].

4.3. Dynamic Anisotropy

The results of our data analysis unambiguously exposed the dynamic anisotropy of RT mixing in realistic environments, such as the experiments in [12], and were in good agreement with the group theory results. The dynamic anisotropy was revealed in the different values of the power-law exponents α and the exponential decay factors β in the spectral shapes $S(k)$ of the kinetic energy fluctuations in distinct velocity components. See Tables 1 and 2, Figures 2a and 5a, Equations (1)–(4) and Ref. [28].

Our data analysis found the following properties of fluctuations spectra of the velocity components v^2, u^2, w^2, see Tables 1 and 2 and Ref. [28]. In these spectra, the self-similar spectral part $\sim k^\alpha$ was the steepest, $\alpha \sim -2$, for the cross-stream velocity v^2 in the direction of the acceleration. Furthermore, it was the most gradual, $\alpha \sim -1$, for the stream-wise velocity u^2 in the direction of the co-flowing streams. The scale-dependent spectral part $\sim \exp(\beta k)$ had the smallest magnitude of the exponential decay factor, $|\beta| = 1.02 \times 10^{-3}$, and the smallest associated length scale, $l^* \sim 7.69 \times 10^{-3}$ [m], for the cross-stream velocity v^2. It had the largest magnitude of the exponential decay factor, $|\beta| = 2.14 \times 10^{-3}$, and the largest associated length scale, $l^* \sim 3.89 \times 10^{-2}$ [m], for the stream-wise velocity u^2. These results were consistent with our theoretical foundation in Equations (1), (3) and (4), and highlighted a number of relevant physics properties of RT mixing [12].

The fluctuations of the cross-stream velocity v^2 in the direction of the acceleration were discussed in detail in our recent work [28]. They are defined by the acceleration g. This is illustrated by the exponent α of the power-law, with the experimental value $\alpha = -2.04$ being within 2% of the theoretical value $\alpha = -2$. This is also illustrated by the characteristic length scale $l^* = 7.69 \times 10^{-3}$ [m], which was set by the exponential decay factor $|\beta| = 1.02$, and which was close, within 30%, to the length scale l_ν set by the acceleration g, and by the kinematic viscosity ν. See Tables 1–3, Equation (1), Section 2.1.2.

The fluctuations of the stream-wise velocity u^2 in the direction of the co-flowing streams were set by the deterministic experimental conditions. This was revealed by the exponent α of the power-law, with the experimental value $\alpha = -1.06$ being close, within 6%, to the theoretical value $\alpha = -1$. This was also revealed by the characteristic scale of length $l^* = 1.61 \times 10^{-2}$ [m], which was defined by the exponential decay factor $|\beta| = 2.14$, and which was within 3% of the length scale $\hat{l}_\nu$, determined by the deterministic noise; the velocity scale $\hat{U}$ and the kinematic viscosity ν. See Tables 1–3, Equation (3).

Recalling that, in the RT mixing experimental setup A1S0, the reported accuracy of the velocity magnitudes of the co-flowing streams, $U = U_1 = U_2$, was within 4%; the data analysis, thus, suggested that even the slightest shear caused by a small deterministic noise ($\hat{U} = 0.01\ U$) was sensed by RT mixing flow at very late times. The strong sensitivity of RT mixing to the deterministic conditions was consistent with the theoretical foundations and was also observed in other experiments [12,24,39].

The fluctuations of the cross-tank velocity w^2 (which was in the direction normal to the acceleration and to the velocity of the co-flowing streams) were set by the combined effects of the acceleration and the deterministic conditions. This was evidenced by the exponent α of the power-law, with experimental value $\alpha = -1.49$ being within 0.07% of $\alpha = -3/2$, which was the mean of the theoretical values of $\alpha = -2$ and $\alpha = -1$ of the cross-stream and stream-wise velocities, respectively. This was further confirmed by the characteristic length scale $l^* = 9.80 \times 10^{-3}$ [m] set by the exponential decay factor $|\beta| = 1.30$, which was within 20% of a length scale 1.18×10^{-2} [m], which was the mean of the length scales $l^* = 7.69 \times 10^{-3}$ [m] and $l^* = 1.61 \times 10^{-2}$ [m] for v^2 and u^2, respectively. See Tables 1–3, Equation (4). Note also that the power-law exponent for the w^2 fluctuations $\alpha = -1.49$ was the closest (within 12%) to the exponent $-5/3$ for the canonical Kolmogorov turbulence [34,35,42].

4.4. *Dynamic Bias*

Our data analysis unambiguously revealed the dynamic bias of RT mixing dynamics in realistic experimental environments; thus, achieving good agreement with the group theory results. See Tables 1 and 2, Figures 2a and 5a, Equations (1)–(4) and Ref. [28].

Particularly, our data analysis exposed the sensitivity of the late time RT dynamics to the deterministic (the initial and the flow) conditions. This property of RT mixing is explained by group theory foundations, which have also explained similar observations in other RT mixing experiments with Reynolds numbers up to 3.2×10^6 for a broad range of setups and conditions [9,11,24,39,41].

While for any initial conditions the RT flow is expected to evolve toward a mixing regime with self-similarly growing amplitude in the acceleration direction, the RT dynamics in the normal directions can be scale-dependent and influenced by the deterministic conditions. We call this sensitivity to deterministic conditions dynamic bias. In our data analysis, the dynamic bias of RT mixing is revealed in the values of the power-law exponent α and the exponential decay factor β of the specific kinetic energy of the stream-wise u^2 velocity component. It is also revealed in the span of scales of fluctuations of the velocities (v^2, u^2, w^2). See Tables 1–3, Figures 2–4.

For the velocity component u^2, the exponent was $\alpha = -1.06$ and the exponential decay factor was $|\beta| = 2.14$, with associated length-scale $l^* = 1.61 \times 10^{-2}$ [m]. These experimental values were in good agreement with the theoretical values set by the velocity scale $\hat{U}$ and the kinematic viscosity ν, with $\alpha = -1$ and $\hat{l}_\nu = 1.50 \times 10^{-2}$ [m]. They evidenced that the stream-wise dynamics in the direction of the co-flowing streams was influenced by the deterministic noise caused by the 1% uncertainty of the velocity magnitude of the co-flowing stream $\hat{U} = 0.01\ U$, $U = U_1 = U_2$. See Tables 1 and 3, Equation (3).

The other indicator of the dynamic bias of RT mixing was the relatively short span of scales. The experimental values of $lg(k_r/k_l)$ for the velocity components (v^2, u^2, w^2) were $(1.60, 1.53, 1.39)$, respectively. This experimental span of scales was substantially, $\sim$ 3-fold, shorter than the span of scales predicted for canonical turbulence. The latter depended on the Reynolds number as $lg(Re^{3/4})$ and was estimated as $lg(Re^{3/4}) \approx 3.40$ for $Re = 3.4 \times 10^4$. The short span of scales diagnosed in the experiments revealed that, in RT mixing, the velocity fluctuations are deterministic rather than stochastic in nature [9,24,30,34,35].

4.5. Analysis Method and Data Interpretation

Since RT mixing is sensitive to deterministic bias and is statistically unsteady, some discussion is required on the effect of the data analysis method on data interpretations [9,10,14,24–27,29,45].

To obtain physics properties of RT mixing in realistic environments, we analyzed raw unprocessed data and we applied compound functions to describe the fluctuations spectra of the diagnosed quantities. In the compound function $S(k) \sim k^\alpha \exp(\beta k)$, the power-law k^α captures the self-invariant part of the spectrum at relatively large length scales, $k \ll k_\nu$, and the exponential $\exp(\beta k)$ captures the scale-dependent part of the spectrum at relatively small length scales, $k \sim k_\nu$.

To precisely evaluate the exponent α (the exponential decay factor β), one needs to grasp accurately, and with ample statistics, the dynamics at relatively small (large) length scales, with $k \sim k_\nu$ ($k \ll k_\nu$). An appropriate choice of the fitting interval $k \in [k_l, k_r]$, and its left (right) boundaries $k_l (k_r)$, is further required to enable a relatively broad span of scales $lg(k_r/k_l)$ of the function $S(k)$, with small enough k_l and with large enough k_r. Hence, on the maps of $\alpha(k_l, k_r)$ and $\beta(k_l, k_r)$, accurate estimates of values α and β are expected to be located in the top left part in the plane (k_l, k_r). This was fully consistent with our results. See Figures 3b and 6b and Ref. [28].

The goodness of fit is the most important aspect of the data analysis. Our previous work quantified goodness of fit by using the standard Kolmogorov–Smirnov (KS) test and the p_{KS} score. Here, we employed the Anderson–Darling test and the parameter p_{AD}, and we compared the residuals with the χ_2^2 distribution. We believe the Anderson–Darling test was more statistically suitable in our case, since the maximum likelihood estimator was computed from the same data set used to verify the fit without re-sampling. The goodness of fit analysis in our present work was consistent with that in the previous work; the dependence of the parameter p_{AD} on the fitting window was qualitatively similar to that of the parameter p_{KS}. See Figures 4b and 7b, Ref. [28] and Refs. [25,36,48,49].

The dependence of the parameter p_{AD} on the fitting interval in the plane (k_l, k_r) suggests that, in order for the residuals to be distributed according to the fitting technique assumptions, the statistical analysis should account for a significant number of modes on the left and on the right ends of the periodogram. Furthermore, in the regions $k \in [k_l, k_r]$, where the parameters α

and β of the fitting function $S(k) \sim k^{\alpha} \exp(\beta k)$ were precisely identified, the fit was excellent and the goodness-of-fit parameter value was high, $p_{AD} > 0.5$. A high precision and a high goodness of fit appear closely interrelated. We made a remarkable observation that, in statistically unsteady RT mixing, similarly to statistically steady processes, an *old-school* approach to quantify a value with small relative errors and with ample statistics works well. To our knowledge, such observations have not been discussed before. See Figures 2–7 and Ref. [28].

4.6. Summary of Properties of RT Mixing

Our analysis of the data of fluctuations of each of the three velocity components identified important physics properties of RT mixing in realistic environments: the anomalous scaling, the dynamic anisotropy, and the dynamic bias. Specifically, the following were identified: (i) In RT mixing the fluctuations spectra differ substantially from the spectra of canonical Kolmogorov turbulence; (ii) In RT mixing, the fluctuations of the specific kinetic energy unambiguously exhibit dynamic anisotropy; (iii) RT mixing dynamics remain sensitive to deterministic conditions, even at very late times. These outcomes are consistent with group theory foundations and with other experiments. See Tables 1 and 2, Figures 2–7, Equations (1)–(4) [9,15,17,23,24,29,30,37,39,44–46,51].

4.7. Analysis Outcomes for Design of Experiments

We, herein, discuss the relevance of our data analysis to the design of experiments on RT mixing in fluids [10,12,19,24,39,41,52].

Traditionally, analysis of data is viewed as a tool for establishing fidelity of a theory by comparing theoretical predictions with experiments. The underlying assumptions are such that the theory is mathematically rigorous and physically accurate, whereas in the experiment, the conditions and parameters closely resemble theoretical assumptions, the diagnostics are accurate and non-intrusive, and the data have ample statistics [9,30,42].

The phenomena of Rayleigh–Taylor instability and RT mixing open new venues for data science, well beyond traditional considerations. Since RT dynamics is highly sensitive to deterministic (the initial and the flow) conditions, in addition to fidelity of the theory, one can employ the analysis of data for systematic control of experimental conditions, and for methodical improvement of experimental design [9,10,24].

For instance, the experiments [12,53] are designed to study the coupling of RT and KH dynamics and the late stages of RT mixing in a large domain. The pure RT dynamics is achieved in the setup A1S0, when the velocities of the co-flowing streams are the same, within a few percent, as $U_1 = U_2 = U$ with $|\hat{U}/U| \approx 10^{-2}$, where $|\hat{U}| \approx |U_1 - U_2|$. The slight distinctions of the velocities of the co-flowing streams are assumed to have negligible effect on the late time RT mixing [12,53,54].

Our data analysis results unambiguously found that even that tiny a distinction may strongly influence the flow characteristics. Particularly, it may lead to significant, in several folds, departures in the parameters of the spectral shapes of the velocity components. This was illustrated by the values of α and β in the spectral shapes $S(k) \sim k^{\alpha} e^{\beta k}$ for the fluctuations of the cross-stream and stream-wise velocity components, with $\alpha = -2.06$, $\beta = -1.06 \times 10^{-3}$ for v^2 and with $\alpha = -1.06, \beta = -2.14 \times 10^{-3}$ for u^2; see Table 1 and Figure 2 and Ref. [28]. The strong sensitivity of RT dynamics to the deterministic noise is consistent with the group theory results [9,26,29] and with experiments [24,41]. It also indicates a need in new approaches for experimental design [10].

Particularly, accurate probing and quantification of properties of RT mixing requires non-intrusive diagnostics of fast events, and ultra-high performance in space–time resolution, bandwidth, and data-acquisition rate, along with tight control of the initial and experimental conditions [10,17,24]. These requirements can be implemented in integrated experimental systems. The systems should incorporate state-of-the-art technology in motion control, precision mechanics, optical imaging, image processing, and digital signal processing, as well as data analysis methods [10,28]. For simultaneous measurements in space and in time of the interface dynamics and the flow fields, the experimental diagnostics may include, for example, particle image velocimetry, planar laser-induced fluorescence and holographic particle image velocimetry. By employing these diagnostic capabilities, one can accurately implement and diagnose RT mixing and can provide data suitable for a direct comparison with the group theory approach and numerical simulations [9,10,17,24,55–57].

4.8. Analysis Outcome for Numerical Simulations

Rayleigh–Taylor mixing is successfully modeled in numerical simulations employing various methods [55,56,58–60]. Traditionally, visual inspection is used to analyze smoothed post-processed numerical data, but our data analysis methodology can provide numerical simulations with a number of advantages, expanding the predictive modeling capabilities of simulations.

Our methodology can be applied to analyze both raw and processed numerical data. It can employ a broad class of fitting functions (e.g., a power-law, an exponential, a compound function) to study fluctuations spectra, structure functions, amplitude growth and growth rate, and other quantities. It can identify the fitting function parameters, including their mean values, relative errors and goodness-of-fit. It can investigate the dependence of the results on the fitting window and self-consistently identify the best fit interval. It can be used for both in-fly diagnostics and in-depth analysis of numerical data. We urge that high quality numerical simulations consider these advantages and incorporate our methodology [55,56,58–60].

4.9. Spectral Shapes in Turbulence and in RT Mixing

Compound functions $k^{\alpha}e^{\beta k}$ are regularly employed in analysis of data from a broad range of realistic turbulent processes [13–17,32,33]. Success was found in the seminal works [13,14], which investigated the fluctuations spectra of isotropic, homogeneous turbulence at large, finite Reynolds numbers. With regards to physics, compound functions $k^{\alpha}e^{\beta k}$ describe processes in which self-similar k^{α} and scale-dependent $e^{\beta k}$ dynamics interplay.

Rayleigh–Taylor mixing is anisotropic, inhomogeneous, non-local, and statistically unsteady, and is sensitive to deterministic conditions. Its properties depart from those of isotropic, homogeneous, local, statistically steady and stochastic Kolmogorov turbulence [9,24,40,41,44]. Rayleigh–Taylor mixing possesses a complex interplay of self-similar and scale-dependent dynamics. Hence, compound functions can be applied to analyze RT spectra, as in other physics processes.

The present work finds that, in addition to the shape $k^{\alpha}e^{\beta k}$ of fluctuations spectra, a thorough analysis of data is expected to include estimates of relative errors of fitting function parameters, evaluations of goodness of fit, and investigations of effects of fitting intervals on results. We urge numerical simulations and experiments of realistic turbulent processes and of Rayleigh–Taylor mixing to consider these findings and incorporate our methodology.

5. Discussion

We analyzed the fluctuations of three velocity components measured in a RT experiment, by a hot-three-wire anemometer, using a group theory-guided data analysis method. The analysis elucidated some key features of the fluctuations spectra in realistic Rayleigh–Taylor instabilities.

We expounded the theoretical framework underpinning the aforementioned statistical method for diagnosing the physics and mathematical properties of RT mixing. The method employs a maximum likelihood estimate to estimate the best fit of the parameters and the standard errors associated with the fit parameters. To evaluate the goodness-of-fit, the Anderson–Darling test [25,36] is employed, including a new test statistic [28], (Sections 4.1 and 4.2). By applying the method over all potential left and right cut-offs of the fitting interval, the effects of the fitting interval, left and right cut-offs, and the noise level on the model fit are investigated (Section 4.3). Group theory establishes the foundation for the method (Sections 2.1.1–2.1.3), justifying the functional form of the power density spectrum $S(k) \sim k^{\alpha} \exp(\beta k)$ we have, herein, applied to analyze the time series (Section 2.1.4, Equation (5). Such forms have been applied in works on classical turbulence [13,14,26,29].

By optimizing each of the fit conditions, including the relative errors of the fit parameters, the goodness-of-fit, and the span of scales, we have identified the optimal fitting interval and determined the associated fit parameters (Section 5). We have further identified the resultant properties of RT mixing (Tables 1 and 2, Figures 5–7, Equations (1)–(4)).

The results of the data analysis indicate that RT mixing may depart substantially from canonical Kolmogorov turbulence, exhibiting dynamic isotropy, anomalous scaling properties, vastly different spectral properties, and the dependence of late-time dynamics on the deterministic conditions [9,11,13–17,24,26,27,32,34,35,37,39,41,44,51].

The data analysis method employed, herein, and the results of our investigation are applicable to future research in fluids, plasmas, and other materials which exhibit RT mixing, or other turbulence-adjacent phenomena. They may assist in informing new research at various scales [8,22].

6. Conclusions

This work investigated the properties of fluctuations in Rayleigh–Taylor mixing, based on analysis of experimental data. For fluctuations of the kinetic energy of each of three velocity components, we studied the fluctuations' properties at large and at small scales. We found, with statistical confidence, that the spectral shapes of the fluctuations spectra in RT mixing experiments are at least as complex as compound functions, represented by a product of a power law and an exponential. The properties of the spectral shape parameters unambiguously identify the dynamic anisotropy and dynamic bias of Rayleigh–Taylor mixing in realistic environments, in agreement with group theory foundations, in conformity with experiments on high Reynolds number RT mixing and on accelerated turbulent flows, and in consistency with the anomalous properties of turbulent flows in realistic environments.

Author Contributions: Conceptualization, S.I.A.; Methodology, K.C.W. and S.I.A.; Software, K.C.W. and S.I.A.; Investigation, K.C.W. and S.I.A.; Data curation, K.C.W.; Writing—original draft, K.C.W. and S.I.A.; Supervision, S.I.A. All authors have read and agreed to the published version of the manuscript.

Funding: The authors appreciate the support of the National Science Foundation (USA) (Award No. 1404449), the Australian Research Council (AUS) (Award No. LE220100132), the Australian Government Research Training Program.

Institutional Review Board Statement: Not applicable.

Informed Consent Statement: Not applicable.

Data Availability Statement: The methods, the results and the data presented in this work are available to the readers in the paper.

Acknowledgments: The authors thank D. Ranjan for making the experimental data available for analysis, P. Suchandra for providing the information on the diagnostics and the resolution, and D. Pfefferle for permitting the use of the maximum likelihood estimator code. The authors deeply thank K.R. Sreenivasan for insightful comments and inspiring discussions.

Conflicts of Interest: The authors declare no conflict of interest.

References

1. Herring, J.; Rilley, J.; Patterson, G.; Kraichnan, R. Growth of uncertainty in decaying isotropic turbulence. *J. Atmos. Sci.* **1973**, *30*, 997–1006. [CrossRef]
2. Herring, J.; Orszag, S.; Kraichnan, R.; Fox, D. Decay of two-dimensional homogeneous turbulence. *J. Fluid Mech.* **1974**, *66*, 417–444. [CrossRef]
3. Kraichnan, R.; Herring, J. A strain-based Lagrangian-history turbulence theory. *J. Fluid Mech.* **1978**, *88*, 355–367. [CrossRef]

4. Gotoh, T.; Rogallo, J.; Herring, J.; Kraichnan, R. Lagrangian velocity correlations in homogeneous isotropic turbulence. *Phys. Fluids A Fluid Dyn.* **1993**, *5*, 2846–2864. [CrossRef]
5. Chen, S.; Doolen, G.; Herring, J.; Kraichnan, R.; Orszag, S.; She, Z. Far-dissipation range of turbulence. *Phys. Rev. Lett.* **1993**, *70*, 3051–3054. [CrossRef] [PubMed]
6. Kerr, R.; Herring, J. Prandtl number dependence of Nusselt number in direct numerical simulations. *Phys. Scr.* **2000**, *419*, 325–344. [CrossRef]
7. Herring, J.; Kimura, Y. Some issues and problems of stably stratified turbulence. *Phys. Scr.* **2013**, *T155*, 014031. [CrossRef]
8. Abarzhi, S.; Gauthier, S.; Sreenivasan, K. Turbulent mixing and beyond: Non-equilibrium processes from atomistic to astrophysical scales I. *Philos. Trans. R. Soc. A Math. Phys. Eng. Sci.* **2013**, *371*, 20120436. [CrossRef] [PubMed]
9. Abarzhi, S. Review of theoretical modelling approaches of Rayleigh-Taylor instabilities and turbulent mixing. *Philos. Trans. R. Soc. A Math. Phys. Eng. Sci.* **2010**, *368*, 1809–1828. [CrossRef] [PubMed]
10. Orlov, S.; Abarzhi, S.; Oh, S.; Barbastathis, G.; Sreenivasan, K. High-performance holographic technologies for fluid-dynamics experiments. *Philos. Trans. R. Soc. A Math. Phys. Eng. Sci.* **2010**, *368*, 1705–1737. [CrossRef] [PubMed]
11. Anisimov, S.; Drake, R.; Gauthier, S.; Meshkov, E.; Abarzhi, S. What is certain and what is not so certain in our knowledge of Rayleigh-Taylor mixing? *Philos. Trans. R. Soc. A Math. Phys. Eng. Sci.* **2013**, *371*, 20130266. [CrossRef] [PubMed]
12. Akula, B.; Suchandra, P.; Mikhaeil, M.; Ranjan, D. Dynamics of unstably stratified free shear flows: An experimental investigation of coupled Kelvin–Helmholtz and Rayleigh–Taylor instability. *J. Fluid Mech.* **2017**, *816*, 619–660. [CrossRef]
13. Kraichnan, R. The structure of isotropic turbulence at very high Reynolds numbers. *J. Fluid Mech.* **1959**, *5*, 497–543. [CrossRef]
14. Sreenivasan, K. On the scaling of the turbulence energy dissipation rate. *Phys. Fluids* **1984**, *27*, 1048–1051. [CrossRef]
15. Sreenivasan, K.; Antonia, R. The phenomenology of small-scale turbulence. *Annu. Rev. Fluid Mech.* **1997**, *28*, 435–472. [CrossRef]
16. Pouquet, A.; Mininni, P. The interplay between helicity and rotation in turbulence: Implications for scaling laws and small-scale dynamics. *Philos. Trans. R. Soc. A* **2010**, *368*, 1635–1662. [CrossRef] [PubMed]
17. Sreenivasan, K. Turbulent mixing: A perspective. *Proc. Natl. Acad. Sci. USA* **2019**, *116*, 18175–18183. [CrossRef]
18. Strutt, R.J.L. Investigation of the Character of the Equilibrium of an Incompressible Heavy Fluid of Variable Density. *Proc. Lond. Math. Soc.* **1883**, *s1-14*, 170–177. [CrossRef]
19. Davies, R.; Taylor, G. The mechanics of large bubbles rising through extended liquids and through liquids in tubes. *Proc. R. Soc. London. Ser. A Math. Phys. Sci.* **1950**, *200*, 375–390. [CrossRef]
20. Arnett, W.D. *Supernovae and Nucleosynthesis: An Investigation of the History of Matter, from the Big Bang to the Present*; Princeton Series in Astrophysics; Princeton University Press: Princeton, NJ, USA, 1996.
21. Haan, S.W.; Lindl, J.D.; Callahan, D.A.; Clark, D.S.; Salmonson, J.D.; Hammel, B.A.; Atherton, L.J.; Cook, R.C.; Edwards, M.J.; Glenzer, S.; et al. Point design targets, specifications, and requirements for the 2010 ignition campaign on the National Ignition Facility. *Phys. Plasmas* **2011**, *18*, 051001. [CrossRef]
22. Abarzhi, S.; Goddard, W. Interfaces and mixing: Nonequilibrium transport across the scales. *Proc. Natl. Acad. Sci. USA* **2019**, *116*, 18171–18174. [CrossRef] [PubMed]
23. Remington, B.; Park, H.S.; Casey, D.; Cavallo, R.; Clark, D.; Huntington, C.; Kuranz, C.; Miles, A.; Nagel, S.; Raman, K.; et al. Rayleigh-Taylor instabilities in high-energy density settings on the National Ignition Facility. *Proc. Natl. Acad. Sci. USA* **2019**, *116*, 18223–18228. [CrossRef] [PubMed]
24. Meshkov, E.; Abarzhi, S. Group theory and jelly's experiment of Rayleigh–Taylor instability and Rayleigh–Taylor interfacial mixing. *Fluid Dyn. Res.* **2019**, *51*, 065502. [CrossRef]
25. Pfefferlé, D.; Abarzhi, S. Whittle maximum likelihood estimate of spectral properties of Rayleigh-Taylor interfacial mixing using hot-wire anemometry experimental data. *Phys. Rev. E* **2020**, *102*, 053107. Erratum in **2022**, *106*, 019901. [CrossRef] [PubMed]
26. Abarzhi, S. Self-similar interfacial mixing with variable acceleration. *Phys. Fluids* **2021**, *33*, 122110. [CrossRef]
27. Abarzhi, S.; Sreenivasan, K. Self-similar Rayleigh-Taylor mixing with accelerations varying in time and in space. *Proc. Natl. Acad. Sci. USA* **2022**, *119*, e2118589119. [CrossRef]
28. Williams, K.; Abarzhi, S. Fluctuations spectra of specific kinetic energy, density, and mass flux in Rayleigh–Taylor mixing. *Phys. Fluids* **2023**, *34*, 122188. [CrossRef]
29. Abarzhi, S.I. On fundamentals of Rayleigh-Taylor turbulent mixing. *EPL Europhysics Lett.* **2010**, *91*, 35001. [CrossRef]
30. Sreenivasan, K. Fluid turbulence. *Rev. Mod. Phys.* **1999**, *71*, S383–S395. [CrossRef]
31. Bershadskii, A. Distributed chaos and turbulence in Bénard-Marangoni and Rayleigh-Bénard convection. *arXiv* **2019**, arXiv:1903.05018.
32. Klewicki, J.; Chini, G.; Gibson, J. Prospectus: Towards the development of high-fidelity models of wall turbulence at large Reynolds number. *Philos. Trans. R. Soc. A* **2017**, *375*, 20160092. [CrossRef]
33. Yakhot, V.; Donzis, D. Emergence of multiscaling in a random-force stirred fluid. *Phys. Rev. Lett.* **2017**, *119*, 044501. [CrossRef]
34. Kolmogorov, A. The Local Structure of Turbulence in Incompressible Viscous Fluid for Very Large Reynolds' Numbers. *Dokl. Akad. Nauk SSSR* **1941**, *30*, 299–303.
35. Kolmogorov, A. On the degeneracy of isotropic turbulence in an incompressible viscous fluid. *Dokl. Akad. Nauk SSSR* **1941**, *31*, 538–541.
36. Anderson, T.; Darling, D. A Test of Goodness-of-Fit. *J. Am. Stat. Assoc.* **1954**, *49*, 765–769. [CrossRef]
37. Robey, H.F.; Zhou, Y.; Buckingham, A.C.; Keiter, P.; Remington, B.; Drake, R. The time scale for the transition to turbulence in a high Reynolds number, accelerated flow. *Phys. Plasmas* **2003**, *10*, 614–622. [CrossRef]

38. Dalziel, S.; Linden, P.; Youngs, D. Self-similarity and internal structure of turbulence induced by Rayleigh–Taylor instability. *J. Fluid Mech.* **1999**, *399*, 1–48. [CrossRef]
39. Meshkov, E.E. *Studies of Hydrodynamic Instabilities in Laboratory Experiments*; FGUC-VNIIEF: Sarov, Russia, 2006; ISBN 5-9515-0069-9. (In Russian)
40. Swisher, N.; Kuranz, C.; Arnett, W.; Hurricane, O.; Robey, H.; Remington, B.; Abarzhi, S. Rayleigh-Taylor mixing in supernova experiments. *Phys. Plasmas* **2015**, *22*, 102707. [CrossRef]
41. Meshkov, E. Some peculiar features of hydrodynamic instability development. *Philos. Trans. R. Soc. A Math. Phys. Eng. Sci.* **2013**, *371*, 20120288. [CrossRef]
42. Landau, L.; Lifshitz, E. *Course of Theoretical Physics*; Elsevier Science: Amsterdam, The Netherlands, 1987; Volume I–X.
43. Abarzhi, S.; Gorobets, A.; Sreenivasan, K. Rayleigh–Taylor turbulent mixing of immiscible, miscible and stratified fluids. *Phys. Fluids* **2005**, *17*, 081705. [CrossRef]
44. Abarzhi, S.; Bhowmick, A.; Naveh, A.; Pandian, A.; Swisher, N.; Stellingwerf, R.; Arnett, W. Supernova, nuclear synthesis, fluid instabilities, and interfacial mixing. *Proc. Natl. Acad. Sci. USA* **2019**, *116*, 18184–18192. [CrossRef] [PubMed]
45. Abarzhi, S.; Hill, D.; Williams, K.; Wright, C. Buoyancy and drag in Rayleigh-Taylor and Richtmyer-Meshkov linear, nonlinear and mixing dynamics. *Appl. Math. Lett.* **2022**, *31*, 108036. [CrossRef]
46. Abarzhi, S.; Hill, D.; Williams, K.; Li, J.; Remington, B.; Martinez, D.; Arnett, W. Fluid dynamics mathematical aspects of supernova remnants. *Phys. Fluids* **2023**, *35*, 034106. [CrossRef]
47. Whittle, P. Curve and Periodogram Smoothing. *J. R. Stat. Soc.* **1957**, *19*, 38–63. [CrossRef]
48. Kolmogorov, A.N. Sulla Determinazione Empirica di una Legge di Distribuzione, *Giorn. Dell'Inst. Ital. Attuari.dell'Ist. Ital. Attuari* **1933**, *4*, 83–91.
49. Smirnov, N. Table for Estimating the Goodness of Fit of Empirical Distributions. *Ann. Math. Statist.* **1948**, *19*, 279–281. [CrossRef]
50. Stephens, M.A. EDF Statistics for Goodness of Fit and Some Comparisons. *J. Am. Stat. Assoc.* **1987**, *69*, 730–737. [CrossRef]
51. Narasimha, R.; Sreenivasan, K. Relaminarization in highly accelerated turbulent boundary layers. *J. Fluid Mech.* **1973**, *61*, 417–447. [CrossRef]
52. Read, K. Experimental investigation of turbulent mixing by Rayleigh–Taylor instability. *Phys. D* **1984**, *12*, 45–58. [CrossRef]
53. Kraft, W.; Banerjee, A.; Andrews, M. On hot-wire diagnostics in Rayleigh–Taylor mixing layers. *Exp. Fluids* **2009**, *47*, 49–68. [CrossRef]
54. Akula, B.; Ranjan, D. Dynamics of buoyancy-driven flows at moderately high Atwood numbers. *J. Fluid Mech.* **2016**, *795*, 313–355. [CrossRef]
55. Reynolds number effects on Rayleigh–Taylor instability with possible implications for type Ia supernovae. *Nat. Phys.* **2006**, *2*, 562–568. [CrossRef]
56. Schilling, O. Self-similar Reynolds-averaged mechanical–scalar turbulence models for Rayleigh–Taylor, Richtmyer–Meshkov, and Kelvin–Helmholtz instability-induced mixing in the small Atwood number limit. *Phys. Fluids* **2021**, *33*, 085129. [CrossRef]
57. Schumacher, J.; Sreenivasan, K. Colloquium: Unusual dynamics of convection in the Sun. *Rev. Mod. Phys.* **2022**, *92*, 041001. [CrossRef]
58. Ristorcelli, J.; Clark, T. Rayleigh–Taylor turbulence: Self-similar analysis and direct numerical simulations. *J. Fluid Mech.* **2004**, *507*, 213–253. [CrossRef]
59. Glimm, J.; Sharp, D.; Kaman, T.; Lim, H. New directions for Rayleigh-Taylor mixing. *Philos. Trans. R. Soc. A Math. Phys. Eng. Sci.* **2013**, *371*, 20120183. [CrossRef] [PubMed]
60. Grinstein, F.; Saenz, J.; Germano, M. Coarse grained simulations of shock-driven turbulent material mixing. *Phys. Fluids* **2021**, *33*, 035131. [CrossRef]

atmosphere

Article

Modal Projection for Quasi-Homogeneous Anisotropic Turbulence

Ying Zhu † and Claude Cambon *

Université de Lyon, Laboratoire de Mécanique des Fluides et d'Acoustique, UMR 5509, Ecole Centrale de Lyon, CNRS, UCBL, INSA, F-69134 Ecully CEDEX, France; yzhu@unice.fr
* Correspondence: claude.cambon@ec-lyon.fr
† Current address: Institut de Physique de Nice, Université Côte d'Azur, CNRS, 17 Rue Julien Lauprêtre, 06200 Nice, France.

Abstract: This article, or essay, addresses the anisotropic structure and the dynamics of quasi-homogeneous, incompressible turbulence. Modal projection and expansions in terms of spherical harmonics in three-dimensional Fourier space are in line with a seminal study by Jack Herring, around the so-called Craya–Herring frame of reference, with a large review of the related approaches to date. The research part is focused on structure and dynamics of rotating sheared turbulence, including a description of both directional and polarization anisotropy with a minimal number of modes. Effort is made to generalize expansions in terms of scalar spherical harmonics (SSHs) to vector spherical harmonics (VSHs). Looking at stochastic fields, for possibly intermittent vector fields, some directions are explored to reconcile modal projection, firstly used for smooth vector fields, and multifractal approaches for internal intermittency but far beyond scalar correlations, such as structure functions. In order to illustrate *turbulence from Earth to planets, stars, and galaxies*, applications to geophysics and astrophysics are touched upon, with generalization to coupled vector fields (for kinetic, magnetic, and potential energies), possibly dominated by waves (Coriolis, gravity, and Alfvén).

Keywords: turbulence; anisotropy; Fourier harmonics

Citation: Zhu, Y.; Cambon, C. Modal Projection for Quasi-Homogeneous Anisotropic Turbulence. *Atmosphere* **2023**, *14*, 1215. https://doi.org/10.3390/atmos14081215

Academic Editors: Boris Galperin, Annick Pouquet and Peter Sullivan

Received: 20 June 2023
Accepted: 19 July 2023
Published: 28 July 2023

1. Introduction

The so-called 'Craya–Herring' frame of reference is very popular, from the seminal article of Jack Herring in 1974 [1]. Among all the scholar contributions of Herring to the physics of fluids, the modal projection is apparently a very narrow topic. Still, we show in this article that such a theme merits a large survey, with the addition of some unpublished studies. Decomposition in terms of spherical harmonics (SHs) was also touched upon by Jack Herring, so various applications of SH to scalar (SSH) and vector (VSH) fields will be discussed and compared.

Modal projections and SH decompositions can be expressed in physical space as well as in Fourier space. It is useful to compare both of them. For instance, the so-called 'wave–vortex' projection by Riley et al. (1981) [2] of a solenoidal (divergence-free) fluctuating velocity field in 3D physical space has a very simple counterpart in 3D Fourier space, where it is purely algebraic and very close to the Herring's projection. More generally, there are strong analogies between a vector field expanded on the surface of a sphere in physical space and its expansion in Fourier space. In this sense, the 3D Fourier counterpart of a solenoidal velocity field is of two components in the surface of the sphere (see Figure 1) since its radial component is zero. Such a velocity field seen in 3D Fourier space is *sphéricole* (from the flat animals living on the surface of a sphere, as imaged by Henri Poincaré).

Our article is organized as follows. Section 2 is devoted to a review of the related proposals for modal projection by Craya (1957) [3], Herring (1974) [1], Riley et al. (1981) [2], Cambon and Jacquin (1989) [4], and Waleffe (1992) [5], often in different contexts, with more recent analyses, mainly for linear dynamics. Applications to statistical moments, such as spectral tensors related to two-point second-order vector correlations, are addressed in Section 3. Rotating shear turbulence is investigated in Section 4 by extending the recent model by Zhu et al. (2019) [6]. In Section 5, a modal projection of poloidal/toroidal type, more general than the one related to Herring's one, is investigated for a vector field, such as a smooth divergence-free velocity field. Promising extensions towards the stochastic modeling of possibly intermittent turbulent fields are presented from a recent project by Daniel Schertzer and followers. Section 6 is devoted to the conclusion and perspectives. Since the special issue in honor of Jack Herring is dedicated to *Turbulence from Earth to Planets, Stars, and Galaxies*, we end up with a list of examples of applications to geophysics and astrophysics of the technical formalism addressed in our article.

2. The Craya–Herring Frame of Reference and Beyond

Antoine Craya is known for using the eponymous frame of reference, thanks to Jack Herring (1974) [1]. Later recognized as a spectral counterpart of a general decomposition in terms of simplified toroidal/poloidal/dilatational modes, this frame leads to expressing the spectral tensors of correlation with a minimal number of scalar (or pseudo-scalar) descriptors without loss of information and for arbitrary anisotropy. Craya provided us with a special angle of attack of the so-called RDT (rapid distortion theory) and quasi-normal (QN) closures, even if he did not work directly on them. Since the original Ph.D. report from Craya (1957) is only in French and not easy to access, we recommend a recent essay with an updated survey and recent progress by the second author [7].

2.1. Related Modal Projections, Analogies, and Differences

Only a solenoidal velocity field is considered in the following. Simplified toroidal–poloidal decomposition (for a solenoidal smooth vector field $\boldsymbol{u}$) follows from the "vortex–wave" decomposition by Riley et al. (1981) [2] with application to stably stratified turbulence:

$$\boldsymbol{u}(\boldsymbol{r}) = \boldsymbol{\nabla} \times (s^{(to)}(\boldsymbol{r})\boldsymbol{n}) + \boldsymbol{\nabla} \times \left(\boldsymbol{\nabla} \times s^{(po)}(\boldsymbol{r})\boldsymbol{n}\right), \tag{1}$$

where $\boldsymbol{n}$ is a fixed unit vector, e.g., aligned with the direction of mean stratification, and anti-parallel to the gravitational acceleration.

Leaving aside the particular application to stably stratified turbulence, this relationship is rewritten from the original reference for an easier comparison with the Craya/Herring one, from a physical space to spectral one.

The counterpart of the previous relationship in 3D Fourier space is

$$\hat{\boldsymbol{u}}(\boldsymbol{k}) = \imath\hat{s}^{(to)}(\boldsymbol{k} \times \boldsymbol{n}) - \hat{s}^{(po)}\boldsymbol{k} \times (\boldsymbol{k} \times \boldsymbol{n}). \tag{2}$$

From the relationship in physical space, it appears that the two components are individually divergence free, with a corresponding, simple orthogonality condition, in 3D Fourier space. The dependency on $r_{\parallel} = \boldsymbol{r} \cdot \boldsymbol{n}$ of both $s^{(to)}$ and $s^{(po)}$, and not only on transverse variables $r_{\perp}$, is essential. If $s^{(to)}$ was independent of the axial variable $r_{\parallel}$, the toroidal component of velocity would be the same as for a two-dimensional (2D) two-component (2C) velocity field generated by a streamfunction, whereas the decomposition here is really three dimensional. A very important part of the flow, however, is zero: it was coined VSHF (vertically sheared horizontal flow) by Smith and Waleffe (2002) [8], or $\boldsymbol{u} = \boldsymbol{u}_{\perp}(r_{\parallel})$, when $\boldsymbol{r} \parallel \boldsymbol{n}$. Similarly, there is a hole in the representation in Fourier space for $\hat{\boldsymbol{u}}$, when $\boldsymbol{k} \parallel \boldsymbol{n}$.

A previous search towards a complete decomposition was proposed as "potosh" (poloidal/toroidal/shear) by Galmiche and Hunt [9] but it was not mathematically correct and will no longer be considered here.

The Craya–Herring decomposition was very similar to Equation (2) as

$$\hat{\boldsymbol{u}}(\boldsymbol{k}) = u^{(1)}(\boldsymbol{k})\boldsymbol{e}^{(1)}(\boldsymbol{\alpha}) + u^{(2)}(\boldsymbol{k})\boldsymbol{e}^{(2)}(\boldsymbol{\alpha}), \tag{3}$$

where $(e^{(1)}, e^{(2)}, \alpha = k/k)$ form a direct orthonormal frame of reference for a given orientation of the wave vector k as shown in Figure 1. At this stage, the case $k_\perp = 0$ is puzzling and ought to be treated as a singular point. The unit vector $e^{(1)}$, defined as

$$e^{(1)} = \frac{k \times n}{\mid k \times n \mid},$$

with $k_\perp = \mid k \times n \mid$, is not defined for $k_\perp = 0$, or, more precisely, it must be defined as a limit that depends on the angle ϕ in polar-spherical coordinates, following a given meridian towards the pole of the sphere. In short, the VSHF is accounted for in the Craya–Herring frame, but it corresponds to $k_\perp = 0$ so that it is concerned by the possible polar multi-definitions of the local frame.

The orthonormal frame is recovered by Ying et al. (2019) [6] for the gradient operator in k-space, passing from Cartesian coordinates to polar-spherical ones, with

$$k\frac{\partial}{\partial k_n} = \alpha_n k\frac{\partial}{\partial k} + e_n^{(2)}\frac{\partial}{\partial \theta} - \frac{e_n^{(1)}}{\sin\theta}\frac{\partial}{\partial \phi}, \tag{4}$$

where θ is the polar angle (on Figure 1) and ϕ the azimuthal angle.

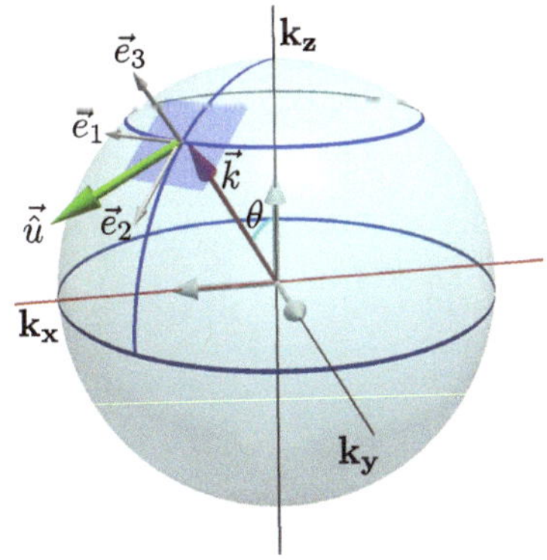

Figure 1. Craya–Herring frame of reference.

Since the $\hat{u}$ vector related to a solenoidal (divergence-free) vector in physical space is of two components in the plane normal to the wave vector, it is interesting to use complex-valued base vectors to project $\hat{u}$, or

$$N(s\alpha) = e^{(2)}(\alpha) - s\imath e^{(1)}(\alpha), \quad s = \pm 1. \tag{5}$$

Introduced by Cambon and Jacquin (1989) [4] in order to study nonlinear rotating turbulence, they were coined *helical modes* by Waleffe (1992). They are eigenmodes of the Curl operator, as $N(\pm k)\,\mathbf{exp}(\pm\imath k \cdot x)$. They simplify the basic equations, even without basic rotation, from the expression of the Lamb vector in 3D Fourier space $\widehat{\omega \times u}$, and allow to ensure the solenoidal constraint $(\nabla \cdot u = 0, k \cdot \hat{u} = 0)$, in avoiding *the byzantine use of projectors* (as jokingly said by Leaf Turner, Los Alamos). In addition, they diagonalize the operator of inertial waves in a rotating frame. The helical decomposition can be written as

$$\hat{u}(k) = \sum_{s=\pm 1} u_s(k) N(s\alpha), \tag{6}$$

for the velocity fluctuation.

2.2. Including Other Coupled Fields, Buoyancy, Magnetic Field

It is possible to include the buoyancy scalar b in stratified turbulence in a new 3D-3C vector as

$$\boldsymbol{w}(\boldsymbol{k}) = \hat{\boldsymbol{u}}(\boldsymbol{k}) + \imath \frac{\hat{b}}{N} \boldsymbol{\alpha}, \tag{7}$$

where N is the Brunt–Wäisälä frequency so that b/N has the dimension of a velocity. The Hermitian symmetry holds with the factor $\imath$, and $(1/2)\hat{b}^* \cdot \hat{b}/N^2$ is related to the density of potential energy as $(1/2)\hat{\boldsymbol{u}}^* \cdot \hat{\boldsymbol{u}}$ for the kinetic energy.

As for MHD flows, a few independent variables can generate both the velocity and the magnetic fields, both being divergence-free vectors.

3. Application to Two-Point Second Order Statistics for Strong Anisotropy

In arbitrary incompressible HAT (homogeneous anisotropic turbulence), the spectral tensor $\hat{R}_{ij}(\boldsymbol{k})$ is the 3D Fourier transform of the two-point second-order correlation tensor

$$R_{ij}(\boldsymbol{r}) = \langle u_i(\boldsymbol{x}) u_j(\boldsymbol{x} + \boldsymbol{r}) \rangle. \tag{8}$$

Its general form calls into play three contributions [4]:

$$\hat{R}_{ij}(\boldsymbol{k}) = \mathcal{E}(\boldsymbol{k}) P_{ij}(\boldsymbol{\alpha}) + \Re(Z(\boldsymbol{k}) N_i N_j) + \imath \epsilon_{ijn} \alpha_n \mathcal{H}(\boldsymbol{k}), \quad P_{ij}(\boldsymbol{\alpha}) = \delta_{ij} - \alpha_i \alpha_j. \tag{9}$$

The relationship (9) involves two scalars (energy and helicity spectra) and one complex-valued pseudo-scalar Z for polarization anisotropy. The energy spectrum is related to the trace, or $\mathcal{E} = (1/2)\hat{R}_{ii}$, and the helicity spectrum $k\mathcal{H}(\boldsymbol{k})$ is related to the purely imaginary and antisymmetric part of $\hat{R}_{ij}$. Last but not least, the real and symmetric deviatoric contribution from polarization, $R_{ij}^{(\mathrm{pol})} = \Re(ZN_iN_j)$ is much less known: It is generated by Z, using the helical modes or directly extracted from the spectral tensor in Cartesian coordinates, from

$$\hat{R}_{ij}^{(\mathrm{pol})}(\boldsymbol{k}) = \frac{1}{2}(P_{im}P_{jn} + P_{in}P_{jm} - P_{ij}P_{mn})\hat{R}_{mn}(\boldsymbol{k}). \tag{10}$$

Of course, the latter equation, which also corresponds to $Z(\boldsymbol{k}) = (1/2)\hat{R}_{mn}N_m^*N_n^*$, is tautological. Our goal is to replace $\hat{R}_{mm}$ in the latter equations with a simpler tensor, to which classical or modified SH expansions may apply.

The first term of Equation (9) can be split into a purely 3D isotropic part and a part that reflects directional anisotropy, or

$$\mathcal{E}(\boldsymbol{k}) P_{ij}(\boldsymbol{\alpha}) = \frac{E(k)}{4\pi k^2} P_{ij}(\boldsymbol{\alpha}) + \hat{R}_{ij}^{(\mathrm{dir})}(\boldsymbol{k}),$$

with

$$\hat{R}_{ij}^{(\mathrm{dir})}(\boldsymbol{k}) = \left(\mathcal{E}(\boldsymbol{k}) - \frac{E(k)}{4\pi k^2}\right) P_{ij}(\boldsymbol{\alpha}). \tag{11}$$

In the presence of the advected scalar field b, an extended spectral tensor can be defined from Equation (7) as

$$\langle w_i^*(\boldsymbol{p}) w_j(\boldsymbol{k}) \rangle = W_{ij}(\boldsymbol{k}) \delta^3(\boldsymbol{k} - \boldsymbol{p}). \tag{12}$$

From the augmented $\mathbf{W}$ tensor vs. $\hat{\mathbf{R}}$, additional spectra for potential energy and for toroidal and poloidal buoyancy fluxes are readily defined.

Similarly for MHD flows, the $(\mathcal{E}, Z, \mathcal{H})$ set for the kinetic flow is complemented by a magnetic set $(\mathcal{E}^M, Z^M, \mathcal{H}^M)$ with the same structure, whereas magnetic–kinetic co-spectra involve the cross helicity, the vector field of the electromotive force, and two unnamed cross-polarization terms. (see [10], Chapter 12).

3.1. Decomposition in Terms of Scalar Spherical Harmonics (SSH)

Regarding the smooth statistical terms $\mathcal{E}, \mathcal{H}$, they are true scalar terms, invariant to any change of the orthonormal frame so that they can be expanded in terms of SSH. The same property is valid for the modulus $\mid Z \mid$ of the polarization term but not for Z as a whole. For instance, a SO(3)-type expansion holds for the scalar $\mathcal{E}$ as

$$\mathcal{E}(\boldsymbol{k}) = \frac{E(k)}{4\pi k^2}\left(1 + H^{2\,(\mathrm{dir})}_{mn}(k)\alpha_m\alpha_n + H^{4\,(\mathrm{dir})}_{mnpq}(k)\alpha_m\alpha_n\alpha_p\alpha_q + \cdots\right). \tag{13}$$

This expansion was established in several papers, quoted in [10]. A very useful identity is

$$H^{2(\mathrm{dir})}_{ij} = -15H^{(\mathrm{dir})}_{ij}, \tag{14}$$

where $2E(k)H^{(\mathrm{dir})}_{ij}(k)$ is the spherical integral (integral on the surface of a sphere of radius $k = \mid \boldsymbol{k} \mid$) of $\hat{R}^{(\mathrm{dir})}_{ij}$ in Equation (11). Nevertheless, it is difficult to extend a practical expansion beyond degree 2 (degree 4 by Rubinstein et al. (2015) [11] and Briard (2017) [12]).

Accordingly, the classical expansion in terms of scalar spherical harmonics is much more practical, especially when the degree increases:

$$\mathcal{E}(\boldsymbol{k}) = \frac{E(k)}{4\pi k^2}\left(1 + \sum_{n=1}^{N}\sum_{m=-2n}^{2n} e^m_{2n}(k)Y^m_{2n}(\theta,\phi)\right), \tag{15}$$

in which $Y^m_n(\theta,\phi)$ are expressed in terms of extended Legendre polynomials $P^m_n(\theta)$ via

$$Y^m_n(\theta,\phi) = P^m_n(\theta)\exp(\imath m\phi). \tag{16}$$

$2n$ is called the degree, only even here, up to a maximum $2N$, and m is the order that is limited by $2n$. In contrast with the expansion in terms of tensors, the properties of orthogonality are obvious. The basis depends on the choice of the polar axis but not the degree so that at any given degree, there are simple linear relationships to pass from $Y^m_n(\theta,\phi)$ to $Y'^m_n(\theta',\phi')$ from a system of polar-spherical coordinates to another one.

Only even degrees are relevant, from the Hermitian symmetry restricted to a purely real term. Note that the number of degrees of freedom is recovered from the tensorial decomposition to the scalar spherical one: at degree 2, there are five $e^m_2(k)$ descriptors, with $m = -2, -1, 0, 1, 2$, and five independent components for the symmetric traceless tensor $H^{(\mathrm{dir})}_{ij}(k)$.

3.2. Angular Harmonics for the Polarization Term

The decomposition in terms of spherical harmonics was touched upon by Jack Herring for homogeneous axisymmetric turbulence with mirror symmetry. In this case, the two-point second order spectral tensor reduces to two components only:

$$\Phi^{11}(k,\theta) = \hat{R}_{ij}e^{(1)}_i e^{(1)}_j \quad \text{and} \quad \Phi^{22}(k,\theta) = \hat{R}_{ij}e^{(2)}_i e^{(2)}_j. \tag{17}$$

For this particular symmetry, it was proposed to expand both Φ^{11} and Φ^{22} in terms of the simple Legendre polynomials $P^0_n(\theta)$. This was correct for the spectral energy, or $(1/2)(\Phi^{11} + \Phi^{22})$, but not for the polarization counterpart, or $(1/2)(\Phi^{22} - \Phi^{11})$, which reduces to Z. In effect, Z may vanish at the pole in the axisymmetric case. This suggests restricting the expansion in terms of the simple Legendre polynomials to $\mathcal{E}$ and to $\tilde{Z}$ with $Z = \sin^2\theta\tilde{Z}$.

From the simple example of axisymmetric turbulence, it is clear that the SSH expansion cannot be applied directly to Z. Setting aside a possible direct application of VSH, in Section 5, a simpler method is first proposed. Its goal is to solve, beyond the axisymmetric case, the problem of the special definition or multiple definition of a vector field in the Craya–Herring frame of reference that implies a similar problem for Z.

For this purpose, one recovers the decomposition by CC and Teissèdre (1985) [13] as

$$Z(\boldsymbol{k}) = \sin^2\theta_k \tilde{Z}(\boldsymbol{k}) - \left(\frac{1+\cos\theta_k}{2}\right)^2 \exp(2\imath\phi_k)\Psi(k) - \left(\frac{1-\cos\theta_k}{2}\right)^2 \exp(-2\imath\phi_k)\Psi^*(k), \quad (18)$$

in which Ψ derives from the value of $\hat{R}_{ij}(k, \boldsymbol{n})$ exactly at the pole $\boldsymbol{\alpha} = \boldsymbol{n}$:

$$\Psi = \frac{1}{2}\left(\hat{R}_{22} - \hat{R}_{11} + \imath(\hat{R}_{12} + \hat{R}_{21})\right). \quad (19)$$

This equation derives from the calculation of $(1/2)N_i^* N_j^* \Psi_{ij}(\boldsymbol{k})$: The choice $\Psi_{ij}(\boldsymbol{k}) = \hat{R}_{ij}(k, \boldsymbol{n}) + \sin^2\theta_k \tilde{\Psi}_{ij}(\boldsymbol{k})$ yields Equation (18). This equation expresses explicitly the *multidefinition* of Z at the exact pole, with $Z \to -\exp(2\imath\phi_k)\Psi$. Accordingly, Ψ in Equations (18) and (19) is the continuous limit of Z at $\phi_k = \pi/2$, but this is not the case in following any other meridian line (fixed ϕ_k) when converging towards the pole. In the whole spectral domain, this equation can be considered as exact provided that $\hat{R}_{ij}(\boldsymbol{k}) - \hat{R}_{ij}(k, \boldsymbol{n})$ behaves as $\sin^2\theta_k$. The particular axisymmetric case corresponds to $\Psi = 0$, $\tilde{Z} = \tilde{Z}(k, \theta_k)$.

Z cannot be expanded in terms of Y_n^m, but Equation (18) suggests transferring the decomposition in terms of scalar spherical harmonics from Z to $\tilde{Z}$, and the preliminary results of CC and Teissèdre (1985) were encouraging [13]. In this case, Ψ gives the polarization of the spectral tensor exactly at the pole, and a correct convergence to this polar value is ensured, with

$$\tilde{Z} = \sum_{n=0}^{N} \sum_{m=n}^{n} z_n^m(k) Y_n^m(\theta, \phi),$$

with both even and odd degrees. Coefficients with odd degrees are imaginary.

As another good property, Equation (18) and the latter is consistent with the degree 2, or

$$Z(\boldsymbol{k}, t) = \frac{5}{2} N_i^*(\boldsymbol{\alpha}) N_j^*(\boldsymbol{\alpha}) H_{ij}^{(\text{pol})}(k, t),$$

with $\Psi(k) = \frac{5}{2}(H_{22}^{(\text{pol})} - H_{11}^{(\text{pol}} + 2\imath H_{12}^{(\text{pol})})$ and $\tilde{Z}(k) = \frac{5}{2}(2H_{33}^{(\text{pol})} - H_{11}^{(\text{pol})} - H_{22}^{(\text{pol}})$.

Because Z represents the tensor of polarization, it is interesting to go beyond SSH, and to look at vectorial spherical harmonics (VSH); if a decomposition is valid for a vector $\boldsymbol{V}$, it should apply to a tensor, forming $\boldsymbol{V} \otimes \boldsymbol{V}$.

Even if physical data for the anisotropic helicity spectrum are missing in HAT, an SSH decomposition can be proposed, similarly to (13) and (15) but with additional, purely imaginary, terms of odd degrees.

Going back to the SO(3)-type expansion, a general one can be proposed as

$$Z(\boldsymbol{k}) = \frac{1}{2}\frac{E(k)}{4\pi k^2}\left(H_{ij}^{2\,(\text{pol})}(k) + \imath H_{ijm}^{3(\text{pol})}(k)\alpha_m + H_{ijmn}^{4\,(\text{pol})}\alpha_m\alpha_n + \cdots\right) N_i^*(\boldsymbol{\alpha}) N_j^*(\boldsymbol{\alpha}). \quad (20)$$

Note that the terms with odd degrees yield imaginary contributions from generating k-modulus tensors. As for the energy spectrum, the identity

$$H_{ij}^{2(\text{pol})}(k) = 5H_{ij}^{(\text{pol})}(k), \quad (21)$$

holds, in which $2E(k)H_{ij}^{(\text{pol})}(k)$ is the spherical integral of $\hat{R}_{ij}^{(\text{pol})}(\boldsymbol{k})$. In addition, terms of degree 3 and 4 were investigated by Briard (2017) [12] but without a practical and systematic way to reach higher degrees (see also Rubinstein et al. (2015) [11]).

4. SSH Decomposition in Rotating Shear Turbulence

In this section, let us concentrate on rotating shear flows. We begin by reviewing the ZCG model and the numerical simulations based on it, with typical combinations of mean shear and system rotation. Using the simulation data, we evaluate the equivalency of the tensorial expansion of the SO(3) type and the SSH decomposition on the $\mathcal{E}$ field. Then, by expanding $\mathcal{E}$ and $|Z|$ to high degrees using SSH, we explore the anisotropy-generating mechanism by the linear operator and anisotropy damping induced by nonlinear terms of ZCG.

4.1. Recalling ZCG Model

ZCG is a statistical spectral model developed for homogeneous anisotropic turbulence by the author Y. Zhu, C. Cambon, and our collaborator F. S. Godeferd. For all relevant formulas, numerical details, and simulation results presented in this section, part of them are published in Zhu et al. (2019) [6], and the other unpublished part is in the Ph.D. thesis of the first author [14].

4.1.1. Spectral Models for Homogeneous Anisotropic Turbulence

Consider an extensional mean flow with a spatially uniform mean-velocity gradient $A_{ij} = S_{ij} + \frac{1}{2}\epsilon_{imj}\mathcal{W}_m$, under a solid body rotation of the frame with the angular velocity $\boldsymbol{\Omega}$, The governing equation for $(\mathcal{E}, Z)$ ($\mathcal{H}$ is neglected) reads

$$(k\dot{\mathcal{E}}) + 2\nu k^3\mathcal{E} + \Re\Big(kZ(\boldsymbol{k},t)S_{ij}N_i(\boldsymbol{\alpha})N_j(\boldsymbol{\alpha})\Big) = kT^{(\mathcal{E})}(\boldsymbol{k},t), \quad (22a)$$

$$(k\dot{Z}) + 2\nu k^3 Z + k\mathcal{E}(\boldsymbol{k},t)S_{ij}N_i(-\boldsymbol{\alpha})N_j(-\boldsymbol{\alpha}) - \imath kZ(\boldsymbol{k},t)((\boldsymbol{\mathcal{W}} + 4\boldsymbol{\Omega})\cdot\boldsymbol{\alpha} - 2\Omega_E) = kT^{(Z)}(\boldsymbol{k},t). \quad (22b)$$

The overdot denotes the advection operator due to the presence of the mean flow, namely $(\dot{...}) = \frac{\partial}{\partial t} - A_{mn}k_m\frac{\partial}{\partial k_n}$. ν is the kinetic viscosity. Ω_E is the rotation rate induced by the advection operator, which can be removed by applying special $\boldsymbol{A}$ and $\boldsymbol{n}$. Equation (22) derives from the Navier–Stokes equations (NSE). The left-hand side (LHS) of Equation (22) represents the linear effects of the mean flow as in viscous linear spectral theory (SLT), which was originally developed as RDT (see [15,16]). Note that the LHS of (22a) and (22b) contains viscous terms and similar terms from strain $\boldsymbol{S}$ (the symmetric part of the mean-velocity gradients), but the antisymmetric part of the mean-velocity gradients (the mean vorticity $\boldsymbol{\mathcal{W}}$) only affects the equation for polarization through a combination with the system vorticity $4\boldsymbol{\Omega}$ (the 'stropholysis' effect coined by [17]). The right-hand side (RHS) of Equation (22) gathers the contribution from two-point third-order correlations mediated by the quadratic nonlinearity of basic NSE that should be closed properly depending on the flow regime and available computational resources.

Previous studies developed a series of sophisticated and successful models based on the high-order closures using the eddy damped quasi-normal Markovian (EDQNM) technique. The EDQNM closure was first built by Orszag for homogeneous isotropic turbulence (HIT) (see [18]), later extended to shear-driven flow and flows driven by buoyancy or coupled fields, such as magneto-hydrodynamics (see review in [10,19]). Concerning shear-driven flows, various formulations and resolution methods can be chosen. For example, EDQNM-1—the model derived for turbulent flows when linear terms are associated with energy production in double correlation equations—keeps the full angular dependence in (22). The model of EDQNM-1 type for unstably stratified homogeneous turbulence in [20] compares well with direct numerical simulation (DNS), but EDQNM-1 for a purely anisotropic velocity field has not been implemented because of the complicated and computationally expensive 3D convolutions involved with the $T^{(\mathcal{E})}$ and $T^{(Z)}$ terms. A simplified model from EDQNM-1—MCS (see [21])—gives the spherically averaged descriptors that depend on radius k by retaining only the first two degrees of the angular harmonics of $\mathcal{E}(\boldsymbol{k})$ and $Z(\boldsymbol{k})$ in (13) and (20). Although validated for different flows, the MCS model shows that it is much less adapted to the representation of angular variations for linear terms than for nonlinear terms based on long-term comparisons with SLT and DNS.

The ZCG model was designed to restore the full angular dependence of linear terms in the equations for $\mathcal{E}(\boldsymbol{k},t)$ and $Z(\boldsymbol{k},t)$, as in EDQNM-1, and to restrict to nonlinear transfer terms $T^{(\mathcal{E})}$ and $T^{(Z)}$ the use of low-degree expansion inspired by MCS. Additionally, to include the effect from the anisotropy higher than degree 2, ZCG added return-to-isotropy (RTI) terms to $T^{(\mathcal{E})}$ and $T^{(Z)}$ as

$$\begin{aligned} T^{(\mathcal{E})}(\boldsymbol{k},t) &= T^{(\mathcal{E})2}(\boldsymbol{k},t) - \varphi^{(\mathrm{RTI})}(k,t)\left(\mathcal{E}(\boldsymbol{k},t) - \mathcal{E}^{(2)}(\boldsymbol{k},t)\right) \\ T^{(Z)}(\boldsymbol{k},t) &= T^{(Z)2}(\boldsymbol{k},t) - \varphi^{(\mathrm{RTI})}(k,t)\left(Z(\boldsymbol{k},t) - Z^{(2)}(\boldsymbol{k},t)\right). \end{aligned} \tag{23}$$

$T^{(\mathcal{E})2}(\boldsymbol{k},t)$ and $T^{(Z)2}(\boldsymbol{k},t)$ are reconstructed from spherically averaged nonlinear descriptors $\tilde{S}_{ij}^{\mathrm{NL(dir)}}(k,t)$ and $\tilde{S}_{ij}^{\mathrm{NL(pol)}}(k,t)$ given by MCS, as an analogy to (13) and (20), truncated at degree 2. $\mathcal{E}^{(2)}(\boldsymbol{k},t)$ and $Z^{(2)}(\boldsymbol{k},t)$ are the exact first two-degree components of (13) and (20), respectively. $\varphi^{(\mathrm{RTI})}(k,t)$ is borrowed from Weinstock's model [22,23], in which only isotropic EDQNM is employed for the nonlinear closure, other than MCS.

Solving the advection operators in (22) is the main challenge for numerically simulating ZCG. In the commonly used approach of SLT, as well as in fully nonlinear DNS by Rogallo (1981) [24] and Lesur and Longaretti (2005) [25], the scheme amounts to following the characteristic lines in terms of $\boldsymbol{k}(t)$ related to the mean Lagrangian trajectories in physical space. As a result, the wave vector is time dependent so that the time evolution of a statistical quantity Φ appears as $\dot{\Phi}(\boldsymbol{k}(t),t) = \dfrac{\partial\Phi}{\partial t} + \dfrac{\partial\Phi}{\partial k_i}\dfrac{\mathrm{d}k_i}{\mathrm{d}t}$ with $\dot{k}_i = -A_{ji}k_j$ the eikonal equation. To avoid the distorted grids in the conventional method, which are difficult to couple with nonlinear models based on spherically averaged descriptors, the ZCG model uses a finite-difference scheme for evaluating the $\partial/\partial k_i$ derivatives (4), with a discretization of the wave vector consistent with the polar-spherical coordinates presented in Figure 1. To fix the multi-definition of $Z(\boldsymbol{k},t)$ at the pole, we look backward to the $\hat{\mathbf{R}}(\boldsymbol{k}.t)$ equation in a small region around the pole. Readers can find the special treatment of the pole zone in Appendix A.

4.1.2. Numerical Simulations for Rotating Shear Flows

Among various combinations of mean-flow gradients and system rotation, the rotating shear flow with the mean plane shear rotating in the spanwise direction is of great interest for its widespread applications from engineering—e.g., turbomachinery or hydroelectric power generation—to geophysics and astrophysics. Previous studies with SLT (spectral linear theory), e.g., ref. [16] for rotating shear turbulence, ref. [26] for stratified shear flows and [27] for rotating stratified shear turbulence, show the global relevance of the Bradshaw number B (see [28]) for characterizing the stability of rotating shear flow: $B = R(R+1)$, in which the Rossby number $R = 2\Omega/-S$ is the ratio of system vorticity Ω to shear-induced vorticity $-S$. Cases with $B < 0$ or $-1 < R < 0$ correspond to the exponential growth of turbulent kinetic energy, and $B > 0$ to exponential decay. Neutral cases are found for $R = 0$ (no background rotation) and $R = -1$ (zero absolute vorticity). Accordingly, four typical rotating shear flows (with various values of R) are numerically simulated in [6] by ZCG, namely using (22) and (23): $R = -5$ corresponds to a stabilizing, anticyclonic case; $R = -1$ for a neutral case with zero absolute vorticity, as encountered in the central region of a rotating channel; $R = -1/2$ for a maximum destabilization, anticyclonic case, as in the pressure side of a rotating channel; and $R = 0$ with no rotation. The mean shear and system rotation are chosen as $A_{ij} = S\delta_{i1}\delta_{j2}$ and $\boldsymbol{\Omega} = \Omega\delta_{i3}$, respectively, to vanish the term of Ω_E in Equation (22).

We first carried out the simulations in the linear limit (namely setting the RHS of Equation (22) zero), with $\nu = 0$ (the inviscid linear limit) and $\nu \neq 0$ (the viscous linear limit), respectively. The ZCG model showed an almost perfect coincidence of the results, compared to their corresponding ones obtained by the STL for all four values of R and compared to the analytical results for the pure shear case when $R = 0$. Then, we turned on the RHS of Equation (22), i.e., simulating the fully nonlinear ZCG model, and compared the data to those obtained by Sahli et al. [29] using DNS. Great agreement of the results between ZCG and DNS for all the four typical values of R firmly validated our ZCG model. It is interesting that although MCS failed in most cases, it succeeded in generating exponentially increasing energy in the maximum destabilization case ($R = -1/2$). Figure 2 presents the time evolution of the kinetic energy $\mathcal{K}(t) = \int_0^\infty E(k,t)dk$ by simulating ZCG, with various Rossby numbers, in the three different limits.

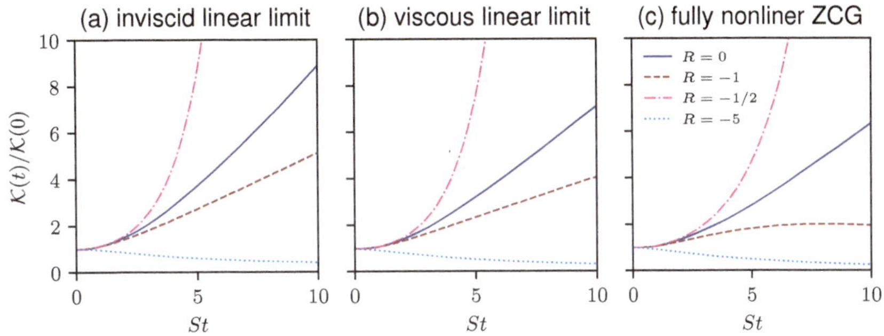

Figure 2. Time evolution of turbulent kinetic energy. Comparisons of results from the cases with typical values of R: (**a**) in the inviscid linear limit; (**b**) in the viscous linear limit, and (**c**) with fully nonlinear ZCG.

Figure 3 exhibits the according time evolution of the deviatoric part of the Reynolds stress tensor b_{12}, which is the Reynolds stress tensor component $R_{12} = \langle u_1(\boldsymbol{x})\, u_2(\boldsymbol{x})\rangle$ non-dimensionalized by $2\mathcal{K}$, and plays an important role in the energy injection by the mean shear. The most remarkable achievement in [6] is that the fully nonlinear ZCG accomplished the clearly steady limit of $b_{12} = -0.14$ in the pure shear case (see Figure 3c), with a constant value very close to the classically expected one, in the range $[-0.16; -0.1]$ (see table p. 443 in [10]), and thereby provided the exponential re-growth of energy (see Figure 2c), for the first time as a spectral model.

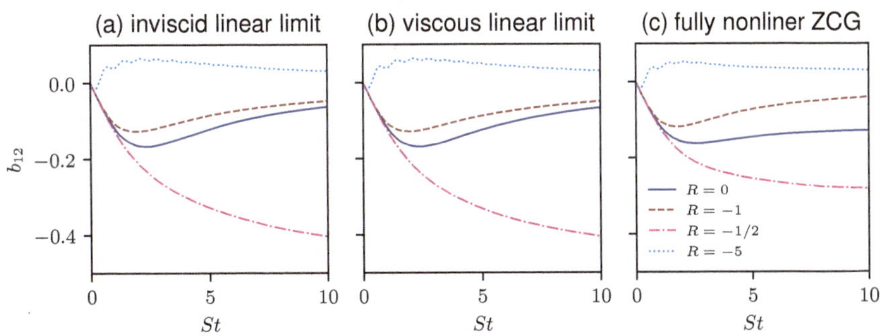

Figure 3. Time evolution of the deviatoric part of the Reynolds stress tensor b_{12}. Comparisons of results from the cases with typical values of R: (**a**) in inviscid linear limit; (**b**) in viscous linear limit; and (**c**) with fully nonlinear ZCG.

4.2. Numerical Validation of the Equivalency of SO(3) Expansion and SSH Decomposition

Here, we use the simulation data introduced above by the ZCG model to validate the equivalency of the tensorial expansion (of SO(3) type) and the SSH decomposition of a scalar field, with their applications on $\mathcal{E}(\boldsymbol{k})$ firstly.

Rubinstein et al. [11] pointed out some properties of the coefficient tensors in the SO(3)-type expansion (13): They can be assumed symmetric under any interchange of indices and also trace free in the extended sense that the contraction of any two indices vanishes identically. There are $2n+1$ linearly independent tensors with this property for each degree $2n$. Note that degree zero corresponds to the isotropic part $\frac{E(k)}{4\pi k^2}$, and the second degree is found as $H_{ij}^{2(\mathrm{dir})} = -15H_{ij}^{\mathrm{dir}}$, which is applied on MCS. Rubinstein et al. [11] extended a practical expansion for degree 4, as

$$E(k)H_{mnpq}^{4(\mathrm{dir})} = \iint_{S_k} \mathcal{E}(\boldsymbol{k})P_{mnpq}(\boldsymbol{\alpha})\mathrm{d}^2\boldsymbol{k}\,, \tag{24}$$

with

$$\begin{aligned} P_{ijpq}(\boldsymbol{\alpha}) =& \alpha_i\alpha_j\alpha_p\alpha_q - \frac{1}{7}(\delta_{ij}\alpha_p\alpha_q + \delta_{ip}\alpha_j\alpha_q + \delta_{iq}\alpha_j\alpha_p \\ &+ \delta_{jp}\alpha_i\alpha_q + \delta_{jq}\alpha_i\alpha_p + \delta_{pq}\alpha_i\alpha_j) + \frac{1}{35}(\delta_{ij}\delta_{pq} + \delta_{ip}\delta_{jq} + \delta_{iq}\delta_{pj})\,. \end{aligned} \tag{25}$$

It is unrealistic to extend to higher expression for its complexity.

To compare the results, we denote

$$\mathcal{E}^0(\boldsymbol{k}) = \&\mathcal{E}^{0(\mathrm{T})}(\boldsymbol{k}) = \mathcal{E}^{0(\mathrm{S})}(\boldsymbol{k}) = \frac{E(k)}{4\pi k^2}\,, \tag{26}$$

for the 0-th degree component of $\mathcal{E}(\boldsymbol{k})$, where'T' and 'S' stand for "tensorial" and "SSH", respectively. For other high-degree components ($2n > 0$), the definitions come from (13) and (15) as

$$\mathcal{E}^{2\mathrm{n}(\mathrm{T})}(\boldsymbol{k}) = \&\frac{E(k)}{4\pi k^2}H_{l_1l_2\ldots l_{2n}}^{2n(\mathrm{dir})}(k)a_{l_1}a_{l_2}\ldots a_{l_{2n}}\,, \tag{27}$$

$$\mathcal{E}^{2\mathrm{n}(\mathrm{S})}(\boldsymbol{k}) = \&\frac{E(k)}{4\pi k^2}\sum_{m=-2n}^{m=2n} e_{2n}^m(k)Y_{2n}^m(\theta_k,\phi_k)\,. \tag{28}$$

The coefficients $e_{2n}^m(k)$ can be found simply by the integrals:

$$e_{2n}^m(k) = 4\pi\iint_{S_k}\frac{\mathcal{E}(\boldsymbol{k})}{E(k)}Y_{2n}^m(\theta,\phi)\mathrm{d}^2\boldsymbol{k}\,. \tag{29}$$

Without loss of generality, we plot $\mathcal{E}^{2(\mathrm{T})}(\boldsymbol{k})$, $\mathcal{E}^{2(\mathrm{S})}(\boldsymbol{k})$, $\mathcal{E}^{4(\mathrm{T})}(\boldsymbol{k})$ and $\mathcal{E}^{4(\mathrm{S})}(\boldsymbol{k})$ divided by $\mathcal{E}^0(\boldsymbol{k})$, for $R = -1/2$, at the middle dimensionless time $St = 5$, for the characteristic wave number $k_\lambda = 1/\lambda$ corresponding to the Taylor scale λ, in Figure 4.

The results in Figure 4 (and other results at different wave numbers and for various values of R in [14]) show clearly an almost perfect coincidence between the SO(3) type decomposition and the SSH expansion.

Further, to observe the scale effects of anisotropy in different degrees, we define a normalized spectrum for $\mathcal{E}$ in degree $2n$ as shown below:

$$a^{2n(\mathcal{E})}(k) = \frac{1}{E(k)}\iint_{S_k}|\mathcal{E}^{2n}(\boldsymbol{k})|\mathrm{d}^2\boldsymbol{k}\,. \tag{30}$$

One always obtains 0 if naively removing the absolute value operator $|\cdot|$ from the integrand for any $n > 0$. Trivially, we have $a^{0(\mathcal{E})}(k) = 1$, and for isotropic turbulence $a^{2n(\mathcal{E})}(k) = 0$ for $n > 0$. We can roughly regard the directional anisotropy of $\hat{\boldsymbol{R}}(\boldsymbol{k})$ in degree $2n > 0$ at k as a negligible component when $a^{2n(\mathcal{E})}(k) \ll 1$; otherwise, it is considerable when $a^{2n(\mathcal{E})}(k) \sim 1$. Figure 5 presents $a^{2n(\mathcal{E})}(k)$ obtained by the SO(3) expansion and the SSH decomposition, respectively, for the maximum destabilization case, at $St = 5$, in three different limits. All of the figures indicate that the spherical harmonics decomposition agrees with the tensorial expansion very well with high accuracy in both degree 2 and degree 4, either with or without a nonlinear mechanism, at any length scale.

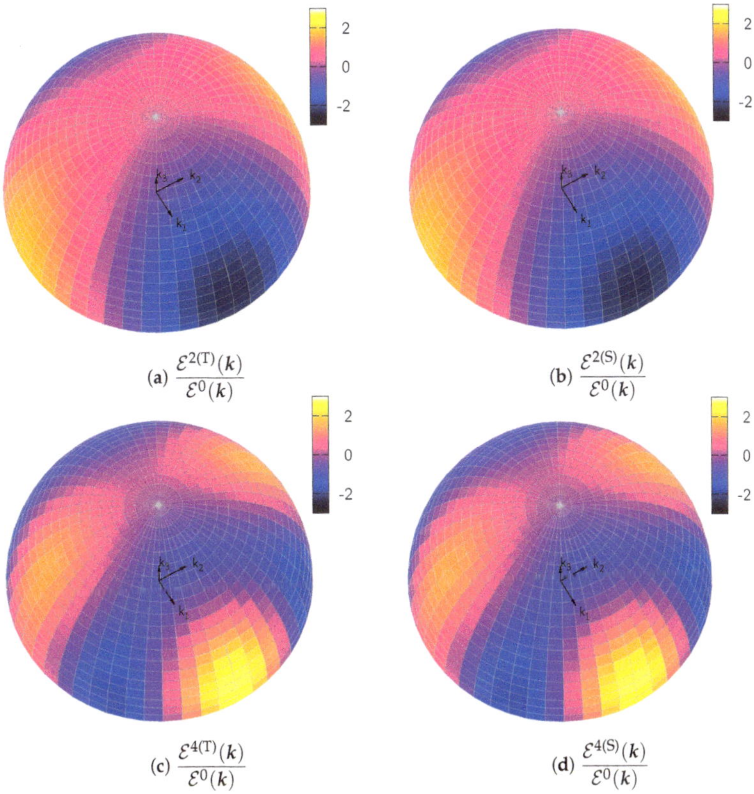

(a) $\frac{\mathcal{E}^{2(\mathrm{T})}(\boldsymbol{k})}{\mathcal{E}^0(\boldsymbol{k})}$ (b) $\frac{\mathcal{E}^{2(\mathrm{S})}(\boldsymbol{k})}{\mathcal{E}^0(\boldsymbol{k})}$

(c) $\frac{\mathcal{E}^{4(\mathrm{T})}(\boldsymbol{k})}{\mathcal{E}^0(\boldsymbol{k})}$ (d) $\frac{\mathcal{E}^{4(\mathrm{S})}(\boldsymbol{k})}{\mathcal{E}^0(\boldsymbol{k})}$

Figure 4. Spherical distributions of $\mathcal{E}^2(\boldsymbol{k})$ and $\mathcal{E}^4(\boldsymbol{k})$ in viscous linear limit with $R = -1/2$ at $St = 5$ at the wavenumber $k = k_\lambda$. Comparison of results obtained by the SO(3)-type expansion and SSH decomposition, respectively.

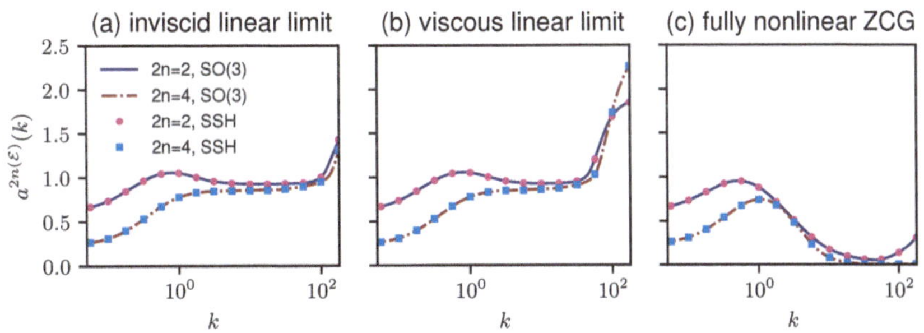

Figure 5. Spherically averaged anisotropy spectra $a^{2n(\mathcal{E})}(k)$ for $\mathcal{E}(\boldsymbol{k})$ in degree $2n = 2$ and $2n = 4$ at $St = 5$. Comparisons of the results obtained respectively by tensorial expansion (SO(3) type) and SSH decomposition: (**a**) in the inviscid linear limit; (**b**) in the viscous linear limit; and (**c**) with fully nonlinear ZCG.

4.3. Analysis on High-Degree Anisotropy

As introduced before, the dynamics of high-degree anisotropy are of great importance in modeling anisotropic turbulence. The SSH decomposition that can be easily extended to high degrees permits us to explore the generating and damping of high-degree anisotropy in rotating shear flows. To consider the polarization anisotropy, we decompose frame-invariant $|Z(\boldsymbol{k})|$ by the SSH, for $Z(\boldsymbol{k})$ cannot be decomposed by spherical harmonics directly since it is singular at the pole. Analogous to (15), one finds

$$|Z(\boldsymbol{k})| = z_0\left(1 + \sum_{n=1}^{\infty}\sum_{m=-2n}^{2n} z_{2n}^{\prime m}(k)Y_{2n,m}(\theta_k,\phi_k)\right), \tag{31}$$

with

$$z_0 = \frac{1}{4\pi k^2}\iint_{S_k}|Z(\boldsymbol{k})|\mathrm{d}^2\boldsymbol{k}, \tag{32}$$

Also, we define

$$Z^{2n}(\boldsymbol{k}) = \sum_{m=-2n}^{m=2n} z_{2n}^{\prime m}(k)Y_{2n}^{m}(\theta_k,\phi_k), \quad n = 1,2,3\ldots, \tag{33}$$

and

$$a^{2n(Z)}(k) = \frac{1}{4\pi k^2 z_0}\iint_{S_k}|Z^{2n}(\boldsymbol{k})|\mathrm{d}^2\boldsymbol{k}. \tag{34}$$

The decomposition allows us to obtain high degrees of polarization anisotropy.

First, we investigate $a^{2n(\mathcal{E})}(k)$ and $a^{2n(Z)}(k)$ for $2n = 2,4,6,8$ in the inviscid limit (Figure 6) and viscous linear limit (Figure 7), for various values of R, at $St = 5$.

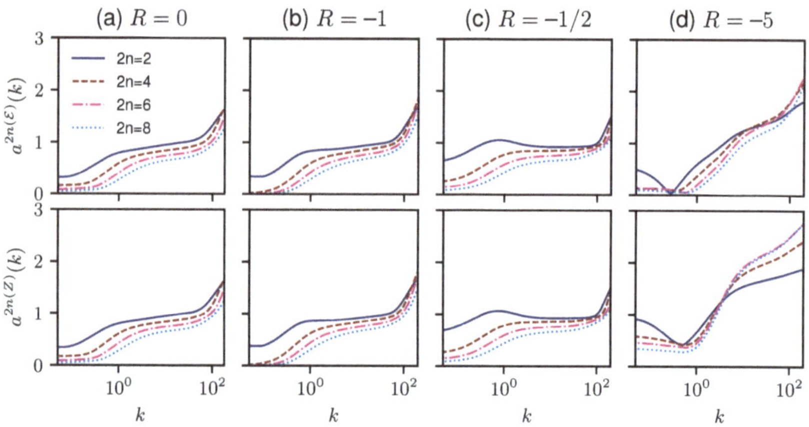

Figure 6. Spherically integral anisotropy $a^{2n(\mathcal{E})}(k)$ for $\mathcal{E}(\boldsymbol{k})$ (top) and $a^{2n(Z)}(k)$ for Z (bottom) in degree $2n = 2,4,6,8$, in inviscid linear limit at $St = 5$ for various values of R.

Overall, for the same case, the distribution of directional anisotropy in terms of degree and wave number looks similar to its corresponding one for polarization anisotropy. The anisotropy of degree 2 generated by the linear operator dominates in most scale ranges and in all cases. Note that $a^{2n(\mathcal{E})}(k)$ and $a^{2n(Z)}(k)$ are the anisotropy divided by the corresponding zero degree values, so they are actually the relative anisotropy in degree $2n$. Accordingly, all the simulations show stronger relative high-degree anisotropy at the largest k's than at the smallest k's. Comparing Figure 7 to Figure 6, not surprisingly, we find that the viscosity acts basically in the largest k range (roughly $k > 100$), namely, the dissipating scale: the viscosity weakens the anisotropy in degree 2 but enhances other higher-degree components for the largest k values. The stabilizing case contributes the most different behavior, which shows quite small anisotropy around the integral length scale (around $k = 1$). In this case, the anisotropy of degree 2 does not dominate in the inertial range (around $k = 10$) nor in the dissipation range.

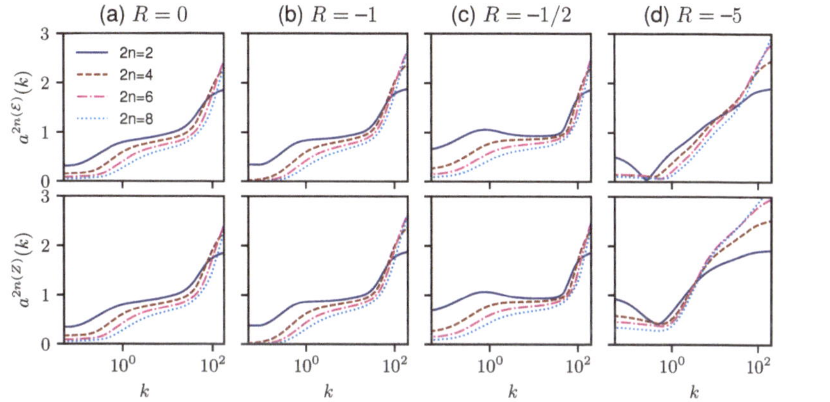

Figure 7. Spherically integral anisotropy $a^{2n(\mathcal{E})}(k)$ for $\mathcal{E}(\boldsymbol{k})$ (top) and $a^{2n(Z)}(k)$ for Z (bottom) in degree $2n = 2, 4, 6, 8$, in viscous linear limit at $St = 5$ for various values of R.

The neologism 'stropholysis' was coined by Kassinos and Reynolds in the different context of improved single-point closure models in introducing new tensors with respect to the Reynolds stress tensor, and better modeling the effects of background rotation. In our straightforward spectral analysis, the corresponding effect breaks the mirror symmetry in the polarization term, with a direct impact on the imaginary part of Z only. In the present case of the rotating shear, the dynamical effect of stropholysis is mediated by the term $\mathcal{W} + 4\boldsymbol{\Omega}$ in Equation (22b). Accordingly, the related effect vanishes in the case of maximum destabilization, and it is expected to damp, in average, the polarization term in all other cases.

The 'stropholysis' term is the only explicitly different term for all cases with various R and the same mean shear rate S: $\mathcal{W} + 4\boldsymbol{\Omega} = (0, S, 0)$ for the pure shear case; $\mathcal{W} + 4\boldsymbol{\Omega} = (0, 0, 0)$ vanishes in the maximum destabilization case; $\mathcal{W} + 4\boldsymbol{\Omega} = (0, -S, 0)$ for the neutral case, with the same net mean vorticity but opposite sign to $R = 0$; and $\mathcal{W} + 4\boldsymbol{\Omega} = (0, -9S, 0)$ for the stabilizing case $R = -5$ with relatively large net mean vorticity.

We present in Figure 8 the results obtained by the fully nonlinear ZCG. In all cases, the anisotropy in all the degrees reduces remarkably in the inertial and dissipation ranges, especially for $2n > 2$. This indicates that the RTI term used in the ZCG model works efficiently for anisotropy higher than degree 2 as we expected.

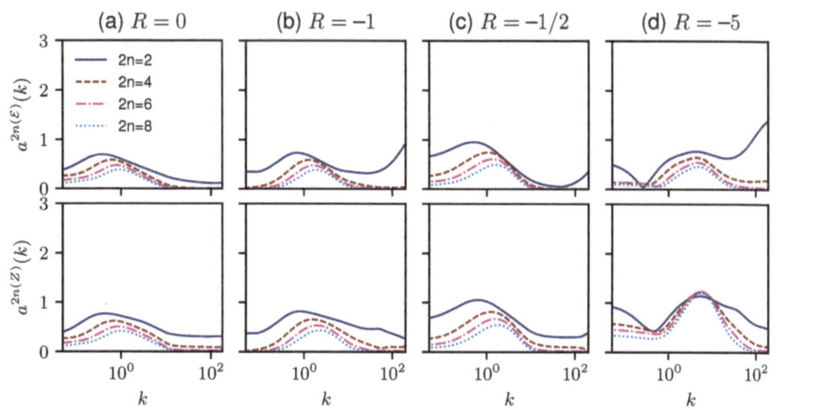

Figure 8. Spherically integral anisotropy $a^{2n(\mathcal{E})}(k)$ for $\mathcal{E}(\boldsymbol{k})$ (top) and $a^{2n(Z)}(k)$ for Z (bottom) in degree $2n = 2, 4, 6, 8$, with fully nonlinear ZCG, at $St = 5$ for various values of R.

5. Towards a More General Decomposition for a Vector Field

This section is motivated by recent studies by Daniel Schertzer, Ioula Chiriginskaia, and coworkers, who mentioned in [30] the following: *For instance, it seems ironic that multifractals have been mostly restricted to scalar-valued fields, whereas cascades were first invoked for the wind velocity.*

In the past, the important context of internal intermittency and its scaling was mainly investigated using *scalar structure functions* for moments of velocity increments in physical space. For models developed in spectral space, shell models were essentially restricted to isotropic turbulence.

Recent attempts to describe really anisotropic turbulent flows in terms of vector fields are thereby reviewed in this section. We first address the case of smooth vector fields, with a possible linkage to a modal projection, more complex than the Craya/Herring one, and related expansions in terms of VSHs. In the second part, intermittent vector fields are touched upon with the multifractal approach of Schertzer's group.

5.1. Using a General Toroidal/Poloidal Decomposition

Meetings around the DFGA theme (dynamics of geophysical and astrophysical fluids) have shown promising and truly multidisciplinary perspectives. For instance, the CNRS International Colloquium—Geophysical and Astrophysical Flows—took an interdisciplinary approach (organized by Philippe Fraunié, on October 16–18, in Paris, France), with several talks that addressed the use of VSH, from smooth to intermittent fields.

Decomposition in terms of VSH (vectorial spherical harmonics) was used by Rieutord (1987) [31] in order to solve linear operators of rotating flow on spheres in physical space. The essential difference with the simplified toroidal/poloidal decomposition in physical space, closely related to that of Craya in Fourier space, is a more complex definition of the toroidal mode as

$$\boldsymbol{u}^{(to)}(\boldsymbol{r}) = \boldsymbol{\nabla} \times \left(s^{(to)}(\boldsymbol{r}) \frac{\boldsymbol{r}}{r} \right), \quad (35)$$

following Chandrasekkhar (1981) [32]. This amounts to the substitution of the unit radial vector $\boldsymbol{r}/r$ to the polar axis $\boldsymbol{n}$, and the problem of some singular points on the surface of a sphere is not restricted to the pole of a polar-spherical system of coordinates.

Looking at the decomposition in terms of spherical harmonics, the (new) toroidal mode is expanded as

$$\boldsymbol{u}^{(to)}(r, \theta_r, \phi_r) = \sum_{n=0}^{N} \sum_{m=-n}^{n} u_n^m(r) \boldsymbol{\nabla} \times \left(Y_n^m(\theta_r, \phi_r) \frac{\boldsymbol{r}}{r} \right)$$

ref. [31].

On the other hand, the similarity between the representation in physical space and the one in Fourier space is conserved, but the counterpart in Fourier space is no longer purely algebraic. A spectral surrogate of Equation (35) is

$$\hat{\boldsymbol{u}}^{(to)}(\boldsymbol{k}) = \boldsymbol{k} \times \left(\frac{\partial s^{(to+)}(\boldsymbol{k})}{\partial \boldsymbol{k}} \right). \quad (36)$$

Interestingly, the change of the operator from physical to spectral space is very similar to the one for the advection operator. For the advection term, one has

$$x_n \frac{\partial}{\partial x_m} \quad \rightarrow \quad -k_m \frac{\partial}{\partial k_n}$$

As inferred from the new toroidal/poloidal (or spheroidal) modal projection in physical space, from Equation (35), we propose

$$\hat{\boldsymbol{u}}(\boldsymbol{k}) = \boldsymbol{\alpha} \times \left(k\frac{\partial}{\partial k}\left(s^{(to+)f}(\boldsymbol{k})\right)\right) + \boldsymbol{\alpha} \times \left(\boldsymbol{\alpha} \times \left(k\frac{\partial}{\partial k}\left(s^{(po+)f}(\boldsymbol{k})\right)\right)\right), \tag{37}$$

where a classical SSH decomposition holds for $s^{(to+)}$ and $s^{(po+)}$, with

$$s^{(to+)}(\boldsymbol{k}) = \sum_{n=0}^{N}\sum_{m=-n}^{n} t_n^m(k)Y_n^m(\theta_k, \phi_k), \quad s^{(po+)}(\boldsymbol{k}) = \sum_{n=0}^{N}\sum_{m=-n}^{n} p_n^m(k)Y_n^m(\theta_k, \phi_k). \tag{38}$$

Finally, the helical modes in Equation (6) are expressed as

$$u_s(\boldsymbol{k}) = \frac{1}{2}\hat{\boldsymbol{u}}(\boldsymbol{k}) \cdot \boldsymbol{N}(-s\boldsymbol{\alpha})$$

$$= -\frac{1}{2}\sum_{n=0}^{N}\sum_{m=-n}^{n} (p_n^m - s\imath t_n^m)\left((Y_n^m)_{,\theta} - s\imath\frac{1}{\sin\theta_k}(Y_n^m)_{,\phi}\right), \quad s = \pm 1. \tag{39}$$

It is, therefore, possible to reconstruct the spectral tensors from the above-mentioned VSH expansion of the helical modes, but this task is very complex. Looking at the two-point second-order spectral tensor, one can recall that the implicit use of VSH should only concern the polarization term Z.

5.2. Recent Progresses towards Stochastic Fields and Multifractal Approach

With the viewpoint of working with intermittent vector fields, is it possible to use Fourier space beyond the conventional shell models for isotropic turbulence? A first answer was given by Chiriginskaya et al. (1998) [33], from a seminal model by Chiriginskaia and Schertzer (1997) [34], who introduced a cascade of gyroscopes. With respect to the very old analogy of the cascade of eddies as the fragmentation of (scalar) grains, with, for instance, the log-normal distribution inspired by Kolmogorov and Obhukov, this approach is based on a dyadic network of triads of vector interactions between a parent eddy and two children eddies, from the largest scale down to the dissipation (in the molecular sense). Each triad forms a (really, 3D) gyroscope. This is a highly seductive approach, and it is based on the formal analogy of the Euler equation for fluids and the Euler equation for a rigid body as put forward by Arnold. This analogy was perhaps pushed to its best form by Waleffe [5] with his principle of triad instability, where the Euler equations for fluid are expanded in terms of helical modes [4]. Needless to say, it seems to be worthwhile to continue in this direction, as well as to clarify the mutual links with the general approach defining multifractal operators acting on vector fields [33].

6. Conclusions, Perspectives of Turbulence from Earth to Planets, Stars, and Galaxies

The present article, started from seminal studies of Jack Herring for modal projection and expansions in terms of spherical harmonics, is essentially technical. We investigate how to project solenoidal fields, mainly the velocity field, on a reduced number of modes. Projection is systematically carried out in physical space and in spectral one for the fluctuating field(s), with comparison, using 3D, direct and inverse, integral Fourier transform. From a random velocity field (possibly augmented from a buoyancy field and from a magnetic field), corresponding expansions of spectral tensors are derived. For the latter statistical applications, we emphasize the two-point second-order spectral tensor of fluctuating velocity. As far as possible, we emphasize the mathematical treatment of vectors, for instance, using VSH, in addition to SSH for scalar fields. In this sense, the expansion of the second-order spectral tensor $\hat{\mathbf{R}}$ can directly result from the VSH expansion of $\hat{\boldsymbol{u}}$ in building $\hat{\boldsymbol{u}}^* \otimes \hat{\boldsymbol{u}}$. This possibility is briefly explored in Section 5.1 but appears as a formidable task. On the other hand, the possible underlying vectorial aspect is reduced to only the polarization part Z of the spectral tensor $\hat{\mathbf{R}}$ since $\mathcal{E}$, $\mathcal{H}$, and even the Z-modulus can be treated as true scalars. Accordingly, a special way of expanding Z is investigated, forcing its correct convergence towards the pole of the system of the polar-spherical system of coordinates, which is puzzling for applying the Craya–Herring projection for non-axisymmetric statistical configurations. The particular example of rotating sheared turbulence is revisited from [6], with new unpublished results, in Section 4. We validate the equivalency of the tensorial SO(3)-type expansion and the SSH decomposition using the data obtained by the ZCG model. The analysis on high-degree anisotropy (higher than in ZCG) is first limited to the linear limit. Also, we verify that the nonlinear terms of the ZCG model damp the high-degree anisotropy efficiently as expected. In addition to the investigation of the anisotropy of the second-order correlations, for quasi-homogeneous turbulence subject to more general mean velocity gradients, the numerical treatment of the convection term by the mean flow in the ZCG model could be applied to DNSs.

New ways for describing intermittent turbulent flows are discussed, using vector fields and thereby going beyond the conventional use of scalar statistics for, for example, structure functions.

Last, but not least, applications to geophysics and astrophysics are expected from the title of the special issue that includes *from Earth to planets, stars, and galaxies*. To which extent are the above-mentioned technical developments useful for these issues? Some recent reviews are in [10,35]. The mathematical structure of the anisotropic turbulent flows is investigated. Still, their governing equations are hardly shown and discussed, except for rotating shear flows in Section 4, in the presence of body forces and/or basic large-scale flow: such a dynamical aspect is essential for geophysics and astrophysics and is discussed as follows.

6.1. Linear and Quasi-Linear Models from Geophysics to Accretion Discs in Astrophysics

If the linear mechanisms that are responsible for turbulence production and generation of various wave regimes are first investigated, we can move from linear stability analyses to the so-called "rapid distortion theory" and beyond. The rapid evolution of developed turbulence can be investigated in neglecting the explicit nonlinear terms but in initializing turbulence, such as for large Reynolds numbers. Linear responses can be applied to both initial data or to a given force representing nonlinearity as an impulsional response. A modern way towards rather complex inhomogeneous flows is in progress in the 'resolvent approach'.

The linear dynamics is useful for geophysical and astrophysical flows that are subjected to various dispersive waves: inertial waves in rotating turbulence, gravity waves in stably stratified flows, Alfvén waves in MHD, and various combinations of them. A recent analysis [36] shows, for instance, how destabilizing the resonances are between these waves in an elliptic core.

Application to the dynamics of accretion discs can be started from simple models for quasi-incompressible, quasi-homogeneous sheared turbulence [37], using spatial Fourier harmonics (or mean-flow-advected Fourier modes). In these studies, the nonlinearity is implicit and contributes to the regeneration of active modes, for which the linear mechanisms of *non-normal algebraic growth* are essential. Additional effects of mean stratification are analyzed in [38].

6.2. Models with Explicit Nonlinearity for Cascades

Wave turbulence theory is a nonlinear theory, even if it is reduced to weak interactions of nonlinear waves (see, e.g., ref. [39] and the references therein). One of the most complete analyses of inertial wave turbulence theory was developed in [40] for the asymptotic dynamics of rotating turbulence. Equations for the fluctuating field and for spectral tensors until the third order were expanded in terms of helical modes, which are the eigenmodes of inertial waves. Close analogies with anisotropic EDQNM, beyond [4], were used. It appears that when the turbulent flow is dominated by dispersive waves, the nonlinearity is drastically reduced and only survives at very large times via the nonlinear resonances of waves. In this asymptotic limit, a quasi-normal assumption, or zero value of cumulants (fourth order if the basic nonlinearity is triadic), becomes exact.

Of course, the cascade process for strong turbulence essentially needs an explicit model of nonlinearity. We mentioned EDQNM (anisotropic multimodal), as for the ZCG model in Section 4. In this case, where the dispersive waves are not dominant, the QN assumption can be a dead end, and ought to be complemented by a QN ingredient, as an empirical damping of fourth-order cumulants. Note that the toroidal mode in stably stratified turbulence is a 3D non-propagating mode, which renders such turbulence outside the scope of conventional wave–turbulence theory. Consequently, a new theory of weak turbulence is in progress for rotating stably stratified flows (Scott and Cambon, just submitted to JFM.)

Some applications to stratified turbulence were carried out with a multi-fractal model, supported by lidar observations [41,42]. Anisotropic scale invariance in atmospheric flows was addressed in [43,44].

Author Contributions: Conceptualization, Y.Z. and C.C.; methodology, Y.Z. and C.C.; software, Y.Z.; validation, Y.Z. and C.C.; formal analysis, Y.Z. and C.C.; investigation, Y.Z. and C.C.; data curation, Y.Z.; writing—original draft preparation, C.C.; writing—review and editing, Y.Z. and C.C.; visualization, Y.Z.; supervision, C.C. All authors have read and agreed to the published version of the manuscript.

Funding: This research received no external funding.

Institutional Review Board Statement: Not applicable.

Informed Consent Statement: Not applicable.

Data Availability Statement: Not applicable.

Acknowledgments: We thank Daniel Schertzer for reviewing the paper, especially for his extension, improvements, and correctness on the multifractal approach in Section 5.2.

Conflicts of Interest: The authors declare no conflicts of interest.

Appendix A. Degenerated Craya Equations Exactly at the Pole

A special definition of the Craya–Herring frame is needed in the pole, when the direction of the wave vector exactly coincides with the polar axis, or $\boldsymbol{\alpha} = \pm \boldsymbol{n}$. We consider half a space, taking into account the Hermitian symmetry, so that we focus on the vicinity of $\boldsymbol{\alpha} = \boldsymbol{n}$. For instance, the Craya frame is replaced by the Cartesian frame at this point. The spectral tensor $\hat{R}_{ij}$ again reduces to four non-zero components because of incompressibility, $\hat{R}_{ij}(k,\boldsymbol{n})n_j = \hat{R}_{ij}(k,\boldsymbol{n})n_i = 0$, say $\hat{R}_{\alpha\beta}(k,\boldsymbol{n})$, with Greek indices restricted to 1, 2 and $n_i = \delta_{i3}$. The degenerated Craya equation is

$$\left(\frac{\partial}{\partial t} + 2\nu k^2\right)\hat{R}_{\alpha\beta}(k,\boldsymbol{n},t) + (A_{\alpha\gamma} + 2\epsilon_{\alpha m\gamma}\Omega_m)\hat{R}_{\gamma\beta} + (A_{\beta\gamma} + 2\epsilon_{\beta m\gamma}\Omega_m)\hat{R}_{\alpha\gamma} = T_{\alpha\beta}. \quad \text{(A1)}$$

Drastic simplifications come from the zero contribution of linear pressure–strain terms, and from zero contribution from the differential operator $-A_{mn}k_m\frac{\partial}{\partial k_n}$ because $A_{mn}n_m = 0$. We can assume that the latter condition is fulfilled for any subsequent choice of A_{ij}, beyond the case of rotating shear flow under consideration. Let us not forget the additional Coriolis force, but its contribution is null if $\boldsymbol{\Omega}$ is chosen perpendicular to $\boldsymbol{n}$. The simplification of the viscous term holds in the same conditions. Incidentally, following characteristic lines, with a time-dependent wave vector (as in conventional RDT, in DNS by Rogallo 1981 [24] or, more explicitly, by Lesur 2005 [25]), the time dependence of $\boldsymbol{k}$ should disappear at the pole.

Accordingly, it is possible to work with only three real quantities, $\hat{R}_{11}$ $\hat{R}_{22}$, $\hat{R}_{12}$, or equivalently with $\mathcal{E}(k,\boldsymbol{n}) = \frac{1}{2}(\hat{R}_{11} + \hat{R}_{22})$ (no polar specificity), and with

$$\Psi(k) = \frac{1}{2}(\hat{R}_{22} - \hat{R}_{11}) + \imath\hat{R}_{12}, \quad \text{(A2)}$$

as the polar surrogate of Z.

Now, we need the polar form of third-order contributions T_{11}, T_{22}, T_{12}, or equivalently $T^{(\mathcal{E})}$ and T^{Ψ}. Without using the helical mode $\boldsymbol{N}$, the MCS (or CR06) expansion of T_{ij}, as the one of $\hat{R}_{ij}$, is

$$T_{ij} = \frac{T}{4\pi k^2} P_{ij}\left(1 - 15\tilde{S}^{\mathrm{NL\,(dir)}}_{mn}\alpha_m\alpha_n\right) +$$
$$+5\frac{T}{4\pi k^2}\left(P_{im}P_{jn} - \frac{1}{2}P_{ij}P_{mn}\right)\tilde{S}^{NL(pol)}_{mn}.$$

For $\alpha_i = n_i = \delta_{i3}$, this yields

$$T_{\alpha\beta} = \frac{T}{4\pi k^2}\left(\delta_{\alpha\beta}\left(1 - 15\tilde{S}^{\mathrm{NL\,(dir)}}_{33}\right) + 5\left(\tilde{S}^{NL(pol)}_{\alpha\beta} + \frac{1}{2}\delta_{\alpha\beta}\tilde{S}^{NL(pol)}_{33}\right)\right). \tag{A3}$$

Equivalently, one has

$$T^{(\mathcal{E})} = \frac{T}{4\pi k^2}\left(1 - 15\tilde{S}^{\mathrm{NL\,(dir)}}_{33}\right), \quad T^{\Psi} = 5\frac{T}{4\pi k^2}\left(\frac{1}{2}\left(\tilde{S}^{NL(pol)}_{22} - \tilde{S}^{NL(pol)}_{11}\right) + \imath\tilde{S}^{NL(pol)}_{12}\right)$$

References

1. Herring, J.R. Approach of axisymmetric turbulence to isotropy. *Phys. Fluids* **1974**, *17*, 859–872. [CrossRef]
2. Riley, J.J.; Metcalfe, R.W.; Weissman, M.A. Direct numerical simulations of homogeneous turbulence in density-stratified fluids. *AIP Conf. Proc.* **1981**, *76*, 79–112.
3. Craya, A. Contribution à l'analyse de la Turbulence Associée à des Vitesses Moyennes. Ph.D. Thesis, Université de Grenoble, Grenoble, France, 1957.
4. Cambon, C.; Jacquin, L. Spectral approach to non-isotropic turbulence subjected to rotation. *J. Fluid Mech.* **1989**, *202*, 295–317. [CrossRef]
5. Waleffe, F. The nature of triad interactions in homogeneous turbulence. *Phys. Fluids A Fluid Dyn.* **1992**, *4*, 350–363. [CrossRef]
6. Zhu, Y.; Cambon, C.; Godeferd, F.; Salhi, A. Nonlinear spectral model for rotating sheared turbulence. *J. Fluid Mech.* **2019**, *866*, 5–32. [CrossRef]
7. Cambon, C. Multipoint turbulence structure and modelling: The legacy of Antoine Craya. *Comptes Rendus Méc.* **2017**, *345*, 627–641. [CrossRef]
8. Smith, L.M.; Waleffe, F. Generation of slow large scales in forced rotating stratified turbulence. *J. Fluid Mech.* **2002**, *451*, 145–168. [CrossRef]
9. Galmiche, M.; Hunt, J.; Thual, O.; Bonneton, P. Turbulence-mean field interactions and layer formation in a stratified fluid. *Eur. J. Mech.-B/Fluids* **2001**, *20*, 577–585. [CrossRef]
10. Sagaut, P.; Cambon, C. *Homogeneous Turbulence Dynamics*, 2nd ed.; Springer: Cham, Switzerland, 2018. [CrossRef]
11. Rubinstein, R.; Kurien, S.; Cambon, C. Scalar and tensor spherical harmonics expansion of the velocity correlation in homogeneous anisotropic turbulence. *J. Turbul.* **2015**, *16*, 1058–1075. [CrossRef]
12. Briard, A.; Iyer, M.; Gomez, T. Anisotropic spectral modeling for unstably stratified homogeneous turbulence. *Phys. Rev. Fluids* **2017**, *2*, 044604. [CrossRef]
13. Cambon, C.; Teissèdre, C.; Jeandel, D. Etude d'effets couplés de rotation et de déformation sur une turbulence homogène. *J. Mec. Theor. Appl.* **1985**, *5*, 629.
14. Zhu, Y. Modelling and Calculation for Shear-Driven Rotating Turbulence, with Multiscale and Directional Approach. Ph.D. Thesis, Université de Lyon, Lyon, France, 2019.
15. Batchelor, G.; Proudman, I. The effect of rapid distortion of a fluid in turbulent motion. *Q. J. Mech. Appl. Math.* **1954**, *7*, 83–103. [CrossRef]
16. Salhi, A.; Cambon, C. An analysis of rotating shear flow using linear theory and DNS and LES results. *J. Fluid Mech.* **1997**, *347*, 171–195. [CrossRef]
17. Kassinos, S.C.; Reynolds, W.C.; Rogers, M.M. One-point turbulence structure tensors. *J. Fluid Mech.* **2001**, *428*, 213–248. [CrossRef]
18. Orszag, S.A. Analytical theories of turbulence. *J. Fluid Mech.* **1969**, *41*, 363–386. [CrossRef]
19. Cambon, C.; Mons, V.; Gréa, B.J.; Rubinstein, R. Anisotropic triadic closures for shear-driven and buoyancy-driven turbulent flows. *Comput. Fluids* **2017**, *151*, 73–84. [CrossRef]
20. Burlot, A.; Gréa, B.J.; Godeferd, F.S.; Cambon, C.; Griffond, J. Spectral modelling of high Reynolds number unstably stratified homogeneous turbulence. *J. Fluid Mech.* **2015**, *765*, 17–44. [CrossRef]
21. Mons, V.; Cambon, C.; Sagaut, P. A spectral model for homogeneous shear-driven anisotropic turbulence in terms of spherically averaged descriptors. *J. Fluid Mech.* **2016**, *788*, 147–182. [CrossRef]
22. Weinstock, J. Theory of the pressure–strain rate. Part 2. Diagonal elements. *J. Fluid Mech.* **1982**, *116*, 1–29. [CrossRef]

23. Weinstock, J. Analytical theory of homogeneous mean shear turbulence. *J. Fluid Mech.* **2013**, *727*, 256–281. [CrossRef]
24. Rogallo, R.S. Numerical Experiments in Homogeneous Turbulence. *NASA STI/Recon Tech. Rep. N* **1981**, *81*, 31508.
25. Lesur, G.; Longaretti, P.Y. On the relevance of subcritical hydrodynamic turbulence to accretion disk transport. *Astron. Astrophys.* **2005**, *444*, 25–44. [CrossRef]
26. Hanazaki, H.; Hunt, J.C.R. Structure of unsteady stably stratified turbulence with mean shear. *J. Fluid Mech.* **2004**, *507*, 1–42. [CrossRef]
27. Salhi, A.; Cambon, C. Stability of rotating stratified shear flow: An analytical study. *Phys. Rev. E* **2010**, *81*, 026302. [CrossRef]
28. Bradshaw, P. The analogy between streamline curvature and buoyancy in turbulent shear flow. *J. Fluid Mech.* **1969**, *36*, 177–191. [CrossRef]
29. Salhi, A.; Jacobitz, F.G.; Schneider, K.; Cambon, C. Nonlinear dynamics and anisotropic structure of rotating sheared turbulence. *Phys. Rev. E* **2014**, *89*, 013020. [CrossRef]
30. Schertzer, D.; Tchiguirinskaia, I. A century of turbulent cascades and the emergence of multifractal operators. *Earth Space Sci.* **2020**, *7*, e2019EA000608. [CrossRef]
31. Rieutord, M. Linear theory of rotating fluids using spherical harmonics part I: Steady flows. *Geophys. Astrophys. Fluid Dyn.* **1987**, *39*, 163–182. [CrossRef]
32. Chandrasekhar, S. *Hydrodynamic and Hydromagnetic Stability*; Dover Publications: New York, USA, 1981.
33. Chigirinskaya, Y.; Schertzer, D.; Lovejoy, S. An alternative to shell-models: More complete and yet simple models of intermittency. In Proceedings of the Advances in Turbulence VII: Proceedings of the Seventh European Turbulence Conference, Saint-Jean Cap Ferrat, France, 30 June–3 July 1998; pp. 263–266.
34. Chigirinskaya, Y.; Schertzer, D. Cascade of Scaling Gyroscopes: Lie Structure, Universal Multifractals and Self-Organized Criticality in Turbulence. In *Stochastic Models in Geosystems*; Molchanov, S.A., Woyczynski, W.A., Eds.; Springer: New York, NY, USA, 1997; pp. 57–81. [CrossRef]
35. Cambon, C.; Laguna, A.A.; Zhou, Y. CFD for turbulence: From fundamentals to geophysics and astrophysics. *Comptes Rendus Méc.* **2022**, *350*, 1–20. [CrossRef]
36. Salhi, A.; Cambon, C. Magneto-gravity-elliptic instability. *J. Fluid Mech.* **2023**, *963*, A9. [CrossRef]
37. Chagelishvili, G.; Zahn, J.P.; Tevzadze, A.; Lominadze, J. On hydrodynamic shear turbulence in Keplerian disks: Via transient growth to bypass transition. *Astron. Astrophys.* **2003**, *402*, 401–407. [CrossRef]
38. Salhi, A.; Lehner, T.; Godeferd, F.; Cambon, C. Wave–vortex mode coupling in astrophysical accretion disks under combined radial and vertical stratification. *Astrophys. J.* **2013**, *771*, 103. [CrossRef]
39. Nazarenko, S. *Wave Turbulence*; Springer Science & Business Media: Berlin/Heidelberg, Germany, 2011; Volume 825.
40. Bellet, F.; Godeferd, F.; Scott, J.; Cambon, C. Wave turbulence in rapidly rotating flows. *J. Fluid Mech.* **2006**, *562*, 83–121. [CrossRef]
41. Radkevich, A.; Lovejoy, S.; Strawbridge, K.; Schertzer, D. The elliptical dimension of space-time atmospheric stratification of passive admixtures using lidar data. *Phys. A Stat. Mech. Its Appl.* **2007**, *382*, 597–615. [CrossRef]
42. Lilley, M.; Lovejoy, S.; Strawbridge, K.; Schertzer, D. 23/9 dimensional anisotropic scaling of passive admixtures using lidar data of aerosols. *Phys. Rev. E* **2004**, *70*, 036307. [CrossRef] [PubMed]
43. Lewis, G.; Lovejoy, S.; Schertzer, D.; Pecknold, S. The scale invariant generator technique for quantifying anisotropic scale invariance. *Comput. Geosci.* **1999**, *25*, 963–978. [CrossRef]
44. Lazarev, A.; Schertzer, D.; Lovejoy, S.; Chigirinskaya, Y. Unified multifractal atmospheric dynamics tested in the tropics: Part II, vertical scaling and generalized scale invariance. *Nonlinear Process. Geophys.* **1994**, *1*, 115–123. [CrossRef]

atmosphere

MDPI

Review

The Surprising Roles of Turbulence in Tropical Cyclone Physics

Kerry Emanuel *, Martin Velez-Pardo and Timothy W. Cronin

Program in Atmospheres, Oceans, and Climate, Massachusetts Institute of Technology, Cambridge, MA 02139, USA; martinvp@mit.edu (M.V.-P.); twcronin@mit.edu (T.W.C.)
* Correspondence: emanuel@mit.edu

Abstract: Tropical cyclones have long been known to be powered by turbulent enthalpy fluxes from the ocean's surface and slowed by turbulent momentum fluxes into the surface. Here, we review evidence that the development and structure of these storms are also partially controlled by turbulence in the outflow near the storm's top. Finally, we present new research that shows that tropical cyclone-like, low-aspect-ratio vortices are most likely in systems in which the bottom heat flux is controlled by mechanical turbulence, and the top boundary is insulating.

Keywords: tropical cyclone; hurricane; turbulence; convection; Rayleigh–Bénard

Citation: Emanuel, K.; Velez-Pardo, M.; Cronin, T.W. The Surprising Roles of Turbulence in Tropical Cyclone Physics. *Atmosphere* **2023**, *14*, 1254. https://doi.org/10.3390/atmos14081254

Academic Editors: Boris Galperin, Annick Pouquet and Peter Sullivan

Received: 22 April 2023
Revised: 19 May 2023
Accepted: 1 August 2023
Published: 7 August 2023

1. Introduction

Some 90 tropical cyclones develop globally each year, producing among the strongest winds and storm tides of any atmospheric phenomenon. Here, we review the critical, multiple roles that turbulence plays in tropical cyclone physics. We dedicate this paper to the memory of Jack Herring, whose own work on turbulence inspired parts of what we present here.

Tropical cyclones are driven by turbulent enthalpy fluxes from the ocean, as shown long ago by Riehl [1] and Kleinschmidt [2]. More recently, numerical simulations of dry tropical cyclones demonstrated that phase change of water is not necessary, though it does affect the structure and intensity of the vortices [3–6]. In the real-world moist case, most of the turbulent enthalpy flux is in the form of latent heat, and so the air–sea temperature differential is small. This, coupled with very strong surface winds, implies that the turbulence driving the surface fluxes is almost entirely mechanical. Likewise, the power dissipation at the sea's surface increases with the cube of the surface wind speed. An upper bound on the wind speed in a steady, axisymmetric tropical cyclone can be derived by equating the power generated by the thermal cycle of the storm with that dissipated by turbulence; this upper bound has been shown to be an excellent predictor of maximum surface wind speed in numerical simulations [7] and bounds the probability distribution of storm lifetime maximum winds in nature [8]. In Section 2, we review the important role of boundary layer turbulence and the challenge of estimating surface fluxes in the highly perturbed sea states in intense storms.

A complete solution of the equations governing a steady, axisymmetric tropical cyclone demands an upper boundary condition as well as the surface boundary conditions described above. As the outflow is constrained by both rotation and ambient stratification, an appropriate upper boundary condition is not evident from first principles. We argued that the outflow of tropical cyclones is self-stratifying, maintaining a constant, critical Richardson Number [9]. In Section 3, we review this idea and its implications for tropical cyclone structure and intensity and present new field observations supporting the idea of self-stratifying outflow.

Research on geophysical convection and on idealized Rayleigh–Bénard convection has proceeded on largely different paths for many decades, coming to quite different conclusions about the nature of convection. In particular, the Rayleigh–Bénard formalism predicts and laboratory experiments confirm that fluxes from the boundaries continue to depend on molecular diffusivities even in the limit of infinite Rayleigh Number, whereas convective boundary layer theory, supported by field experiments, shows no explicit dependence on molecular diffusivity. In rotating Rayleigh–Bénard convection, the horizontal spacing of vortices scales with the fluid depth, whereas tropical cyclones in idealized experiments are spaced at much greater distances. In Section 4, we show evidence that these discrepancies result from very different assumptions about the physical natures of the horizontal boundaries in the two cases.

2. The Critical Role of Boundary Layer Turbulence in Tropical Cyclone Physics

As air spirals into the core of a mature tropical cyclone, seawater evaporates into it, and turbulent kinetic energy is dissipated into heat. These two processes raise the specific moist entropy of the air, defined as [10]

$$s = (c_p + c_l q_t)\ln(T) + \frac{L_v q}{T} - (1 + q_t)R_d \ln(p) - R_v q_t T \ln(H), \tag{1}$$

where c_p is the heat capacity at constant pressure of dry air, c_l is the heat capacity of liquid water, q_t is the mass concentration of water in all its phases, T is absolute temperature, q is the mass concentration of water vapor, L_v is the latent heat of vaporization, R_d is the universal gas constant divided by the mean molecular weight of dry air, p is total air pressure, R_v is the universal gas constant divided by the molecular weight of water, and H is the relative humidity.

The air arrives in the eyewall with increased entropy and then ascends through the eyewall, conserving its entropy in a nearly adiabatic expansion. The excess entropy is lost at very low temperatures (usually less than −70 °C in the upper troposphere (15–18 km above the surface) through a combination of asymmetric export and infrared radiative cooling. Eventually, air descends back into the boundary layer, following the moist adiabatic temperature profile typical of the unperturbed tropical atmosphere. (Although this air descends unsaturated and is cooled radiatively, it follows the same thermodynamic pathway as a saturated sample, in which evaporation of condensed water takes place. A hurricane in which phases changes of water occur reversibly has been successfully simulated [6]).

These four legs of the airflow can be described as a nearly ideal Carnot cycle [11], with isothermal expansion as air flows along the surface into the low-pressure core while the enormous heat bath provided by the ocean keeps it at nearly constant temperature, followed by moist adiabatic expansion, followed by nearly isothermal compression, and, finally, followed by compression along a moist adiabat.

In Figure 1a, we sketch two adjacent Carnot circuits in the radius–height plane: A-B-C-D-A, and A-B′-C′-D-A. As the gradients near points A and D are very weak, we ignore the differences between the thermal properties of air of the two circuits at these points. These cycles are illustrated in thermodynamic phase space in Figure 1b. The ordinate is absolute temperature, and the abscissa is the entropy of saturated air s^*. In [7], we developed integral relationships for both circuits by combining the first law of thermodynamics with the kinetic energy equation and integrating the result around the circuits. By subtracting one the circuit integral of the first circuit from the second, we arrived at

$$\oint_{inner} T\frac{ds}{dt} = -\oint_{inner} \mathbf{V}\cdot\mathbf{F} - \oint_{inner} \frac{dq_t}{dt}\left[\frac{1}{2}|\mathbf{V}|^2 + gz\right], \tag{2}$$

where *"inner"* denotes the circuit B′-B-C-C′-B′. The left side of (2) is the total thermodynamic energy available around the circuit, whereas the first term on the right is the power dissipation (dot product of the vector velocity and frictional slowing, **F**). The last term in Equation (2) represents the work used to accelerate the water substance to the free stream velocity and to lift it against gravity g, where z is altitude. This term can be shown to be small [7].

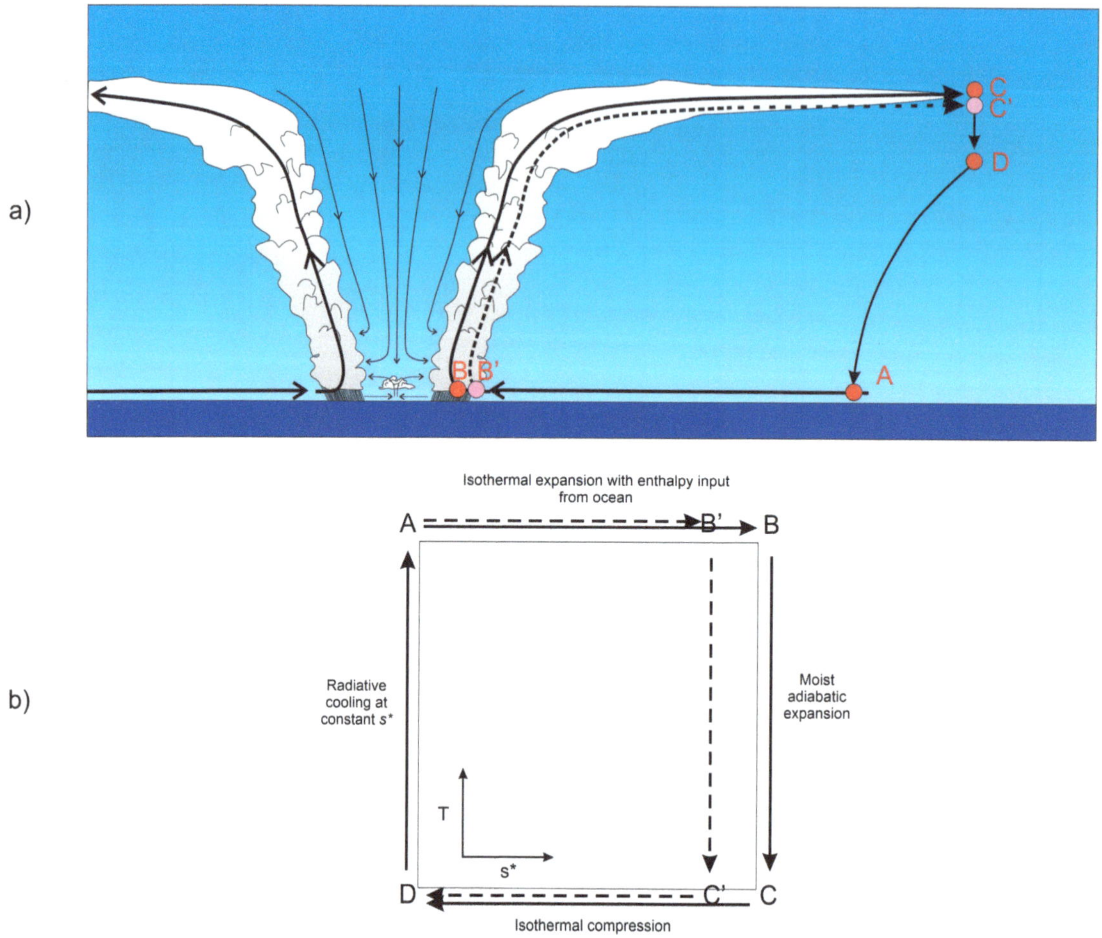

Figure 1. (**a**) Radius-height sketch through a mature, axisymmetric tropical cyclone. The differential Carnot cycle is based on the circuits A-B-C-D and A-B′-C′-D. (**b**) Cycles of (**a**) in thermodynamic space in which saturation entropy is on the abscissa and temperature is on the ordinate. See text for explanation.

Therefore, Equation (2) may be approximated by

$$\oint_{inner} T\frac{ds}{dt} = -\oint_{inner} \mathbf{V}\cdot\mathbf{F}. \tag{3}$$

We take points B and B′ to lie on either side of the radius of maximum surface wind speed. We take entropy and angular momentum to be conserved on the legs B-C and B′-C′. Entropy increases from B′ to C′, and the same quantity is lost between C and C′. Angular momentum is diminished from B′ to B, and the same amount is regained by turbulent exchange with the environment between C and C′. However, the right hand side of Equation (3) evaluated between C and C′ is small compared to the contribution along the leg B′-B, provided the point C is not too far from the storm center [11], and we neglect it here. We model the turbulent exchanges between the air and the surface according to the neutral aerodynamic flux formulae. With these assumptions, and taking the path length B′-B to be vanishingly small, Equation (3) becomes

$$\frac{T_s - T_{out}}{T_s}\left[C_{k10}|\mathbf{V}_{10}|(k_0^* - k_{10}) + C_{D10}|\mathbf{V}_{10}|^3\right] = C_{D10}|\mathbf{V}_{10}|^3, \tag{4}$$

where $|\mathbf{V}_{10}|$ is the 10 m wind speed at the radius of maximum winds, the drag coefficient C_{D10} and enthalpy exchange coefficient C_{k10} pertain to 10 m altitude, T_s is the surface temperature, T_{out} is the temperature at C-C′, k_0^* is the saturation enthalpy of the sea surface, and k_{10} is the enthalpy at 10 m. (The altitude 10 m is used here only by convention).

Collecting terms in Equation (4) results in an expression for the 10 m wind speed:

$$|\mathbf{V}_{10}|^2 = \frac{C_{k10}}{C_{D10}}\frac{T_s - T_{out}}{T_{out}}(k_0^* - k_{10}). \tag{5}$$

The surface wind speed calculated this way has become known as the potential intensity. Numerical simulations undertaken in [7] confirm the validity of Equation (5), and observations show that Equation (5) provides a useful upper bound on real cyclone wind speeds [8]. (But note that Equation (5) cannot be used directly to estimate potential intensity from environmental quantities because both k_0^* and k_{10} must be evaluated at the radius of maximum winds. In particular, k_0^* increases inward owing to falling surface pressure; this pressure fall is in turn proportional to $|\mathbf{V}_{10}|_2$. This represents an important positive feedback in the amplification of the cyclone, and at high sea surface temperatures can run away, resulting in a phenomenon known as a hypercane [12]).

An important effect that has been neglected in this derivation is radial diffusion of momentum in the turbulent boundary layer near the radius of maximum winds. In classical boundary layer theory e.g., [13], radial turbulent fluxes are neglected. But, hurricane dynamics leads to very strong frontogenesis at the eyewall [14], which can only result in strong radial turbulent fluxes. So far, the effect of these fluxes on storm intensity has not been accounted for theoretically, though radial turbulent fluxes are represented in comprehensive numerical models of hurricanes and are known empirically to have a large effect on storm intensity [15].

Note that the potential intensity depends only on the ratio of the two exchange coefficients, not their individual values. The rate of intensification and decay of these storms does, however, depend on the individual values of the coefficients [16]. The important point here is that tropical cyclones are largely controlled by turbulent fluxes of enthalpy and momentum between the atmosphere and ocean.

Over fixed surfaces, the neutral exchange coefficients are functions of the roughness scale of the landscape. But, over water, the roughness is itself a function of wind speed and is controlled mostly by shorter wind-driven waves [17,18]. At low wind speeds, the drag coefficient increases with wind speed, whereas the available evidence suggests that the enthalpy exchange coefficient is nearly constant.

At wind speeds above about 30 ms^{-1}, typical of hurricanes, the character of the sea surface begins to change. Spray begins to form [19], and at yet larger wind speeds may become the rate-limiting process for air–sea fluxes [20,21]. Observations [22] and laboratory measurements [23,24] suggest that the drag coefficient ceases to rise with wind speed and may even decline. What happens to the enthalpy flux coefficient at these extreme wind speeds remains a vexing question, and field measurements and laboratory inferences show very large scatter in the estimates. Recently, it has become possible to directly simulate the formation of spray droplets e.g., [25], as shown in Figure 2.

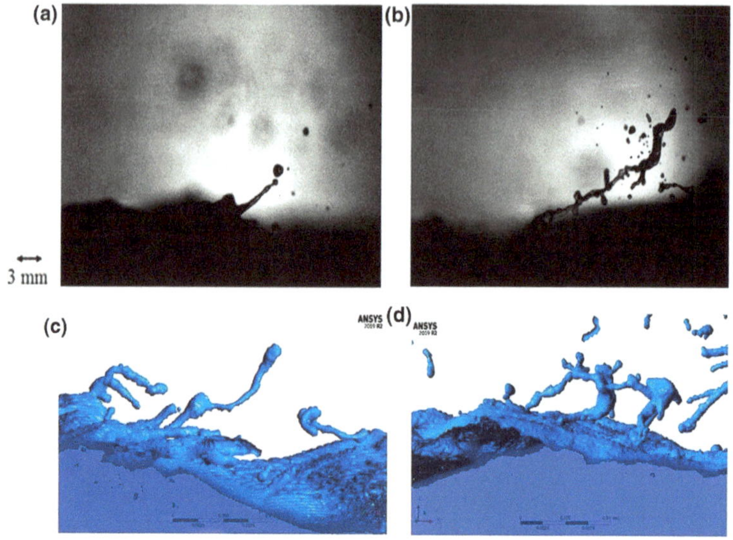

Figure 2. Formation of spray droplets in a laboratory apparatus (**a**,**b**) and in direct numerical simulations (**c**,**d**), under tropical cyclone-like conditions (10-m wind speed of 40 ms^{-1}). Panels (**a**) and (**c**) pertain to a pure water surface, whereas a surfactant has been added in (**b**) and (**d**). From [25].

For a simple system consisting of a semi-infinite body of pure, unstratified water surmounted by a semi-infinite, unstratified atmosphere driven by an imposed, constant horizontal pressure gradient, just three non-dimensional parameters have been identified as controlling the system dynamics [26]. Two of these depend only on gravity and the molecular properties of the two fluids. The third is defined as

$$R_u \equiv \frac{u_*^4}{\sigma g}, \tag{6}$$

where u_* is the friction velocity, σ is the kinematic surface tension, and g is the acceleration of gravity. In this simple system, the wind dependence of the character of the system should depend only on R_u, which may be interpreted as the ratio of the Charnock length to the scale of the largest stable droplets in spray. For example, R_u should control the fraction of the sea surface covered by whitecaps in equilibrium. In [26] it was hypothesized that all nondimensional quantities, such as the drag coefficient, should become constant in the limit of infinite R_u. This hypothesis has not been tested experimentally. A comprehensive review of spray-mediated turbulent fluxes may be found in [27].

What is clear at present is that boundary layer turbulence near the radius of maximum winds is a *mare incognitum* of tropical cyclone physics. We know from theory and numerical simulations that radial turbulent diffusion is important but we lack a quantitative theoretical model for it. Moreover, the all-important surface enthalpy and momentum fluxes are likely to be strongly affected and are perhaps rate-limited by sea spray. Research on these issues remains a critical and challenging issue in tropical cyclone physics.

3. The Outflow Layer

Observations, modeling, and theory all show that outside the hurricane's eye the free tropical troposphere is nearly in a state of "slantwise convective adjustment" in which the temperature lapse rate is moist adiabatic along the vortex lines of the balanced flow [28].

In particular, this is true in tropical cyclone eyewalls and approximately in the outer regions [29]; here, the balanced vortex lines lie on surfaces of absolute angular momentum:

$$M = rV + \Omega r^2, \tag{7}$$

where r is radius, V is the azimuthal velocity, assumed to be in gradient wind balance above the boundary layer, and Ω is the earth's angular velocity vector projected onto the local vertical. In an inviscid axisymmetric flow, M is conserved following air motion. Moreover, M increases monotonically with radius; indeed, the criterion for inertial instability is that angular momentum decrease with radius. Observations and numerical simulations confirm that M increases with radius, although in the outflow layer the increase can be very slow.

The condition that the temperature lapse rate follows a moist adiabat along an angular momentum surface is equivalent to holding the saturation entropy constant on each M surface, where saturation entropy is defined as the entropy air would have if water were saturated at the same pressure and temperature. From Equation (1), this quantity may be written as

$$s^* = \left(c_p + c_l q_t\right)\ln(T) + \frac{L_v q^*}{T} - (1 + q_t)R_d \ln(p), \tag{8}$$

where q^* is the mass concentration of water vapor at saturation. Convective neutrality demands that $s^* = s^*(M)$; saturation entropy is only a function of angular momentum (and time, in slowly evolving flows that maintain convective neutrality). Also, since s^* is a function of pressure and temperature alone, air density can be expressed as a function of pressure and s^* alone (neglecting the direct effect of water substance on air density). From this, and from gradient wind and hydrostatic balance, one can form a thermal wind relation in angular momentum coordinates [2,11,30]:

$$\frac{V}{r} = \frac{V_t}{r_t} - (T - T_t)\frac{ds^*}{dM}, \tag{9}$$

where the subscript t denotes evaluation at a specific point along the angular momentum surface. As one travels up an M surface from the top of the turbulent boundary layer, the azimuthal velocity V vanishes at some particular altitude, or temperature, which serves here as a proxy for altitude. It is convenient to define the point t as that point, with radius r_t and temperature T_t. If, furthermore, we evaluate the expression at the top of the boundary layer, Equation (9) becomes

$$\frac{V_b}{r_b} = -(T_b - T_t)\frac{ds^*}{dM}, \tag{10}$$

where the subscript b denotes evaluation at the top of the turbulent boundary layer. To solve Equation (10) for the azimuthal wind at the boundary layer top, we need two boundary conditions: one that determines ds^*/dM and the other that determines T_t. The first is effectively a lower boundary condition and can be derived from conservation equations for entropy and angular momentum in the boundary layer. We assume that the boundary layer is well mixed in the vertical (technically, along angular momentum surfaces) and that actual entropy in the boundary layer is equal to s^* just above the top of the boundary layer; this is equivalent to asserting that a sample of air in the boundary layer is neutrally buoyant when lifted moist adiabatically along angular momentum surfaces. Then the entropy and angular momentum conservation equations integrated vertically through the boundary layer are, respectively,

$$hT_b\frac{ds}{dt} = C_k|\mathbf{V}_{10}|(k_0^* - k_{10}) + C_D|\mathbf{V}_{10}|^3, \tag{11}$$

And

$$h\frac{dM}{dt} = -C_D r|\mathbf{V}_{10}|V_{10}, \tag{12}$$

where h is the local depth of the boundary layer and we have approximated the temperature at the surface by T_b and the radius of angular momentum surfaces in the boundary layer by r_b. We next take the ratio of Equations (11) to (12), which gives ds/dM $(= ds^*/dM)$ in the steady state, and substitute that result into (10) to give

$$V_{10}V_b = \frac{T_b - T_t}{T_b}\left[\frac{C_k}{C_D}(k_0^* - k_{10}) + |\mathbf{V}_{10}|^2\right]. \tag{13}$$

Comparing Equation (13) to the Carnot derivation result Equation (5), we see that the two are equivalent only if $V_{10}\ V_b = |\mathbf{V}_{10}|^2$. The product of the gradient wind and the azimuthal wind component at the reference altitude equals the square of the magnitude of the wind (including its radial component) at the reference altitude. (Since $V_{10} < |\mathbf{V}_{10}|$, it follows that $V_b > |\mathbf{V}_{10}|$: the reference surface wind speed must be less than the gradient wind. Hurricane forecasters have long known this to be the case). Hereafter, we will make use of $V_{10}\ V_b = |\mathbf{V}_{10}|^2$ and also make the approximation that $|\mathbf{V}_{10}| = V_b$.

We could use Equation (13) with this approximation to predict the radial distribution of gradient wind if we knew how the temperature T_t varied with angular momentum. Empirically, from detailed numerical simulations, we know that this "outflow temperature" increases as angular momentum increases [9]. In a coordinate system in which absolute temperature serves as an altitude proxy and angular momentum serves as a radial proxy, the actual situation resembles the sketch displayed in Figure 3.

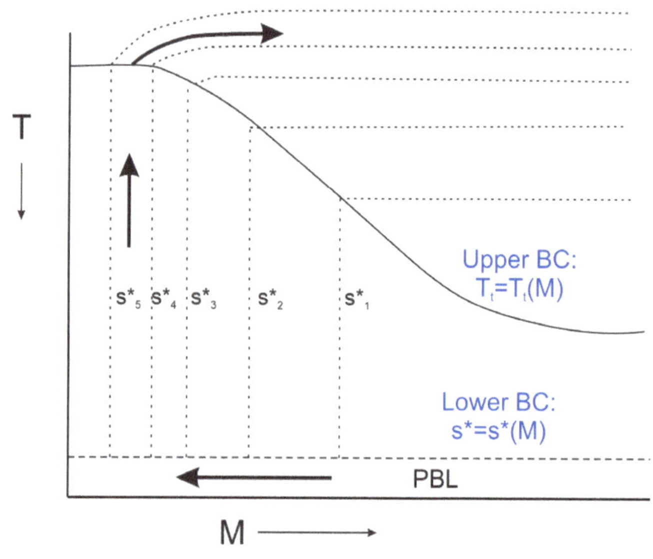

Figure 3. Sketch of distribution of various quantities in a steady, axisymmetric tropical cyclone. Angular momentum increases along the abscissa, whereas temperature decreases along the ordinate. Dotted lines show surfaces of constant saturation entropy, and the thick arrows indicate airflow. The thin solid curve shows the surface along which the azimuthal component of the gradient wind vanishes.

The temperature at the location of vanishing azimuthal velocity increases with angular momentum. Why? The curve connecting the loci of vanishing azimuthal wind is quite certainly not the tropopause. Almost all the storm outflow lies above this curve but below or near the tropopause. In [9], we proposed that the flow near this curve was turbulent and self-stratifying, keeping the local gradient Richardson Number near its critical value for Kelvin–Helmholtz instability. We demonstrated both that the outflow is indeed quite turbulent, and the Richardson Number is near unity in a high-resolution, axisymmetric numerical simulation of a tropical cyclone. Small-scale turbulence is parameterized in this model, and production of turbulence kinetic energy shuts down when the Richardson Number exceeds unity.

How turbulent is the outflow in real tropical cyclones? Almost all the direct measurements we have in tropical cyclones come from reconnaissance aircraft, whose maximum flight altitudes are well below the outflow of these storms. A strong desire to measure hurricane outflow was part of the motivation behind an experiment mounted by the Office of Naval Research in 2015, which deployed, among other platforms, a WB-57 aircraft capable of flying at 18 km altitude, thus overflying most tropical cyclones. The WB-57 deployed dropwindsondes, packages of in situ instruments that measure temperature, pressure, humidity, and wind speed and direction (through GPS tracking). These are parachuted down to the surface and telemeter data back to the parent aircraft.

As it happens, one of the tropical cyclones measured in the 2015 field campaign was eastern North Pacific Hurricane Patricia, which produced the highest wind speeds ever measured in a tropical cyclone. Calculations of (inverse) Richardson Number from one of the sondes that fell through Patricia's outflow is shown in Figure 4.

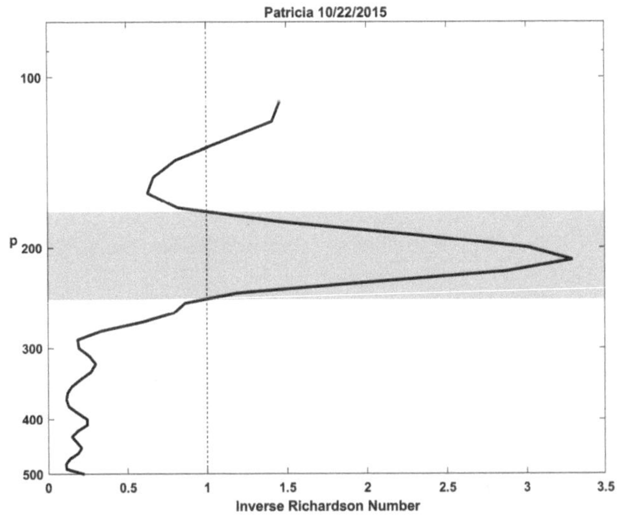

Figure 4. Inverse Richardson number as a function of pressure as measured by a dropsonde deployed from a WB-57 aircraft flying above Hurricane Patricia on 22 October 2015. This dropsonde fell through Patricia's outflow about 450 km north of the storm center. Shading shows region of Richardson Number < 1.

The Richardson number in the outflow was quite small (large inverse Richardson Number), suggesting that the outflow was indeed turbulent.

In [16] we showed that the assumption of a critical Richardson number and taking the outflow temperature T_t to be the undisturbed tropopause temperature at the storm center closes the problem and yields the complete axisymmetric structure of the steady-state tropical cyclone. For the special case of neglecting isothermal expansion effects and dissipative heating the results is

$$V_b = V_{max}\frac{r_{max}}{r}\left[\frac{2\left(\frac{r}{r_{max}}\right)^2}{2-\frac{C_k}{C_D}+\frac{C_k}{C_D}\left(\frac{r}{r_{max}}\right)^2}\right]^{\frac{1}{2-C_k/C_D}}-\Omega r, \tag{14}$$

where

$$V_{max}^2 = \left(\frac{1}{2}\frac{C_k}{C_D}\right)^{\frac{\frac{C_k}{C_D}}{2-\frac{C_k}{C_D}}} \frac{C_k}{C_D}(T_b - T_t)({s_0}^* - {s_e}^*), \tag{15}$$

and

$$r_{max} \cong \left(\frac{1}{2}\right)^{\frac{1}{2}} \frac{\Omega r_o^2}{\sqrt{(T_b - T_{trop})({s_0}^* - {s_e}^*)}}. \tag{16}$$

Here ${s_0}^*$ is the saturation entropy of the sea surface (approximated as constant here) and ${s_e}^*$ is the saturation entropy of the unperturbed troposphere, another constant. The quantity r_o is the "outer radius", the distance from the storm center beyond which there is no storm-related circulation, while r_{max} is the radius at which the surface winds attain their maximum value, V_{max}. Note that (16) only provides a relationship between these two radii; one must specify one or the other. The theory, in its present form, does not yield an absolute radial dimension of the vortex. In nature, there is a broad range of tropical cyclone diameters that follow a log normal distribution [31,32], but we have no theory for either the median of that distribution or that predicts a log-normal distribution.

The predictions entailed by the above equations were found to describe the intensities of numerically simulated tropical cyclones quite well [9]. The predicted structure was also a good fit to numerical simulations of dry cyclones as well as those in which condensed water was not permitted to fall out of the system [6], but it failed to capture the slow decline of winds outside the storm core in moist, precipitating cyclones. In this case, the wind profile was governed by the requirement that radiative subsidence match Ekman suction at the top of the boundary layer [33].

The surprising conclusion of this analysis, to the extent that it is correct, is that turbulence in tropical cyclone outflow has a strong effect on both storm structure and intensity. It would appear that turbulence in both the boundary layer and high up in the storm circulation is essential for the existence and nature of tropical cyclones. But what, if anything, does this have to do with idealized convection between two parallel plates, about which there is an extensive literature going back to Bénard's original 1900 paper [34]? We will explore this issue in the next section.

4. Tropical Cyclones in Parallel Plate Convection

At first glance, the preceding development bears little relationship to classical studies of parallel-plate convection. We treated a mature, steady-state vortex in isolation, with no discussion of development or as part of a systematic spatial pattern of convection. Although the theoretical development does not identify any absolute horizontal scales, natural tropical cyclones have very low aspect ratios (~1/50), whereas classical parallel-plate convection has aspect ratios of order unity. Can tropical-cyclone-like vortices develop in parallel-plate convection, and, if so, under what conditions? The first of these questions was answered as affirmative in [4], in which patterns of vortices developed in doubly periodic domains simulating rotating radiative–convective equilibrium without any phase change in water. In this case, the surface fluxes were calculated using aerodynamic flux formulae, as in the preceding development, but the compensating cooling was distributed through the interior of the fluid rather than at an upper boundary. The vortices that developed behaved much like real tropical cyclones, with low aspect ratios (but not as low as their moist counterparts).

The conditions under which rotating parallel plate convection organizes into tropical cyclone-like vortices are explored in [5]. They performed direct numerical simulations of idealized, parallel-plate convection in rotating, doubly periodic domains, exploring sensitivities to both nondimensional parameters and the type of boundary condition used. In particular, they applied four types of thermal boundary conditions: constant temperature, constant flux, zero flux, and wind-dependent flux (to mimic the aerodynamic fluxes found over rough surfaces). No-slip conditions were applied on velocities, and the square domain horizontal scale was 16 times the domain depth. The governing nondimensional parameters are a flux Rayleigh number, the Prandtl Number, and a convective Rossby number, defined as

$$Ro_c \equiv \frac{1}{2\Omega}\left(\frac{F_B}{H^2}\right)^{\frac{1}{3}}, \tag{17}$$

where Ω is the domain angular velocity, F_B is the buoyancy flux, and H is the domain depth. In all the simulations, the Prandtl Number was fixed at 1.0, and for most of the simulations the flux Rayleigh Number was set to 10^9, which permits moderately turbulent flows while keeping computational costs low enough to perform many simulations.

The simulations with fixed boundary temperatures produced convective patterns resembling those found in classical studies of rotating parallel-plate convection with fixed temperature at both boundaries. The aspect ratio of the convection in these cases is of order unity. With one or two constant flux or wind-dependent flux boundary conditions, larger scale structures appear, consistent with previous studies of convection with constant flux boundary conditions. But if the rotation is sufficiently slow ($Ro_c \gtrsim 1$), tropical cyclone-like vortices appear for certain combinations of thermal boundary conditions, as shown in Figure 5.

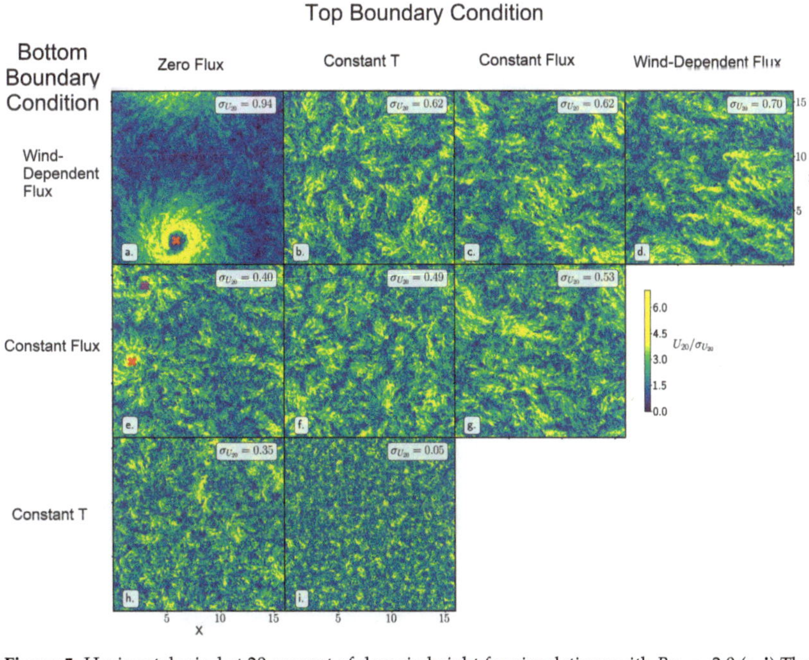

Figure 5. Horizontal wind at 20 percent of domain height for simulations with $Ro_c = 2.0$ (**a–i**) These snapshots were taken after 10 rotational periods. Magnitudes are normalized by the standard deviation in each domain, and color shades saturate at 3.5 times the standard deviation. Rows represent different bottom thermal boundary conditions, and columns are top boundary condition. Locations of near-surface pressure minima corresponding to persistent cyclonic vortices are indicated with "x" markers. (This is Figure 3 of reference [5]. ©American Meteorological Society. Used with permission. This preliminary version has been accepted for publication in the *Journal of the Atmospheric Sciences* and may be fully cited. The final typeset copyedited article will replace the Early Online Release when it is published.).

5. Summary

Turbulence plays a critical role in tropical cyclone physics. Shear-driven turbulence near the sea surface controls the rates at which enthalpy and momentum are transferred between the ocean and atmosphere, ultimately determining the intensities of tropical cyclones that are otherwise uninfluenced by environmental interactions such as wind shear. Solutions of the equations governing steady, axisymmetric tropical cyclones also require upper boundary conditions, and, although there are no widely accepted conditions, turbulence in the outflow layer may dictate the density stratification of the outflow, which would close the solution. Such a solution has been found and is in reasonably good accord with the results of numerical simulations.

As tropical cyclones are driven by boundary heat fluxes, phase changes of water are not necessary either to simulate or to understand the phenomenon, though they do influence the intensities and structures of the resulting vortices [4]. They emerge in simulations of parallel-plate convection if the heat sink is distributed through the fluid, and the bottom thermal boundary condition is either one of constant or wind-dependent flux. We encourage those exploring parallel-plate convection to consider applying aerodynamically rough boundaries and/or distributed heat sinks/sources in the fluids.

Author Contributions: K.E. performed the research described in Sections 2 and 3 and wrote the first draft of this article. M.V.-P. and T.W.C. performed the research described in Section 4. All authors have read and agreed to the published version of the manuscript.

Funding: Kerry Emanuel was supported in part by the National Science Foundation under grant NSF AGS-1906768.

Institutional Review Board Statement: Not applicable.

Informed Consent Statement: Not applicable.

Data Availability Statement: No new data were generated for this paper. Data sources and availability for the research reviewed in this article may be found in the original references cited.

Conflicts of Interest: The authors declare no conflict of interest.

References

1. Riehl, H. A Model for Hurricane Formation. *J. Appl. Phys.* **1950**, *21*, 917–925. [CrossRef]
2. Kleinschmidt, E., Jr. Grundlagen Einer Theorie Der Tropischen Zyklonen. *Arch. Meteorol. Geophys. Bioklimatol. Ser. A* **1951**, *4*, 53–72. [CrossRef]
3. Mrowiec, A.A.; Garner, S.T.; Pauluis, O.M. Axisymmetric Hurricane in a Dry Atmosphere: Theoretical Framework and Numerical Experiments. *J. Atmos. Sci.* **2011**, *68*, 1607–1619. [CrossRef]
4. Cronin, T.W.; Chavas, D.R. Dry and Semidry Tropical Cyclones. *J. Atmos. Sci.* **2019**, *76*, 2193–2212. [CrossRef]
5. Velez-Pardo, M.; Cronin, T.W. Convective Organization and Dry Tropical Cyclones in Direct Numerical Simulations of Idealized Fluid Setups. *J. Atmos. Sci.* 2023, *in press*.
6. Wang, D.; Lin, Y. Size and Structure of Dry and Moist Reversible Tropical Cyclones. *J. Atmos. Sci.* **2020**, *77*, 2091–2114. [CrossRef]
7. Rousseau-Rizzi, R.; Emanuel, K. An Evaluation of Hurricane Superintensity in Axisymmetric Numerical Models. *J. Atmos. Sci.* **2019**, *76*, 1697–1708. [CrossRef]
8. Emanuel, K.A. A Statistical Analysis of Tropical Cyclone Intensity. *Mon. Wea. Rev.* **2000**, *128*, 1139–1152. [CrossRef]
9. Emanuel, K.; Rotunno, R. Self-Stratification of Tropical Cyclone Outflow. Part I: Implications for Storm Structure. *J. Atmos. Sci.* **2011**, *68*, 2236–2249. [CrossRef]
10. Iribarne, J.V.; Godson, W.L. *Atmospheric Thermodynamics*; D. Reidel: Dordrecht, The Netherlands, 1973; 222p.
11. Emanuel, K.A. An Air-Sea Interaction Theory for Tropical Cyclones. Part I: Steady State Maintenance. *J. Atmos. Sci.* **1986**, *43*, 585–605. [CrossRef]
12. Emanuel, K.A.; Speer, K.; Rotunno, R.; Srivastava, R.; Molina, M. Hypercanes: A Possible Link in Global Extinction Scenarios. *J. Geophys. Res.* **1995**, *100*, 13755–13765. [CrossRef]
13. Stull, R.B. *An Introduction to Boundary Layer Meteorology*; Kluwer Academic Publishers: Boston, MA, USA, 1988.
14. Emanuel, K.A. Some Aspects of Hurricane Inner-Core Dynamics and Energetics. *J. Atmos. Sci.* **1997**, *54*, 1014–1026. [CrossRef]
15. Bryan, G.H. Effects of Surface Exchange Coefficients and Turbulence Length Scales on the Intensity and Structure of Numerically Simulated Hurricanes. *Mon. Weather Rev.* **2012**, *140*, 1125–1143. [CrossRef]
16. Emanuel, K. Self-Stratification of Tropical Cyclone Outflow: Part II: Implications for Storm Intensification. *J. Atmos. Sci.* **2012**, *69*, 988–996. [CrossRef]

17. Charnock, H. Wind Stress on a Water Surface. *Q. J. R. Meteorol. Soc.* **1955**, *81*, 639–640. [CrossRef]
18. Large, W.G.; Pond, S. Sensible and Latent Heat Flux Measurements over the Ocean. *J. Phys. Ocean.* **1982**, *12*, 464–482. [CrossRef]
19. Andreas, E.L.; Decosmo, J. The Signature of Sea Spray in the HEXOS Turbulent Heat Flux Data. *Bound. Layer Meteorol.* **2002**, *103*, 303–333. [CrossRef]
20. Andreas, E.L.; Emanuel, K. Effects of Sea Spray on Tropical Cyclone Intensity. *J. Atmos. Sci.* **2001**, *58*, 3741–3751. [CrossRef]
21. Sroka, S.; Emanuel, K. Sensitivity of Sea-Surface Enthalpy and Momentum Fluxes to Sea Spray Microphysics. *J. Geophys. Res. Ocean.* **2022**, *127*, e2021JC017774. [CrossRef]
22. Powell, M.D.; Vickery, P.J.; Reinhold, T.A. Reduced Drag Coefficients for High Wind Speeds in Tropical Cyclones. *Nature* **2003**, *422*, 279–283. [CrossRef]
23. Donelan, M.A.; Haus, B.K.; Reul, N.; Plant, W.J.; Stiassnie, M.; Graber, H.C.; Brown, O.B.; Saltzman, E.S.D. On the Limiting Aerodynamic Roughness of the Ocean in Very Strong Winds. *Geophys. Res. Lett.* **2004**, *31*, L18306. [CrossRef]
24. Troitskaya, Y.; Kandaurov, A.; Ermakova, O.; Kozlov, D.; Sergeev, D.; Zilitinkevich, S. Bag-Breakup Fragmentation as the Dominant Mechanism of Sea-Spray Production in High Winds. *Sci. Rep.* **2017**, *7*, 1614. [CrossRef] [PubMed]
25. Vanderplow, B.; Soloviev, A.V.; Dean, C.W.; Haus, B.K.; Lukas, R.; Sami, M.; Ginis, I. Potential Effect of Bio-Surfactants on Sea Spray Generation in Tropical Cyclone Conditions. *Sci. Rep.* **2020**, *10*, 19057. [CrossRef]
26. Emanuel, K. A Similarity Hypothesis for Air-Sea Exchange at Extreme Wind Speeds. *J. Atmos. Sci.* **2003**, *60*, 1420–1428. [CrossRef]
27. Sroka, S.; Emanuel, K. A Review of Parameterizations for Enthalpy and Momentum Fluxes from Sea Spray in Tropical Cyclones. *J. Phys. Oceanogr.* **2021**, *51*, 3053–3069. [CrossRef]
28. Emanuel, K.A. On Assessing Local Conditional Symmetric Instability from Atmospheric Soundings. *Mon. Weather Rev.* **1983**, *111*, 2016–2033. [CrossRef]
29. Bryan, G.H.; Rotunno, R.D. Evaluation of an Analytical Model for the Maximum Intensity of Tropical Cyclones. *J. Atmos. Sci.* **2009**, *66*, 3042–3060. [CrossRef]
30. Shutts, G.J. Hurricane Structure and the Zero Potential Vorticity Approximation. *Mon. Weather Rev.* **1981**, *109*, 324–329. [CrossRef]
31. Dean, L.; Emanuel, K.; Chavas, D.R. On the Size Distribution of Atlantic Tropical Cyclones. *Geophy. Res. Lett.* **2009**, *36*, L14803. [CrossRef]
32. Chavas, D.R.; Emanuel, K.A. A QuickSCAT Climatology of Tropical Cyclone Size. *Geophys. Res. Lett.* **2010**, *37*, L18816. [CrossRef]
33. Chavas, D.R.; Lin, N. A Model for the Complete Radial Structure of the Tropical Cyclone Wind Field. Part II: Wind Field Variability. *J. Atmos. Sci.* **2016**, *73*, 3093–3113. [CrossRef]
34. Bénard, H. Étude Expérimentale Des Courants de Convection Dans Une Nappe Liquide—Régime Permanent: Tourbillons Cellulaires. *J. Phys. Theor. Appl.* **1900**, *9*, 513–524. [CrossRef]

atmosphere

MDPI

Article

Exact Intermittent Solutions in a Turbulence Multi-Branch Shell Model

Ben Ajzner and Alexandros Alexakis *

Laboratoire de Physique de l'Ecole Normale Supérieure, Université PSL, Centre National de la Recherche Scientifique, Sorbonne Université, Université de Paris, F-75005 Paris, France
* Correspondence: alexakis@phys.ens.fr

Abstract: Reproducing complex phenomena with simple models marks our understanding of the phenomena themselves, and this is what Jack Herring's work demonstrated multiple times. In that spirit, this work studies a turbulence shell model consisting of a hierarchy of structures of different scales ℓ_n such that each structure transfers its energy to two substructures of scale $\ell_{n+1} = \ell_n/\lambda$. For this model, we construct exact inertial range solutions that display intermittency, i.e., absence of self-similarity. Using a large ensemble of these solutions, we investigate how the probability distributions of the velocity modes change with scale. It is demonstrated that, while velocity amplitudes are not scale-invariant, their ratios are. Furthermore, using large deviation theory, we show how the probability distributions of the velocity modes can be re-scaled to collapse in a scale-independent form. Finally, we discuss the implications the present results have for real turbulent flows.

Keywords: turbulence; intermittency

Citation: Ajzner, B.; Alexakis, A. Exact Intermittent Solutions in a Turbulence Multi-Branch Shell Model. *Atmosphere* **2023**, *14*, 1316. https://doi.org/10.3390/atmos14081316

Academic Editors: Boris Galperin, Annick Pouquet and Peter Sullivan

Received: 26 July 2023
Revised: 12 August 2023
Accepted: 16 August 2023
Published: 20 August 2023

1. Introduction

Constructing simple models that reproduce the phenomenologically complex behaviour of fluid flows has always been a driving force in turbulence research and is a direction in which Jack Herring's work excelled. There are numerous works in his career explaining complex phenomena in fluid dynamics with simplified models [1–13]. In particular, the energy cascade in scale space is a phenomenon that has met various modeling approaches in the literature, such as direct interaction approximation [1,5,14–16], eddy damping quasi-normal Markovian models [17–20] energy diffusion models [21,22], and shell models [23–26]. Such models have led to predictions about the direction of cascade, the power-law exponents of the energy spectra, and intermittency. Intermittency that still escapes a firm quantitative understanding manifests itself as a deviation from self-similarity and from the prediction obtained on purely dimensional grounds. In particular, shell models have been used to study intermittency for many years now. Their simplicity has enabled examining asymptotically large Reynolds numbers and merits various rigorous analytical studies [27–32]. Recent reviews can be found in [33–35]. Typically, shell models quantify all structures of a given scale ℓ by a single real or complex amplitude U_ℓ. As such, spatial intermittency that is linked to the appearance of rare but extremely intense structures cannot be captured this way. Nonetheless, the temporal variation of the modes U_l does display intermittency, as has been demonstrated by many models [24–26,36]. This type of intermittency has been recently linked to the fluctuation dissipation theorem [37]. Furthermore, a solvable (but not energy-conserving) model was also derived and studied in [38].

In the spirit discussed in the first paragraph, we here construct and study a binary tree shell model for turbulence that displays intermittency. In this model, energy at each scale is split between multiple different structures. Each structure transfers its energy into two smaller-scale structures, building a binary tree structure as shown in Figure 1. In this way, the number of structures increases exponentially as smaller scales are reached. Such models with binary structure were introduced in the 1990s but have not been investigated extensively [39–41]. Here, we follow a similar analysis as in [42], where stationary solutions of non-binary shell models were investigated. We demonstrate that such analysis allows the construction of exact stationary solutions that display intermittency.

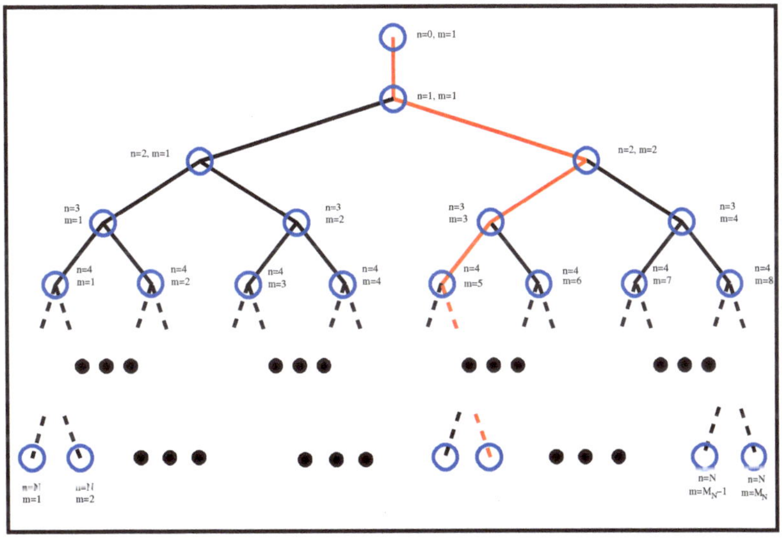

Figure 1. A sketch of the two-branch ($\mu = 2$) shell model. Each node marked by blue circles represents one dynamical mode of amplitude $U_{n,m}$ marked by the two indexes n, m, where n characterises the scale $\ell_n = \ell_0 \lambda^n$ and m characterises the number of the mode in that scale. In each scale n, there are $M_n = \mu^{n-1}$ modes. The red line indicates one possible path on which energy cascades.

2. Multi-Branch Shell Models

We consider the evolution of a turbulent flow modeled by the real amplitudes $U_{n,m}$ of structures of scale $\ell_n = 1/k_n$ where

$$k_n = \lambda^n k_0 \quad \text{or} \quad \ell_n = \ell_0/\lambda^n \tag{1}$$

and $1 < \lambda$. At scale ℓ_1, there is one structure whose amplitude is given by $U_{1,1}$; this structure will transfer its energy to $\mu \in \mathbb{N}$ structures of scale ℓ_2, each one of which will transfer its energy to μ structures of scale ℓ_3 and so on, as shown in Figure 1 for $\mu = 2$. The volume of each structure is given by $V_i = \ell_n^D$, where D is the spatial dimension. If the cascade process is space-filling, the number of substructures μ is related to λ and D by

$$\lambda^D = \mu. \tag{2}$$

Accordingly, at energy scale ℓ_i, we have $M_n = \mu^{n-1}$ (with $M_0 = 1$) structures so that if we consider N such scales we have a total of

$$Z = 1 + \sum_{n=1}^{N} M_n = \frac{\mu^N - 1}{\mu - 1} + 1 \tag{3}$$

structures. The energy of every structure is given by

$$E_{n,m} = \frac{1}{2}\rho U_{n,m}^2 V_n = \frac{1}{2}\rho U_{n,m}^2 \ell_n^D \tag{4}$$

so the total energy is given by

$$E = \frac{1}{2}\sum_{n=0}^{N}\frac{1}{M_n}\sum_{m=1}^{M_n} U_{n,m}^2 \tag{5}$$

where ρ is from now on taken to be unity.

In the Desnianskii and Novikov model [23], structures of scale ℓ_n interact with only structures of scale ℓ_{n+1} and ℓ_{n-1}, and there is no branching $\mu = 1$. The amplitudes U_n then follow the following dynamical equation:

$$\dot{U}_n = ak_n[U_{n-1}U_{n-1} - \lambda U_n U_{n+1}] + bk_n[U_n U_{n-1} - \lambda U_{n+1}U_{n+1}] - \nu k^2 U_n + F_n \tag{6}$$

For $\nu = 0$ and $F_n = 0$, this system conserves the energy (5) (with $M_n = 1$) for any value of a, b. The flux of energy across a scale ℓ_n is given by:

$$\Pi_n = ak_n U_n U_{n-1} U_{n-1} + bk_n U_n U_n U_{n-1}. \tag{7}$$

The Desnianskii and Novikov model [23] is the simplest energy-conserving model one can consider. Since its construction, more complex models have been designed that include more distant mode couplings and complex amplitude modes. The newer models display chaotic dynamics and also conserve more invariants than just the energy. The most popular ones are the SABRA and the GUY model [24–26,36]. A comparison of the two can be found in [34]. Although these models are more realistic, here, we are going to keep the structure of the Desnianskii and Novikov model [23] because its simplicity allows for analytical treatment.

Expanding on the Desnianskii and Novikov model, allowing each structure to branch out to two ($\mu = 2$) smaller-scale structures $U_{n,m}$ results in the following dynamical equation:

$$\begin{aligned}\dot{U}_{n,m} = \; & ak_n\left[U_{n-1,m^*}U_{n-1,m^*} - \frac{\lambda}{2}(U_{n,m}U_{n+1,m'} + U_{n,m}U_{n+1,m'+1})\right] + \\ & bk_n\left[U_{n,m}U_{n-1,m^*} - \frac{\lambda}{2}(U_{n+1,m'}U_{n+1,m'} + U_{n+1,m'+1}U_{n+1,m'+1})\right] \\ & -\nu k_n^2 U_{n,m} + F_{n,m}\end{aligned} \tag{8}$$

where ν is the viscosity, $F_{n,m}$ is the forcing, and a, b are again free parameters. The branching diagram for the model given in (9) is provided in Figure 1. The integers m' and $m' + 1$ correspond to the index of scales ℓ_{n+1}, with which the mode $U_{n,m}$ is linked where m' is explicitly given by $m' = 2m - 1$ and m^* corresponds to the index of scale ℓ_{n-1} linked to $U_{n,m}$ given by $m^* = \text{Int}[(m+1)/2]$, as illustrated in the left panel of Figure 2. For $\nu = 0$, $F_{n,m} = 0$, and, for any value of a, b, the system conserves the energy (5), where now $M_n = 2^{n-1}$. The energy flux $\Pi_{n,m}$ through a scale ℓ_n and structure (n, m) expressing the rate energy from the large scales $(i < n)$ is lost to the smaller scales $(i \geq n)$ through the structure m due to the non-linearity, which is given by

$$\Pi_{n,m} = ak_n U_{n,m} U_{n-1,m^*} U_{n-1,m^*} + bk_n U_{n,m} U_{n,m} U_{n-1,m^*} \tag{9}$$

The total flux through scale ℓ_n is then given by

$$\Pi_n = \frac{1}{M_n}\sum_{m=1}^{M_n} \Pi_{n,m} \tag{10}$$

Conservation of energy by the non-linear terms implies that, at scales smaller than the forcing scale and larger than the dissipation scale (ℓ_ν), the flux Π_n is constant and equal to the energy injection/dissipation ϵ

where $\theta_{n,m}$ is arbitrary. Equations (18) and (19) form a recurrence relation out of which, given $r_{1,1}$ and a choice of $\theta_{n,m}$, one can construct all $r_{n,m}$. Then, given $r_{n,m}$, one can obtain $U_{n,m}$ based on (14) as

$$U_{n,m} = U_{1,1}\, r_{1,1}\, r_{2,m_1}\, r_{3,m_2} \,\cdots\, r_{n,m} \tag{20}$$

where $m_1, m_2, \ldots$ are the m one crosses along the path from (1,1) to (n, m) as shown by the red line in Figure 1. This recurrence relation, however, does not always lead to bounded solutions of $r_{n,m}$. For some values of $\theta_{n,m}$, the resulting x, y can be negative or zero. Negative values can lead to un-physical solutions with negative flux from the small to the large scales. If the flux $\Pi_{n,m}$ is negative for some (n,m), then the sum of the flux of the "daughter" nodes $\Pi_{n+1,m'} + \Pi_{n+1,m'+1}$ has to be negative and equal to $\Pi_{n,m}$ and so on for their "daughter" nodes. There would then exist at least one descendant at each scale with negative flux, and this can only be realised if there is a source of energy at the very small scales, which is un-physical. Negative flux stationary solutions are thus not accepted. Furthermore, if x or y is zero, it means that the particular branch is zero for all subsequent values (i.e., all descendants). We need thus to limit the choice of θ so that positive and finite $r_{n,m}$ are obtained.

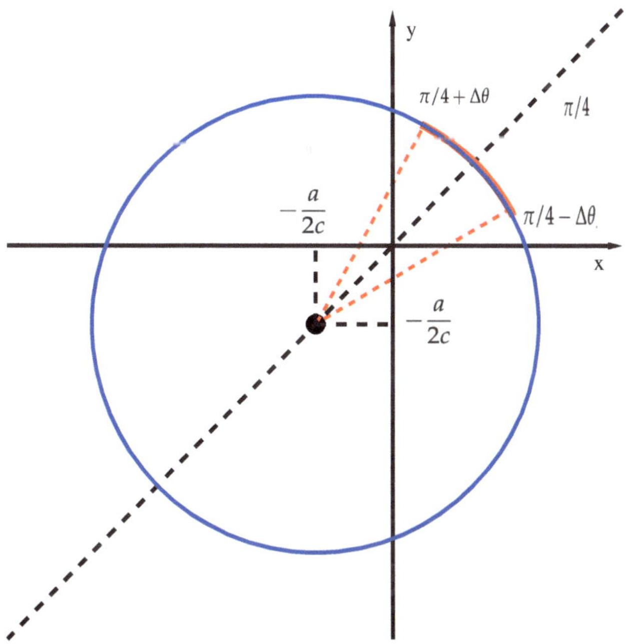

Figure 4. A plot of all admissible solutions x, y of (16). The red line shows the ones chosen for a given choice of $\Delta\theta$.

The simplest case is obtained by choosing $\theta_{n,m} = \pi/4$. It corresponds to an equal part of energy being distributed to the left and the right branch and leads to the Kolmogorov solution $r_{n,m} = 1$ or in terms of the velocity $U_{n,m} = \lambda^{n/3}$) (where $U_{1,1} = 1$ is assumed). It corresponds to a finite flux non-intermittent (self-similar) solution.

Intermittency, however, can manifest itself if we chose $\theta_{n,m} \neq \pi/4$ so that energy is not equally distributed in the left and right branch. Here, we will chose $\theta_{n,m}$ randomly with uniform distribution in the range $\theta_{\min} = \pi/4 - \Delta\theta < \theta_{n,m} < \pi/4 + \Delta\theta = \theta_{\max}$ for a given $\Delta\theta < \pi/4$, as shown in red in Figure 4. Then, it can be shown that, for $c > a$, there exists $r_{\max} > r_{\min} > 0$ such that, for all $r \in (r_{\max}, r_{\min})$, both $x \in (r_{\max}, r_{\min})$ and $y \in (r_{\max}, r_{\min})$. For $c \leq a$, the recurrence relation converges either to $r_{n,m} = 0$ or $r_{n,m} = \infty$ and we are going to limit ourselves only to the $c > a$ case here. To obtain $r_{\max}, r_{\min}$, one needs to note that, from the recurrence relation (19), the largest value of $r_{n+1,m'} = r_{max}$ is obtained when $\theta = \theta_{max}$ and $r_{n,m} = r_{min}$, while the smallest value of $r_{n+1,m'} = r_{min}$ is obtained when $\theta = \theta_{min}$ and $r_{n,m} = r_{max}$. This leads to the following relations

$$r_{\max} = -\frac{a}{2c} + \sqrt{2}\cos(\theta_{\min})\sqrt{\frac{a}{cr_{\min}^2} + \frac{1}{r_{\min}} + \frac{a^2}{4c^2}} \tag{21}$$

$$r_{\min} = -\frac{a}{2c} + \sqrt{2}\cos(\theta_{\max})\sqrt{\frac{a}{cr_{\max}^2} + \frac{1}{r_{\max}} + \frac{a^2}{4c^2}}. \tag{22}$$

We arrive at exactly the same relations if we examine Equation (19).

We solved Equations (21) and (22) numerically and the results are shown in the left panel of Figure 5 for three different values of c/a. For $\Delta\theta = 0$, only the Kolmogorov solution is allowed with $r_{max} = r_{min} = 1$. As $\Delta\theta = 0$ increases, $r_{n,m}$ cover a wider range of values up until a critical value of $\Delta\theta = \Delta\theta_c$ for which r_{min} becomes zero and r_{max} diverges. The value of this critical angle as a function c/a is shown in the right panel of the same figure. $\Delta\theta_c$ is zero for $c/a = 1$ and grows for larger values approaching $\Delta\theta_c = \pi/4$ as $c/a \to \infty$ (not demonstrated here).

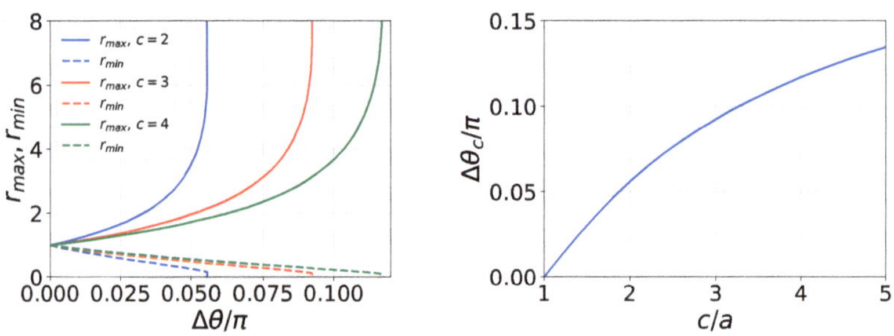

Figure 5. Left: The minimum and maximum value of r as a function of $\Delta\theta$. **Right**: The critical value of $\Delta\theta$ for which r_{max} diverges as a function of c/a.

For any given choice of $\Delta\theta < \Delta\theta_c$, we can construct an ensemble of exact solutions of the present model by following the recurrence relations (18) and (19), picking each time randomnly $\theta_{n,m} \in (\pi/4 - \Delta\theta, \pi/4 + \Delta\theta)$ and reconstructing $U_{n,m}$ by Equation (20). We note that, other than c/a, the only other parameter that controls the ensemble of solutions considered is $\Delta\theta/\Delta\theta_c$, which provides a measure of how much our ensemble deviates from the Kolmogorov solution $\Delta\theta = 0$. This process has direct links with the random cascade models studied in the past [43–45]; however, we need to note that, unlike the random cascade models, the solutions found here are energy-conserving.

4. Statistical Behaviour and Intermittency

In this section, we examine a large ensemble of the exact solutions shown in the previous section and investigate their properties. For our investigation, we have set $c/a = 2$ and we consider only a single path (as the one shown in red in Figure 1) and not the full tree. The differences in the statistics between the two choices (single path and full tree) lie in the cross correlations between different modes that are not captured in the single path. As an example, we mention that the flux Π_n in Equation (10) is identically equal to ϵ for every realisation, while the flux $\Pi_{n,m}$ given in Equation (9) fluctuates and only its mean value is equal to ϵ

$$\langle \Pi_{n,m} \rangle = \Pi_n = \epsilon.$$

Along such path, we consider three different ensembles for $\Delta\theta/\Delta\theta_c = 0.1, 0.5, 0.9$, each one composed of 10^7 different solutions. The solutions were constructed by picking randomly $\theta_{n,m}$ for each node examined, from a uniform distribution between $\pi/4 - \Delta\theta$ and $\pi/4 + \Delta\theta$. The value of n varied from $n = 1$ to $n = 200$. We note that, if the full tree was investigated instead of a single path for such large value n, it would require to solve for 2^{200} degrees of freedom, which is computationally unattainable.

In the top panels of Figure 6, we plot the probability distribution function (PDF) $P_U(U_{n,m})$ of the variable $U_{n,m}$ for the three values of $\Delta\theta/\Delta\theta_c = 0.1, 0.5, 0.9$ (from left to right) and different values of n. The PDFs of different values of n do not seem to overlap, although the x-axis has been normalised by the Kolmogorov prediction $\lambda^{-1/3}$. Instead, as large values of n are reached, the PDFs display larger tails, reaching values of $U_{n,m}$ much larger and much smaller than their mean values. The closer $\Delta\theta$ is to the critical value $\Delta\theta_c$, the larger this deviation is. On the other hand, the PDFs $P_r(r_{n,m})$ of the ratios $r_{n,m}$ that are displayed in the lower panels of Figure 6 do not display such widening. For sufficiently large n, all PDFs collapse to the same functional form that depends only on the choice of $\Delta\theta_c$. This implies that, while $U_{n,m}$ are not self-similar under scale transformations, their ratios $r_{n,m}$ are!

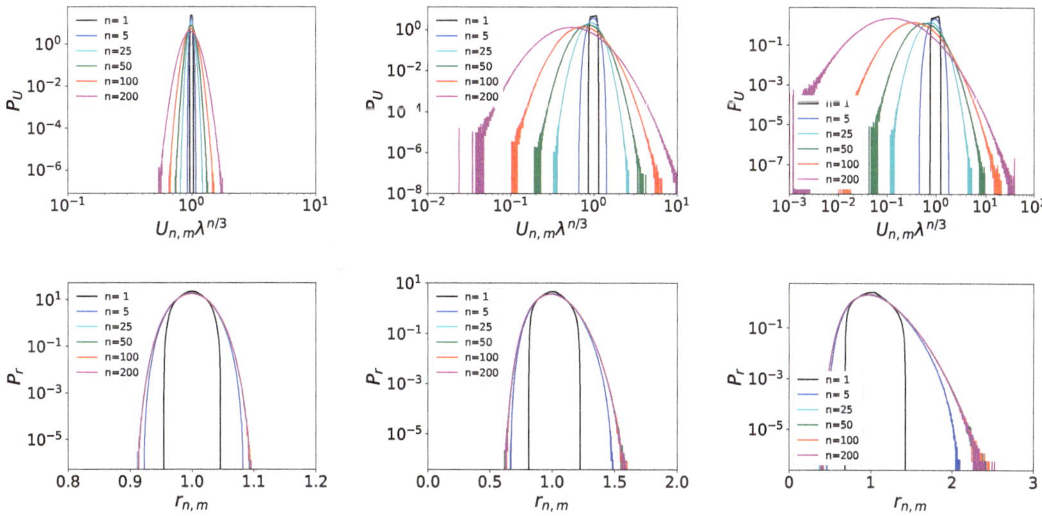

Figure 6. Top panels: PDFs $P_U(U_{n,m})$ of the velocity modes $U_{n,m}$ for the three different ensembles (left $\Delta\theta/\Delta\theta_c = 0.1$, center $\Delta\theta/\Delta\theta_c = 0.5$, and right $\Delta\theta/\Delta\theta_c = 0.9$) for different values of n. **Bottom panels:** PDFs $P_r(r_{n,m})$ of the velocity ratios $r_{n,m}$ for the same ensembles and the same n.

The same behaviour can be observed for the energy fluxes $\Pi_{n,m}$. In the top panels of Figure 7, we plot the PDFs P_Π of $\Pi_{n,m}$ for the same values of $\Delta\theta$ and n as in Figure 6. As with the velocity amplitudes $U_{n,m}$, as n is increased, the PDFs of $\Pi_{n,m}$ widen without collapsing to an n-independent form. In the lower panel of the same figure, we plot the the PDFs $P_\pi(\pi_{n,m})$ of the flux ratio $\pi_{n,m}$. It is defined as

$$\pi_{n,m} = \frac{\Pi_{n,m}}{\Pi_{n-1,m^*}} \tag{23}$$

after a little algebra and using (9) and (18) leads to

$$\pi_{n+1,m'} = 1 + f(r_{n,m})\cos(2\theta_{n,m}) \tag{24}$$

where $f(r) = 1 + (a/2c)^2 r^2/(a/c + r)$. The flux ratio, much like the velocity ratio $r_{n,m}$, does converge to an n-independent PDF as large values of n are reached. Furthermore, the functional form of this PDF appears to be flat, limited by a minimum and a maximum value of $\pi_{n,m}$. This appears to be so because $f(r)$ in (24) varies little with r for small variations in r and the variations in $\pi_{n,m}$ are mostly controlled by the variations in $\theta_{n,m}$.

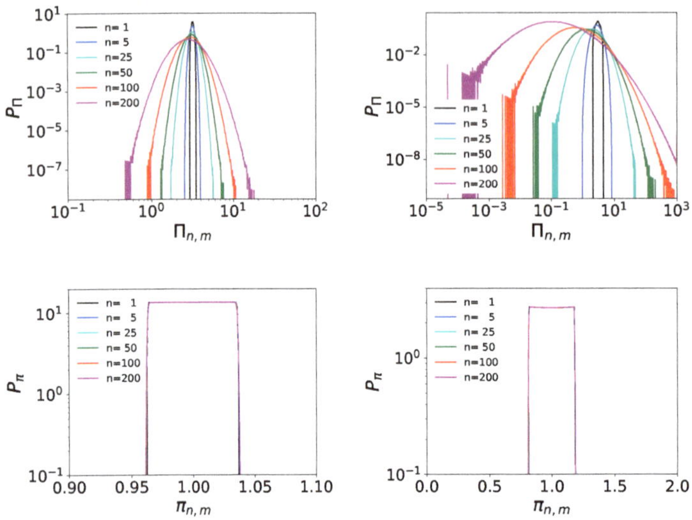

Figure 7. **Top panels:** PDFs $P_\Pi(\Pi_{n,m})$ of the fluxes $\Pi_{n,m}$ for the three different ensembles (left $\Delta\theta/\Delta\theta_c = 0.1$, center $\Delta\theta/\Delta\theta_c = 0.5$, and right $\Delta\theta/\Delta\theta_c = 0.9$) for different values of n. **Bottom panels:** PDFs $P_\pi(\pi_{n,m})$ of the velocity ratios $\pi_{n,m}$ for the same ensembles and the same n.

Given that the PDFs of $r_{n,m}$ and $\pi_{n,m}$ arrive at an n-independent form, a large n has some implications for the evolution in n of the PDFs P_U, P_Π. Both $U_{n,m}$ and $\Pi_{n,m}$ can be written as a product of all $r_{n',m}$ and $\pi_{n',m}$ with $n' \leq n$. As a result, the logarithms of $U_{n,m}$ and $\Pi_{n,m}$ can be written as

$$\ln(U_{n,m}) = \ln(U_{1,1}) + nL_U, \quad \ln(\Pi_{n,m}) = \ln(\Pi_{1,1}) + nL_\Pi \tag{25}$$

where L_U and L_Π stand for the mean value of the logarithms of $r_{n,m}$ and $\pi_{n,m}$, respectively:

$$L_U = \frac{1}{n}\sum_{n'=1}^{n} \ln(r_{n',m}), \quad \text{and} \quad L_\Pi = \frac{1}{n}\sum_{n'=1}^{n} \ln(\pi_{n',m}). \tag{26}$$

The properties of $U_{n,m}$ and $\Pi_{n,m}$ remind of the random cascades studied in the past [43–45]. However, while the random cascade models were not conserving energy, in the present model, energy is conserved exactly. Another important difference here is that $r_{n',m}$ and $\pi_{n,m}$ are not independent but each one depends on the value of the previous one. Nonetheless, we can proceed assuming such independence, although not entirely correct. In that case, P_U and P_Π can be reconstructed using large deviation theory [46]. In this framework, L_U and L_Π follow for large n a distribution of the form

$$P_{L_U}(L_U) \propto \exp[-nI_U(L_U)], \quad \text{and} \quad P_{L_\Pi}(L_\Pi) \propto \exp[-nI_\Pi(L_\Pi)] \tag{27}$$

where I_U and I_Π are called the rate functions that can in principle be obtained from P_r and P_π using the Legendre–Fenchel transform [46]. Here, we limit ourselves in noting that, if

P_{L_U} and P_{L_Π} follow the form of Equation (27), then the distributions of $U_{n,m}$ and $\Pi_{n,m}$ that are linked to L_U and L_Π by (25) should take the form

$$P_U(U_{n,m}) \propto \exp\left[-nI_U\left(\frac{1}{n}\ln\left(\frac{U_{n,m}}{U_{1,1}}\right)\right)\right], \quad P_\Pi(\Pi_{n,m}) \propto \exp\left[-nI_\Pi\left(\frac{1}{n}\ln\left(\frac{\Pi_{n,m}}{\Pi_{1,1}}\right)\right)\right] \tag{28}$$

where only the largest terms in n are kept. To test this prediction, we plot in Figure 8 $(P_U/P_U^*)^{1/n}$ as a function $(U_{n,m}/U^*)^{1/n}$ (top panels) and $(P_\Pi/P_\Pi^*)^{1/n}$ as a function $(\Pi_{n,m}/\Pi^*)^{1/n}$, where U^* and Π^* correspond to the value at which the probability obtains its maximum P_U^*, P_Π^*. With this normalization, the PDFs both for $U_{n,m}$ and for $\Pi_{n,m}$ collapse, indicating that the large deviation principle works well for this model.

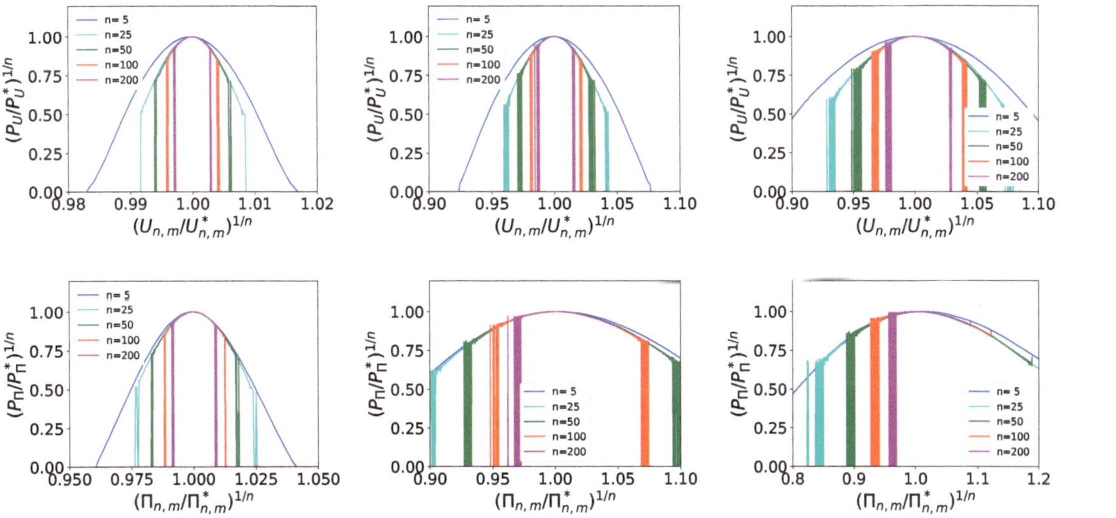

Figure 8. Top panels: PDFs $P_U(U_{n,m})$ for the different cases examined normalised using the predictions of large deviation theory (left $\Delta\theta/\Delta\theta_c = 0.1$, center $\Delta\theta/\Delta\theta_c = 0.5$, and right $\Delta\theta/\Delta\theta_c = 0.9$). **Bottom panels:** The same for the PDFs $P_\Pi(\Pi_{n,m})$.

As a final look at the intermittency problem, we display in the top panels of Figure 9 the first ten structure functions $S_p(\ell_p)$ defined as

$$S_p(\ell_n) = \left\langle U_{n,m}^p \right\rangle \tag{29}$$

where the angular brackets stand for ensemble average. The structure functions have been normalised by the Kolmogorov scaling to emphasise the differences. The structure functions are fitted to power laws

$$S_p(\ell_n) \propto \ell_n^{\zeta_p} \tag{30}$$

and the measured exponents ζ_p are plotted in the lower panels of Figure 9. The exponents show similar behaviour, with real turbulence displaying larger values for $p < 3$ and smaller values for $p > 3$, while the exact result $\zeta_3 = 1$ is satisfied. It is worth noting that the deviations from the Kolmogorov scaling are not universal but depend on our choice of ensemble, which is controlled by $\Delta\theta$.

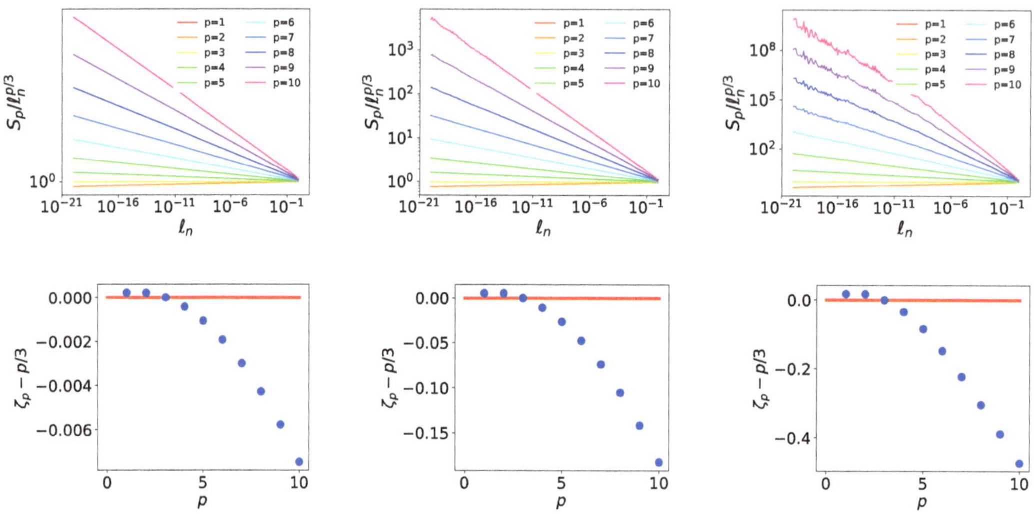

Figure 9. Top panels: Structure functions up to 10th order for the three different values of $\Delta\theta$ examined (left $\Delta\theta/\Delta\theta_c = 0.1$, center $\Delta\theta/\Delta\theta_c = 0.5$, and right $\Delta\theta/\Delta\theta_c = 0.9$). **Bottom panels:** The resulting exponents ζ_p.

5. Discussion and Conclusions

One can argue that the exact stationary solutions obtained in this work have little to do with real turbulence that displays chaotic spatio-temporal dynamics. This may be true and multi-branch models with two neighbour interactions as in [39–41] that display chaotic dynamics should be investigated instead. The present results, however, do point to a clear instructive demonstration of how intermittency can appear in realistic flows and how it can be modeled. Furthermore, it leads to a series of clear messages that are described below that are of great use in future turbulence research and can guide measurements in numerical simulations and experiments.

First, we note that intermittency appearing in stationary fields found here comes in contrast with the typical shell model studies in single-branch models for which intermittency comes from the temporal dynamics alone as only a single structure exists for each scale ℓ_n. In the latter case, intermittency has been linked to the temporal dynamics through the fluctuation dissipation theorem [37]. In reality, both temporal and spatial dynamics contribute to the presence of intermittency and their role needs to be clarified.

In the present model, randomness comes from our choice of $\theta_{n,m}$ and the resulting intermittency depends on that choice. In reality (or in more complex shell models), such randomness comes from local chaotic dynamics that need to studied in order to clarify which processes lead to enhanced cascade and with what probability. Multi-branch models based on the more complex GOY or SABRA as proposed in [39–41] can help in this direction. The additional coupling terms introduced in these works avoid phase synchronization and lead to chaotic dynamics. Chaos can remove the arbitrariness of the choice of θ that should ideally be self-imposed by the dynamics.

Perhaps the most interesting implication of this work is that it suggests new ways to plot data from experimental and numerical simulations. One way suggested by this work is, instead of focusing on the PDFs of velocity differences, experimental or numerical data could focus on the PDFs of ratios of velocity differences. The latter are shown in this work to become scale-independent and could lead to more precise measurements. An alternative way is to re-scale the PDFs of velocity differences using the large deviation prediction (27), as was conducted in Figure 8. Of course, realistic data $n \propto \ln(L/\ell_n)$ are not precisely defined and an optimal choice should be searched for.

A good model of a complex phenomenon, in the authors' opinion, is not one that quantitatively reproduces experimental measurements through parameter fitting but rather one that unravels the processes involved. In that respect, we believe that the present model and results are very fruitful. We only hope that this work comes close to the standards set by Jack Herring. A.A. met Jack Herring during his ASP post doc in 2004–2006. Jack is fondly remembered stopping by the offices of post docs just to see if they are OK. He will be greatly missed.

Author Contributions: Conceptualization, A.A.; methodology, A.A. and B.A.; software, B.A.; validation, B.A.; formal analysis, A.A. and B.A.; investigation, A.A. and B.A.; resources, A.A.; data curation, A.A.; writing—original draft preparation, A.A.; writing—review and editing, A.A.; visualization, A.A. and B.A.; supervision, A.A.; project administration, A.A.; funding acquisition, A.A. All authors have read and agreed to the published version of the manuscript.

Funding: This work was also supported by the Agence nationale de la recherche (ANR DYSTURB project No. ANR-17-CE30-0004).

Institutional Review Board Statement: Not applicable.

Informed Consent Statement: Not applicable.

Data Availability Statement: No new data were created.

Acknowledgments: A.A. would also like to acknowledge the help of the Pétrélis brothers, Francois and Nicolas, for their help with the large deviation theory. We would also like to thank Annick Pouquet for inviting us to write an article in this special issue dedicated to Jack Herring.

Conflicts of Interest: The authors declare no conflict of interest. The funders had no role in the design of the study; in the collection, analyses, or interpretation of data; in the writing of the manuscript; or in the decision to publish the results.

References

1. Herring, J. Self-consistent-field approach to turbulence theory. *Phys. Fluids* **1965**, *8*, 2219–2225. [CrossRef]
2. Herring, J. Approach of axisymmetric turbulence to isotropy. *Phys. Fluids* **1974**, *17*, 859–872. [CrossRef]
3. Herring, J. Theory of two-dimensional anisotropic turbulence. *J. Atmos. Sci.* **1975**, *32*, 2254–2271. [CrossRef]
4. Herring, J.; Orszag, S.; Kraichnan, R.; Fox, D. Decay of two-dimensional homogeneous turbulence. *J. Fluid Mech.* **1974**, *66*, 417–444. [CrossRef]
5. Herring, J.R. On the statistical theory of two-dimensional topographic turbulence. *J. Atmos. Sci.* **1977**, *34*, 1731–1750. [CrossRef]
6. Newman, G.R.; Herring, J.R. A test field model study of a passive scalar in isotropic turbulence. *J. Fluid Mech.* **1979**, *94*, 163–194. [CrossRef]
7. Herring, J.R. Statistical theory of quasi-geostrophic turbulence. *J. Atmos. Sci.* **1980**, *37*, 969–977. [CrossRef]
8. Herring, J.R. Theoretical calculations of turbulent bispectra. *J. Fluid Mech.* **1980**, *97*, 193–204. [CrossRef]
9. Herring, J.; Schertzer, D.; Lesieur, M.; Newman, G.; Chollet, J.; Larcheveque, M. A comparative assessment of spectral closures as applied to passive scalar diffusion. *J. Fluid Mech.* **1982**, *124*, 411–437. [CrossRef]
10. Herring, J.; McWilliams, J. Comparison of direct numerical simulation of two-dimensional turbulence with two-point closure: The effects of intermittency. *J. Fluid Mech.* **1985**, *153*, 229–242. [CrossRef]
11. Chen, H.; Herring, J.R.; Kerr, R.M.; Kraichnan, R.H. Non-Gaussian statistics in isotropic turbulence. *Phys. Fluids A Fluid Dyn.* **1989**, *1*, 1844–1854. [CrossRef]
12. Herring, J. Subgrid scale modeling—An introduction and overview. In Proceedings of the Turbulent Shear Flows I: Selected Papers from the First International Symposium on Turbulent Shear Flows, The Pennsylvania State University, University Park, Pennsylvania, PA, USA, 18–20 April 1977; Springer: Berlin/Heidelberg, Germany, 1979; pp. 347–352.
13. Herring, J.R.; Métais, O. Numerical experiments in forced stably stratified turbulence. *J. Fluid Mech.* **1989**, *202*, 97–115. [CrossRef]
14. Kraichnan, R.H. The structure of isotropic turbulence at very high Reynolds numbers. *J. Fluid Mech.* **1959**, *5*, 497–543. [CrossRef]
15. Kraichnan, R.H. Decay of Isotropic Turbulence in the Direct-Interaction Approximation. *Phys. Fluids* **1964**, *7*, 1030–1048. [CrossRef]
16. Herring, J. Self-Consistent-Field Approach to Nonstationary Turbulence. *Phys. Fluids* **1966**, *9*, 2106–2110. [CrossRef]
17. Orszag, S.A. Analytical theories of turbulence. *J. Fluid Mech.* **1970**, *41*, 363–386. [CrossRef]
18. Leith, C. Atmospheric predictability and two-dimensional turbulence. *J. Atmos. Sci.* **1971**, *28*, 145–161. [CrossRef]

19. Fouquet, A.; Lesieur, M.; André, J.; Basdevant, C. Evolution of high Reynolds number two-dimensional turbulence. *J. Fluid Mech.* **1975**, *72*, 305–319. [CrossRef]
20. Lesieur, M.; Ossia, S. 3D isotropic turbulence at very high Reynolds numbers: EDQNM study. *J. Turbul.* **2000**, *1*, 007. [CrossRef]
21. Leith, C. Diffusion approximation to inertial energy transfer in isotropic turbulence. *Phys. Fluids* **1967**, *10*, 1409–1416. [CrossRef]
22. Leith, C.E. Diffusion approximation for two-dimensional turbulence. *Phys. Fluids* **1968**, *11*, 671–672. [CrossRef]
23. Desnianskii, V.; Novikov, E. Simulation of cascade processes in turbulent flows: PMM vol. 38, n = 3, 1974, pp. 507–513. *J. Appl. Math. Mech.* **1974**, *38*, 468–475. [CrossRef]
24. Gledzer, E. Eb gledzer, dokl. akad. nauk sssr 200, 1043. *Dokl. Akad. Nauk SSSR* **1973**, *200*, 1043.
25. Yamada, M.; Ohkitani, K. Lyapunov spectrum of a chaotic model of three-dimensional turbulence. *J. Phys. Soc. Jpn.* **1987**, *56*, 4210–4213. [CrossRef]
26. L'vov, V.S.; Podivilov, E.; Pomyalov, A.; Procaccia, I.; Vandembroucq, D. Improved shell model of turbulence. *Phys. Rev. E* **1998**, *58*, 1811. [CrossRef]
27. Constantin, P.; Levant, B.; Titi, E.S. Analytic study of shell models of turbulence. *Phys. D Nonlinear Phenom.* **2006**, *219*, 120–141. [CrossRef]
28. Constantin, P.; Levant, B.; Titi, E.S. Regularity of inviscid shell models of turbulence. *Phys. Rev. E* **2007**, *75*, 016304. [CrossRef]
29. Falkovich, G.; Kadish, Y.; Vladimirova, N. Multimode correlations and the entropy of turbulence in shell models. *Phys. Rev. E* **2023**, *108*, 015103. [CrossRef]
30. Vincenzi, D.; Gibbon, J.D. How close are shell models to the 3D Navier–Stokes equations? *Nonlinearity* **2021**, *34*, 5821. [CrossRef]
31. Benzi, R.; Levant, B.; Procaccia, I.; Titi, E.S. Statistical properties of nonlinear shell models of turbulence from linear advection models: Rigorous results. *Nonlinearity* **2007**, *20*, 1431. [CrossRef]
32. Ditlevsen, P.; Giuliani, P. Cascades in helical turbulence. *Phys. Rev. E* **2001**, *63*, 036304. [CrossRef]
33. Biferale, L. Shell models of energy cascade in turbulence. *Annu. Rev. Fluid Mech.* **2003**, *35*, 441–468. [CrossRef]
34. Ditlevsen, P.D. *Turbulence and Shell Models*; Cambridge University Press: Cambridge, UK, 2010.
35. Alexakis, A.; Biferale, L. Cascades and transitions in turbulent flows. *Phys. Rep.* **2018**, *767*, 1–101.
36. Jensen, M.; Paladin, G.; Vulpiani, A. Intermittency in a cascade model for three-dimensional turbulence. *Phys. Rev. A* **1991**, *43*, 798. [CrossRef] [PubMed]
37. Aumaitre, S.; Fauve, S. Intermittency as a consequence of a stationarity constraint on the energy flux. 2023, *in press*.
38. Mailybaev, A.A. Solvable intermittent shell model of turbulence. *Commun. Math. Phys.* **2021**, *388*, 469–478. [CrossRef]
39. Aurell, E.; Frick, P.; Shaidurov, V. Hierarchical tree-model of 2D-turbulence. *Phys. D Nonlinear Phenom.* **1994**, *72*, 95–109. [CrossRef]
40. Aurell, E.; Dormy, E.; Frick, P. Binary tree models of high-Reynolds-number turbulence. *Phys. Rev. E* **1997**, *56*, 1692. [CrossRef]
41. Benzi, R.; Biferale, L.; Tripiccione, R.; Trovatore, E. (1+1)-dimensional turbulence. *Phys. Fluids* **1997**, *9*, 2355–2363. [CrossRef]
42. Biferale, L.; Blank, M.; Frisch, U. Chaotic cascades with Kolmogorov 1941 scaling. *J. Stat. Phys.* **1994**, *75*, 781–795. [CrossRef]
43. Novikov, E.; Stewart, R. Turbulent intermittency and the spectrum of fluctuations of energy dissipation. *Izv. Akad. Nauk. SSSR Geofiz.* **1964**, *3*, 408–413.
44. Yaglom, A. Effect of fluctuations in energy dissipation rate on the form of turbulence characteristics in the inertial subrange. *Dokl. Akad. Nauk SSSR* **1966**, *166*, 49–52.
45. Mandelbrot, B.B.; Mandelbrot, B.B. Intermittent turbulence in self-similar cascades: Divergence of high moments and dimension of the carrier. In *Multifractals and 1/f Noise: Wild Self-Affinity in Physics (1963–1976)*; Springer: Berlin/Heidelberg, Germany, 1999; pp. 317–357.
46. Touchette, H. The large deviation approach to statistical mechanics. *Phys. Rep.* **2009**, *478*, 1–69. [CrossRef]

atmosphere

MDPI

Article

Intermittency Scaling for Mixing and Dissipation in Rotating Stratified Turbulence at the Edge of Instability

Annick Pouquet [1,2,*,†], Duane Rosenberg [3,†], Raffaele Marino [4,†] and Pablo Mininni [5,6,†]

1 Laboratory for Atmospheric and Space Physics, University of Colorado, Boulder, CO 80309, USA
2 National Center for Atmospheric Research, P.O. Box 3000, Boulder, CO 80307, USA
3 Cooperative Institute for Research in the Atmosphere (CIRA), NOAA/OAR Global Systems Laboratory, Colorado State University, 325 Broadway Boulder, Fort Collins, CO 80305, USA; duaner62@gmail.com
4 CNRS, École Centrale de Lyon, INSA Lyon, Université Claude Bernard Lyon 1, Laboratoire de Mécanique des Fluides et Acoustique, UMR5509, Université de Lyon, F-69134 Ecully, France; raffaele.marino@ec-lyon.fr
5 Facultad de Ciencias Exactas y Naturales, Departamento de Fisica, Universidad de Buenos Aires, Ciudad Universitaria, Buenos Aires 1428, Argentina; mininni@df.uba.ar
6 Instituto de Fisica Interdisciplinaria y Aplicada (INFINA), CONICET—Universidad de Buenos Aires, Ciudad Universitaria, Buenos Aires 1428, Argentina
* Correspondence: pouquet@ucar.edu
† These authors contributed equally to this work.

Abstract: Many issues pioneered by Jackson Herring deal with how nonlinear interactions shape atmospheric dynamics. In this context, we analyze new direct numerical simulations of rotating stratified flows with a large-scale forcing, which is either random or quasi-geostrophic (QG). Runs were performed at a moderate Reynolds number Re and up to 1646 turn-over times in one case. We found intermittent fluctuations of the vertical velocity w and temperature θ in a narrow domain of parameters as for decaying flows. Preliminary results indicate that parabolic relations between normalized third- and fourth-order moments of the buoyancy flux $\propto \langle w\theta \rangle$ and of the energy dissipation emerge in this domain, including for passive and active scalars, with or without rotation. These are reminiscent of (but not identical to) previous findings for other variables and systems such as oceanic and atmospheric flows, climate re-analysis data, fusion plasmas, the Solar Wind, or galaxies. For QG forcing, sharp scaling transitions take place once the Ozmidov length scale ℓ_{Oz} is resolved—ℓ_{Oz} being the scale after which a turbulent Kolmogorov energy spectrum likely recovers at high Re.

Keywords: rotating stratified turbulence; intermittency; dissipation; buoyancy flux; kurtosis

Citation: Pouquet, A.; Rosenberg, D.; Marino, R.; Mininni, P. Intermittency Scaling for Mixing and Dissipation in Rotating Stratified Turbulence at the Edge of Instability. *Atmosphere* **2023**, *14*, 1375. https://doi.org/10.3390/atmos14091375

Academic Editor: Stephan De Wekker

Received: 9 August 2023
Revised: 24 August 2023
Accepted: 26 August 2023
Published: 31 August 2023

1. Introduction

Much progress has been made in our understanding of turbulence, developing and analyzing experiments, observations, direct numerical simulations (DNSs), as well as numerous theoretical and modeling techniques, among them the direct interaction approximation (DIA) stemming from a renormalization-group (RG) procedure [1], the test-field model (TFM), and related second-order closures, often Markovianized such as the Eddy-Damped Quasi-Normal Markovian (EDQNM), or the quasi-normal scale elimination [2] (see [3–5] for recent reviews, and this volume). These flows are complex, due to the large number of eddies interacting non-linearly, as in homogeneous isotropic fully developed turbulence (FDT), and due to their inherent lack of predictability [6–8]. Flow structures change in the presence of waves, stemming for example from imposed gravity, rotation, or a strong uniform magnetic field and, specifically, their behaviors differ according to how fast the waves are relative to the eddy turn-over time $\tau_{NL} = L_{int}/U_{rms}$ with U_{rms}, L_{int} the *r.m.s.* velocity and the integral length scale. One of the governing parameters of the resulting dynamics, beyond the Reynolds number, is the ratio $G = \tau_P/\tau_{NL}$ with τ_P the characteristic period of the waves; G is the Rossby number in the presence of an imposed rotation, or it is the Froude number in the presence of stratification.

Accurate direct numerical simulations with pseudo-spectral methods were developed in particular at MIT and NCAR in the early 70s and were immediately put to use by Jack Herring to study turbulence, as e.g., in the context of the growth of errors in numerical modeling of the atmosphere [8], of two-dimensional (2D), axisymmetric or QG turbulence [9–11], for comparing closures to DNS [12,13], including for non-Gaussian statistics and intermittency in the far dissipation range [14,15]. Jack Herring was central to this introduction of models and DNS to complement experimental and theoretical studies of FDT, and he then moved on to study stratified turbulence [16,17].

Characteristic of turbulence is the presence of localized extreme dissipation in many settings [18,19]. In the ocean, satellite altimetry allows to study the interactions of large-scale planetary or (sub-)synoptic-scale waves with shear structures, as observed for example in the Gulf Stream and the Kuroshio Current [20,21]. Turbulence leads to filamentation, as seen in phytoplankton dynamics forced by hurricanes [22], as well as to energy transfer and dissipative events (e.g., [23,24]). There is also recent radar data indicative of extreme vertical drafts in the (strongly stratified) polar summer mesosphere [25]. Furthermore, waves can change the energy spectra, as demonstrated in weak turbulence [3], detected in the atmosphere [26] and in DNS [27], and as shown in magnetohydrodynamic (MHD) as well [28,29]. Atmospheric data can be related to numerical results for strongly stratified flows showing quantitatively that turbulence interacting with fast waves can be more intermittent than FDT [30,31] in a narrow domain of parameters (see [32] for Solar Wind observations, and [33] for energy dissipation in reconnection events leading also to particle acceleration in the magnetotail).

In this context, we present preliminary results using DNS on relationships between normalized third and fourth order moments taken as proxies to characterize the intermittency and dissipative properties of rotating strongly stratified turbulence, contrasting behaviors for high *vs.* low kurtosis of the vertical velocity with varying Froude and Reynolds numbers (see next Section for definitions). Similar relationships have been found in the turbulent atmosphere and ocean [34–38], climate data re-analysis and glaciology [39,40], and in fusion plasmas [41–43], the magnetosheath, the interplanetary magnetic field or the cosmos [44–48]. Equations, definitions and our numerical methodology are given in the next section. Some of these properties for several field variables for the new runs performed herein are presented in Section 3 for Navier–Stokes and stratified flows. Rotating stratified turbulence is analyzed in somewhat more detail in Section 4. An overview is given in Section 5, and Section 6 presents a discussion and conclusions.

2. Numerical Set-Up

2.1. Equations and Definitions

We perform several sets of numerical simulations for an incompressible ($\nabla \cdot \mathbf{u} = 0$) velocity field $\mathbf{u} = (u, v, w) = (\mathbf{u}_\perp, w)$ with rotation and stratification. The active scalar θ (called in the following *temperature*) is in velocity units; strictly speaking, θ and the temperature have opposite signs, but they are linearly related through a thermal expansion coefficient. With ν, κ the kinematic viscosity and thermal diffusivity, taken equal, p the pressure, $N = [-g\partial_z\bar{\theta}/\theta_0]^{1/2}$ the Brunt-Väisälä frequency, and $f_0 = 2\Omega$, $\mathbf{\Omega} = \Omega\mathbf{z}^\star$ an imposed rotation in the vertical direction of unit vector $\mathbf{z}^\star$, the Boussinesq equations are:

$$\partial_t\mathbf{u} + (\mathbf{u}\cdot\nabla)\mathbf{u} = -\nabla p - N\theta\mathbf{z}^\star + 2\mathbf{u}\times\mathbf{\Omega} + \nu\nabla^2\mathbf{u} + \mathbf{F}_u\,, \quad (1)$$

$$\partial_t\theta + (\mathbf{u}\cdot\nabla)\theta = Nw + \kappa\nabla^2\theta + F_\theta\,. \quad (2)$$

Two types of forcing for $\mathbf{F}_u, F_\theta$ have been implemented, both set in the large scales at a wavenumber $k_F \approx 2.5$. The first one has random phases and is white noise in time. The second forcing, for the rotating stratified turbulence (RST) case, ensures a quasi-geostrophic balance at all times in the forcing modes (see [11]), following the procedure described in detail in [49] for QG initial conditions with zero vertical velocity. Pressure balance with the Coriolis force in the horizontal is enforced, as well as hydrostatic momentum balance due to gravity (see [50] for a different implementation in a study of balanced dynamics). One might note that the time-scale associated with the forcing, in particular in the case of white noise, may matter in the interactions between waves and eddies, with more waves for short correlation times, in particular at low Froude number for which N becomes comparable to $1/\Delta t$, where Δt is the time-step [27]. The buoyancy flux B_f, the kinetic energy dissipation ϵ_{du} and the potential vorticity P_V (point-wise invariant in the absence of dissipation) are:

$$B_f = N\langle w\theta\rangle \; ; \; \epsilon_{du} = \nu\left\langle |\omega|^2 \right\rangle \; ; \; P_V = f\partial_z\theta - N\omega_z + \omega \cdot \nabla\theta \; , \tag{3}$$

with $\omega = \nabla \times \mathbf{u}$ the vorticity. B_f is an energy exchange term with the same physical dimension as ϵ_{du}. It leads to local changes in density in the presence of gravity waves which need to be modeled in large-eddy simulations. Characteristic scales can be defined as the Taylor micro-scale λ_T that factors in the dissipation in the inertial range, the dissipation scale η_K based on a Kolmogorov spectrum, the buoyancy scale L_B measuring the effect of stratification in the large scales, and the Ozmidov scale ℓ_{Oz}:

$$\lambda_T = \left(\frac{\langle u^2\rangle}{\langle \omega^2\rangle}\right)^{1/2}, \; \eta_K = \left(\frac{\epsilon_{du}}{\nu^3}\right)^{-1/4}, \; L_B = \frac{U_{rms}}{N}, \; \ell_{Oz} = \left(\frac{\epsilon_{du}}{N^3}\right)^{1/2} \; ; \; \frac{\ell_{Oz}}{\eta_K} = \left(\frac{L_B}{\lambda_T}\right)^{3/2} . \tag{4}$$

The last expression indicates the multi-scale link in the dynamics of the flow between the development of turbulence and its stratification. The dimensionless parameters governing the fluid behavior are the Reynolds, Rossby and Froude numbers, Re, Ro, Fr, with the Prandtl number $Pr \equiv \nu/\kappa = 1$ here:

$$Re = \frac{U_{rms}L_{int}}{\nu} \; , \; Ro = \frac{U_{rms}}{fL_{int}} \; , \; Fr = \frac{U_{rms}}{NL_{int}} \; ; \; R_\lambda = \frac{U_{rms}\lambda_T}{\nu} \; , \; R_B = ReFr^2 \; , \tag{5}$$

with R_λ, R_B the Taylor and buoyancy Reynolds numbers. For the purely stratified runs, the forcing scale $L_f = 2\pi/k_F$ is used in the evaluation of the parameters instead of the integral scale L_{int}. One also defines the interaction parameter R_{IB} and, when involving gravity through N, the Richardson number Ri based on an overall vertical gradient of $\mathbf{u}_\perp$, as well as a local gradient Richardson number Ri_g:

$$R_{IB} = \frac{\omega_{rms}^2}{N^2} = \frac{\epsilon_{du}}{\nu N^2} = \left(\frac{L_B}{\lambda_T}\right)^2 ; \; Ri = \frac{N^2}{\langle \partial_z u_\perp^2 \rangle} \; , \; Ri_g = \frac{N(N - \partial_z\theta)}{[(\partial_z u)^2 + (\partial_z v]^2]} \; , \; I_{Ri} = \langle Ri_g \rangle \; . \tag{6}$$

The phenomenological evaluation of kinetic energy dissipation, $\epsilon_D = U_{rms}^3/L_{int}$, is based on expressing that it occurs in an eddy turn-over time, irrespective of the (small) viscosity. When ϵ_{du} becomes comparable to ϵ_D, as in FDT, one has $R_B = R_{IB}$. Note also that Ri, Ri_g, in the absence of uniform shear, are based on large-scale shear resulting from the overall nonlinear dynamics. One can also define a Taylor Froude number F_λ and relate it to the other parameters. After simple manipulations, one has:

$$F_\lambda \equiv \frac{U_{rms}}{N\lambda_T} = \frac{L_B}{\lambda_T} = \frac{ReFr}{R_\lambda} = R_{IB}^{1/2} = \left(\frac{\ell_{Oz}}{\eta_K}\right)^{2/3} . \tag{7}$$

The RST runs cover the three regimes identified in [51]: wave and viscous-dominated at small $[Fr, R_B]$, eddy-wave interactions at intermediate values, and the eddy-dominated regime with a resolved turbulent inertial range beyond the Ozmidov scale ($\ell_{Oz} > \eta_K$). These three regimes can likely be related to the classification proposed in [52] for the nocturnal planetary boundary layer (PBL) into weakly stable, transitional, and very stable regimes: when increasing the Froude number, there is a drop in the Richardson number, particularly marked in the transition regime, and a sharp increase in eddy diffusivities. Another classification of stratified flows can be found in [53] (see also [54,55]) in which the role of viscous dissipation, governed by a small Reynolds number, is particularly emphasized. This is in contrast to the role of fast waves compared to nonlinear eddies, as governed by a small Froude number; the competition between these two effects can be encompassed in the parameter $R_B = ReFr^2$, or in R_{IB}, but not necessarily uniquely [56]. At low R_B, anisotropy comes strongly into play. Previous studies (see e.g., [57]) have shown that when the system is forced three-dimensionally, this results in a significant excitation of wave modes which, as a result of viscous effects, can excite less vortical modes. However, their interactions with waves are still present [58], and we call this regime wave dominated since no turbulence range is excited. We also note that, at a given (high) buoyancy Reynolds number, all three regimes can be identified in the energy spectra in a given flow, provided L_B, ℓ_{Oz} and η_K are resolved (see [59] for an analysis of these regimes using a turbulence closure). Different values for R_{IB} can be found when using the energy injection rate $\langle \mathbf{F}_u \cdot \mathbf{u} \rangle$ instead of ϵ_{du}, the two equilibrating only on average. This is particularly relevant when there are strong intermittent fluctuations in the system [60,61], and this balance can also be broken in the presence of an inverse energy cascade. Finally, the skewness and excess kurtosis of a random variable V, with averages taken over three-dimensional (3D) space and with $S_V^G = 0$, $K_V^G = 0$ for a Gaussian, are defined as usual, with the following parabolic relation to be tested in this paper, e.g., for $V = B_f$:

$$S_V = \left\langle V^3 \right\rangle / \left\langle V^2 \right\rangle^{3/2}, \; K_V = \left\langle V^4 \right\rangle / \left\langle V^2 \right\rangle^2 - 3 \, ; \;\; K_V(S_V) = a_V S_V^2 + b_V \, . \tag{8}$$

2.2. Overview of the Direct Numerical Simulations

We use a variety of DNS (see Table 1). The runs labeled H are for the hydrodynamic case (Navier–Stokes, $N \equiv 0, \Omega \equiv 0$); the $S_{i=1,6}$ runs deal with the Boussinesq equations with $\Omega \equiv 0$. The R, Q runs are RST with random or QG forcing; all these runs have $N/f_0 \approx 5$, leading to $0.28 \leq Ro \leq 3.7$. Smaller Rossby numbers would imply the development of a strong inverse energy cascade (see e.g., [62] and references therein), a cascade that would be altered by the large-scale forcing when there is insufficient scale separation as here. The dependence of kurtosis on skewness for various fields in the presence of both inverse and direct energy cascades is left for future work.

The linear dimension of the cubic grid is n_p, varying from 128 to 512. The code is a versatile pseudo-spectral formulation of the time-integration of a large set of partial differential equations for fluid and plasma turbulence, with efficient hybrid parallelism implemented in a periodic box [63]. There is a version of the code in non-cubic geometry, as well as with non-periodic boundary conditions in one direction for the incompressible case [64]. T_M is the extension in terms of turn-over time τ_{NL} of the segment of the simulation used to build the statistics, with $41 \leq T_M \leq 1646$ across all runs. Data is gathered roughly four times every τ_{NL}, but for purely stratified runs, statistics of the dimensionless parameters are taken around the first peak of dissipation in order to avoid the effect of energy accumulation in the slow-mode leading to vertically sheared horizontal winds [65,66]. An estimate of the error on the skewness, and thus of its departure from Gaussianity, is $\sqrt{6/N_S}$, and twice that for the kurtosis [38,67], N_S being the number of independent samples. For our runs, the worst-case gives $S > 0.38$ for a flow to be determined to be non-Gaussian, and $\approx$0.06 in the best case. All Taylor Reynolds numbers are above $\approx$27 and should thus suffice to determine small-scale intermittent properties of the flow [68,69] when properly averaged over long times. In fact, local quasi-singular behavior in the intensity and curvature (morphology) of the vorticity and strain tensor can already be discerned at moderate R_λ when accurately evaluating the geometrical properties and anisotropy of the flow close to such structures through a so-called sparseness scale [70–72].

Finally, in Table 1, K_w^M is the maximum over time of the kurtosis of the vertical velocity, a signature of large-scale intermittency and strong dissipation in the transitional regime [30,31,73,74]. Most of the RST runs are at low to intermediate R_B with high K_w^M, as for the stratified runs. How well the runs are resolved is measured through $Res_N = k_{max}\eta_K$, with $k_{max} = n_p/3$ using a standard dealiasing method. $Res_N \geq 1$ implies that the dissipation scale is resolved by the numerical grid; a, b are the parameters when performing

$aS^2 + b$ fits analyzing $K(S)$ for the buoyancy flux; r.m.s. errors are between 1.8 and 6.3% for all runs for a, and between 0.8 and 2.3% for b, with respective averages for the eight QG runs of 4.27% and 1.44%. The extent of the turbulent range beyond the Ozmidov scale is $Res_T = \ell_{Oz}/\eta_K = F_\lambda^{3/2}$; when greater than unity, a sharp transition occurs in the $K(S)$ scaling parameters for both B_f and the kinetic energy dissipation ϵ_{du} (see Section 4 below).

Table 1. Characteristics of the hydrodynamics (H), stratified (S) and RST runs with random (R) or QG (Q) forcing. K_w^M is the temporal maximum kurtosis of vertical velocity, T_M the final time in units of τ_{NL}, and $Res_N = k_{max}\eta_K$ the numerical resolution for FDT. $[a_B, b_B]$ are fit coefficients for the parabolic law of B_f, $Res_T = \ell_{Oz}/\eta_K = F_\lambda^{3/2}$ is the extent of the fully turbulent range, and $I_{Ri} = \langle Ri_g \rangle$ is the averaged gradient Richardson number. Res_T and I_{Ri} mark a clear scaling transition.

ID	n_p	N	Fr	Re	R_λ	R_B	Ro	K_w^M	T_M	Res_N	a_B	b_B	Res_T	I_{Ri}
H1	128	0	–	306	28	–	–	0.1	142	2.0	2.8	6.2	–	0.82
H2	512	0	–	821	53	–	–	0.06	150	3.7	3.9	7.0	–	0.34
S1	512	16.	0.032	3223	162	3.3	–	5.06	42	1.96	2.49	7.96	0.51	7.1
S2	512	14.	0.036	3209	162	4.2	–	3.88	64	1.97	1.06	8.51	0.61	5.4
S3	512	11.8	0.042	3158	155	5.6	–	5.6	59	1.97	5.52	12.	0.79	3.7
S4	512	8.	0.060	3017	134	11.	–	15.3	428	1.97	13.8	29.4	1.6	1.6
S5	512	5.	0.089	2792	113	22.	–	6.09	98	1.90	14.6	10.3	3.3	0.72
S6	512	2.95	0.145	2692	93.3	56.	–	4.09	41	1.71	5.31	8.61	8.6	0.28
R1	128	4.78	0.056	454	43	1.4	0.28	8.7	148	2.3	7.2	8.2	0.45	4.1
R2	128	3	0.1	345	31	3.3	0.49	5.5	136	2.2	3.1	8.8	1.2	2.7
R3	128	1.4	0.24	299	27	16.6	1.2	1.2	134	2.1	3.7	5.6	4.3	1.3
R4	512	1.4	0.36	2231	94	293	1.8	0.37	51	6.2	3.1	4.0	25.	0.3
Q1	128	4.78	0.07	631	53	3.2	0.36	10.8	197	2.5	3.4	16.5	0.76	3.1
Q2	128	3	0.12	450	38	6	0.6	5.6	169	2.0	3.6	9.9	1.7	0.69
Q4	256	3.5	0.11	942	63	11.3	0.55	4.16	91	3.0	4.4	10.3	2.1	1.8
Q3	128	1.4	0.29	370	31	30.5	1.4	1.6	1646	1.7	1.58	5.5	6.4	0.9
Q5	256	1.4	0.36	694	48	87.5	1.8	1.0	89	3.1	1.45	6.2	11.9	0.6
Q7	512	1.4	0.36	1063	62	140	1.8	0.69	183	3.8	1.6	5.8	15.3	0.5
Q8	512	1.4	0.40	2641	102	385	1.9	0.8	96	3.1	1.7	6.0	33.3	0.3
Q6	256	0.7	0.73	636	44	343	3.7	0.22	86	3.2	1.4	3.8	34.3	0.2

Many studies of stratified flows have been performed [27,53,54,75–80]. Small-scale intermittency is present once $\ell_{Oz}/\eta_K > 1$, whereas at large-scale it is observed in a restricted range of values of Ri_g strongly peaked for shear instabilities (as well as in a narrow range of Fr, R_B, Res_T), without rotation [30], or with it as shown in this paper. An increase in the amplitude of nonlinear P_V when going to higher buoyancy Reynolds number is also observed [74,75]. Note also that, in the presence of a strong Coriolis force, a quasi-geostrophic balance is obtained that can lead not only to an inverse energy cascade to large scales (Ref. [62] and references therein), but in fact to a dual constant-flux energy cascade to both large and small scales [81–83].

3. Large-Scale and Small-Scale Intermittency in Turbulence

Intermittency, measured by non-Gaussian statistics [14], can modify passive scalar advection such as pollutants, and thus the chemistry within turbulent flows, e.g., through segregation as in the convective boundary layer, in particular when strong shear is present [84] (see [48,85] in the interstellar medium and galactic contexts). These baroclinic instabilities can be abrupt and result in turbulent bursts, reminiscent of large-scale *on-off* intermittency (LSI). There are in fact several such findings of LSI, starting with the probability distribution function (PDF) of the temperature which can be non-Gaussian (e.g., Métais and Herring [16], Rorai et al. [73]). One of the striking new results in [31] is that dissipation can be stronger locally than for FDT, as quantified by measuring the minimum flow volume needed to reach some level of energy dissipation, say 50%. Close to the peak of intermittency of w, θ, this volume can be as small as 12%, compared to 15% for FDT at a similar Reynolds number, and less than 4% for the active scalar, possibly indicating than a log-normal distribution is not adequate for the statistics of these flows [86] (see [87] for the FDT case). The waves can enhance dissipation in hot spots even when, globally, the flow is constrained by the large-scale, smooth dynamics. Extreme events are observed frequently in the atmosphere, e.g., in terms of w, θ [88], of the rate of energy dissipation [89] as well as for extreme precipitation [90,91]. A possible role in the formation of such structures is through the precession resonance of Rossby waves which was analyzed e.g., in [92,93], leading to their breaking as well as to their interactions with potential vorticity [94,95]. Similarly, in the ocean, wave breaking leads to efficient mixing processes [77–79], and observations of double-signed skewness can be the signature of a front, modeled in a dichotomic way for relative vorticity [67] as for the Agulhas current.

A simple non-Gaussian model is to express the PDF for field V as the sum of a normal distribution and its square, the relative weight of the two being governed by an open parameter. When the quadratic term dominates, one finds $K_V = 3S_V^2/2 + b$ [96]. In fusion plasmas, similar relationships are obtained [41,42,97,98], with different quadratic laws followed by different experiments [43]. The properties of the high-variability parts of turbulent flows, such as shear layers and their relaxation, seem universal and can be described by the model in [34], such shear layers developing at the boundaries [97], or as internal structures due to the turbulence. In fact, high kurtosis for the velocity and temperature at the edge of a shear layer have been found in a DNS of a Kelvin-Helmholtz instability [99], and singular shear layers can lead to intrinsic stochasticity [100]. Moreover, interactions with large scales in fluid turbulence represent roughly 20% of the kinetic energy flux at moderate Re. For structure functions beyond the energy level, including in the absence of waves, the nonlinear interactions are found to be mostly non-local, involving the integral scale [101], but with a tendency towards locality as $Re \to \infty$. Large-scale extreme events can arise as well in climatological studies, for example from quasi-resonant amplifications of planetary waves [92]. One also recovers a quadratic $K(S)$ law by adding three Gaussian PDFs, a model suitable for climate temperature variability [102], with the result that knowledge of skewness becomes a predictor of kurtosis. In fact, (S, K) mapping can be taken as an indicator of the turbulent dynamics in all these flows [44,46,67,96].

3.1. Is There a Skewness–Kurtosis Relationship for Fluid Turbulence with a Passive Scalar?

Curiously perhaps, FDT (with a passive scalar) has barely been analyzed in the context of $K(S)$ laws. This problem was studied in Herring and Kerr [12] in the decaying context and for $Pr = 0.5$. The skewness of the velocity was compared with experiments and with predictions of turbulence closures at R_λ comparable to what we compute in this paper, but here in the forced case. The flatness evaluated from the DNS was also contrasted with that given by closures (DIA and TFM). Good agreement was observed up to $R_\lambda \approx 30$. At high R_λ (above 800 or so), the experimental and numerical data reviewed in [103] for fluid turbulence does lend some credence to a relationship $K \sim S^j$, $j \gtrapprox 2.5$ for the velocity derivative, as noted in [43] (see also [104], Figure A6, for a recent estimate of K, S for numerical and experimental fluid turbulence with R_λ in excess of 6000). Furthermore, a statistical study of passive scalars in the atmosphere under many physical conditions gives a parabolic relationship, with in some cases rather high values of both S_P and K_P [36,37]. Several PDFs have been invoked to justify a quadratic relationship between K and skewness S, depending on the tails of such distributions [105], and exponential wings in PDFs of the horizontal vorticity for RST flows have been found [74]. Thus, we briefly discuss first the kurtosis and skewness data for FDT and a passive scalar with $Pr = 1$.

Figure 1 (top) displays $K(S)$ for run H2 for the vertical velocity, temperature and ω_z (a–c). At the bottom, $K(S)$ is shown for the buoyancy flux B_f (we are not considering here the joint w, θ distribution), for runs H1 (d) and H2 (e), with least-square fits for $[a, b]$ given in Table 1. Note that, for a unimodal and symmetric distribution, one can derive estimates that improve on the Cauchy-Schwarz inequality, namely $K \geq S^2 - 6/5$ [106], thus limiting potential closures [107]. In all the $K(S)$ figures, two parabolae are drawn, $K(S) = 1.5S^2 - 1.1$ and $K(S) = 3(S^2 + 2)$ (using $n = 5$ in [108], Equation (9b)). The first formulation appears in a variety of data, as described earlier. It should also be mentioned that it is quite close to the approximation developed in [109] (who also considers joint distributions), namely $K \approx 16S^2/9$, emanating from an expansion of a PDF assumed to be quasi-Gaussian and first analyzed on data of coastal waves [110] (see e.g., [111] in the context of the Weibull distribution for a model of sea surface wind velocity, and [112] for recent data analysis of strong waves on a beach with a sloping bottom). The second $K(S)$ formulation arises from the analysis of stochastically-generated skewed distributions (SGS) for which general $K(S)$ relations are obtained. These SGS distributions can be related to Markov processes, thereby shedding some light on the underlying stochastic systems [108]. As expected, no measurable intermittency is seen in Figure 1 in the large-scale field w, with both S and K close to zero, but θ, ω_z display measurable non-zero values that are at times higher than those given in Herring and Kerr [12], the forcing likely allowing for wider excursions. In plasmas, the $K \sim S^2$ law obtains for both quiet and active regions and, in that sense, the relationship itself may not be related to the development of strong turbulence, but rather to intrinsic multi-scale properties of the fields themselves [41]; ϵ_{du} (not shown) is also close to a quadratic law, but with higher coefficients. Taking ω_z as a (skewed) velocity gradient, the values of S, K are in fact consistent with data in [113], but higher Reynolds numbers will have to be explored to see whether a parabolic law actually appears.

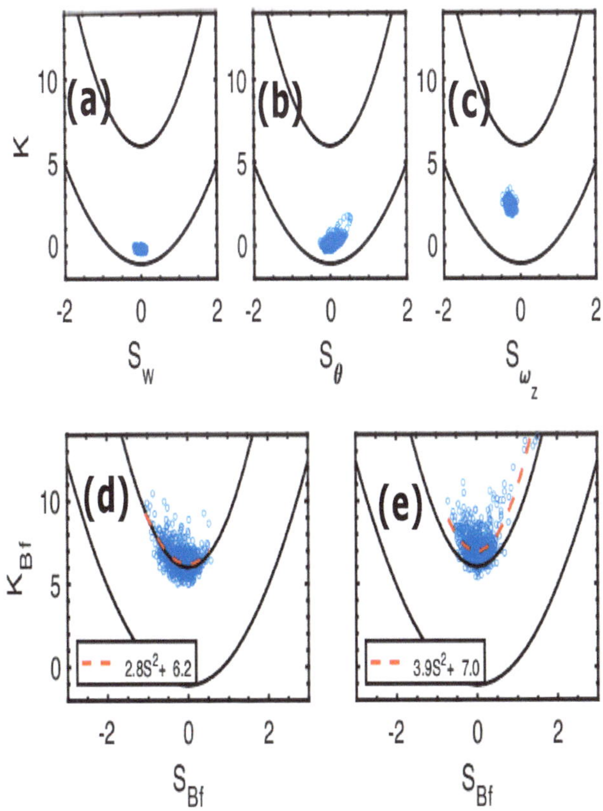

Figure 1. Kurtosis vs. skewness for the vertical velocity w (**a**), temperature θ (**b**) and vertical vorticity (**c**) for run H2. Bottom: $K(S)$ for $\langle w\theta \rangle$ for runs H1 (**d**) and H2 (**e**). All lower solid lines follow $K = 3S^2/2 - 1.1$ and upper ones, $K = 3S^2 + 6$ (see text). Red dashed lines are fits as given in insets.

Many models describe the interactions between widely separated scales from energy-containing to dissipative eddies. The combination of (weak) fast additive noise and large-scale multiplicative noise is known to produce PDFs with power-law tails [114]. Further examples are *on-off* intermittency in FDT close to instability [115], the scale-dependent creation of intermittency [116], or the magnetic dynamo problem [117]. Chaotic dynamics can also be seen as quasi-linear behavior forced by non-Gaussian processes whose events in the tail can trigger bifurcations in the system [118]. In this context, let us mention the climate model of Hasselmann [119] which begins with a linear stochastic (Langevin) equation for the temperature anomaly Θ, namely $\partial_t \Theta = -\lambda\Theta + \eta$; here, λ represents the heat flux and η is a noise term modeling small-scale fast fluctuations. One can in fact write $\lambda = \bar{\lambda} + \lambda'$ with $\bar{\lambda}$ an average, making the noise multiplicative [38] (see [120] for a recent review). Indeed, small-scale eddies are in large numbers, have low energy compared to that of the large scales and are intermittent, leading to complex behavior. This makes it plausible to model them through a stochastic linear noise acting on the slower large scale eddies (see also [121] for another Langevin approach, and [122] for stochastic turbulence models). One can then derive the parabolic relationship in Equation (8) with the expectation that $a \geq 3/2,\ b \geq 0$ [38], a finding corroborated by several analyses of climate data. The long-term (monthly onward) climate statistics are Gaussian, but the daily averages reflect the stochasticity of small-scale eddies [38] which also interact with waves. Therefore, on some intermediate temporal scale for the large structures, these small eddies can be viewed as noise, although there exist nonlocal (in scale) interactions between small and large scales. This leads to enhanced dissipation at the level that is needed in order that, in its adimensionalized form and at least in the absence of waves, it be of order unity, both for fluid turbulence [19,123] and for fusion and space plasmas [124,125].

3.2. $K(S)$ Relationships for the Buoyancy Flux in the Purely Stratified Case

For strong stratification, the vertical velocity intermittency can be significant, like in the nocturnal PBL [18,34,52], or for DNS in a narrow range of Froude number [30]. Does a $K(S)$ relationship ensue in that regime? To answer this question, we examine the results of a series of purely stratified runs with $0.032 \leq Fr \leq 0.145$, and with high $Re \approx 3000$. At left in Figure 2 is $K(S)$ for w for the run with the highest kurtosis of w [30,31]. We observe that there is a large number of data points with large excess kurtosis, but the skewness of w remains close to zero. On the other hand, such is not the case for the buoyancy flux B_f shown at right in Figure 2 for all the S runs. In the inset, three parabolae are given (dash; dots; and dash-dots) corresponding to different [a,b] fits. The first remark is that, for stratified turbulence, a parabolic $K(S)$ law seems to emerge as well for B_f (and for ϵ_{du}, not shown). Note that practically all runs lie on the lower parabola marked with dash, and showing as an orange cloud of points in the plot. Superimposed on this, for two runs close to the maximum of K_w, there is an explosion in the values taken by the kurtosis (blue triangles and black squares, runs S3 and S4). This confirms the analysis performed in [31] that dissipation in this Fr range takes high local values; it corresponds to a transitional phenomenon with high intermittency of vertical velocity, flux and dissipation itself, and thus leading to enhanced mixing as we move from regime I (wave dominated) to regime II (eddy-wave interactions), likely similar to what happens in the nocturnal PBL [52]. The a coefficient in front of the quadratic term in the $K(S)$ law for the vertical velocity in the PBL is thought to depend on the stability of the layer through Ri_g [54,126]. Indeed, strong vertical drafts are clearly correlated with $Ri_g \approx 1/4$ (see [30], Figure 3). It may be that $K(S)$ laws are emerging here from the flow being at the edge of instability [127,128]. An analysis similar to that of Vieillefosse [129] yields nontrivial relations between high-order moments of the field gradients near the onset of convective instabilities [130]. An inverse scaling, on more than two orders of magnitude, between two a-priori unrelated parameters, namely one governing stratification (the gradient Richardson number) and one governing the turbulence in the presence of stratification (the buoyancy Reynolds number)– was observed in DNS of rotating stratified flows [51]. One could thus infer that the a parameter just mentioned would depend in a similar fashion on R_B. This inverse $Ri_g \sim R_B^{-1}$ relationship is significant because it implies a strong dynamical effect of vertical shear, and a balance between vertical gradients (of the turbulent horizontal velocity field) and the turbulent isotropic eddies beyond the Ozmidov scale [51], as hinted in fact by several authors over the years in varied contexts.

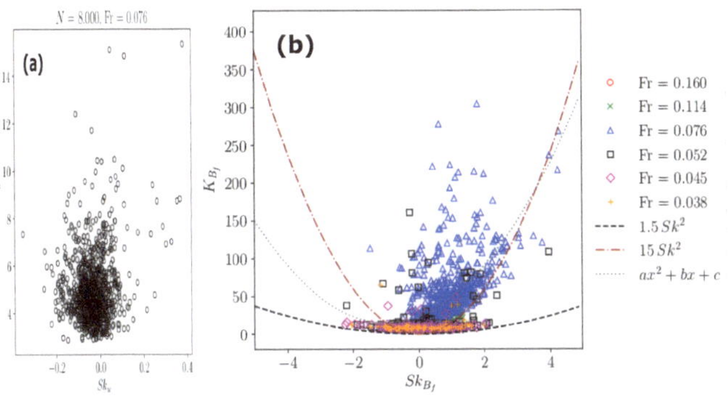

Figure 2. (**a**): Kurtosis vs. skewness of the vertical velocity for run S3 at the peak of K_w [30,31]. Note the narrow range of skewness allowing to focus on small S_w, high K_w values. (**b**): $K_D(S_D)$ for the buoyancy flux B_f for all stratified runs; parabolic fits and color symbols are given at right.

4. The Case of Rotating Stratified Turbulence (RST)

Intermittent vertical velocities are also found in RST, for decaying flows [74] and, as shown here, in forced runs. We look first at global dynamics. In Figure 3a is given as a function of time the total kinetic energy for run H1 as a base, and for three runs with varying Fr and $0.36 \leq Ro \leq 3.7$. The higher level of saturation for the RST runs is due to the fact that on the one hand, there is a lesser dissipation due to the strong effect of waves, and on the other hand, $Ro \approx 0.36$ for run Q1, which is low enough that an inverse transfer of energy to the large scales is plausible, although it cannot be distinguished directly on energy spectra which are still peaked at $k \approx k_F$ (see Figure 3d). Higher Ro may correspond to the environment of tropical storms in which turbulence also plays a dual role [131,132]. In Figure 3b we plot K_w (dotted magenta line), the kinetic (solid blue), potential (black dashed) and total energy dissipation rates for run R1, with the time expressed in units of the turnover time τ_{NL}. We see that K_w can take large values, as in the purely stratified case [30]. There are also discernible excursions in the dissipation rates. The high-K values are associated with local instabilities and overturning regions; these excursions in K_w are also present in several other runs. We recover here the behavior analyzed in detail in [30,31] for purely stratified flows: narrow peaks in K_w as a function of Fr and R_B occur, compatible with observed strong updrafts in the DNS. The temporal evolution of the kinetic energy E_V and enstrophy $\langle|\omega|^2\rangle$ are given for the total length of that run in (c), showing a persistence in the amplitude of fluctuations. The sporadic energetization of the flow has its signature in energy spectra shown in Figure 3d for run R1; indeed, the kinetic and potential spectra gain close to a factor of ten at the smallest scales for an excited state (with a peak in K_w) when compared to quiet times (with a trough in K_w), whereas the spectra have comparable amplitudes at the largest scales. Similar differences in small-scale spectra occur for the QG runs. Note also that, the forcing being centered in the large scales, no inverse cascade of energy that would occur in the presence of strong rotation can be observed in these runs (see [81,82] for DNS resolving both the inverse and direct energy cascades).

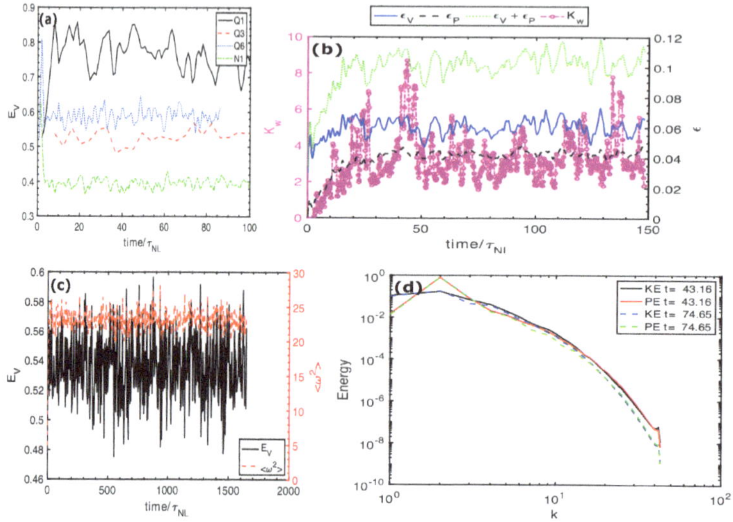

Figure 3. Time variation of (**a**) kinetic energy for runs (see inset) up to $t/\tau_{NL} \approx 100$; (**b**) kinetic, potential and total dissipation (scale at right), and K_w, run R1; (**c**): kinetic energy (black) and enstrophy (red, scale at right) for the total length of Q3. (**d**): Kinetic (solid black/dash blue) and potential (solid red/dash green) energy spectra at *resp.* maximum and minimum kurtosis of w, run R1.

In Figure 4, we give $K(S)$ for w for runs R1–R3 (a–c) , and for B_f for runs Q1, Q3, and Q8 (d–f), with buoyancy Reynolds numbers varying from 1.4 to 385. At top, the skewness is typically bounded by ± 1 or less, and there is a clear decrease in the kurtosis of the vertical velocity as R_B increases from left to right. We note that, for all these runs, and similarly to the FDT and purely stratified cases, the vertical velocity has basically no skewness but its kurtosis can take moderate values, especially so at intermediate R_B (here, the first plot), although no parabolic law emerges from the data. For the buoyancy flux, there is a marked transition towards a standard scaling ($\approx 1.5S^2 + b$), once the Kolmogorov regime is sufficiently resolved, i.e., for high enough R_B, Ri_g, and with a classical FDT dissipation taking place (Figure 4e,f). Note that the intermittency uncovered in this paper likely deals more with intrinsic properties of dissipation than local accelerations that might be observed in a boundary layer. A different approach, following [109], is to expand the PDF in terms of its cumulant under the assumption of closeness to a Gaussian distribution. In the simplest case, a relationship is obtained, namely $K = [4S/3]^2$, close to the previous expression ($a \approx 1.5$ for small b). The $K(S)$ plots for the temperature (Figure 5a,b) and the vertical component of the vorticity (c,d plots in Figure 5) are similar as well, as shown for runs Q2 (a,c) and Q7 (b,d). In all cases, the skewness stays close to zero and there is a decrease of kurtosis as the flows become more dynamical.

A last issue is how much mixing occurs, estimated through B_f, ϵ_{du} taken as proxies [79,86]. Thus, we give in Figure 6, $K(S)$ for ϵ_{du} for runs Q1, Q2 (a,b) with $Res_T \approx 1$, and Q5, Q6 (d,e, $Res_T > 10$). Again, a transition is observed, with steeper scaling in the latter case; similar results obtain for FDT, with a best fit for ϵ_{du} (run H2) of $K = 2.6S^2 + 7.2$ (not shown). Note that, for the components of the strain tensor itself, $S_{ij} = [\partial_i u_j + \partial_j u_i]/2$, we have $K \geq 21S^2/4$ in the incompressible case [133]. Figure 6 (right) gives the PDFs of ϵ_{du} for run Q3 for times with a high (c) and low (f) K_w; a ≈ -3 power law is plausible, with a larger extension when the intermittency of w is stronger; note that a shallower scaling for ϵ_{du} has been observed in oceanic data, together with substantially higher values of K, S [134]. In the fits, $\alpha = \sqrt{1 + S^2_{\epsilon_{du}}/4}$ (see [108]).

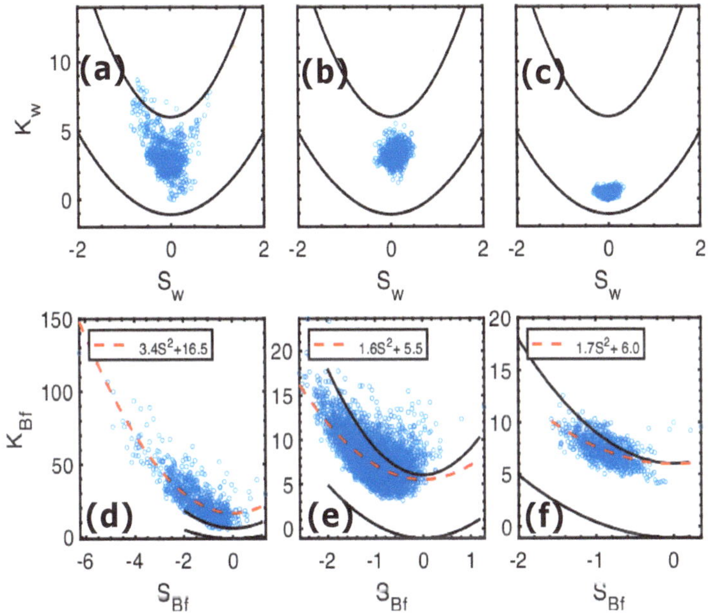

Figure 4. $K_w(S_w)$ of vertical velocity for RST runs (**a**) R1, (**b**) R2 and (**c**) R3. $K_{B_f}(S_{B_f})$ of buoyancy flux for QG runs (**d**) Q1, (**e**) Q3 and (**f**) Q8. Insets give best fits indicated by dashed red lines.

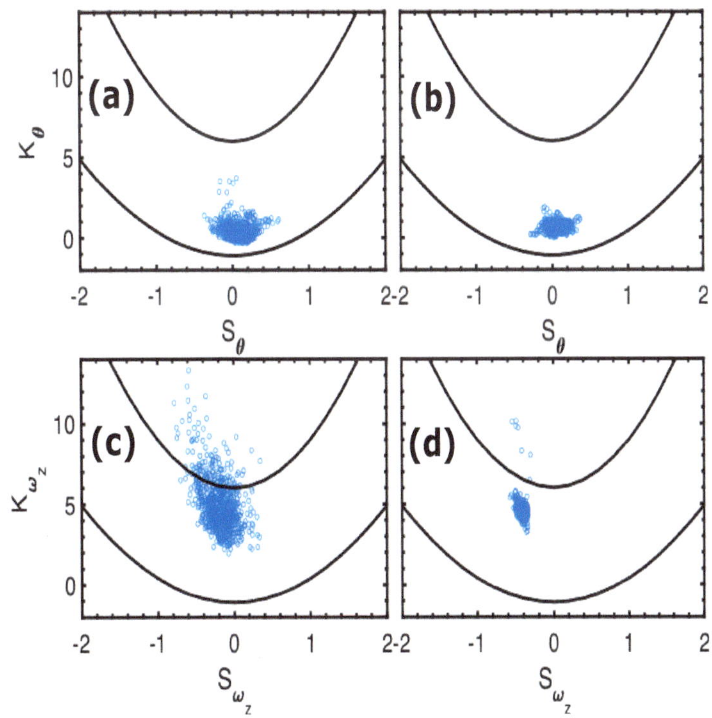

Figure 5. $K(S)$ for θ and the z−component of vorticity for runs Q2 (left, (**a**,**c**)) and Q8 (right, (**b**,**d**)).

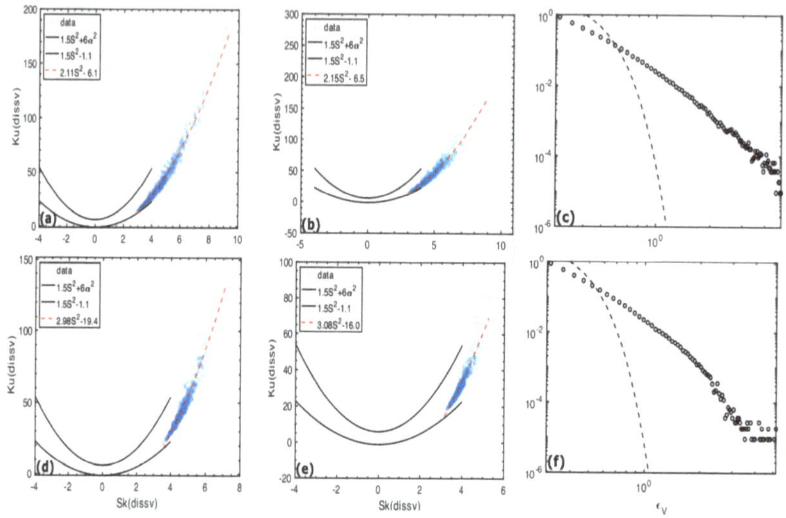

Figure 6. $K(S)$ for kinetic energy dissipation ϵ_{du} for (**a**,**b**) Q1, Q2, and (**d**,**e**) Q5, Q6. At right, (**c**,**f**) give the PDFs of ϵ_{du} at different times (see text).

5. Possible Frameworks for K(S) Relationships

It is not clear what the link of the above results is with the analysis of complex turbulent flows using the concept of avalanches and of self-organized criticality [127,135]. In that latter case, the dissipative structures at small scale propagate in some fashion their intermittency to the large scales, just as small avalanches become large through merging. Saying it another way, dissipative structures near strain and vortices, or in current sheets and filaments, exist at the large scale characteristic of the flow and at the small scale corresponding to their thickness, of the order of the dissipation length. Therefore, the governing parameter may not be Re or R_λ, but could be related to the ability of the small-scale structures to organize into larger clusters, or in other words to a transition at the local (in scale) Reynolds number of unity in the vicinity of the Kolmogorov scale in fluid turbulence. The occurrence of a quadratic $K(S)$ law, albeit with varying coefficients, has been noted before. One should analyze large-enough samples [108,136]), but the non-Gaussianity of turbulence is ascertained, for non-vanishing dissipation, by the exact laws stemming from the conservation of quadratic invariants, as for kinetic energy in fluid turbulence [137,138], under several conditions:

$$\left\langle \delta u_L^3(r) \right\rangle = -\frac{4}{5}\epsilon_{du} r \ , \quad \left\langle \delta u_L(r) \Sigma^i \delta u_i^2(r) \right\rangle = -\frac{4}{3}\epsilon_{du} r \ , \tag{9}$$

with u_L the longitudinal velocity along the distance $|r|$, and δ expressing velocity differences. In MHD, cross-field correlations emerge [139] (see [140] for a review), which are central to the structure of reconnection events [33,47,141]. Inertia-gravity waves, which exchange energy between the kinetic and potential modes, will not alter the balance at the level of total energy dissipation. However, when writing an equation for the temporal evolution of B_f, it occurs at the scale of the turn-over time since it deals with large-scale variables, whereas the equipartition deficiency $N\langle[\theta^2 - w^2]\rangle$ due to waves evolves on a fast time for small Fr and can be averaged to be modeled as a stochastic forcing for B_f, ϵ_{du}; this makes their temporal evolutions akin to systems already discussed herein (see e.g., [38,119]).

Another remark is that, obtaining quadratic $K(S)$ laws for both weak and strong turbulence, diffusive or not, and dispersive or not, could lead one to conclude that the issue is more of a mathematical constraint than a physical one. However, the $K(S)$ relationship is important in limiting possible closures applied to turbulence, as for example in the context of solar convection (see [107] for recent developments of closures beyond the quasi-normal framework). It has also been remarked that these systems have in common a convective nonlinearity, however weak or intense [97]. The specific transition in the scaling of the $K(S)$ laws obtained here is linked to the fact that waves become progressively more strongly coupled to nonlinear eddies, or in other words steepen significantly, thus leading to enhanced dissipation and to the onset of a Kolmogorov energy spectrum at small scale. The existence of $K \sim S^2$ laws at high and low R_λ leads to another possible origin for it, namely the presence of intermittency in the dissipation range common to all such flows. As pointed out in [15,142], this range can be prone to extreme intermittency because of a faster than algebraic decrease of the intensity of turbulence at high wavenumber, a spectral domain that exists even in the near absence of a nonlinear inertial range at low R_λ. This likely leaves only a few shear structures dominating the overall statistics. In this context, studies of Lagrangian turbulence have been able to show the link between the dissipation range and large scales, with both short and long-range correlations [143].

Computations performed at moderate R_λ identify small-scale intermittency with numerous large-scale structures [15], as opposed to the more familiar vortex filaments [144–146]. Such nonlocal structures could be detected in rotating stratified flows using specific algorithms (see [135] for the identification of current sheets at the onset of the dissipation range of plasmas in the MHD regime). In that spirit, it would be of interest to consider as well the case of quasi-geostrophy, that of cubic nonlinearities like for compressible flows, or for solitons (such as for the Korteweg–deVries or the cubic nonlinear Schrödinger (NLS) equations and their multi-dimensional extensions), all in the presence of forcing and dissipation (see also [147] where the $K(S)$ behavior for the NLS equation is associated with *on-off* intermittency). The large-scale behavior linked with the presence of Rossby waves and their possible nonlinear coupling and breaking may also reinforce or even alter large-scale intermittency.

Finally, a link between inertial and dissipative ranges is consistent with a linear stochastic model for the fast small scales which is successful in giving $K(S) \sim S^2$ for climate data [38,39]. In decaying experiments at both low and high R_λ using sulfur hexafluoride, the inertial range does not follow a strict power-law, but rather may display a logarithmic correction that is independent of Reynolds number and that may arise from the dissipative range, putting into question their independence [148]. Nonlinearities do not seem to be central to the establishment of non-Gaussian statistics beyond their (necessary) mode-coupling role [149], and it is rather the susceptibility of the system being close to criticality, linking all scales through dissipation that is central to turbulence dynamics and intermittency. This can manifest itself as avalanches in granular media, or in the development of shear instabilities in (e.g., RST) flows close to the critical Richardson number ([74,75,89,127,150] and references therein).

6. Discussion, Conclusions, and Perspectives

In the nocturnal planetary boundary layer, " ... turbulence is expected to become more intermittent in very stable conditions" (Mahrt et al. [52]). Relating to these observations, we are likely dealing in this paper with two phenomena, possibly superposed and intertwined. On the one hand, we have the increased intermittency, as diagnosed by a high kurtosis of vertical velocity and temperature, if in a narrow domain of Froude and buoyancy Reynolds number [30,74], and signaling the onset of small-scale turbulence interacting with the larger-scale (inertia)-gravity waves. The signature in physical space consists of a high local vertical velocity and hence a high local Reynolds number, allowing for instabilities to develop until the system evolves (in a turn-over time) back to a state dominated by the fast waves [31,127,130,151], and the cycle can restart. Thus, it leads to strong local energy dissipation in an otherwise wave-dominated rather placid system, and as such these isolated hot spots are a critical factor in controlling the overall dissipation. This dynamics can be modeled e.g., through the temporal evolution of the velocity and temperature gradient matrices, taking into account the anisotropy of the system and using a closure for the pressure Hessian [130]. High kurtosis values for the vertical velocity are narrowly centered around a peak in Froude number (and in R_{IB}), although it is not known at this point how much the peak depends on the Reynolds number itself. They can be associated for example with jet meandering in the ocean and modeled through stochastic parametrization. In fact, it has been hypothesized – and studied in depth over the years– that shear flows are central to the critical instability of turbulence, a point further developed recently in a global fluid context [152–155], and applied to stratified fluids as well [127] (see [128] for RST, and [156] for plasmas). This transition could also be linked to a role-change in potential vorticity P_V at small scales due to the breaking, at the Ozmidov scale onward, of gravity waves [80].

On the other hand, different $K(S)$ laws emerge with different coefficients for the buoyancy flux, the kinetic energy dissipation as well as for mixing as measured through the ratio B_f/ϵ_{du}. This second phenomenon of a transition between regimes for $K(S)$ is new to our knowledge, certainly in the specific context of rotating stratified flows, but of course long computations at higher Re, R_{IB} with a resolved range beyond ℓ_{Oz} are needed. The scaling corresponds to a shift, in the small scales and possibly as well in the early dissipation range, in the nature of an intermittency which becomes stronger as the nonlinear eddies get stronger, and with a $K(S)$ scaling for B_f which seems to approach that found for other turbulent systems [38,41,46]. This is linked to the well-known transition at the Ozmidov scale of an energy spectrum becoming progressively more Kolmogorov-like. Many models have been proposed for these $K(S)$ parabolic laws, as partially reviewed all along in the paper. More work will be needed to sort out the different possibilities, and to reach higher buoyancy Reynolds numbers corresponding to conditions pertaining to those in atmospheric and oceanic flows. In order to obtain dissipative structures at higher R_B for long-time runs, a computation in a non-cubic geometry will help tremendously. For example, it was shown in [157] that in a flat box with a 1:8 aspect ratio, the process of front and filament formation can be very efficient in sheared flows [130]. The evolution of extreme events in such flows with comparable geometries was also considered in [158] for their effect on the mixing of neutrally buoyant particles (see also [159] for a substantially higher aspect ratio together with using hyper-viscosity). Similarly, peak vertical vorticity can be close to two orders of magnitude larger than the imposed rotation, as shown in [160] within a Large Eddy Simulation (LES) framework. Another feature of these flows is their persistent anisotropy [59,128,161,162], associated, for example, with seasonal oceanic variability [163]. Thus, the precise formulation of the small-scale parametrization and of the dissipation of resolved scales in numerical codes may affect the mesoscale dynamical properties such as energy spectra [164]. Related evaluations for plasma disruptions and fusion, from MHD to Hall-MHD and beyond, might also be of interest to understand better– and possibly control to some extent– these strong small-scale disruptive features.

One of the several models of dissipative events leading to a quadratic $K(S)$ relationship not yet mentioned here is in the context of so-called $1/f^n$ noise as a signature of long-time correlations in turbulent hydrodynamic and MHD flows [165] where it is shown that the transition from Gaussian statistics in the inviscid case to a non-Gaussian one in the presence of dissipation can be associated with varying properties of energy transfer through scales. Are there several different classes of universality, though? In this context, we have already noted that, for the Navier–Stokes equations, there is a clear departure from standard coefficients for the law $K(S) = 3S^2/2 - 1.1$, as well as for low Froude numbers in the case of stratified turbulence and RST.

Another potential issue is the form of the dissipation operator. When using either eddy viscosity or hyper-viscosity as a model , or LES, a sub-grid scale algorithm, or the α-model Lagrangian description for fluids and MHD, do similar quadratic $K(S)$ relations arise? What about for the cyclostrophic balance in helical stratified flows [166]? Or in the ideal ($\nu = 0$) case in its pseudo-dissipative range stemming from the eddy viscosity due to the small-scale equilibrium range [167], or in 2D MHD or even in 1D (allowing for the long temporal integrations which are needed in this intermediate inviscid but turbulence-like phase)? In other words, does modifying the way energy is dissipated in turbulence alter the $K(S)$ relation for velocity or dissipation, therefore suggesting that it is a crucial feature for dissipative events? We already know that the PDFs of velocity gradients in the NS case are affected [168], but it is not clear whether the $K(S)$ law would be. Such renormalized dissipation arise at high Re, whereas a $K \sim S^2$ relation is present as well at moderate Re, for which the viscosity is directly active. If a change in the $K(S)$ law occurs, beyond the one identified in this paper and related to the onset of a Kolmogorov range beyond ℓ_{Oz}, what functional form does it take, and are there consequences for accuracy in an evaluation of the large-scale dynamics that weather and climate codes seek to reproduce faithfully while using techniques such as LES (see, e.g.,, the discussion in [108])?

We have presented preliminary results, limited in Reynolds number, and much remains to be done. One issue is to examine the dual constant-flux energy cascade when forcing is at smaller scale [81,82]. Other questions arise, such as the persistence of large-scale intermittency of vertical velocity in RST in the presence of a resolved turbulence range, or of strong helicity (see e.g., [169,170] on intermittency and helicity in FDT). Nonlinear interactions $\propto \mathbf{u} \times \boldsymbol{\omega}$ are weakened in vortex tubes, potentially allowing waves to play a more prominent role in driving the flow closer to the critical state in which strong vertical drafts and ensuing dissipation occur. A wave-vortex decomposition, such as that studied e.g., in [171,172], could be of help to distinguish between these different regimes. One can also ask how anisotropy in RST flows is affected by these strong updrafts. Large-scale intermittency has also been linked to the role of helicity (velocity-vorticity correlations) [173], and the eigenvectors of strain can play an important role as well in intermittent energy transfer, as shown in a simple model in which alignment of vorticity with strain eigenvector corresponding at first to its intermediate eigenvalue, and featuring the combination of a shear flow with a horizontal straining field [174] (see also [175]), and yet anisotropy is essential in possibly stopping the formation of singular structures [70].

We wish to end this text with a direct quote from a review article on waves and coherent flows in the tropics (Stephan et al. [176], see page 2619): "Theory provides the qualitative framework—the causal narrative—within which the quantitative information from models and observations can be meaningfully compared and usefully combined." The research of Jack Herring was on the idealized, theoretical side of the study of the dynamics of the atmosphere, pioneering the inclusive use of theoretical and numerical approaches. As such, it has provided invaluable information, and causal accounts, for the idealized behavior of turbulence and the central role played by nonlinear effects in such—and other geophysical, astrophysical as well as engineering, chemical, biological and climatological—complex systems.

Author Contributions: All authors contributed equally on all aspects of this paper. All authors have read and agreed to the published version of the manuscript.

Funding: Duane Rosenberg acknowledges support from award NOAA/OAR/NA19OAR4320073. Raffaele Marino acknowledges support from the project EVENTFUL (ANR-20-CE30-0011), funded by the French Agence Nationale de la Recherche-ANR through the program AAPG-2020.

Data Availability Statement: Data is available for all runs from the authors upon request.

Acknowledgments: We all would like to express our gratitude to Jack Herring, and acknowledge many enlightening discussions with him over our years at the Geophysical Turbulence Program which Jack helped create at NCAR, and which was extremely successful in the turbulence community at large. His interest in pursuing and encouraging colleagues to pursue a theoretical understanding of turbulence in atmospheric and geophysical flows in general never waned, nor did his kindness to others and his open-mindedness. Finally, Annick Pouquet is thankful to LASP (University of Colorado at Boulder), and in particular to Bob Ergun.

Conflicts of Interest: The authors declare no conflict of interest.

References

1. Kraichnan, R. The structure of isotropic turbulence at very high Reynolds numbers. *J. Fluid Mech.* **1959**, *5*, 497–543. [CrossRef]
2. Sukoriansky, S.; Galperin, B.; Staroselsky, I. A quasinormal scale elimination model of turbulent flows with stable stratification. *Phys. Fluids* **2005**, *17*, 085107. [CrossRef]
3. Nazarenko, S. *Wave Turbulence*; Lecture Notes in Physics; Springer: Berlin/Heidelberg, Germany, 2011; Volume 825.
4. Zhou, Y. Turbulence theories and statistical closure approaches. *Phys. Rep.* **2021**, *935*, 1–117.
5. Marston, J.B.; Tobias, S.M. Recent Developments in Theories of Inhomogeneous and Anisotropic Turbulence. *Ann. Rev. Fluid Mech.* **2022**, *55*, 1–29. [CrossRef]
6. Lorenz, E.N. The predictability of a flow which possesses many scales of motion. *Tellus* **1969**, *21*, 289–307. [CrossRef]
7. Leith, C.; Kraichnan, R. Predictability of turbulent flows. *J. Atmos. Sci.* **1972**, *29*, 1041–1058. [CrossRef]
8. Herring, J.R.; Riley, J.J.; Patterson, G.S., Jr.; Kraichnan, R.H. Growth of Uncertainty in Decaying Isotropic Turbulence. *J. Atm. Sci.* **1973**, *30*, 997–1006. [CrossRef]
9. Herring, J.R. Approach of axisymmetric turbulence to isotropy. *Phys. Fluids* **1974**, *17*, 859–872. [CrossRef]
10. Schumann, U.; Herring, J. Axisymmetric homogeneous turbulence: A comparison of direct spectral simulations with the direct-interaction approximation. *J. Fluid Mech.* **1976**, *76*, 755–782. [CrossRef]
11. Herring, J.R. Statistical theory of quasi-geostrophic turbulence. *J. Atmos. Sci.* **1980**, *37*, 969–977. [CrossRef]
12. Herring, J.R.; Kerr, R.M. Comparison of direct numerical simulations with predictions of two-point closures for isotropic turbulence convecting a passive scalar. *J. Fluid Mech.* **1982**, *118*, 205–219. [CrossRef]
13. Herring, J.; McWilliams, J. Comparison of direct numerical simulations of two-dimensional turbulence with two-point closure: The effects of intermittency. *J. Fluid Mech.* **1985**, *153*, 229–242. [CrossRef]
14. Chen, H.; Herring, J.R.; Kerr, R.M.; Kraichnan, R.H. Non-Gaussian statistics in isotropic turbulence. *Phys. Fluids* **1989**, *A1*, 1844–1855. [CrossRef]
15. Chen, S.; Doolen, G.; Herring, J.R.; Kraichnan, R.H.; Orszag, S.A.; She, Z.S. Far-Dissipation Range of Turbulence. *Phys. Rev. Lett.* **1993**, *70*, 3051–3054. [CrossRef]
16. Métais, O.; Herring, J. Numerical simulations of freely evolving turbulence in stably stratified fluids. *J. Fluid Mech.* **1989**, *202*, 117–148. [CrossRef]
17. Kimura, Y.; Herring, J. Diffusion in stably stratified turbulence. *J. Fluid Mech.* **1996**, *328*, 253–269. [CrossRef]
18. Champagne, F. The fine-scale structure of the turbulent velocity field. *J. Fluid Mech.* **1978**, *86*, 67–108. [CrossRef]
19. Buaria, D.; Bodenschatz, E.; Pumir, A. Vortex stretching and enstrophy production in high Reynolds number turbulence. *Phys. Rev. F* **2020**, *5*, 104602. [CrossRef]
20. Klymak, J.M.; Pinkel, R.; Rainville, L. Direct Breaking of the Internal Tide near Topography: Kaena Ridge, Hawaii. *J. Phys. Oceano.* **2008**, *38*, 380–399. [CrossRef]
21. Belonenko, T.V.; Sandalyuk, N.V.; Gnevyshev, V.G. Interaction of Rossby waves with the Gulf Stream and Kuroshio using altimetry in a framework of a vortex layer model. *Adv. Space Res.* **2023**, *71*, 2384–2393. [CrossRef]
22. Davis, A.Y.; Yan, X.H. Hurricane forcing on chlorophyll-a concentration off the northeast coast of the U.S. *Geophys. Res. Lett.* **2004**, *31*, L17304. [CrossRef]
23. Grisouarda, N.; Leclair, M.; Gostiaux, L.; Staquet, C. Large-scale energy transfer from an internal gravity wave reflecting on a simple slope. *Proc. IUTAM* **2013**, *8*, 119–128. [CrossRef]
24. McWilliams, J.C. A survey of submesoscale currents. *Geosci. Lett.* **2019**, *6*, 3. [CrossRef]
25. Chau, J.; Marino, R.; Feraco, F.; Urco, J.; Baumgarten, G.; Lubken, F.; Hocking, W.; Schult, C.; Renkwitz, T.; Latteck, R. Radar observation of extreme vertical drafts in the polar summer mesosphere. *Geophys. Res. Lett.* **2021**, *48*, e2021GL094918. [CrossRef]
26. Nastrom, G.D.; Gage, K. A climatology of atmospheric wavenumber spectra of wind and temperature observed by commercial aircraft. *J. Atmos. Sci.* **1985**, *42*, 950–960. [CrossRef]
27. Waite, M.L. Random forcing of geostrophic motion in rotating stratified turbulence. *Phys. Fluids* **2017**, *29*. [CrossRef]
28. Mininni, P.D.; Pouquet, A. Energy spectra stemming from interactions of Alfvén waves and turbulent eddies. *Phys. Rev. Lett.* **2007**, *99*, 254502. [CrossRef]
29. Matthaeus, W. Turbulence in space plasmas: Who needs it? *Phys. Plasmas* **2021**, *28*, 032306. [CrossRef]
30. Feraco, F.; Marino, R.; Pumir, A.; Primavera, L.; Mininni, P.; Pouquet, A.; Rosenberg, D. Vertical drafts and mixing in stratified turbulence: Sharp transition with Froude number. *Eur. Phys. Lett.* **2018**, *123*, 44002. [CrossRef]
31. Marino, R.; Feraco, F.; Primavera, L.; Pumir, A.; Pouquet, A.; Rosenberg, D. Turbulence generation by large-scale extreme drafts and the modulation of local energy dissipation in stably stratified geophysical flows. *Phys. Rev. F* **2022**, *7*, 033801. [CrossRef]
32. Osman, K.; Matthaeus, W.H.; Wand, M.; Rapazzo, A. Intermittency and local heating in the solar wind. *Phys. Rev. Lett.* **2012**, *108*, 261102. [CrossRef] [PubMed]
33. Ergun, R.E.; Goodrich, K.A.; Wilder, F.D.; Ahmadi, N.; Holmes, J.C.; Eriksson, S.; Stawarz, J.E.; Nakamura, R.; Genestreti, K.J.; Hesse, M.; et al. Magnetic Reconnection, Turbulence, and Particle Acceleration: Observations in the Earth's Magnetotail. *Geophys. Res. Lett.* **2018**, *45*, 3338–3347. [CrossRef]
34. Lenschow, D.; Mann, J.; Kristensen, L. How long is long enough when measuring fluxes and other turbulence statistics? *J. Atm. Oc. Tech.* **1994**, *11*, 661–673. [CrossRef]

35. Mole, N.; Clarke, E. Relationships between higher moments of concentration and of dose in turbulent dispersion. *Bound. Layer Met.* **1995**, *73*, 35–52. [CrossRef]
36. Lewis, D.; Chatwin, P.; Mole, N. Investigation of the collapse of the skewness and kurtosis exhibited in atmospheric dispersion data. *Nuovo Cimento* **1997**, *20*, 385–397.
37. Schopflocher, L.; Sullivan, P.J. The relationship between skewness and kurtosis of a diffusing scalar. *Bound. Layer Met.* **2005**, *115*, 341–358. [CrossRef]
38. Sura, P.; Sardeshmukh, P.D. A global view of non-Gaussian SST variability. *J. Phys. Oceano.* **2008**, *38*, 639–647. [CrossRef]
39. Sardeshmukh, P.D.; Sura, P. Reconciling Non-Gaussian Climate Statistics with Linear Dynamics. *J. Climate* **2009**, *22*, 1193–1207. [CrossRef]
40. Kaur, S.; Lukovich, J.V.; Ehn, J.K.; Barber, D.G. Higher-order statistical moments to analyse Arctic sea-ice drift patterns. *Ann. Glaciology* **2021**, *61*, 464–471. [CrossRef]
41. Labit, B.; Furno, I.; Fasoli, A.; Diallo, A.; Müller, S.; Plyushchev, G.; Podestà, M.; Poli, F. Universal Statistical Properties of Drift-Interchange Turbulence in TORPEX Plasmas. *Phys. Rev. Lett.* **2007**, *98*, 255002. [CrossRef]
42. Krommes, J.A. The remarkable similarity between the scaling of kurtosis with squared skewness for TORPEX density fluctuations and sea-surface temperature fluctuations. *Phys. Plasmas* **2008**, *15*, 030703. [CrossRef]
43. Sattin, F.; Agostini, M.; Cavazzana, R.; Serianni, G.; Scarin, P.; Vianello, N. About the parabolic relation between the skewness and the kurtosis in time series of experimental data. *Phys. Scrip.* **2009**, *79*, 045006. [CrossRef]
44. Burkhart, B.; Stanimirovic, S.; Lazarian, A.; Kowal, G. Characterizing magnetohydrodynamic turbulence in the small Magellanic cloud. *Astrophys. J.* **2010**, *708*, 1204–1210. [CrossRef]
45. Hamza, A.; Meziane, K. On turbulence in the quasi-perpendicular bow shock. *Planet. Space Sci.* **2011**, *59*, 475–481. [CrossRef]
46. Osmane, A.; Dimmock, A.; Pulkkinen, T. Universal properties of mirror mode turbulence in the Earth magnetosheath. *Geophys. Res. Lett.* **2015**, *42*, 3085–3092. [CrossRef]
47. Miranda, R.A.; Schelin, A.B.; Chian, A.C.L.; Ferreira, J.L. Non-Gaussianity and cross-scale coupling in interplanetary magnetic field turbulence during a rope-rope magnetic reconnection event. *Ann. Geophys.* **2018**, *36*, 497–507. [CrossRef]
48. Maier, E.; Elmegreen, B.G.; Hunter, D.A.; Chien, L.H.; Hollyday, G.; Simpson, C.E. The Nature of Turbulence in the LITTLE THINGS Dwarf Irregular Galaxies. *Astrophys. J.* **2017**, *153*, 163. [CrossRef]
49. Rosenberg, D.; Pouquet, A.; Marino, R.; Mininni, P. Evidence for Bolgiano-Obukhov scaling in rotating stratified turbulence using high-resolution direct numerical simulations. *Phys. Fluids* **2015**, *27*, 055105. [CrossRef]
50. Kafiabad, H.; Bartello, P. Balance dynamics in rotating stratified turbulence. *J. Fluid Mech.* **2016**, *795*, 914–949. [CrossRef]
51. Pouquet, A.; Rosenberg, D.; Marino, R.; Herbert, C. Scaling laws for mixing and dissipation in unforced rotating stratified turbulence. *J. Fluid Mech.* **2018**, *844*, 519–545. [CrossRef]
52. Mahrt, L.; Sun, J.; Blumen, W.; Delany, T.; Oncley, S. Nocturnal boundary-layer regimes. *Bound. Lay. Met.* **1998**, *88*, 255–278. [CrossRef]
53. Brethouwer, G.; Billant, P.; Lindborg, E.; Chomaz, J.M. Scaling analysis and simulation of strongly stratified turbulent flows. *J. Fluid Mech.* **2007**, *585*, 343–368. [CrossRef]
54. Laval, J.P.; McWilliams, J.; Dubrulle, B. Forced stratified turbulence: Successive transitions with Reynolds number. *Phys. Rev. E* **2003**, *68*, 036308. [CrossRef] [PubMed]
55. Legaspi, J.D.; Waite, M.L. Prandtl number dependence of stratified turbulence. *J. Fluid Mech.* **2020**, *903*, A12. [CrossRef]
56. Venayagamoorthy, S.; Koseff, J. On the flux Richardson number in stably stratified turbulence. *J. Fluid Mech.* **2016**, *798*, R1–R10. [CrossRef]
57. Oks, D.; Mininni, P.; Marino, R.; Pouquet, A. Inverse cascades and resonant triads in rotating and stratified turbulence. *Phys. Fluids* **2017**, *29*, 111109. [CrossRef]
58. Lelong, M.P.; Riley, J. Internal wave-vortical mode interactions in strongly stratified flows. *J. Fluid Mech.* **1991**, *232*, 1–19. [CrossRef]
59. Sukoriansky, S.; Galperin, B. An analytical theory of the buoyancy-Kolmogorov subrange transition in turbulent flows with stable stratification. *Phil. Trans. Roy. Soc. A* **2013**, *371*, 20120212. [CrossRef]
60. Falcon, E.; Aumaitre, S.; Falcon, C.; Laroche, C.; Fauve, S. Fluctuations of Energy Flux in Wave Turbulence. *Phys. Rev. Lett.* **2008**, *100*, 064503. [CrossRef]
61. Lee, J.S.; Kwon, C.; Park, H. Everlasting initial memory threshold for rare events in equilibration processes. *Phys. Rev. E* **2013**, *87*, 020104. [CrossRef]
62. Sukoriansky, S.; Galperin, B. QNSE theory of turbulence anisotropization and onset of the inverse energy cascade by solid body rotation. *J. Fluid Mech.* **2016**, *805*, 384–421. [CrossRef]
63. Rosenberg, D.; Mininni, P.D.; Reddy, R.; Pouquet, A. GPU Parallelization of a Hybrid Pseudospectral Geophysical Turbulence Framework Using CUDA. *Atmosphere* **2020**, *11*, 178. [CrossRef]
64. Fontana, M.; Bruno, O.P.; Mininni, P.D.; Dmitruk, P. Fourier continuation method for incompressible fluids with boundaries. *Comp. Phys. Comm.* **2020**, *256*, 107482. [CrossRef]
65. Godeferd, F.; Cambon, C. Detailed investigation of energy transfers in homogeneous stratified turbulence. *Phys. Fluids* **1994**, *6*, 2084–2100. [CrossRef]

66. Rorai, C.; Mininni, P.; Pouquet, A. Stably stratified turbulence in the presence of large-scale forcing. *Phys. Rev. E* **2015**, *92*, 013003. [CrossRef] [PubMed]
67. Hughes, C.W.; Thompson, A.F.; Wilson, C. Identification of jets and mixing barriers from sea level and vorticity measurements using simple statistics. *Ocean Mod.* **2010**, *32*, 44–57. [CrossRef]
68. Yakhot, V.; Donzis, D.A. Anomalous exponents in strong turbulence. *Phys. D* **2018**, *384*, 12. [CrossRef]
69. Khurshid, S.; Donzis, D.A.; Sreenivasan, K.R. Emergence of universal scaling in isotropic turbulence. *Phys. Rev. E* **2023**, *107*, 045102. [CrossRef]
70. Bradshaw, Z.; Farhat, A.; Grujic, Z. An Algebraic Reduction of the Scaling Gap in the Navier-Stokes Regularity Problem. *Arch. Ration. Mech. Anal.* **2019**, *231*, 1983–2005. [CrossRef]
71. Rafner, J.; Grujic, Z.; Bach, C.; Bærentzen, J.A.; Gervang, B.; Jia, R.; Leinweber, S.; Misztal, M.; Sherson, J. Geometry of turbulent dissipation and the Navier-Stokes regularity problem. *Sci. Rep.* **2021**, *11*, 8824. [CrossRef]
72. Buaria, D.; Pumir, A.; Bodenschatz, E. Generation of intense dissipation in high Reynolds number turbulence. *Phil. Trans. A* **2022**, *380*, 20210088. [CrossRef] [PubMed]
73. Rorai, C.; Mininni, P.; Pouquet, A. Turbulence comes in bursts in stably stratified flows. *Phys. Rev. E* **2014**, *89*, 043002. [CrossRef] [PubMed]
74. Rosenberg, D.; Pouquet, A.; Marino, R. Correlation between buoyancy flux, dissipation and potential vorticity in rotating stratified turbulence. *Atmosphere* **2021**, *12*, 157. [CrossRef]
75. Pouquet, A.; Rosenberg, D.; Stawarz, J.; Marino, R. Helicity Dynamics, Inverse, and Bidirectional Cascades in Fluid and Magnetohydrodynamic Turbulence: A Brief Review. *Earth Space Sci.* **2019**, *6*, 1–19. [CrossRef]
76. Tuckerman, L.S.; Chantry, M.; Barkley, D. Patterns in Wall-Bounded Shear Flows. *Ann. Rev. Fluid Mech.* **2020**, *52*, 343–367. [CrossRef]
77. McWilliams, J.C. Submesoscale currents in the ocean. *Proc. Roy. Soc. A* **2016**, *472*, 20160117. [CrossRef]
78. Legg, S. Mixing by Oceanic Lee Waves. *Ann. Rev. Fluid Mech.* **2021**, *53*, 173–201. [CrossRef]
79. Caulfield, C.P. Layering, Instabilities, and Mixing in Turbulent Stratified Flows. *Ann. Rev. Fluid Mech.* **2021**, *53*, 113–145. [CrossRef]
80. Waite, M.L.; Richardson, N. Potential Vorticity Generation in Breaking Gravity Waves. *Atmosphere* **2023**, *14*, 881. [CrossRef]
81. Pouquet, A.; Marino, R. Geophysical turbulence and the duality of the energy flow across scales. *Phys. Rev. Lett.* **2013**, *111*, 234501. [CrossRef]
82. Marino, R.; Pouquet, A.; Rosenberg, D. Resolving the paradox of oceanic large-scale balance and small-scale mixing. *Phys. Rev. Lett.* **2015**, *114*, 114504. [CrossRef] [PubMed]
83. Faranda, D.; Lembo, V.; Iyer, M.; Kuzzay, D.; Chibbaro, S.; Daviaud, F.; Dubrulle, B. Computation and Characterization of Local Subfilter-Scale Energy Transfers in Atmospheric Flows. *J. Atmos. Sci.* **2018**, *75*, 2175–2186. [CrossRef]
84. Wang, Y.; Brasseur, G.P.; Wang, T. Segregation of Atmospheric Oxidants in Turbulent Urban Environments. *Atmosphere* **2022**, *13*, 315. [CrossRef]
85. Falgarone, E.; Pety, J.; Hily-Blandt, J. Intermittency of interstellar turbulence: Extreme velocity-shears and CO emission on milliparsec scale. *Astron. Astrophys.* **2009**, *507*, 355–368. [CrossRef]
86. Cael, B.; Mashayek, A. Log-Skew-Normality of Ocean Turbulence. *Phys. Rev. Lett.* **2021**, *126*, 224502. [CrossRef]
87. Elsinga, G.E.; Ishihara, T.; Hunt, J.C. Extreme dissipation and intermittency in turbulence at very high Reynolds number. *Proc. Roy. Soc. A* **2020**, *476*, 20200591. [CrossRef]
88. Moeng, C.; Rotunno, R. Vertical velocity skewness in the buoyancy-driven boundary layer. *J. Atmos. Sci.* **1990**, *47*, 1149–1162. [CrossRef]
89. Kit, E.; Hocut, C.M.; Liberzon, D.; Fernando, H.J.S. Fine-scale turbulent bursts in stable atmospheric boundary layer in complex terrain. *J. Fluid Mech.* **2017**, *833*, 745–772. [CrossRef]
90. de Vries, A.J. A global climatological perspective on the importance of Rossby wave breaking and intense moisture transport for extreme precipitation events. *Weather Clim. Dyn.* **2021**, *2*, 129–161. [CrossRef]
91. White, R.H.; Kai Kornhuber, O.M.; Wirth, V. From Atmospheric Waves to Heatwaves: A Waveguide Perspective for Understanding and Predicting Concurrent, Persistent, and Extreme Extratropical Weather. *BAMS* **2022**, *103*, E923–E935. [CrossRef]
92. Petoukhov, V.; Rahmstorf, S.; Petri, S.; Schellnhuber, H.J. Quasiresonant amplification of planetary waves and recent Northern Hemisphere weather extremes. *Proc. Nat. Acad. Sci. USA* **2013**, *110*, 5336–5341. [CrossRef] [PubMed]
93. Raphaldini, B.; Peixoto, P.S.; Teruya, A.S.W.; Raupp, C.F.M.; Bustamante, M.D. Precession resonance of Rossby wave triads and the generation of low-frequency atmospheric oscillations. *Phys. Fluids* **2022**, *34*, 076604. [CrossRef]
94. Wirth, V.; Riemer, M.; Chang, E.K.; Martius, O. Rossby Wave Packets on the Midlatitude Waveguide—A Review. *Month. Wea. Rev.* **2018**, *146*, 1965–2001. [CrossRef]
95. Antokhina, O.; Antokhin, P.; Gochakov, A.; Zbirannik, A.; Gazimov, T. Atmospheric Circulation Patterns Associated with Extreme Precipitation Events in Eastern Siberia and Mongolia. *Atmosphere* **2023**, *14*, 480. [CrossRef]
96. Lenschow, D.H.; Lothon, M.; Mayor, S.D.; Sullivan, P.P.; Canut, G. A Comparison of Higher-Order Vertical Velocity Moments in the Convective Boundary Layer from Lidar with in Situ Measurements and Large-Eddy Simulation. *Bound. Lay. Met.* **2012**, *143*, 107–123. [CrossRef]

97. Sandberg, I.; Benkadda, S.; Garbet, X.; Ropokis, G.; Hizanidis, K.; del Castillo-Negrete, D. Universal Probability Distribution Function for Bursty Transport in Plasma Turbulence. *PRL* **2009**, *103*, 165001. [CrossRef]
98. Guszejnov, D.; Lazanyi, N.; Bencze, A.; Zoletnik, S. On the effect of intermittency of turbulence on the parabolic relation between skewness and kurtosis in magnetized plasmas. *Phys. Plasma* **2013**, *20*, 112305. [CrossRef]
99. Fritts, D.C.; Werne, J. Stratified shear turbulence: Evolution and statistics. *Geophys. Res. Lett.* **2009**, *26*, 439–442.
100. Thalabard, S.; Bec, J.; Mailybaev, A.A. From the butterfly effect to spontaneous stochasticity in singular shear flows. *Comm. Phys.* **2020**, *122*, 3. [CrossRef]
101. Mininni, P.; Alexakis, A.; Pouquet, A. Nonlocal interactions in hydrodynamic turbulence at high Reynolds numbers: The slow emergence of scaling laws. *Phys. Rev. E* **2008**, *77*, 036306. [CrossRef]
102. Tamarin-Brodsky, T.; Hodges, K.; Hoskins, B.J.; Sheperd, T.G. A Simple Model for Interpreting Temperature Variability and Its Higher-Order Changes. *J. Clim.* **2022**, *35*, 387–403.
103. Sreenivasan, K.; Antonia, R.A. The phenomenology of small-scale turbulence. *Ann. Rev. Fluid Mech.* **1997**, *29*, 435–472. [CrossRef]
104. Schroder, M.; Batge, T.; Bodenschatz, E.; Wilczek, M.; Bagheri, G. Estimating the turbulent kinetic energy dissipation rate from one-dimensional velocity measurements in time. *Atm. Measur. Tech.* **2023**, *preprint*.
105. Sura, P.; Hannachi, A. Perspectives of Non-Gaussianity in Atmospheric Synoptic and Low-Frequency Variability. *J. Clim.* **2015**, *28*, 5091–6014. [CrossRef]
106. Klaassen, C.A.; Mokveld, P.J.; van Es, B. Squared skewness minus kurtosis bounded by 186/125 for unimodal distributions. *Stat. Prob. Lett.* **2000**, *50*, 131–135. [CrossRef]
107. Gryanik, V.M.; Hartmann, J. On a Solution of the Closure Problem for Dry Convective Boundary Layer Turbulence and Beyond. *J. Atmos. Sci.* **2022**, *79*, 1405–1428. [CrossRef]
108. Sardeshmukh, P.; Compo, G.P.; Penland, C. Need for Caution in Interpreting Extreme Weather Statistics. *J. Clim.* **2015**, *28*, 9166–9185. [CrossRef]
109. Longuet-Higgins, M. The effect of non-linearities on statistical distributions in the theory of sea waves. *J. Fluid Mech.* **1963**, *17*, 459–480. [CrossRef]
110. Ochi, M.K.; Wang, W.C. Non-Gaussian characteristics of coastal waves. *International Conference of Coastal Engineering*; ASCE: New York, NY, USA, 1984; Volume 19, pp. 516–531.
111. Monahan, A.H. The Probability Distribution of Sea Surface Wind Speeds. Part I: Theory and SeaWinds Observations. *J. Clim.* **2006**, *19*, 497–516. [CrossRef]
112. Abroug, I.; Matar, R.; Abcha, N. Spatial Evolution of Skewness and Kurtosis of Unidirectional Extreme Waves Propagating over a Sloping Beach. *J. Marine Sci. Eng.* **2022**, *10*, 1475. [CrossRef]
113. Schumacher, J.; Scheel, J.D.; Krasnova, D.; Donzis, D.A.; Yakhot, V.; Sreenivasan, K.R. Small-scale universality in fluid turbulence. *Proc. Nat. Acad. Sci. USA* **2014**, *111*, 10961–10965. [CrossRef] [PubMed]
114. Nakao, H. Asymptotic power law of moments in a random multiplicative process with weak additive noise. *Phys. Rev. E* **1998**, *58*, 1591–1600. [CrossRef]
115. van Kan, A.; Alexakis, A.; Brachet, M.E. Levy on-off intermittency. *Phys. Rev. E* **2021**, *103*, 052115. [CrossRef] [PubMed]
116. Laval, J.; Dubrulle, B.; Nazarenko, S. Nonlocality and intermittency in three-dimensional turbulence. *Phys. Fluids* **2001**, *13*, 1995–2002. [CrossRef]
117. Alexakis, A.; Ponty, Y. Effect of the Lorentz force on on-off dynamo intermittency. *Phys. Rev. E* **2008**, *77*, 056308. [CrossRef]
118. Brunton, S.L.; Brunton, B.W.; Proctor, J.L.; Kaiser, E.; Kutz, J.N. Chaos as an intermittently forced linear system. *Nat. Comm.* **2017**, *8*, 1–9. [CrossRef]
119. Hasselmann, K. Stochastic climate models. Part I. Theory. *Tellus* **1976**, *28*, 473–484.
120. Berner, J.; Achatz, U.; Batté, L.; Bengtsson, L.; de la Cámara, A.; Christensen, H.M.; Colangeli, M.; Coleman, D.R.B.; Crommelin, D.; Dolaptchiev, S.I.; et al. Stochastic Parameterization: Toward a New View of Weather and Climate Models. *Bull. Am. Met. Soc.* **2017**, *98*, 565–588. [CrossRef]
121. Farago, J. Injected Power Fluctuations in Langevin Equation. *J. Stat. Phys.* **2002**, *107*, 781–803. [CrossRef]
122. Farrell, B.; Ioannou, P. A Theory of Baroclinic Turbulence. *J. Atmos. Sci.* **2009**, *66*, 2444–2454. [CrossRef]
123. Djenidi, L.; Lefeuvre, N.; Kamruzzaman, M.; Antonia, R. On the normalized dissipation C_ϵ in decaying turbulence. *J. Fluid Mech.* **2017**, *817*, 61–79. [CrossRef]
124. Mininni, P.; Pouquet, A. Finite dissipation and intermittency in MHD. *Phys. Rev. E* **2009**, *80*, 025401. [CrossRef] [PubMed]
125. Bandyopadhyay, R.; Oughton, S.; Wan, M.; Matthaeus, W.H.; Chhiber, R.; Parashar, T.N. Finite Dissipation in Anisotropic Magnetohydrodynamic Turbulence. *Phys. Rev. X* **2018**, *8*, 041052. [CrossRef]
126. Maurizi, A. On the dependence of third- and fourth-order moments on stability in the turbulent boundary layer. *Nonlin. Process. Geophys.* **2013**, *13*, 119–123. [CrossRef]
127. Smyth, W.; Nash, J.; Moum, J. Self-organized criticality in geophysical turbulence. *Sci. Rep.* **2019**, *9*, 3747. [CrossRef]
128. Pouquet, A.; Rosenberg, D.; Marino, R. Linking dissipation, anisotropy and intermittency in rotating stratified turbulence. *Phys. Fluids* **2019**, *31*, 105116. [CrossRef]
129. Vieillefosse, P. Internal motion of a small element of fluid in an inviscid flow. *Phys. A* **1984**, *125*, 150–162. [CrossRef]
130. Sujovolsky, N.; Mininni, P. From waves to convection and back again: The phase space of stably stratified turbulence. *Phys. Rev. Fluids* **2020**, *5*, 064802. [CrossRef]

131. Chien, M.H.; Pauluis, O.M.; Almgren, A.S. Hurricane-Like Vortices in Conditionally Unstable Moist Convection. *JAMES* **2022**, *14*, e2021MS002846. [CrossRef]
132. Emanuel, K.; Velez-Pardo, M.; Cronin, T.W. The Surprising Roles of Turbulence in Tropical Cyclone Physics. *Atmosphere* **2023**, *14*, 1254. [CrossRef]
133. Betchov, R. An inequality concerning the production of vorticity in isotropic turbulence. *J. Fluid Mech.* **1956**, *1*, 497–504. [CrossRef]
134. Lozovatsky, I.; Fernando, H.; Planella-Morato, J.; Liu, Z.; Lee, J.; Jinadasa, S. Probability Distribution of Turbulent Kinetic Energy Dissipation Rate in Ocean: Observations and Approximations. *J. Geophys. Res.* **2017**, *122*, 8293–8308. [CrossRef]
135. Uritsky, V.; Pouquet, A.; Rosenberg, D.; Mininni, P.; Donovan, E. Structures in magnetohydrodynamic turbulence: Detection and scaling. *Phys. Rev. E* **2010**, *82*, 056326. [CrossRef]
136. Celikoglu, A.; Tirnakli, U. Comment on Universal relation between skewness and kurtosis in complex dynamics. *Phys. Rev. E* **2015**, *92*, 066801. [CrossRef]
137. Kolmogorov, A.N. Dissipation of energy in locally isotropic turbulence [English translation in Proc. Roy. Soc. London A 434 (1991) 15–17]. *Dokl. Akad. Nauk SSSR* **1941**, *32*, 19–21.
138. Antonia, R.; Ould-Rouis, M.; Anselmet, F.; Zhu, Y. Analogy between predictions of Kolmogorov and Yaglom. *J. Fluid Mech.* **1997**, *332*, 395–409. [CrossRef]
139. Politano, H.; Pouquet, A. Dynamical length scales for turbulent magnetized flows. *Geophys. Res. Lett.* **1998**, *25*, 273–276. [CrossRef]
140. Marino, R.; Sorriso-Valvo, L. Scaling laws for the energy transfer in space plasma turbulence. *Phys. Rep.* **2023**, *1006*, 1–144. . [CrossRef]
141. Pouquet, A.; Sulem, P.L.; Meneguzzi, M. Influence of velocity-magnetic field correlations on decaying magnetohydrodynamic turbulence with neutral X-points. *Phys. Fluids* **1988**, *31*, 2635–2643. [CrossRef]
142. Kraichnan, R.H. Intermittency in the Very Small Scales of Turbulence. *Phys. Fluids* **1967**, *10*, 2080–2082. [CrossRef]
143. Mordant, N.; Delour, J.; Lévêque, E.; Arnéodo, A.; Pinton, J.F. Long Time Correlations in Lagrangian Dynamics: A Key to Intermittency in Turbulence. *Phys. Rev. Lett.* **2002**, *89*, 254502. [CrossRef] [PubMed]
144. Ishihara, T.; Gotoh, T.; Kaneda, Y. Study of High Reynolds Number Isotropic Turbulence by Direct Numerical Simulation. *Ann. Rev. Fluid Mech.* **2009**, *41*, 165–180. [CrossRef]
145. Yeung, P.K.; Zhai, X.M.; Sreenivasan, K.R. Extreme events in computational turbulence. *Proc. Nat. Acad. Sci. USA* **2015**, *112*, 12633–12638. [CrossRef] [PubMed]
146. Faller, H.; Geneste, D.; Chaabo, T.; Cheminet, A.; Valori, V.; Ostovan, Y.; Cappanera, L.; Cuvier, C.; Daviaud, F.; Foucaut, J.M.; et al. On the nature of intermittency in a turbulent von Karman flow. *J. Fluid Mech.* **2021**, *914*, A2. [CrossRef]
147. Galuzio, P.; Lopes, S.; dos Santos Lima, G.; Viana, R.; Benkadda, M. Evidence of determinism for intermittent convective transport in turbulence processes. *Phys. A* **2014**, *402*, 8–13. [CrossRef]
148. Küchler, C.; Bewley, G.P.; Bodenschatz, E. Universal velocity statistics in decaying turbulence. *Phys. Rev. Lett.* **2023**, *131*, 024001. [CrossRef]
149. Bramwell, S.T.; Christensen, K.; Fortin, J.Y.; Holdsworth, P.C.W.; Jensen, H.J.; Lise, S.; Lopez, J.M.; Nicodemi, M.; Pinton, J.F.; Sellitto, M. Universal Fluctuations in Correlated Systems. *Phys. Rev. Lett.* **2000**, *84*, 3744–3747. [CrossRef]
150. Chini, G.P.; Michel, G.; Julien, K.; Rocha, C.B.; Caulfield, C.P. Exploiting self-organized criticality in strongly stratified turbulence. *J. Fluid Mech.* **2022**, *933*, A22. [CrossRef]
151. Watanabe, T.; Riley, J.R.; de Bruyn-Kops, S.M.; Diamessis, P.J.; Zhou, Q. Turbulent/non-turbulent interfaces in wakes in stably stratified fluids. *J. Fluid Mech.* **2016**, *797*, R1. [CrossRef]
152. Barkley, D. Theoretical perspective on the route to turbulence in a pipe. *J. Fluid Mech.* **2016**, *803*, 1–80. [CrossRef]
153. Sano, M.; Tamai, K. A universal transition to turbulence in channel flow. *Nat. Phys.* **2016**, *12*, 249–254. [CrossRef]
154. Shih, H.Y.; Hsieh, T.L.; Goldenfeld, N. Ecological collapse and the emergence of travelling waves at the onset of shear turbulence. *Nat. Phys.* **2016**, *12*, 245–248. [CrossRef]
155. Manneville, P. Laminar-Turbulent Patterning in Transitional Flows. *Entropy* **2017**, *19*, 316. [CrossRef]
156. Gurcan, O.D.; Diamond, P.H. Zonal flows and pattern formation. *J. Phys. A* **2015**, *48*, 293001. [CrossRef]
157. Sujovolsky, N.; Mininni, P.; Pouquet, A. Generation of turbulence through frontogenesis in sheared stratified flows. *Phys. Fluids* **2018**, *30*, 086601. [CrossRef]
158. Reartes, C.; Mininni, P. Dynamical regimes and clustering of small neutrally buoyant inertial particles in stably stratified turbulence. *Phys. Rev. Fluids* **2023**, *8*, 054501. [CrossRef]
159. Kafiabad, H.A.; Bartello, P. Spontaneous imbalance in the non-hydrostatic Boussinesq equations. *J. Fluid Mech.* **2018**, *847*, 614–643. [CrossRef]
160. Sullivan, P.P.; McWilliams, J.C. Langmuir Turbulence and filament frontogenesis in the oceanic surface boundary layer. *J. Fluid Mech.* **2019**, *879*, 512–553. [CrossRef]
161. Sagaut, P.; Cambon, C. *Homogeneous Turbulence Dynamics*; Cambridge University Press: Cambridge, UK, 2008.
162. Kit, E.; Fernando, H. Small-scale anisotropy in stably stratified turbulence; Inferences based on katabatic flows. *Atmosphere* **2023**, *14*, 918. [CrossRef]
163. Galperin, B.; Sukoriansky, S.; Qiu, B. Seasonal oceanic variability on meso- and submesoscales: A turbulence perspective. *Ocean Dyn.* **2021**, *71*, 475–489. [CrossRef]

164. Wang, J.W.A.; Sardeshmukh, P. Inconsistent Global Kinetic Energy Spectra in Reanalyses and Models. *J. Atmos. Sci.* **2021**, *78*, 2589–2604.
165. Bertin, E.; Holdsworth, P.C.W. Dissipation induced non-Gaussian energy fluctuations. *Eur. Phys. J.* **2013**, *102*, 50004. [CrossRef]
166. Rorai, C.; Rosenberg, D.; Pouquet, A.; Mininni, P. Helicity dynamics in stratified turbulence in the absence of forcing. *Phys. Rev. E* **2013**, *87*, 063007. [CrossRef] [PubMed]
167. Cichowlas, C.; Bonaïti, P.; Debbasch, F.; Brachet, M. Effective Dissipation and Turbulence in Spectrally Truncated Euler Flows. *Phys. Rev. Lett.* **2005**, *95*, 264502. [CrossRef] [PubMed]
168. Spyksma, K.; Magcalas, M.; Campbell, N. Quantifying effects of hyperviscosity on isotropic turbulence. *Phys. Fluids* **2012**, *24*, 125102. [CrossRef]
169. Choi, Y.; Kim, B.; Lee, C. Alignment of velocity and vorticity and the intermittent distribution of helicity in isotropic turbulence. *Phys. Rev. E* **2009**, *80*, 017301. [CrossRef]
170. Buaria, D.; Pumir, A.; Bodenschatz, E. Self-attenuation of extreme events in Navier-Stokes turbulence. *Nat. Comm.* **2020**, *11*, 5852. [CrossRef]
171. Bartello, P. Geostrophic adjustment and inverse cascade in rotating stratified turbulence. *J. Atmos. Sci.* **1995**, *52*, 4410–4428. [CrossRef]
172. Herbert, C.; Marino, R.; Pouquet, A.; Rosenberg, D. Waves and vortices in the inverse cascade regime of rotating stratified turbulence with or without rotation. *J. Fluid Mech.* **2016**, *806*, 165–204. [CrossRef]
173. Baj, P.; Portela, F.A.; Carter, D.W. On the simultaneous cascades of energy, helicity, and enstrophy in incompressible homogeneous turbulence. *J. Fluid Mech.* **2022**, *952*, A20. [CrossRef]
174. Brachet, M.; Meneguzzi, M.; Vincent, A.; Politano, H.; Sulem, P. Numerical evidence of smooth self-similar dynamics and possibility of subsequent collapse for 3D ideal flows. *Phys. Fluids* **1992**, *A4*, 2845–2854. [CrossRef]
175. Hamlington, P.E.; Schumacher, J.; Dahm, W.J. Direct assessment of vorticity alignment with local and nonlocal strain rates in turbulent flows. *Phys. Fluids* **2008**, *20*, 111703. [CrossRef]
176. Stephan, C.C.; Žagar, N.; Shepherd, T.G. Waves and coherent flows in the tropical atmosphere: New opportunities, old challenges. *Quart. J. Roy. Met. Soc.* **2021**, *147*, 2597–2624. [CrossRef]

atmosphere

Review

Vectorial EM Propagation Governed by the 3D Stochastic Maxwell Vector Wave Equation in Stratified Layers

Bryce M. Barclay *, Eric J. Kostelich and Alex Mahalov

School of Mathematical and Statistical Sciences, Arizona State University, Tempe, AZ 85287, USA; kostelich@asu.edu (E.J.K.); mahalov@asu.edu (A.M.)
* Correspondence: bryce.barclay@asu.edu

Abstract: The modeling and processing of vectorial electromagnetic (EM) waves in inhomogeneous media are important problems in physics and engineering, and new methods need to be developed to incorporate novel vector sensor technology. Vectorial phenomena of EM waves in stratified atmospheric layers can be incorporated into governing equations by retaining the gradient of the refractive index when deriving the Maxwell Vector Wave Equation (MVWE) from Maxwell's equations. The MVWE, as opposed to the scalar wave, Helmholtz, and paraxial equations, couples the EM field components in inhomogeneous media and thus captures important physics phenomena such as depolarization. Here, recent developments are reviewed on using sensor time series data to reconstruct electromagnetic waves that propagate through stratified media. In modern applications, often many sensors can be deployed simultaneously to observe electromagnetic waves. These networks of sensors can be used to improve the quality of the reconstructed EM wave fields and cross-validate the observed sensor time series. Lastly, the effects of noisy current densities on sensor time series are considered. Generally, as the sensor observes for longer periods of time, the variance of estimates of the wave field obtained from sensor time series data increases. As a result, longer sensor observation times do not always result in better estimates of the EM wave fields, and an optimal observation time can be obtained.

Keywords: Maxwell's equations; vectorial electromagnetic waves; stratified atmospheric layers; stochastic electrodynamics; sensor networks

Citation: Barclay, B.M.; Kostelich, E.J.; Mahalov, A. Vectorial EM Propagation Governed by the 3D Stochastic Maxwell Vector Wave Equation in Stratified Layers. *Atmosphere* **2023**, *14*, 1451. https://doi.org/10.3390/atmos14091451

Academic Editors: Boris Galperin, Annick Pouquet and Peter Sullivan

Received: 11 August 2023
Revised: 14 September 2023
Accepted: 15 September 2023
Published: 18 September 2023

1. Introduction

With the development of modern vectorial EM sensor technologies, new vectorial methods for modeling and processing EM waves, derived from fundamental Maxwell's equations, need to be created. New methods also need to be developed for processing signals observed from a network of sensors. For example, CubeSats are inexpensive and can be quickly deployed, making it practical to utilize networks of antennas and sensing equipment to detect electromagnetic waves. The National Research Council stated that the deployment of satellite constellations was a priority for the coming decades [1]. Satellite constellations will greatly advance our scientific understanding of the interactions of atmospheric layers through the collection of environmental data. Placement conditions to optimize the design of CubeSat constellations for communication networks are discussed in [2]. Additionally, CubeSat missions are being developed with vector sensor technology [3,4]. New imaging techniques have also been developed to remotely capture 3D structures in the ionosphere from the ground [5].

In this review, we discuss how the Maxwell vector wave equation (MVWE), an equation modeling vectorial EM propagation derived directly from Maxwell's equations, has been utilized in recent advancements to more accurately model EM wave propagation in inhomogeneous and random media. The MVWE, as opposed to the scalar wave, Helmholtz, and paraxial equations, models vectorial EM waves by retaining the gradient of the refractive index, and, as a result, the MVWE captures important physics phenomena associated with the coupling of electric field components such as depolarization, which occurs in EM waves propagating through inhomogeneous media.

Application of the MVWE has been demonstrated for the important case of stratified media. Examples of stratified media include the ionosphere and sharp layers such as air–sea and Rayleigh–Taylor interfaces [6–8]. For a broad overview of Rayleigh–Taylor instabilities and turbulence, see [9]. In these cases, the permittivity gradient is large and affects propagating EM waves. When waves propagate to the Karman line (at 100 km above the Earth's surface), a portion of the wave penetrates into the ionosphere and another portion reflects back into the mesosphere, and as a result of the permittivity gradient, the electric fields are coupled in this region and EM waves can become depolarized. This phenomenon is related to the temperature gradient ∇T and the angle between the electric field $\mathbf{E}$ and the gradient. The thermodynamic structure of the atmosphere in the vertical direction affects communication and the propagation of EM waves [10]. NASA's Ionospheric Connection Explorer (ICON) mission, launched in 2019, seeks to understand how space weather interacts with the weather on Earth. Radio and GPS signals that propagate through the ionosphere can be disrupted by dense pockets of charged particles [11,12]. Fast winds in this region power electric fields that can affect satellites [11,13]. The ICON mission provided direct measurements of the Karman layer.

This paper is dedicated to Dr. Jackson Rea Herring. Jack was a Senior Scientist at the Mesoscale and Microscale Meteorology Division at the National Center for Atmospheric Research (NCAR) at Boulder, Colorado, a position he had held since 1978. Jack made fundamental contributions to research on stratified turbulence [14,15]. The Craya–Herring decomposition is widely used by researchers in the field. Stably stratified flows appear in many areas of physics and engineering and are fundamental to the study of geophysics. Stable stratification occurs in the Earth's atmosphere and planetary flows more generally. He was a leader in the Geophysical Turbulence Program (GTP) at NCAR [16]. Over the years, our scientific community benefited greatly from Jack's generosity and the GTP scientific activities including workshops and a special collection of articles on geophysical turbulence which he edited [17,18]. Jack Herring contributed to atmospheric and oceanic flows, and his work influenced many publications in stratified flows more generally [19,20].

The present review is narrower in scope, focusing on electromagnetic wave propagation through inhomogeneous stratified media and processing of sensor observation time series in random media. Sensor time series data can be used with an eigenfunction decomposition of the EM wave field obtained from general PDE theory to reconstruct the EM field at any other location (the procedure is described in Section 3). Sensor networks can also be utilized to obtain better estimates of EM fields.

This paper is organized as follows. In Section 2, the Maxwell vector wave equation is derived for electromagnetic wave propagation in inhomogeneous media and the important case of a stratified layered medium is discussed. In Section 3, work on electromagnetic wave field reconstruction from sensor network data is reviewed. In Section 4, the stochastic Maxwell vector wave equation is introduced and studied. Finally, concluding remarks are made on wave propagation through stratified layers.

2. Vectorial Electromagnetic Wave Propagation through Stratified Layers

In their most fundamental form, electromagnetic waves are described by Maxwell's equations. Often, approximations are made to simplify these equations to decoupled scalar wave equations. This approximation, however, neglects important vectorial phenomena such as depolarization. Thus, it is important to study the vectorial wave equation derived directly from Maxwell's equations, which is referred to here as the Maxwell vector wave equation (MVWE). This equation can be derived from the differential form of Faraday's law of induction and Ampére's law:

$$\nabla \times \mathbf{E} = -\frac{\partial \mathbf{B}}{\partial t} \tag{1}$$

$$\nabla \times \mathbf{H} = \mathbf{J} + \frac{\partial \mathbf{D}}{\partial t} \tag{2}$$

with the constitutive relations $\mathbf{D} = \epsilon(\mathbf{x})\mathbf{E}$ and $\mathbf{H} = \mu^{-1}\mathbf{B}$, where $\mathbf{J}$ is the current density, $\mathbf{E}$ is the electric field, $\mathbf{H}$ is the magnetic field, $\mathbf{D}$ is the electric displacement, $\mathbf{B}$ is the magnetic induction, $\mathbf{x} = (x_1, x_2, x_3)$ represents the Cartesian spatial coordinates, $\epsilon(\mathbf{x})$ is the spatially varying permittivity, and μ is the permeability constant. Differentiating Ampére's law with respect to time gives

$$\nabla \times \left(\mu^{-1}\frac{\partial \mathbf{B}}{\partial t}\right) = \frac{\partial \mathbf{J}}{\partial t} + \epsilon(\mathbf{x})\frac{\partial^2 \mathbf{E}}{\partial t^2}. \tag{3}$$

Finally, using Faraday's law gives the Maxwell vector wave equation

$$\epsilon(\mathbf{x})\frac{\partial^2 \mathbf{E}}{\partial t^2} + \nabla \times \left(\mu^{-1}\nabla \times \mathbf{E}\right) = -\frac{\partial \mathbf{J}}{\partial t}. \tag{4}$$

For constant μ, the vector identity $\nabla \times \nabla \times \mathbf{E} = \nabla(\nabla \cdot \mathbf{E}) - \Delta\mathbf{E}$ can be applied to obtain the alternate form

$$\frac{\partial^2 \mathbf{E}}{\partial t^2} - c^2(\mathbf{x})\Delta\mathbf{E} + c^2(\mathbf{x})\nabla(\nabla \cdot \mathbf{E}) = -[\epsilon(\mathbf{x})]^{-1}\frac{\partial \mathbf{J}}{\partial t} \tag{5}$$

where $c^2(\mathbf{x}) = (\mu\epsilon(\mathbf{x}))^{-1}$. This equation clearly resembles a standard wave equation with the additional term $c^2(\mathbf{x})\nabla(\nabla \cdot \mathbf{E})$. To reduce Equation (5) to a standard wave equation, it is often assumed that $\nabla \cdot \mathbf{E} = 0$ using Gauss's Law $\nabla \cdot \mathbf{D} = 0$. This assumption, however, neglects vectorial and depolarization effects of electromagnetic propagation in inhomogeneous media since

$$\nabla \cdot \mathbf{D} = \epsilon(\mathbf{x})\nabla \cdot \mathbf{E} + \nabla\epsilon(\mathbf{x}) \cdot \mathbf{E} = 0, \tag{6}$$

which implies that

$$\nabla \cdot \mathbf{E} = -[\epsilon(\mathbf{x})]^{-1}\nabla\epsilon(\mathbf{x}) \cdot \mathbf{E} \neq 0. \tag{7}$$

Thus, Equation (5) can alternatively be written in terms of the parameter $\gamma(\mathbf{x}) := [\epsilon(\mathbf{x})]^{-1}\nabla\epsilon(\mathbf{x}) = 2\nabla\log(n(\mathbf{x}))$, where n is the refractive index, to represent the coupling of the components of the electric field via the permittivity gradient:

$$\frac{\partial^2 \mathbf{E}}{\partial t^2} - c^2(\mathbf{x})\Delta\mathbf{E} - c^2(\mathbf{x})\nabla(\gamma \cdot \mathbf{E}) = -[\epsilon(\mathbf{x})]^{-1}\frac{\partial \mathbf{J}}{\partial t}. \tag{8}$$

The inverse of the magnitude of γ, i.e., $L_\epsilon = |\gamma|^{-1}$, is the length scale that characterizes the inhomogeneity of the medium. Without the coupling of the electric field components, depolarization effects are neglected.

Simpler models than Equation (8) that preserve coupling effects can be obtained in special cases. One important such case is when the medium is vertically stratified, which can be represented by a permittivity $\epsilon(x_3)$ that depends only on the vertical coordinate x_3. In practice, a vertically stratified permittivity can be obtained via ensemble averages of permittivity data observed at different times, horizontal averages of permittivity data observed at different horizontal locations, or using a standard profile (for example, see [21]).

The gradient of the permittivity points in the x_3 direction, and therefore, $\gamma = (0, 0, \gamma_3(x_3))$. The coupled Equations (4) become an upper-triangular system:

$$\frac{\partial^2 E_1}{\partial t^2} - c^2(x_3)\Delta E_1 - c^2(x_3)\gamma_3(x_3)\frac{\partial E_3}{\partial x_1} = -[\epsilon(x_3)]^{-1}\frac{\partial J_1}{\partial t} \tag{9}$$

$$\frac{\partial^2 E_2}{\partial t^2} - c^2(x_3)\Delta E_2 - c^2(x_3)\gamma_3(x_3)\frac{\partial E_3}{\partial x_2} = -[\epsilon(x_3)]^{-1}\frac{\partial J_2}{\partial t} \tag{10}$$

$$\frac{\partial^2 E_3}{\partial t^2} - c^2(x_3)\Delta E_3 - c^2(x_3)\frac{\partial(\gamma_3(x_3)E_3)}{\partial x_3} = -[\epsilon(x_3)]^{-1}\frac{\partial J_3}{\partial t}. \tag{11}$$

In this system, the field component parallel to the direction of stratification acts as a forcing term for the orthogonal field components. The orthogonal field components solve forced scalar wave equations. As a result, a wave source that is initially linearly polarized parallel to the direction of stratification will depolarize and activate orthogonal components of the electric field. Depolarization can be defined, as in [22], as the ratio of the complex electric field that is orthogonally polarized to the transmitted wave ($E_\perp$) to the field that is copolarized to the transmitted wave ($E_\parallel$):

$$\text{depolarization} = E_\perp / E_\parallel. \tag{12}$$

Depolarization effects have been studied in the context of evaporation ducts at the air–sea interface [23]. Evaporation ducts are refracting layers where humidity sharply decreases near water–air interfaces such as boundaries of clouds, the marine boundary layer, and the air–sea interface. At the air–sea interface, the permittivity changes sharply, which results in depolarized electromagnetic waves. The effect of evaporation ducts on radio frequency (RF) propagation was studied in the Coupled Air–Sea Processes and Electromagnetic Ducting Research (CASPER) project [24]. The effect of random, inhomogeneous media on wave propagation has also been studied using a simpler scalar wave equation approach [25–28].

Other types of layers encountered in electromagnetic applications are discussed in the conclusion. Jack Herring made fundamental contributions to the study of stratified layered media characterized by the Brunt–Väisälä frequency, $N^2 = g\theta^{-1}\frac{d\theta}{dx_3}$, where $\theta(x_3)$ is the potential temperature and g is the acceleration of gravity [14,15].

3. Robust Reconstruction of Wave Dynamics Using Data from Noisy Sensors and Heterogeneous Sensor Networks

In this section, the results for the wave equation

$$\frac{\partial^2 u}{\partial t^2} - c^2(\mathbf{x})\Delta u = F(\mathbf{x}, t) \tag{13}$$

are reviewed. Equation (13) has the form of (9) and (10) in the stratified upper-triangular system and is also a fundamental equation in acoustics. In [29], Equation (13) is studied from a signal processing perspective. An important problem in signal processing is understanding how and when to use information obtained from multiple sources. When an approximation of an electric wave field in some region of space is needed, a network of electric field sensors can be utilized. In [29], the problem of constructing wave fields from a network of sensors is analyzed. Each sensor obtains a time series of observations $R_\mathbf{s}(t)$ of the wave field at a fixed spatial location $\mathbf{s}$, i.e., $R_\mathbf{s}(t) = u(\mathbf{s}, t)$, where u is the wave field. The sensor location $\mathbf{s}$ is represented by the fixed Cartesian coordinates (s_1, s_2, s_3) relative to the frame of reference. Natural questions to ask are (1) how large the sensor network should be to construct the wave field and (2) what the level of error in such approximations is.

The solutions of (13) can be described in terms of eigenfunctions Φ of the weighted Laplace operator

$$c^2(\mathbf{x})\Delta\Phi + \lambda\Phi = 0 \tag{14}$$

with suitable boundary conditions along with initial conditions

$$u(\mathbf{x},0) = f(\mathbf{x}) = \sum_n \hat{f}_n \Phi_n(\mathbf{x}) \tag{15}$$

$$\frac{\partial u}{\partial t}(\mathbf{x},0) = g(\mathbf{x}) = \sum_n \hat{g}_n \Phi_n(\mathbf{x}). \tag{16}$$

The general solution can then be written as a sum

$$u(\mathbf{x},t) = \sum_n A_n(t)\Phi_n(\mathbf{x}) \tag{17}$$

$$A_n(t) = \hat{f}_n \cos(\sqrt{\lambda_n} t) + \frac{\hat{g}_n}{\sqrt{\lambda_n}} \sin(\sqrt{\lambda_n} t) \tag{18}$$

which is written here and throughout in the real, as opposed to complex, form. The eigenfunctions $\Phi_n(\mathbf{x})$ can be thought of more generally as "features" of the problem (13) together with boundary and initial data. For (13), the features are eigenfunctions of the weighted Laplace operator, and for the Maxwell vector wave Equation (4), the features are eigenfunctions of the weighted curl-squared operator (discussed in the next section). Importantly, the only requirement for the analysis that follows is that the features be orthogonal. Thus, different boundary conditions and differential operators can be substituted whenever the eigenfunctions are orthogonal. The general features framework is used in many other areas of research, including machine learning, compressed sensing, and dynamic mode decomposition [30,31].

The sensor time series $R_\mathbf{s}(t)$ is almost periodic in time, i.e., a sum of trigonometric functions

$$R_\mathbf{s}(t) = \sum_n c_n \cos(\omega_n t) + d_n \sin(\omega_n t). \tag{19}$$

Although each term is periodic, the sum is not periodic, because the frequencies ω_n are noncommenserate, which is the case in the processing of signals in practice. The space of almost periodic functions is a Hilbert space with the inner product

$$(h(t), k(t)) = \lim_{T\to\infty} \frac{1}{T} \int_0^T h(t)k(t)\, dt. \tag{20}$$

Therefore, the frequencies ω_n and coefficients c_n and d_n can be obtained using the inner products:

$$\alpha := (R_\mathbf{s}(t), \cos(\omega t)) \tag{21}$$

$$\beta := (R_\mathbf{s}(t), \sin(\omega t)). \tag{22}$$

When every eigenvalue uniquely identifies an eigenfunction (i.e., every eigenvalue has multiplicity one), the wave field u can be fully reconstructed from one sensor.

Given a sensor time series $R_\mathbf{s}(t)$ at spatial location $\mathbf{s}$, the time series of the wave field at any other location $\mathbf{x}_0$ can be obtained via [29]

$$R_{\mathbf{x}_0}(t) = \sum_n \frac{\Phi_n(\mathbf{x}_0)}{\Phi_n(\mathbf{s})} c_n \cos(\omega_n t) + \frac{\Phi_n(\mathbf{x}_0)}{\Phi_n(\mathbf{s})} d_n \sin(\omega_n t) \tag{23}$$

when $F = 0$. Thus, in this framework, the form of the solution (17) obtained from general PDE theory can be used together with the time series data $R_\mathbf{s}(t)$ of the wave field at a fixed spatial location $\mathbf{s}$ to obtain the wave field solution at any other location $\mathbf{x}_0$. In general, the single time series $R_\mathbf{s}(t)$ does not provide enough information to construct u because one frequency ω may correspond to multiple eigenfunctions Φ_n. Instead, the information obtained is a sum over all eigenfunctions Φ_i that correspond to the same eigenvalue λ_j:

$$\alpha_j = \sum_{i \in I_j} \Phi_i(\mathbf{s}) \frac{\hat{f}_i}{2}. \tag{24}$$

Thus, a single time series $R_{\mathbf{s}}(t)$ cannot generally be used to obtain the amplitudes $\hat{f}_i$ and one needs a network of sensors. The data obtained from a network of sensors give a linear system:

$$2\begin{pmatrix} \alpha_j^{\mathbf{s}_1} \\ \vdots \\ \alpha_j^{\mathbf{s}_n} \end{pmatrix} = \begin{pmatrix} \Phi_{i_1}(\mathbf{s}_1) \dots \Phi_{i_{m_j}}(\mathbf{s}_1) \\ \vdots \\ \Phi_{i_1}(\mathbf{s}_n) \dots \Phi_{i_{m_j}}(\mathbf{s}_n) \end{pmatrix} \begin{pmatrix} \hat{f}_{i_1} \\ \vdots \\ \hat{f}_{i_{m_j}} \end{pmatrix} \tag{25}$$

and a method such as ordinary least squares must be chosen to obtain an estimate of the coefficients $\hat{f}_i$. This linear system can be written more succinctly as $\alpha_J = \Phi_J \hat{f}_J$. In practice, only finitely many eigenfunctions are of interest and the signals are bandlimited, but some eigenvalues will still correspond to many eigenfunctions. In Figure 1, least squares reconstruction is applied to the systems (25) for a two-dimensional case. The spatial domain is $\mathcal{D} = [0, \pi] \times [0, 1]$, and the boundary conditions are periodic. In the case of periodic boundary conditions, multiple sensors are always required because there are eigenvalues with a multiplicity greater than one, i.e., in one dimension:

$$\Phi_n(x) = \sin(n\pi x/L) \quad \text{and} \quad \Phi_{-n}(x) = \cos(n\pi x/L). \tag{26}$$

In Figure 1, a network of seven sensors is used to construct approximations of the system. In the first panel, the initial state $f(\mathbf{x})$ of the system, which is an off-center Gaussian bump, is displayed and the location of two additional sensors is shown in red. In the second and third panels of Figure 1, the time series of the two additional sensors are approximated. In both cases, the frequencies and amplitudes of the signal are recovered.

When the method of least squares is used, the error in the estimates $\tilde{f}_J$ of the coefficients $\hat{f}_J$ can be written in terms of the matrix Φ_J. Suppose the system is contaminated with Gaussian noise so that $\alpha_J = \Phi_J \hat{f}_J + \epsilon_J$ where $\epsilon_J \sim \mathcal{N}_n(0, \sigma^2 I_n)$ for some $\sigma > 0$. The best linear unbiased estimate $\tilde{f}_J$ of $\hat{f}_J$ and its covariance are

$$\tilde{f}_J = (\Phi_J^T \Phi_J)^{-1} \Phi_J^T \alpha_J \tag{27}$$

$$\sigma_J^2 = \mathbb{E}[(\tilde{f}_J - \hat{f}_J)(\tilde{f}_J - \hat{f}_J)^T | \Phi_J] = (\Phi_J^T \Phi_J)^{-1} \sigma^2. \tag{28}$$

To see how the error depends explicitly on the number of noisy sensors n in the network, the entries of the matrix Φ_J must be given a distribution. In the simplest case where $[\Phi_{i_1}(\mathbf{s}_k), \dots, \Phi_{i_{m_j}}(\mathbf{s}_k)]$ is normally distributed $\mathcal{N}(0, \Sigma)$, then

$$(\Phi_J^T \Phi_J)^{-1} \sim \mathcal{W}_{m_j}^{-1}(\Sigma^{-1}, n) \tag{29}$$

where $\mathcal{W}^{-1}$ denotes the inverse Wishart distribution. Alternatively, if the eigenfunctions are known, then a distribution is given to the sensor placement such as

$$\mathbf{s} \sim \text{Unif}(\mathcal{D}). \tag{30}$$

In Figure 2, the variance (28) is shown on a log–log plot for different distributions of $\Phi_i(\mathbf{s}_k)$. For the first three, the sensor is placed uniformly at random in the domain, and the eigenfunctions

$$\Phi_1(x) = \sin(4x) \quad \mathcal{D} = [-\pi, \pi] \tag{31}$$

$$\Phi_2(\mathbf{x}) = \sin(7x)\cos(12\pi y) \quad \mathcal{D} = [-\pi, \pi] \times [-1, 1] \tag{32}$$

$$\Phi_3(\mathbf{x}) = \cos(16x)\cos(23\pi y)\sin(28\pi z) \quad \mathcal{D} = [-\pi, \pi] \times [-1, 1] \times [-1, 1] \tag{33}$$

are used. The variance (28) is also plotted for the case where $\Phi_i(\mathbf{s}_k) \sim \mathcal{N}(0, 1)$. On this log–log plot, the slope of each line is approximately -1, meaning that the variance in each case decays as $\mathcal{O}(n^{-1})$, where n is the number of sensors. Figure 3 shows how the relative error depends on the number of sensors and the noise level σ of the sensors. In this figure, the error is the $L^2([0, 100])$ error in approximating the time series of an additional sensor.

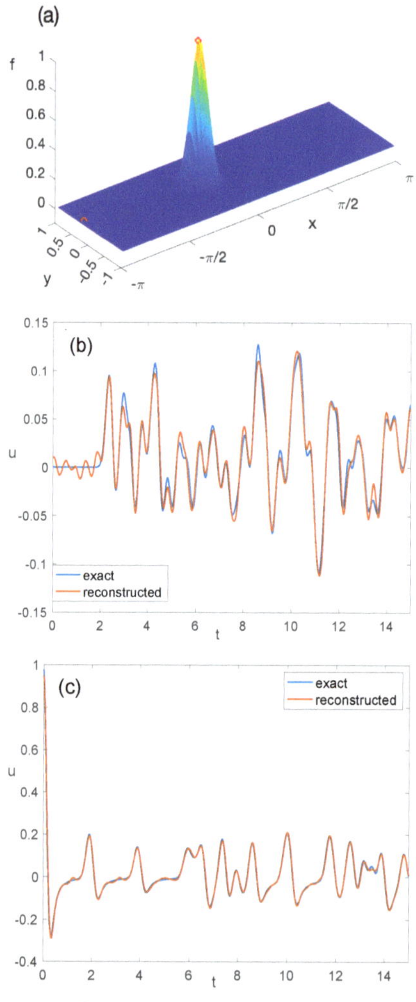

Figure 1. Time series data from a network of seven sensors are used to approximate the wave field u by solving the systems (25). (**a**) The locations of an eighth and ninth sensor in red plotted on the initial state of the system (15). (**b**,**c**) Comparisons of time series of the eighth and ninth sensors to the reconstructions of the time series at those locations using the network of seven sensors.

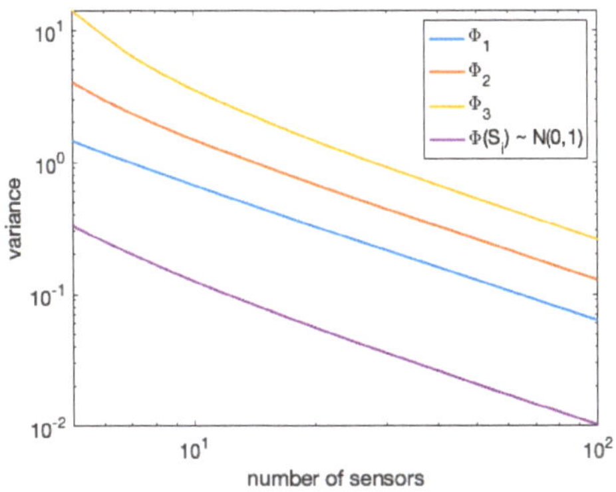

Figure 2. The variance of the least squares estimate (28) of a coefficient $\hat{f}_i$ plotted against the number of sensors used. Each line displays the variance for a different assumption for how the data $\Phi_i(S_k)$ are obtained.

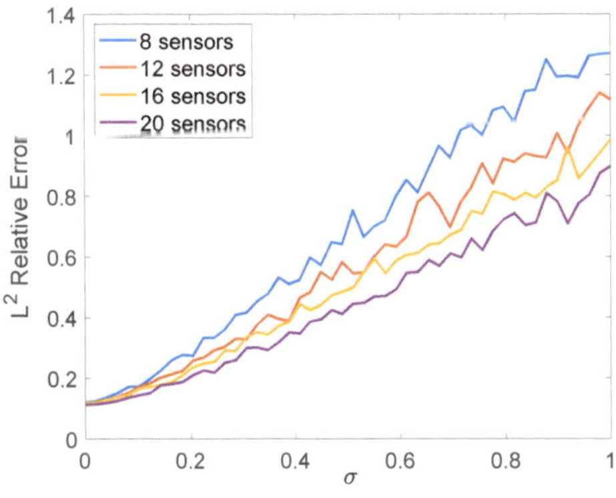

Figure 3. The $L^2([0,\tau])$ relative error in approximating the time series of a sensor plotted against the sensor noise level σ and the number of sensor time series used for approximation. The final time is $\tau = 100$.

4. 3D Stochastic Maxwell Vector Wave Equation and Optimal Observation Time

In this section, the signal processing of vectorial electromagnetic waves in the presence of a noisy current density $\mathbf{J}(\mathbf{x}, t)$ is discussed (see [32], chapter 3). In this context, the medium parameters, permittivity $\epsilon(\mathbf{x})$ and permeability $\mu(\mathbf{x})$, are general functions of space or are stratified and the current density is stochastic.

The electric field $\mathbf{E}$ satisfies the Maxwell vector wave equation:

$$\epsilon(\mathbf{x})\frac{\partial^2 \mathbf{E}}{\partial t^2} + \nabla \times \left([\mu(\mathbf{x})]^{-1}\nabla \times \mathbf{E}\right) = -\frac{\partial \mathbf{J}}{\partial t}. \tag{34}$$

The electric field solutions can be written in the form

$$\mathbf{E}(\mathbf{x}, t) = \sum_n A_n(t)\mathbf{\Phi}_n(\mathbf{x}) \tag{35}$$

where $\mathbf{\Phi}(\mathbf{x})$ are eigenfunctions of the weighted curl-squared operator:

$$\nabla \times \left([\mu(\mathbf{x})]^{-1}\nabla \times \mathbf{\Phi}\right) - \lambda\epsilon(\mathbf{x})\mathbf{\Phi} = 0. \tag{36}$$

Stochasticity is introduced to the problem via the current density

$$\frac{\partial \mathbf{J}}{\partial t} := \epsilon(\mathbf{x})\sqrt{Q}\frac{\partial \mathcal{W}}{\partial t} \tag{37}$$

where $\mathcal{W}$ is a cylindrical Wiener process and Q is a self-adjoint trace-class operator defined by $Q\mathbf{\Phi}_n = \sigma_n^2\mathbf{\Phi}_n$ (see [33,34] for more details). The Wiener process $\sqrt{Q}\mathcal{W}$ can be written in terms of independent one-dimensional Wiener processes $\mathcal{W}_n(t)$:

$$\sqrt{Q}\mathcal{W} = \sum_n \sigma_n \mathcal{W}_n(t)\mathbf{\Phi}_n(\mathbf{x}). \tag{38}$$

Using the orthonormality of the eigenfunctions $\mathbf{\Phi}_n(\mathbf{x})$, a sequence of ODEs for the $A_n(t)$ terms is obtained:

$$A_n''(t) + \lambda_n A_n(t) = \sigma_n \frac{\partial \mathcal{W}_n}{\partial t}(t). \tag{39}$$

The solution is

$$A_n(t) = \hat{f}_n \cos(\sqrt{\lambda_n} t) + \frac{\hat{g}_n}{\sqrt{\lambda_n}} \sin(\sqrt{\lambda_n} t) + \int_0^t \frac{\sigma_n}{\sqrt{\lambda_n}} \sin(\sqrt{\lambda_n}(t-r))\, d\mathcal{W}_n(r)$$

and its variance grows linearly with time:

$$\mathrm{VAR}[A_n(t)] = \frac{\sigma_n^2}{4\lambda_n^{3/2}}\left(2\sqrt{\lambda_n}t - \sin(2\sqrt{\lambda_n}t)\right). \tag{40}$$

In the framework of the Maxwell vector wave equation, as in the last section, the inner products (21) can be utilized to reconstruct the electric field from sensor time series data; in this context, however, the time series data are bombarded with noise from the stochastic current density. Because the variance of the solution grows with time as a result of noise and a minimum-noise reconstruction is desired, sensor time series observed for a shorter time interval $[0, \tau]$ can provide a better reconstruction of the noiseless electric field.

This can be made explicit by calculating the mean squared error (MSE) of each coefficient $\hat{f}_j$ as a function of the sensor observation time τ:

$$\mathrm{MSE}(\tau; j) = [\mathrm{BIAS}(\tau; j)]^2 + \mathrm{VAR}(\tau; j). \tag{41}$$

The error resulting from the noisy current density is then described in terms of the variance term VAR, and the deterministic error is described in terms of the squared-bias term $[\mathrm{BIAS}]^2$. The variance can be computed explicitly in simple cases and grows linearly with respect to the length of the time series τ, i.e., $\mathrm{VAR}(\tau; j) \sim \mathcal{O}(\tau)$. As a result, there is a bias–variance tradeoff. In Figure 4, the mean squared error of a coefficient $\hat{f}_j$ is displayed. The MSE initially decreases as a result of the decreasing squared bias. After reaching a minimum, the MSE begins to grow roughly linearly as a result of the variance term. The observation time τ that minimizes the MSE is the "optimal" observation time. In Figure 4, the dominant variance term $\mathrm{VAR}_{\mathrm{res}}$ which can be calculated analytically is displayed along with the squared bias and MSE. This optimal observation time τ can be utilized to estimate each of the coefficients $\hat{f}_j$ and reconstruct the electric field $\mathbf{E}$.

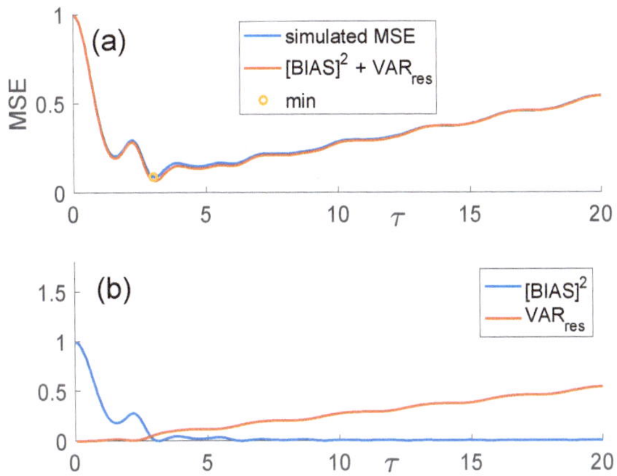

Figure 4. The decomposition of the relative mean squared error (41) of a coefficient $\hat{f}_j$ into squared-bias and variance terms. (**a**) The mean squared error (41) versus sensor observation time τ. (**b**) The squared-bias and variance terms versus τ.

5. Concluding Remarks

In this review, we discussed recent advancements in the modeling and processing of vectorial electromagnetic signals propagating in inhomogeneous, stratified media. With new developments in vector sensor technology and the development of cheaper satellites, corresponding advancements in signal processing need to be made. The Maxwell vector wave equation, the governing equation for vectorial EM waves derived directly from Maxwell's equations, incorporates relevant fundamental physics phenomena that are absent in scalar wave equation approaches. In applications involving EM wave propagation through media with sharp refractive index gradients, the modeling of vectorial effects is essential. Beyond the incorporation of vectorial effects, we reviewed new results on the importance of sensor networks for the accurate reconstruction of electromagnetic signals as well as modeling noisy environments via stochastic currents.

Author Contributions: Conceptualization, B.M.B., E.J.K. and A.M.; methodology, B.M.B. and A.M.; software, B.M.B.; formal analysis, B.M.B.; investigation, B.M.B.; writing—original draft preparation, B.M.B., E.J.K. and A.M.; visualization, B.M.B.; funding acquisition, A.M. All authors have read and agreed to the published version of the manuscript.

Funding: This work was partially supported by the Air Force Office of Scientific Research under award number FA9550-23-1-0177.

Institutional Review Board Statement: Not applicable.

Informed Consent Statement: Not applicable.

Data Availability Statement: The data used in this article are available upon request from the authors.

Conflicts of Interest: The authors declare no conflict of interest.

References

1. *Solar and Space Physics: A Science for a Technological Society*; National Academies Press: Washington, DC, USA, 2013.
2. Kak, A.; Akyildiz, I.F. Designing large-scale constellations for the Internet of Space Things with CubeSats. *IEEE Internet Things J.* **2020**, *8*, 1749–1768. [CrossRef]

3. Lind, F.D.; Erickson, P.J.; Hecht, M.; Knapp, M.; Crew, G.; Volz, R.; Swoboda, J.; Robey, F.; Silver, M.; Fenn, A.J.; et al. AERO & VISTA: Demonstrating HF radio interferometry with vector sensors. In Proceedings of the 33rd Annual AIAA/USU Conference on Small Satellites, Logan, UT, USA, 20 August 2019.
4. Knapp, M.; Robey, F.; Volz, R.; Lind, F.; Fenn, A.; Morris, A.; Silver, M.; Klein, S.; Seager, S. Vector antenna and maximum likelihood imaging for radio astronomy. In Proceedings of the 2016 IEEE Aerospace Conference, Big Sky, MT, USA, 5–12 March 2016; pp. 1–17.
5. Loi, S.T.; Murphy, T.; Cairns, I.H.; Menk, F.W.; Waters, C.L.; Erickson, P.J.; Trott, C.M.; Hurley-Walker, N.; Morgan, J.; Lenc, E.; et al. Real-time imaging of density ducts between the plasmasphere and ionosphere. *Geophys. Res. Lett.* **2015**, *42*, 3707–3714. [CrossRef]
6. Abarzhi, S.I. Review of theoretical modelling approaches of Rayleigh–Taylor instabilities and turbulent mixing. *Philos. Trans. R. Soc. A Math. Phys. Eng. Sci.* **2010**, *368*, 1809–1828. [CrossRef] [PubMed]
7. Zhou, Y.; Williams, R.J.; Ramaprabhu, P.; Groom, M.; Thornber, B.; Hillier, A.; Mostert, W.; Rollin, B.; Balachandar, S.; Powell, P.D.; et al. Rayleigh–Taylor and Richtmyer–Meshkov instabilities: A journey through scales. *Phys. D Nonlinear Phenom.* **2021**, *423*, 132838. [CrossRef]
8. Mahalov, A. Multiscale modeling and nested simulations of three-dimensional ionospheric plasmas: Rayleigh–Taylor turbulence and nonequilibrium layer dynamics at fine scales. *Phys. Scr.* **2014**, *89*, 098001. [CrossRef]
9. Zhou, Y. Rayleigh–Taylor and Richtmyer–Meshkov instability induced flow, turbulence, and mixing. I. *Phys. Rep.* **2017**, *720–722*, 1–136.
10. Tatarski, V.I. *Wave Propagation in a Turbulent Medium*; McGraw-Hill: Columbus, OH, USA, 1961.
11. Immel, T.J.; Harding, B.J.; Heelis, R.A.; Maute, A.; Forbes, J.M.; England, S.L.; Mende, S.B.; Englert, C.R.; Stoneback, R.A.; Marr, K.; et al. Regulation of ionospheric plasma velocities by thermospheric winds. *Nat. Geosci.* **2021**, *14*, 893–898. [CrossRef] [PubMed]
12. Sahba, S.; Sashidhar, D.; Wilcox, C.C.; McDaniel, A.; Brunton, S.L.; Kutz, J.N. Dynamic mode decomposition for aero-optic wavefront characterization. *Opt. Eng.* **2022**, *61*, 013105. [CrossRef]
13. Tran, L. Strong Winds Power Electric Fields in the Upper Atmosphere, NASA's ICON Finds. 2021. Available online: https://www.nasa.gov/feature/goddard/2021/strong-winds-power-electric-fields-in-upper-atmosphere-icon (accessed on 4 August 2023).
14. Herring, J.; Kimura, Y. Some issues and problems of stably stratified turbulence. *Phys. Scr.* **2013**, *2013*, 014031. [CrossRef]
15. Kimura, Y.; Herring, J. Energy spectra of stably stratified turbulence. *J. Fluid Mech.* **2012**, *698*, 19–50. [CrossRef]
16. Herring, J. A Brief History of the Geophysical Turbulence Program at NCAR. In *IUTAM Symposium on Developments in Geophysical Turbulence*; Springer: Berlin/Heidelberg, Germany, 2000; pp. 1–4.
17. Charbonneau, P.; Herring, J.; Milliff, R. Preface. *Theoret. Comput. Fluid Dyn.* **1998**, *11*, 137–138.
18. Babin, A.; Mahalov, A.; Nicolaenko, B. On nonlinear baroclinic waves and adjustment of pancake dynamics. *Theor. Comput. Fluid Dyn.* **1998**, *11*, 215–235. [CrossRef]
19. Mahalov, A. Turbulence and waves in the upper troposphere and lower stratosphere. In *Aviation Turbulence: Processes, Detection, Prediction*; Sharman, R., Lane, T., Eds.; Springer: Berlin/Heidelberg, Germany, 2016; pp. 407–424.
20. Fernando, H.J.S. Turbulent mixing in stratified fluids. *Annu. Rev. Fluid Mech.* **1991**, *23*, 455–493. [CrossRef]
21. Kelley, M.C. *The Earth's Ionosphere: Plasma Physics and Electrodynamics*; Academic Press: Burlington, MA, USA, 2009.
22. Cox, D. Depolarization of radio waves by atmospheric hydrometeors in earth-space paths: A review. *Radio Sci.* **1981**, *16*, 781–812. [CrossRef]
23. Shaffer, S.R.; Mahalov, A. Permittivity gradient induced depolarization: Electromagnetic propagation with the Maxwell vector wave equation. *IEEE Trans. Antennas Propag.* **2020**, *69*, 1553–1559. [CrossRef]
24. Wang, Q.; Alappattu, D.P.; Billingsley, S.; Blomquist, B.; Burkholder, R.J.; Christman, A.J.; Creegan, E.D.; De Paolo, T.; Eleuterio, D.P.; Fernando, H.J.S.; et al. CASPER: Coupled air–sea processes and electromagnetic ducting research. *Bull. Am. Meteorol. Soc.* **2018**, *99*, 1449–1471. [CrossRef]
25. McDaniel, A.; Mahalov, A. Lensing effects in a random inhomogeneous medium. *Opt. Express* **2017**, *25*, 28157–28166. [CrossRef]
26. McDaniel, A.; Mahalov, A. Coupling of paraxial and white-noise approximations of the Helmholtz equation in randomly layered media. *Phys. D Nonlinear Phenom.* **2020**, *409*, 132491. [CrossRef]
27. McDaniel, A.; Mahalov, A. Stochastic mirage phenomenon in a random medium. *Opt. Lett.* **2017**, *42*, 2002–2005. [CrossRef]
28. Mahalov, A.; McDaniel, A. Long-range propagation through inhomogeneous turbulent atmosphere: Analysis beyond phase screens. *Phys. Scr.* **2019**, *94*, 034003. [CrossRef]
29. Barclay, B.M.; Kostelich, E.J.; Mahalov, A. Sensor Placement Sensitivity and Robust Reconstruction of Wave Dynamics from Multiple Sensors. *SIAM J. Appl. Dyn. Syst.* **2022**, *21*, 2297–2313. [CrossRef]
30. Erichson, N.B.; Mathelin, L.; Kutz, J.N.; Brunton, S.L. Randomized dynamic mode decomposition. *SIAM J. Appl. Dyn. Syst.* **2019**, *18*, 1867–1891. [CrossRef]
31. Hashemi, A.; Schaeffer, H.; Shi, R.; Topcu, U.; Tran, G.; Ward, R. Generalization bounds for sparse random feature expansions. *Appl. Comput. Harmon. Anal.* **2023**, *62*, 310–330. [CrossRef]
32. Barclay, B. Stochastic Maxwell's Equations: Robust Reconstruction of Wave Dynamics from Sensor Data and Optimal Observation Time. Ph.D. Thesis, Arizona State University, Tempe, AZ, USA, 2023.

33. Da Prato, G.; Zabczyk, J. *Stochastic Equations in Infinite Dimensions*, 2nd ed.; Cambridge University Press: Cambridge, UK, 2014.
34. Flandoli, F.; Mahalov, A. Stochastic three-dimensional rotating Navier–Stokes equations: Averaging, convergence and regularity. *Arch. Ration. Mech. Anal.* **2012**, *205*, 195–237. [CrossRef]

atmosphere

MDPI

Article

Mean Flow from Phase Averages in the 2D Boussinesq Equations

Beth A. Wingate [1,*], Juliane Rosemeier [1] and Terry Haut [2]

[1] Department of Mathematics and Statistics, University of Exeter, Exeter EX4 4QF, UK; j.rosemeier@exeter.ac.uk
[2] Lawrence Livermore National Laboratory, Livermore, CA 94550, USA; haut3@llnl.gov
* Correspondence: b.wingate@exeter.ac.uk

Abstract: The atmosphere and ocean are described by highly oscillatory PDEs that challenge both our understanding of their dynamics and their numerical approximation. This paper presents a preliminary numerical study of one type of phase averaging applied to mean flows in the 2D Boussinesq equations that also has application to numerical methods. The phase averaging technique, well-known in dynamical systems theory, relies on a mapping using the exponential operator, and then an averaging over the phase. The exponential operator has connections to the Craya–Herring basis pioneered by Jack Herring to study the fluid dynamics of oscillatory, nonlinear fluid dynamics. In this paper, we perform numerical experiments to study the effect of this averaging technique on the time evolution of the solution. We explore its potential as a definition for mean flows. We also show that, as expected from theory, the phase-averaging method can reduce the magnitude of the time rate of change in the PDEs, making them potentially suitable for time stepping methods.

Keywords: mean flow formulation; 2D Boussinesq equations; exponential of linear operator; averaging

Citation: Wingate, B.A.; Rosemeier, J.; Haut, T. Mean Flow from Phase Averages in the 2D Boussinesq Equations. *Atmosphere* **2023**, *14*, 1523. https://doi.org/10.3390/atmos14101523

Academic Editors: Boris Galperin, Annick Pouquet and Peter Sullivan

Received: 20 August 2023
Revised: 13 September 2023
Accepted: 20 September 2023
Published: 30 September 2023

1. Introduction

The mean flow in the present study is a time-average of a system of equations that has been mapped using the exponential operator. The idea is that oscillatory dynamics, once they are mapped into a new 'moving frame', can be composed of a slowly varying mean and fast departures from that mean. This idea has a long history in studying the time evolution of the dynamics and in the numerical approximation of oscillatory ODEs and PDEs, which we discuss in the next paragraph. The technique explored in this paper has also been proposed as a way of time stepping that uses a serial prediction step and a time-parallel correction step. The time stepping technique shows promise but has only been applied to ODEs and 1D PDEs. In this paper, we use the mapping and averaging technique as a diagnostic in numerical experiments to (1) study the impact of this mapping and averaging on the time-evolution of the 2D Boussinesq equations, (2) to provide an initial assessment of whether it might be a useful definition of a mean flow, and (3) to assess its potential usefulness in predictor–corrector-type numerical methods.

The technique of mapping and averaging, now a classic method in dynamical systems analysis, is discussed and developed in [1], where they also give a brief history of the averaging method since the lectures of Jacobi, which date to 1842. One of their key points is that the oscillatory system of equations needs to be mapped into their 'standard form'. This mapping is equivalent to our use of the exponential map in this paper. Further, the eigen-basis used for the exponential map in this paper has a relationship to the Craya–Herring basis pioneered by Jack Herring in [2]. Furthermore, in several papers, including [3,4], Herring and his collaborators used this basis as a means to understand the dynamics of waves relative to vortical aspects of the flow. In this paper we explore decompositions of this nature, but for constructing phase averages of the nonlinear dynamics. Ref. [1] gives a demonstration (page 28) of why straightforward averaging without mapping, which they call naive averaging, gives approximation errors. The technique in this paper also plays a role in the theory of fast singular limits [5–10]. The mapping concept without averaging has also been useful in the study of geophysical fluid dynamics, e.g., [11–13], to study nonlinear resonances.

Another way of understanding this technique, for equations in the form of (1), is to think of it as nonlinear phase averaging because, once the PDE has been mapped using the exponential operator (2), all the phase information arises in the nonlinear term (see Equation (3)). Recent papers that explicitly discuss phase averaging in geophysical fluid dynamics include [14–16].

From a PDEs point of view, the averaging technique in this paper is related to the concept of 'cancellation of oscillations' [7,10] to treat hyperbolic PDEs. The authors of [8] refer to this technique and treat geophysical applications, that is, the Rotating Boussinesq Equations and the Rotating Shallow Water Equations. In this context, an exponential operator and infinite time averaging leads to limiting equations for the 'slow'dynamics. This type of averaging was further developed in [17,18] for geophysical applications. In [18,19], the analysis was extended to the case when more than one fast time scale is present.

The present study, like those mentioned in the previous paragraph, uses an exponential operator as a mapping. We emphasize that using the exponential operator mitigates the linear oscillations in the problem, but the oscillations are not eliminated—there are oscillations that develop from the nonlinear terms. Therefore, as an additional ingredient, we evaluate an integral with the scaled kernel function $\mathcal{K}$, that is, we compute a convolution. It is this technique that potentially can describe general time-mean flows and that can potentially be used in numerical time stepping methods. It differs from the technique used in the study of fast singular limits in that the averaging is over a finite window and uses a smooth integration kernel over just a few oscillations, rather than an infinite time average.

Numerically, the classical method of averaging is described in [1]. In geophysical fluid dynamics applications, a pioneering paper [20] directly applies the method of [5,6] to the shallow water equations; they found that the method worked well when the oscillations were fast, but became less accurate as the frequency of the oscillations decreased. The phase-averaged mean flows studied in this paper use a technique proposed by [21–23] as a coarse time stepping algorithm for a numerical time-stepping method. This technique has also been used for time-stepping on the sphere [16].

The oscillatory equations we study in this paper are the 2D Boussinesq equations given in (6)–(8); before we discuss those, we first describe the phase-averaging method we use in this paper. We therefore study the more general form of oscillatory equations:

$$\frac{d\mathbf{u}}{dt} + \frac{1}{\varepsilon}\mathcal{L}\mathbf{u} = \mathcal{N}(\mathbf{u}). \tag{1}$$

The linear operator $\mathcal{L}$ has purely imaginary eigenvalues and makes the problems stiff. Additionally, the non-linearity $\mathcal{N}$ is bi-linear in fluid applications. One possibility is to eliminate the linear term in the time evolution problem (1) by using the transformation

$$\mathbf{w}(t) = \exp\left(\frac{\mathcal{L}}{\varepsilon}t\right)\mathbf{u}(t). \tag{2}$$

This leads to the following problem:

$$\frac{d\mathbf{w}}{dt} = \exp\left(\frac{\mathcal{L}}{\varepsilon}t\right)\mathcal{N}\left(\exp\left(-\frac{\mathcal{L}}{\varepsilon}t\right)\mathbf{w}\right). \tag{3}$$

Applying averaging to further mitigate the oscillatory stiffness leads to a new problem:

$$\frac{d\tilde{\mathbf{w}}}{dt} = \frac{1}{\eta}\int_{-\eta/2}^{\eta/2} \mathcal{K}\left(\frac{s}{\eta}\right)\exp\left(\frac{\mathcal{L}}{\epsilon}(s+t)\right)\mathcal{N}\left(\exp\left(-\frac{\mathcal{L}}{\epsilon}(s+t)\right)\tilde{\mathbf{w}}(t)\right)ds. \tag{4}$$

Solving Equation (4) can be very useful for numerical time-stepping, such as in the Parareal method, because the time rate-of-change of the unknowns becomes less oscillatory. This path was extensively studied in [21–23].

From a fluid dynamics point of view, this type of equations may be interpreted as the mean flow. In the present study, we want to step in a new, less-explored direction by applying the transformation and averaging to a given solution $\mathbf{u}(t)$ of problem (1) as a diagnostic tool and investigating if such formulations provide a suitable formulation for mean flows. In this case, we evaluate

$$\overline{\mathbf{w}}(t) = \frac{1}{\eta}\int_{-\eta/2}^{\eta/2} \mathcal{K}\left(\frac{s}{\eta}\right)\mathbf{w}(t+s)ds, \tag{5}$$

where $\mathbf{w}$ is given by Relation (2). The difference between Equations (4) and (5) is that, in the first equation, the averaging procedure is applied to the right-hand side of a differential equation, whereas, in the second equation, it is applied to a given solution of that differential equation. Thus, in the first case, averaging is applied to make a time evolution problem more well-behaved for numerical computations; in the second case, it serves to extract information from data. The averaging is discussed in more detail in Section 2.3.

Already, in [8], it was suggested to use exponential mapping for the formulation of the mean flow for geophysical fluid dynamics applications. However, the approach described there differs from the ideas in the present expositions in the following ways. First, the exponential in this exposition is applied to data representing a solution to the Boussinesq equations. We do not use it to derive analytical expressions for leading order solutions when the method of multiple scales is applied. Second, for the formulation of the mean flow, the authors of [8] proposed to consider only modes which satisfy the condition for resonant triads. In the present exposition, this constraint is relaxed and not only resonant modes, but also other slow modes are admitted for the formulation of the mean flow.

In the present study, the focus is the 2D Boussinesq equations. In particular, we study the case of an initial buoyancy distribution that decays with time:

The Boussinesq equations are

$$u_t + uu_x + wu_z + \frac{\partial}{\partial x}\left(\Delta^{-1}\left(-\nabla\cdot(\boldsymbol{u}\cdot\nabla\boldsymbol{u}) - N\frac{\partial\rho'}{\partial z}\right)\right) = \nu\Delta u, \tag{6}$$

$$w_t + uw_x + ww_z + \frac{\partial}{\partial z}\left(\Delta^{-1}\left(-\nabla\cdot(\boldsymbol{u}\cdot\nabla\boldsymbol{u}) - N\frac{\partial\rho'}{\partial z}\right)\right) + N\,\rho' = \nu\Delta w, \tag{7}$$

$$\frac{\partial\rho'}{\partial t} + u\rho'_x + w\rho'_z - N\,w = \kappa\Delta\rho', \tag{8}$$

where the the Brunt–Väisälä frequency is $N = \sqrt{\frac{bg}{\rho_0}}$. We assume periodic boundary conditions with the vector of unknowns $\mathbf{u} = (\mathbf{v}, \rho')$. The divergence constraint is given by $div\,\mathbf{v} = 0$, where $\mathbf{v} = (u, w)^T$. The pressure obeys a Poisson equation (see [24]). We study the averaging technique for oscillating and decaying dynamics.

In [8], the following identity was derived for a leading order solution of the rotating Boussinesq equations:

$$\mathbf{u}^0 = e^{-\tau\mathcal{L}}\overline{\mathbf{u}}, \tag{9}$$

where $\overline{u}$ satisfies an averaged equation, which contains the non-linearity and is given by

$$\frac{d\mathbf{u}}{dt} = -\lim_{\tau\to\infty}\frac{1}{\tau}\int_0^{\tau} e^{s\mathcal{L}}\mathcal{B}(e^{s\mathcal{L}}\mathbf{u}(t), e^{s\mathcal{L}}\mathbf{u}(t))ds. \tag{10}$$

Note that we take $\tau \to \infty$ in Equation (10). The integrand in Equation (10) can be represented using a Fourier decomposition in space. Applying the Riemann–Lebesque lemma, one sees that only the modes which form direct resonances are retained by the intregration process. However, we relax this constraint and allow some slow oscillations too. This provides a new tool to define mean flows, where we allow selected oscillations to be in the formulation. The new averaging procedure, which differs from the one given by Equation (10), is of central significance to decide which oscillations remain in the solution and is explained in more detail in Section 2.3.

In the remainder of the article, we describe the details of the diagnostic we use to compute the time-mean flow. In the results section, we (1) present preliminary numerical results that show the behavior of the mapping and averaging on the time-evolution of the buoyancy, (2) characterize the mean flow that results from the mapping and averaging, including exploring the effect of changing the averaging window, and (3) compute the time rate-of-change of the buoyancy, to assess whether it can mitigate the oscillatory stiffness for use in numerical methods.

2. Methods

We compute a temporal mean flow for solutions to the Boussinesq Equations. For that, the following procedure is pursued. First, a solution to the Boussinesq Equations, denoted as $\mathbf{u}$, is computed. Particulars can be found in Section 2.1. Then, the exponential of the linear operator is applied to the solution and we obtain the mapped solution $\mathbf{w}$. The computation of the exponential is described in Section 2.2 and Appendix A. In the next step, the mapped solution is averaged according to Equation (5). Thus, we compute $\overline{\mathbf{w}}$. The mapped solution $\mathbf{w}$ can be represented in terms of spatial Fourier modes. When applying the averaging from Equation (5), we can choose, which modes shall be averaged or remain in the solution, by adjusting η, denoted as the averaging window. If the period of a mode is larger than η, the mode will still be present in the solution—see Section 2.3 for details. In the last step, we apply the inverse of the transformation (2).

Performing the previously described procedure, we create several data sets:

1. The unchanged data from the numerical solution of the 2D Boussinesq equations $\mathbf{u}$;
2. The data after applying the transformation, yielding the mapped solution $\mathbf{w}$;
3. The data after applying the transformation and the averaging procedure, which give the averaged solution $\overline{\mathbf{w}}$;
4. The data after applying the inverse transformation $\overline{\mathbf{u}}$.

The above items are illustrated in Figure 1. Comparison of the data sets helps to assess how well the proposed procedure is suited to describe mean flows and potentially to be used for time-stepping methods.

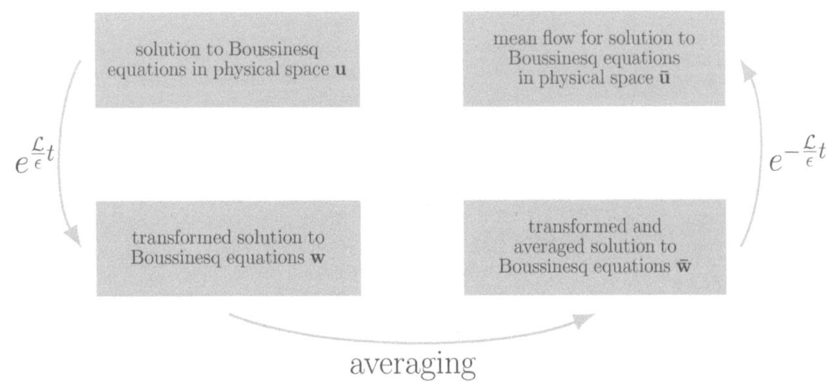

Figure 1. The figure illustrates how the data sets for the mean flow representation are created. The gray boxes represent the data sets. The arrows show the operations which are applied to the data sets.

2.1. Numerical Computation of Solutions to the Boussinesq Equations

We compute numerical solutions using the Dedalus software [25]. We use a Fourier discretization in space in a 2π periodic 2D domain and 3rd-order 4-stage Runge–Kutta Method (RK433) time-stepper. For all the simulations, the time step is $\Delta t = 0.001$, the spatial grid is (N_x, N_z) with $Nx = Nz = 128$, the viscosity is $\nu = 0.0001$, and the dissipation for buoyancy is $\kappa = 0.0001$. For the initial condition, $u(0, \mathbf{x}) = w(0, \mathbf{x}) = 0$, while the initial buoyancy is

$$\rho'(0, \mathbf{x}) = \sqrt{\epsilon_f}\sin(3z)\sin(3x), \tag{11}$$

with $\epsilon_f = 25$. The energy oscillates between kinetic and potential energy in time while it decays. The higher the value of N, the more frequent the oscillations. Figure 2 shows an example of the dynamics in decay as it oscillates between kinetic and potential energy. At some time after the simulation begins, higher frequency oscillations are excited, providing a challenge for studying oscillatory mean flows.

In this study, we use 3 different values of N: $N = 1, 10$, and 20. We define the Froude number to be a function of the character of the initial condition, using similar ideas to [12], who studied the forced case

$$Fr = 2\frac{(\epsilon_f k_f^2)^{\frac{1}{3}}}{N}, \tag{12}$$

which gives us equivalent Froude numbers of $Fr = 10.6, 1.06$, and 0.53.

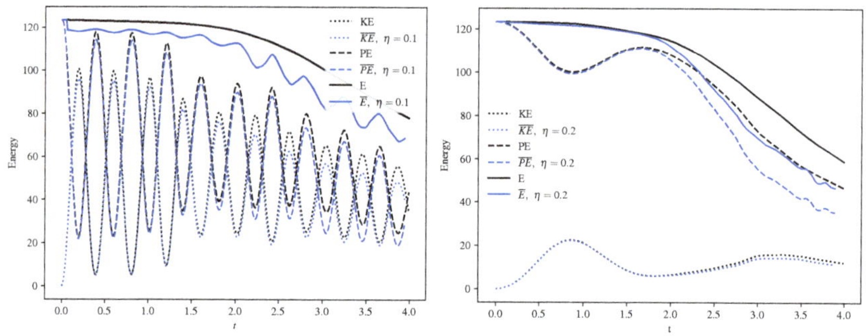

Figure 2. The total, kinetic, and potential energy (black lines) compared to their mean values (blue lines). The panel on the left shows the case for $N = 10$ and η = 0 .1. The panel on the right shows the case for less frequent oscillations when $N = 1$ and η = 0.2. Solid lines are the total energy, dotted lines are the kinetic energy, and dashed lines are the potential energy. The solutions oscillate between kinetic and potential energy as they decay. The principal effect of the averaging window on the mapped and averaged solution is to track the frequency of the energetic exchanges between kinetic and potential energy and to reduce the magnitude of the oscillations. The case $N = 10$ on the left suggests that the phase-averaged mean flow shows an oscillatory total energy.

2.2. Fourier Analysis and the Computation of the Exponential of $\mathcal{L}$ with the Divergence Constraint

For the implementation of the exponential of the linear operator, we apply a Fourier expansion in space to the solutions of the Boussinesq equations, i.e., we have

$$\mathbf{u}(t, \mathbf{x}) = \sum_{k=-N}^{N} \widehat{\mathbf{u}}(t)e^{i\mathbf{k}\cdot\mathbf{x}}. \tag{13}$$

Each mode $e^{i\mathbf{k}\cdot\mathbf{x}}$ is applied to the linear operator, which is given by

$$L\mathbf{u} = \begin{pmatrix} -\frac{\partial}{\partial x}\left(\Delta^{-1}N\frac{\partial\rho'}{\partial z}\right) \\ -\frac{\partial}{\partial z}\left(\Delta^{-1}N\rho'\right) + N\rho' \\ -Nw. \end{pmatrix} \tag{14}$$

Consequently, a 3×3 matrix is obtained for each mode. Then, the matrix exponentials for the different modes are computed. To compute the exponential of the linear operator, it is sufficient to know the exponential for each mode $e^{i\mathbf{k}\cdot\mathbf{x}}$. Due to linearity, we have

$$e^{\frac{\mathcal{L}}{\epsilon}t}u(t,\mathbf{x}) = \sum_{k=-N}^{N} e^{\frac{\mathcal{L}}{\epsilon}t}e^{i\mathbf{k}\cdot\mathbf{x}}\hat{\mathbf{u}}(t) = \sum_{k=-N}^{N} B_k\hat{\mathbf{u}}(t)e^{i\mathbf{k}\cdot\mathbf{x}}, \tag{15}$$

where $B_k e^{i\mathbf{k}\cdot\mathbf{x}} = e^{\frac{\mathcal{L}}{\epsilon}t}e^{i\mathbf{k}\cdot\mathbf{x}}$. Applying the exponential, we obtain $\mathbf{w}$, given by (2).

Here, we briefly describe the computation of the matrix exponential. Further details of the computations can be found in Appendix A. The horizontal wave number is denoted as k_1. With k_3, we refer to the vertical wave number. First, the eigenvalues are computed. Then, the corresponding eigenvectors are identified. With the eigenvectors, we know the transformation matrix T_k and can also compute its inverse T_k^{-1}. In the computations, we distinguish between four different cases (first: $k_1 = 0, k_3 = 0$, second: $k_1 = 0, k_3 \neq 0$; third: $k_1 \neq 0, k_3 = 0$; fourth: $k_1 \neq 0, k_3 \neq 0$). The matrices are diagonalizable, except for the case $k_1 = 0, k_3 \neq 0$. For that case, we compute the Jordan normal form. To compute the exponential of the matrix B_k, where B_k comes from the spatial discretization of the linear operator and depends on the wave numbers, we use the identity $\exp(B_k) = T_k \exp(M_k) T_k^{-1}$. In the relation, the matrix M_k is the diagonal matrix or the matrix in Jordan normal form with the computed eigenvalues. Finally, we enforce the divergence constraint.

2.3. Averaging with a Convolution Integral

A transformed solution of the Boussinesq equations $\mathbf{w}$ is convolved by applying a scaled filter function $\mathcal{K}$:

$$\mathcal{K}\left(\frac{s}{\eta}\right) = \frac{1}{K_0}\exp\left(\frac{1}{(s/\eta - 1/2)(s/\eta + 1/2)}\right), \tag{16}$$

where the constant K_0 is chosen so that integrating the kernal function along the real line gives 1. We scale the filter function by choosing an averaging window η. The idea behind the choice of the averaging window η is that it must mitigate the fast oscillations while leaving the coarser behavior of the solution unaffected. The bigger the averaging window, the more oscillations are mitigated. The smaller the averaging window, the more oscillations are resolved. Temporal averaging was also investigated in other contexts, like ordinary differential equations [1], heterogeneous multiscale methods [26,27], and PDE analysis [8].

To further illustrate the averaging, let us suppose that a slow function is superimposed by a fast periodic function with zero mean. When we compute the integral of the fast function over an interval of length η, assuming that η is as large as a few times the period of the fast periodic function, the positive and negative contributions cancel each other. However, the integrand is weighted by the scaled kernel function $\mathcal{K}$ with compact support, which decays fast close to the boundary of the compact support. In addition, we do not assume that the length of the exact period is known. Thus, in the general case, the oscillations are not exactly canceled, but rather they are mitigated, as can be seen in Figure 3. Technical details of the averaging can be found in [27] in Lemma 2.2.

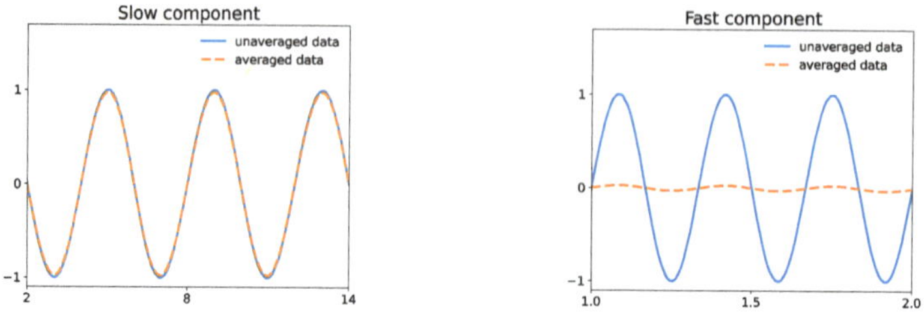

Figure 3. This figure demonstrates the effect of the averaging window on oscillatory functions. In the left panel, the averaging window η is smaller than the period of the oscillations. The data are barely altered. In the right panel, the averaging window η is larger than the period of the oscillations. Therefore, the oscillations are damped and the mean value is between the peaks and valleys of the signal.

3. Results and Discussion

We begin with a description of the time evolution of the dynamics (unaveraged) for the specific case studied in this work. In all cases, the initial condition is given by (11) and the velocity components are zero initially.

Figure 2 shows the evolution of total, potential, and kinetic energy for $N = 10$ on the left and $N = 1$ on the right. The data are shown in black lines for the unaveraged case with total energy: solid, kinetic energy: dotted, and potential energy: dashed. These graphs show that the dynamics decay as it evolves, and that the case with the larger Brunt–Väisälä frequency has more frequent energy exchanges between potential and kinetic energy compared to the $N = 1$ case.

These more frequent energy exchanges in potential and kinetic energy can also be seen by examining the time history of $\mathbf{u}[3](x_p, z_p, t) = \rho'(x_p, z_p, t)$ at a single, randomly chosen point $(x_p, z_p) = (\frac{2\pi\, 60}{128}, \frac{2\pi\, 60}{128})$. Other choices of the location of the point show similar results. This will be explored in the next section where we examine vertical slices, which show the time evolution for all the values of z at a fixed x_p. Compare the solid black line in the upper right panel of Figure 4 for $N = 10$ to the same style of line in Figure 4 for the $N = 1$ data. For the $N = 10$ case, the solution has more frequent oscillations than the $N = 1$ case. For both values of N, the solution breaks up into higher oscillations as the solution decays. This feature is used to illustrate the effect of the averaging in the following sections.

Finally, before we go on to study the effect of the averaging, we look at the time evolution of the single point data in the mapped domain $\mathbf{w}[3](x_p, z_p, t)$. These data are computed by applying the matrix exponential, as in (2), to the vector $\mathbf{u}$. This has the effect of showing what the time evolution of the solution looks like in the moving frame and allows us to observe the effect of the 'phase scrambling' in the nonlinear term, as in (3). The evolution in the moving frame is shown in the solid black line in the left panels of Figure 4. For $N = 20$ and $N = 10$, there are more frequent oscillations in the moving frame than in the ordinary frame shown in the right panel. In contrast, for the $N = 1$ case, the difference between the solution in the moving frame $\mathbf{w}$ and the ordinary frame $\mathbf{u}$ is not as different as the other cases, which is because the moving frame is slow for $N = 1$.

In the next three subsections, we examine the effect of the mapping and averaging on these oscillatory, decaying solutions.

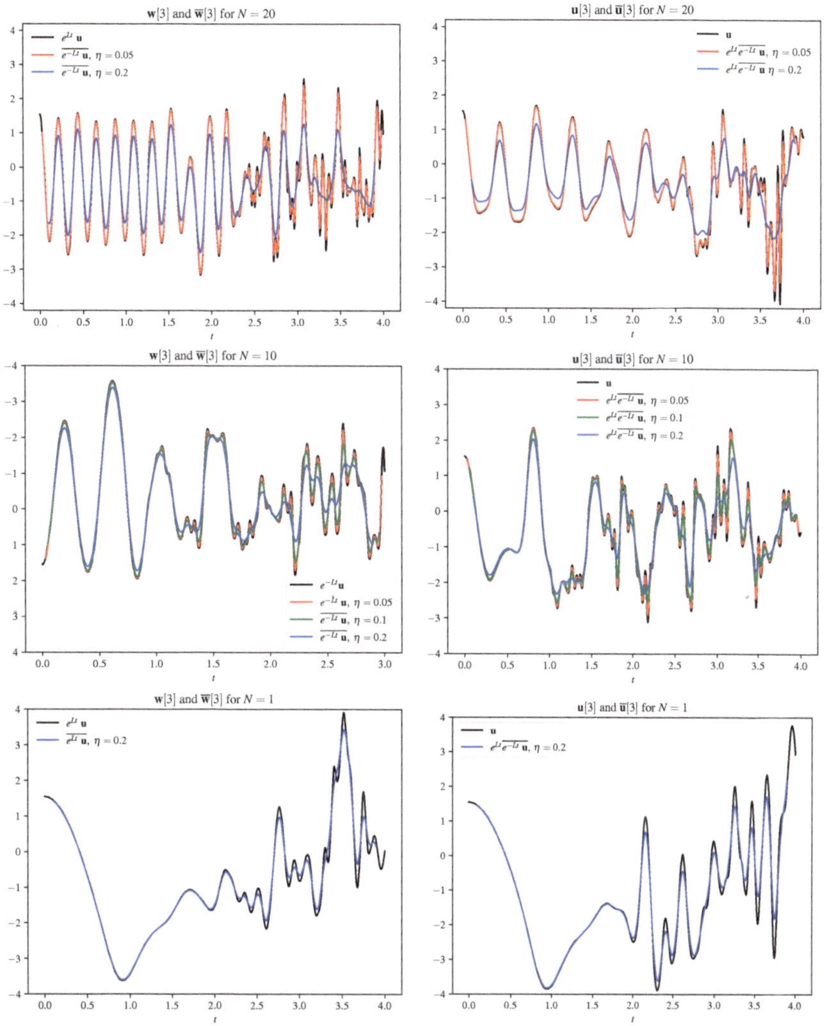

Figure 4. Time series at a single point for the 3rd component of each vector field. The top row is for $N = 20$, middle for $N = 10$, and bottom for $N = 1$. Panels on the left show the time evolution of the data in the moving frame, while, on the right, they depict the evolution in the ordinary frame. Within each graph, colored lines showing the effect of changing the averaging window. As N decreases, there is less difference between the dynamics in the moving frame and the ordinary frame, an effect of the moving frame oscillating at a slower rate over the time scale of the simulation. There are more oscillations in the moving frame, where the averaging takes place, than in the ordinary frame.

3.1. Impact of Mapping and Averaging

We now describe the impact of the mapping and averaging technique on the time evolution of the ρ' dynamics by examining time series for three different values of N in the context of the four data sets $(\mathbf{u}, \mathbf{w}, \overline{\mathbf{w}}, \overline{\mathbf{u}})$ described in the second paragraph of Section 2. Because we focus on ρ', when $\mathbf{u}, \mathbf{w}, \overline{\mathbf{w}}, \overline{\mathbf{u}}$ appear in a figure, these are the third component of the vector field.

Figure 4 presents the time series at a single point, with each row of the figure corresponding to a different value of N. In all three rows, the left panel shows the time evolution in the moving frame $\mathbf{w}$ and its average $\overline{\mathbf{w}}$. In contrast, the right panel shows the ordinary variable $\mathbf{u}$ and the average in the moving frame mapped back to the ordinary frame $\overline{\mathbf{u}}$. The significance of the left panels is that they show the result of how the oscillations are being 'scrambled' and evolved by the nonlinear term in (2) as the solution undergoes decay from its initial condition. A first observation is that, for the two higher values of N studied, the oscillations in $\mathbf{w}$, the moving frame variables, are faster than those observed in $\mathbf{u}$. For some parameter regimes in 2D Boussinesq dynamics, one might expect the dynamics of $\mathbf{w}$ to be of lower frequency than $\mathbf{u}$, not higher, so further studies of the case with forcing, where low frequency buoyancy layers are formed, could be interesting (see [12]).

A second observation is that, for the higher values of N at later times, when the faster oscillations appear, the averaging from the nonlinear phase scrambler in the moving frame, $\overline{\mathbf{w}}$, tracks the mean value, as expected. This is in contrast to the panel on the right where, for example, for $N = 20$ in the range $3.5 < t < 4$, when the average is mapped back to the ordinary frame, the average does not track the mean value. This is also as expected because, in this domain, the average is modulated by the oscillations. As mentioned in the introduction, ref. [1] gives an example of why straightforward time averages, in the unmapped equations, do not give accurate results. The comparison that we just described, where the modulated mean, $\overline{\mathbf{u}}$, does not necessarily track the mean of $\mathbf{u}$, is a characteristic feature of averages taken in the modulated domain. We discuss this more in the next section where we consider the mean-flow possibilities.

Finally, we make a few observations about the effect of the averaging window η on the mean of the time series. In Figure 4, on the left panels, the effect of varying the averaging window is shown. The longer the averaging window, the more the amplitudes are damped. At early times, for $N = 20$, the main effect is to reduce the magnitude of the oscillations, but, at later times, the red line, for which a small averaging window $\eta = 0.05$ was used, shows little difference from the unaveraged case. However, when the largest averaging window is applied, for $\eta = 0.2$, the blue line for $3 < t < 4$ shows that the mapping and averaging tracks the mean of the oscillations. This similar behavior can be seen in the unmapped domain on the right, where, as pointed out earlier, the blue line, with the larger averaging window, is slower, but does not follow the mean of the oscillations. For the middle panel, when $N = 10$, we show three different averaging window lengths ($\eta = 0.05$, 0.1, and 0.2). For the largest averaging window (blue line), the mean tracks the mean of the oscillations. This is true in the right panel as well. Finally, for the lowest value of N, $N = 1$, the choice of averaging window has a similar effect, except that, at early times, it does not have that much of an effect, while, at later times, when the faster oscillations begin, it appears to track through the mean value.

Figures 5 and 6 show the time evolution of four data sets for every point in the z plane for a fixed $x = x_p$, for $N = 10$ and $N = 1$, respectively. Rather than grouping $\mathbf{w}$ and $\overline{\mathbf{w}}$ together and $\mathbf{u}$ and $\overline{\mathbf{u}}$ together, as in the previous paragraphs, this time we show a panel for each data set in the same counterclockwise order as shown in Figure 1. The upper left graph shows the time evolution of $\mathbf{u}$, the lower left shows the time evolution in the moving frame $\mathbf{w}$, and the lower right panel shows the average in the moving frame $\overline{\mathbf{w}}$. For $N = 10$, $\mathbf{u}$ oscillates between positive and negative values of the buoyancy and then we see the creation of higher frequency waves as the solution decays. We observe that, even when the higher frequency oscillations appear, larger scale (in time) patterns can also be seen. The lower right panel shows the effect of averaging in the moving frame, where the larger structures in time now look smoother. For the case when $N = 1$, the oscillatory pattern between positive and negative values are less frequent, since N is smaller. Comparing the two cases $N = 10$ and $N = 1$, the graphs show that $N = 1$ has an overall large-scale (in time) pattern of oscillations, whereas the $N = 10$ case has frequent temporal oscillations.

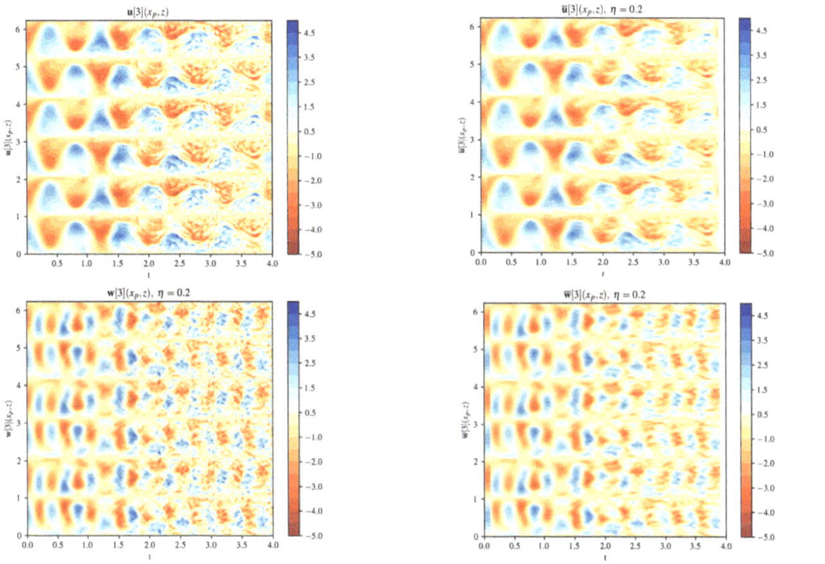

Figure 5. Time evolution of the 3rd component of the 4 data sets described in Figure 1 for $N = 10$. The top left panel shows $\mathbf{u}$ in the ordinary frame. The oscillatory pattern here is also seen in the potential energy shown in Figure 2. The large-scale pattern of oscillations between positive and negative values can be seen even as higher frequency waves appear. There are approximately 2 times more oscillations in $\mathbf{w}$ than in the ordinary domain. The appearance of the waves is more noticeable starting at approximately $t = 1$. The lower right panel shows the time evolution of the average in the moving frame $\overline{\mathbf{w}}$, which smooths the higher frequency oscillations. Finally, mapping back to the ordinary frame, the upper right panel shows the result of the mapping and averaging on the variables in the ordinary frame, $\overline{\mathbf{u}}$.

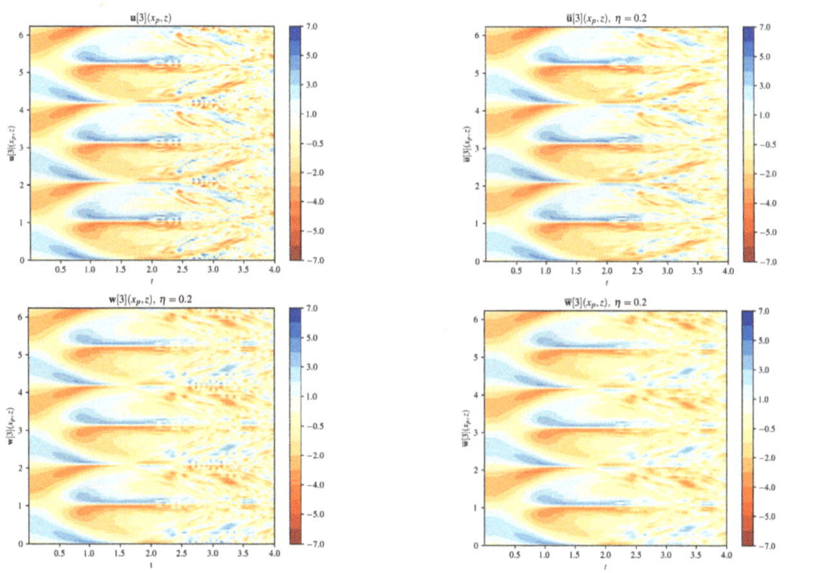

Figure 6. Time evolution of the 3rd component of the 4 data sets described in Figure 1 for $N = 1$. The difference between the ordinary frame (**top left** panel) and mapped frame (**bottom left** frame) is

less pronounced than the corresponding frames for $N = 10$ (Figure 5) because the oscillations in the exponential operator for $N = 1$ are less frequent. The higher frequencies that appear later in the simulation are smoothed by the averaging, with the result in the top right panel, which shows the result of the mapping and averaging in the ordinary domain $\overline{\mathbf{u}}$.

Finally, we examine the effect of the mapping and averaging on the energy. Figure 2 shows the effect of the averaging on the kinetic, potential, and total energy for the $\mathbf{u}$ and $\overline{\mathbf{u}}$ data sets for $N = 10$ on the left and $N = 1$ on the right. For $N = 10$, the mean tracks the oscillation between kinetic and potential, and shows an oscillation in the overall mean flow, which is discussed a bit more in the next section. For $N = 1$, the mean total energy decays faster than total energy, due to the average over the oscillations.

3.2. *Definition of a Mean Flow*

For a preliminary assessment of whether the mapping and averaging procedure could be a suitable definition of a mean flow for the 2D Boussinesq equations, we break the unknowns into its mean and fluctuation:

$$\mathbf{u} = \overline{\mathbf{u}} + \mathbf{u}', \tag{17}$$

and in the moving frame:

$$\mathbf{w} = \overline{\mathbf{w}} + \mathbf{w}'. \tag{18}$$

Integration over the oscillations would give us zero as positive and negative components cancel each other. From a mean flow, we would expect that only the mean behavior without the departures is retained.

Revisiting Figure 4, we make the following two observations: (1) The fastest variations in the solution are not visible anymore; (2) The amplitudes are damped when the solutions have steep gradients. Therefore, the component which is not retained contains fast oscillations and has positive and negative components.

Figure 7 shows the point-wise departure from the mean in the moving frame, $\mathbf{w}'$, for three different averaging windows. The main variations from the mean occurs when the filtering window is larger, as we expect, and also where there are higher oscillations at later times in the simulation. We also observe that, once the higher frequency oscillations have begun, the variation from the mean is also nearly periodic. Variation from the mean for the ordinary frame $\mathbf{u}'$ (not shown) shows a different detailed pattern with a similar periodicity.

Additionally, the choice of the averaging window allows us to determine which oscillations should be filtered and how strong the damping should be. For large averaging windows, more oscillations are filtered and the damping is stronger than for small averaging windows. This means that the definitions of the mean flow and the departures from the mean depend on the choice of the length of the averaging window. In both the moving frame and the ordinary frame, periodic patterns can be seen in the deviation from the mean.

Finally, we point out that total energy in Figure 2 for $N = 10$ shows an oscillation in the mean and that the the distance from the mean to the instantaneous energy decay is the energy from the variations from the mean. This oscillatory pattern in the mean flow is a potential area to explore in further work.

Our observation in the last section, that the mapping and averaging tracks the mean in the moving frame but not in the ordinary frame, will also be interesting in the context of fast singular limits, where it is expected that low-frequency structures will appear over time, rather than only high frequencies. To obtain a more complete understanding a wider range of behaviors, such as the creation of low frequencies, would be useful.

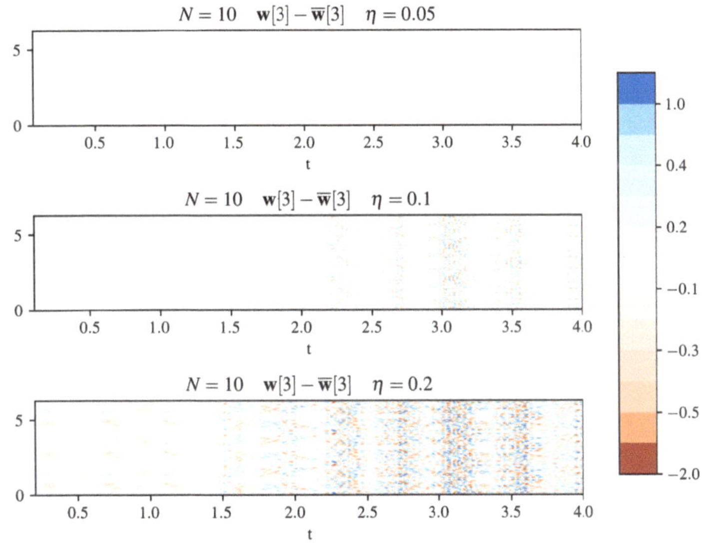

Figure 7. This figure shows the departure from the mean for three different averaging windows, η, for the case when $N = 10$, in the moving frame. As the averaging window, η, is increased, the excursions from the mean flow also increase. The excursions also show a periodic pattern.

3.3. Potential for Use in Numerical Methods

As mentioned in the introduction, the technique explored in this paper has been proposed for use as a prediction step for time-parallel predictor–corrector methods used to solve PDEs of type (1). An error bound for a predictor-step algorithm using this method can be found in [22] and does not require that ϵ be small. As a part of the error estimate, the error bound for the approximation of the time rate of change of the unknowns depends on the averaging window, η. We therefore are interested in whether the time rate of change of the variable examined in this paper has this property. To study this, we apply a first-order finite difference method to the moving frame $\mathbf{w}$ and $\overline{\mathbf{w}}$ data sets. Figure 8 shows the time rate of change for the $N = 10$ case depicted in the time series of Figure 4. This shows that the nonlinear averaging process reduces the amplitude and frequency of the oscillations, making it possible that this method could provide a good predictor for numerical methods. However, it is clear that the filter width matters, in line with the error estimate, and that it may need to be dynamically adjusted. For example, the solid blue line in Figure 8 shows that it has fewer large oscillations than the unaveraged case, but whether this is adequate to take larger time steps is unclear.

We have also pointed out, in previous sections, that, for the simulations in this study, the actual mean tracks $\overline{\mathbf{w}}$ rather than $\overline{\mathbf{u}}$. It would be of interest to explore this more fully in future work.

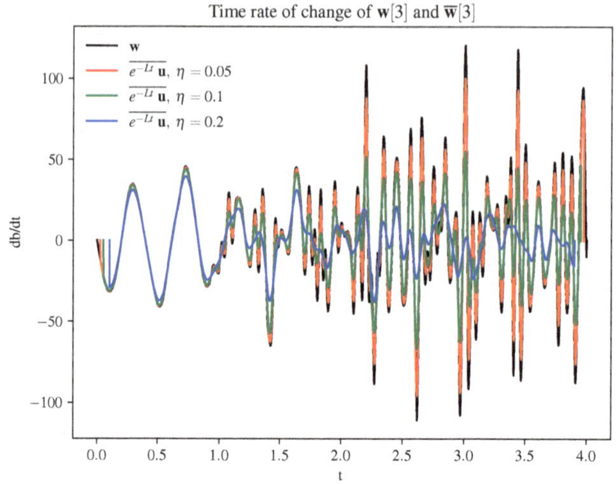

Figure 8. This figure shows the time rate of change of the 3rd component of $\mathbf{w}$ and $\overline{\mathbf{w}}$. The wider the value of η, the smaller the amplitude of the time rate of change, and, in some cases, the less frequent the oscillations. While this could be useful for taking large time steps, the degree of regularity depends strongly on the value of η.

4. Conclusions

In this work, we have presented numerical studies of the oscillating and decaying 2D Bousinnesq equations and developed a diagnostics to study a nonlinear averaging procedure. The goal was to present a preliminary study of the impact of the mapping and averaging technique by using it as a diagnostic, then assess whether it could make a good definition of a mean flow and potentially help mitigate oscillatory stiffness in numerical methods.

We made the following observations:

1. The nonlinear averaging procedure depends strongly on the value of η, but could potentially provide a reasonable representation of the time-mean of the solutions, whether or not N is large. Further, the average 'modulates' the waves, as in (2). This has the effect that the mean $\overline{\mathbf{u}}$ depends on how fast the oscillations are in the exponential operator (2), which could be interesting for further studies. Further work is needed to assess whether this would be a good definition of a mean flow in fluid dynamics. For example, it would be interesting to examine a case that admits a low-frequency solution such as the 2D Boussinesq equations with forcing, as in [12] and examples shown in [8];
2. We also observed that the method could potentially be used to take larger time steps, as has been shown in simple examples [21], and that the ability to do so critically depends on the value of η.

Author Contributions: Conceptualization, B.A.W., T.H. and J.R.; methodology, B.A.W., T.H. and J.R.; software, B.A.W. and J.R.; validation, B.A.W. and J.R.; writing—original draft preparation, B.A.W. and J.R.; writing—review and editing, B.A.W. and J.R.; funding acquisition, B.A.W., T.H., and J.R. All authors have read and agreed to the published version of the manuscript.

Funding: Beth Wingate's research was funded by the UK Engineering Physical Science Research Council (EPSRC) grant number EP/R029628/1, Leverhulme Trust Research Fellowship RF-2022-013. Juliane Rosemeier's research was funded by the Deutsche Forschungsgemeinschaft (DFG, German Research Foundation)—463179503. Terry Haut's research was funded by the US Department of Energy.

Institutional Review Board Statement: Not applicable.

Informed Consent Statement: Not applicable.

Data Availability Statement: Not applicable.

Acknowledgments: We thank the University of Exeter for hosting Juliane Rosemeier's DFG Walter Benjamin Fellowship, and the US Department of Energy Lawrence Livermore National Laboratory for hosting Juliane Rosemeier and Beth Wingate as visitors.

Conflicts of Interest: The funders had no role in the design of the study; in the collection, analyses, or interpretation of data; in the writing of the manuscript; or in the decision to publish the results.

Abbreviations

The following abbreviations are used in this manuscript:

PDE	partial differential equation
ODE	ordinary differential equation

Appendix A. Computational Details for the Exponential

Appendix A.1. The Eigenvalues and Eigenvectors

First case: $k_1 = 0, k_3 = 0,$
eigenvalues: $\omega_1 = 0, \quad \omega_2 = -iN, \quad \omega_3 = iN$

$$\text{eigenvectors: } v_1 = \begin{pmatrix} 1 \\ 0 \\ 0 \end{pmatrix} \quad v_2 = \frac{1}{\sqrt{2}} \begin{pmatrix} 0 \\ i \\ 1 \end{pmatrix} \quad v_3 = \frac{1}{\sqrt{2}} \begin{pmatrix} 0 \\ -i \\ 1 \end{pmatrix}$$

Second case: $k_1 = 0, k_3 \neq 0,$
eigenvalues: $\omega_1 = \omega_2 = \omega_3 = 0$

$$\text{eigenvectors: } v_1 = \begin{pmatrix} 1 \\ 0 \\ 0 \end{pmatrix} \quad v_2 = \begin{pmatrix} 0 \\ 0 \\ 1 \end{pmatrix}$$

$$\text{generalized eigenvector: } \tilde{v} = \begin{pmatrix} 0 \\ -\frac{1}{N} \\ 0 \end{pmatrix}$$

Third case: $k_1 \neq 0, k_3 = 0,$
In this case, we have the same eigenvalues and eigenvectors as in the the first case.
Fourth case: $k_1 \neq 0, k_3 \neq 0,$
eigenvalues: $\omega_1 = 0, \quad \omega_2 = iN|k_1|/\|\mathbf{k}\|, \quad \omega_3 = -iN|k_1|/\|\mathbf{k}\|$

$$\text{eigenvectors: } v_1 = \begin{pmatrix} 1 \\ 0 \\ 0 \end{pmatrix} \quad v_2 = \frac{1}{\sqrt{2}} \begin{pmatrix} i\frac{k_1 k_3}{|k_1|\|\mathbf{k}\|} \\ -i\frac{|k_1|}{\|\mathbf{k}\|} \\ 1 \end{pmatrix} \quad v_3 = \frac{1}{\sqrt{2}} \begin{pmatrix} -i\frac{k_1 k_3}{|k_1|\|\mathbf{k}\|} \\ i\frac{|k_1|}{\|\mathbf{k}\|} \\ 1 \end{pmatrix}$$

Appendix A.2. Diagonal Matrix or Jordan Normal Form

First case: $k_1 = 0, k_3 = 0,$

$$D = \begin{pmatrix} 0 & 0 & 0 \\ 0 & -iN & 0 \\ 0 & 0 & iN \end{pmatrix}$$

Second case: $k_1 = 0, k_3 \neq 0,$

$$J = \begin{pmatrix} 0 & 0 & 0 \\ 0 & 0 & 1 \\ 0 & 0 & 0 \end{pmatrix}$$

Third case: $k_1 \neq 0, k_3 = 0,$
In this case, we have the diagonal matrix as in the the first case.
Fourth case: $k_1 \neq 0, k_3 \neq 0,$

$$D = \begin{pmatrix} 0 & 0 & 0 \\ 0 & iN\frac{|k_1|}{\|\mathbf{k}\|} & 0 \\ 0 & 0 & -iN\frac{|k_1|}{\|\mathbf{k}\|} \end{pmatrix}$$

Appendix A.3. Transformation Matrix and Inverse

First case: $k_1 = 0, k_3 = 0,$

$$T = \begin{pmatrix} 1 & 0 & 0 \\ 0 & \frac{i}{\sqrt{2}} & -\frac{i}{\sqrt{2}} \\ 0 & \frac{1}{\sqrt{2}} & \frac{1}{\sqrt{2}} \end{pmatrix}, \qquad T^{-1} = \begin{pmatrix} 1 & 0 & 0 \\ 0 & -\frac{i}{\sqrt{2}} & \frac{1}{\sqrt{2}} \\ 0 & \frac{i}{\sqrt{2}} & \frac{1}{\sqrt{2}} \end{pmatrix}$$

Second case: $k_1 = 0, k_3 \neq 0,$

$$T = \begin{pmatrix} 1 & 0 & 0 \\ 0 & 0 & -\frac{1}{N} \\ 0 & 1 & 0 \end{pmatrix}, \qquad T^{-1} = \begin{pmatrix} 1 & 0 & 0 \\ 0 & 0 & 1 \\ 0 & -N & 0 \end{pmatrix}$$

Third case: $k_1 \neq 0, k_3 = 0,$

The transformation matrix and inverse are the same as in the the first case.

Fourth case: $k_1 \neq 0, k_3 \neq 0,$

$$T = \begin{pmatrix} 1 & \frac{i}{\sqrt{2}}\frac{k_1 k_3}{|k_1|\|\mathbf{k}\|} & -\frac{i}{\sqrt{2}}\frac{k_1 k_3}{|k_1|\|\mathbf{k}\|} \\ 0 & -\frac{i}{\sqrt{2}}\frac{k_1}{\|\mathbf{k}\|} & \frac{i}{\sqrt{2}}\frac{k_1}{\|\mathbf{k}\|} \\ 0 & \frac{1}{\sqrt{2}} & \frac{1}{\sqrt{2}} \end{pmatrix}, \qquad T^{-1} = i\frac{\|\mathbf{k}\|}{k_1} \begin{pmatrix} -i\frac{k_1}{\|\mathbf{k}\|} & -i\frac{k_1 k_3}{|k_1|\|\mathbf{k}\|} & 0 \\ 0 & \frac{1}{\sqrt{2}} & -\frac{i}{\sqrt{2}}\frac{k_1}{\|\mathbf{k}\|} \\ 0 & -\frac{1}{\sqrt{2}} & -\frac{i}{\sqrt{2}}\frac{k_1}{\|\mathbf{k}\|} \end{pmatrix}$$

Appendix A.4. The Matrices with Divergence Constraint

First case: $k_1 = 0, k_3 = 0,$

$$\begin{pmatrix} 1 & 0 & 0 \\ 0 & \frac{1}{2}(e^{-iNt} + e^{iNt}) & \frac{1}{2}(e^{-iNt} - e^{iNt}) \\ 0 & \frac{1}{2}(-e^{-iNt} + e^{iNt}) & \frac{1}{2}(e^{-iNt} + e^{iNt}) \end{pmatrix}$$

Second case: $k_1 = 0, k_3 \neq 0,$

$$\begin{pmatrix} 0 & 0 & 0 \\ 0 & \frac{1}{2}(e^{-iNt} + e^{iNt}) & \frac{1}{2}(e^{-iNt} - e^{iNt}) \\ 0 & \frac{1}{2}(-e^{-iNt} + e^{iNt}) & \frac{1}{2}(e^{-iNt} + e^{iNt}) \end{pmatrix}$$

Third case: $k_1 \neq 0, k_3 = 0,$

$$\begin{pmatrix} 1 & 0 & 0 \\ 0 & 0 & 0 \\ 0 & 0 & 1 \end{pmatrix}$$

Fourth case: $k_1 \neq 0, k_3 \neq 0,$

$$\begin{pmatrix} 0 & -\frac{k_3}{2k_1}\left(e^{iN\frac{k_1}{|k|}t} + e^{-iN\frac{k_1}{|k|}t}\right) & i\frac{k_3|k_1|}{2|\mathbf{k}|k_1}\left(-e^{iN\frac{k_1}{|k|}t} + e^{-iN\frac{k_1}{|k|}t}\right) \\ 0 & \frac{1}{2}\left(e^{iN\frac{k_1}{|k|}t} + e^{-iN\frac{k_1}{|k|}t}\right) & -i\frac{|k_1|}{2|\mathbf{k}|}\left(-e^{iN\frac{k_1}{|k|}t} + e^{-iN\frac{k_1}{|k|}t}\right) \\ 0 & i\frac{\mathbf{k}}{2k_1}\left(e^{iN\frac{k_1}{|k|}t} - e^{-iN\frac{k_1}{|k|}t}\right) & \frac{1}{2}\left(e^{iN\frac{k_1}{|k|}t} + e^{-iN\frac{k_1}{|k|}t}\right) \end{pmatrix}$$

References

1. Verhulst, S.F.; Murdock, J. *Averaging Methods in Nonlinear Dynamical Systems*, 2nd ed.; Springer: New York, NY, USA, 2007. [CrossRef]
2. Herring, J.R. Approach of axisymmetric turbulence to isotropy. *Phys. Fluids* **1974**, *17*, 859–872. [CrossRef]
3. Herring, J.R.; Métais, O. Numerical experiments in forced stably stratified turbulence. *J. Fluid Mech.* **1989**, *202*, 97–115. [CrossRef]
4. Métais, O.; Herring, J.R. Numerical simulations of freely evolving turbulence in stably stratified fluids. *J. Fluid Mech.* **1989**, *202*, 117–148. [CrossRef]
5. Babin, A.; Mahalov, A.; Nicolaenko, B. Global splitting, integrability and regularity of 3D Euler and Navier-Stokes equations for uniformly rotating fluids. *Eur. J. Mech. B Fluids* **1996**, *15*, 291–300.
6. Babin, A.; Mahalov, A.; Nicolaenko, B.; Zho, Y. On the asymptotic regimes and the strongly stratified limit of rotating Boussinesq equations. *Theor. Comput. Fluid Dyn.* **1997**, *9*, 223–251. [CrossRef]

7. Bogoliubov, N.; Mitropolsky, Y. *Asymptotic Methods in the Theory of Nonlinear Oscillations*; Gordon and Breach: New York, NY, USA, 1961.
8. Embid, F.; Majda, J. Averaging over fast gravity waves for geophysical flows with arbitrary potential vorticity. *Commun. Partial. Differ. Equ.* **1996**, *21*, 619–658. [CrossRef]
9. Klainerman, S.; Majda, A.J. Singular limits of quasilinear hyperbolic systems with large parameters and the incompressible limit of compressible fluids. *Commun. Pure Appl. Math.* **1981**, *34*, 481–524. [CrossRef]
10. Schochet, S. Fast Singular Limits of Hyperbolic PDEs. *J. Differ. Equ.* **1994**, *114*, 476–512. [CrossRef]
11. Newell, A. Rossby wave packet interactions. *J. Fluid Mech.* **1969**, *35*, 255–271. [CrossRef]
12. Smith, L.M. Numerical study of two-dimensional stratified turbulence. *Contemp. Math.* **2001**, *283*, 91–106.
13. Smith, L.; Waleffe, F. Generation of slow large scales in forced rotating stratified turbulence. *J. Fluid Mech.* **2002**, *451*, 145–168. [CrossRef]
14. Kafiabad, H.A.; Vanneste, J.; Young, W.R. Wave-averaged balance: A simple example. *J. Fluid Mech.* **2021**, *911*, R1.
15. Wagner, G.L.; Young, W.R. Available potential vorticity and wave-averaged quasi-geostrophic flow. *J. Fluid Mech.* **2015**, *785*, 401–424.
16. Yamazaki, H.; Cotter, C.; Wingate, B. Time parallel integration and phase averaging for the nonlinear shallow water equations on the sphere. *Q. J. R. Meteorol. Soc.* **2023**, *accepted*. [CrossRef]
17. Majda, A.; Embid, P. Averaging over Fast Gravity Waves for Geophysical Flows with Unbalanced Initial Data. *Theoret. Comput. Fluid Dyn.* **1998**, *11*, 155–169. [CrossRef]
18. Whitehead, J.P.; Haut, T.; Wingate, B. The effect of two distinct fast time scales in the rotating, stratified Boussinesq equations: Variations from quasi-geostrophy. *Theor. Comput. Fluid Dyn.* **2018**, *32*, 713–732. [CrossRef]
19. Mu, P.; Ju, Q. Three-scale singular limits of the rotating stratified Boussinesq equations. *Appl. Anal.* **2021**, *100*, 2405–2417. [CrossRef]
20. Jones, D.A.; Mahalov, A.; Nicolaenko, B. A Numerical Study of an Operator Splitting Method for Rotating Flows with Large Ageostrophic Initial Data. *Theor. Comput. Fluid Dyn.* **1999**, *13*, 143. [CrossRef]
21. Haut, T.; Wingate, B. An asymptotic parallel-in-time method for highly oscillatory pdes. *SIAM J. Sci. Comput.* **2014**, *36*, A693–A713. [CrossRef]
22. Peddle, A.G.; Haut, T.; Wingate, B. Parareal convergence for oscillatory pdes with finite time-scale separation. *SIAM J. On Scientific Comput.* **2019**, *41*, A3476–A3497. [CrossRef]
23. Rosemeier, J.; Haut, T; Wingate B. Multi-level Parareal algorithm with Averaging for Oscillatory Problems. *SIAM J. Sci. Comput.* **2023**, *submitted*.
24. Majda, A. *Introduction to P.D.E.'s and Waves for the Atmosphere and Ocean*; Courant Lecture Notes; New York University; Courant Institute of Mathematical Sciences and American Mathematical Society: 2002 ; Volume 9.
25. Burns, K.J.; Vasil, G.M.; Oishi, J.S.; Lecoanet, D.; Brown, B.P. Dedalus: A Flexible Framework for Numerical Simulations with Spectral Methods. *Phys. Rev. Res.* **2020**, 2, 023068. [CrossRef]
26. Weinan, E; Engquist, B. Multiscale modeling and computation. *Not. Am. Math. Soc.* **2003**, *50*, 1062–1070.
27. Engquist, B.; Tsai, R. Heterogeneous multiscale methods for stiff ordinary differential equations. *Math. Comput.* **2005**, *74*, 1707–1742. [CrossRef]

atmosphere

MDPI

Article

Evolution of a Stratified Turbulent Cloud under Rotation

Tianyi Li [1,2], Minping Wan [1,3,*] and Shiyi Chen [1,3,4,*]

1 Guangdong Provincial Key Laboratory of Turbulence Research and Applications, Department of Mechanics and Aerospace Engineering, Southern University of Science and Technology, Shenzhen 518055, China; tianyi.li@roma2.infn.it

2 Department of Physics and INFN, University of Rome "Tor Vergata", Via della Ricerca Scientifica 1, 00133 Rome, Italy

3 Guangdong-Hong Kong-Macao Joint Laboratory for Data-Driven Fluid Mechanics and Engineering Applications, Southern University of Science and Technology, Shenzhen 518055, China

4 Eastern Institute for Advanced Study, Ningbo 315201, China

* Correspondence: wanmp@sustech.edu.cn (M.W.); chensy@sustech.edu.cn (S.C.)

Abstract: Localized turbulence is common in geophysical flows, where the roles of rotation and stratification are paramount. In this study, we investigate the evolution of a stratified turbulent cloud under rotation. Recognizing that a turbulent cloud is composed of vortices of varying scales and shapes, we start our investigation with a single eddy using analytical solutions derived from a linearized system. Compared to an eddy under pure rotation, the stratified eddy shows the physical manifestation of a known potential vorticity mode, appearing as a static stable vortex. In addition, the expected shift from inertial waves to inertial-gravity waves is observed. In our numerical simulations of the turbulent cloud, carried out at a constant Rossby number over a range of Froude numbers, stratification causes columnar structures to deviate from vertical alignment. This deviation increases with increasing stratification, slowing the expansion rate of the cloud. The observed characteristics of these columnar structures are consistent with the predictions of linear theory, particularly in their tilt angles and vertical growth rates, suggesting a significant influence of inertial-gravity waves. Using Lagrangian particle tracking, we have identified regions where wave activity dominates over turbulence. In scenarios of milder stratification, these inertial-gravity waves are responsible for a significant energy transfer away from the turbulent cloud, a phenomenon that attenuates with increasing stratification.

Keywords: localized turbulence; stratification; rotation; inertial-gravity waves

Citation: Li, T.; Wan, M.; Chen, S. Evolution of a Stratified Turbulent Cloud under Rotation. *Atmosphere* **2023**, *14*, 1590. https://doi.org/10.3390/atmos14101590

Academic Editor: Doris Folini

Received: 11 September 2023
Revised: 12 October 2023
Accepted: 20 October 2023
Published: 22 October 2023

1. Introduction

Localized turbulence is common in both atmospheric and oceanic flows, which are strongly influenced by the Earth's rotation and stable density stratification [1–6]. Understanding the evolution of these turbulent patches under the combined effects of rotation and stratification is crucial for gaining deeper insights into various geophysical phenomena. A key mechanism behind the generation of such turbulence is the local breakdown of internal waves [7,8]. In particular, the resulting turbulence often exhibits horizontal scales that are significantly larger than their vertical counterparts. Empirical observations from deep waters and the equatorial pycnocline also reveal pronounced differences in horizontal and vertical scales of turbulent patches [2,3]. Furthermore, mixing induced by internal wave breaking tends to weaken local stratification. Such reduced stratification is also observed in turbulent regions within the ocean mixed layer [9]. With these characteristics in mind, our present study is concerned with a horizontally extended turbulent cloud within a vertically stably stratified, rotating fluid, where the effects of rotation are stronger than or comparable to the stratification. Our focus is on understanding the formation of flow structures and analyzing the associated energy dispersion patterns.

While numerous studies have highlighted the dominance of waves in spatially uniform and stationary scenarios of rotating/stratified turbulence [10–15], the dynamics of localized turbulent patches has also attracted considerable attention. In particular, under rotation, these turbulent patches often give rise to columnar vortices parallel to the axis of rotation, associated with inertial waves [16–19]. For example, in their oscillating-grid turbulence experiments, Dickinson and Long [17] found that, for a Rossby number close to unity, the boundary between perturbed and quiescent regions moves at speeds related to certain wave velocities. Similarly, Davidson et al. [18] showed experimentally that the leading edges of columnar vortices under rotation align with inertial wave group velocities. Using direct numerical simulations (DNS), Ranjan and Davidson [19] observed these columnar structures, attributed their formation to inertial waves, and emphasized their role in significant energy dissipation from localized turbulence.

Analogous to turbulent patches under rotation, the presence of internal wave radiation is clearly evident in localized turbulence within stratified fluids [20–23], such as wakes caused by object motion [24,25]. The flow structures in these scenarios are predominantly wave-dominated. In addition, these waves contribute significantly to energy reorganization, highlighting their fundamental role in the atmosphere and oceans [26]. Anisotropic structures and wave propagation are observed in both rotating and stratified turbulent flows. These phenomena highlight the complex interplay between rotation and stratification in geophysical flows [27–29]. While many studies have investigated the turbulent patch behavior in rotating and stratified fluids, the emphasis has often been on aspects such as lateral intrusions, Thorpe scales, and the vertical expansion of the patch [30–33]. However, the waves emitted by these disturbances often receive less attention. One notable exception [34] used DNS to study a buoyant cloud subjected to vertical rotation and horizontal gravity, reminiscent of equatorial conditions. It was confirmed that the formation of columnar structures was due to inertial waves arising from the buoyant cloud.

Building on previous work, we explore an under-explored area: the evolution of a horizontal turbulent cloud under vertical rotation and stratification. This has significant relevance in geophysical contexts such as the high latitudes of the atmosphere and oceans [35] and the deep Arctic waters [36,37]. In particular, our study examines scenarios where rotation effects outweigh stratification, in contrast to many typical geophysical flows where stratification predominates. In this context, we address three key questions: First, how does stratification affect the resulting flow structures, as opposed to purely rotating scenarios? Second, are these structures influenced by inertial-gravity waves, and how does different stratification affect their formation? Third, what fraction of the cloud's energy is transferred to waves at different stratification levels?

2. The Evolution of a Single Stratified Eddy under Rotation

In this section, we study the evolution of a single stratified eddy under rotation. This is motivated by the fact that a turbulent cloud can be considered as a sea of randomly oriented vortex blobs with different scales. First, we introduce inertial-gravity waves under the Boussinesq approximation. Second, we solve analytically the initial value problem of a compact vortex blob in a rapidly rotating stratified environment. Finally, a numerical simulation is performed to validate the analytical solution. It is shown that as the eddy evolves, part of the energy is retained in the eddy by the potential vorticity (PV) mode, while the other part is carried away from the eddy by inertial-gravity waves.

2.1. Inertial-Gravity Waves

The Boussinesq set of equations for a linearly stratified fluid under system rotation can be written as:

$$\nabla \cdot \boldsymbol{u} = 0, \tag{1}$$

$$\frac{\partial \boldsymbol{u}}{\partial t} + \boldsymbol{u} \cdot \nabla \boldsymbol{u} = -\frac{1}{\rho_0} \nabla p + 2\boldsymbol{u} \times \boldsymbol{\Omega} - N\phi \boldsymbol{e}_z + \nu \nabla^2 \boldsymbol{u}, \tag{2}$$

$$\frac{\partial\phi}{\partial t} + \boldsymbol{u}\cdot\boldsymbol{\nabla}\phi = Nu_z + \kappa\nabla^2\phi, \tag{3}$$

where $\boldsymbol{u}$ is the velocity vector, u_z is the z-component of the velocity, p is the modified pressure incorporating a centrifugal term, $\boldsymbol{\Omega} = \Omega\boldsymbol{e}_z$ represents the rotation vector, N is the Brunt–Väisälä frequency, and ν and κ are the kinematic viscosity and the diffusion coefficient, respectively. We define $\phi = (g/\rho_0 N)\rho'$, which has the dimension of the velocity, and ρ' is the density perturbation from the ambient density.

Inertial-gravity waves can be obtained after a linearization of the Boussinesq equations. Neglecting the molecular diffusion and the second-order terms of $\boldsymbol{u}$, $\boldsymbol{\omega} = \boldsymbol{\nabla}\times\boldsymbol{u}$ and ϕ one obtains:

$$\frac{\partial\boldsymbol{\omega}}{\partial t} + \boldsymbol{\nabla}\times(2\boldsymbol{\Omega}\times\boldsymbol{u}) = -N\boldsymbol{\nabla}\phi\times\boldsymbol{e}_z, \tag{4}$$

$$\frac{\partial\phi}{\partial t} = Nu_z, \tag{5}$$

from which one finds (see Ref. [38], p. 56)

$$\frac{\partial^2}{\partial t^2}\nabla^2 u_z + 4\Omega^2\frac{\partial^2 u_z}{\partial z^2} + N^2\left(\frac{\partial^2 u_z}{\partial x^2} + \frac{\partial^2 u_z}{\partial y^2}\right) = 0. \tag{6}$$

This equation admits inertial-gravity waves with the dispersion relation given as

$$\varpi = \pm\sqrt{N^2k_h^2 + 4\Omega^2k_z^2}/k, \tag{7}$$

where ϖ is the frequency, while $k_h = \sqrt{k_x^2 + k_y^2}$ and $k = \sqrt{k_x^2 + k_y^2 + k_z^2}$ are the horizontal wavenumber and the total wavenumber, respectively. The group velocity is $\boldsymbol{c}_g = \boldsymbol{\nabla}\varpi$, whose z-component is

$$c_{g,z} = \frac{\partial\varpi}{\partial k_z} = \pm\frac{(4\Omega^2 - N^2)k_h^2k_z}{\sqrt{N^2k_h^2 + 4\Omega^2k_z^2}k^3}. \tag{8}$$

2.2. Analytical Study of a Single Eddy

Following Davidson et al. [18], we consider the evolution of an eddy in a stratified fluid under rotation, assuming that the process is axisymmetric with respect to the eddy axis. In this context, in a cylindrical coordinate system (r, θ, z), an axisymmetric velocity field $\boldsymbol{u}$ can be decomposed into azimuthal and poloidal components

$$\boldsymbol{u} = (\Gamma/r)\boldsymbol{e}_\theta + \boldsymbol{\nabla}\times[(\psi/r)\boldsymbol{e}_\theta], \tag{9}$$

where $\Gamma = u_\theta r$ is the angular momentum and ψ is the Stokes streamfunction. We substitute Equation (9) into the linearized inviscid vorticity Equation (4) and the linearized non-diffusive Equation (5) to obtain

$$\frac{\partial\Gamma}{\partial t} = 2\Omega\frac{\partial\psi}{\partial z}, \quad \frac{\partial}{\partial t}(r\omega_\theta) = 2\Omega\frac{\partial\Gamma}{\partial z} + Nr\frac{\partial\phi}{\partial r}, \quad \frac{\partial\phi}{\partial t} = \frac{N}{r}\frac{\partial\psi}{\partial r}, \tag{10}$$

where $\nabla_*^2\psi = (r\partial/\partial r)(r^{-1}\partial\psi/\partial r) + \partial^2\psi/\partial z^2 = -r\omega_\theta$, ∇_*^2 being the Stokes operator. Combining the equations in (10) yields

$$\frac{\partial}{\partial t}\left(\frac{\partial^2}{\partial t^2}\nabla_*^2\Gamma + \left[(2\Omega)^2\frac{\partial^2}{\partial z^2} + N^2\left(r\frac{\partial}{\partial r}\right)\left(\frac{\partial}{r\partial r}\right)\right]\Gamma\right) = 0. \tag{11}$$

Equation (11) can be readily solved by using the Hankel-cosine transform

$$\hat{u}_\theta = \frac{1}{2\pi^2}\int_0^\infty\int_0^\infty r u_\theta J_1(k_r r)\cos(k_z z)\mathrm{d}r\mathrm{d}z, \tag{12}$$

where J_1 is the Bessel function of the first kind of order 1. Here, k_r and k_z represent the wavenumbers in the r-direction and the z-direction, respectively, with $0 \le k_r, k_z < \infty$. Given the initial conditions $\hat{u}_\theta = \hat{u}_\theta^{(0)}$, $\psi = 0$ and $\phi = 0$, we can derive that

$$\hat{u}_\theta = \left[\hat{u}_\theta^{(0)}/\left(N^2k_r^2+(2\Omega)^2k_z^2\right)\right]\left[N^2k_r^2+(2\Omega)^2k_z^2\cos\left(\sqrt{N^2k_r^2+(2\Omega)^2k_z^2}\,t/k\right)\right], \tag{13}$$

from which we obtain

$$\begin{aligned} u_\theta = & \underbrace{4\pi\int_0^\infty\int_0^\infty\left(N^2k_r^3\hat{u}_\theta^{(0)}/\alpha^2\right)J_1(k_r r)\cos(k_z z)\mathrm{d}k_r\mathrm{d}k_z}_{\text{PV mode}} \\ & \underbrace{+2\pi\int_0^\infty\int_0^\infty\left[(2\Omega)^2k_rk_z^2\hat{u}_\theta^{(0)}/\alpha^2\right]J_1(k_r r)[\cos(k_z z-\alpha t/k)+\cos(k_z z+\alpha t/k)]\mathrm{d}k_r\mathrm{d}k_z,}_{\text{inertial-gravity waves}} \end{aligned} \tag{14}$$

where we introduce $\alpha = \sqrt{N^2k_r^2+(2\Omega)^2k_z^2}$ for conciseness. In the absence of stratification (i.e., when $N = 0$), Equations (13) and (14) are consistent with the results presented in Davidson et al. [18], which consider only the rotation of the system. However, the inclusion of stratification introduces a new time-independent term and changes the frequencies of the previously identified wave-like components. Note that the phase velocities of the wave-like components, given by $\pm\alpha/k$, closely match those of the inertial-gravity waves according to Equation (7), provided we replace k_r with k_h. This alignment strongly suggests that these wave-like components are indeed manifestations of the inertial-gravity waves. In a spatially uniform rotating stratified flow with periodic boundary condition, Smith and Waleffe [39] identified the linear eigenmodes as two inertial-gravity waves and a PV mode, suggesting that the time-independent term in (14) is consistent with the zero-frequency PV mode. However, our solution shows how these PV mode and inertial-gravity waves manifest in the flow evolution from a spatially localized initial condition—an aspect not addressed in previous studies. Later in Section 2.3, we will explore the associated flow structures and elucidate their role in energy dispersion.

2.3. Numerical Validation of the Analytical Results

To assess the applicability of the analytical results in Section 2.2 and to develop a better understanding of the evolution of a single eddy in a rotating stratified fluid, a simple initial velocity field of the Gaussian-eddy form is chosen:

$$\boldsymbol{u} = \Lambda r\exp\left(-\frac{r^2+z^2}{\delta^2}\right)\boldsymbol{e}_\theta, \tag{15}$$

for which Davidson et al. [18] have given an analytical solution when only the rotation is considered. In this equation, Λ is the characteristic angular rotation rate and δ is the characteristic size. The corresponding Rossby and Froude numbers are defined by $Ro = \Lambda/(2\Omega)$ and $Fr = \Lambda/N$, respectively. To validate our analytical solutions, we chose parameters $Ro = 0.02$ and $Fr = 0.16$, corresponding to a vortex characterized by $\Lambda = 0.422$ and $\delta = 0.125$. This choice corresponds to a parameter set from our turbulent cloud studies in Section 3. We note that the validation of our analytical solutions remains robust across different parameter sets in these studies. The analytical solution is obtained by substituting (15) into (14). We then performed a DNS to study the evolution of this Gaussian eddy. Equations (1)–(3) are solved using a parallelized pseudo-spectral code in a 512^3

periodic box of size $l_{box} = 2\pi$, where the fourth-order Runge–Kutta time-stepping scheme is employed [40]. The linear terms caused by rotation and stratification, together with the viscous and diffusive terms, are integrated exactly by using an integrating factor technique. A combination of phase-shifting and truncation is used to de-alias the nonlinear terms (see e.g., Ref. [41]).

Figure 1a,b show the contours of u_θ^2 (normalized by its maximum value) in the x-z plane passing through the axis of the Gaussian eddy at $t/t_f = 13.5$. These are results from the analytical solution and the DNS, respectively. Here $t_f = 1/(2\Omega)$ denotes the rotation time scale. The DNS result, which considers the full Navier–Stokes equations including viscous and nonlinear effects, is still in close agreement with the analytical counterpart. Although there are discrepancies due to these effects, the overall trend and structure are consistent between the two. A prominent region of u_θ^2 near the origin stands out, which is absent in the unstratified scenario (Ref. [19], Figure 3). This region is indicative of the PV mode, while the other intense regions represent inertial-gravity waves.

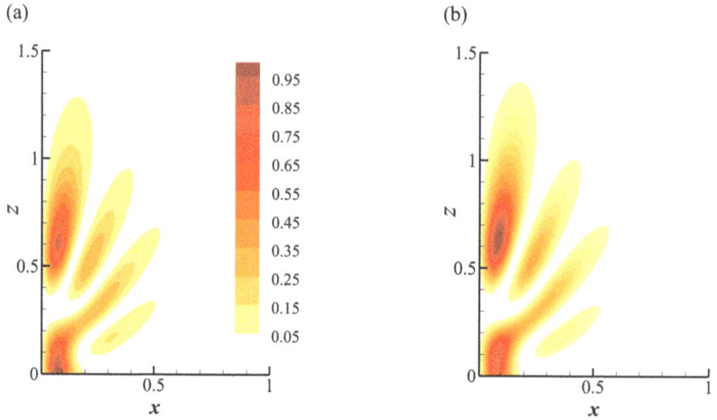

Figure 1. Contours of $u_\theta^2/(u_\theta^2)_{max}$ in the x-z plane through the axis of the Gaussian eddy at $t/t_f = 13.5$. The results are from (**a**) the analytical solution (14) and (**b**) the DNS.

To illustrate the evolution of the flow structures and their corresponding energy dispersion, Figure 2 shows iso-surfaces of the vertical vorticity, $\omega_z = 0.1\Lambda$, color-coded with ϕ, at times $t/t_f = 13.5$ and 16.9. As the system begins to evolve, a central vortex appears at the origin, flanked by two vortices, one above and one below (Figure 2a). Over time, the central vortex undergoes minor changes, while the flanking vortices become smaller and their centers move further away from the origin (Figure 2b). This behavior is in stark contrast to the purely rotating case [34], where only two columnar structures emerge, progressively moving away from each other as they elongate. The shifts in the flow structures can be attributed to the interplay between the Coriolis and buoyancy forces. In Figure 2a, all three cyclonic vortices ($w_z > 0$) are influenced by the Coriolis force, causing them to stretch horizontally. For the central vortex, ϕ has negative values at the top and positive ones at the base. This distribution results in a vertical stretch from the buoyancy force, which causes a horizontal contraction of the vertex due to the incompressibility of the fluid, counteracting the Coriolis force. As a result, this vortex remains relatively stable, maintaining a nearly consistent energy distribution over time. For each flanking vortex, ϕ retains its sign, showing larger absolute values for smaller magnitudes of $|z|$. This implies that the buoyancy force is acting to move these vortices away from the origin and to compress them vertically. The interaction of the Coriolis and buoyancy forces causes these vortices to recede and decrease in size, resulting in energy dispersion and potential-kinetic energy exchange. It is worth noting that a similar phenomenon to the central vortex has been documented in fully developed rotating stratified turbulence, where it correlates with a pronounced exchange between kinetic and potential energy [40].

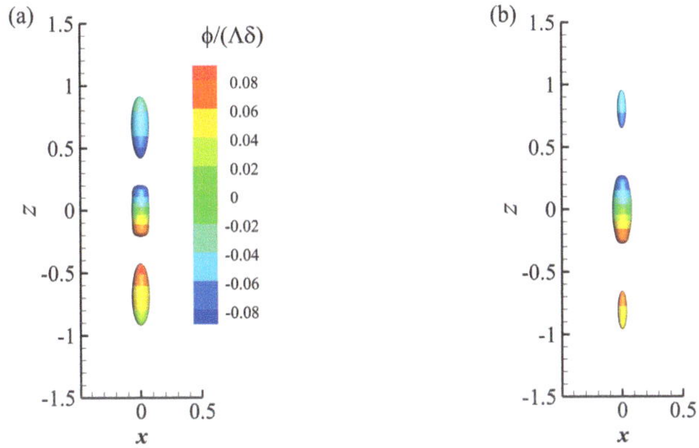

Figure 2. The iso-surfaces of $\omega_z = 0.1\Lambda$ (front view) color-coded with ϕ obtained from the DNS at (**a**) $t/t_f = 13.5$ and (**b**) $t/t_f = 16.9$.

3. A Stratified Turbulent Cloud under Rotation

In this section, we move from the study of a single eddy to the nonlinear evolution of a turbulent cloud. We observe characteristics compatible with inertial-gravity waves in DNS by studying the flow structures. Finally, we calculate the energy exchanged from the turbulent cloud by these waves to understand their role in energy transfer for localized turbulent patches.

3.1. DNS of a Turbulent Cloud

Following the procedure described by Ranjan and Davidson [19], we performed DNS on fully developed homogeneous isotropic turbulence to obtain the initial velocity field for the turbulent cloud study. We then selected a velocity field corresponding to a specific moment characterized by an integral length scale of $l_0 \sim \int_0^\infty k^{-1}E(k)\mathrm{d}k / \int_0^\infty E(k)\mathrm{d}k \approx 0.04 l_{box}$ and a Reynolds number $Re_0 = u_0 l_0/\nu \approx 169$, where $E(k)$ is the energy spectrum and u_0 is the r.m.s. velocity. To numerically generate a horizontal turbulent cloud, the resulting velocity field was spatially filtered with

$$f(z) = \exp\left(-10^a |z - \pi|^m\right), \tag{16}$$

as shown in Figure 3a. Here we set $a = 4$ and $m = 9$, resulting in a cloud thickness of $l_c = 0.13 l_{box}$, approximately three times the integral length scale l_0. The velocity field is then projected to be divergence-free. The buoyancy field was initialized with $\phi = 0$ throughout the computational domain, which ensures a uniformly linear density profile, indicating no initial mixing within the cloud. Figure 3b shows the initial turbulent cloud as indicated by the velocity module, and its spectrum in the x-y plane, averaged over its central part in the z-direction, is shown in Figure 3c.

To investigate the effect of the stratification on the evolution of a turbulent cloud under rotation, we conducted DNS with a constant Rossby number, $Ro = 0.11$, and five different Froude numbers: $Fr = \infty$ (corresponding to $N = 0$, a purely rotating scenario), 0.88, 0.44, 0.22 and 0.11. All simulations started from the same initial condition with no external forcing (i.e., the decaying case). The simulation parameters are given in Table 1, where $Ro = u_0/(2\Omega l_0)$ and $Fr = u_0/(N l_0)$. With the exception of the purely rotating case R0.11, the simulations are denoted by their Froude numbers, e.g., the case F0.88 corresponds to $Fr = 0.88$. For practical reference, we have also listed the Burger number, $Bu = (Ro/Fr)^2$ [42], for each case in Table 1. This number indicates the balance between stratification and rotation in the flows studied.

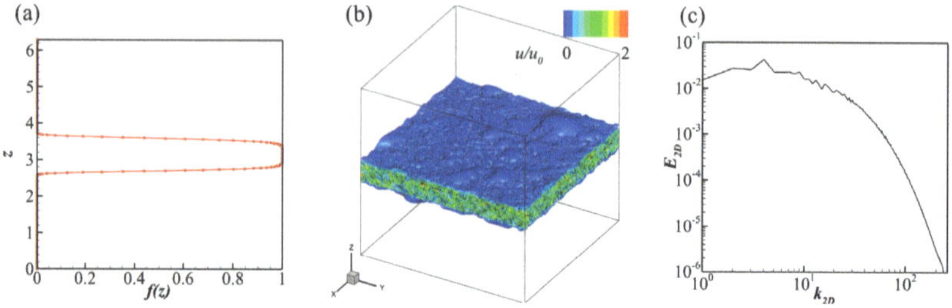

Figure 3. (**a**) The spatial filter given by $f(z) = \exp(-10^a|z-\pi|^m)$ with $a = 4$ and $m = 9$. (**b**) Initial turbulent cloud visualized by the velocity module. (**c**) Energy spectrum in the x-y plane, averaged over different z values around the center of the cloud.

Table 1. Simulation parameters for the DNS of a turbulent cloud. In all cases, we have set $\nu = \kappa = 0.001$.

Case	Resolution	Ro	2Ω	Fr	N	Bu	Re_0	l_c/l_{box}	l_c/l_0
R0.11	512^3	0.11	21.1	∞	0.0	0.00	169	0.13	3
F0.88	512^3	0.11	21.1	0.88	2.6	0.02	169	0.13	3
F0.44	512^3	0.11	21.1	0.44	5.3	0.06	169	0.13	3
F0.22	512^3	0.11	21.1	0.22	10.6	0.25	169	0.13	3
F0.11	512^3	0.11	21.1	0.11	21.1	1.00	169	0.13	3

3.2. Evolution of Flow Structures

To show the evolution of the turbulent cloud structures, Figure 4 shows iso-surfaces of u_z for all simulations at different times. For the case R0.11 under pure rotation, Figure 4a shows the emergence of vertical columnar structures from the turbulent cloud, which extend into the quiescent region, in agreement with previous experimental results [18] and numerical study [19]. For the weakly stratified case F0.88, Figure 4b shows subtle differences in the flow structures at $t/t_f = 8.4$ (middle panels) and more pronounced variations at $t/t_f = 12.7$ (right panels) compared to the case R0.11.

For the moderately stratified cases F0.44 and F0.22, Figure 4c,d show that the columnar structures tend to deviate from a purely vertical orientation as they extend into the quiescent region. This tilt becomes more pronounced with increasing stratification. A comparison of Figure 4a–e shows that, at equivalent time intervals, the vertical growth of these structures decreases with decreasing Fr. This suggests that strong density stratification limits the vertical expansion of the cloud, consistent with the observation in experiments [31].

In the case F0.11 (Figure 4e), the rotation frequency and the Brunt–Väisälä frequency are identical. As a result, no structures emerge from the turbulent cloud, suggesting minimal energy emission from the cloud. This behavior mirrors the case without both rotation and stratification, as documented in Ref. [19]. A plausible interpretation is that the stratification strength is sufficient to completely inhibit the vertical expansion of the turbulent cloud in the presence of system rotation. It is noteworthy that Equation (7) for this scenario yields $\varpi = 2\Omega = N$, implying uniform frequencies of inertial-gravity waves in all directions. This monochromatic behavior is consistent with systems in which both rotation and stratification are absent.

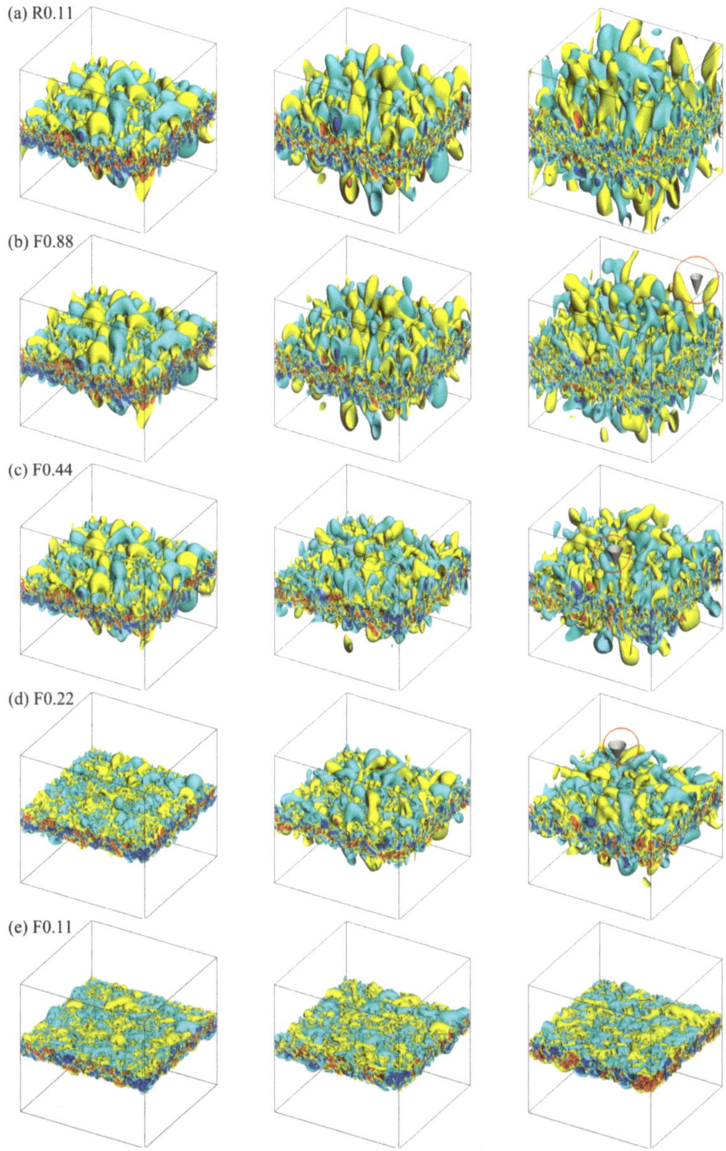

Figure 4. The evolution of the iso-surfaces of u_z for the cases: (**a**) R0.11, (**b**) F0.88, (**c**) F0.44, (**d**) F0.22 and (**e**) F0.11 at t/t_f = 4.2 (left), 8.4 (middle) and 12.7 (right). The color scheme is as follows: $0.24u_0$ in yellow, $0.50u_0$ in red, $-0.24u_0$ in cyan, and $-0.50u_0$ in blue. Cones (highlighted by red circles) are plotted to indicate the predicted tilt angle of flow structures from the linear theory.

3.3. *Are the Flow Structures Formed by Inertial-Gravity Waves?*

Columnar structures from a turbulent cloud under rotation have been identified as inertial waves [19], while the large-scale pancake structures from a stratified turbulent cloud are recognized as internal gravity waves [26]. This section examines whether the flow structures discussed in Section 3.2 from the stratified turbulent cloud under rotation can be attributed to inertial-gravity waves.

In all simulations, the growth of flow structures in different directions leads to the expansion of turbulent clouds in the vertical direction, which can be quantified by tracking the boundary of the $|u_z|$ iso-surfaces with time. The reason we choose u_z to indicate structure is that it satisfies the wave equation as described by (6). In particular, given a threshold $|u_z^b|$, the upper and lower boundaries of the cloud can be obtained from the positions where $|u_z| = |u_z^b|$. We can therefore calculate the vertical extent of the cloud, $h(x, y, t)$, for each point on the x-y plane and time t. The mean wave-cloud thickness $\overline{h}(t)$ is defined as the average of $h(x, y, t)$ over x and y. Figure 5a shows the mean wave-cloud thickness for all the simulations where we choose $|u_z^b| = 0.35u_0$. The results remain robust when the threshold varies around this value, and this threshold is used consistently throughout the analysis in this paper. For the purely rotating case R0.11, the mean wave-cloud thickness shows a linear growth with time, resulting from the inertial waves with $k_z = 0$ traveling at the group velocity $c_g = 2\Omega/k_h$ [19]. For the case F0.88 with weak stratification, the mean wave-cloud thickness is close to that in the case R0.11 at $t \leq 0.2$, indicating that the stratification does not affect the cloud before $t = 0.2$. This is because the linear time scale corresponding to the rotation, $t_f = 1/(2\Omega)$, is smaller than that corresponding to the stratification, $t_N = 1/N$. When $t > 0.2$, the mean wave-cloud thickness shows a smaller linear growth rate. For the cases F0.44 and F0.22 with stronger stratification, the behavior of the mean wave-cloud thicknesses can be interpreted as the superposition of a linear growth and an oscillation with a fixed frequency over time. In addition, the frequency is larger in the case F0.22. The mean wave-cloud thickness in the case F0.11 is almost constant with time, which is consistent with the observation in Figure 4e.

To investigate whether the flow structures are closely related to the inertial-gravity waves, we compare the results from the DNS with those predicted by the linear theory. Since the initial turbulent cloud can be viewed as a distribution of random vortex blobs of different shapes and sizes, we consider the linear evolution of a Gaussian vortex satisfying Equation (15) under the same Rossby and Froude numbers. We set $\Lambda = 2.31$, $\delta = 0.125$, and utilize $|u_z^b| = 0.002\Lambda\delta$ to determine the vertical extent of the eddy, h_e. This extent is defined as the largest vertical distance between the positions where $|u_z| = |u_z^b|$. Figure 5b shows that the results for h_e align qualitatively with those of $\overline{h}$ in Figure 5a, suggesting that the expansion of the turbulent cloud may be prominently influenced by the linear mechanism. Note that the difference for the case F0.11, where the curve appears flat in Figure 5a and wave-like in Figure 5b, arises because $\overline{h}$ represents an average over various x and y, whereas h_e illustrates the results for a single eddy.

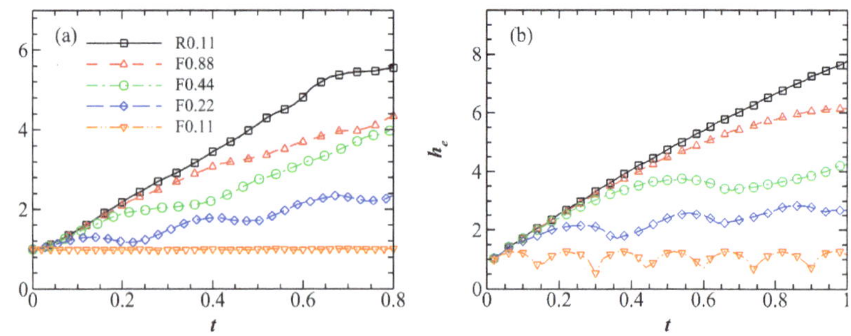

Figure 5. (**a**) The mean wave-cloud thickness $\overline{h}$ of turbulent clouds in the DNS. (**b**) The vertical extent h_e of a Gaussian eddy evolving under the same Ro and Fr as those in the DNS.

We now show quantitatively that the flow structures are closely related to inertial-gravity waves. For a vortex blob in the initial turbulent cloud, we suppose that it is characterized by a length scale of $1/k$. Then its rescaled group velocity in the vertical direction is

$$\tilde{c}_{g,z} = kc_{g,z} = \pm \frac{2\Omega(1-(N/(2\Omega))^2)(k_z/k_h)}{\sqrt{(N/(2\Omega))^2 + (k_z/k_h)^2(1+(k_z/k_h)^2)}}. \quad (17)$$

Given the ratio of the Brunt–Väisälä frequency to the rotation frequency $N/(2\Omega)$, it is determined that when

$$\left(\frac{k_z}{k_h}\right)^2 = \frac{1}{4}\left[-(N/(2\Omega))^2 + \sqrt{(N/(2\Omega))^4 + 8(N/(2\Omega))^2}\right], \quad (18)$$

$\tilde{c}_{g,z}$ reaches the maximum $\tilde{c}_{g,z}^{max}$, which is the vertical growth rate of the turbulent cloud predicted by the linear inertial-gravity waves. Furthermore, the corresponding value of $k_z/k_h = (k_z/k_h)_0$ characterizes the tilt angle of the flow structure. The vertical growth rate of the turbulent cloud in the DNS can be determined from the growth rate of the mean wave-cloud thickness, $\overline{h}(t)$, as shown in Figure 5a. The rate is calculated using a linear fitting method. In Figure 6a we present two vertical growth rates: α_{DNS} derived from DNS and α_{LT} derived from linear theory. To facilitate comparison, both rates are normalized by their values at $N/2\Omega = 0$. Figure 6b shows the angle between the flow structure and the vertical direction, $\theta = \arctan[(k_z/k_h)_0]$, predicted by linear theory for different $N/(2\Omega)$. The results are consistent with Figure 4, where the angle becomes larger as $N/(2\Omega)$ increases. For quantitative comparison, we also plot cones in the third columns of Figure 4b–d, where the angle between the cone surface and the vertical axis is equal to θ. It is notable that there is a strong agreement between the DNS results and the predictions of linear theory. However, for the case F0.22, both α and θ are slightly underestimated by the linear theory. This discrepancy could be due to the interactions between waves and turbulence, especially given the small vertical expansion of the cloud in this case. It is important to note that in scenarios where both rotation and stratification are present, the frequency of the inertial-gravity waves (7) tied to $\tilde{c}_{g,z}^{max}$ is not zero. This explains the oscillatory component observed in $\overline{h}$ for the cases F0.44 and F0.22. When only rotation or stratification is present, the maximum group velocity always coincides with inertial (or gravity) waves of zero frequency. Under these circumstances, tools such as two-dimensional energy spectrum analysis can be used effectively, given the consistent phase of u_z [19] or u_x [26] at the edge of the turbulent cloud.

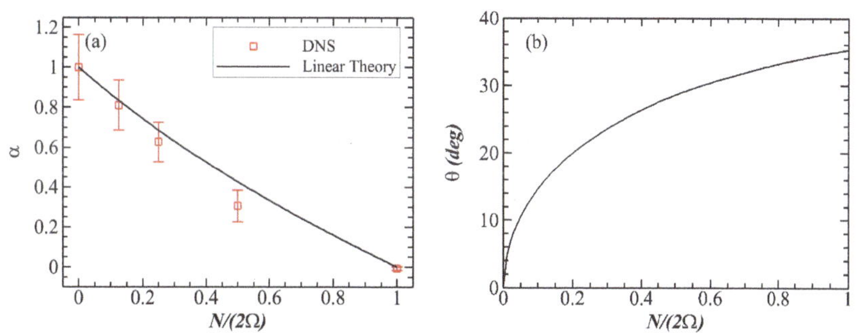

Figure 6. (**a**) Normalized vertical growth rates of the turbulent cloud, α, obtained from the DNS results and the linear theory. (**b**) The angle θ between the flow structure and the vertical direction as predicted by the linear theory.

3.4. *Wave-Dominated and Turbulence-Dominated Regions*

In Section 3.3 it is highlighted that the flow structures emerging from the turbulent cloud show a strong association with inertial-gravity waves. In stratified turbulence, PV has been used to identify turbulent/non-turbulent interfaces [26,43]. Our study, however, uses Lagrangian particle tracking to measure the vertical extent of turbulent advection [19]. Lagrangian particles act as passive tracers, with their position $\boldsymbol{r}$ at time t given by $\mathrm{d}\boldsymbol{r}/\mathrm{d}t = \boldsymbol{u}(\boldsymbol{r}, t)$. In this relationship, $\boldsymbol{u}$ represents the velocity of the fluid. To determine $\boldsymbol{u}$ at the particle positions, we use a 6th-order Lagrange interpolation. Furthermore, the positions of these particles are integrated using a 2nd-order Adams–Bashforth scheme.

Tracer particles with a number $N_p = 65536$ are initialized randomly from a uniform distribution throughout the initial turbulent cloud, the vertical range of which is denoted as $[z_0, z_1]$. Let N_p^{out} be the number of particles outside of the range of the initial turbulent cloud. The r.m.s. distance of these particles to the initial turbulent cloud can be calculated as

$$d_{rms} = \sqrt{\frac{1}{N_p^{out}} \sum_{p=1}^{N_p^{out}} d_p^2}, \tag{19}$$

where

$$d_p = \begin{cases} z_p - z_1, & z_p > z_1 \\ z_0 - z_p, & z_p < z_0 \end{cases} \tag{20}$$

represents the distance of the particle p to the initial turbulent cloud. We consider $[z_0 - d_{rms}, z_1 + d_{rms}]$ as the vertical range of the turbulence-dominated region and the other domain as the wave-dominated region. Figure 7 shows the ratio of particles outside the initial turbulent cloud, N_p^{out}/N_p, and d_{rms} normalized with the initial cloud thickness l_c. For the case R0.11, both N_p^{out}/N_p and d_{rms}/l_c increase with time, while for the cases F0.88 and F0.44 both quantities first increase and then decrease after reaching the maximum values, likely approaching a stationary state due to the stable stratification. However, in the cases F0.22 and F0.11, where the stratification is strong, both quantities immediately reach stationary states around small values.

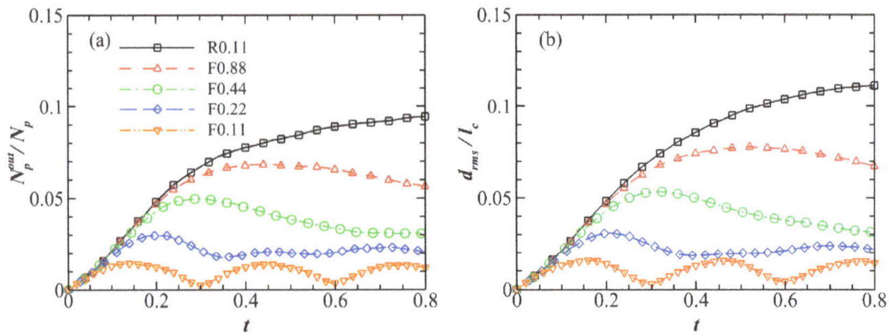

Figure 7. (**a**) The ratio of particles outside the initial turbulent cloud. (**b**) The r.m.s. distance of particles outside the initial turbulent cloud to its boundary.

Table 2 presents the maximum values of N_p^{out}/N_p and d_{rms}/l_c for all cases, where $\max(d_{rms}/l_c)$ can also be interpreted as the maximum vertical expansion of the initial turbulent cloud. It is observed that both quantities become smaller with increasing stratification, indicating the limitation of vertical turbulent advection by stratification.

Table 2. Maximum proportion of particles located outside the initial turbulent cloud, and the maximum vertical cloud expansion for the different cases.

Case	R0.11	F0.88	F0.44	F0.22	F0.11
$\max(N_p^{out}/N_p)$	9.45%	6.86%	4.98%	2.99%	1.44%
$\max(d_{rms}/l_c)$	11.16%	7.77%	5.33%	3.08%	1.57%

Once the regions dominated by waves and turbulence are identified, we can determine the energy of the turbulent cloud carried away by the inertial-gravity waves, denoted as

E^{wave}. This energy is equal to the energy within the wave-dominated region. Figure 8a shows that for cases where $Ro < Fr$, the ratio E^{wave}/E exhibits a monotonic increase with time. In particular, for a given time point, a decrease in Fr results in smaller values of E^{wave}/E. This suggests that under stronger stratification, inertial-gravity waves contain a smaller fraction of the energy. In contrast, for the scenario where $Ro = Fr$, the ratio E^{wave}/E oscillates with a fixed pattern. Its values fluctuate between nearly zero and a peak of about 15%, indicating that the inertial-gravity waves only sporadically extract a small fraction of energy from the turbulent cloud. Thus, while inertial waves account for a significant fraction of the energy in a system dominated by rotation alone, this energy contribution diminishes with the introduction of stratification. Figure 8b shows the ratio of potential energy to total energy in the wave-dominated region, E_P^{wave}/E^{wave}. As stratification increases, potential energy becomes increasingly dominant and varies more frequently with time. This highlights the increasing importance of the kinetic-potential energy exchange and suggests that the dominant inertial-gravity waves exhibit greater phase velocities under increased stratification. It is worth noting that a similar kinetic-potential energy exchange phenomenon has been reported in fully developed rotating stratified turbulence [44].

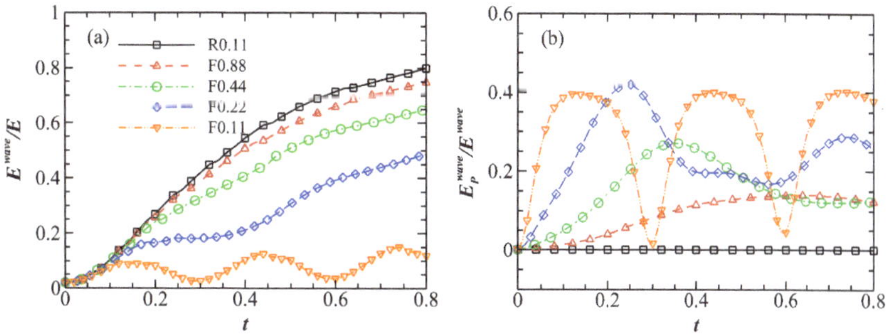

Figure 8. (**a**) The energy ratio within the inertial-gravity waves. (**b**) The ratio of potential energy to total energy in the wave-dominated region.

4. Discussion and Conclusions

In this study, we investigate the evolution of a stratified turbulent cloud influenced by rotation. By analyzing the behavior of a single eddy under both rotation and stratification, we observe clear deviations from the purely rotating scenario. In particular, within the stratified environment, we detect a spatial manifestation of a known PV mode, characterized as a stable vortex at the origin. Furthermore, the system transitions from exhibiting inertial waves to producing inertial-gravity waves, whose propagation relations depend on both rotation and stratification.

Numerical simulations have been performed to investigate the nonlinear evolution of the turbulent cloud influenced by different degrees of stratification under rotation. In cases where $Ro < Fr$, columnar structures emerge spontaneously from the turbulent cloud. These structures are vertically oriented in the case of pure rotation, but deviate from the vertical as stratification is introduced. As Fr decreases, the tilt angle increases and the rate of cloud expansion decreases. Conversely, in the $Ro = Fr$ system, the turbulent cloud remains devoid of distinct structures, reminiscent of scenarios without rotation and stratification. The flow structures are observed to have characteristics consistent with inertial-gravity waves. This observation is supported by the agreement between DNS and linear theory, which gives consistent predictions for the vertical growth rates of the turbulent cloud and the tilt angles of these structures. Using Lagrangian particle tracking, we distinguish between wave-dominated and turbulence-dominated regions. It is found that the inertial gravity waves emitted by the cloud can transport a substantial fraction of the energy to the neighboring quiescent fluid, which decreases with increasing stratification. This suggests the central role of inertial-gravity waves in energy transfer processes within atmospheres and oceans, particularly around regions of localized turbulence.

The present study elucidates the effects of stratification on localized turbulence under system rotation. First, compared to the purely rotating case, the presence of stratification introduces the zero-frequency PV mode, whose energy distribution barely changes with time. While this mode is not new, our finding is its distinct spatial manifestation within a non-uniformly turbulent background. Second, as the stratification intensifies, the dominant inertial-gravity waves exhibit increased tilting and reduced vertical group velocity. In combination, these phenomena limit the energy emitted by the initial turbulent cloud. It should be noted that this study focuses on the parameter regime where $Ro < Fr$ (the corresponding Bu ranges from 0 to 1). While the $Ro > Fr$ regime is thought to dominate in the Earth's atmosphere and oceans [7], the $Ro < Fr$ regime is relevant for several geophysical flows, such as in the deep sea where the influence of stratification is weak but rotation effects are prominent [45–49].

There are still many open questions that need to be addressed. Specifically, the potential role of the Ro/Fr ratio in influencing the flow dynamics and the effect of the Reynolds number warrant further investigation [50,51]. Previous research has shown that the absence of triadic resonance in the parameter range $1/2 \leq N/(2\Omega) \leq 2$ is associated with an enhanced inverse cascade [39,50]. The particular significance of this range requires further investigation. In addition, the pronounced enhancement of vertical velocity, termed "vertical drafts", is a common observation in many DNS studies of stratified flows [52,53] and in the oceans [54]. We did not detect such vertical drafts in our simulations, which may be due to the initialization from homogeneous isotropic turbulence rather than rotating stratified turbulence, or the Froude numbers considered are not favorable for such a phenomenon. Given its importance, this phenomenon is worthy of systematic investigation for the development of turbulent clouds in the future. Furthermore, the study of the corresponding Lagrangian statistics is another interesting research direction [55].

Author Contributions: Conceptualization, T.L., M.W. and S.C.; methodology, T.L.; formal analysis, T.L.; investigation, T.L.; resources, M.W. and S.C.; data curation, T.L.; writing—original draft preparation, T.L.; writing—review and editing, M.W. and S.C.; visualization, T.L.; supervision, M.W. and S.C.; project administration, M.W. and S.C.; funding acquisition, M.W. and S.C. All authors have read and agreed to the published version of the manuscript.

Funding: This research was funded by NSFC Basic Science Center Program, grant number 11988102; NSFC, grant number 12225204; Department of Science and Technology of Guangdong Province, grant number 2023B1212060001; Shenzhen Science and Technology Innovation Commission, grant number KQTD20180411143441009; and European Research Council (ERC) under the European Union's Horizon 2020 research and innovation programme, grant agreement number 882340.

Institutional Review Board Statement: Not applicable.

Data Availability Statement: The data and code used in this study are available on request from the corresponding author. The data and code are not publicly available because we want to understand the specific usage by those requesting access.

Acknowledgments: Numerical simulations were supported by the Center for Computational Science and Engineering of Southern University of Science and Technology. T.L. extends gratitude to Vikrant Gupta for proofreading the manuscript. M.W. acknowledges the support from Centers for Mechanical Engineering Research and Education at MIT and SUSTech.

Conflicts of Interest: The authors declare no conflict of interest.

Abbreviations

The following abbreviations are used in this manuscript:

DNS	Direct numerical simulation
PV	Potential vorticity

References

1. Grant, H.; Moilliet, A.; Vogel, W. Some observations of the occurrence of turbulence in and above the thermocline. *J. Fluid Mech.* **1968**, *34*, 443–448. [CrossRef]
2. Nasmyth, P.W. Oceanic Turbulence. Ph.D. Thesis, University of British Columbia, Vancouver, BC, Canada, 1970.
3. Wijesekera, H.W.; Dillon, T.M. Internal waves and mixing in the upper equatorial Pacific Ocean. *J. Geophys. Res. Ocean.* **1991**, *96*, 7115–7125. [CrossRef]
4. Nash, J.; Alford, M.; Kunze, E.; Martini, K.; Kelly, S. Hotspots of deep ocean mixing on the Oregon continental slope. *Geophys. Res. Lett.* **2007**, *34*, L01605. [CrossRef]
5. Davidson, P. *Turbulence: An Introduction for Scientists and Engineers*; Oxford University Press: Oxford, UK, 2015.
6. Yang, C.F.; Chi, W.C.; van Haren, H. Deep-sea turbulence evolution observed by multiple closely spaced instruments. *Sci. Rep.* **2021**, *11*, 3919. [CrossRef] [PubMed]
7. Riley, J.J.; Lelong, M.P. Fluid motions in the presence of strong stable stratification. *Annu. Rev. Fluid Mech.* **2000**, *32*, 613–657. [CrossRef]
8. Sutherland, B.R. *Internal Gravity Waves*; Cambridge University Press: Cambridge, UK, 2010.
9. Thorpe, S.A. *The Turbulent Ocean*; Cambridge University Press: Cambridge, UK, 2005.
10. Yarom, E.; Sharon, E. Experimental observation of steady inertial wave turbulence in deep rotating flows. *Nat. Phys.* **2014**, *10*, 510–514. [CrossRef]
11. Buzzicotti, M.; Aluie, H.; Biferale, L.; Linkmann, M. Energy transfer in turbulence under rotation. *Phys. Rev. Fluids* **2018**, *3*, 034802. [CrossRef]
12. Alexakis, A.; Biferale, L. Cascades and transitions in turbulent flows. *Phys. Rep.* **2018**, *767*, 1–101. [CrossRef]
13. Di Leoni, P.C.; Alexakis, A.; Biferale, L.; Buzzicotti, M. Phase transitions and flux-loop metastable states in rotating turbulence. *Phys. Rev. Fluids* **2020**, *5*, 104603. [CrossRef]
14. Savaro, C.; Campagne, A.; Linares, M.C.; Augier, P.; Sommeria, J.; Valran, T.; Viboud, S.; Mordant, N. Generation of weakly nonlinear turbulence of internal gravity waves in the Coriolis facility. *Phys. Rev. Fluids* **2020**, *5*, 073801. [CrossRef]
15. Monsalve, E.; Brunet, M.; Gallet, B.; Cortet, P.P. Quantitative experimental observation of weak inertial-wave turbulence. *Phys. Rev. Lett.* **2020**, *125*, 254502. [CrossRef] [PubMed]
16. Hopfinger, E.; Browand, F.; Gagne, Y. Turbulence and waves in a rotating tank. *J. Fluid Mech.* **1982**, *125*, 505–534. [CrossRef]
17. Dickinson, S.C.; Long, R.R. Oscillating-grid turbulence including effects of rotation. *J. Fluid Mech.* **1983**, *126*, 315–333. [CrossRef]
18. Davidson, P.; Staplehurst, P.; Dalziel, S. On the evolution of eddies in a rapidly rotating system. *J. Fluid Mech.* **2006**, *557*, 135–144. [CrossRef]
19. Ranjan, A.; Davidson, P. Evolution of a turbulent cloud under rotation. *J. Fluid Mech.* **2014**, *756*, 488–509. [CrossRef]
20. Thorpe, S. On the layers produced by rapidly oscillating a vertical grid in a uniformly stratified fluid. *J. Fluid Mech.* **1982**, *124*, 391–409. [CrossRef]
21. Browand, F.; Guyomar, D.; Yoon, S.C. The behavior of a turbulent front in a stratified fluid: Experiments with an oscillating grid. *J. Geophys. Res. Ocean.* **1987**, *92*, 5329–5341.
22. De Silva, I.; Fernando, H. Some aspects of mixing in a stratified turbulent patch. *J. Fluid Mech.* **1992**, *240*, 601–625. [CrossRef]
23. De Silva, I.; Fernando, H.J. Experiments on collapsing turbulent regions in stratified fluids. *J. Fluid Mech.* **1998**, *358*, 29–60. [CrossRef]
24. Gilreath, H.; Brandt, A. Experiments on the generation of internal waves in a stratified fluid. *AIAA J.* **1985**, *23*, 693–700. [CrossRef]
25. Rowe, K.; Diamessis, P.; Zhou, Q. Internal gravity wave radiation from a stratified turbulent wake. *J. Fluid Mech.* **2020**, *888*, A25. [CrossRef]
26. Maffioli, A.; Davidson, P.; Dalziel, S.; Swaminathan, N. The evolution of a stratified turbulent cloud. *J. Fluid Mech.* **2014**, *739*, 229–253. [CrossRef]
27. Veronis, G. The analogy between rotating and stratified fluids. *Annu. Rev. Fluid Mech.* **1970**, *2*, 37–66. [CrossRef]
28. Pedlosky, J. *Geophysical Fluid Dynamics*; Springer: New York, NY, USA, 2013.
29. Vallis, G.K. *Atmospheric and Oceanic Fluid Dynamics*; Cambridge University Press: Cambridge, UK, 2017.
30. Manins, P. Intrusion into a stratified fluid. *J. Fluid Mech.* **1976**, *74*, 547–560. [CrossRef]
31. Davies, P.A.; Fernando, H.J.; Besley, P.; Simpson, R.J. Generation and spreading of a turbulent mixed layer in a rotating, stratified fluid. *J. Geophys. Res. Ocean.* **1991**, *96*, 12567–12585. [CrossRef]
32. Folkard, A.M.; Davies, P.A.; Fernando, H.J. Measurements in a turbulent patch in a rotating, linearly-stratified fluid. *Dyn. Atmos. Ocean.* **1997**, *26*, 27–51. [CrossRef]
33. Wells, J.R.; Helfrich, K.R. A laboratory study of localized boundary mixing in a rotating stratified fluid. *J. Fluid Mech.* **2004**, *516*, 83–113. [CrossRef]
34. Ranjan, A.; Davidson, P. DNS of a Buoyant Turbulent Cloud under Rapid Rotation. In *Advances in Computation, Modeling and Control of Transitional and Turbulent Flows*; World Scientific: Singapore, 2016; pp. 452–460.
35. Emery, W.; Lee, W.; Magaard, L. Geographic and seasonal distributions of Brunt-Väisälä frequency and Rossby radii in the North Pacific and North Atlantic. *J. Phys. Ocean.* **1984**, *14*, 294–317. [CrossRef]
36. Jones, E.; Rudels, B.; Anderson, L. Deep waters of the Arctic Ocean: Origins and circulation. *Deep. Sea Res. Part I Oceanogr. Res. Pap.* **1995**, *42*, 737–760. [CrossRef]

37. Woodgate, R.A.; Aagaard, K.; Muench, R.D.; Gunn, J.; Björk, G.; Rudels, B.; Roach, A.; Schauer, U. The Arctic Ocean boundary current along the Eurasian slope and the adjacent Lomonosov Ridge: Water mass properties, transports and transformations from moored instruments. *Deep. Sea Res. Part I Oceanogr. Res. Pap.* **2001**, *48*, 1757–1792. [CrossRef]
38. Lesieur, M. *Turbulence in Fluids: Stochastic and Numerical Modelling*; Springer Science & Business Media: Dordrecht, The Netherlands, 1987.
39. Smith, L.M.; Waleffe, F. Generation of slow large scales in forced rotating stratified turbulence. *J. Fluid Mech.* **2002**, *451*, 145–168. [CrossRef]
40. Li, T.; Wan, M.; Wang, J.; Chen, S. Flow structures and kinetic-potential exchange in forced rotating stratified turbulence. *Phys. Rev. Fluids* **2020**, *5*, 014802. [CrossRef]
41. Canuto, C.; Hussaini, M.Y.; Quarteroni, A.; Thomas, A., Jr. *Spectral Methods in Fluid Dynamics*; Springer Science & Business Media: Berlin/Heidelberg, Germany, 2012.
42. Cushman-Roisin, B.; Beckers, J.M. *Introduction to Geophysical Fluid Dynamics: Physical and Numerical Aspects*; Academic Press: Waltham, MA, USA, 2011.
43. Watanabe, T.; Riley, J.J.; de Bruyn Kops, S.M.; Diamessis, P.J.; Zhou, Q. Turbulent/non-turbulent interfaces in wakes in stably stratified fluids. *J. Fluid Mech.* **2016**, *797*, R1. [CrossRef]
44. Li, T.; Wan, M.; Wang, J.; Chen, S. Spectral energy transfers and kinetic-potential energy exchange in rotating stratified turbulence. *Phys. Rev. Fluids* **2020**, *5*, 124804. [CrossRef]
45. Wingate, B.A.; Embid, P.; Holmes-Cerfon, M.; Taylor, M.A. Low Rossby limiting dynamics for stably stratified flow with finite Froude number. *J. Fluid Mech.* **2011**, *676*, 546–571. [CrossRef]
46. Heywood, K.J.; Garabato, A.C.N.; Stevens, D.P. High mixing rates in the abyssal Southern Ocean. *Nature* **2002**, *415*, 1011–1014. [CrossRef]
47. Van Haren, H.; Millot, C. Gyroscopic waves in the Mediterranean Sea. *Geophys. Res. Lett.* **2005**, *32*, L24614. [CrossRef]
48. Timmermans, M.L.; Melling, H.; Rainville, L. Dynamics in the deep Canada Basin, Arctic Ocean, inferred by thermistor chain time series. *J. Phys. Oceanogr.* **2007**, *37*, 1066–1076. [CrossRef]
49. Timmermans, M.L.; Rainville, L.; Thomas, L.; Proshutinsky, A. Moored observations of bottom-intensified motions in the deep Canada Basin, Arctic Ocean. *J. Mar. Res.* **2010**, *68*, 625–641. [CrossRef]
50. Marino, R.; Mininni, P.D.; Rosenberg, D.; Pouquet, A. Inverse cascades in rotating stratified turbulence: Fast growth of large scales. *Europhys. Lett.* **2013**, *102*, 44006. [CrossRef]
51. Rosenberg, D.; Pouquet, A.; Marino, R.; Mininni, P.D. Evidence for Bolgiano-Obukhov scaling in rotating stratified turbulence using high-resolution direct numerical simulations. *Phys. Fluids* **2015**, *27*, 055105. [CrossRef]
52. Feraco, F.; Marino, R.; Pumir, A.; Primavera, L.; Mininni, P.D.; Pouquet, A.; Rosenberg, D. Vertical drafts and mixing in stratified turbulence: Sharp transition with Froude number. *Europhys. Lett.* **2018**, *123*, 44002. [CrossRef]
53. Marino, R.; Feraco, F.; Primavera, L.; Pumir, A.; Pouquet, A.; Rosenberg, D.; Mininni, P.D. Turbulence generation by large-scale extreme vertical drafts and the modulation of local energy dissipation in stably stratified geophysical flows. *Phys. Rev. Fluids* **2022**, *7*, 033801. [CrossRef]
54. D'Asaro, E.A.; Lien, R.C.; Henyey, F. High-frequency internal waves on the Oregon continental shelf. *J. Phys. Oceanogr.* **2007**, *37*, 1956–1967. [CrossRef]
55. Buaria, D.; Pumir, A.; Feraco, F.; Marino, R.; Pouquet, A.; Rosenberg, D.; Primavera, L. Single-particle Lagrangian statistics from direct numerical simulations of rotating-stratified turbulence. *Phys. Rev. Fluids* **2020**, *5*, 064801. [CrossRef]

atmosphere

Article

Dimensional Transitions in Turbulence: The Effects of Rotation and Stratification

Guido Boffetta

Department of Physics and INFN, University of Torino, via Pietro Giuria 1, 10125 Torino, Italy; guido.boffetta@unito.it

Abstract: The transition from two-dimensional to three-dimensional turbulence is a fascinating problem which finds applications in the study of geophysical flows. This paper briefly reviews the research in this field with emphasis on the role of rotation and stratification, two important ingredients of geophysical flows at large scales. By means of direct numerical simulations of the Navier–Stokes equations, the conditions for the emergence of a split cascade, with a simultaneous cascade of energy to both the large and the small scales, are discussed.

Keywords: turbulence; two-dimensional turbulence; rotating turbulence

Citation: Boffetta, G. Dimensional Transitions in Turbulence: The Effects of Rotation and Stratification. *Atmosphere* **2023**, *14*, 1688. https://doi.org/10.3390/atmos14111688

Academic Editors: Boris Galperin, Annick Pouquet and Peter Sullivan

Received: 10 October 2023
Revised: 7 November 2023
Accepted: 10 November 2023
Published: 15 November 2023

1. Introduction

Geophysical flows at large scales can often be considered as quasi-two-dimensional (2D). While this is evident from a geometrical point of view, the effects of confinement on the dynamics of the flow are less obvious, since the confinement is much larger than the dissipative viscous scales and the flow is intrinsically three-dimensional (3D). Strictly speaking, no real flow is two-dimensional. Nonetheless, when one considers a very thin layer of fluid, one can safely assume that both the vertical velocity and the vertical dependence of the fields are suppressed and the flow can be described by the 2D version of the Navier–Stokes equation. We recall that the main difference between the 3D and the 2D versions is the absence, for the latter, of the *vortex stretching* term which is responsible for the non-conservation (and the amplification) of vorticity [1]. As a consequence, in two dimensions the enstrophy (average square vorticity) is an inviscid invariant together with the kinetic energy (average square velocity), and this prevents the viscous dissipation of kinetic energy at small scales, which therefore flows to the large scales, producing the inverse cascade.

Turbulent flows in quasi-2D geometries exhibit a rich and interesting phenomenology which has stimulated several studies. Numerical simulations [2–4] and laboratory experiments [5,6] have shown that, by changing the confinement, 2D and 3D phenomenologies coexist in a range of parameters with the simultaneous presence of an inverse and a direct cascade of energy. The key parameter which controls the relative flux of the two cascades is the ratio S between the confining scale and the characteristic forcing scale at which energy is injected. In particular, it has been found that there exists a critical ratio S_c above which the inverse cascade vanishes and the turbulent flow fully recovers the 3D phenomenology [2]. We remark that in particular conditions, such as the presence of a large-scale vortex, this transition can become more complex and displays bistability and hysteresis [7,8].

The dimensional transition, and in particular the critical value S_c, is affected by other physical ingredients, including rotation [9,10] and stratification, the presence of a magnetic field, or more exotic features [11]. A rotation of the flow in the direction of confinement in general favors bidimensionalization, i.e., boosts the inverse flux with respect to the non-rotating case at fixed S, and increases the critical value S_c. Conversely, the presence of a stable stratification of the density field in the Boussinesq approximation reduces the inverse energy flux and therefore decreases the critical value S_c. In this paper, we will review the dimensional transition in turbulent flows both induced by spatial confinement alone and with the presence of rotation or stratification. The results presented are obtained on the basis of extensive direct numerical simulations of the Navier–Stokes–Boussinesq model.

Jack Herring made fundamental and pioneering contributions in the study of two-dimensional turbulence by comparing the statistical theories with direct numerical simulations and recognizing the physical origin of the discrepancies observed in the direct cascade of enstrophy [12] and its non-universality with respect to the forcing statistics [13], highlighting one important difference between 2D and 3D turbulence. Remarkably, in an important paper in 1989, Herring also observed the emergence of an inverse cascade in a forced, stably stratified three-dimensional turbulent flow [14], thus anticipating by many years the systematic studies of dimensional transition in turbulence which is the subject of this contribution.

The remaining part of the paper is organized as follows. Section 2 introduces the mathematical model and the relevant dimensionless number of the problem, while in Section 3, we report the results of numerical simulations. Section 4 is devoted to the conclusions.

2. Mathematical Models

We consider, in general, an incompressible flow in the presence of a stable stratification in the (vertical) direction of the acceleration $\boldsymbol{g} = (0, 0, -g)$ with a constant density gradient γ and in a frame of reference rotating with constant angular velocity $\boldsymbol{\Omega} = (0, 0, \Omega)$. The flow is confined in a box of volume $V = L_x \times L_y \times L_z$, and, in order to preserve homogeneity, periodic boundary conditions are imposed in all the directions. In the Boussinesq approximation, the equation of motion for the velocity field $\boldsymbol{u}(\boldsymbol{x}, t)$ and the density $\phi(\boldsymbol{x}, t)$ are, respectively,

$$\partial_t \boldsymbol{u} + \boldsymbol{u} \cdot \boldsymbol{\nabla} \boldsymbol{u} + 2\boldsymbol{\Omega} \times \boldsymbol{u} = -\frac{1}{\rho_0}\boldsymbol{\nabla} p + \nu \nabla^2 \boldsymbol{u} - N\phi \boldsymbol{e}_3 + \boldsymbol{f} \tag{1}$$

$$\partial_t \phi + \boldsymbol{u} \cdot \boldsymbol{\nabla} \phi = \kappa \nabla^2 \phi + N \boldsymbol{e}_3 \cdot \boldsymbol{u} \tag{2}$$

together with the incompressibility condition $\boldsymbol{\nabla} \cdot \boldsymbol{u} = 0$. In (1), ν is the kinematic viscosity, κ is the scalar diffusivity, and $N = \sqrt{\gamma g / \rho_0}$ is the Brunt–Väisäla frequency associated with the stratification. The field ϕ has the dimension of a velocity and is proportional to the deviation of the density field from the linear stable profile, $\rho = \rho_0 + \gamma(\phi/N - z)$. The flow is sustained by the external forcing $\boldsymbol{f}(\boldsymbol{x}, t)$, which is a stochastic Gaussian, white in time noise, active only on the horizontal components u_x, u_y of the velocity, and depends on the horizontal components x, y only. The forcing is localized in Fourier space in a narrow band of wave numbers around $k_f = 2\pi / L_f$ and injects energy into the system at a fixed rate ε_I [15].

The relevant dimensionless parameters are the *thickness numbers*

$$S = \frac{L_z}{L_f} \tag{3}$$

which quantify the vertical confinement with respect to the forcing scale, the *Froude number*

$$Fr = \frac{(k_f^2 \varepsilon_I)^{1/3}}{N} \tag{4}$$

ratio of the forcing frequency to the Brunt–Väisäla frequency, and the *Rossby number*

$$Ro = \frac{(k_f^2 \varepsilon_I)^{1/3}}{\Omega} \tag{5}$$

which is the ratio of the characteristic frequency at the forcing scale to the rotation rate. We remark that, since the flow has periodic boundary conditions, the effect of a finite L_z is that just of reducing the number of modes available in the z direction. Different boundary conditions, such as those present in laboratory experiments, can change quantitatively the results [16]. We also remark that the horizontal scales L_x and L_y are assumed to be much larger than the other scales, and they will not play any role in the dynamics.

In the absence of forcing and dissipation, (1) and (2) conserve the total energy density, given by the sum of kinetic and potential contributions

$$E = E_k + E_p = \frac{1}{2}\langle|\boldsymbol{u}|^2\rangle + \frac{1}{2}\langle\phi^2\rangle \tag{6}$$

where $\langle\ldots\rangle \equiv \frac{1}{V}\int d^3x$ represents the average over the domain of volume V. From (1) and (2), the balances of kinetic and potential energy can be written as

$$\frac{dE_k}{dt} \equiv \varepsilon_\alpha = \varepsilon_f - \varepsilon_\nu - \varepsilon_x \tag{7}$$

$$\frac{dE_p}{dt} = \varepsilon_x - \varepsilon_\kappa \tag{8}$$

where $\varepsilon_\nu = \nu\langle\partial_i u_j \partial_i u_j\rangle$ (summation over repeated indices is assumed) is the viscous dissipation rate and $\varepsilon_\kappa = \kappa\langle\partial_i\phi\partial_i\phi\rangle$ is the diffusive dissipation rate. By introducing the vorticity field $\boldsymbol{\omega} = \boldsymbol{\nabla}\times\boldsymbol{u}$, the viscous dissipation can be written ad $\varepsilon_\nu = 2\nu Z$, where $Z = \langle|\boldsymbol{\omega}|^2\rangle/2$ is the enstrophy. The term $\varepsilon_x = N\langle u_3\phi\rangle$ represents the exchange rate between kinetic and potential energies. Although the sign of this exchange rate is not defined a priori, the analysis of the Kármán–Howarth–Monin equations shows that it is positive on average [17]. This indicates that part of the kinetic energy is converted to produce potential energy in the flow, which is eventually dissipated by the last term in (8). In the presence of an inverse cascade, the flow displays a long non-stationary transient with the energy being transferred to large scales. Eventually the energy reaches the largest scale (the box size) and starts to accumulate until the amplitude of this mode saturates by the dissipation induced by viscosity [18,19]. Since in a turbulent flow the Reynolds number (inverse of viscosity) is large, this effect appears for a very long time only, and the rate ε_α of energy growth during this long transient is therefore used as a measure of the strength of the inverse cascade.

Direct numerical simulations of the models (1) and (2) are performed in a box of dimension $L_x = L_y$ and L_z discretized on a uniform grid at resolution $N_x \times N_y \times N_z$ by a fully parallel pseudospectral code, with a standard 2/3 dealiasing scheme. Most of the results are obtained at a resolution $N_x = N_y = 512$ and $N_z = SN_x/8$ with $S = L_z/L_f$ ranging from $S = 0.25$ to $S = 8.0$ and forcing wavenumber $k_f = 8$. For the non-rotating, non-stratified case we use also data from a simulation at $N_x = N_y = 4069$ and $N_z = 64$ with $k_f = 16$. For rotating turbulence, we analyze six simulations at Rossby number $Ro = 0.1, 0.2, 0.67, 1.0, 1.33, 2.0$, while for the stratified flow we have eight different Froude numbers $Fr = 0.15, 0.2, 0.25, 0.3, 0.4, 0.5, 0.6, 0.75$. In all the simulations, the viscous and the diffusive terms are replaced by a hyperviscous damping scheme of higher order [20]. More details on the numerical simulations can be found in [17,20,21].

3. Numerical Results

3.1. *Transition in the Absence of Rotation and Stratification*

The simplest case is when rotation and stratification can be neglected. Formally, this corresponds to having $Ro^{-1} = Fr^{-1} = 0$ (in terms of the physical parameters $\Omega = 0$ and $N = 0$). In this limit, (2) decouples from (1): ϕ becomes a passive scalar which can be neglected since it does not affect the dynamics. The thickness scale controls the fate of the energy injected in the flow at the scale L_f. Extensive numerical simulations [2,4,20] complemented with laboratory experiments [5,6] show that when the thickness is large

($S > 1$), the flow develops a 3D motion at the injection scale and energy is transferred to small scales by a direct cascade, as in the usual 3D turbulence. In the limit of small $S \ll 1$, the flow becomes essentially two-dimensional (i.e., vertical motions are suppressed) and an inverse cascade of energy is observed [22]. Remarkably, for intermediate values of the thickness number, one observes a *split cascade* with a fraction of the injected energy going to the large scales and the remaining energy flowing to small scales. In this regime, the fraction of energy flowing to large/small scales is controlled by the parameter S.

Figure 1 shows the fraction of energy transferred to large scales (i.e., the rate of kinetic energy growth), which quantifies the strength of the inverse cascade as a function of S from direct numerical simulations of a thin layer of fluid. As shown in similar works [4] for $S \geq S_c \simeq 1/2$, the inverse cascade disappears and all the energy flows to small scales where it is dissipated by viscosity. As shown by Figure 1, for $S = 1/8$, about one-half of the kinetic energy is transferred to the large scales.

We remark that the threshold $S_c = 1/2$ is not expected to be a universal value as it depends on the details of the forcing and also on the precise definition of L_f. Indeed, different forcing schemes lead to different values of S_c at which the inverse flux vanishes [2,23].

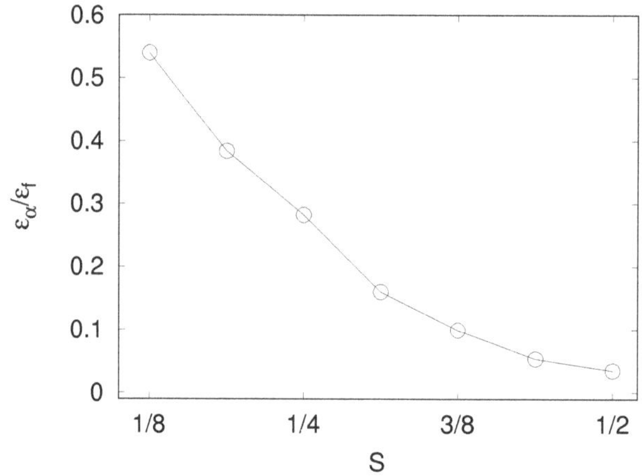

Figure 1. Growth rate of kinetic energy ε_α normalized with the energy input ε_f as a function of the thickness number $S = L_z/L_f$. Simulations at resolution $N_x = N_y = 512$, $N_z = 64S$ data from [17].

The kinetic energy spectrum in the split cascade regime at $S = 1/4$ is shown in Figure 2. The spectrum is computed at a late time when the inverse cascade is well developed and we are able to observe the presence of three different scaling laws. Only the two horizontal components of the velocity are used to compute the spectrum, and the wavenumber is restricted to the horizontal wavenumber $k = (k_x^2 + k_y^2)^{1/2}$ on the plane $k_z = 0$. This is equivalent to the spectrum of the velocity field averaged over the vertical direction z.

At large scales, $k < k_f$, we clearly see the inverse energy cascade with Kolmogorov scaling $k^{-5/3}$ [24]. At intermediate scales, $k_f < k < k_z$, we observe a steeper spectrum, compatible with a direct enstrophy cascade. Finally, for $k > k_f$, the flow becomes three-dimensional, and a direct energy cascade, again with Kolmogorov scaling, is observed. We remark that the intermediate, enstrophy cascade, spectrum is expected in a narrow range of wavenumbers with ratio $k_z/k_f = 1/S$. Figure 2 also shows the spectrum of the vertical component of the velocity. This is strongly suppressed at large scales, but it reaches the horizontal spectrum at $k \geq k_z$, an indication of the recovery of isotropy at small scales.

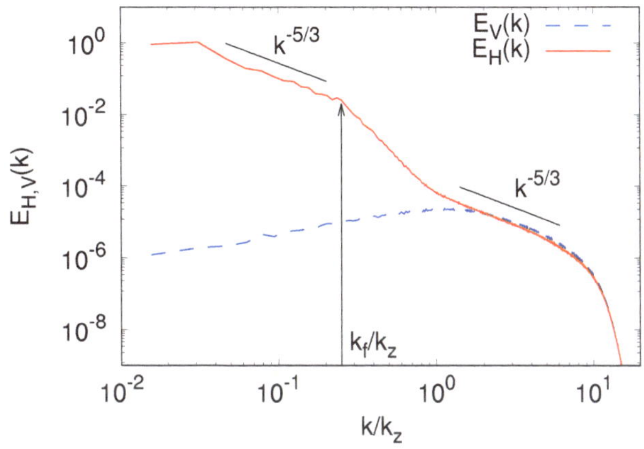

Figure 2. Two-dimensional energy spectra of the vertically averaged 2*D* horizontal velocity field (red, solid line) and of the vertically averaged vertical velocity field (blue, dashed line) for a simulation at $S = 1/4$. Resolution $N_x = N_y = 4096$, $N_z = 64$. Data from [20].

In order to better understand the phenomenology of the split cascade, it is useful to consider the vorticity equation, obtained by taking the curl of (1):

$$\partial_t \omega + \boldsymbol{u} \cdot \boldsymbol{\nabla} \omega = \omega \cdot \boldsymbol{\nabla} \boldsymbol{u} + \nu \nabla^2 \omega f_\omega \tag{9}$$

where $f_\omega = \boldsymbol{\nabla} \times \boldsymbol{f}$. As discussed in the Introduction, in two-dimensions the vortex stretching term in (9) identically vanishes, and therefore, in that case the enstrophy is a second inviscid invariant. This is the main difference with respect to 3D turbulence where the vortex stretching term is responsible for the increase in enstrophy which dissipates the energy flowing to small scales.

In the case considered here, the flow is three-dimensional, and therefore enstrophy is not expected to be conserved. Nonetheless, the appearance of a 2D-like phenomenology suggests that, at large scales, the vortex stretching terms are dynamically suppressed and therefore the "large scale enstrophy" is conserved. In order to quantify this effect, we compute the different contributions to the enstrophy flux. By taking the Fourier transform of (9) and multiplying by $\hat{\omega}^*$ (the conjugate of the Fourier transform), we define

$$\Pi_Z(k) = \int_{|q| \le k} \hat{\omega}^*(\boldsymbol{q}) (\widehat{\boldsymbol{u} \cdot \boldsymbol{\nabla} \omega})(\boldsymbol{q}) d\boldsymbol{q} \tag{10}$$

as the enstrophy flux and

$$\Sigma_Z(k) = \int_{|q| \le k} \hat{\omega}^*(\boldsymbol{q}) (\widehat{\omega \cdot \boldsymbol{\nabla} \boldsymbol{u}})(\boldsymbol{q}) d\boldsymbol{q} \tag{11}$$

as the spectral enstrophy production. While $\Pi_Z(k)$ represents the transport of enstrophy across the wavenumber k by the nonlinear (advection) term, $\Sigma_Z(k)$ is the generation of enstrophy at wavenumbers smaller than k due to the vortex stretching mechanism. The results from the simulation are shown in Figure 3. For wavenumber $k < k_f$, in the region of the inverse energy cascade, both the enstrophy flux and production are close to zero, as in the 2D case. In the intermediate range of wavenumber $k_f < k < k_z$, the production of enstrophy (i.e., the vortex stretching term) is still negligible while the flux is constant, as in a 2D direct enstrophy cascade. At large wavenumbers $k > k_z$, the production of enstrophy becomes significant, and therefore the flux is not constant anymore but grows following the production term.

The fact that enstrophy behaves at large scales as a quasi-invariant is remarkable since it indicates that the development of the inverse energy cascade is not caused by the two-dimensionalization of the flow but by the presence of a second sign-definite conserved quantity at large scales. The development of an inverse energy cascade due to the presence of a second inviscid invariant has been observed also in other turbulent systems, e.g., in a homogeneous isotropic 3D flow with mirror symmetry broken such that helicity has a well-defined sign at all wave numbers [25].

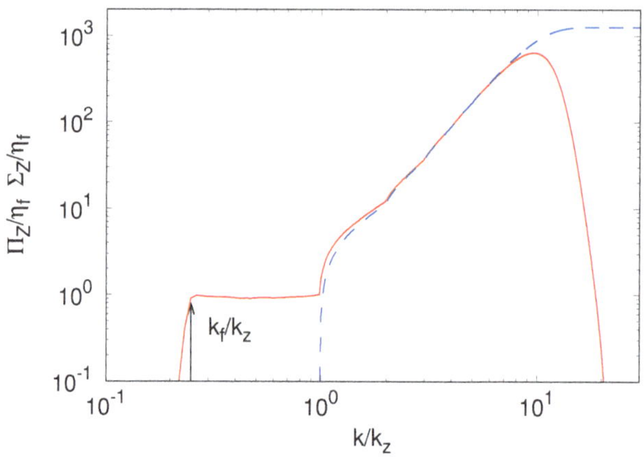

Figure 3. Spectral enstrophy flux $\Pi_Z(k)$ (red, solid line) and spectral enstrophy production $\Sigma_Z(k)$ (blue, dashed line) for a simulation at $S = 1/4$. All quantities are normalized with the enstrophy input η_f. Resolution $N_x = N_y = 4096$, $N_z = 64$, data from [20].

3.2. Dimensional Transition in Rotating Turbulence

We now consider how the phenomenology described in the previous section changes in the presence of rotation. To this aim we consider the case of finite Rossby number Ro in the limit of infinite Froude number Fr. This corresponds to integrating (9) with $N = 0$ and $\Omega > 0$. The results of this section are to a large extent presented in [21].

It is well known that rotation in general favors the two-dimensionalization of the flow, and the Taylor–Proudman theorem supports this idea. In turbulent flows, the onset of an inverse cascade in the presence of rotation has been observed in laboratory experiments [26,27]. However, it is not obvious how, starting from a 3D turbulent flow, the actual two-dimensionalization would occur. Different mechanisms have been proposed, including (nearly) resonant triad interactions [28,29] and weak turbulence theory [30,31].

In the presence of confinement, it has been observed that a moderate rotation enhances the fraction of energy which is transferred to large scales and that the split cascade survives also for $S > 1/2$ [21]. Figure 4 displays the fraction of energy transferred to large scale for different Rossby numbers. We observe that indeed ε_α increases by decreasing Ro and that, for the strongest rotation (smallest Ro), a residual inverse cascade is present even for $S = 10$. We remark that this result is consistent with the observation, based on asymptotic techniques, that for sufficiently rapid rotation, the dimensional transition persists for large S [10]. Moreover, in the limit of strong rotation it can be shown that 2D turbulence is stable to vertically dependent perturbations [32,33]. In the case of fast rotation, it has been shown that the inverse cascade is also sustained by the interaction among triads with the same sign of helicity [34]. From the data in Figure 4, it is evident that S_c is a decreasing function of Ro.

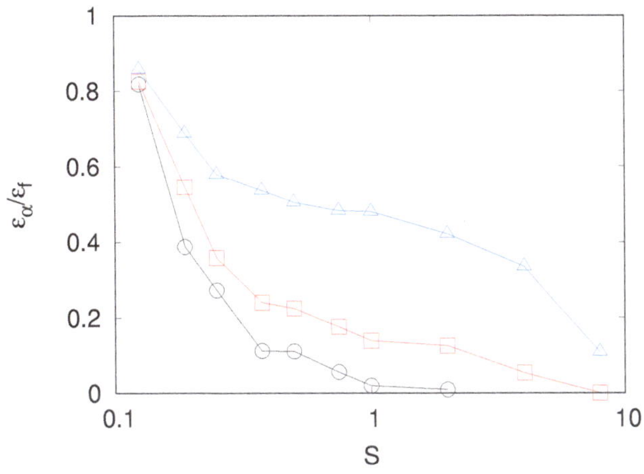

Figure 4. Relative strength of the inverse cascade ε_α as a function of the confinement number S for different values of the Rossby number. $Ro = 0.67$ (blue triangles), $Ro = 1.0$ (red squares), and $Ro = 1.33$ (black circles). Simulations with $N_x = N_y = 512$ and $N_z = 64S$, data from [21].

Figure 5 shows the spectral enstrophy flux and the enstrophy production for a non-rotating ($Ro \to \infty$) simulation at $S = 0.188$ and a simulation with $Ro = 2/3$ and $S = 4$. By comparing Figures 1 and 4, we see that the two flows have approximatively the same ratio between the inverse and direct energy fluxes, and therefore it is interesting to compare whether this is due to the same mechanism. Figure 5 shows that this is not the case. As discussed in the previous section, in the non-rotating case (black lines) the vortex stretching contribution to the flux is completely suppressed for $k < k_z$, and a direct cascade with an almost constant flux of enstrophy is observed for $k_f < k < k_z$. In the case of rotating turbulence (red lines), we observe that the suppression of the vortex stretching term is less sharp, and, consequently, there is no evidence of a direct cascade of enstrophy with a constant flux. This suggests that a different mechanism for the split cascade is at work in this case. In spite of these differences, it is interesting to observe that at small scales the enstrophy production and the enstrophy flux of the two cases become indistinguishable.

3.3. Cyclonic-Anticyclonic Asymmetry in a Thin Layer

One of the distinctive features of rotating turbulence is the breakdown of the symmetry between cyclonic (i.e., rotating in the direction of $\mathbf{\Omega}$) and anticyclonic vortices, i.e., an asymmetric distribution of the vorticity ω_z with a predominance of positive values. This feature has been observed in many experiments and numerical simulations [35–40], both in forced and decaying turbulence. Nonetheless, the origin of the asymmetry is still not fully understood, and different types of arguments have been proposed to explain the phenomenon [41,42]. A simple, natural way to quantify the asymmetry of ω_z is by computing the skewness of its distribution

$$Sk = \frac{\langle \omega_z^3 \rangle}{\langle \omega_z^2 \rangle^{3/2}} \, . \tag{12}$$

It is interesting to study how this asymmetry, due to the background rotation of the flow, is influenced by the confinement in the direction of rotation. Indeed, in the limit of a two-dimensional flow, $S \to 0$, rotation cannot induce an asymmetry in the distribution of ω_z since in this case the Coriolis term $\mathbf{\Omega} \times \boldsymbol{u}$ disappears in the 2D version of (9). The dependence of Sk on the confinement S for a turbulent flow rotating with $Ro = 2/3$ is shown in Figure 6. We observe that indeed $Sk \to 0$ at small S, as expected for a purely 2D flow. For $S > 1$, the numerical indication is that the skewness saturates at a value $Sk \simeq 1$, at least in the range of confinement studied. We remark that for all these values of S, an inverse cascade is present in the flows, as shown by Figure 4.

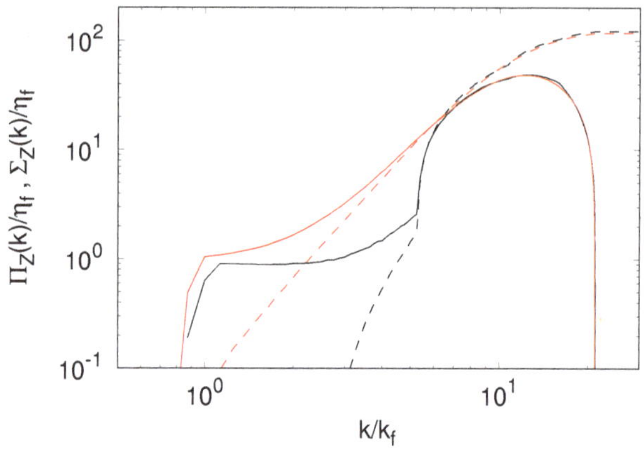

Figure 5. Spectral enstrophy fluxes $\Pi_Z(k)$ (solid lines) and enstrophy productions $\Sigma_Z(k)$ (dashed lines) for two simulations: $Ro \to \infty$ and $S = 0.188$ (lower, black lines) and $Ro = 2/3$ and $S = 4$ (upper, red lines). All quantities are normalized with the enstrophy input. Simulations at $N_x = N_y = 512$, $N_z = 64S$. Data from [21].

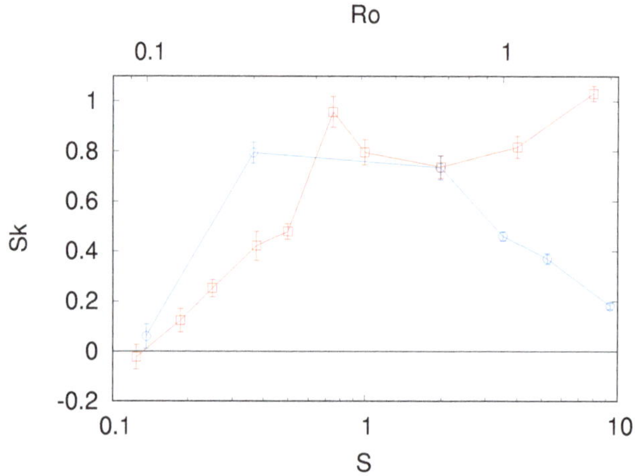

Figure 6. Skewness of the distribution of the vertical vorticity ω_z as a function of the thickness number S for Rossby number $Ro = 2/3$ (red squares) and as a function of the Rossby number Ro for the thickness number $S = 2$ (blue circles). The error bars represent the error on the mean over the duration of the simulation. Simulations at $N_x = N_y = 512$, $N_z = 64S$. Data from [21].

Figure 6 also shows the dependence of Sk on the Rossby number. We observe that in this case, the behavior is non-monotonic with a maximum of asymmetry for intemediate values of Ro.

3.4. Dimensional Transition in Stably Stratified Flows

In this section, we consider the effect of a stable density stratification on the dimensional turbulent transition in a thin layer. We recall that stably stratified flows occur naturally in many instances of geophysical flows, both in the ocean and in the atmosphere.

Therefore, many laboratory experiments and numerical simulations have been devoted to the study of stratified turbulence [43–45]. In particular, we remark that early numerical simulations of stratified turbulence reported the existence of an inverse cascade [14], while other simulations have observed a direct cascade [46,47], in particular in the limit of strong stratification [48,49].

From these considerations, it is therefore interesting to study how stratification affects the dimensional transition in a thin turbulent layer and in particular how critical thickness S_c observed in non-stratified flows is affected. This problem has been addressed on the basis of extensive direct numerical simulations of the models (1) and (2) with $\mathbf{\Omega} = 0$ and for different values of the Froude number Fr and of the thickness number S. We remark that the two limits of thin layer $S \to 0$ and of strong stratification $Fr \to 0$ are not expected to commute. In particular, if the limit $Fr \to 0$ is taken first, one cannot expect that the 2D limit is recovered when $S \to 0$ [17].

For each simulation at given values of the parameters (Fr, S), the growth rate of kinetic energy is measured and used, according to (7), as a proxy for the inverse energy flux ε_α. As shown in Figure 7, for each value of the stratification parameter Fr, the ratio $\varepsilon_\alpha/\varepsilon_f$ is a decreasing function of the thickness number S. This indicates that the presence of a stable stratification in the flow reduces the intensity of the inverse cascade.

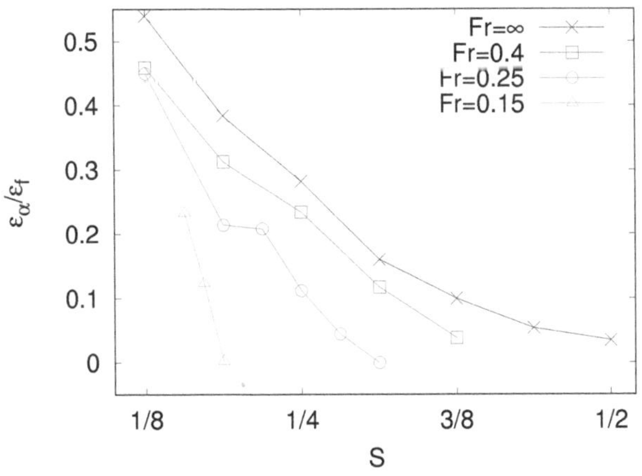

Figure 7. Relative inverse flux $\varepsilon_\alpha/\varepsilon_f$ as a function of the thickness number $S = L_z/L_f$ for different values of the Froude number Fr. Simulations at resolutions $N_x = N_y = 512$, $Nz = 64S$. Data from [17].

The inverse flux is found to vanish at the critical thickness $S_c(Fr)$, which becomes smaller as the stratification increases. The critical value is bounded by the value in the absence of stratification, $S_c(Fr) \leq S_c(\infty) \simeq 1/2$. Figure 8 shows the critical value S_c as a function of the Froude number. For strong stratification, we observe approximatively a linear behavior $S_c \sim Fr$. A geometrical interpretation of this behavior is the following [17]. By introducing the characteristic vertical scale of the layered structures which characterize a stratified flow $L_v \simeq \varepsilon_f^{1/3} L_f^{1/3}/N$ [49], one has

$$Fr = \frac{\varepsilon_f^{1/3} L_f^{-2/3}}{N} = \frac{L_v}{L_f} = \frac{L_v}{L_z} S \tag{13}$$

The behavior seen in Figure 8 $S_c \simeq Fr$ corresponds to $L_v \simeq L_z$ and therefore suggests that the inverse cascade is suppressed when the typical thickness of the layered structure becomes small enough to fit in the fluid layer. In this case, the kinetic energy injected at L_f

is converted into potential energy by the term ε_x in (7) and is eventually dissipated at small scales according to (8) [17].

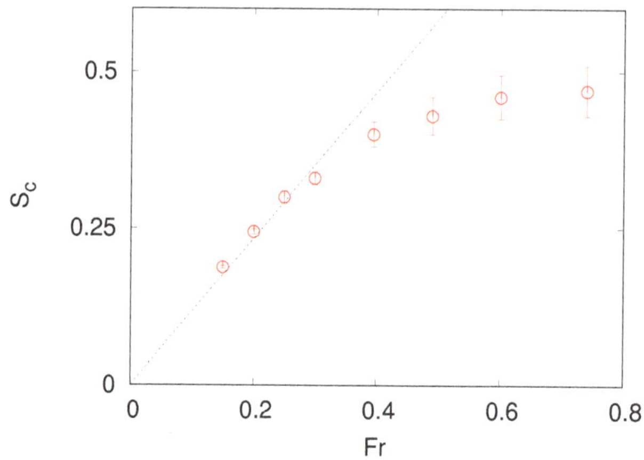

Figure 8. Critical thickness number S_c for the suppression of the inverse cascade as a function of the Froude number Fr. The line represents the fit $S_c = 1.17Fr$ over the first four points. Simulations at resolutions $N_x = N_y = 512$, $Nz = 64S$. Data from [17].

The analysis of spectral fluxes of both kinetic and potential energy confirms this picture. Indeed, by increasing the stratification at fixed S, the inverse flux of kinetic energy is reduced (as shown in Figure 7), while the direct flux of energy is enhanced in the range of wavenumber $k_f < k < k_z$ (remember that, since $S < 1$, $k_z > k_f$). For $k > k_z$, the analysis of fluxes shows the development of a cascade of potential energy which subtracts kinetic energy at large wavenumbers. This process is activated when the thickness of the vertical structures L_v becomes smaller than L_z, and therefore the suppression of the inverse flux is expected to be proportional to Fr, as shown in Figure 8. It is interesting to observe that the scaling shown in Figure 8 for small Fr can be extended to the fluctuations of the vertical velocity. Indeed, by incompressibility, one can write $u_z \simeq u_x L_v / L_x$, and therefore (13) implies that $u_z \propto Fr$ for small Froude numbers. This scaling was already observed in a remarkable paper by Herring on numerical simulations of stratified turbulence [14].

4. Conclusions

This paper briefly reviewed the recent studies on the transition between 2D and 3D turbulence in a confined thin fluid layer in the presence of rotation or a stable density stratification. It has been shown that confinement and rotation produce an inverse cascade of energy by a similar mechanism which suppresses enstrophy production at large scales. The presence of a density stratification, on the contrary, reduces the intensity of the inverse cascade and the critical thickness at which the inverse flux vanishes.

A natural extension of the studies reported here is to include both rotation and stratification in the fluid layer as it occurs in the oceans and in the atmosphere. Extensive studies in this direction have shown that stratification helps rotation to transfer energy to the two-dimensional modes [47,50] and, in particular conditions, can develop a Kolmogorov-like spectrum [51]. Moreover, numerical simulations of the 3D Boussinesq equation in the presence of both rotation and a stable stratification have shown that a small-scale energy flux can coexist with a large scale dynamics dominated by quasigeostrophic motion [52]. The key parameter which controls the ratio between large-scale and small-scale fluxes is the ratio Ro/Fr [44,53], and this ratio is found to be close to that estimated for the ocean when using realistic values of the parameters [54]. These, together with other studies in recent years [55–58], have achieved at least a partial understanding of the complex phenomenology which emerges when turbulence is coupled with the fundamental physical ingredients of geophysical flows (confinement, rotation, and stratification). These results are therefore an important guide for the analysis of more complex models for the atmosphere and the oceans.

Funding: This research received no external funding.

Institutional Review Board Statement: Not applicable.

Informed Consent Statement: Not applicable.

Data Availability Statement: The data used in this study are available on request from the corresponding author. Data are not publicly available because the author wants to understand the specific usage by those requesting access.

Conflicts of Interest: The author declare no conflict of interest.

References

1. Frisch, U. *Turbulence: The Legacy of AN Kolmogorov*; Cambridge University Press: Cambridge, UK, 1995.
2. Smith, L.; Chasnov, J.; Waleffe, F. Crossover from Two- to Three-Dimensional Turbulence. *Phys. Rev. Lett.* **1996**, *77*, 2467–2470. [CrossRef] [PubMed]
3. Ngan, K.; Straub, D.; Bartello, P. Aspect ratio effects in quasi-two-dimensional turbulence. *Phys. Fluids* **2005**, *17*, 125102. [CrossRef]
4. Celani, A.; Musacchio, S.; Vincenzi, D. Turbulence in more than two and less than three dimensions. *Phys. Rev. Lett.* **2010**, *104*, 184506. [CrossRef] [PubMed]
5. Xia, H.; Byrne, D.; Falkovich, G.; Shats, M. Upscale energy transfer in thick turbulent fluid layers. *Nature Phys.* **2011**, *7*, 321–324. [CrossRef]
6. Byrne, D.; Xia, H.; Shats, M. Robust inverse energy cascade and turbulence structure in three-dimensional layers of fluid. *Phys. Fluids* **2011**, *23*, 95109. [CrossRef]
7. De Wit, X.M.; Van Kan, A.; Alexakis, A. Bistability of the large-scale dynamics in quasi-two-dimensional turbulence. *J. Fluid Mech.* **2022**, *939*, R2. [CrossRef]
8. van Kan, A.; Nemoto, T.; Alexakis, A. Rare transitions to thin-layer turbulent condensates. *J. Fluid Mech.* **2019**, *878*, 356–369. [CrossRef]
9. Pestana, T.; Hickel, S. Regime transition in the energy cascade of rotating turbulence. *Phys. Rev. E* **2019**, *99*, 053103. [CrossRef]
10. van Kan, A.; Alexakis, A. Critical transition in fast-rotating turbulence within highly elongated domains. *J. Fluid Mech.* **2020**, *899*, A33. [CrossRef]
11. Alexakis, A.; Biferale, L. Cascades and transitions in turbulent flows. *Phys. Rep.* **2018**, *767*, 1–101.
12. Herring, J.; Orszag, S.; Kraichnan, R.; Fox, D. Decay of two-dimensional homogeneous turbulence. *J. Fluid Mech.* **1974**, *66*, 417. [CrossRef]
13. Herring, J.; McWilliams, J. Comparison of direct numerical simulation of two-dimensional turbulence with two-point closure: The effects of intermittency. *J. Fluid Mech.* **1985**, *153*, 229–242. [CrossRef]
14. Herring, J.; Métais, O. Numerical experiments in forced stably stratified turbulence. *J. Fluid Mech.* **1989**, *202*, 97–115. [CrossRef]
15. Lindborg, E.; Brethouwer, G. Stratified turbulence forced in rotational and divergent modes. *J. Fluid Mech.* **2007**, *586*, 83–108. [CrossRef]
16. Boffetta, G.; Musacchio, S.; Mazzino, A.; Rosti, M. Transient inverse energy cascade in free surface turbulence. *Phys. Rev. Fluids* **2023**, *8*, 034601. [CrossRef]
17. Sozza, A.; Boffetta, G.; Muratore-Ginanneschi, P.; Musacchio, S. Dimensional transition of energy cascades in stably stratified forced thin fluid layers. *Phys. Fluids* **2015**, *27*, 035112. [CrossRef]
18. van Kan, A.; Alexakis, A. Condensates in thin-layer turbulence. *J. Fluid Mech.* **2019**, *864*, 490–518. [CrossRef]
19. Musacchio, S.; Boffetta, G. Condensate in quasi-two-dimensional turbulence. *Phys. Rev. Fluids* **2019**, *4*, 022602. [CrossRef]
20. Musacchio, S.; Boffetta, G. Split energy cascade in turbulent thin fluid layers. *Phys. Fluids* **2017**, *29*, 111106. [CrossRef]
21. Deusebio, E.; Boffetta, G.; Lindborg, E.; Musacchio, S. Dimensional transition in rotating turbulence. *Phys. Rev. E* **2014**, *90*, 023005. [CrossRef] [PubMed]
22. Benavides, S.J.; Alexakis, A. Critical transitions in thin layer turbulence. *J. Fluid Mech.* **2017**, *822*, 364–385. [CrossRef]
23. Poujol, B.; van Kan, A.; Alexakis, A. Role of the forcing dimensionality in thin-layer turbulent energy cascades. *Phys. Rev. Fluids* **2020**, *5*, 064610. [CrossRef]
24. Boffetta, G.; Ecke, R. Two-Dimensional Turbulence. *Ann. Rev. Fluid Mech.* **2012**, *44*, 427. [CrossRef]
25. Biferale, L.; Musacchio, S.; Toschi, F. Inverse Energy Cascade in Three-Dimensional Isotropic Turbulence. *Phys. Rev. Lett.* **2012**, *108*, 164501. [CrossRef]
26. Yarom, E.; Vardi, Y.; Sharon, E. Experimental quantification of inverse energy cascade in deep rotating turbulence. *Phys. Fluids* **2013**, *25*, 85105. [CrossRef]

27. Campagne, A.; Gallet, B.; Moisy, F.; Cortet, P.P. Direct and inverse energy cascades in a forced rotating turbulence experiment. *Phys. Fluids* **2014**, *26*, 125112. [CrossRef]
28. Smith, L.; Waleffe, F. Transfer of energy to two-dimensional large scales in forced, rotating three-dimensional turbulence. *Phys. Fluids* **1999**, *11*, 1608–1622. [CrossRef]
29. Chen, Q.; Chen, S.; Eyink, G.; Holm, D. Resonant interactions in rotating homogeneous three-dimensional turbulence. *J. Fluid Mech.* **2005**, *542*, 139–164. [CrossRef]
30. Cambon, C.; Rubinstein, R.; Godeferd, F.S. Advances in wave turbulence: Rapidly rotating flows. *New J. Phys.* **2004**, *6*, 73. [CrossRef]
31. Scott, J.F. Wave turbulence in a rotating channel. *J. Fluid Mech.* **2014**, *741*, 316–349. [CrossRef]
32. Gallet, B. Exact two-dimensionalization of rapidly rotating large-Reynolds-number flows. *J. Fluid Mech.* **2015**, *783*, 412–447. [CrossRef]
33. Seshasayanan, K.; Gallet, B. Onset of three-dimensionality in rapidly rotating turbulent flows. *J. Fluid Mech.* **2020**, *901*, R5. [CrossRef]
34. Buzzicotti, M.; Aluie, H.; Biferale, L.; Linkmann, M. Energy transfer in turbulence under rotation. *Phys. Rev. Fluids* **2018**, *3*, 034802. [CrossRef]
35. Moisy, F.; Morize, C.; Rabaud, M.; Sommeria, J. Decay laws, anisotropy and cyclone–anticyclone asymmetry in decaying rotating turbulence. *J. Fluid Mech.* **2011**, *666*, 5–35. [CrossRef]
36. Bourouiba, L.; Bartello, P. The intermediate Rossby number range and two-dimensional-three-dimensional transfers in rotating decaying homogeneous turbulence. *J. Fluid Mech.* **2007**, *587*, 139. [CrossRef]
37. Praud, O.; Sommeria, J.; Fincham, A.M. Decaying grid turbulence in a rotating stratified fluid. *J. Fluid Mech.* **2006**, *547*, 389–412. [CrossRef]
38. van Bokhoven, L.J.A.; Cambon, C.; Liechtenstein, L.; Godeferd, F.S.; Clercx, H.J.H. Refined vorticity statistics of decaying rotating three-dimensional turbulence. *J. Turbul.* **2008**, *9*, N6. [CrossRef]
39. Gallet, B.; Campagne, A.; Cortet, P.P.; Moisy, F. Scale-dependent cyclone-anticyclone asymmetry in a forced rotating turbulence experiment. *Phys. Fluids* **2014**, *26*, 35108. [CrossRef]
40. Boffetta, G.; Toselli, F.; Manfrin, M.; Musacchio, S. Cyclone–anticyclone asymmetry in rotating thin fluid layers. *J. Turb.* **2021**, *22*, 242–253. [CrossRef]
41. Bartello, P.; Metais, O.; Lesieur, M. Coherent structures in rotating three-dimensional turbulence. *J. Fluid Mech.* **1994**, *273*, 1–30. [CrossRef]
42. Sreenivasan, B.; Davidson, P.A. On the formation of cyclones and anticyclones in a rotating fluid. *Phys. Fluids* **2008**, *20*, 085104. [CrossRef]
43. Métais, O.; Bartello, P.; Garnier, E.; Riley, J.; Lesieur, M. Inverse cascade in stably stratified rotating turbulence. *Dyn. Atmos. Ocean.* **1996**, *23*, 193–203. [CrossRef]
44. Smith, L.M.; Waleffe, F. Generation of slow large scales in forced rotating stratified turbulence. *J. Fluid Mech.* **2002**, *451*, 145–168. [CrossRef]
45. Billant, P.; Chomaz, J.M. Experimental evidence for a new instability of a vertical columnar vortex pair in a strongly stratified fluid. *J. Fluid Mech.* **2000**, *418*, 167–188. [CrossRef]
46. Waite, M.L.; Bartello, P. Stratified turbulence dominated by vortical motion. *J. Fluid Mech.* **2004**, *517*, 281–308. [CrossRef]
47. Marino, R.; Mininni, P.D.; Rosenberg, D.; Pouquet, A. Inverse cascades in rotating stratified turbulence: Fast growth of large scales. *Europhys. Lett.* **2013**, *102*, 44006. [CrossRef]
48. Brethouwer, G.; Billant, P.; Lindborg, E.; Chomaz, J.M. Scaling analysis and simulation of strongly stratified turbulent flows. *J. Fluid Mech.* **2007**, *585*, 343–368. [CrossRef]
49. Lindborg, E. The energy cascade in a strongly stratified fluid. *J. Fluid Mech.* **2006**, *550*, 207–242. [CrossRef]
50. Brunet, M.; Gallet, B.; Cortet, P.P. Shortcut to geostrophy in wave-driven rotating turbulence: The quartetic instability. *Phys. Rev. Lett.* **2020**, *124*, 124501. [CrossRef]
51. Marino, R.; Mininni, P.D.; Rosenberg, D.L.; Pouquet, A. Large-scale anisotropy in stably stratified rotating flows. *Phys. Rev. E* **2014**, *90*, 023018. [CrossRef]
52. Pouquet, A.; Marino, R. Geophysical turbulence and the duality of the energy flow across scales. *Phys. Rev. Lett.* **2013**, *111*, 234501. [CrossRef] [PubMed]
53. Waite, M.L.; Bartello, P. The transition from geostrophic to stratified turbulence. *J. Fluid Mech.* **2006**, *568*, 89–108. [CrossRef]
54. Marino, R.; Pouquet, A.; Rosenberg, D. Resolving the paradox of oceanic large-scale balance and small-scale mixing. *Phys. Rev. Lett.* **2015**, *114*, 114504. [CrossRef] [PubMed]
55. Bartello, P. Geostrophic adjustment and inverse cascades in rotating stratified turbulence. *J. Atmos. Sci.* **1995**, *52*, 4410–4428. [CrossRef]
56. Herbert, C.; Marino, R.; Rosenberg, D.; Pouquet, A. Waves and vortices in the inverse cascade regime of stratified turbulence with or without rotation. *J. Fluid Mech.* **2016**, *806*, 165–204. [CrossRef]

57. Pouquet, A.; Marino, R.; Mininni, P.D.; Rosenberg, D. Dual constant-flux energy cascades to both large scales and small scales. *Phys. Fluids* **2017**, *29*, 111108. [CrossRef]
58. Van Kan, A.; Alexakis, A. Energy cascades in rapidly rotating and stratified turbulence within elongated domains. *J. Fluid Mech.* **2022**, *933*, A11. [CrossRef]

atmosphere

MDPI

Article

Log-Lattices for Atmospheric Flows

Quentin Pikeroen , Amaury Barral, Guillaume Costa and Bérengère Dubrulle *

University Paris-Saclay, CEA, CNRS, SPEC, 91191 Gif-sur-Yvette, France; quentin.pikeroen@lsce.ipsl.fr (Q.P.); guillaume.costa@cea.fr (G.C.)

* Correspondence: berengere.dubrulle@cea.fr

Abstract: We discuss how the projection of geophysical equations of motion onto an exponential grid allows the determination of realistic values of parameters at a moderate cost. This allows us to perform many simulations over a wide range of parameters, thereby leading to general scaling laws of transport efficiency that can then be used to parametrize the turbulent transport in general climate models for Earth or other planets. We illustrate this process using the equation describing heat transport in a dry atmosphere to obtain the scaling laws for the onset of convection as a function of rotation. We confirm the theoretical scaling of the critical Rayleigh number, $\mathrm{Ra}_c \sim E^{-4/3}$, over a wide range of parameters. We have also demonstrated the existence of two regimes of convection: one laminar regime extending near the convection onset, and one turbulent regime occurring as soon as the vertical Reynolds number reaches a value of 10^4. We derive general scaling laws for these two regimes, both for the transport of heat and the dissipation of kinetic energy, and values of anisotropy and temperature fluctuations.

Keywords: convection; rotation; turbulence

Citation: Pikeroen, Q.; Barral, A.; Costa, G.; Dubrulle, B. Log-Lattices for Atmospheric Flows. *Atmosphere* **2023**, *14*, 1690. https://doi.org/10.3390/atmos14111690

Academic Editors: Boris Galperin, Annick Pouquet and Peter Sullivan

Received: 26 September 2023
Revised: 10 November 2023
Accepted: 13 November 2023
Published: 15 November 2023

1. Foreword by B. Dubrulle

I met Jack in 1999, when I came to the MMM division of NCAR for a one-year sabbatical. I had been attracted there by Jack's reputation from conversations with Annick Pouquet, Uriel Frisch, and Maurice Meneguzzi. Being a theoretician of turbulence interested in geophysical applications, I then knew that I would find in Jack a very good interlocutor and benefit greatly from his physical insights, his broad knowledge about turbulence and geophysical flows, and his open mind. I met Jack regularly during my stay, and we spoke about all sorts of topics. It was after a discussion with them that I started to investigate Rayleigh–Bénard flows and heat transfer properties—I published two papers on this topic that year. About Jack, I keep the memory of a true "gentleman of science", very kind to junior scientists (as I still was at that time), with a great sense of humor and an immense knowledge that he was keen to share. It is certainly thanks to Jack that I jumped into the modeling of geophysical flows, and I will always be grateful to him for this.

2. Introduction

Ultra-high Reynolds number flows are ubiquitous in geosciences due to small viscosities, large dimensions or velocity. They are described by the Navier–Stokes equations (NSEs). A natural control parameter of the NSEs is the Reynolds number $Re = LU/\nu$, built using the viscosity ν and characteristic length L and velocity U. Classical turbulent flows are thought to be described by NSEs, with $Re \gg 1$. In 1941, Kolmogorov [K41] used such equations to predict the shape of the energy spectrum $E(k)$ derived from the Fourier transform of the velocity correlation function for isotropic and homogeneous turbulence stirred at a constant rate ϵ. He found that it should scale like $E(k) \sim \epsilon^{2/3}k^{-5/3}$ in the range $1/L \ll k \ll \nu^{-3/4}\epsilon^{1/4}$, where k is the wavenumber. This prediction was verified in 1962 on data from a turbulent flow in Seymour Narrows [1] and appears to be one of the most robust laws of turbulence [2,3], being independent of the boundaries or the stirring process.

At large wavenumbers, viscous processes take over, and the spectrum decays very fast, so that the energy contained in wavenumbers greater than $k_d = \nu^{-3/4}\epsilon^{1/4}$ is negligible. The overall behavior of $E(k)$ can be used to infer the typical number of dynamically active modes as $N = (k_d L)^3$. In geophysical flows, such a number can be quite large; for example, for the atmosphere, $L = 10^3$ km (the size of the large typhoons or cyclones), while k_d^{-1} is smaller than 10 mm, resulting in $k_d L > 10^8$. Direct numerical simulations of the Navier–Stokes equations for such flows are thus impossible, as the total computational cost of reading-in/writing-out and coupling all these modes exceeds the capability of current most powerful computers by many orders of magnitude.

If one wishes to simulate ultra-large Reynolds numbers like the atmosphere, one has no other way out than to empirically decimate modes via a clever selection of grid points or modes. Simulating viscous flows with just as many scales as needed to "get the physics right" has been and still is the holy grail of all researchers in the computational fluid community. If a well-established theory of turbulence was available, including a deep understanding of all interactions between scales, the quest would probably be over by now. In the absence of such a complete theory, we need empirical yet clever strategies for mode number reduction. Jack Herring worked many years on such an issue, using, e.g., two-point closure (see, e.g., ref. [4]).

Nowadays, the most popular approach is to use a large Eddy simulation strategy, in which only large scales are simulated. Present climate models have a grid size of 10 km, allowing the handling of data volumes in 2 CPU seconds per time step. However, there is no free lunch; the price to pay for such a mode reduction is the addition of a (sometime very large) damping to avoid accumulation of energy at the smallest simulated scale. In LES-type climate simulations, the damping is the same as if the atmosphere was made of peanut butter and the ocean of honey, so that no fluctuations can develop. This is problematic to capture possible bifurcations, such as those observed in von Karman flows [5].

In this paper, we consider another class of model in which mode reduction is achieved by keeping modes following a geometric progression. Such an approximation leaves out a lot of possible interactions, as we shall see. However, it allows for reaching very small scales with a very small number of modes. In the atmosphere, for example, only 27 modes are necessary to go from $k_0 = 1/L$ to $10^8/L$ by a geometrical progression of step 2 (38 with a step being the golden number). This means that we can perform a "resolved" 3D simulation of an atmospheric flow with less than 2×10^4 modes (respectively, less than 6×10^4), corresponding to the number of modes involved in an LES simulation at a resolution of 37 km. The corresponding models correspond to fluid dynamics on log-lattices, the properties of which were detailed in [6]. For example, they respect the classical and basic properties of the Navier–Stokes equation, such as the constancy of the energy flux in the inertial range. Given the potential of such models to describe ultra-high Reynolds number flow at a low price, we investigate here further properties of such models with respect to one important open problem of atmospheric flows, which was dear to Jack, namely the influence of rotation on convection.

3. Log-Lattice Framework

Consider a velocity field that obeys the Navier–Stokes equation (NSE), with viscosity ν and forcing. Its Fourier transform, denoted $u(t,k)$, is a complex field that obeys the equation:

$$\begin{aligned} \mathbf{i}k_j u_j &= 0, \\ \partial_t u_i + \mathbf{i}k_j u_i * u_j &= -\mathbf{i}k_i p - \nu k^2 u_i + f_i, \\ u(k,t) &= u^*(-k,t). \end{aligned} \tag{1}$$

where $*$ denotes the traditional convolution product, that involves coupling of modes $u(t,p)$ and $u(t,q)$ for each k, such that $k_i = p_i + q_i$ for any ith component. The (constant) density has been set to 1 for convenience, and thus disappears in front of the pressure term in Equation (1). The convolution sum is computed directly, but is less costly than a fast Fourier transform because only a limited amount of local interactions are kept [6].

In traditional spectral simulations of the NSE, the wavenumbers are discretized on a linear grid, and can be written as $k = (m_1, \ldots, m_3)$ with $m_i \in \mathbb{Z}^3$. The convolution is ensured by the condition $m = n + q$ for any $(m, n, q) \in \mathbb{Z}^3$, this allows for a lot of interactions between Fourier modes.

In the log-lattice framework, we now impose that the wavenumbers follow a geometric progression:

$$k = (\lambda^{m_1}, \ldots, \lambda^{m_3}), \tag{2}$$

with $m_i \in \mathbb{Z}^3$, where λ is a parameter yet to be defined. It is imposed by the condition that the convolution product appearing in Equation (1) has some non-zero solution, which is only possible if the equation:

$$\lambda^m = \lambda^p + \lambda^q, \tag{3}$$

has some solutions for any $m, p, q \in \mathbb{Z}^3$. As discussed in [6], Equation (3) can only be achieved provided λ takes some specific values, among which $\lambda = 2$, $\lambda \equiv \phi = (1 + \sqrt{5})/2 \approx 1.618$, the golden mean and $\lambda \equiv \sigma =\approx 1.325$ (the plastic number (σ is defined as the common real root to $\sigma^3 - \sigma - 1 = 0$ and $\sigma^5 - \sigma^4 - 1 = 0$)). Different values of λ correspond to different numbers of coupling between modes; they become more and more numerous and less and less local as $\lambda \to 1$ [6]. In this sense, log-lattices can be seen as a special case of the sparse Fourier model [7], the REWA model [8], or fractal decimated models [9,10], in which the non-linear interactions of the NSE are projectively decreased randomly or in a scale-invariant manner.

The choice of λ is fixed by several considerations: larger values of λ are cheaper, both in computation time and in memory. However, lower values of λ give rise to more interactions, and thus more fluctuations, which is more realistic for simulating a turbulent flow. As discussed in [11], the case $\lambda = 2$ should be avoided when simulating incompressible dynamics because it lacks backscatter. On the other hand, ref. [12] checked whether the scaling law properties, such as the spectral slope or blow-up exponent, do not depend on the value of λ. Therefore, we work below with $\lambda = \phi$, which is the best compromise between all constraints.

3.1. Energy Spectra

It is possible to write the mean energy that measures the scaling properties of the mode at $k_n = \lambda^n$ as:

$$E_m(n) =< ||u(k)||^2 >_{S_n}, \tag{4}$$

where the average is taken over wavenumbers in the shell delimited by spheres with radii λ^{n-1} and λ^n. Specifically:

$$< ||u(k)||^2 >_{S_n}, = \frac{1}{N_k} \sum_{\lambda^{n-1} \leq ||k|| < \lambda^n} \sum_{\alpha=1}^{d} |u_\alpha(k)|^2, \tag{5}$$

where d is the space dimension and $N_k \sim (\log(k))^{d-1}$ is the number of wavenumbers in the shell S_n.

From this quantity, one can define a pseudo-energy spectrum as:

$$E(k) = \frac{\lambda^{-n}}{\lambda - 1} < ||u(k)||^2 >_{S_n}, \tag{6}$$

Examples of such a spectrum in $d = 3$ are shown in Figure 1 for $\lambda = 2$. One sees that it is self-similar and displays a very clear $k^{-5/3}$ law. This spectrum has been obtained by simulating Equation (1) with 20^3 modes, which can easily be performed on a PC. Due to the exponential spacing, it enables reaching resolutions and inertial ranges even larger than those achieved by oceanic measurements, a performance which is out-of-reach of DNS simulations of the NSE. This shows the importance of FDLL in geophysics.

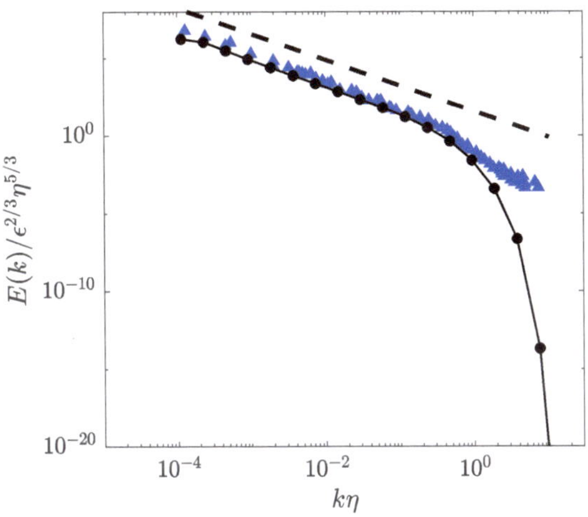

Figure 1. Renormalized energy spectrum $E(k)/\epsilon^{2/3}\eta^{5/3}$ as a function of the renormalized wavenumber $k\eta$ for simulation of the Navier–Stokes equation on a 20^3 log-lattice (black dotted line) and for a turbulent flow in Seymour Narrows (Ocean) (blue squares). The black dashed line corresponds to $E(k) \sim k^{-5/3}$. This figure was drawn by the authors using data extracted from the published graph shown in [1].

3.2. Generalizations

Equations (1)–(3) define the Navier–Stokes equation on a log-lattice. By extension, any equation of fluid dynamics on s log-lattice can be defined by performing the following two steps [6]: (i) Write the equation in the Fourier-space. (ii) Replace any convolution product by the convolution on a log-lattice. This construction guarantees that the resulting equation obeys all the conservation laws and symmetries of the original equations [6].

3.3. Limitations of Log-Lattices

Computations on log-lattices involve a number of limitations. First, due to the local nature of the convolution in Fourier space, log-lattices are unable to describe the so-called "non-local" interactions that are involved, for example, in the shearing of small eddies by large eddies [13]. This may explain why log-lattice simulations do not display intermittency for the structure functions [14]. Due to their sparsity, log-lattices cannot describe dispersive wave resonances, like inertial or gravity waves. This means that they cannot capture dissipative phenomena induced by these waves, and that they can only capture the dissipation due to small-scale turbulent eddies. Finally, due to the spectral nature of the construction, it may appear that the log-lattice framework is only appropriate for homogeneous flows, i.e., far from boundaries. In a recent work, ref. [15] have, however, shown that the extension to flow with boundaries is possible via lattice symmetrization around the boundary and careful treatment of the resulting discontinuity. Despite the high relevance of this situation for geophysical flows, we here concentrate on the simpler case of homogenous flows, and show how log-lattice simulations enable the recovery of some well-known features of homogeneous rotating convections.

4. Homogeneous Rotating Convections on Log-Lattices

4.1. Definitions

We now consider a rotating homogeneous fluid, with coefficient of thermal dilation α, viscosity ν and diffusivity κ, subject to a temperature gradient ΔT over a length H and vertical gravity g. Its dynamics are given by the HRB set of wequations [11,16–18],

$$\partial_t u + u \cdot \nabla u + \frac{1}{\rho_0}\boldsymbol{\nabla} p + 2\boldsymbol{\Omega} \times u = \nu \nabla^2 u + \alpha g \theta \vec{z} - f u, \tag{7}$$

$$\partial_t \theta + u \cdot \boldsymbol{\nabla}\theta = \kappa \nabla^2 \theta + u_z \frac{\Delta T}{H} - f\theta, \tag{8}$$

$$\nabla \cdot u = 0, \tag{9}$$

where u is the velocity, θ is the temperature deviation from the equilibrium profile (where $\theta = 0$), which means by definition that $T = -\Delta T z/H + \theta$, $\boldsymbol{\Omega}$ is the rotation vector, ρ_0 is the (constant) reference density, p is the pressure and f is a (Rayleigh) friction term, accounting for the friction at the boundary layer that cannot be resolved by the present framework. We will assume below that this friction is concentrated only on a large scale. As proven in [11], this friction is mandatory to allow the system to reach well-defined stationary states. Indeed, for periodic boundary conditions, exponential instabilities can grow because there is no wall to stop them. They are characterized by the accumulation of energy on a large scale. Introducing a large scale friction allows damping of the inverse cascade and avoids the concentration of energy on the large scale. This set of equations has to be completed with boundary conditions. In this paper, we consider periodic boundary conditions and focus on the case when the rotation is aligned with the z-axis, $\boldsymbol{\Omega} = \Omega e_z$.

4.2. Non-Dimensional Numbers

We can use five independent non-dimensional numbers to characterize the system:

- The Rayleigh number $\mathrm{Ra} = \alpha g H^3 \Delta T/(\nu\kappa)$, which characterizes the forcing by the temperature gradient.
- The Prandtl number $\mathrm{Pr} = \nu/\kappa$, which is the ratio of the fluid viscosity to its thermal diffusivity.
- The Nusselt number $\mathrm{Nu} = JH/\kappa\Delta T$ that characterizes the mean total heat flux is the z direction is $J = \partial_z < u_z\theta > -\kappa\Delta T$.
- The Ekman number $\mathrm{E} = \nu/(2\Omega H^2)$, measuring the importance of the rotation with respect to the diffusive process.
- The Rossby number $\mathrm{Ro} = \sqrt{\alpha g \Delta T}/(2\Omega\sqrt{H})$, measuring the importance of the rotation with respect to buoyancy. In terms of other variables, we have $\mathrm{Ro} = \mathrm{E}\sqrt{\mathrm{Ra}}/(\sqrt{Pr})$.
- The friction coefficient $F = f\sqrt{H}/\sqrt{\alpha g \Delta T}$, which provides the intensity of the Rayleigh damping.

4.3. Equations on Log-Lattice

The regimes we want to explore are very turbulent regimes where the viscosity and diffusivity do not play any role anymore. Therefore, it is natural to adimensionalize the equation in terms of "inertial quantities", i.e., using the vertical width H as a unit of length, the free fall velocity $U = \sqrt{\alpha g \Delta T H}$ as a unit of velocity, and ΔT as a unit of temperature.

Then, to define the Rotating Homogeneous Rayleigh–Benard (RHRB) equations on log-lattices, we take the Fourier transform of Equation (7) that can be written in non-dimensional form as (with the Einstein convention on summed repeated indices):

$$\begin{aligned} \mathbf{i}k_j u_j &= 0, && (10)\\ \partial_t u_i + \mathbf{i}k_j u_i * u_j &= -\mathbf{i}k_i p + \theta\delta_{i3} - F u_i \delta_{k\approx k_{\min}} \\ &\quad - \sqrt{\frac{\mathrm{Pr}}{\mathrm{Ra}}} k^2 u_i - \frac{1}{\mathrm{Ro}} \epsilon_{i3k} u_k && (11)\\ \partial_t \theta + \mathbf{i}k_j \theta * u_j &= u_z - \frac{k^2\theta}{\sqrt{\mathrm{Ra\,Pr}}} - F\theta\delta_{k\approx k_{\min}}, && (12)\\ u(k,t) &= u^*(-k,t), && (13)\\ \theta(k,t) &= \theta^*(-k,t). && (14) \end{aligned}$$

where the Dirac $\delta_{k\approx k_{\min}}$ filters out the small scales.

In these equations, the convolution product is taken over the log-lattice, see Equation (3).

4.4. Convection Onset

Convection is an instability, so it sets-up at a certain critical value of the parameter Ra_c. Detailed computation of this parameter is performed in [19] for the case $F = 0$. In a nutshell, we assume that we are very near the threshold, so that deviations from the "equilibrium state" $u = \theta = 0$ are small. This will allow us to neglect all non-linear terms in Equation (14). Then, we look for solutions behaving as:

$$\begin{aligned} u(k,t) &= (u(k), v(k), w(k))e^{\sigma t}, \\ p(k,t) &= p(k)e^{\sigma t}, \\ \theta(k,t) &= \theta(k)e^{\sigma t}, \end{aligned} \tag{15}$$

where σ is the growth rate of the instability; if σ has a negative real part, then all perturbation decay, while instability develops when the real part of $\sigma > 0$. Plugging this decomposition into Equation (14) and neglecting non-linear terms, we get:

$$\begin{aligned} ik_x u + ik_y v + ik_z w &= 0, \\ \sigma u - \frac{v}{Ro} &= -ik_x p - \sqrt{\frac{\mathrm{Pr}}{\mathrm{Ra}}} k^2 u, \\ \sigma v + \frac{u}{Ro} &= -ik_y p - \sqrt{\frac{\mathrm{Pr}}{\mathrm{Ra}}} k^2 v, \\ \sigma w &= -ik_z p - \sqrt{\frac{\mathrm{Pr}}{\mathrm{Ra}}} k^2 w + \theta, \\ \sigma\theta &= w - \frac{1}{\sqrt{\mathrm{Ra\,Pr}}} k^2 \theta. \end{aligned} \tag{16}$$

This represents a linear, homogeneous system of equations in the variable (u, v, w, p, θ). If we want this system to have other solutions than $(0,0,0,0,0)$, we must impose the determinant of the system to be zero, which provides us with an expression linking σ, k and the parameters of the system. Taking $Pr = 1$ for simplicity (the general case leads to the same conclusions with different prefactors), we get:

$$\sigma^2 + 2k^2\,\mathrm{Ra}^{-1/2}\,\sigma + \frac{k^4}{\mathrm{Ra}} + (1+\mu^2)^{-1}\left(\frac{\mu^2}{\mathrm{Ro}^2} - 1\right) = 0, \tag{17}$$

where $\mu^2 = k_z^2/(k_x^2 + k_y^2)$. This is a second order equation for σ, there are therefore two solutions. Since the prefactor of σ is positive, it means that the sum of the two solutions is negative. To ensure that there exists at least one solution with a positive real part, we must then ensure that the product of the solution, given by a term independent of σ, is negative. To study the consequences of this condition, we consider two limiting cases.

Onset at Zero Rotation

We first consider the case with no rotation, Ro $\to \infty$. Then, the condition for instability is $\frac{(1+\mu^2)k^4}{\mathrm{Ra}} - 1 < 0$. Since $(1+\mu^2)k^4$ is at a minimum when $k_{\mathrm{min}} = (2\pi, 2\pi, 2\pi)$, where k_{min} is the minimal wavenumber, this is achieved whenever Ra $> \frac{3}{2}k_{\mathrm{min}}^4$. In our simulations, $k_{\mathrm{min}} \sim 2\pi\sqrt{3}$; this gives $\mathrm{Ra}_c \sim 2.1 \times 10^4$.

4.5. Onset at Large Rotation

In the limit of large rotation, $Ro \to 0$ and the condition for instability now reads $\frac{k^4}{\mathrm{Ra}} + (1+\mu^2)^{-1}(\frac{\mu^2}{\mathrm{Ro}^2} - 1) < 0$. Introducing $\alpha = k^4/\,\mathrm{Ra}$, we may further simplify the condition by noting that at the instability threshold, μ is close to 1, and $k_z^2 \ll k_x^2 + k_y^2 \sim$ Ra. The condition for instability then becomes $\mathrm{Ra}^{3/2} > 4\pi^2\,\mathrm{E}^{-2}\frac{1}{(1-\alpha)\sqrt{\alpha}}$. The instability criterion is given by the condition that there exists at least α such that the equality is satisfied, i.e., when the minimum of the right-hand side of the condition of instability is achieved. This is true for $\alpha = 1/3$, resulting in $\mathrm{Ra}^{3/2} > 4\pi^2\,\mathrm{E}^{-2}\frac{3\sqrt{3}}{2}$, giving a critical Rayleigh number

$$\mathrm{Ra}_c = (6\pi^2\sqrt{3})^{2/3}\,\mathrm{E}^{-4/3} \approx 22\,\mathrm{E}^{-4/3}\,. \quad (18)$$

We note that this equation is similar to Equation (184) on p. 106 of [19]. This means that the larger the rotation (the smaller E), the more difficult it is to achieve convection; rotation stabilizes the flow.

4.6. Phenomenology When $F = 0$

In the case when $F = 0$, there are a number of simple physical arguments that provide the scaling laws relation between non-dimensional numbers.

4.6.1. Non-Rotating Case

We first consider the non-rotating case, $\mathrm{E} = \infty$. In this case, the phenomenology of homogeneous convection distinguishes three regimes for the behavior of the heat flux as a function of the forcing [11,16]:

(I): When $\mathrm{Ra} \leq \mathrm{Ra}_c$, we are in the laminar case. The fluid is at rest, $< u_z\theta >= 0$, and the heat flux is only piloted by the Fourier law, so that $J = \kappa\Delta T/H$ and $Nu = 1$.

(II): Above the critical threshold for instability, when $\mathrm{Ra} >\sim \mathrm{Ra}_c$, convection sets in, $< u_z\theta >$ starts becoming positive, and we have $\mathrm{Nu} \sim (\mathrm{Ra} - \mathrm{Ra}_c)^{\chi}$, where χ is an exponent characterizing the (super)-critical transition to convection.

(III): When $\mathrm{Ra} \gg \mathrm{Ra}_c$, the turbulence becomes fully developed, and we are entering an "ultimate" regime (also called the Spiegel regime), in which the heat flux does not depend on the viscosity or the diffusivity any more. In that case, we have $\mathrm{Nu} \sim (\mathrm{Ra}\,\mathrm{Pr})^{1/2}$ [20–22] and $\mathrm{Re} \sim (\mathrm{Ra}\,/\,\mathrm{Pr})^{1/2}$ [23].

This regime is very difficult to observe in DNSs because of the need for high resolution to be able to cope with $\mathrm{Ra} \gg \mathrm{Ra}_c$. In the atmosphere, Ra is typically of the order of $\mathrm{Ra} \sim 10^{18-22}$, so we expect the atmosphere to be in this ultimate regime.

Note that in the case where boundaries are present, there is the possibility of an intermediate regime (the Malkus regime), where the heat flux is piloted by the boundary layers and $\mathrm{Nu} \sim \mathrm{Ra}^{\gamma}$, $\gamma \sim 1/3$. This regime is frequently observed in laboratory experiments, but it does not apply to homogeneous turbulence because of the absence of boundaries.

4.6.2. Rotating Case

Let us consider the case with rotation. As shown by many studies, the rotation has a stabilizing influence on the flow, so that the threshold for instability now increases with increasing rotation. Detailed studies show that $\mathrm{Ra}_c \sim \mathrm{E}^{-4/3}$ [24]. In addition, the rotation modifies the structure of the convective cells, which become aligned with the vertical axis in the case of strong rotation. This profoundly changes the heat transfer. To account for this effect, ref. [25] suggests performing the same phenomenology as in the non-rotating case, using Ra $E^{4/3}$ instead of Ra. The two turbulent regimes in the presence of rotation are now:

(R1) A rotating Malkus regime in the presence of boundaries, where the heat flux is independent of the height of the cell and in which Nu $\sim$ $\mathrm{Ra}^3\,\mathrm{E}^4$. This regime cannot be present in HRB equations.

(R2) A rotating ultimate regime, also called a geostrophic turbulent (GT) regime, at Reynolds numbers larger than the threshold for onset of turbulence in the boundary layers, or when boundaries do not limit the heat flux, like in HRB equations. This regime is then found by stating that the relation between Nu and Ra $\mathrm{E}^{4/3}$ and Pr should be such that the energy flux J is independent of κ and ν, resulting in [25].

$$\mathrm{Nu} \sim \mathrm{Ra}^{3/2}\,\mathrm{E}^2\,\mathrm{Pr}^{-1/2}\,. \tag{19}$$

An interesting property of the geostrophic turbulent regime is that it can be expressed as a universal law, independent of the rotation and the Prandtl number, using the "turbulent" coordinates [25]:

$$\begin{aligned} \mathrm{Nu}_* &= \frac{\mathrm{Nu}\,\mathrm{E}}{\mathrm{Pr}}, \\ \mathrm{Ra}_* &= \frac{\mathrm{Ra}\,\mathrm{E}^2}{\mathrm{Pr}}. \end{aligned} \tag{20}$$

In this case, the relation (19) becomes:

$$\mathrm{Nu}_* \sim \mathrm{Ra}_*^{3/2}\,. \tag{21}$$

In this regime, we further have Re $\sim$ Ra E^2 / Pr [23].

4.6.3. Log-Lattice Simulation Details

To simulate these equations, we can perform log-lattice simulations. The minimum wave vector of the grid is set to $k^{1D}_{\mathrm{min}} = 2\pi$ to match a simulation on a box of size $\tilde{L} = 1$. Note that the absence of wavenumbers ($k = 0$) in our simulation means that we will not be able to capture large-scale vertical structures that are often observed in rotating turbulence. However, as shown below, this does not seem to affect the scaling laws of the transport of heat and momentum that we observe. The grid size N (such that the maximum wavenumber in the x, y or z direction is proportional to λ^N) is then set to reach the dissipative scale for both velocity and temperature. We have verified that the size of the grid for 3D simulations ($N \geq 13$) does not affect the mean value of the observables Nu, Re, ..., which are already converged for grids of size $N \geq 6$. However, the tail of the pdfs does depend on N. Another 3D simulation set at $N = 20$ (not shown here, both vs. Ra and Pr) displays the same scaling laws as the $N = 13$ case, confirming this analysis.

5. Results

5.1. Non-Rotating Case

As reference, we have performed simulations without rotation for various Ra values up to 10^{25} at Pr $= 1$ and Pr $= 0.7$ and for $F = 1$. Note that present direct numerical simulations for these Prandtl numbers are limited to Ra $\approx 10^{15}$ [26]. The resulting heat transfer, Nu, as a function of the Rayleigh number Ra is shown in Figure 2. In that case, we observe that the convection starts as $\mathrm{Ra}_c = 6 \times 10^5$, which is larger than the value predicted by the linear theory at $F = 0$, and those observed in previous log-lattice simulations [11]. This means that friction stabilizes the convection in the non-rotating regime. In addition, we observe a near onset regime Nu $\sim (\mathrm{Ra} - \mathrm{Ra}_c)^{3/2}$, which then turns into an asymptotic regime, where Nu $\sim \sqrt{\mathrm{Ra}}$. This regime itself splits in two distinct regimes, characterized by the same scaling law Nu $\sim \mathrm{Ra}^{1/2}$, but with a very different amplitude: (i) a *laminar regime*, characterized by low fluctuations of the heat transfer. This laminar regime is well fitted by an empirical law

$$\mathrm{Nu} = A(\mathrm{Ra} - \mathrm{Ra}_c)^{3/2} / (\mathrm{Ra} \,/\, \mathrm{Ra}_t + 1), \tag{22}$$

where $A = 7$, $\mathrm{Ra}_c = 5 \times 10^5$ is the critical number for convection onset, and $\mathrm{Ra}_t = 5 \times 10^6$ is the transition number from near-onset to the asymptotic regime. This laminar regime is followed by (ii) a ***turbulent regime***, with much larger fluctuations. The typical time behavior of the heat transfer in the two regimes is displayed in Figure 3, which showcases both regimes. The laminar regime is stable up to $\mathrm{Ra} \approx 10^{12}$, and unstable above; if we wait long enough, the solution jumps from the laminar regime to the turbulent regime, as illustrated in Figure 3. Looking at Nu as a function of time in this regime in Figure 3, we observe that the simulation first follows a rather long evolution along the laminar regime, before suddenly transitioning towards the turbulent regime as instability develops, increasing its energy by several orders of magnitude in the process.

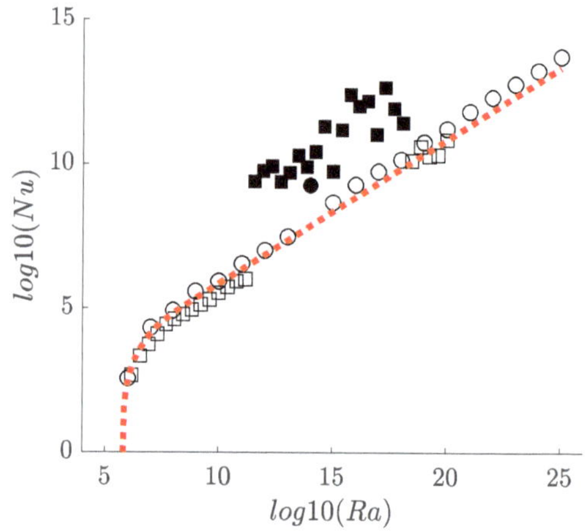

Figure 2. Non-dimensional heat transfer Nu vs. Rayleigh number Ra in 3D for $\mathrm{Pr} = 0.7$ (circle) and $\mathrm{Pr} = 1$ (squares). The dotted line corresponds to the empirical law: $\mathrm{Nu} = 7(\mathrm{Ra} - \mathrm{Ra}_c)^{3/2} / (\mathrm{Ra} \,/\, \mathrm{Ra}_t + 1)$, with $\mathrm{Ra}_c = 6 \times 10^5$ and $\mathrm{Ra}_t = 5 \times 10^6$ that connects the near-convection onset regime to the asymptotic law $\mathrm{Nu} \sim 20\sqrt{\mathrm{Ra}}$ for large Ra, corresponding to asymptotic non-rotating ultimate regime scaling. This regime is itself split into a laminar regime (open symbols) and a turbulent regime (filled symbols).

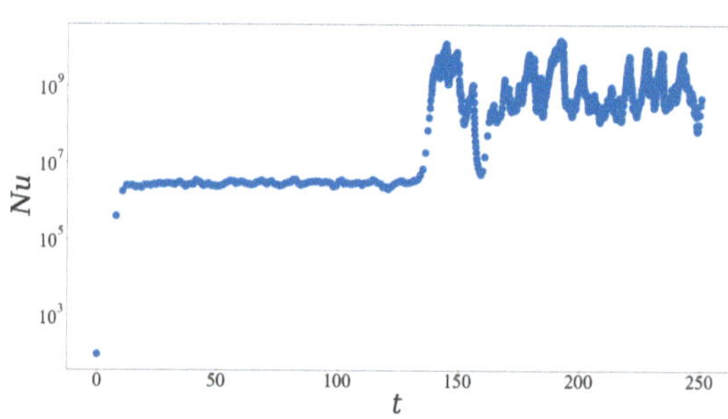

Figure 3. Non-dimensional heat transfer Nu versus time t for $\mathrm{Ra} = 3.46 \times 10^{11}$. The flow starts in a laminar regime, then abruptly transitions to a turbulent regime.

5.2. Rotating Case

Parameter Space and Critical Rayleigh Number

We turn now to the case with rotation. Figure 4 shows the parameter space we have explored, with E ranging from 10^{-9} to 10^{-1} and Ra up to 10^{14}. In this range of parameters, we observe typically three types of behaviors: (i) Conductive behaviors, where both the velocity and temperature fluctuations are zero and $\mathrm{Nu} = 1$. These cases are shown as white symbols in Figure 4. (ii) Transitional regimes, where the velocity and temperature fluctuations are decaying very slowly but steadily along with Nu over the time of the simulation, so that asymptotically, they are likely to converge to the conductive limit. The typical timescale to reach this limit increases as E decays and Ra increases. To save computational time, we stopped the simulation before reaching the limit for $\mathrm{E} \geq 10^{-6}$, but have reported these points as yellow points in Figure 4. (iii) Convective regimes, where both the velocity and RMS velocities reach a stationary state; these points are reported as black points in Figure 4. This representation allows for a clear view of the stabilizing influence of rotation on the convection threshold. In the asymptotic regime of large rotation, $\mathrm{E} \ll 1$, such an influence is very well described by the prediction of the linear theory at $F = 0$, Equation (18), which is reported as a red line in the diagram. This means that, as rotation increases, the friction becomes less important in the dynamics.

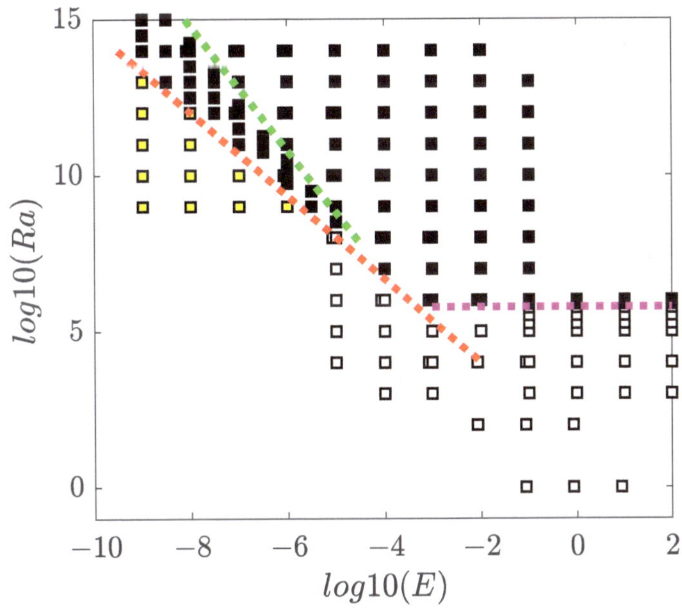

Figure 4. Parameter space covered by our log-lattice simulations at $\mathrm{Pr} = 0.7$. The color of the symbols indicates the three possible regimes: conductive (white), transitional (yellow), and convective (black). The red line is the theoretical asymptotic prediction given by Equation (18). The magenta line is the theoretical convection threshold in the absence of rotation $\mathrm{Ra} = 6 \times 10^5$. The green line has equation $\mathrm{Ra} = 0.06E^{-2}$ and delineates regions of the parameter space where the turbulence is influenced by rotation (below the line) or not influenced by rotation (above the line), as diagnosed by the behavior of the kinetic energy dissipation, see Figure 5. The geostrophic turbulent regime is observed in between the red and the green line.

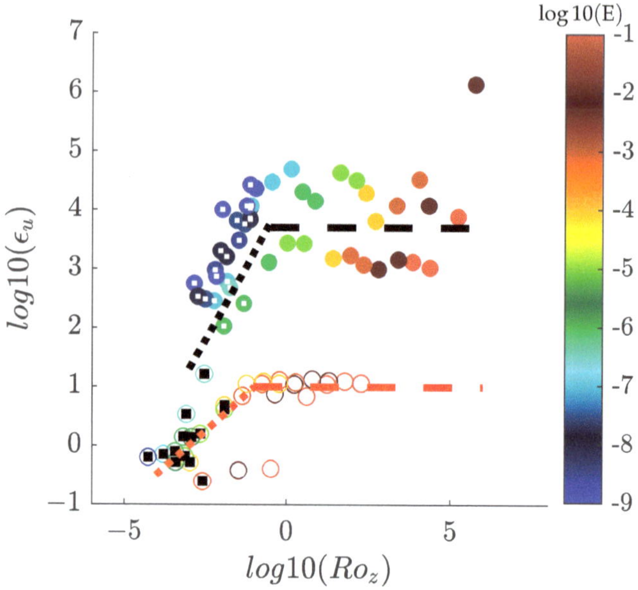

Figure 5. Non-dimensional energy dissipation $\epsilon_u = \nu < (\nabla u)^2 > H/U^3 = \mathrm{Ra\,Nu\,Pr}^{-2}/(\mathrm{Ra\,Pr})^{3/2}$ vs. vertical Rossby number Ro_z in 3D for $\mathrm{Pr} = 0.7$ for rotating HRB simulations on log-lattices. The symbols are colored according to their Ekman number, E. The open symbols trace the laminar regime, while the filled symbols trace the turbulent regime. The rotation-dominated regimes are highlighted by a black (respectively, white) square for the laminar (respectively, turbulent) regime. The black (respectively, red) dotted line corresponds to $\epsilon_u \sim \mathrm{Ro}_z$ (respectively, $\epsilon_u \sim \mathrm{Ro}_z^{1/2}$), while the black (respectively, red) dashed line corresponds to $\epsilon_u = 3.7$ (respectively, $\epsilon_u = 1$).

5.3. *Influence of Friction*

To elucidate this observation, we computed the ratio of the kinetic energy dissipated by friction, FU^2, to the kinetic energy dissipated by in the flow by viscous processes, ϵ_u. This is reported in Figure 6. We indeed observe two regimes: one at a low value of $\mathrm{Ra} < 10^{11}$, where the friction dominates the dissipation, and one at a larger $\mathrm{Ra} > 10^{11}$, where the friction is negligible. We will see below that this induces two different regimes termed "laminar" and "turbulent" regimes. In addition, we observe that as the rotation becomes larger, the influence of friction decreases.

5.3.1. Laminar vs. Turbulent Regime

Figure 7 reports the heat transport Nu as a function of Rayleigh number, Ra, and the Ekman number, E. One can see that larger Ekman numbers correspond to a smaller heat transport at a given Ra. However, rotation does not suppress the existence of a laminar to turbulent transition, which is already present in the non-rotating case. Like in the non-rotating case, it occurs when the vertical Reynolds number exceeds $u_z^{rms} L/\nu = \mathrm{Re}_c = 10^4$, corresponding to $\mathrm{Ra} \sim 10^{12}$, with a bi-stability occurring around $\mathrm{Ra} = 10^{11}$, see Figure 8. This critical value separates two different scaling regimes: one such that $\mathrm{Re} \sim \mathrm{Ra}^{1/2}$, and one such that $\mathrm{Re} \sim \mathrm{Ra}$.

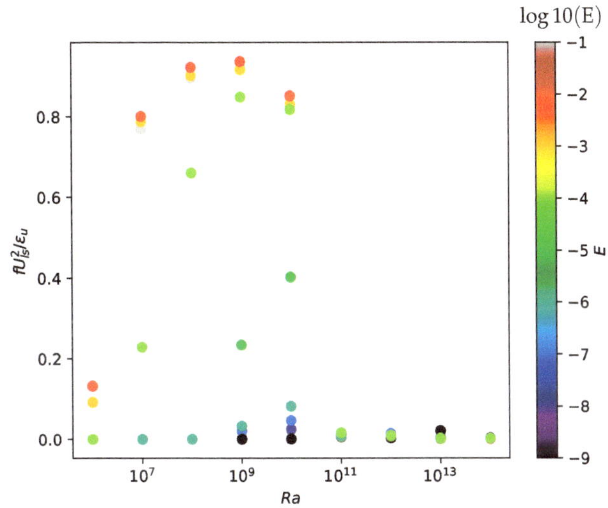

Figure 6. Ratio of the kinetic energy dissipated by friction, ΓU^2, to the kinetic energy dissipated by viscous processes, ϵ_u, as a function of the Rayleigh number, Ra, and the Ekman number E. The points are color-coded by log 10(E). The friction dominates for low-rotation and for Ra $< 10^{11}$.

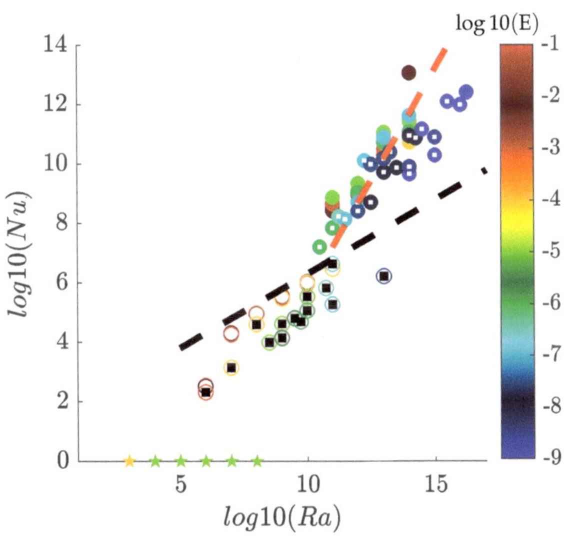

Figure 7. Non-dimensional heat transfer Nu vs. Rayleigh number Ra in 3D for Pr = 0.7 for rotating HRB simulations on log-lattices. The symbols are colored according to their Ekman number E. The stars traces the conductive regime. The open symbols trace the laminar regime, while the filled symbols trace the turbulent regime. The rotation-dominated regimes are highlighted by a black (respectively, white) square for the laminar (respectively, turbulent) regime. The black dashed line is $\mathrm{Nu} = 20\sqrt{\mathrm{Ra}}$, corresponding to asymptotic non-rotating ultimate regime scaling. The red dotted line is $\mathrm{Nu} \sim \mathrm{Ra}^{3/2}$, corresponding to the geostrophic turbulent regime, see Figure 9 for an exact representation of the corresponding scaling law.

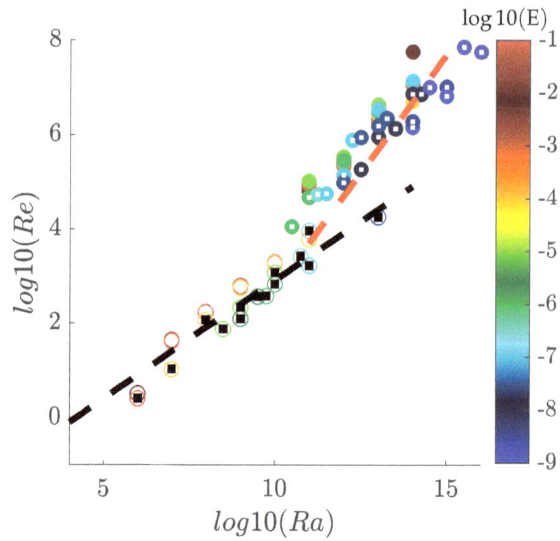

Figure 8. Vertical Reynolds number $\mathrm{Re} = \sqrt{< u_z^2 >} H/\nu$ as a function of Rayleigh number Ra in 3D for $\mathrm{Pr} = 0.7$ for rotating HRB simulations on log-lattices. The symbols are colored according to their Ekman number, E. The open symbols trace the laminar regime, while the filled symbols trace the turbulent regime. The rotation-dominated regimes are highlighted by a black (respectively, white) square for the laminar (respectively, turbulent) regime. The black (respectively, red) dashed line follows the equation $y \sim x^{1/2}$ (respectively, $y \sim x^1$).

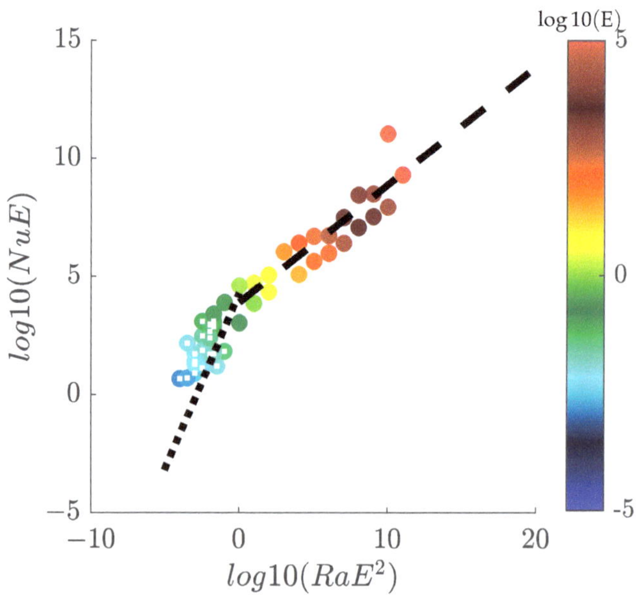

Figure 9. Universal law governing the heat transfer in the turbulent regime Nu E as a function of $\mathrm{Ra\,E}^2$ in 3D for $\mathrm{Pr} = 0.7$ for rotating HRB simulations on log-lattices. The symbols are colored according to their Ekman number E. The symbols tagged by a white square trace the rotation-dominated regime, while the filled symbols trace the rotation-independent regime. The black dotted (respectively, dashed) line follows the equation $y \sim x^{3/2}$ (respectively, $y \sim x^{1/2}$).

5.3.2. Influence of Rotation and Onset of Rotation-Dominated Regimes

The influence of rotation on turbulence is well documented (see, e.g., ref. [27,28]), and is characterized by a *decrease* in the efficiency of the transport properties with respect to the non-rotating case. For the case of heat transport, this is already clear from Figure 7, as already discussed. In the case of the turbulent kinetic energy dissipation $\nu < (\nabla u)^2 >$, it is known that the decrease is proportional to the vertical Rossby number, $\mathrm{Ro}_z = u_z/2H\Omega$, when Ro_z goes to zero [28]. We indeed observe this effect in our simulations, as illustrated in Figure 5, where the non-dimensional turbulent kinetic energy dissipation $\epsilon_u = \nu < (\nabla u)^2 > H/U^3$ is shown as a function of the vertical Rossby number. We see that in both the laminar and turbulent regimes, the energy dissipation indeed decreases for sufficiently low vertical Rossby numbers. In the turbulent regime, the decrease is indeed proportional to Ro_z, indicated by the black dotted line. In the laminar regime, however, the decrease is milder, following $\mathrm{Ro}_z^{1/2}$. In the turbulent case, the rotation-dominated regime starts below $\mathrm{Ro}_z \sim 0.1$, while in the laminar case, it starts at $\mathrm{Ro}_z \sim 0.03$. We can use this change in regime to tag the simulations that are or are not influenced by the rotation. We denote them below by a white dot inside a filled symbol for the turbulent regime, and a black dot inside an open symbol for the laminar regime. Reporting this in Figures 7 and 8, we see that this regime corresponds to a lower heat transport and a smaller vertical Reynolds number, in agreement with the general findings that rotation impedes heat transport and vertical velocities. Note that the vertical Rossby number scales similar to the global Rossby number, following $\mathrm{Ro}_z \sim \mathrm{Ra}^{1/2}\,\mathrm{E}$ in both the laminar and turbulent regimes, see Figure 10.

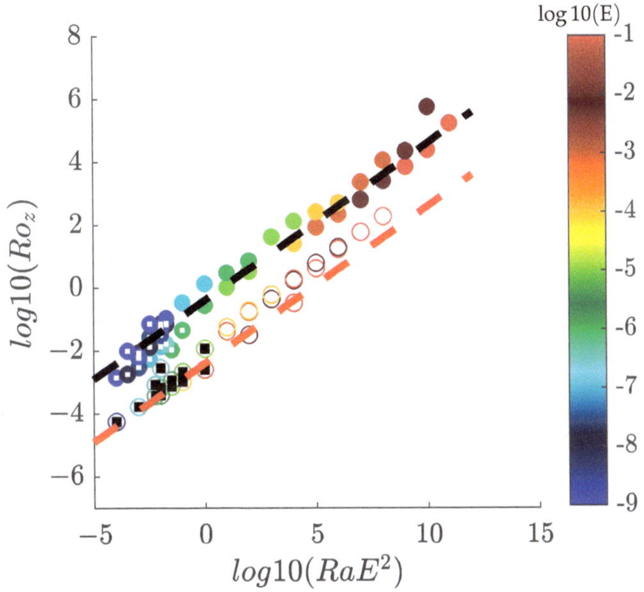

Figure 10. Vertical Rossby number Ro_z vs. "turbulent" Rayleigh number $\mathrm{Ra}\,E^2$ in 3D for $\mathrm{Pr} = 0.7$ for rotating HRB simulations on log-lattices. The symbols are colored according to their Ekman number E. The open symbols trace the laminar regime, while the filled symbols trace the turbulent regime. The rotation-dominated regimes are tagged by a black (respectively, white) square for the laminar (respectively, turbulent) regime. The black (respectively, red) dashed line follows $0.4\,\mathrm{Ra}^{1/2}\,E$ (respectively, $0.04\,\mathrm{Ra}^{1/2}\,E$).

5.3.3. Temperature Fluctuation and Anisotropy

Another interesting indicator of the influence of rotation on convection is given by the behavior of the temperature fluctuations, displayed in Figure 11. In the laminar case, they plateau at a low value (less than 0.01) as Ra increases, showing that the dynamics are indeed laminar. The rotation even tends to decrease the size of the fluctuations. In the laminar case, the temperature fluctuations increase with Ra, showing that convection is increasingly vigorous; even so, large rotations tend to somehow impede the development of excessively large fluctuations. The rotation also influences the anisotropy of the turbulence, as seen in Figure 12. Both in the laminar and turbulent cases, the anisotropy is well above 1/2, which would correspond to a situation where the kinetic energy is split evenly between motions along and perpendicular to the rotation axis. Due to the special nature of our projection, we cannot obtain a meaningful representation of what this result means in the physical space. It might, however, be interpreted as the influence of strong up (respectively, down)-drafts that are observed in direct numerical simulations to carry the heat from the bottom to the top (respectively, the cold fluid from the top to the bottom). The values we obtain here for the anisotropy are much larger to what are usually observed in simulations or experiments of RB convection with boundaries, but they are compatible with observations of radiative convection [29] or convection with no-slip boundary conditions [25], which have observed the formation of extremely extended plumes extending towards the whole bulk of the flow. The decrease in anisotropy observed at decreasing E values may be connected with the stabilizing influence of rotation, which impedes vertical fluctuations [27].

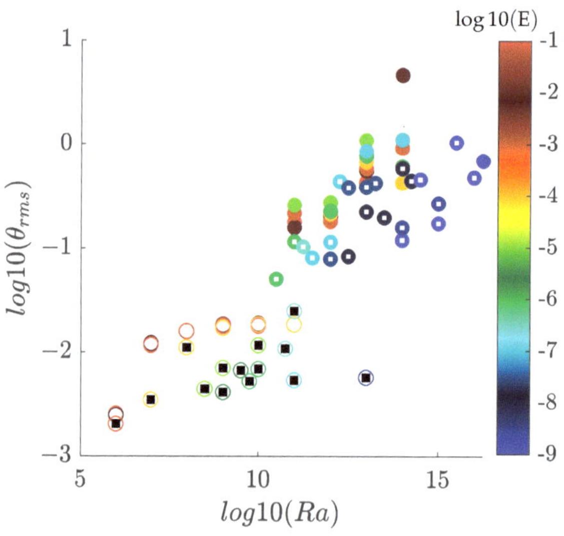

Figure 11. Temperature fluctuations $\theta_{rms} = \sqrt{< \theta^2 >}$ vs. Rayleigh number Ra in 3D for Pr $= 0.7$ for rotating HRB simulations on log-lattices. The symbols are colored according to their Ekman number E. The open symbols trace the laminar regime, while the filled symbols trace the turbulent regime. The rotation-dominated regimes are highlighted by a black (respectively, white) square for the laminar (respectively, turbulent) regime.

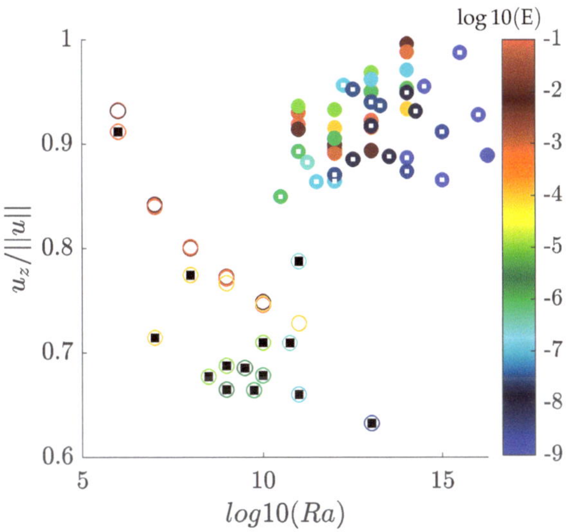

Figure 12. Velocity anisotropy $\sqrt{<u_z^2>}/\sqrt{<u^2>}$ vs. Rayleigh number Ra in 3D for Pr = 0.7 for rotating HRB simulations on log-lattices. The symbols are colored according to their Ekman number E. The open symbols trace the laminar regime, while the filled symbols trace the turbulent regime. The rotation-dominated regimes are highlighted by a black (respectively, white) square for the laminar (respectively, turbulent) regime.

5.3.4. Laminar and Turbulent Scaling Laws and GT Regimes

The laminar regime starts from the convection onset. It is therefore likely that it is influenced by near-onset dynamics. A natural idea is then to try to see whether it also follows the near-onset scaling law of the non-rotating case, Equation (22), albeit with Ra_c being replaced by its rotating value ($22\,\mathrm{E}^{-4/3}$) and Ra_t being replaced by $B\,\mathrm{E}^{-4/3}$, where the constant B needs to be determined. With this hypothesis, we then find that in the laminar regime, $\mathrm{Nu}\,E^{2/3}$ should be a function of $\mathrm{Ra}\,\mathrm{E}^{4/3}$, where the function satisfies Equation (22), with $A = 7$, $\mathrm{Ra}_c = 22$, and $\mathrm{Ra}_t = 5 \times 10^2$, see Figure 13. The turbulent regime corresponds to large Reynolds numbers, in which viscosity and diffusivity should not play a role any more. It is then natural to represent them in the turbulent variables Nu E and $\mathrm{Ra}\,\mathrm{E}^2$ (where we have omitted the Pr dependence, since it a constant in all our data sets), which is shown in Figure 9. This representation indeed collapses the data with two different scaling laws: one with exponent $3/2$ for regimes influenced by rotation—this is the GT regime—and one with exponent $1/2$ corresponding to the turbulent regime not influenced by rotation. The precise location in the parameter space where the GT regime occurs can be computed using the fit of the vertical Rossby number as a function of $\mathrm{Ra}\,\mathrm{E}^2$, see Figure 10. The condition $\mathrm{Ro}_z < 0.1$ then translates into the condition $\mathrm{Ra} \leq 0.06E^{-2}$, which is the green line in Figure 4. The GT regime is then found to be in-between the blue and the green lines, which is the region where we concentrate additional numerical simulations.

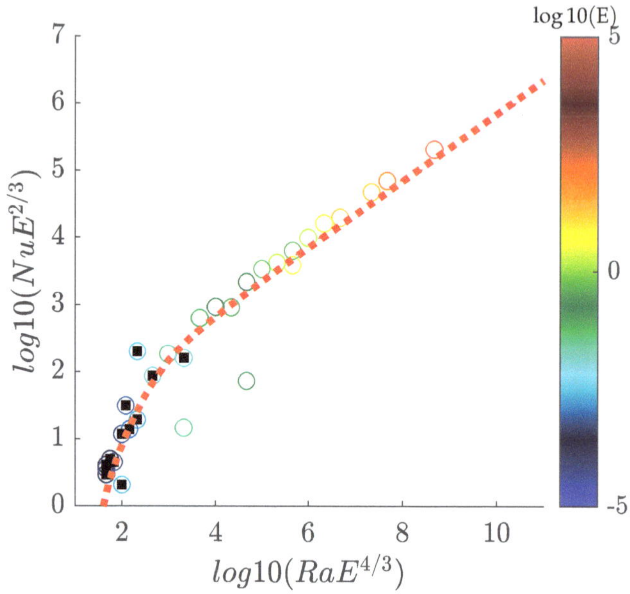

Figure 13. Universal law governing the heat transfer in the laminar regime Nu $E^{2/3}$ as a function of Ra $E^{4/3}$ in 3D for Pr $= 0.7$ for rotating HRB simulations on log-lattices. The symbols are colored according to their Ekman number E. The open symbols containing by a black square trace the rotation-dominated regime, while the open symbols trace the rotation-independent regime. The red dotted line follows Equation (22), with $A = 7$, $\mathrm{Ra}_c = 22$ and $\mathrm{Ra}_t = 5 \times 10^2$.

6. Discussion

We have shown that the projection of geophysical equations of motion allow for the determination of realistic values of parameters at a moderate cost. This allows us to perform many simulations over a wide range of parameters, thereby leading to general scaling laws of transport efficiency that can then be used to parametrize the turbulent transport in general climate models for Earth or other planets. We have illustrated this process using the equation describing the heat transport in a dry atmosphere to obtain the scaling laws for the onset of convection as a function of rotation, and confirmed the theoretical result of $\mathrm{Ra}_c \sim E^{-4/3}$ over a wide range of parameters. We have also demonstrated the existence of two regimes of convection: one laminar regime extending near the convection onset and one turbulent regime occurring as soon as the vertical Reynolds number reaches a value of 10^4. We have derived general scaling laws for these two regimes, both for the transport of heat and the dissipation of kinetic energy, and values of the anisotropy and temperature fluctuations. The set-up we have used here is far from being able to reproduce the full complexity of the atmosphere, as it models the friction at the bottom with a simple law, and ignores moisture dynamics. We plan to include these features in future work. Finally, it is not clear how the projection of dynamics onto log-lattices influences the results we are deriving. It is quite remarkable that the procedure is able to capture scaling laws, as we recover here some results already obtained in experiments [29] and numerical simulations [25] of convection without boundary layers, albeit with different prefactors. This of course has some important implications when translating these scaling laws as parametrizations in models. However, if we believe that the scaling laws themselves are robust, it only takes comparisons with a few direct numerical simulations to recalibrate the constants and turn our laws into useful parametrizations.

Author Contributions: Conceptualization, all authors; methodology, all authors; software, A.B., G.C. and Q.P.; validation, all authors; formal analysis, B.D.; investigation, Q.P.; resources, A.B.; data curation, all authors; writing—original draft preparation, B.D.; writing—review and editing, all authors; visualization, B.D.; supervision, B.D.; project administration, B.D.; funding acquisition, B.D. All authors have read and agreed to the published version of the manuscript.

Funding: This work received funding through the PhD fellowship programs of the Ecole Polytechnique and Ecole Normale Superieure Paris-Saclay, and through the Agence Nationale pour la Recherche, via the grants ANR TILT grant agreement no. ANR-20-CE30-0035 and ANR BANG grant agreement no. ANR-22-CE30-0025.

Institutional Review Board Statement: Not applicable.

Informed Consent Statement: Not applicable.

Data Availability Statement: The data presented in this study are available on request from the corresponding author. The data are not publicly available due to restrictions.

Conflicts of Interest: The authors declare no conflict of interest.

References

1. Grant, H.L.; Stewart, R.W.; Moilliet, A. Turbulence spectra from a tidal channel. *J. Fluid Mech.* **1962**, *12*, 241–268. [CrossRef]
2. Frisch, U. *Turbulence, the Legacy of A. N. Kolmogorov;* Cambridge University Press: Cambridge, UK, 1996.
3. Dubrulle, B. Beyond Kolmogorov cascades. *J. Fluid Mech.* **2019**, *867*, P1. [CrossRef]
4. Herring, J.R. Some Contributions of Two-Point Closure to Turbulence. In *Proceedings of the Frontiers in Fluid Mechanics;* Davis, S.H., Lumley, J.L., Eds.; Springer: Berlin/Heidelberg, Germany, 1985; pp. 68–87.
5. Dubrulle, B.; Daviaud, F.; Faranda, D.; Marié, L.; Saint-Michel, B. How many modes are needed to predict climate bifurcations? Lessons from an experiment. *Nonlinear Process. Geophys.* **2022**, *29*, 17–35. [CrossRef]
6. Campolina, C.S.; Mailybaev, A.A. Fluid dynamics on logarithmic lattices. *Nonlinearity* **2021**, *34*, 4684–4715. [CrossRef]
7. Meneguzzi, M.; Politano, H.; Pouquet, A.; Zolver, M. A Sparse-Mode Spectral Method for the Simulation of Turbulent Flows. *J. Comput. Phys.* **1996**, *123*, 32–44. [CrossRef]
8. Grossmann, S.; Lohse, D.; L'vov, V.; Procaccia, I. Finite size corrections to scaling in high Reynolds number turbulence. *Phys. Rev. Lett.* **1994**, *73*, 432–435. [CrossRef]
9. Frisch, U.; Pomyalov, A.; Procaccia, I.; Ray, S.S. Turbulence in Noninteger Dimensions by Fractal Fourier Decimation. *Phys. Rev. Lett.* **2012**, *108*, 074501. [CrossRef] [PubMed]
10. Lanotte, A.S.; Benzi, R.; Malapaka, S.K.; Toschi, F.; Biferale, L. Turbulence on a Fractal Fourier Set. *Phys. Rev. Lett.* **2015**, *115*, 264502. [CrossRef] [PubMed]
11. Barral, A.; Dubrulle, B. Asymptotic ultimate regime of homogeneous Rayleigh-Bénard convection on logarithmic lattices. *J. Fluid Mech.* **2023**, *962*, A2. [CrossRef]
12. Pikeroen, Q.; Barral, A.; Costa, G.; Campolina, C.; Mailybaev, A.; Dubrulle, B. Tracking complex singularities of fluids on log-lattices. *Nonlinearity* 2023, *submitted* .
13. Laval, J.P.; Dubrulle, B.; Nazarenko, S. Nonlocality and intermittency in three-dimensional turbulence. *Phys. Fluids* **2001**, *13*, 1995–2012. [CrossRef]
14. Costa, G.; Barral, A.; Dubrulle, B. Reversible Navier–Stokes equation on logarithmic lattices. *Phys. Rev. E* **2023**, *107*, 065106. [CrossRef] [PubMed]
15. Campolina, C.S. Fluid Dynamics on Logarithmic Lattices. Ph.D. Thesis, IMPA, Rio de Janeiro, Brazil , 2022.
16. Calzavarini, E.; Lohse, D.; Toschi, F.; Tripiccione, R. Rayleigh and Prandtl number scaling in the bulk of Rayleigh-Bénard turbulence. *Phys. Fluids* **2005**, *17*, 055107. [CrossRef]
17. Calzavarini, E.; Doering, C.R.; Gibbon, J.D.; Lohse, D.; Tanabe, A.; Toschi, F. Exponentially growing solutions in homogeneous Rayleigh-Bénard convection. *Phys. Rev. E* **2006**, *73*, 035301. [CrossRef]
18. Calzavarini, E.; Lohse, D.; Toschi, F. Homogeneous Rayleigh-Bénard Convection. In *Proceedings of the Progress in Turbulence II;* Oberlack, M., Khujadze, G., Günther, S., Weller, T., Frewer, M., Peinke, J., Barth, S., Eds.; Springer: Berlin/Heidelberg, Germany, 2007; pp. 181–184.
19. Chandrasekhar, S. *Hydrodynamic and Hydromagnetic Stability;* Courier Corporation: North Chelmsford, MA, USA, 2013.
20. Spiegel, E.A. A Generalization of the Mixing-Length Theory of Turbulent Convection. *Astrophys. J.* **1963**, *138*, 216. [CrossRef]
21. Kraichnan, R.H. Turbulent Thermal Convection at Arbitrary Prandtl Number. *Phys. Fluids* **1962**, *5*, 1374–1389. [CrossRef]
22. Grossmann, S.; Lohse, D. Scaling in thermal convection: A unifying theory. *J. Fluid Mech.* **2000**, *407*, 27–56. [CrossRef]
23. Ecke, R.E.; Shishkina, O. Turbulent Rotating Rayleigh–Bénard Convection. *Annu. Rev. Fluid Mech.* **2023**, *55*, 603–638. [CrossRef]
24. Cheng, J.S.; Stellmach, S.; Ribeiro, A.; Grannan, A.; King, E.M.; Aurnou, J.M. Laboratory-numerical models of rapidly rotating convection in planetary cores. *Geophys. J. Int.* **2015**, *201*, 1–17. [CrossRef]
25. Plumley, M.; Julien, K. Scaling Laws in Rayleigh-Bénard Convection. *Earth Space Sci.* **2019**, *6*, 1580–1592. [CrossRef]

26. Iyer, K.P.; Scheel, J.D.; Schumacher, J.; Sreenivasan, K.R. Classical 1/3 scaling of convection holds up to Ra = 10^{15}. *Proc. Natl. Acad. Sci. USA* **2020**, *117*, 7594–7598. [CrossRef] [PubMed]
27. Dubrulle, B.; Valdettaro, L. Consequences of rotation in energetics of accretion disks. *Astron. Astrophys.* **1992**, *263*, 387–400.
28. Campagne, A.; Machicoane, N.; Gallet, B.; Cortet, P.P.; Moisy, F. Turbulent drag in a rotating frame. *J. Fluid Mech.* **2016**, *794*, R5. [CrossRef]
29. Bouillaut, V.; Miquel, B.; Julien, K.; Aumaître, S.; Gallet, B. Experimental observation of the geostrophic turbulence regime of rapidly rotating convection. *Proc. Natl. Acad. Sci. USA* **2021**, *118*, e2105015118. [CrossRef] [PubMed]

atmosphere

MDPI

Article

Turbulence Transitions in Kelvin–Helmholtz Instability "Tube" and "Knot" Dynamics: Vorticity, Helicity, and Twist Waves

David C. Fritts [1,2,*], Thomas S. Lund [1], Adam C. Lund [1] and Ling Wang [1]

1 Global Atmospheric Technologies and Sciences (GATS), 3360 Mitchell Lane, Boulder, CO 80301, USA
2 Center for Space and Atmospheric Research (CSAR), Embry-Riddle Aeronautical University, Daytona Beach, FL 32114, USA
* Correspondence: dave@gats-inc.com

Abstract: We address the sources and dynamics of vorticity and helicity and their relations in transitions to turbulence arising due to Kelvin–Helmholtz instability (KHI) "Tube" and "Knot" (T&K) events. Such events are common in the atmosphere and oceans, and initial numerical simulations reveal that T&K dynamics significantly accelerate turbulence transitions and enhance KHI peak and mean energy dissipation rates. KHI T&K events arise where emerging KH billows exhibit varying wavelengths, phases, amplitudes, and/or discontinuities along their axes. As the KH billows intensify, these regions evolve roughly orthogonal billow cores and induced vortex tubes in close proximity. Their mutual advection as they intensify induces large-amplitude Kelvin vortex waves, or "twist waves", that arise where locally uniform vortices are distorted by axial or radial advection. The twist waves propagate along, and fragment, the vortex tubes and billow cores, thus accounting for the emergence of helicity and the down-scale energy, enstrophy, and helicity fluxes within the turbulence inertial range. We describe the results of four direct numerical simulations (DNS) addressing KHI T&K dynamics in large and idealized small domains. The large-domain vorticity fields reveal the character and diversity of KHI T&K dynamics, the emergence of twist waves at larger and smaller scales, and their driving of turbulence transitions. Two small-domain DNS exhibit idealized KHI T&K events arising from KH billows that are mis-aligned and that exhibit phase variability along their axes. A third examines the interactions of two vortex tubes in close proximity. These reveal that twist waves drive the character and evolutions of the vorticity and helicity fields.

Keywords: Kelvin–Helmholtz instability; tube and knot dynamics; vortex dynamics; twist waves; helicity; turbulence cascades

Citation: Fritts, D.C.; Lund, T.S.; Lund, A.C.; Wang, L. Turbulence Transitions in Kelvin–Helmholtz Instability "Tube" and "Knot" Dynamics: Vorticity, Helicity, and Twist Waves. *Atmosphere* **2023**, *14*, 1770. https://doi.org/10.3390/atmos14121770

Academic Editors: Boris Galperin, Annick Pouquet and Peter Sullivan

Received: 30 September 2023
Revised: 3 November 2023
Accepted: 7 November 2023
Published: 30 November 2023

1. Introduction

Evidence of vorticity dynamics driving accelerated and apparently more energetic transitions to turbulence in stratified shear flows has spanned almost four decades. Initial evidence was obtained from laboratory stratified shear flow studies of Kelvin–Helmholtz instabilities (KHIs) by Thorpe [1,2], who recognized their significance and coined the "Tube" and "Knot" (T&K) labels. More recently, Thorpe [3] noted apparent evidence of T&K dynamics in thin tropospheric cloud layers, revealing discontinuous and mis-aligned KHI. However, the first direct evidence including resolved transitional vorticity dynamics was provided by high-resolution imaging of OH airglow and polar mesospheric clouds (PMCs) at ~80–90 km in the atmosphere and accompanying initial "large-eddy simulation" (LES) and "direct numerical simulation" (DNS) modeling [4,5].

Modeling of KHI T&K dynamics to date has focused on the vorticity dynamics driving the energy and enstrophy cascades, relying on descriptions in terms of local, nonlinear versions of Kelvin vortex waves [6], or "twist waves", that are linear analytic perturbations of an idealized uniform vortex. Transient, local twist waves arise where initial uniform vortices are perturbed by local displacements (a) normal to their axes [7], (b) along their axes by an adjacent, roughly orthogonal, vortex yielding axial compression and stretching, or (c) via external shear or strain [8,9]. An equivalent description of these interactions was provided in terms of vortex stretching and twisting [10].

Spectral transfers of energy, enstrophy, and helicity have been described as fluxes driven by global three-component interactions; see, in particular, reviews of helicity cascades and references therein [11–13]. These cascades represent domain averages over all triad interactions, and they successfully capture the expected spectral character for observed down-scale and up-scale cascades implied by individual triad interactions. The inferred spectral fluxes must be consistent with inferences from local dynamics defined by superpositions of all spectral modes. However, the local dynamics in physical, rather than spectral, space provide a very different, and arguably more insightful, understanding of the local vorticity dynamics driving these turbulence transitions and evolutions.

Initial studies of turbulence transitions driven by mis-aligned or varying KH billows [4,5] showed larger-scale vortex knots to emerge via two primary vortex interactions. Sites having mis-aligned initial KH billows where two billows connect to one, or three connect to two, initiate strong links among the co-rotating billow cores that entwine and distort them. These dynamics rapidly induce intense interactions among roughly orthogonal large-scale vortices thereafter. Other sites where adjacent KH billows exhibit roughly parallel phase variations along their axes initiate vortex tubes on the intermediate vortex sheets wrapping under and over the adjacent KH billows. The larger (smaller) KH billow phase variations favor larger (smaller) vortex tubes, respectively. Where emerging vortex tubes are advected and stretched over and under the adjacent KH billows, they "attach" to the billows and also induce roughly orthogonal larger- and smaller-scale vortex alignments in close proximity. These sites likewise lead to vortex knots, but they are typically delayed and/or less intense than those initiated by mis-aligned initial KH billows.

The primary mechanism driving the cascades of energy, enstrophy, and helicity to smaller scales in these studies is the excitation of strongly nonlinear twist waves on vortices in close proximity, and oriented roughly orthogonally [4,5]. Mode-1 twist waves exhibit helical or "corkscrew" shapes that propagate along the vortices, but do not directly drive the cascade to smaller scales. Mode-2 twist waves also easily arise and achieve very large amplitudes that unravel single initial vortices into two intertwined helical vortices at smaller scales outward from their source, as noted above. Mode-3 and higher twist waves are surely also excited, but they are apparently not generated efficiently and have not been identified at observable amplitudes in the various modeling studies to date.

Smaller-scale knots arise thereafter where twist waves or other vortices interact in close proximity. These dynamics accelerate at smaller scales and closer proximity, and rapidly drive a broad turbulence inertial range. The DNS of these KHI T&K dynamics to date reveal the occurrence of knots extending from the initial KH billow scales to ~30 times smaller for the specified initial Richardson and Reynolds numbers, *Ri* and *Re*.

Our purposes here are to explore the implications of KHI T&K dynamics revealed by the vorticity evolution for one large multi-scale DNS and both the vorticity and helicity evolutions for three idealized KHI T&K cases. The first DNS (Case 1) describes a large-domain, primarily three-KH billow event for a minimum initial $Ri = 0.1$ and a moderate $Re = 5000$ initiated with weak initial random noise and exhibiting diverse T&K interactions and both knot types noted above. The second, third, and fourth DNS describe idealized T&K events where (a) one KH billow core links to two where they are initially mis-aligned along their axes (Case 2), (b) adjacent KH billows link via vortex tubes where they exhibit correlated phase variations along their axes (Case 3), and (c) two orthogonal linear vortices having different scales and intensities emerge in close proximity. Cases 2, 3, and 4 are performed to illustrate the vortex dynamics accounting for the emergence of twist waves and helicity at larger and smaller scales and their subsequent evolutions. The spectral model, its initial conditions, and the analysis methods employed are described in Section 2. The Case 1 evolution illustrating multiple observed T&K dynamics [4,5] and their relative importance is described in Section 3. Cases 2, 3, and 4 are described in Section 4. Section 5 discusses these results relative to previous studies addressing similar dynamics. Our summary and conclusions are presented in Section 6.

2. Methods: Direct Numerical Simulations

2.1. Fourier Spectral Model Formulation and Solution Method

The Spectral Atmosphere Model (SAM) formulation, solution methods, boundary and initial conditions, and analysis and visualization methods are described below. The SAM solves the Boussinesq Navier-Stokes equations written in strong conservation law (divergence) form as follows:

$$\partial u_j/\partial x_j = 0 \quad (1)$$

$$\partial u_i/\partial t + \partial(u_i u_j)/\partial x_j = -(1/\rho_0)\,\partial p'/\partial x_i + (\theta'/\theta_0) g\delta_{i3} + \nu_{turb}\partial^2 u_i/\partial x_j \partial x_j \quad (2)$$

and

$$\partial\theta/\partial t + \partial(\theta u_j)/\partial x_j = \kappa_{turb}\partial^2\theta/\partial x_j\partial x_j \quad (3)$$

Here, (u, v, w) are velocity components along (x, y, z); p, ρ, and θ are pressure, density, and potential temperature (and $\theta = T$ for a Boussinesq fluid); ν_{turb} and κ_{turb} are the turbulence kinematic viscosity and thermal diffusivity; primes and "0" subscripts denote perturbation and mean quantities; and successive indices imply summation. The SAM employs a pseudo-spectral algorithm to compute nonlinear products, removal of aliasing errors employing the 2/3rds truncation rule, and a low-storage third-order Runge–Kutta method for time advancement. The SAM is also highly optimized for massively parallel computers [5]. The Case 1 DNS employed up to 3600 processors, resolved 3600 modes, and enabled resolution of 1200 wavelengths along the direction of the shear flow.

Elevated turbulence kinematic viscosity, ν_{turb}, and thermal diffusivity, κ_{turb}, are assumed because (a) we expect the background atmosphere at higher altitudes to exhibit weaker and stronger turbulence nearly continuously in altitude and (b) even a representative $Re = 5000$ yields KHI T&K dynamics and turbulence spanning 3 decades of scales along x, requiring a very computationally costly DNS.

2.2. Boundary and Initial Conditions and Computational Domains

SAM is triply periodic and employs initial streamwise (along x) velocity and stability profiles given by

$$U(z) = U_0 \cos(\pi z/Z) \tanh(z/h) \quad (4)$$

and

$$N^2(z) = (g/\theta_0)\, d\theta/dz = N_0{}^2 + (N_m{}^2 - N_0{}^2)\, sech^2[(z - z_0)/h] \quad (5)$$

We assume $U_0 = 31.25$ m/s, $h = 350$ m, $z = 0$ at the domain center, $N_0{}^2 = 10^{-4}$ s^{-2}, and $N_m{}^2 = 8 \times 10^{-4}$ s^{-2}, yielding an initial buoyancy period $T_b = 222$ s at $z = 0$ (see Figure 1). The chosen U_0 and h are 2.5 times larger than specified for the initial DNS assuming a stratospheric application. They imply a nominal KHI wavelength of λ_h~4–5 km representative of initial KHI T&K dynamics observed in polar mesospheric cloud (PMC) imaging ([5], Figure 4) and seen to arise frequently in multi-scale GW events in the mesosphere. They are also intended to approximate a strong shear layer arising in an environment driven the superposition of several large-amplitude gravity waves (GWs) yielding a maximum dU/dz and minimum Ri at the maximum in N^2.

We chose a Case 1 domain enabling 3 or 4 initial KH billows along a shear of length $X = 3L$ and width $Y = 9L$ for $L = 5$ km. We specified h/L to allow 3 or 4 KH billows along X rather than 1 or 2, or 2 or 3, because fewer billows in a periodic domain favor KH billow pairing, which appears to not be a major component of KHI T&K dynamics in observations to date [4,5]. This domain geometry will be seen below to exhibit additional T&K dynamics and transitions to turbulence arising from a different initial noise seed.

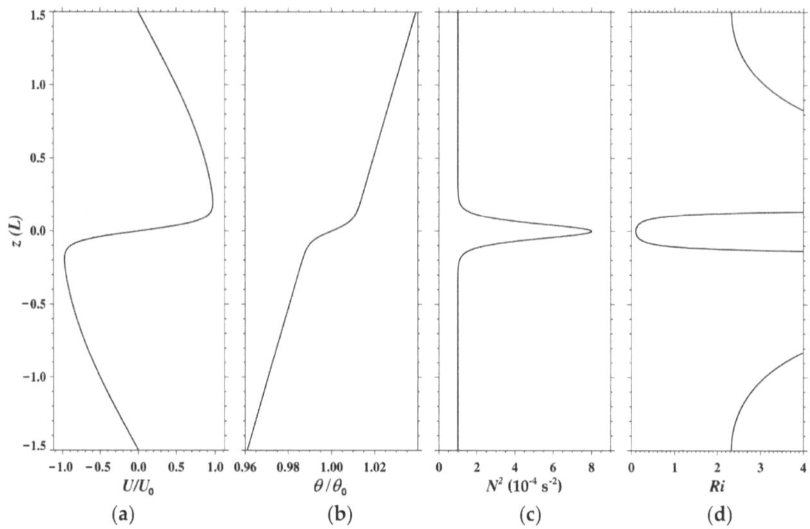

Figure 1. Initial profiles of (**a**) U/U_0, (**b**) θ/θ_0, (**c**) N^2, and (**d**) Ri used for the large domain DNS.

For reference to KHI T&K events seen at the PMC layer, the Case 1 SAM DNS was performed for a turbulence Reynolds number $Re = U_0h/\nu_{turb} = 5000$, a Richardson number $Ri = N^2/(dU/dz)^2 = N^2h^2/U_0{}^2 = 0.1$, corresponding to a Froude number $Fr = Ri^{-0.5} = 3.16$, and a Prandtl number $Pr = 1$. These choices enabled the emergence of the following responses:

(1) larger- and smaller-scale T&K dynamics as noted above;
(2) smaller-scale secondary convective instabilities (CIs) in the outer KH billow cores and secondary KHI on the vortex sheets between, and wrapping around, adjacent billows where the billows exhibit curvature in the x-y plane; and
(3) an extended turbulence inertial range enabling an assessment of energy dissipation rates to small scales.

The Case 1 $Ri = 0.1$ at $z = 0$ and $Re = 5000$ assuming a weak to moderate turbulence viscosity ν_{turb}~0.7 m^2s^{-1} in the upper stratosphere or mesosphere (in order to constrain Re) suggest secondary CI in the outer billow cores, but no secondary KHI in the stratified braids between adjacent KH billows in the absence of local T&K dynamics. For Case 1, the SAM was seeded with nondivergent white noise in velocity at $t = 0$ having $u_{rms} = U_0 \times 10^{-5}$. The smallest scales dissipate quickly, causing the noise spectrum to rapidly acquire a weak inertial range/dissipation range shape. The larger-scale initial noise induces KH billows that exhibit more variable phases along y than modeled previously [4,5]. This noise seed was chosen specifically to include multiple sites where KH billows are mis-aligned along y, at which 2 KH billows link to 1, or 3 billows link to 2 that did not arise for the noise seed in the initial DNS [5].

A smaller domain having horizontal dimensions of $(X, Y, Z) = (2, 10.4, 3.5)L$ was used to induce the primary interactions accompanying initial KH billows that are mis-aligned and exhibit phase variations along y in Cases 2 and 3. These cases were intended to enable the visualization of idealized T&K dynamics and their implications for the emergence and evolution of helicity in the T&K regions. They were performed for smaller $Re = 2000$ and 2500 in order to suppress the very small-scale vorticity structures arising in Case 1.

2.3. Resolution Requirements

Our chosen Re and Ri and the large Case 1 computational domain enabled a diversity of KHI T&K dynamics and demanded major computational resources, as noted above. Our desire to resolve instability dynamics to the smallest scales enabled by our chosen Re and Ri employing DNS required an isotropic resolution of

$$\Delta x < 1.8\eta \tag{6}$$

for a turbulence Kolmogorov scale $\eta = (\nu^3/\varepsilon)^{1/4}$ [14,15], with $\nu = \nu_{turb}$ in our DNS here, and energy dissipation rate

$$\varepsilon = 2\nu <S_{ij}\, S_{ij}> \tag{7}$$

where brackets denote spatial averaging, strain tensor components are given by

$$S_{ij} = \frac{1}{2}\left(\frac{\partial u_i}{\partial x_j} + \frac{\partial u_j}{\partial x_i}\right) \tag{8}$$

and repeated indices imply summation. Mean ε was assessed on all two-dimensional (2-D) model domain planes, and the constraint employed the largest ε on any plane, but typically the (x, y) plane at $z = 0$. This is much stricter than what is typically used because of the very intermittent regions of strong turbulence accompanying initial turbulence transitions and is consistent with our previous usage [5]. Resolution of initial instabilities was judged to be needed to ensure that they had realistic impacts on the subsequent evolutions of Kelvin twist waves and helicity, and these dynamics spanned 3 decades of scales during the most intense turbulence in Case 1. The peak resolution for Case 1 yielded a typical minimum $\Delta x < 1.6\eta$, but at very high computational cost. Cases 2–4 also satisfied this constraint over the intervals shown.

2.4. Three-Dimensional (3-D) Volumetric Visualization

Three-dimensional imaging of the vorticity dynamics underlying the KHI T&K evolutions employs the same methods we have used previously [4,5]. Negative values of the second eigenvalue, λ_2, of the tensor $\boldsymbol{L} = \boldsymbol{R}^2 + \boldsymbol{S}^2$ correspond to features having strong rotation, as opposed to shearing, and its magnitude measures rotational intensity [16].

Components of the strain tensor, $\boldsymbol{S}$, are given in Equation (8); those for the rotation tensor are given by

$$R_{ij} = \frac{1}{2}\left(\frac{\partial u_i}{\partial x_j} - \frac{\partial u_j}{\partial x_i}\right) \tag{9}$$

Negative λ_2 corresponds to the subset of total vorticity primarily driving the turbulence field. This enables tracking the evolution from emerging KH billow cores and vortex tubes to their interactions in knot regions driving twist waves and helicity, the fragmentation of the larger-scale vortices, and similar interactions driving the energy, enstrophy, and helicity cascades deep into the inertial range [5]. These dynamics are illustrated with 3-D imaging of λ_2 in the large SAM domain from the initial to intermediate stages of KHI T&K dynamics in Figures 2–6.

Local helicity $H = \boldsymbol{u}\cdot\boldsymbol{\zeta}$ for vorticity defined as $\boldsymbol{\zeta} = \nabla \times \boldsymbol{V}$, was employed previously for studies of the degree of "knotted-ness" of tangled vortex lines [11,17]. Similarly, our previous studies of KHI T&K dynamics revealed inferred strong helicity in the λ_2 fields at larger and smaller scales accompanying twist wave generation in KHI vortex knots. For our purposes, these dynamics are illustrated with 3-D imaging of λ_2 for the 4 cases described, and of H in Cases 2–4.

3. Large-Domain KHI T&K Imaging

The large-domain Case 1 evolution is shown in Figure 2 from an early stage of KH billow formation (denoted 0 T_b) to a time 2.25 T_b later. The interval spans the dynamics that arise in small domains without T&K influences [18,19] and the T&K dynamics identified to date enabled by large domains noted above. An adaptive λ_2 color scale is employed separately at each time in order to reveal vortex features spanning a wide range of intensities. Specifically, the KH billow cores intensify significantly from 0 to 2.25 T_b, but appear to diminish due to the emergence of much more intense smaller-scale dynamics.

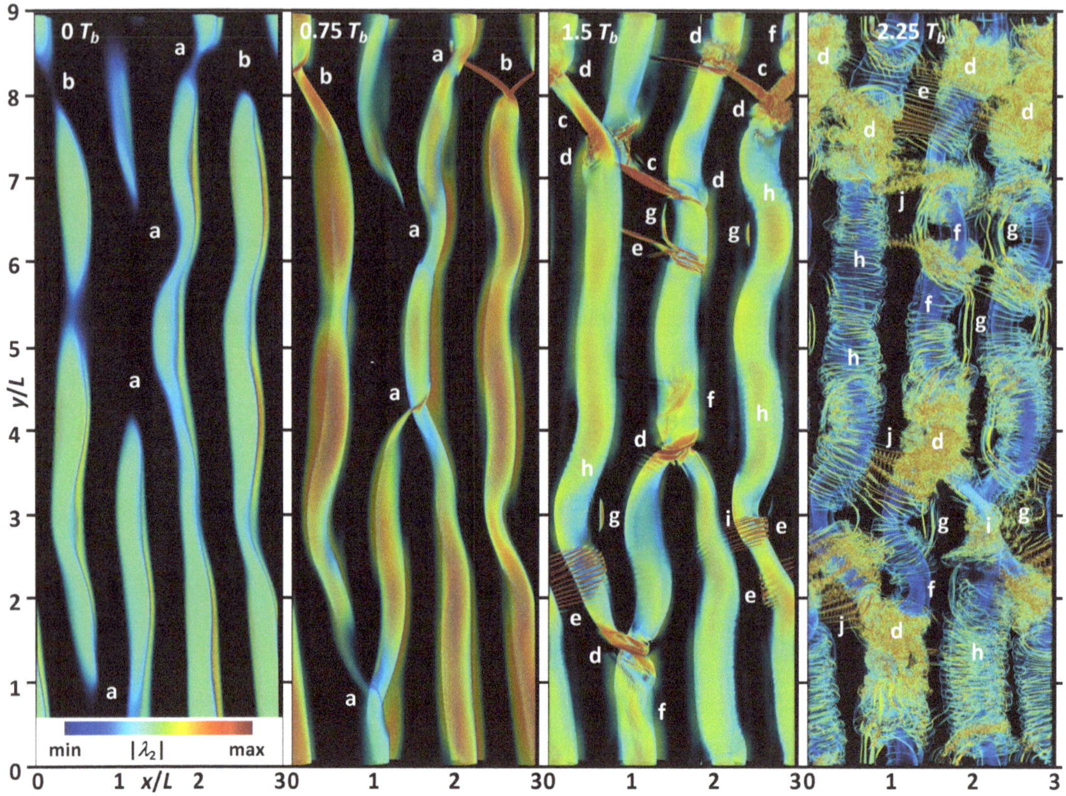

Figure 2. Three-dimensional imaging of λ_2 in the large SAM Case 1 domain viewed from above from the early stages of billow formation. Letters identify features discussed in the text.

These fields reveal a diversity of T&K dynamics at larger and smaller scales, the larger of which are the major drivers of strong turbulence thereafter. All the KHI T&K dynamics emerge and drive strong turbulence transitions within 2.25 T_b, and largely prior to significant transitions by secondary CI and KHI. The T&K dynamics include:

(1) initial links between mis-aligned emerging KH billow "ends" and continuous adjacent billows (sites "a");
(2) additional links of mis-aligned emerging KH billows where one billow "end" links to two billows (sites "b");
(3) large-scale vortex tubes emerging on the intensifying vortex sheets between KH billows where they exhibit significant phase variations along y that link adjacent billows as they intensify (sites "c");
(4) initial large-scale vortex knots driven by links between billow cores and/or large-scale vortex tubes (sites "d");
(5) smaller vortex tubes that form where adjacent billows exhibit weaker phase variations along y (sites "e");
(6) initiation of large-scale twist waves in the billow cores by the emerging knots (sites "f") that propagate along the billow cores thereafter.

Secondary instabilities arising in the presence and/or absence of T&K dynamics by 1.5 T_b include the following:

(7) secondary KHI on the intensifying vortex sheets that are enhanced where billow curvature increases stretching and intensification of the vortex sheets between adjacent KH billows (sites "g");
(8) initial secondary CI in the outer billows where they exhibit rapid intensification with and without apparent T&K enhancements; see sites "h" in the regions having the largest negative λ_2 (red) at 0.75 and 1.5 T_b;
(9) a billow-pairing event, which can also occur in the absence of varying KH billow phases along y (site "i").

Note that vortex tube "ends" and knots discussed here and below are sites where apparently distinct features appear to "attach". However, vortex field lines are continuous and distinct features cannot attach; see the more detailed discussion of Case 3 below.

The Case 1 λ_2 fields at 1.5 and 2.25 T_b in Figure 2 reveal dramatic progressions from an initially laminar flow to emerging and/or intense turbulence arising from diverse KHI dynamics. All the T&K dynamics identified at 1.5 T_b except the billow-scale twist waves yield strong turbulence or its precursors revealed by intense, small-scale λ_2 features at 2.25 T_b. However, the occurrence and character of many features at small scales cannot be seen in Figure 2. Hence, expanded views of the λ_2 fields at smaller and larger y at 2.25 T_b are provided in Figures 3 and 4 and in a subdomain spanning 1.6 L and 2.1 L along x and y, respectively, in Figure 5. These views reveal a diversity of transitional dynamics, several types of responses of which remain laminar in some regions and exhibit initial transitions to turbulence elsewhere. Note that the imaging was restricted to $|z| < 0.2\,L$ because of the very large data files. This enables viewing into the billow cores where upward displacements at their larger displacements toward positive x exclude secondary CI from the volume viewed. Examples seen in Figures 3 and 4, and the zoomed view in Figure 5, include the following:

(1) small-scale vortex tubes between adjacent billows that remain laminar in some regions (sites "e"), but which exhibit links among them that emerge as secondary T&K dynamics on the vortex sheets, and become intense and initiate breakdown as they wrap over the KH billows at larger x (sites "j");
(2) secondary KHIs on the intermediate vortex sheets that are largely laminar, but exhibit mutual interactions driving smaller scale features in multiple regions (sites "g");
(3) secondary CIs at multiple sites that achieve large amplitudes and initiate mutual interactions driving emerging smaller-scale responses (sites "h");
(4) regions in which secondary KHIs and CIs evolve in increasing proximity where the vortex sheets wrap around the KH billows and initiate turbulence transitions among roughly orthogonal vortices (sites "k");
(5) increasing billow core displacements along x that are coherent along y with λ_y~$2L$ (sites "l") denoted a "crankshaft" instability, for which there is now significant observational evidence and are seen to arise in the final field in Figure 2 in the two KH billows at upper right;
(6) emergence of these various dynamics in close proximity that also excite larger- and smaller-scale twist waves on the billow cores, vortex tubes, and smaller-scales vortices arising from larger-scale events (sites "f").

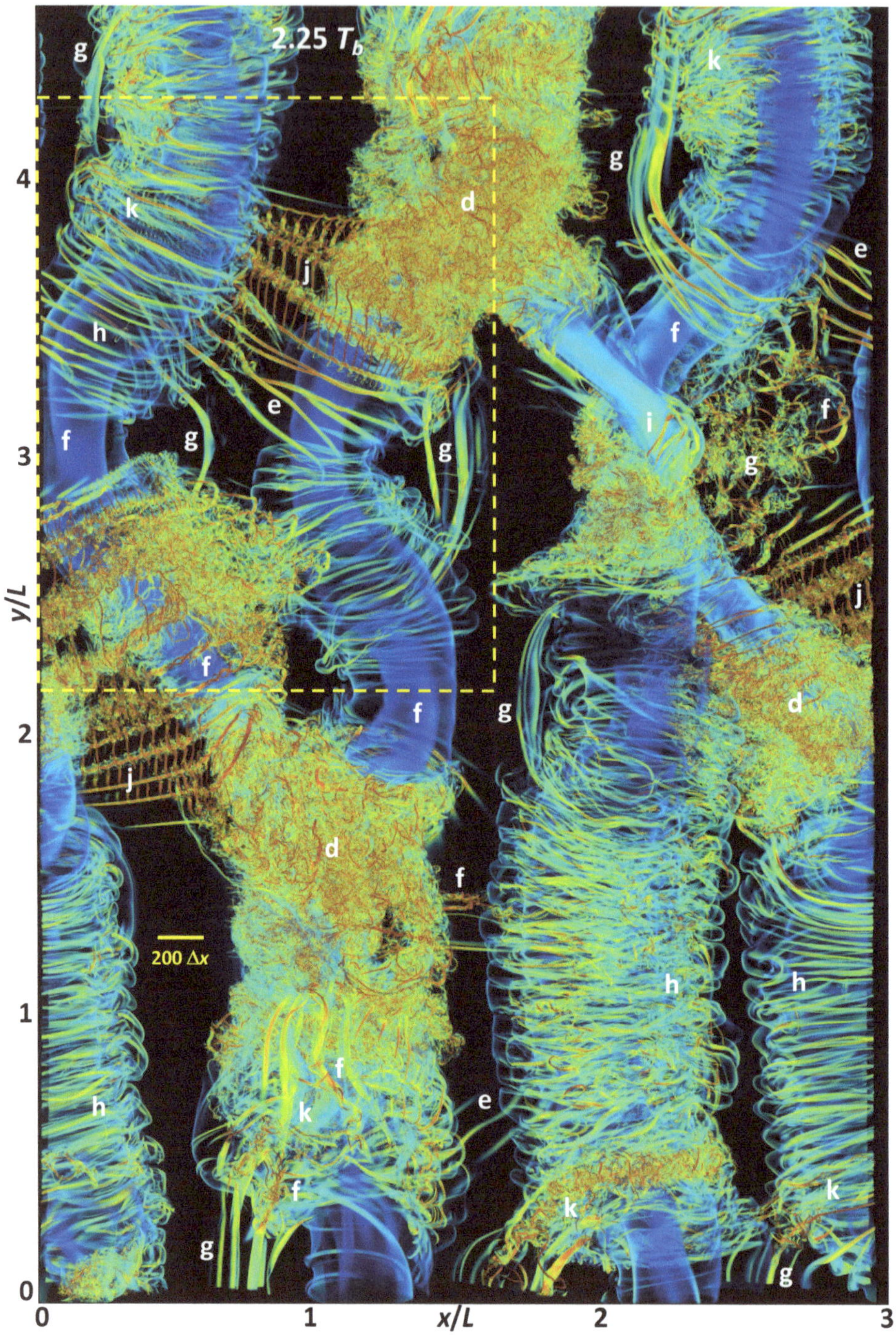

Figure 3. An expanded view of the 3-D imaging in Figure 2 at 2.25 T_b for y/L = 0–4.5. Feature labels are as in Figure 2. Note the yellow bar having a length of 200 Δx for reference. The dashed yellow rectangle indicates the region shown with a zoomed view in Figure 5.

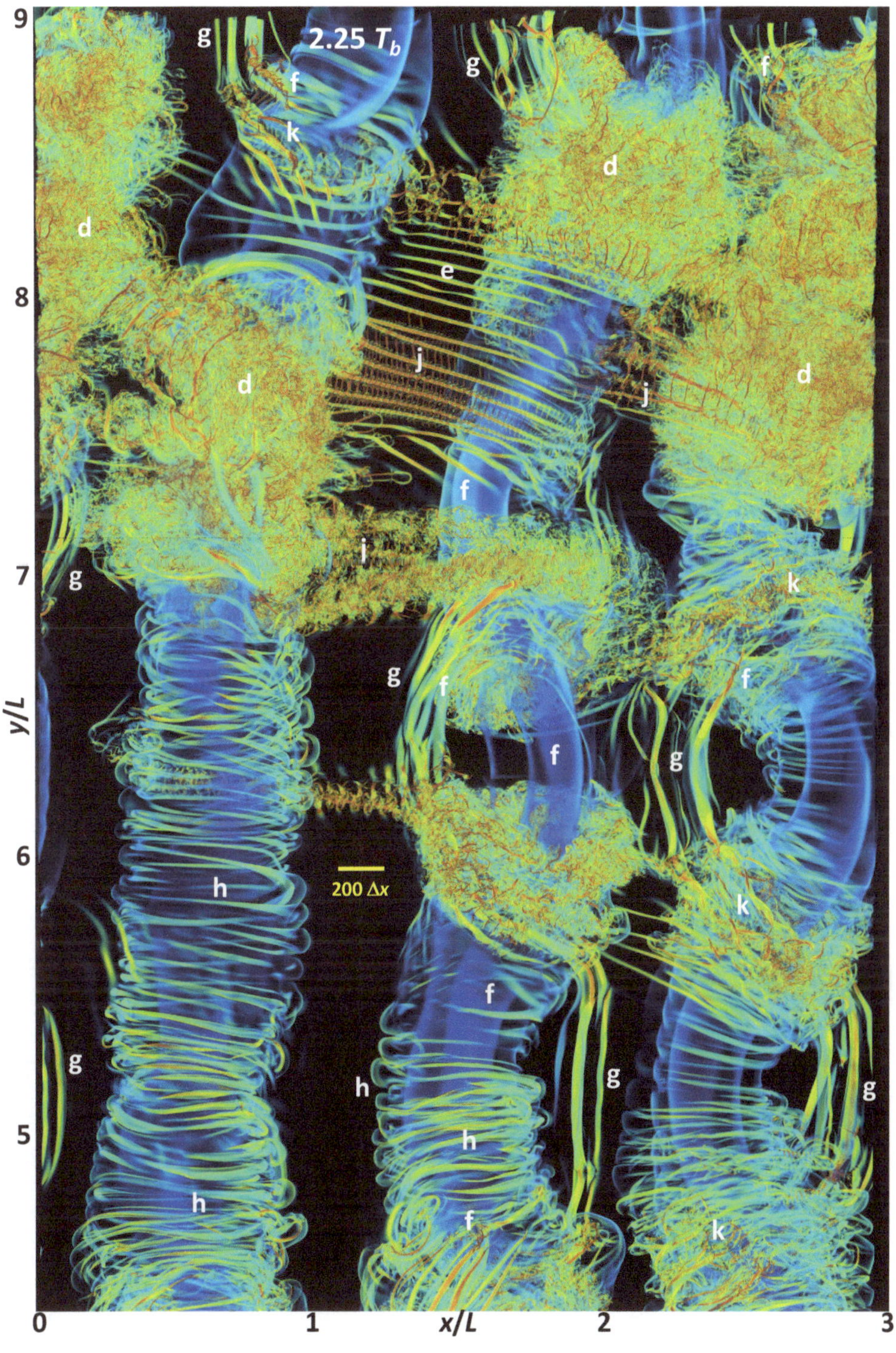

Figure 4. As in Figure 3, for the portion of the domain for y/L = 4.5–9. Note the 200 Δx reference.

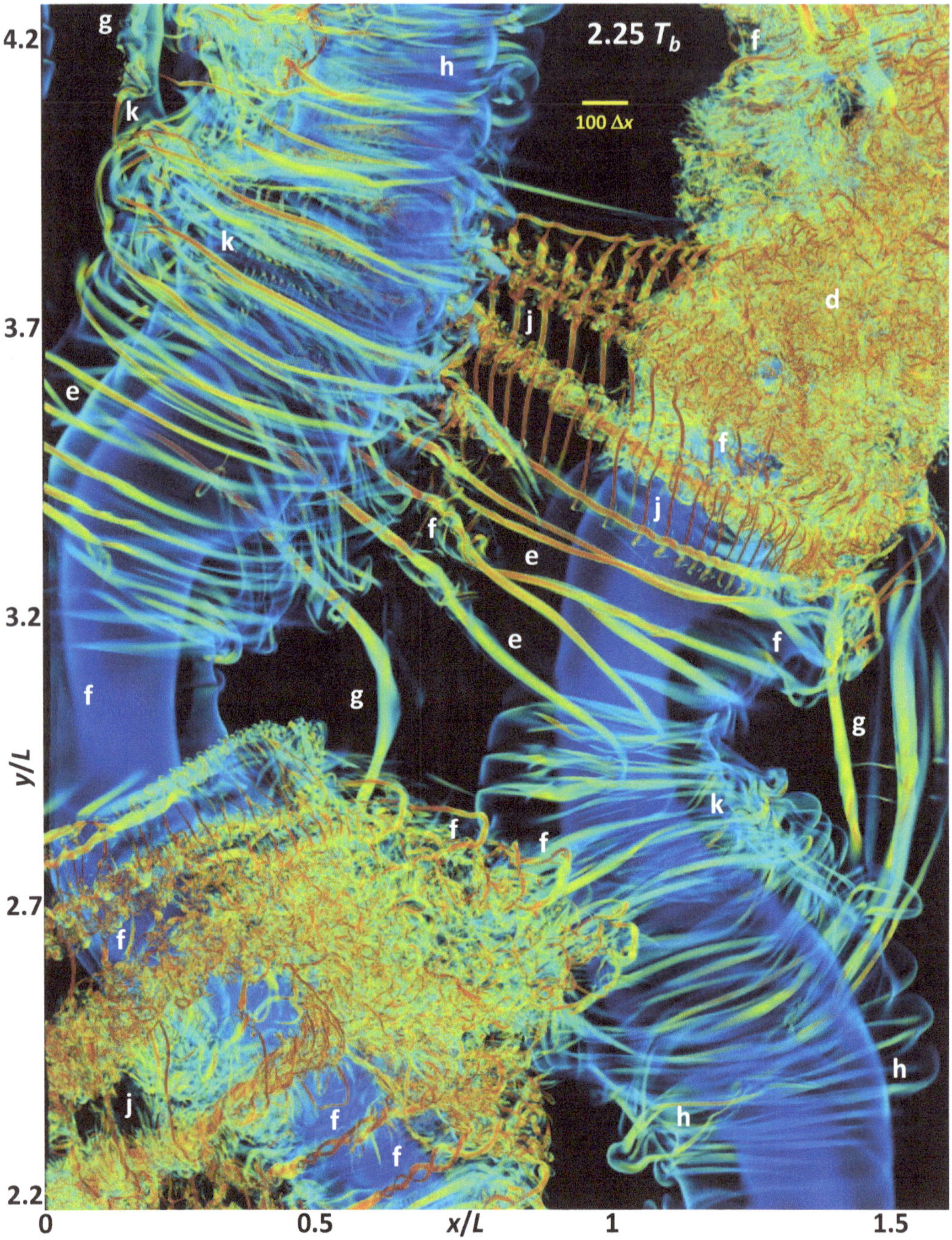

Figure 5. An expanded view of the yellow rectangle in Figure 3, showing smaller-scale features in greater detail. Feature labels are as in Figures 2–4. Note the 100 Δx reference at top.

The importance of various T&K dynamics is assessed qualitatively using their contributions to local energy dissipation rates, ε. We employ ε(*x*,*y*) at *z* = 0 at 2.25 T_b for this purpose because volumetric views, as used for $\lambda_2(x,y)$ in Figures 2–5, mask the peak ε in deep T&K events such as knots by much weaker ε at the event extrema in *z*. Comparisons of λ_2 as shown in Figures 2–5 and $\log_{10}\varepsilon(x,y)$ at *z* = 0 at 2.25 T_b are shown in Figure 6 and reveal the following sites to drive the strongest dynamics and ε arising by 2.25 T_b:

(1) the intense initial knots (sites "d") account for the largest ε spanning the largest regions horizontally;

(2) smaller knot regions arising from the billow pairing (and merging) event (site "i") achieve comparable peak ε, but are more confined spatially;

(3) secondary KHI T&K events along the inclined vortex sheets (sites "j") yield comparable peak ε because this is where the vortex sheets are most intensified by adjacent KH billows having large phase variations along *y*.

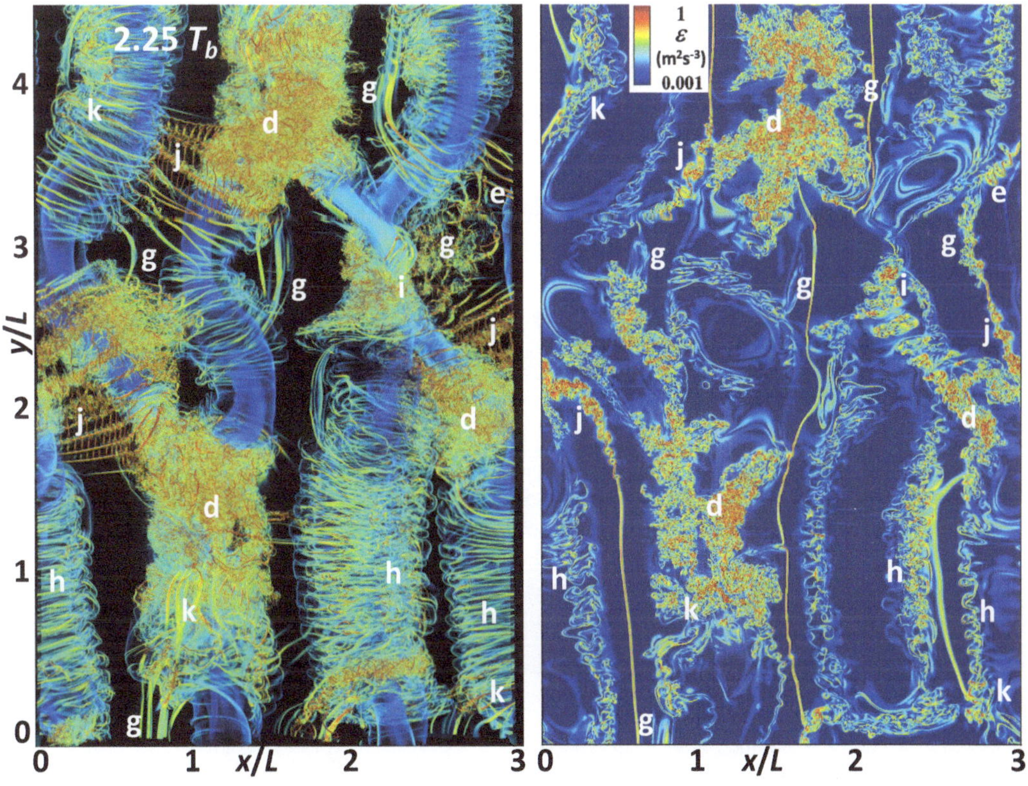

Figure 6. As in Figure 3 (**left**) and the corresponding $\log_{10}\varepsilon(x, y)$ at $z = 0$ and 2.25 T_b (**right**).

Additional sources of emerging instabilities and weaker, but increasing, ε in order of importance include the following:

(4) regions in which secondary KHI are entrained over/under KH billows into close proximity to CIs induce their close interactions and turbulence transitions (sites "k");

(5) intensifying secondary KHIs where they are enhanced by distorted KH billows (site "g" at larger *x* than the billow pairing event);

(6) intensifying and interacting CIs in the billow exteriors inducing initial turbulence transitions (sites "h").

These various dynamics enable several broad conclusions:

(7) large-scale KHI T&K dynamics drive turbulence events that are more intense than those occurring in their absence, consistent with the initial DNS cited above;
(8) large-scale KHI T&K dynamics induce intensification of, and interactions among, smaller-scale secondary CIs and KHIs yielding localized, intense features at small scales without large-scale, energetic precursors;
(9) secondary CIs and KHIs also exhibit smaller-scale, but much weaker, self-interactions without large-scale influences on comparable time scales.

4. Idealized Small-Domain KHI T&K Imaging

We now describe three small DNS of idealized KHI T&K interactions that reveal the transitions to turbulence seen in vorticity, described by λ_2, and helicity H. Case 2 is motivated by sites in the Case 1 DNS seen to drive the initial vortex knots (sites "b" to "d" in Figures 2–6), where two KH billows link to one via strong vortex tubes. Case 3 approximates sites "e" in Case 1 in which two billows link via multiple, smaller vortex tubes where the billows exhibit common phase variations along their axes. Case 4 examines the mutual interactions of two orthogonal linear vortices in close proximity that approximate interactions seen to arise in previous large-eddy simulations (LES) and DNS of turbulence transitions due to KHI T&K [4,5] and similar dynamics in other turbulence transitions.

4.1. Case 2: One KH Billow Linking to Two Billows along Their Axes

Case 2 is an idealized version of the sites labeled "b" at 0.75 T_b in Figure 2. It was performed in a small periodic domain containing sites linking one KH billow to two, one of which is shown here. An initial $Ri = 0.1$ ($Fr = 3.16$) was employed as for Case 1, but a smaller $Re = 2000$ was specified to restrict the evolving dynamics to larger spatial scales. Three-dimensional imaging of λ_2 and H spanning 1.25 T_b is shown in Figure 7, viewed from above and negative x, with positive spanwise vorticity ζ_y to the left. Successive images are expanded to reveal the details as they intensify and cascade to smaller scales. Accompanying videos, Case2a.mp4 and Case2b.mp4 in the Supplementary Materials, provide animations of the of λ_2 and H evolutions shown at left and right in Figure 7.

The first images show a time at which weak initial vortex tubes induced by stretching of the intermediate vortex sheets have linked by wrapping over (under) the billow ends at larger (smaller) y (sites "a" in λ_2 at left) and induced initial mode-2 twist waves in the billow cores. These features and interactions intensify rapidly due to continuing KH billow roll-up and vortex sheet stretching, yielding strong initial mode-2 twist waves in the billow cores by 0.25 T_b (sites "b") and differential vortex stretching and compression (sites "c") where the tubes link to the billow cores by 0.5 T_b. These evolving dynamics act as strong sources of additional, smaller-scale, twist waves on the vortex tube "ends" and on the initial, larger-scale twist waves in the billow cores seen emerging at 0.5 T_b (sites "c" and "d"). They intensify dramatically and drive increasingly smaller twist waves that account for the initial turbulence transitions in the vortex tubes and their links to the billows by 0.75 T_b (sites "e" and "f"). The further evolution seen at 1 and 1.25 T_b reveals extensions of these dynamics along the billow cores (sites "g") and to smaller scales and higher intensities in the turbulence inertial range within the emerging vortex knots (sites "h").

The corresponding Case 2 evolution of helicity, H, is shown on the right in Figure 7 from the same perspective. H is also very weak during the initial links between mis-aligned KH billows because both ζ_y and the initial axial velocities, primarily $u_2 = v$, are very small. These features intensify by 0.25 T_b and exhibit positive (negative) H where the emerging tubes link to the KH billows, implying positive (negative) v at larger (smaller) y for $\zeta_y > 0$. By 0.5 T_b, the vortex tubes have expanded and evolved oppositely signed H sheaths by entraining the vortex sheets from larger and smaller x and y (sites "i") on which the tubes arose.

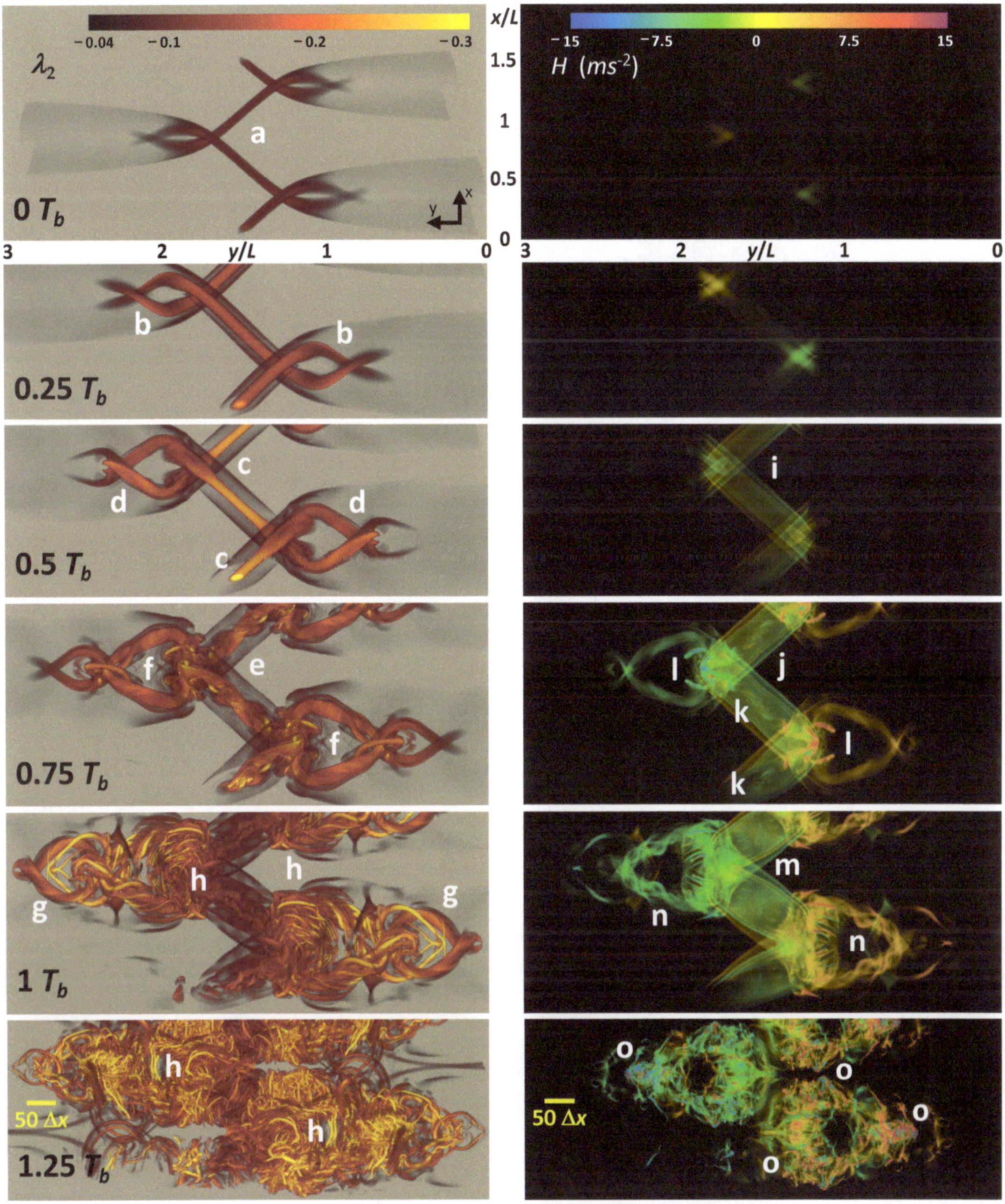

Figure 7. Case 2 λ_2 (**left**) and H (**right**) 3-D imaging spanning 1.25 T_b. As in Case 1, $t = 0$ is defined at a time exhibiting weak initial billow cores. The view is from above and slightly negative x. Feature labels are described in the text. Note the 50 Δx reference at lower left in each panel set.

The vortex tubes expand significantly by 0.75 T_b and exhibit intensifying H sheaths at increasing radii (sites "j") that remain laminar beyond 1 T_b. However, the most significant features arising by 0.75 T_b are strong H signatures of emerging smaller- and larger-scale twist waves on the vortex tubes and on the initial, larger-scale twist wave vortices extending into the billow cores (sites "k" and "l", respectively). In both cases, these reveal finer-scale, more variable features than seen in λ_2 for the opacity scales employed here. The H fields thus enable probing of the interiors of the superposed twist waves in the vortex cores and the evolving vortex knots emerging by 0.75 T_b and intensifying thereafter. H is highly correlated with twist waves on the vortex tubes, as seen in the billow cores at 1 T_b, but its magnitude and sign are determined by the twist wave rotation and larger-scale velocity field along their axes.

Smaller-scale twist waves seen in H in the vortex tubes at 0.75 T_b are seen to break down to smaller, weaker vortex features by 1 T_b (sites "m"). Because these dynamics are largely confined to the vortex tube interiors, the vortex sheaths around the tubes remain intact. In contrast, the larger-scale twist waves in the billow cores emerging by 0.75 T_b exhibit rapid intensification by 1 T_b (sites "n") and break down to very small-scale, but coherent, features by 1.25 T_b (sites "o"). As noted above, these responses also reveal that H imaging identifies smaller-scale, apparently more intense, vortex features in the centers of the vortex knots that are masked by vortex features having smaller H at larger radii.

Intensifying and expanding twist wave interactions in the vortex tubes drive the breakup of the initial H sheaths by 1.25 T_b. These interactions also drive the most intense H within the billow cores where they first arose: at sites "l" at 0.75 T_b, sites "n" at 1 T_b, and sites "o" at 1.25 T_b. Because H magnitudes depend on larger-scale advection, however, λ_2 is a more confident indicator of emerging turbulent vorticity dynamics.

4.2. Case 3: Two KH Billows Having Varying Phases Linking via Smaller Vortex Tubes

Case 3 is an idealized version of the sites labeled "e" at 1.5 T_b in Figure 2. It was performed in the Case 2 domain to explore the λ_2 and H evolutions arising where adjacent billows exhibit common phase variations along their axes. Case 3 employed a smaller $Ri = 0.05$ ($Fr = 4.47$) to enable a more rapid evolution of small-scale vortices than in Case 1 and a smaller $Re = 2500$ to restrict the vorticity dynamics to larger spatial scales. Case 3 imaging of the λ_2 and H fields spanning 0.5 T_b is shown at top and bottom in Figure 8 in a subdomain spanning the region of varying billow phases viewed from above with positive x upward and positive y to the left. Accompanying videos, Case3a.mp4 and Case3b.mp4 in the Supplementary Materials, provide animations of the of λ_2 and H evolutions shown in Figure 8.

The λ_2 image at 0 T_b shows emerging vortex tubes (sites "a") having positive ζ_y and negative ζ_x wrapping under (over) the KH billows at smaller (larger) x having positive ζ_y and ζ_x. The orthogonal alignments of these features, and their co-rotation, induce differential axial stretching and intensification, or compression and weakening, at different sites. The induced motions cause the KH billows to intensify (weaken) at sites "b" ("c") and the vortex tubes between adjacent billows to increase and intensify (sites "d") and "attach" to the outer KH billows by 0.3 T_b (sites "e"), but they remain largely laminar at this time. Increasing vortex tube intensities and proximity beginning by 0.3 T_b drive increasingly rapid and intense vortex knot formations thereafter (sites "f"). These are seen to (a) entrain the KH billow cores and vortex tubes, (b) exhibit rapid cascades to nests of intense twist waves and small-scale vortices, and (c) become largely turbulent by 0.5 T_b.

Our discussion of Case 1 noted that vortex tubes appear to attach to the outer "edges" of the respective KH billows, and these same dynamics are seen at higher resolution in Figure 8 at 0 T_b. This field reveals that the vortex tubes arise on the vortex sheets wrapping over and under adjacent KH billows, hence are distinct because vortex field lines remain continuous and arose at different locations in the initial, undisturbed, shear flow. Thus, they only appear to "connect" to the KH billow, and lead to increasingly entangled "knots" accompanying the cascade to smaller scales.

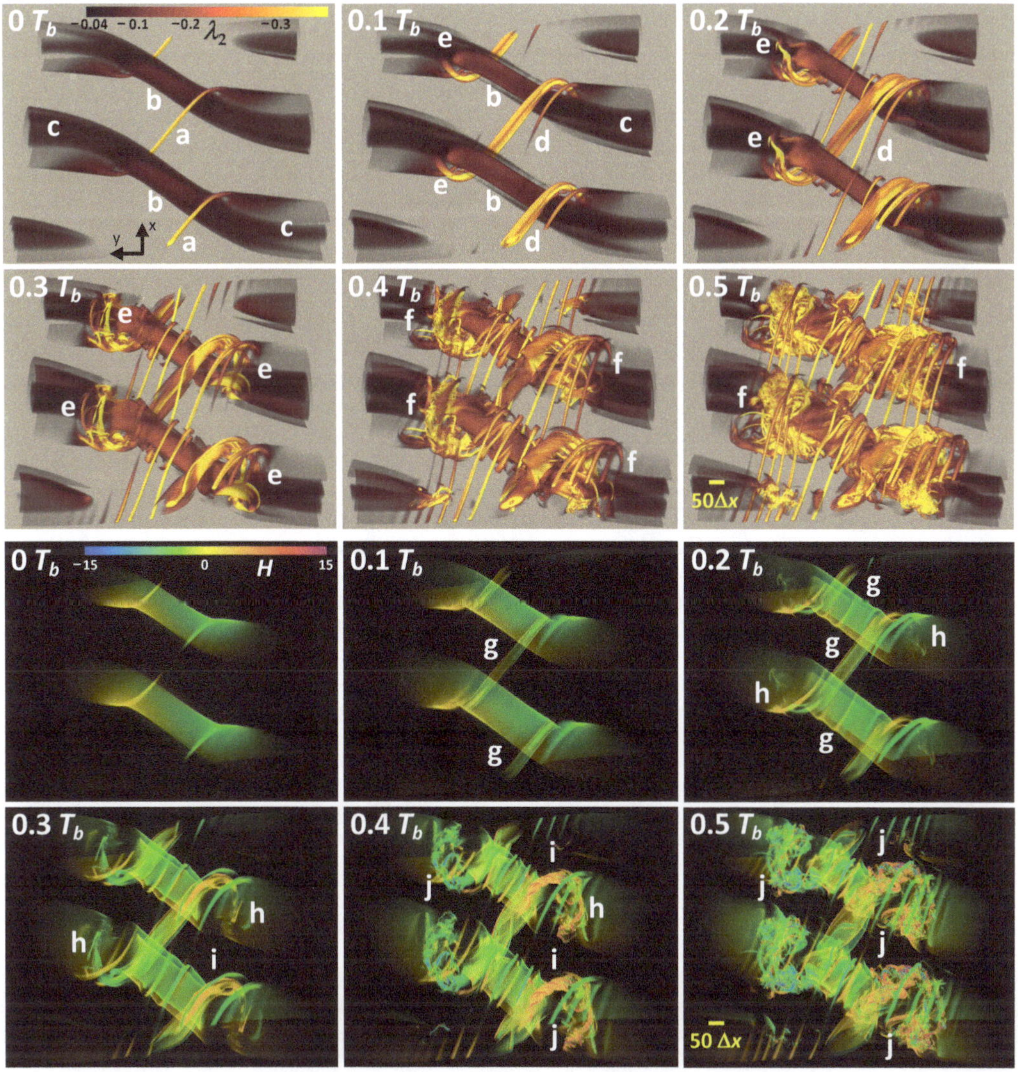

Figure 8. Case 3 λ_2 (**top**) and H (**bottom**) 3-D imaging spanning 0.5 T_b. As in Case 2, t = 0 is defined at a time exhibiting weak initial billow cores. The view is from above and negative x at 45° off vertical. Feature labels are described in the text; the color and opacity scales are specified to best reveal the key features. Note the 50 Δx reference at lower right in each panel set.

Imaging of H in the lower panels of Figure 8 reveals a number of features that are similar to those described in Case 2 above. These include H sheaths of opposite signs arising around both vortex tubes linking adjacent KH billows and the KH billows where they are intensified via axial stretching by the emerging vortex tubes at 0.1 and 0.2 T_b (sites "g"). Small-scale H features also emerge and intensify thereafter where the vortex knots arise at sites "e" in λ_2 imaging in the upper panels of Figure 8 (sites "h"). The H fields reveal the emergence of large-amplitude twist waves from 0.3 to 0.4 T_b, seen in the λ_2 fields but displayed more clearly in H at these times (sites "i"). Additionally, the H imaging reveals the scales and character of strong vortex dynamics within emerging knots that are masked by smaller-scale, but intense, vortex features having strong λ_2 responses at 0.4–0.5 T_b (sites "j").

4.3. Case 4: Orthogonal Linear Vortices in Close Proximity

Case 4 was also performed in a small idealized periodic domain to enable the exploration of vortex dynamics among two orthogonal linear vortices having differing radii, denoted r_0 and $r_1 = 1.5r_0$, the same peak rotational velocities v_0 at these radii, a core separation of 5 r_0, and a Lamb–Oseen radial form [20,21] for the vortex with radius r_0 and circulation Γ_0 given by

$$v(r) = (\Gamma_0/2\pi r)\{1 - \exp[-1.26(r/r_0)^2]\} \tag{10}$$

Case 4 was motivated by the sites labeled "k" at 2.25 T_b in Figures 3–5 that reveal the emergence of roughly orthogonal vortices in close proximity. These dynamics are distinct from those described in Cases 2 and 3 because the vortices exhibit no initial links. The environment was specified to be adiabatic ($N = 0$) to exclude baroclinic vorticity generation. Different vortex radii were specified to illustrate more general responses of roughly orthogonal vortices driven into close proximity by larger-scale advection at larger scales in the turbulence inertial range. As in Cases 2 and 3, a small $Re = r_0v_0/\nu = 2500$ was specified to restrict the vorticity dynamics to resolvable spatial scales. Case 4 nevertheless yielded intense, small-scale vortex dynamics at late times. In order to show the evolution spanning very large λ_2 variations, we employed a color scale varying as $\sqrt{(-\lambda_2)}$ denoted L2.

Imaging spanning 40 T_0 with $T_0 = r_0/v_0$ is shown with wider and zoomed subdomain views at earlier and later times in Figures 9 and 10. Positive ζ_x and ζ_y for each initial vortex are shown with yellow arrows in the upper left panel of Figure 9. These imply converging (diverging) vertical motions in quadrants 1 (4) and downward (upward) motions in quadrants 2 (3). The dominant responses at early times are the distortions of the initial vortices driven by their mutual advection. Vortex distortions induce large-scale twist waves on each. Of these, the stronger distortions by the larger and stronger vortex impose much more dramatic responses on the weaker vortex along x. Induced perturbations to the larger vortex along y remain small until late times. Seen to emerge by 20 T_0 are small-scale vortices that are initiated in quadrant 4 (sites "a"). These arise due to the divergent motions along z and are distorted by, and advected around, the initial vortices. They intensify strongly, but play no roles in the transitions to turbulence in this event at the times shown in Figure 9. Accompanying videos, Case4a.mp4 and Case4b.mp4 in the Supplementary Materials, show the earlier and zoomed later evolutions displayed in Figures 9 and 10.

Advection of the amplifying twist wave on the smaller vortex (with local $\zeta_x > 0$) around the larger initial vortex (with primarily $\zeta_y > 0$) drives these vortices into close proximity by 40 T_0 (site "b"), after which they begin to interact more strongly. The initial larger-scale vortex having $\zeta_y > 0$ is intensified via stretching along y due to the strong curvature of the initial vortex having $\zeta_x > 0$ in close proximity in this region. These vortex alignments drive increasingly strong interactions among adjacent vortex tubes having opposite local ζ_y. Surprisingly, it is the larger, weaker, vortex aligned initially along y that exhibits the most rapid intensification and kinking at this stage. Despite these apparently intense vortex interactions, there is no evidence of transitions to smaller-scale turbulence by 44 T_0.

Turbulence transitions begin with emerging, small-scale, twist waves that arise and link initial vortices where they are most distorted at 45 T_0 (site "d"). These features intensify rapidly and exhibit smaller-scale twist waves that initiate the first turbulence transition by 47 T_0. Entrainment of smaller-scale vortices initiates additional sites driving initial turbulence transitions (sites "e"). However, the strongest transitions are driven by interactions among the intense, larger-scale, initial vortices. A small initial vortex "loop" (site "f") arising by 47 T_0 expands by ~5 and 20 times by 48 and 49 T_0, respectively. These dynamics, and influences by the entraining smaller-scale vortices, drive the emergence of strong turbulence transitions at 47 T_0 (sites "g") that expand throughout the large-scale vortex knot thereafter.

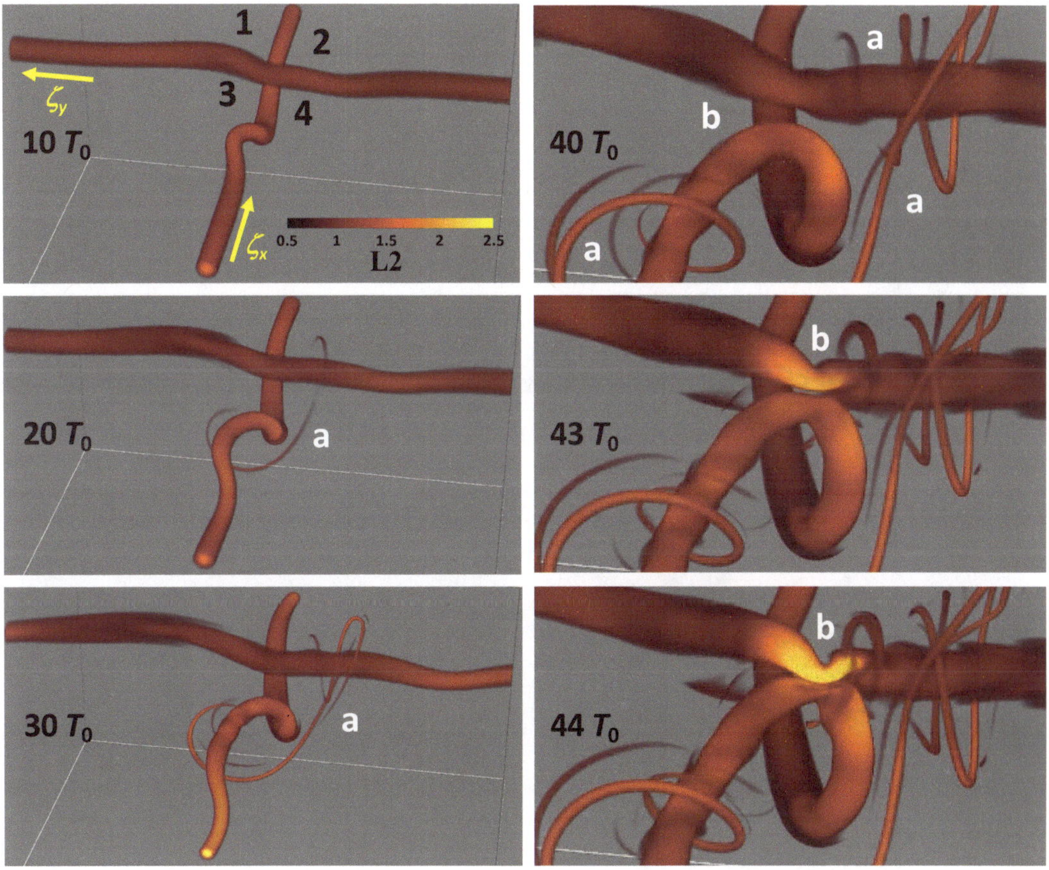

Figure 9. Case 4 λ_2 3-D imaging spanning 34 T_0 showing the evolution from early vortex tube deformations to initial intense interactions. The color scale is at upper left; labels are described in the text.

Importantly, the apparently most rapid and intense transitions are driven by direct interactions among the highly distorted initial vortices seen emerging by 47 T_0. While the larger-scale knot features are largely laminar to 47 T_0, smaller-scale vortices arising by 47 T_0 (sites "g") intensify significantly by 48 T_0 and drive much stronger twist waves and more rapid turbulence transitions thereafter. See the Case4b.mp4 movie of this evolution.

Case 4 H imaging corresponding to that for λ_2 shown in Figures 9 and 10 from 44 to 50 T_0 is shown in Figure 11. Features seen in λ_2 imaging that exhibit similar responses in H are identified in Figure 11 with the same labels at the same times (sites "b", "c", "e", "f", and "g"). The λ_2 and H imaging at 44 T_0 reveal similar larger-scale features exhibiting weak H in the vortex cores, but also smaller-scale features including H sheaths and additional smaller-scale vortices that are not seen in the λ_2 imaging (sites "h"). The H imaging at 46 T_0 reveals the same vortex features at sites "b" and "c", but also somewhat stronger and finer-scale features in the initial vortices that are not revealed by the λ_2 imaging (sites "i"). Similarly, the H imaging at 47 and 48 T_0 reveals several features noted in the λ_2 fields at these times in Figure 10 (sites "e", "f", and "g"), but the majority of significant features seen in H imaging highlight vortex structures occurring on much smaller scales that are not seen clearly in the λ_2 imaging (sites "j" and "k", respectively). The final H imaging at 49 and 50 T_0 reveals the emergence of a diversity of features at multiple sites extending to very small spatial scales, some of which exhibit similar feature scales and orientations common to λ_2 imaging (sites "g"), but the majority of which do not.

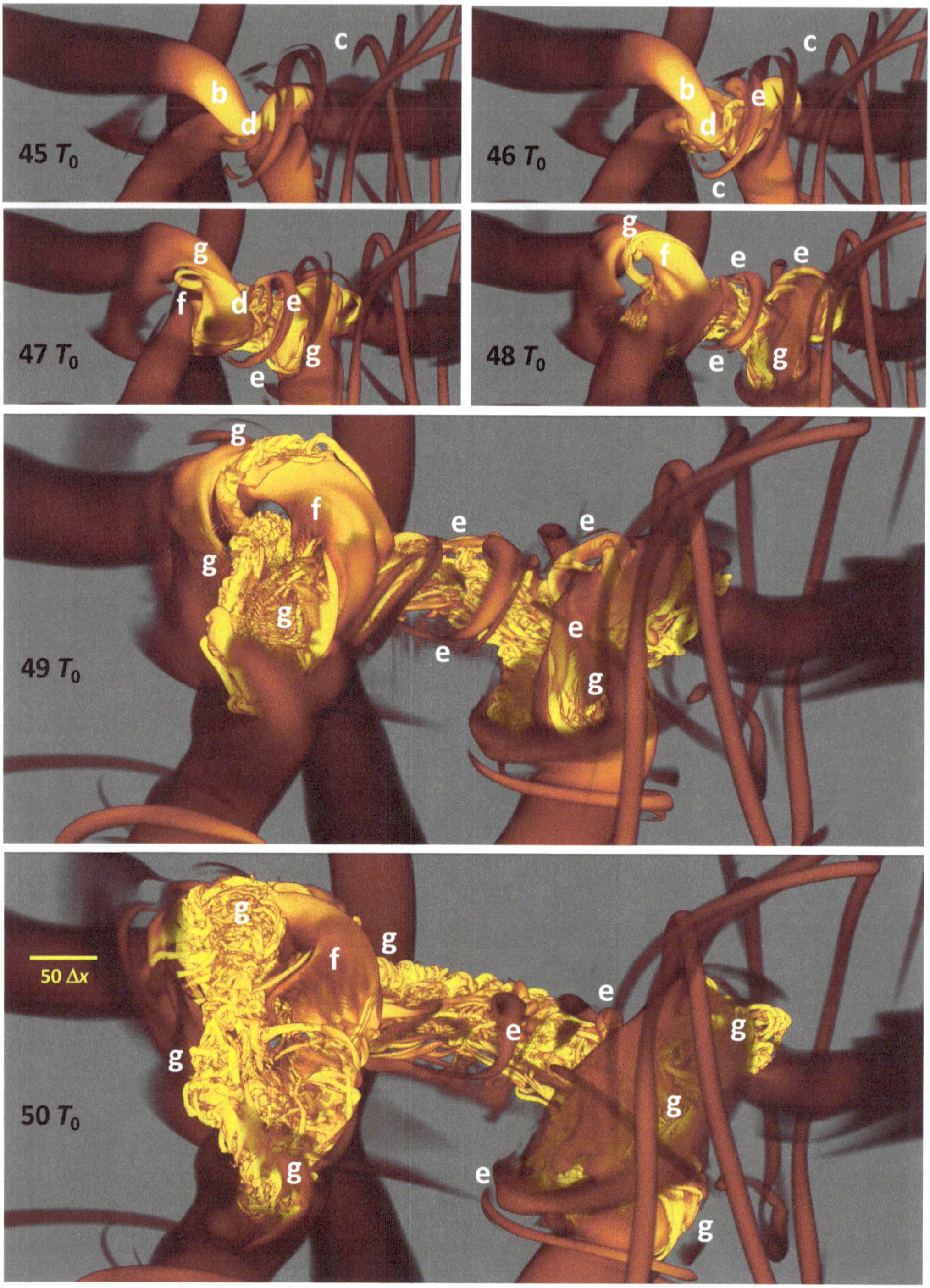

Figure 10. As in Figure 9, showing further zoomed views from 45 to 50 T_0. The λ_2 color scale is as in Figure 9. Note the 50 Δx reference at lower left.

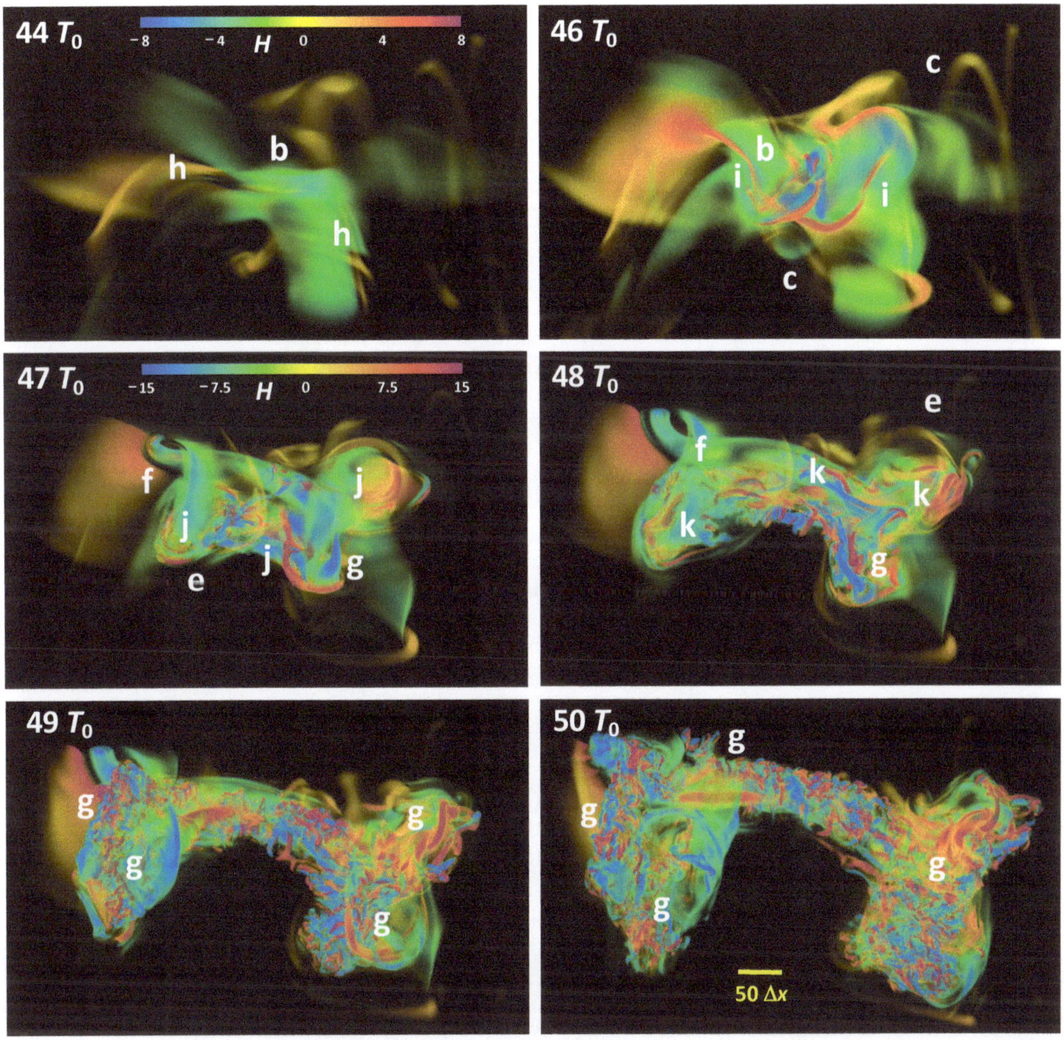

Figure 11. As in Figures 7 and 8, showing zoomed views of H from 44 to 50 T_0. The H color scales for the upper 2 and the lower 4 images are shown in the upper 2 images at left. Features are labeled as in Figures 9 and 10 and described in the text. Note the 50 Δx reference at lower right.

5. Discussion

KHIs have been known for many years to drive turbulence and mixing in the atmosphere, oceans, and lakes. Their potential significance motivated a wide range of observational, laboratory, theoretical, and modeling studies exploring their evolutions, environmental responses, and secondary instabilities driving turbulence transitions for diverse Re, Ri, and Pr. Laboratory studies were the first to reveal KHI T&K dynamics clearly and recognize their significance [1,2,22], but they did not enable their quantitative analysis and interpretation. A wide range of theoretical and modeling studies of KHI spanning over 4 decades identified secondary KHIs and CIs occurring for idealized KHI in narrow domains that precluded T&K dynamics, e.g., [18,19] and references therein. However, it required new, high-resolution imaging of large-scale KHI at altitudes of ~80–90 km that resolved features and evolutions recognized as KHI T&K dynamics to guide initial modeling efforts [4,5].

KHI T&K dynamics now appear to be widespread throughout the atmosphere and likely oceans, lakes, and other stratified and sheared fluids [23]. They also arise readily in multi-scale GW environments [24], and may account for "missing" mixing inferred from global atmosphere modeling [25,26]. They likely also play similar roles in other geophysical and astrophysical environments [13], especially the oceans, where they may have implications for larger-scale stratification and overturning circulations [27,28], but their implications in these environments have yet to be explored.

The high-resolution modeling described here had two primary objectives. The first was to illustrate the more significant types of KHI T&K dynamics identified to date in observations and initial LES and DNS modeling. The second was to reveal specific T&K pathways to turbulence via imaging of feature evolutions in three idealized DNS designed to exclude competing T&K dynamics.

The first objective was addressed using a large, periodic DNS domain employing an initial noise seed that induced three or four initial KH billows at a finite amplitude along x at different y. This yielded (a) the expected secondary instabilities of idealized KHI, specifically secondary CI and KHI, (b) localized KH billow pairing, and (c) initial KHI T&K features driving turbulence transitions for the chosen initial conditions. These include the following in order of decreasing turbulence intensities:

(1) intense vortex knots forming where KH billows are mis-aligned along their axes;
(2) intense vortex knots forming where one or several larger-scale vortex tubes emerge on the intermediate vortex sheets and link adjacent billows where they exhibit larger phase variations along their axes;
(3) delayed and weaker vortex knots that form where multiple, smaller-scale vortex tubes arise on the intermediate vortex sheets and link adjacent billows where they exhibit smaller phase variations along their axes;
(4) secondary KHI T&K dynamics emerging on intensifying vortex sheets initiated by smaller-scale vortex tubes where billow phase variations are smaller;
(5) a localized KH billow pairing event yielding a delayed and weaker knot;
(6) regions where secondary KHIs are advected into close proximity to secondary CIs over and under the KH billows;
(7) stretched and intensified secondary KHIs where billow curvature is large;
(8) secondary CIs where KH billow phases are slowly varying along y.

The second objective was addressed by comparing imaging of (a) the rotational component of vorticity revealed by λ_2 as in our initial KHI T&K studies [4,5] and of (b) H employed previously to assess the degree of "knotted-ness" in turbulent flows [11,17]. In our applications, the λ_2 and H fields highlighted different features of the dynamics driving turbulence transitions. λ_2 imaging emphasized rapid transitions from larger-scale vortex interactions to smaller-scale turbulence where twist waves arose due to emerging strong, roughly orthogonal vortices in close proximity, specifically those initiating vortex knots in Case 1 and described in greater detail in Cases 2–4. These λ_2 fields captured the initial interactions and influences very well, including the emergence of twist waves at larger and smaller scales. As these dynamics progressed, however, smaller-scale vortex dynamics without significant H appear to have masked the larger-scale dynamics driving the emerging turbulence transitions in the vortex knot interiors in some cases.

The H imaging in Cases 2–4 revealed additional flow features that augmented our understanding of the emerging vortex character and the evolving dynamics prior to strong turbulence transitions. Importantly, it also revealed more persistent and coherent features at larger scales in the emerging vortex knot interiors extending to later times that were not revealed in the λ_2 imaging, but must continue to drive additional turbulence transitions extending to later times.

Kelvin vortex waves, or "twist waves" in our terminology, have received significant theoretical attention, e.g., [29], among many others. Multiple studies have revealed a broad spectrum of axial, azimuthal, and radial variations of vorticity and phase, and helical forms arising in varying environments. However, these methodologies cannot address the nonlinear dynamics of interacting vortices driving the turbulence cascade.

Our KHI T&K modeling addressed the vorticity dynamics of evolving multi-scale KHI environments driving the transitions to turbulence at Re = 5000 that is sufficient to enable a significant turbulence inertial range and more than 3 decades of scales resolved along the shear flow in Case 1. It was performed in a large domain enabling multiple KH billows and all of the transitional dynamics revealed in laboratory and theoretical studies, and atmospheric and oceanic imaging and profiling of KHI to date [23]. These dynamics rapidly become highly nonlinear and increasingly complex due to vortex interactions that are not possible to address via other methods. Where emerging KH billows exhibit mis-aligned or varying phases along their axes, they induce additional vortex stretching of the intermediate vortex sheets. These sites yield emerging, smaller-scale, vortex tubes evolving increasingly orthogonal alignments relative to the billow cores as they advect over and under adjacent billows. They appear to connect to the intensifying billow cores and exhibit rapidly intensifying mutual interactions thereafter. These "connections" are only apparent, however, because the interacting KH billow and vortex sheet vortices have their origins in very different regions of the initial shear flow.

The resulting interactions yield increasingly intense, smaller-scale vortices arising on each initial vortex that exhibit helical responses, most typically mode-1 single helix and mode-2 double helix twist waves. These arise due to axial and/or radial deformations or displacements that stretch, compress, strengthen, and/or weaken the initial vortices. Mode-1 twist waves propagate along the initial vortex typically without contributing to its breakup. Mode-2 twist waves, in contrast, "un-wrap" and fragment the initial vortex, driving the cascade of vorticity to smaller scales. Mode-3 and higher twist waves likely also arise, but are challenging to identify among the accelerating, intensifying, and tightly packed dynamics driving the emergence of vortex knots at the initial interaction sites.

Twist wave dynamics arise at successively smaller scales as orthogonal vortex alignments continue to emerge due to larger-scale advection accompanying the continuing KH billow evolutions. Their wavelengths along their parent vortices extend from ~2 L in the billow cores emerging prior to initial turbulence transitions seen at 0.75 T_b in Figure 2 to ~0.01 L in the vortex knots and secondary T&K dynamics at 2.25 T_b in Figure 5. They are also seen having wavelengths as small as ~10 Δx at 50 T_b in Figure 10. Hence, twist waves play central roles in the cascade to smaller scales throughout the turbulence inertial range.

6. Summary and Conclusions

Results presented above reveal the various pathways to enhanced turbulence accompanying KHI T&K dynamics that arise where adjacent KH billows interact due to initial variations or mis-alignments along their axes. They are demonstrably widespread, and perhaps ubiquitous, in the atmosphere [23,24,30–33].

Current parameterizations attribute these responses to breaking GWs accompanying their amplitude growth with increasing altitude. However, more recent observations and DNS modeling suggest that KHI, and especially the expectation of enhanced responses due to their ubiquitous T&K dynamics, may contribute comparable or larger responses. Extensive aircraft measurements of turbulence in the lower stratosphere [34] in a strong mountain wave environment over and around the Drake Passage suggest that KHI, rather than GW breaking, appeared to account for the major turbulence at flight altitudes. A similar conclusion arose from a recent multi-scale GW DNS [24], where turbulence due to KHI was stronger than turbulence associated with GW breaking at later, potentially more realistic, stages of the simulation. Additionally, KHI arise at the most strongly stratified layers in a variable environment, thus potentially contribute more significant mixing. GW breaking, in contrast, arises in the least stable phases of the superposed GW field, hence yields mixing, in part at least, of an already weakly-stratified environment.

Mixing dynamics have been a significant interest of various communities for many years [35,36], and KHI T&K dynamics appear likely to play an important role in accounting for their influences and enabling improved parameterizations and applications.

Supplementary Materials: The following supporting information can be downloaded at: https://www.mdpi.com/article/10.3390/atmos14121770/s1. Supplementary movies of Figures 7–10 include Case2a.mp4, Case2b.mp4, Case3a.mp4, Case3b.mp4, Case4a.mp4, and Case4b.mp4.

Author Contributions: Conceptualization, D.C.F.; methodology, D.C.F., T.S.L., L.W. and A.C.L.; software, T.S.L. and A.C.L.; validation, T.S.L. and L.W.; formal analysis, D.C.F., L.W. and T.S.L.; investigation, D.C.F., L.W. and T.S.L.; resources, D.C.F., L.W. and T.S.L.; data curation, L.W. and T.S.L.; writing—original draft preparation, D.C.F.; writing—review and editing, T.S.L.; visualization, L.W., A.C.L. and T.S.L.; supervision, D.C.F.; project administration, D.C.F.; funding acquisition, D.C.F. All authors have read and agreed to the published version of the manuscript.

Funding: This research was funded by the U.S. National Science Foundation under grants AGS-2017263, AGS-2032678, AGS-2128443, AGS-2129222, AGS-2131350, AGS-2325006 and by the U.S. Air Force Office of Scientific Research under grant FA9550-18-1-0009.

Data Availability Statement: The imaging fields presented in this study can be made available by request to the corresponding author. However, no modeling data are publicly available because the data volumes in the smallest single precision binary format are as large as ~600 GB individually and together exceed ~10 TB. Hence, the authors themselves do not have, and cannot download, the full 3-D data.

Acknowledgments: This paper was prepared as a tribute to Jack Herring, who the lead author was very fortunate to know, interact with, and learn from as a graduate student and then a postdoc at NCAR from 1975 to 1979, and thereafter on subsequent visits. Jack was a good friend and an inspiration to many.

Conflicts of Interest: The authors are employees of GATS. The company had no roles in the design of the study; in the collection, analysis, or interpretation of data; in the writing of the manuscript, or in the decision to publish the articles. The paper reflects the views of the scientists and not the company.

References

1. Thorpe, S.A. Laboratory observations of secondary structures in Kelvin-Helmholtz billows and consequences for ocean mixing, Geophys. Astrophys. *Fluid Dyn.* **1985**, *34*, 175–199. [CrossRef]
2. Thorpe, S.A. Transitional phenomena and the development of turbulence in stratified fluids: A review. *J. Geophys. Res.* **1987**, *92*, 5231–5248.
3. Thorpe, S.A. The axial coherence of Kelvin-Helmholtz billows. *Q. J. R. Meteorol. Soc.* **2002**, *128*, 1529–1542. [CrossRef]
4. Fritts, D.C.; Wieland, S.A.; Lund, T.S.; Thorpe, S.A.; Hecht, J.H. Kelvin-Helmholtz billow interactions and instabilities in the mesosphere over the Andes Lidar Observatory: 2. Modeling and interpretation. *J. Geophys. Res. Atmos.* **2021**, *125*, e2020JD033412. [CrossRef]
5. Fritts, D.C.; Wang, L.; Lund, T.S.; Thorpe, S.A. Multi-scale dynamics of Kelvin–Helmholtz instabilities. Part 1. Secondary instabilities and the dynamics of tubes and knots. *J. Fluid Mech.* **2022**, *941*, A30. [CrossRef]
6. Lord, K. Vibrations of a columnar vortex. *Philos. Mag.* **1880**, *10*, 155–168.
7. Arendt, S.; Fritts, D.C.; Andreassen, O. The initial value problem for Kelvin vortex waves. *J. Fluid Mech.* **1997**, *344*, 181–212. [CrossRef]
8. Moore, D.W.; Saffman, P.G. The instability of a straight vortex filament in a strain field. *Proc. R. Soc. Lond. Ser. A* **1975**, *346*, 413–425.
9. Tsai, C.-Y.; Widnall, S.E. The stability of short waves on a straight vortex filament in a weak externally imposed strain field. *J. Fluid Mech.* **1976**, *73*, 721–733. [CrossRef]
10. Yan, Z.; Li, X.; Yu, C.; Wang, J.; Chen, S. Dual channels of helicity cascade in turbulent flows. *J. Fluid Mech.* **2020**, *20*, R2. [CrossRef]
11. Moffatt, H.; Tsinober, A. Helicity in laminar and turbulent flow. *Annu. Rev. Fluid Mech.* **1992**, *24*, 281–312. [CrossRef]
12. Chen, Q.; Chen, S.; Eyink, G. The joint cascade of energy and helicity in three-dimensional turbulence. *Phys. Fluids* **2003**, *15*, 361–374. [CrossRef]
13. Pouquet, A.; Rosenberg, D.; Stawarz, J.E.; Marino, R. Helicity dynamics, inverse, and bidirectional cascades in fluid and magnetohydrodynamic turbulence: A brief review. *Earth Space Sci.* **2019**, *6*, 351–369. [CrossRef]
14. Moin, P.; Mahesh, K. Direct numerical simulation: A tool in turbulence research. *Annu. Rev. Fluid Mech.* **1998**, *30*, 78. [CrossRef]
15. Pope, S.B. *Turbulent Flows;* Cambridge University Press: Cambridge, UK, 2000. [CrossRef]
16. Jeong, J.; Hussain, F. On the identification of a vortex. *J. Fluid Mech.* **1995**, *285*, 69–94.

17. Moffatt, H.K. The degree of knotted-ness of tangled vortex lines. *J. Fluid Mech.* **1969**, *44*, 705–719. [CrossRef]
18. Mashayek, A.; Peltier, W.R. The "zoo" of secondary instabilities precursory to stratified shear transition. Part 1. Shear aligned convection, pairing, and braid instabilities. *J. Fluid Mech.* **2012**, *708*, 5–44. [CrossRef]
19. Mashayek, A.; Peltier, W.R. The "zoo" of secondary instabilities precursory to stratified shear transition. Part 2. The influence of stratification. *J. Fluid Mech.* **2012**, *708*, 45–70. [CrossRef]
20. Oseen, C.W. Uber die Wirbelbewegung in einer reibenden Flussigkeit. *Ark. Mat. Astro. Fys.* **1912**, *7*, 14–26.
21. Batchelor, G.K. *An Introduction to Fluid Dynamics*; Cambridge University Press: Cambridge, UK, 1967.
22. Caulfield, C.P.; Yoshida, S.; Peltier, W.R. Secondary instability and three-dimensionalization in a laboratory accelerating shear layer with varying density differences. *Dyn. Atmos. Oceans* **1996**, *23*, 139–153. [CrossRef]
23. Fritts, D.C.; Baumgarten, G.; Pautet, P.-D.; Hecht, J.H.; Williams, B.P.; Kaifler, N.; Kaifler, B.; Kjellstrand, C.B.; Wang, L.; Taylor, M.J.; et al. Kelvin Helmholtz Instability "Tube" & "Knot" Dynamics, Part I: Expanding Observational Evidence of Occurrence and Environmental Influences. *J. Atmos. Sci.* **2023**, *60*, 2419–2437. [CrossRef]
24. Fritts, D.C.; Wang, L. Kelvin Helmholtz Instability "Tube" & "Knot" Dynamics, Part II: KHI T&K Dynamics in a Multi-Scale Gravity Wave Direct Numerical Simulation. *J. Atmos. Sci.* **2023**, *80*, 2439–2457. [CrossRef]
25. Garcia, R.R.; López-Puertas, M.; Funke, B.; Marsh, D.R.; Kinnison, D.E.; Smith, A.K.; González-Galindo, F. On the distribution of CO_2 and CO in the mesosphere and lower thermosphere. *J. Geophys. Res. Atmos.* **2014**, *119*, 5700–5718. [CrossRef]
26. Liu, H.-L. Effective vertical diffusion by atmospheric gravity waves. *Geophys. Res. Lett.* **2021**, *48*, e2020GL091474. [CrossRef]
27. Mashayek, A.; Ferrari, R.; Nikurashin, M.; Peltier, W.R. Influence of enhanced abyssal diapycnal mixing on ocean stratification and overturning circulation. *J. Phys. Oceanogr.* **2015**, *45*, 2580–2597. [CrossRef]
28. Ferrari, R.; Mashayek, A.; McDougall, T.J.; Nikurashin, M.; Campin, J.M. Turning Ocean Mixing Upside Down. *J. Phys. Oceanogr.* **2016**, *46*, 2239–2261. [CrossRef]
29. Fabre, D.; Sipp, D.; Jacquin, L. Kelvin waves and the singular modes of the Lamb–Oseen vortex. *J. Fluid. Mech.* **2006**, *551*, 235–274. [CrossRef]
30. Eaton, F.; McLaughlin, S.A.; Hines, J.R. A new frequency-modulated continuous wave radar for studying planetary boundary layer morphology. *Radio Sci.* **1995**, *30*, 75–88. [CrossRef]
31. Fernando, H.J.S.; Weil, J.C. Whither the Stable Boundary Layer? *Bull. Am. Meteorol. Soc.* **2010**, *91*, 1475–1484. [CrossRef]
32. Luce, H.; Hassenpflug, G.; Yamamoto, M.; Fukao, S.; Sato, K. High-Resolution Observations with MU Radar of a KH Instability Triggered by an Inertia–Gravity Wave in the Upper Part of a Jet Stream. *J. Atmos. Sci.* **2008**, *65*, 1711–1718. [CrossRef]
33. Chau, J.L.; Urco, J.M.; Avsarkisov, V.; Vierinen, J.P.; Latteck, R.; Hall, C.M.; Tsutsumi, M. Four-dimensional quantification of Kelvin-Helmholtz instabilities in the polar summer mesosphere using volumetric radar imaging. *Geophys. Res. Lett.* **2020**, *47*, e2019GL086081. [CrossRef]
34. Dörnbrack, A.; Bechtold, P.; Schumann, U. High-resolution aircraft observations of turbulence and waves in the free atmosphere and comparison with global model predictions. *J. Geophys. Res. Atmos.* **2021**, 127. [CrossRef]
35. Caulfield, C.P. Layering, instabilities, and mixing in turbulent stratified flows. *Annu. Rev. Fluid Mech.* **2021**, *53*, 113–145. [CrossRef]
36. Gregg, M.C.; D'Asaro, E.A.; Riley, J.J. Mixing efficiency in the ocean. *Annu. Rev. Mar. Sci.* **2018**, *10*, 443–473. [CrossRef]

atmosphere

Article

Multi-Scale Reconstruction of Turbulent Rotating Flows with Generative Diffusion Models

Tianyi Li [1], Alessandra S. Lanotte [2,*], Michele Buzzicotti [1], Fabio Bonaccorso [1] and Luca Biferale [1]

[1] Department of Physics and INFN, University of Rome "Tor Vergata", Via della Ricerca Scientifica 1, 00133 Rome, Italy; tianyi.li@roma2.infn.it (T.L.); michele.buzzicotti@roma2.infn.it (M.B.); fabio.bonaccorso@roma2.infn.it (F.B.); biferale@roma2.infn.it (L.B.)

[2] Istituto di Nanotecnologia, CNR NANOTEC and INFN, Via per Monteroni, 73100 Lecce, Italy

* Correspondence: alessandrasabina.lanotte@cnr.it

Abstract: We address the problem of data augmentation in a rotating turbulence set-up, a paradigmatic challenge in geophysical applications. The goal is to reconstruct information in two-dimensional (2D) cuts of the three-dimensional flow fields, imagining spatial gaps present within each 2D observed slice. We evaluate the effectiveness of different data-driven tools, based on diffusion models (DMs), a state-of-the-art generative machine learning protocol, and generative adversarial networks (GANs), previously considered as the best-performing method both in terms of point-wise reconstruction and the statistical properties of the inferred velocity fields. We focus on two different DMs recently proposed in the specialized literature: (i) RePaint, based on a heuristic strategy to guide an unconditional DM for flow generation by using partial measurements data, and (ii) Palette, a conditional DM trained for the reconstruction task with paired measured and missing data. Systematic comparison shows that (i) DMs outperform the GAN in terms of the mean squared error and/or the statistical accuracy; (ii) Palette DM emerges as the most promising tool in terms of both point-wise and statistical metrics. An important property of DMs is their capacity for probabilistic reconstructions, providing a range of predictions based on the same measurements, enabling uncertainty quantification and risk assessment.

Keywords: flow reconstruction; machine learning; rotating turbulence

Citation: Li, T.; Lanotte, A.S.; Buzzicotti, M.; Bonaccorso, F.; Biferale, L. Multi-Scale Reconstruction of Turbulent Rotating Flows with Generative Diffusion Models. *Atmosphere* **2024**, *15*, 60. https://doi.org/10.3390/atmos15010060

Academic Editors: Annick Pouquet, Boris Galperin and Peter Sullivan

Received: 29 November 2023
Revised: 27 December 2023
Accepted: 30 December 2023
Published: 31 December 2023

1. Introduction

In atmospheric and oceanic forecasting, the accurate estimation of systems from incomplete observations is a challenging task [1–5]. These environments, often characterized by turbulent dynamics, require effective reconstruction techniques to overcome the common problem of temporally or spatially gappy measurements. The challenge arises from factors such as instrument sensitivity, the natural sparsity of observational data, and the absence of direct information, for example, in the case of deeper ocean layers [6–10]. Established data assimilation techniques, such as variational methods [11,12] and ensemble Kalman filters [13,14], effectively merge time-series observations with model dynamics to attack the inverse problem. When measurements are limited to a single time point, gappy proper orthogonal decomposition (POD) [15] and extended POD [16] deal with spatially incomplete data by exploiting pretrained statistical relationships between measurements and missing information for the data augmentation goal. These POD-based methods are widely used in fluid mechanics [17–19] and geophysical fluid dynamics [20,21] to reconstruct flow fields.

POD-based methods are fundamentally linear, yielding reconstructions with smooth flow properties, associated with few leading POD modes. In the context of turbulent flows, this implies that POD-like methods primarily emphasize large-scale structures [22,23]. In recent years, machine learning has led to an increasing number of successful applications in reconstruction tasks for simple and idealized fluid mechanics problems (see [24] for a brief review). We refer to super-resolution applications (i.e., finding high-resolution flow fields from low-resolution data) [25–27], inpainting (i.e., reconstructing flow fields having spatial damages) [23,28], and inferring volumetric flows from surface or two-dimensional (2D)-section measurements [29–31]. However, much remains to be clarified concerning benchmarks and challenges, and this is even more important for realistic turbulent set-up and at increasing flow complexity, e.g., for increasing Reynolds numbers. When dealing with turbulent systems, the quality of reconstruction tasks must be judged according to two different objectives: (i) the point-wise error, given by the success in filling gappy or damaged regions of the instantaneous fields with data close to the ground truth configuration by configuration; (ii) statistical error, by reproducing statistical multi-scale and multi-point properties, such as the probability distribution functions (PDFs), spectra, etc., of the system.

To move from proof-of-concept to quantitative benchmarks, in a previous work [23], we systematically compared POD-based methods with generative adversarial networks (GANs) [32] using both point-wise and statistical reconstruction objectives for fully developed rotating turbulent flows, accounting for different gap sizes and geometries. GANs belong to the large family of generative models, i.e., machine learning algorithms that produce data according to a probability distribution optimized to resemble that of the data used in the training. The learning task is produced by two networks that compete with each other: First, a generative network is used to predict the data in the gap from the input measurement to obtain a good point-wise reconstruction. Second, to overcome the lack of expressiveness in the multi-scale (with low energetic content) flow structures, a second adversarial network, called the discriminator, is used to optimize the statistical properties of the generated data. Contrary to expectations, despite their non-linearity, GANs only matched the best linear POD techniques in point-wise reconstruction. However, GANs showed superior performance in capturing the statistical multi-scale non-Gaussian fluctuation characteristics of three-dimensional (3D) turbulent flow [23].

From our previous comparative study, we also observed that GANs pose many challenges in the training processes, due to the presence of instability and the necessity for hyper-parameter fine-tuning to achieve a suitable compromise in the multi-objective task. Furthermore, a common limitation of our GANs and POD-based methods is that they provide *only a deterministic* reconstruction solution. This singular output contrasts with the intrinsic nature of turbulence reconstruction, which is a one-to-many problem with multiple plausible solutions. The ability to generate an ensemble of possible reconstructions is critical for practical atmospheric and oceanic forecasting, e.g., in relation to uncertainty quantification and the risk assessment of rare, high-impact events [33–35].

More recently, diffusion models (DMs) [36] have emerged as a powerful generative tool, showing exceptional success in domains such as computer vision [36–38], audio synthesis [39], and natural language processing [40], particularly outperforming GANs in image synthesis [38]. Their applications have also extended to fluid dynamics for super-resolution [41], flow prediction [42], and Lagrangian trajectory generation [43]. By introducing Markov chains to effectively generate data samples (see Section 2.2), the implementation of DMs eliminates the need to resort to the less stable adversarial training of GANs, making DMs generally more stable in the training stage. Another characteristic of DMs is their inherent stochasticity in the generation process, which allows them to produce multiple outputs that adhere to the learned distribution conditioned on the same input.

This study is the first attempt to use state-of-the-art generative DMs for the reconstruction of 2D velocity fields of rotating turbulence, a complex system characterized by both large-scale vortices and highly non-Gaussian and intermittent small-scale fluctuations [44–48]. This novel application of DMs provides a quantitative estimation of their effectiveness in the context of turbulent flows. Our objectives are twofold: first, we aim to make comprehensive comparisons with the best-performing GAN methods from our previous research, and second, we aim to investigate the effectiveness of DMs in probabilistic reconstruction tasks. The statistical description of rotating and/or stratified turbulence is a fundamental problem for atmospheric and oceanic flows, particularly when the rotation is strong enough and the quasi-geostrophic approximation can be applied. Jack Herring made key contributions in this field, by pioneering not only the application of closure theories to understand the behavior of turbulent motions in the two ranges of scales smaller and larger than those of the energy source [49,50], but also numerical simulations where the inverse energy transfer towards large scales was observed [51]. Our study deals with the complex description and reconstruction of the interplay of 3D almost-isotropic small-scale fluctuations with an approximate 2D large-scale flow.

The paper is organized as follows: In Section 2, we introduce the system under consideration and the two adopted strategies for flow reconstruction using DMs. The first is a heuristic conditioning method applied to an unconditional DM designed for flow generation, as demonstrated by RePaint [52]. The second strategy uses a supervised approach, training a DM conditioned on measurements, similar to the Palette method [53,54]. In Section 3, we discuss the performance of the two DMs in point-wise and statistical property reconstruction, in comparison with the previously analyzed GAN method [22]. In Section 4, we study the probabilistic reconstruction capacity of the DMs. We end with some comments in Section 5.

2. Methods

2.1. Problem Setup and Data Preparation

In this study, we focus on rotating turbulence [55,56], which is of considerable physical interest to geophysical and astrophysical problems [57,58]. Under strong rotation, the flow tends to become quasi-2D, characterized by the formation of large-scale coherent vortical structures parallel to the rotation axis [47]. We adopt the same experimental framework as our previous work [23], and explore possible improvements from DMs. We set up a mock field-measurement, anticipating being able to obtain data from a gappy 2D slice of the original 3D volume of rotating turbulence, orthogonal to the axis of rotation. The full 2D image is denoted as (I), the support of the measured domain as (S), and the support of the gap where we miss the data as (G). Here, (G) represents a centrally located square gap of variable size, as shown in Figure 1a. We use the TURB-Rot database [59] obtained from direct numerical simulation (DNS) of the incompressible Navier–Stokes equations for rotating fluid in a 3D periodic domain, which can be written as

$$\frac{\partial \boldsymbol{u}}{\partial t} + \boldsymbol{u} \cdot \boldsymbol{\nabla} \boldsymbol{u} + 2\boldsymbol{\Omega} \times \boldsymbol{u} = -\frac{1}{\rho}\boldsymbol{\nabla} \tilde{p} + \nu \Delta \boldsymbol{u} + \boldsymbol{f}, \tag{1}$$

where $\boldsymbol{u}$ is the incompressible velocity, $\boldsymbol{\Omega} = \Omega \hat{\boldsymbol{x}}_3$ is the rotation vector, and $\tilde{p}$ represents the pressure modified by a centrifugal term. The regular, cubic grid has $N^3 = 256^3$ points. The statistically homogeneous and isotropic forcing $\boldsymbol{f}$ acts at large scales around $k_f = 4$, and it is the solution of a second-order Ornstein–Uhlenbeck process [60,61]. In the stationary state, with $\Omega = 8$, the Rossby number is $Ro = \sqrt{\mathcal{E}}/(\Omega k_f) \approx 0.1$, where $\mathcal{E}$ represents the kinetic energy. The viscous dissipation $\nu \Delta \boldsymbol{u}$ is replaced by a hyperviscous term $\nu_h \Delta^2 \boldsymbol{u}$ to increase the inertial range, while a large-scale linear friction term $\alpha \Delta^{-1} \boldsymbol{u}$ is added to the r.h.s. of Equation (1) to reduce the formation of a large-scale condensate [46], associated with the inverse energy cascade well-developed at this Rossby number. The Kolmogorov dissipative wavenumber, $k_\eta = 32$, is chosen as the scale at which the energy spectrum begins to decay exponentially. An effective Reynolds number is defined as $Re_{eff} = (k_0/k_\eta)^{-3/4} \approx 13$, with the smallest wavenumber $k_0 = 1$. The integral length scale is $L = \mathcal{E} / \int kE(k)\, \mathrm{d}k \approx 0.15 L_0$, where $L_0 = 2\pi$ is the domain length, and the integral time scale is $T_L = L_0/\mathcal{E}^{1/2} \approx 0.185$. For further details of DNS, see [59]; a sketch of the original 2D spectrum is also shown in Figure 1b.

Data were extracted from the DNS by sampling the full 3D velocity field (Figure 1a) during the stationary stage at intervals of $\Delta t_s = 5.41 T_L$ to reduce temporal correlation. We collected 600 early snapshots for training and 160 later snapshots for testing, with the two collections separated by over $3400 T_L$ to ensure independence. To manage the data volume while preserving complexity, the resolution of the sampled fields was reduced from 256^3 to 64^3 using Galerkin truncation in Fourier space, with the truncation wavenumber set to k_η. We then selected x_1-x_2 planes at different x_3-levels and augmented them by random shifts with the periodic boundary conditions, resulting in a train/test split of 84,480/20,480 samples.

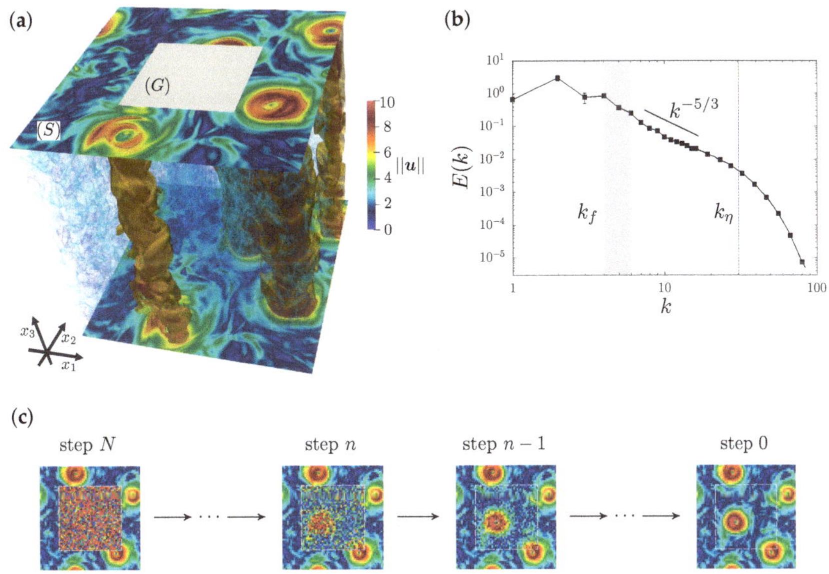

Figure 1. (**a**) Visualization of the velocity magnitude from a three-dimensional (3D) snapshot extracted from our numerical simulations. The two velocity planes (in the x_1-x_2 directions) at the top and bottom of the integration domain show the velocity magnitude. In the 3D volume, we visualize a rendering of the small-scale velocity filaments developed by the 3D dynamics. The gray square on the top level is an example of the damaged gap area, denoted as (G), while the support where we assume to have the measurements is denoted as (S), and their union defines the full 2D image, $(I) = (S) \cup (G)$. A velocity contour around the most intense regions ($\|\boldsymbol{u}\| > 6.35$) highlights the presence of the quasi-2D columnar structures (almost constant along x_3-axis), due to the effect of the Coriolis force induced by the frame rotation. (**b**) Energy spectra averaged over time. The range of scales where forcing is active is indicated by the gray band. The dashed vertical line denotes the Kolmogorov dissipative wavenumber. The reconstruction of the gappy area is based on a downsized image on a grid of 64^2 collocation points, which corresponds to a resolution of the order of $1/k_\eta$. (**c**) Sketch illustration of the reconstruction protocol of a diffusion model (DM) in the backward phase (see later), which uses a Markov chain to progressively generate information through a neural network.

For a baseline comparison, we use the best-performing GAN tailored for this setup in [23], which showed point-wise error close to the best POD-based method and good multi-scale statistical properties. In our analyses, we focus only on the velocity magnitude, $u(x_1, x_2) = \|\boldsymbol{u}(x_1, x_2)\|$. Briefly, the GAN framework consists of two competing convolutional neural networks: the first network is a generator that transforms input measurements into predictions for the missing or damaged data; the second is a discriminator that works to discriminate between generated data and real fields. The training of the generator minimizes a loss function consisting of the mean squared error (MSE) and an adversarial loss provided by the discriminator, optimizing point-wise accuracy and statistical fidelity, respectively. A more detailed description of the GAN can be found in [23].

2.2. DM Framework for Flow Field Generation

Before moving to the more difficult task to inpaint a gap conditioned on some partial measurements of each given image, we need to define how to generate unconditional flow realizations. Unlike GANs, which map input noise to outputs in a single step, DMs use a Markov chain to incrementally denoise and generate information through a neural network (see Figure 1c for a qualitative visual example of one generation event). This finer-grained framework, coupled with an explicit log-likelihood training objective, tends to yield more stable training than the tailored loss functions of GANs, but still has the capability of generating realistic samples. Another feature of DMs is their inherent stochasticity in the generation process, which allows them to produce multiple outputs that adhere to the learned distribution conditioned on the same input.

In this section, we introduce the DM framework for flow field generation. The velocity magnitude field on the full 2D domain (I) is denoted by $\mathcal{V}_I = \{u(\boldsymbol{x})|\boldsymbol{x} \in I\}$, and the distribution of this field is represented as $p(\mathcal{V}_I)$. In order to train the model, we need first to produce a set of images with larger and larger noise. To do that, the DM framework defines a *forward process* or *diffusion process* that incrementally adds Gaussian noise to the data until they become indistinguishable from white noise after N diffusion steps (Figure 2).

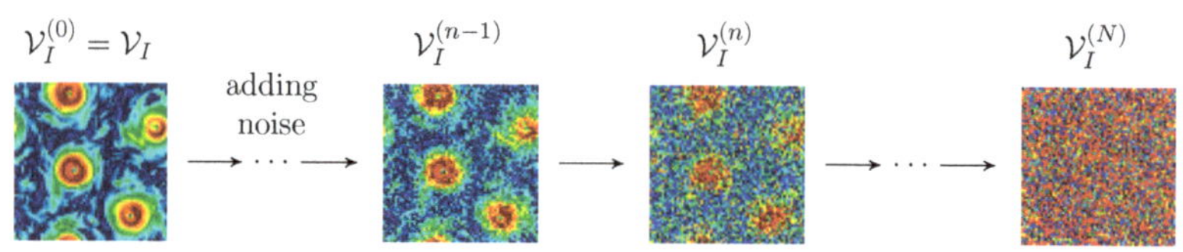

Figure 2. Diagram of the forward process in the DM framework. Starting with the original field $\mathcal{V}_I^{(0)} = \mathcal{V}_I$, Gaussian noise is incrementally added over N diffusion steps, transforming the original 64^2 image into white noise on the same resolution grid, $\mathcal{V}_I^{(N)}$.

This set of *diffused* images is used for training a network to perform a *backward* denoising process, starting from the set of pure i.i.d. Gaussian-noise 2D realizations and seeking to reproduce the set of images in the training dataset. Once the training is accomplished, the parameters of the network are frozen and used to generate brand new images by sampling from any realization of pure random images in the input (see Figure 3a for a sketch summary). The forward diffusion process is expressed in terms of a sequence of N steps, conditioned on the original set of images, i.e., for each image in the training dataset, we produce N noisy copies with an increasing amount of diffusion:

$$q\left(\mathcal{V}_I^{(1:N)}|\mathcal{V}_I^{(0)}\right) := \prod_{n=1}^{N} q\left(\mathcal{V}_I^{(n)}|\mathcal{V}_I^{(n-1)}\right), \tag{2}$$

where $\mathcal{V}_I^{(0)} = \mathcal{V}_I$ is the initial magnitude field and $\mathcal{V}_I^{(N)} \sim \mathcal{N}(\boldsymbol{0}, \boldsymbol{I})$ represents the final white-noise state, an ensemble of Gaussian images made of uncorrelated pixels with zero mean and unit variance. The notation $\mathcal{V}_I^{(1:N)}$ is used to denote the entire sequence of generated noisy fields, $\mathcal{V}_I^{(1)}, \ldots, \mathcal{V}_I^{(N)}$.

Each step, $n = 1, \ldots, N$, of the forward process can be directly obtained as

$$q\left(\mathcal{V}_I^{(n)}|\mathcal{V}_I^{(n-1)}\right) \to \mathcal{V}_I^{(n)} \sim \mathcal{N}\left(\sqrt{1-\beta_n}\mathcal{V}_I^{(n-1)}, \beta_n \boldsymbol{I}\right), \tag{3}$$

which implies sampling from a Gaussian distribution where the mean of $\mathcal{V}_I^{(n)}$ is given by $\sqrt{1-\beta_n}\mathcal{V}_I^{(n-1)}$ and the variance is $\beta_n \boldsymbol{I}$. The variance schedule $\beta_1, \ldots, \beta_N$ is predefined to allow a continuous transition to the pure Gaussian state. For more details on the variance schedule and other aspects of the DMs used in this study, see Appendix B.

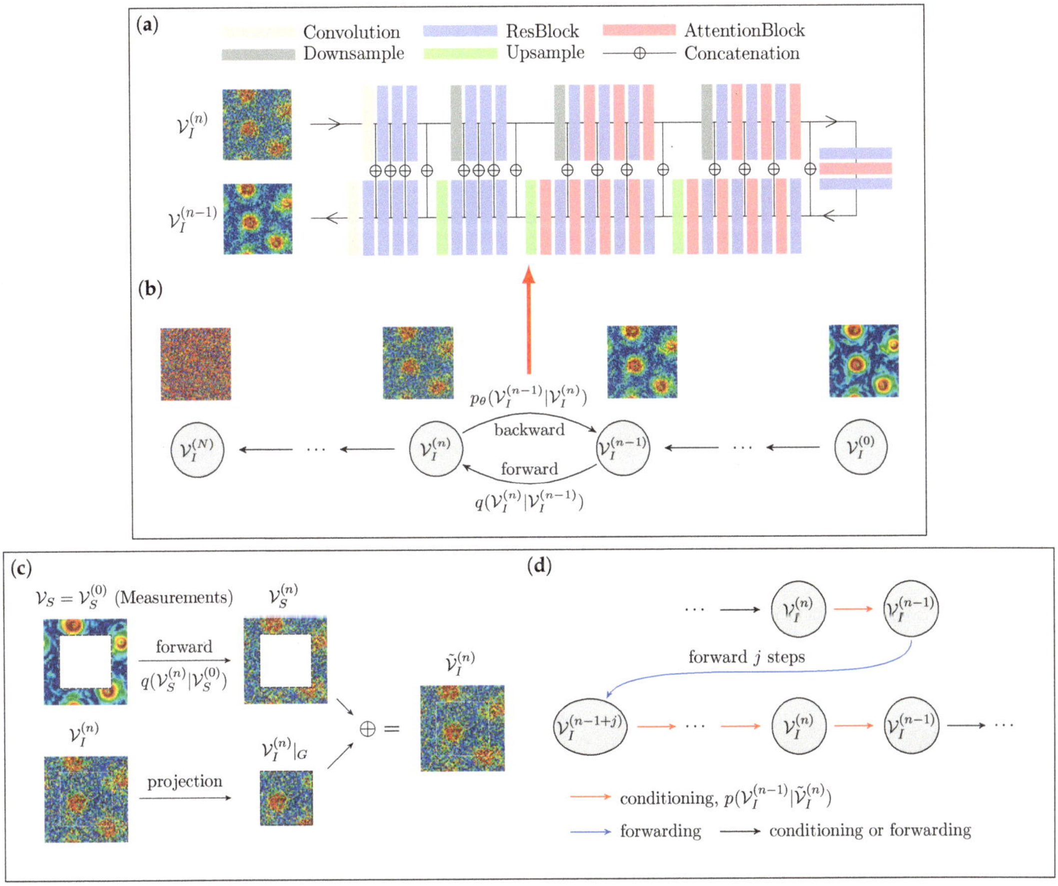

Figure 3. Schematic representation of the DM flow field generation framework used by RePaint for flow reconstruction. (**a**,**b**) Training stage: (**a**) the neural network architecture, U-Net [62], that takes a noisy flow field as input at step n and predicts a denoised field at step $n-1$; (**b**) The scheme of the forward and backward diffusion Markov processes. The forward process (from right to left) incrementally adds noise over N steps, while the backward process (from left to right), modeled by the U-Net, iteratively reconstructs the flow field by denoising the noisy data. More details on the network architecture can be found in Appendix B. (**c**,**d**) Reconstruction stage starting from a *damaged* field with a square mask. (**c**) Conditioning the backward process with the measurement, $\mathcal{V}_S$, involves projecting the noisy state of the entire 2D field, $\mathcal{V}_I^{(n)}$, onto the gap region, $\mathcal{V}_I^{(n)}|_G$, and combining it with the noisy measurement, $\mathcal{V}_S^{(n)}$, obtained from the forward process up to the corresponding step. In this way, we obtain $\tilde{\mathcal{V}}_I^{(n)}$ and then enforce the conditioning by inputting it into a backward step. (**d**) A resampling approach is used to deal with the incoherence in $\tilde{\mathcal{V}}_I^{(n)}$ introduced by the 'rigid' concatenation. First, we perform a backward step to obtain $\mathcal{V}_I^{(n-1)}$, and then some noise is added by j forward steps (blue arrow). Finally, the field is resampled backwards by the same number of iterations, going back to the original step.

The DM trains a neural network to approximate the reverse process of Equation (3), denoted as $p_\theta\left(\mathcal{V}_I^{(n-1)}|\mathcal{V}_I^{(n)}\right)$. This approximation allows the generation of new velocity magnitude fields from Gaussian noise, $p\left(\mathcal{V}_I^{(N)}\right) = \mathcal{N}(\mathbf{0}, \boldsymbol{I})$, through a *backward process* (see Figure 3a) described by

$$p_\theta\left(\mathcal{V}_I^{(0:N)}\right) := p\left(\mathcal{V}_I^{(N)}\right) \prod_{n=1}^{N} p_\theta\left(\mathcal{V}_I^{(n-1)} | \mathcal{V}_I^{(n)}\right), \tag{4}$$

where it is important to notice that the stochasticity in the process allows for the production of different final images even when starting from the same noise. In the continuous diffusion limit, characterized by sequences of small values of β_n, the backward process has a functional form identical to that of the forward process, as discussed in [63,64]. Consequently, the neural network is tasked with predicting the mean $\mu_\theta(\mathcal{V}_I^{(n)}, n)$ and covariance $\Sigma_\theta(\mathcal{V}_I^{(n)}, n)$ of a Gaussian distribution:

$$p_\theta\left(\mathcal{V}_I^{(n-1)} | \mathcal{V}_I^{(n)}\right) \rightarrow \mathcal{V}_I^{(n-1)} \sim \mathcal{N}\left(\mu_\theta(\mathcal{V}_I^{(n)}, n), \Sigma_\theta(\mathcal{V}_I^{(n)}, n)\right). \tag{5}$$

The neural network is optimized to minimize an upper bound of the negative log likelihood,

$$\mathbb{E}_{q(\mathcal{V}_I^{(0)})}[-\log(p_\theta(\mathcal{V}_I^{(0)}))]. \tag{6}$$

This training objective tends to result in more stable training compared to the tailored loss functions used in GANs. For a detailed derivation of the loss function and insights into the training details, please refer to Appendix A.

2.3. *Flow Field Data Augmentation with DMs: RePaint and Palette Strategies*

REPAINT. The RePaint approach aims to reconstruct missing information in the flow field using a DM that has been trained to generate the full 2D flow field from Gaussian noise, as described in the section above, without any conditioning on the measured data, and without relying on any further model training. To achieve the correct reconstruction, RePaint aims to ensure the conditioning on the measurements only by redesigning an ad hoc generation protocol [52]. As discussed above, during training, DM learns to approximate the backward transition probability to step on a sample $\mathcal{V}_I^{(n-1)}$, only from the knowledge of the sample obtained in the previous step $\mathcal{V}_I^{(n)}$; hence, DM models the one-step backward transition probability, $p_\theta\left(\mathcal{V}_I^{(n-1)} | \mathcal{V}_I^{(n)}\right)$. The goal of RePaint is to set up a generative process where this backward probability is also conditioned on some measured data, denoted as $\mathcal{V}_S$. In this way, each new sample in the backward direction is generated from the one-step backward conditioned probability, defined as $p_\theta\left(\mathcal{V}_I^{(n-1)} | \mathcal{V}_I^{(n)}, \mathcal{V}_S\right)$. To achieve this goal, RePaint substitutes the DM model input, $\mathcal{V}_I^{(n)}$, with another 2D field, $\tilde{\mathcal{V}}_I^{(n)}$, which is given by the union of $\mathcal{V}_I^{(n)}$ projected only to have support inside the gap (G) and the measured data on the support (S) propagated at step n according to the forward process, namely, $\tilde{\mathcal{V}}_I^{(n)} = \mathcal{V}_I^{(n)}|_G \oplus \mathcal{V}_S^{(n)}$. In summary, at any generic backward step n, RePaint approximates the conditional backward probability as follows:

$$p_\theta\left(\mathcal{V}_I^{(n-1)} | \mathcal{V}_I^{(n)}, \mathcal{V}_S\right) \approx p_\theta\left(\mathcal{V}_I^{(n-1)} | \tilde{\mathcal{V}}_I^{(n)}\right), \quad \text{where} \quad \tilde{\mathcal{V}}_I^{(n)} = \mathcal{V}_I^{(n)}|_G \oplus \mathcal{V}_S^{(n)}. \tag{7}$$

Here, $\mathcal{V}_I^{(n)}|_G$ represents the projection of the sample generated by the backward process at step n projected inside the gap region (the central square), while $\mathcal{V}_S^{(n)}$ is the noisy version of the measured data (outside the square gap) that are obtained by a forward propagation up to step n of the measurements. At this point, $\tilde{\mathcal{V}}_I^{(n)}$, replacing $\mathcal{V}_I^{(n)}$, is given as input to the model and it is used to obtain the next sample at step $n-1$, $\mathcal{V}_I^{(n-1)}$ (see Figure 3c).

The propagation of information from the measurements into the gap happens thanks to the application of the non-linear (and non-local) function approximated by the U-Net employed in the DM. Hence, the output of the U-Net, describing the probability of moving from step n to $n-1$, is the result of non-local convolutions mixing information in the two regions (S) and (G). In this way, the model mitigates the discontinuities generated across the gap by merging the generated and the measured data. Furthermore, to allow a deeper propagation of information, improving correlations between the measurements and the generated data, RePaint employs a resampling strategy [52]. The idea of resampling, as shown schematically in Figure 3d, is that each sample at step $n-1$, extracted from the conditioned probability, $p_\theta\left(\mathcal{V}_I^{(n-1)}|\tilde{\mathcal{V}}_I^{(n)}\right)$, is not directly used as input to move backward at step $n-2$, but, instead, it is first propagated forward for j steps (by adding more noise) before returning according to the conditioned backward process at step $n-1$. This operation gives the U-Net model the opportunity to iterate the propagation of information from the measured region inside the gap. Resampling can be applied at different steps multiple times, resulting in a back-and-forth progression during the generation process, as opposed to a monotonic backward progression from $n = N$ to $n = 0$. Further details, such as the network architecture and other parameters, can be found in Appendix B. As demonstrated in computer vision applications [52,65–67], this strategy has the advantage of being easily generalizable to diverse tasks, such as free-form inpainting with arbitrary mask shapes. However, this introduces several new challenges in the design of such a convoluted generation protocol, which is neither trivial in its optimization nor in its implementation.

PALETTE. An alternative approach to perform flow field reconstruction is to train the DM directly to learn the backward probability distribution conditioned on the measured data, previously introduced as $p_\theta\left(\mathcal{V}_G^{(n-1)}|\mathcal{V}_G^{(n)},\mathcal{V}_S\right)$. This method, called Palette, has been successfully used in various computer vision applications, such as image-to-image translation tasks [53,54]. The idea is to train a U-Net using the same strategy as any unconditioned DM, but giving the network as input the additional information coming from the measurements, at any step during the diffusion process. This allows the model to learn during training how to use information from the available data to achieve optimal reconstruction inside the gap. In addition, unlike the RePaint method, Palette always uses the measured data without adding noise. In this way, the forward process can be defined as for the pure generation case, but it takes place only within the gap region, while the data on the support, $\mathcal{V}_S$, are frozen throughout the diffusion process and serve as an additional input to the model. A schematic summary of the Palette approach is shown in Figure 4. Once the DM model is trained, since the reconstruction process is Markovian as in the standard generative DM, the conditional probability of the reconstructed field, $p_\theta\left(\mathcal{V}_G^{(0)}|\mathcal{V}_S\right)$, can be determined through the following iterative process:

$$p_\theta\left(\mathcal{V}_G^{(0)}|\mathcal{V}_S\right) = p\left(\mathcal{V}_G^{(N)}\right)\prod_{n=1}^{N} p_\theta\left(\mathcal{V}_G^{(n-1)}|\mathcal{V}_G^{(n)},\mathcal{V}_S\right), \tag{8}$$

starting from any Gaussian noise $\mathcal{V}_G^{(N)}$. To facilitate the comparison with the GAN model implemented in our previous works [23,28], we trained a separate Palette model for each fixed mask size. Let us stress that both methods are capable of training on a free-form mask [68]. More details on Palette are provided in Appendix B.

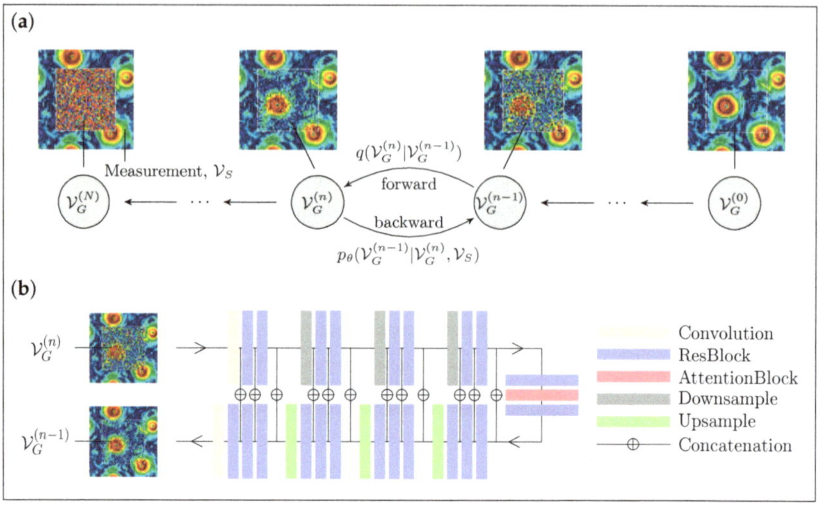

Figure 4. Schematic of the DM Palette protocol. (**a**) In the backward process (from left to right), we start from pure noise in the gap, $\mathcal{V}_G^{(N)}$, combined with the measurements in the frame, $\mathcal{V}_S$, to progressively denoise the missing information using the U-Net architecture described in (**b**). (**b**) A sketch of the U-Net integrating the measurement, $\mathcal{V}_S$, and the noisy data within the gap, $\mathcal{V}_G^{(n)}$, for a backward step.

3. Comparative Analysis of DMs and the GAN in Flow Reconstruction

To provide a systematic comparison between DMs and the GAN in flow reconstruction, we focus on cases where the 2D velocity magnitude fields have a central square gap of variable size, spanning $0.1 < l/l_0 < 1$, where l_0 is the size of the whole flow domain. In this section, only one reconstruction realization is performed for all the image data in the testing ensemble, i.e., we do not further explore the possibility of assessing the robustness of the prediction by sampling over the ensemble of predicted images (see next section). The initial evaluation focuses on the reconstructed single-point velocity magnitude, which is strongly influenced by large-scale coherent structures. Then, we analyze the reconstruction process from a multi-scale perspective by examining the statistical properties of the gradient of the reconstructed velocity magnitude and by looking at other scale-dependent statistics in both real and Fourier space.

3.1. Large-Scale Information

To quantify the reconstruction error between the predicted velocity magnitude, $u_G^{(p)}$, and the true velocity magnitude, $u_G^{(t)}$, within the gap region, we introduce the normalized MSE as follows:

$$\mathrm{MSE}(u_G) = \langle \Delta_{u_G} \rangle / E_{u_G} \,. \tag{9}$$

Here, Δ_{u_G} represents the spatially averaged L_2 error in the central, gappy region for a single-flow configuration, and it is calculated as

$$\Delta_{u_G} = \frac{1}{A_G} \int_G |u_G^{(p)}(\boldsymbol{x}) - u_G^{(t)}(\boldsymbol{x})|^2 \mathrm{d}\boldsymbol{x} \,, \tag{10}$$

where A_G denotes the area of the gap. Averaging $\langle \cdot \rangle$ is performed over the test dataset. The normalization factor, E_{u_G}, is defined as the product of the standard deviations of the predicted and true velocity magnitudes within the gap:

$$E_{u_G} = \sigma_G^{(p)} \sigma_G^{(t)} \,, \tag{11}$$

where

$$\sigma_G^{(p)} = \frac{1}{A_G^{1/2}} \int_G \langle (u_G^{(p)})^2 \rangle^{1/2} \mathrm{d}\mathbf{x} \tag{12}$$

and $\sigma_G^{(t)}$ is similarly defined. This choice for the normalization term, E_{u_G}, ensures that predictions with significantly low or high energy levels will result in a large MSE.

In our analysis, we use the Jensen–Shannon (JS) divergence to assess the distance between the PDF of a predicted quantity and the PDF of the true data. Specifically, the JS divergence applied to two distributions $P(x)$ and $Q(x)$ defined on the same sample space is

$$D_{\mathrm{JS}}(P \parallel Q) = \frac{1}{2} D_{\mathrm{KL}}(P \parallel M) + \frac{1}{2} D_{\mathrm{KL}}(Q \parallel M), \tag{13}$$

where $M = \frac{1}{2}(P + Q)$ and

$$D_{\mathrm{KL}}(P \parallel Q) \equiv \int_{-\infty}^{\infty} P(x) \log\left(\frac{P(x)}{Q(x)}\right) \mathrm{d}x \tag{14}$$

is the Kullback–Leibler (KL) divergence. As the two distributions get closer, the value of the JS divergence becomes smaller, with a value of zero indicating that P and Q are identical.

Figure 5a shows the MSE(u_G) as a function of the normalized gap size, l/l_0. It shows that Palette achieves a comparable MSE with respect to GAN for most gap sizes. Only for the largest gap size, $l/l_0 = 62/64$, is the MSE of Palette significantly better than that of GAN. On the other hand, RePaint has a larger MSE for all sizes compared to the other two methods, demonstrating the limitations of the RePaint approach in enforcing correlations between the measurements and generated data without being specifically trained on a reconstruction problem, as in the other two approaches. The red baseline, derived from predictions using randomly shuffled test data, represents the case where the predictions guess the exact statistical properties, $\langle u_G^{(p)} \rangle = \langle u_G^{(t)} \rangle$ and $\langle (u_G^{(p)})^2 \rangle = \langle (u_G^{(t)})^2 \rangle$, but lose all correlation with the measurements, $\langle u_G^{(p)} u_G^{(t)} \rangle = \langle u_G^{(p)} \rangle \langle u_G^{(t)} \rangle$.

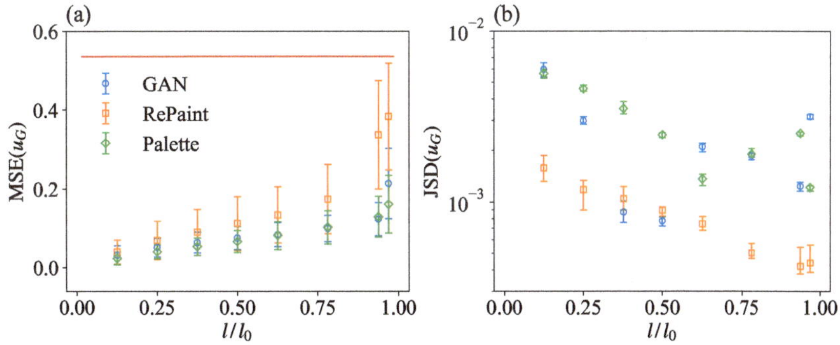

Figure 5. (**a**) The mean squared error (MSE) between the true and the generated velocity magnitude, as obtained from GAN, RePaint, and Palette, for a square gap with variable size. Error bars indicate the standard deviation. The red horizontal line represents the uncorrelated baseline MSE, ≈ 0.54. (**b**) The Jensen–Shannon (JS) divergence between the probability density functions (PDFs) for the true and generated velocity magnitude. The mean and error bars represent the average and range of variation of the JS divergence across 10 batches, each with 2048 samples.

We now examine the velocity magnitude PDFs as predicted by the different methods and compare them with the true data. In Figure 5b, we present the JS divergence between the predicted and true velocity magnitudes, denoted as JSD(u_G) $= D_{\mathrm{JS}}(\mathrm{PDF}(u_G^{(p)}) \parallel \mathrm{PDF}(u_G^{(t)}))$. First of all, it is important to highlight that all the JSD(u_G) values are well below 10^{-2}, suggesting that there is always a close match between the PDFs of the reconstructed and of the true velocity magnitude. The agreement between the different PDFs is also shown in Figure 6, where one can see the extremely good performance of all models in closely matching the PDFs of the generated velocity magnitude with the ground truth one. Going back to the results presented in Figure 5b, it is possible to note that in the small gap region, $l/l_0 \leq 0.4$, there is a monotonic behavior of the JS divergence, which tends to decrease as the gap increases. This can be interpreted by the fact that the main contribution to the JS divergence is due to statistical fluctuations in the PDF tails which are less accurately estimated when the gap is small. This behavior is clearly visible in the results of the RePaint approach, which shows a monotonic decrease in the JS divergence over the whole range of gaps analyzed. The same effect is not visible in the other approaches in the range above $l/l_0 = 0.5$. The reason is probably that both GAN and Palette rely on different training to reconstruct different gap sizes, and the fluctuations due to the training convergence could be underestimated. The non-monotonicity is much more pronounced in the GAN results as this approach is known to be less stable during training. The analysis shows that, compared to the other two approaches, RePaint trained on the pure generation without any conditioning is the best method to obtain a statistical representation of the true data.

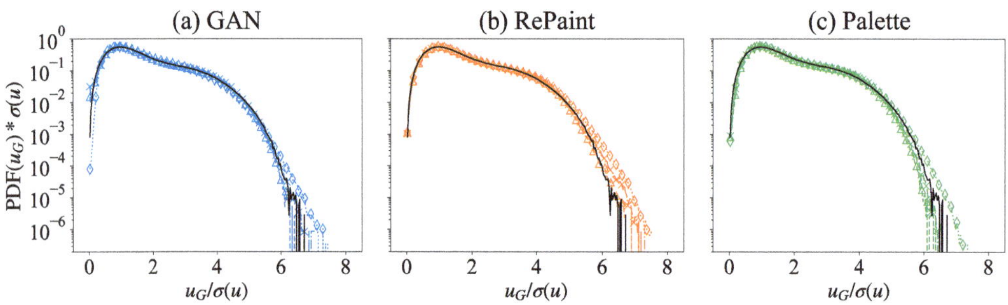

Figure 6. PDFs of the velocity magnitude in the missing region obtained from (**a**) GAN, (**b**) RePaint, and (**c**) Palette for a square gap of variable size $l/l_0 = 24/64$ (triangle), 40/64 (cross), and 62/64 (diamond). The PDF of the true data over the whole region is plotted for reference (solid black line) and $\sigma(u)$ is the standard deviation of the original data over the full domain.

In Figure 7, we compare the PDFs of the spatially averaged L_2 error, Δ_{u_G}, for different flow configurations. For small and medium gap sizes (Figure 7a,b), the PDFs of GAN and Palette closely match, whereas the PDF of RePaint, although similar in shape, exhibits a range with larger errors. For the largest gap size (Figure 7c), Palette is clearly the most accurate, predicting the smallest errors. Again, RePaint performs the worst, characterized by a peak at high error values and a broad error range.

Finally, Figure 8 provides a visual qualitative idea of the reconstruction capabilities of the instantaneous velocity magnitude field using the three adopted models. While all methods generally perform well in locating vortex structures within smaller gaps and produce realistic turbulent reconstructions, RePaint is a notable exception. In particular, for the largest gap (in Figure 8c), RePaint's performance lags significantly behind the other two methods, failing to accurately predict vortex positions and resulting in a significantly larger MSE.

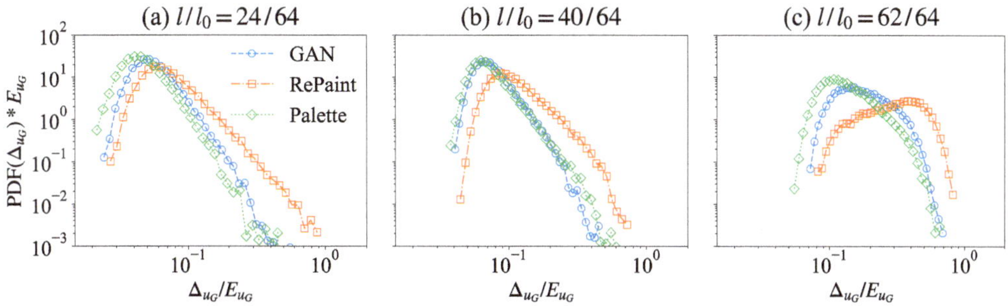

Figure 7. The PDFs of the spatially averaged L_2 error for a single flow configuration obtained from GAN, RePaint, and Palette models. The gap size changes from (**a**) $l/l_0 = 24/64$, to (**b**) $40/64$ and (**c**) $62/64$.

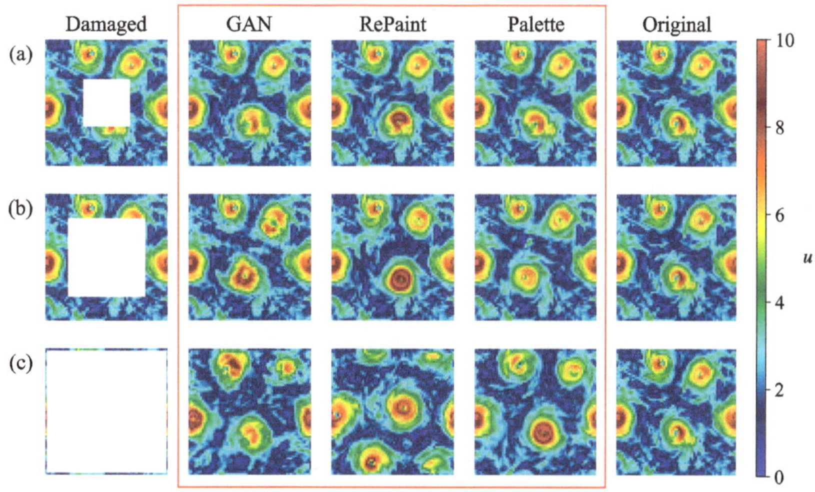

Figure 8. Examples of reconstruction of an instantaneous field (velocity magnitude) for a square gap of size (**a**) $l/l_0 = 24/64$, (**b**) $l/l_0 = 40/64$ and (**c**) $l/l_0 = 62/64$. The damaged fields are shown in the first column, while the second to fourth columns, circled by a red rectangle, show the reconstructed fields obtained from GAN, RePaint, and Palette. The ground truth is shown in the fifth column.

3.2. Multi-Scale Information

This section presents a quantitative analysis of the multi-scale information reconstructed by the different methods. We begin by examining the gradient of the reconstructed velocity magnitude in the missing region, denoted as $\partial u_G / \partial x_1$. Figure 9a shows the MSE of this gradient, $\text{MSE}(\partial u_G / \partial x_1)$, defined similarly to Equation (9). The results show that Palette consistently achieves the lowest MSE. GAN's performance is comparable for most gap sizes, but deteriorates significantly at the extremely large gap size. In contrast, while RePaint has larger point-wise reconstruction errors for the gradient, it maintains the smallest JS divergence, $\text{JSD}(\partial u_G / \partial x_1)$, as shown in Figure 9b, indicating its robust statistical properties.

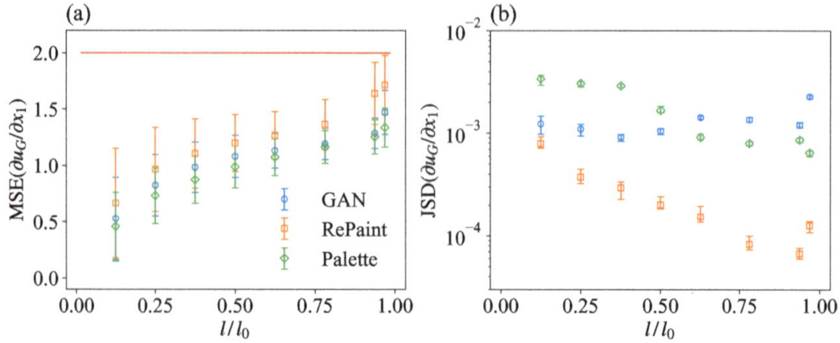

Figure 9. (**a**) MSE and (**b**) JS divergence between the PDFs for the gradient of the original and generated velocity magnitude, as obtained from GAN, RePaint, and Palette, for a square gap with variable size. The red horizontal line in (**a**) represents the uncorrelated baseline, equal to 2. Error bars are obtained in the same way as in Figure 5.

For small gap sizes, Palette has larger JS divergence than GAN, while the situation is reversed at higher gap values (Figure 9b). It is worth noting that, like the velocity module PDFs, the reconstructed gradient PDFs are very well matched by all three methods. In contrast to velocity, the GAN reconstruction is less accurate for very large gaps in the case of gradient statistics, as can be seen by comparing the PDFs in the different panels of Figure 10. This last observation limits the applications of the GAN to the modeling of small-scale turbulent observables.

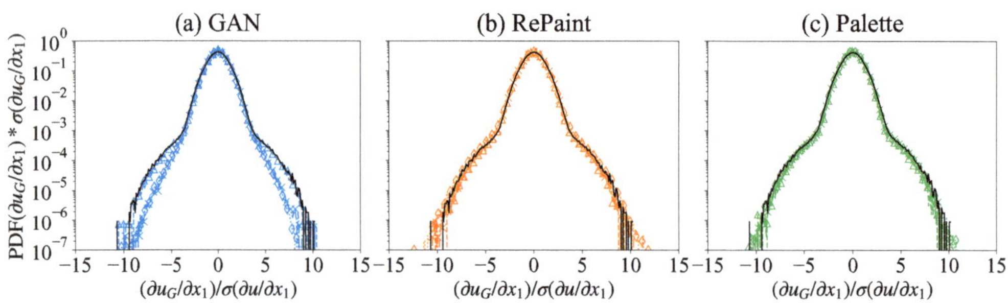

Figure 10. The PDFs of the gradient of the reconstructed velocity magnitude in the missing region obtained from (**a**) GAN, (**b**) RePaint, and (**c**) Palette, for a square gap of variable size $l/l_0 = 24/64$ (triangle), $40/64$ (cross), and $62/64$ (diamond). The PDF of the true data over the whole region is plotted for reference (solid black line) and $\sigma(\partial u/\partial x_1)$ is the standard deviation of the original data over the full domain.

The results of these methods can be more directly visualized by examining the gradient of the reconstruction samples, as shown in Figure 11. For small gap sizes (Figure 11a), all three methods produce realistic predictions that correlate well with the original structure. However, for medium and large gap sizes (Figure 11b,c), only Palette is able to generate gradient structures that are well correlated with the ground truth.

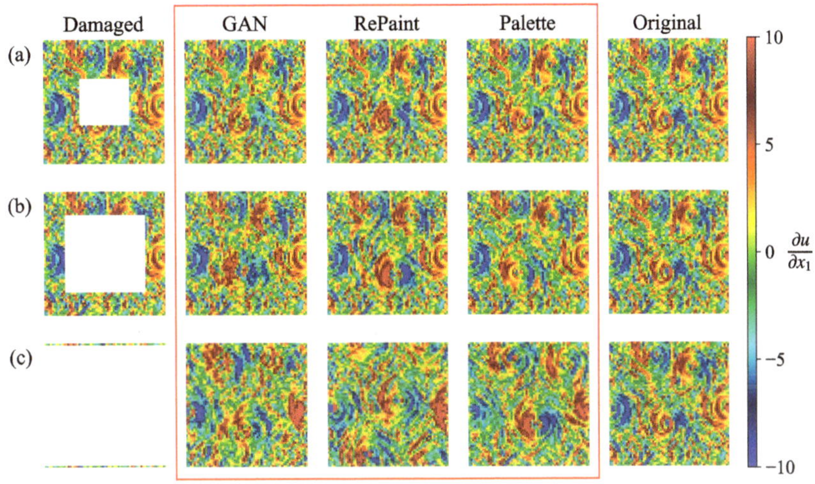

Figure 11. The gradient of the velocity magnitude fields shown in Figure 8. The first column shows the damaged fields with a square gap of size (**a**) $l/l_0 = 24/64$, (**b**) $l/l_0 = 40/64$ and (**c**) $l/l_0 = 62/64$. Note that for the case $l/l_0 = 62/64$, the gap extends almost to the borders, leaving only a single vertical velocity line on both the left and right sides, where the original gradient field is missing. The gradient of the reconstructions from GAN, RePaint, and Palette, shown in the second to fourth columns, is surrounded by a red rectangle for emphasis, while the fifth column shows the ground truth.

The better performance of DMs in capturing the statistical properties is further demonstrated by a scale-by-scale analysis of the 2D energy spectrum obtained from the reconstructed fields,

$$E(k) = \sum_{k \le \|\boldsymbol{k}\| < k+1} \frac{1}{2} \langle \hat{u}(\boldsymbol{k}) \hat{u}^*(\boldsymbol{k}) \rangle. \tag{15}$$

Here, $\boldsymbol{k} = (k_1, k_2)$ denotes the horizontal wavenumber, $\hat{u}(\boldsymbol{k})$ is the Fourier transform of the velocity magnitude, and $\hat{u}^*(\boldsymbol{k})$ is its complex conjugate. Direct comparison of the spectra are shown in Figure 12a–c, for three gap sizes. In Figure 12d–f, we plot the ratio of the reconstructed to the original spectra, denoted as $E(k)/E^{(t)}(k)$. Deviations from unity in this ratio better highlight the wavenumber regions where the reconstruction is less accurate. While all the methods produce satisfactory energy spectra, a closer examination of the ratio to the original energy spectrum shows that RePaint and Palette maintain uniformly good correspondence across all scales and for all gap sizes. Conversely, GAN performs well at small gap sizes, but exhibits poorer performance at large wavenumbers for medium and large gap sizes.

Consistent results are observed when examining the flatness of the velocity magnitude increments:

$$F(r) = \langle (\delta_r u)^4 \rangle / \langle (\delta_r u)^2 \rangle^2, \tag{16}$$

where $\delta_r u = u(\boldsymbol{x} + \boldsymbol{r}) - u(\boldsymbol{x})$ and $\boldsymbol{r} = (r, 0)$, with $\langle \cdot \rangle$ denoting the average over the test data and over $\boldsymbol{x}$, for points $\boldsymbol{x}$ and $\boldsymbol{x} + \boldsymbol{r}$, where only one, or both of them, are within the gap. The flatness calculated over the entire region of the original field is also shown for comparison. In Figure 13, the flatness results further confirm that RePaint and Palette consistently maintain their high-quality performance across all scales. In contrast, while GAN is effective at small gap sizes, it faces challenges in maintaining similar standards at small scales for medium and large gap sizes.

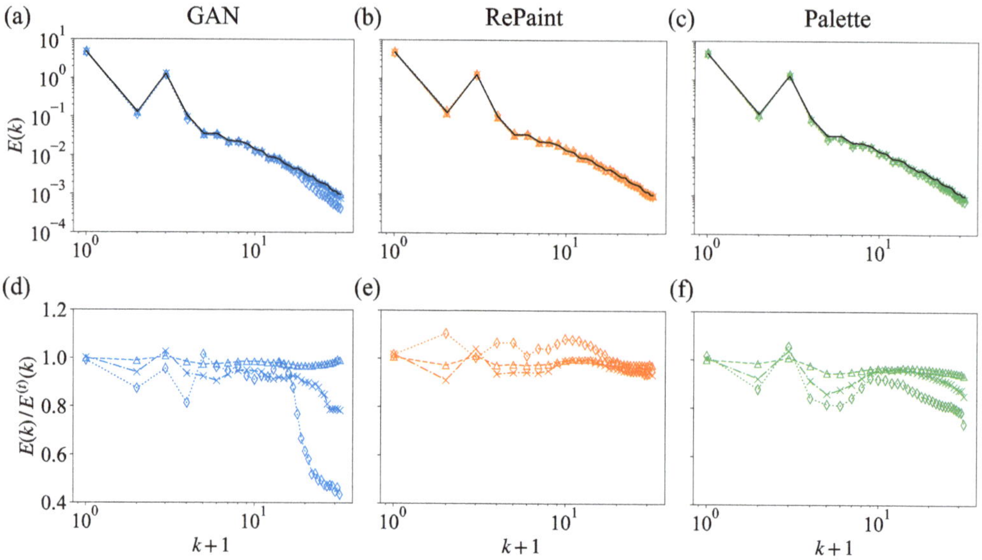

Figure 12. Energy spectra of the original velocity magnitude (solid black line) and the reconstructions obtained from (**a**) GAN, (**b**) RePaint, and (**c**) Palette for a square gap of sizes $l/l_0 = 24/64$ (triangle), $40/64$ (cross), and $62/64$ (diamond). The corresponding $E(k)/E^{(t)}(k)$ is shown in (**d**–**f**), where $E(k)$ and $E^{(t)}(k)$ are the spectra of the reconstructed fields and the ground truth, respectively.

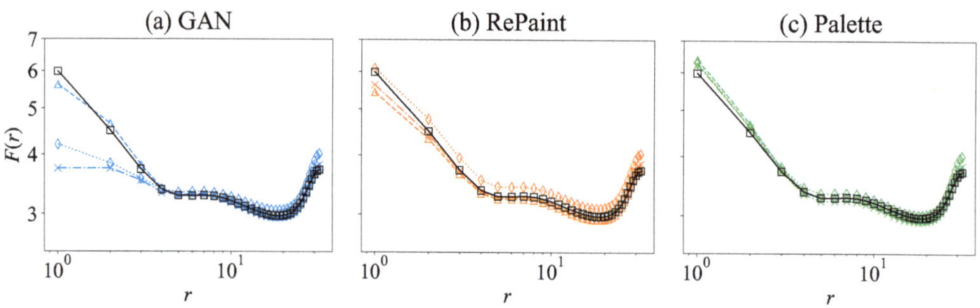

Figure 13. The flatness of the original field (solid black line) and the reconstructions obtained from (**a**) GAN, (**b**) RePaint, and (**c**) Palette for a square gap of sizes $l/l_0 = 24/64$ (triangle), $40/64$ (cross), and $62/64$ (diamond).

4. Probabilistic Reconstructions with DMs

So far, we have analyzed the performances of the three models in the reconstruction of the velocity magnitude itself and its statistical properties. In this section, we explore the probabilistic reconstruction capabilities of DMs, i.e., the fact that DMs provide us with many possible reconstructions that we can quantify in terms of a mean error and a variance. This is a significant advantage over the GAN architecture we have used in this work. It is worth noting that the implementation of stochastic GANs is also possible, although not of interest for our analysis. Focusing on a specific gap size, we select two flow configurations: the first is a configuration for which the discrepancy between the reconstructed fields and the true data–as quantified by the mean L_2 error—is small and comparable across GAN, Palette, and RePaint; the second is a more complex situation for reconstruction, as all models display large discrepancies from the true data. For each of these two configurations, we performed 20,480 reconstructions using RePaint and Palette.

Figure 14a displays the PDFs of the spatially averaged L_2 errors across different reconstruction realizations, compared to GAN's unique reconstruction error indicated by a blue dashed line. The comparison shows that Palette achieves a lower mean L_2 error than GAN, along with a smaller variance, indicating high model confidence for this case. In contrast, RePaint tends to produce higher errors with a wider variance. The comparison is more evident in Figure 14b, where it appears that GAN provides a realistic reconstruction with accurate vortex positioning. As for RePaint, it sometimes inaccurately predicts the vortex positions (Figure 14c (L)) or fails to accurately represent the energy distribution, even when the position is correct (Figure 14c (S) and (M)), leading to larger errors. Conversely, Figure 14d shows that Palette consistently predicts the correct position of vortex structures, with variations in vortex shape or energy distribution being the primary factors affecting the narrow reconstruction error distribution.

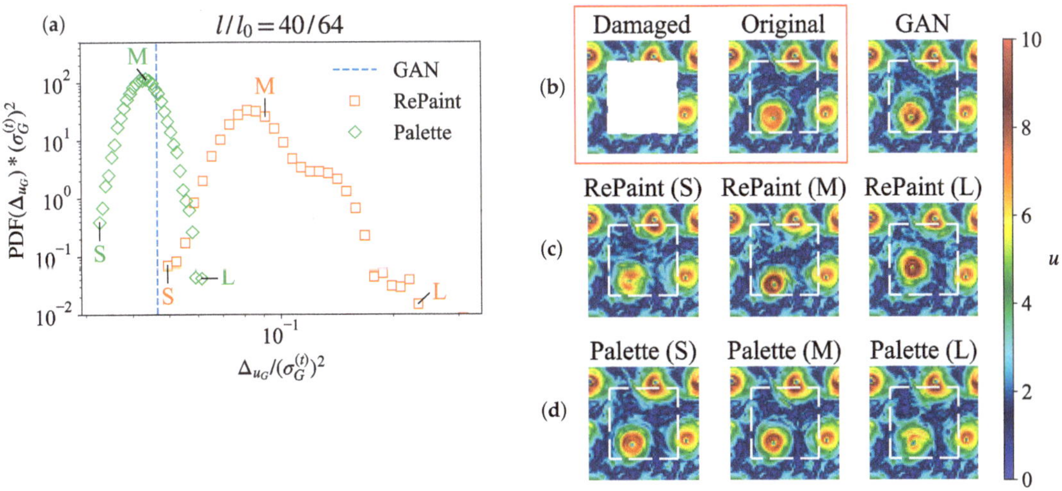

Figure 14. Probabilistic reconstructions from DMs for a fixed measurement outside a square gap with size $l/l_0 = 40/64$ for a configuration where all models give quite small reconstruction errors. (**a**) PDFs of the spatially averaged L_2 error over different reconstructions obtained from RePaint and Palette. The blue vertical dashed line indicates the error for the GAN case. (**b**) The damaged measurement and ground truth, inside a red box, and the prediction from GAN. (**c**) The reconstructions from RePaint with a small L_2 error (S), the mean L_2 error (M), and with a large L_2 error (L). (**d**) The reconstructions from Palette corresponding to a small L_2 error (S), the mean L_2 error (M), and a large L_2 error (L).

Figure 15 presents the same evaluations for the configuration where all models produce large errors. As shown in Figure 15a, both RePaint and Palette show significant variance in errors, with their mean errors exceeding that of GAN. The ground truth, examined in Figure 15b, highlights the inherent difficulty of this reconstruction scenario. In particular, an entire vortex structure is missing, and the proximity of two strong vortices suggests a potential transient state, possibly involving vortex merging or vortex breakdown. These situations may be rare in the training data, leading to a complete failure of GAN to accurately predict the correct vortex position, as shown in Figure 15b. For RePaint, the challenge of this reconstruction is reflected in the different predictions of the vortex positions. While some of these predictions are more accurate than GAN's, RePaint also tends to produce incoherence around the gap boundaries (Figure 15c). Conversely, Figure 15d shows that Palette's predictions are not only more consistent with the measurements, but also provide a range of reconstructions with different vortex positions.

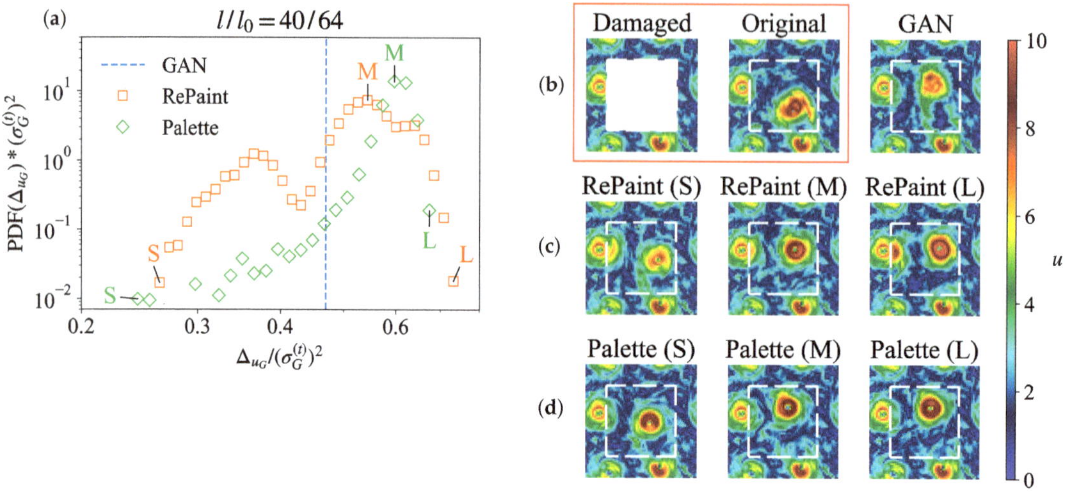

Figure 15. Similar to Figure 14, but for a flow configuration chosen for its large reconstruction errors from GAN, RePaint, and Palette. The red box contains the damaged measurement and ground truth.

5. Conclusions and Discussion

In this study, we investigated the data augmentation ability of DMs for damaged measurements of 2D snapshots of a 3D rotating turbulence at moderate Reynolds numbers. The Rossby number is chosen such as to produce a bidirectional energy cascade at both large and small scales. Two DM reconstruction methods are investigated: RePaint, which uses a heuristic strategy to guide an unconditional DM in the flow generation, and Palette, a conditional DM trained with paired measurements and missing information. As a benchmark, we compared these two DMs with the best-performing GAN method on the same dataset. We showed that there exists a trade-off between obtaining a reliable L_2 error and good statistical reconstruction properties. Typically, models that are very good for the former are less accurate for the latter. Overall, according to our analysis, Palette seems to be the most promising tool considering both metrics. Indeed, our comparative study shows that while RePaint consistently exhibits superior statistical reconstruction properties, it does not achieve small L_2 errors. Conversely, Palette achieves the smallest L_2 errors along with very good statistical results. Moreover, we observe that GAN fails to provide statistical properties that are as accurate as the DMs at small scales for medium and large gaps.

Concerning probabilistic reconstructions, a crucial feature for turbulent studies and uncertainty quantification for both theoretical and practical applications, we evaluated the effectiveness of the two DM methods on two specific configurations of different complexity. For the configuration with sufficient information in the measurement, Palette shows errors that tend to be smaller than GAN and exhibits a small variance, indicating high model confidence. However, RePaint faces challenges in accurately predicting large-scale vortex positions and struggles to achieve an accurate energy distribution. This difficulty partly stems from RePaint's heuristic conditioning strategy, which cannot effectively guide the generative process using the measurement. In a more complex scenario characterized by the presence in the gappy region of an entire large-scale structure, GAN completely fails to predict the correct vortex position, while both DMs can localize it with higher precision by taking advantage of multiple predictions, although RePaint shows incoherence around the gap boundaries.

In summary, this study establishes a new state-of-the-art method for the 2D snapshot reconstruction of 3D rotating turbulence using conditional DMs, surpassing the previous GAN-based approach. The better performance of DMs over GANs stems from their iterative, denoising construction process, which builds up the prediction scale-by-scale, resulting in better performance across all scales. The inherent stochasticity of this iterative process yields a probabilistic set of predictions conditioned on the measurement, in contrast to the unique prediction of the GAN implemented here. Our study opens the way to further applications for risk assessment of extreme events and in support of various data assimilation methods. It is important to note that DMs are significantly more computationally expensive than GANs due to the iterative inference steps. Despite this, many efforts in the computer vision field have been devoted to accelerating this process [69,70].

A promising avenue for future studies could focus on flows at higher Reynolds numbers and Rossby numbers, close to the critical transition, leading to the inverse energy cascade, a very complex turbulent situation where both 3D and 2D physics coexist in a multi-scale environment. In addition, it would be interesting to test our method in practical scenarios, such as wind farm flows [71–73], where it could potentially help to synthesize inflow conditions and reconstruct the associated dynamics. The present work focuses on cases where only instantaneous measurements are available, without temporal sequences. In the future, machine learning methods, such as long short-term memory (LSTM) [74] and transformer [75], could be used to exploit temporal correlations.

Author Contributions: All authors contributed to the conceptualization and to the writing of the results. T.L. and M.B. contributed to the development of the methodology, A.S.L. and L.B. to the formal analysis, and F.B. to the curation of the data and software. All authors have read and agreed to the published version of the manuscript.

Funding: This research was funded by the European Research Council (ERC) under the European Union's Horizon 2020 research and innovation programme, grant agreement number 882340 and by the EHPC-REG-2021R0049 grant. ASL received partial financial support from ICSC—Centro Nazionale di Ricerca in High Performance Computing, Big Data, and Quantum Computing, funded by the European Union—NextGenerationEU. LB and MB received partial funding from the program FARE-MUR R2045J8XAW. MB acknowledges the hospitality of the Eindhoven University of Technology TUe, where part of this work was carried out.

Institutional Review Board Statement: Not applicable.

Informed Consent Statement: Not applicable.

Data Availability Statement: The 2D snapshots of velocity data from rotating turbulence used in this study are openly available on the open-access Smart-TURB portal (http://smart-turb.roma2.infn.it) (access date: 14 June 2020), under the TURB-Rot repository [59]. The codebase for the two DM reconstruction methods, RePaint and Palette, is available at https://github.com/SmartTURB/repaint-turb (access date: 23 November 2023) and https://github.com/SmartTURB/palette-turb (access date: 16 November 2023), respectively.

Conflicts of Interest: The authors declare no conflicts of interest.

Abbreviations

The following abbreviations are used in this manuscript:

2D	Two-dimensional
DM	Diffusion model
GAN	Generative adversarial network
POD	Proper orthogonal decomposition
PDF	Probability density function
3D	Three-dimensional
DNS	Direct numerical simulation
MSE	Mean squared error
JS	Jensen–Shannon
KL	Kullback–Leibler

Appendix A. Training Objective of DM for Flow Field Generation

A notable property of the forward process is that it allows closed-form sampling of $\mathcal{V}_I^{(n)}$ at any given diffusion step n [76]. With definitions of $\alpha_n := 1 - \beta_n$ and $\bar{\alpha}_n := \Pi_{i=1}^n \alpha_i$, we have

$$q(\mathcal{V}_I^{(n)}|\mathcal{V}_I^{(0)}) \rightarrow \mathcal{V}_I^{(n)} \sim \mathcal{N}(\sqrt{\bar{\alpha}_n}\mathcal{V}_I^{(0)}, (1-\bar{\alpha}_n)\boldsymbol{I}). \tag{A1}$$

Specifically, for any real flow field $\mathcal{V}_I^{(0)}$, we can directly evaluate its state after n diffusion steps using

$$\mathcal{V}_I^{(n)} = \sqrt{\bar{\alpha}_n}\mathcal{V}_I^{(0)} + \sqrt{1-\bar{\alpha}_n}\epsilon, \tag{A2}$$

where $\epsilon \sim \mathcal{N}(\boldsymbol{0}, \boldsymbol{I})$.

To optimize the negative log likelihood, $\mathbb{E}_{q(\mathcal{V}_I^{(0)})}[-\log(p_\theta(\mathcal{V}_I^{(0)}))]$, which is numerically intractable, we focus on optimizing its usual variational bound:

$$L := \mathbb{E}_{q(\mathcal{V}_I^{(0)})}\mathbb{E}_{q(\mathcal{V}_I^{(1:N)}|\mathcal{V}_I^{(0)})}\left[-\log\frac{p_\theta(\mathcal{V}_I^{(0:N)})}{q(\mathcal{V}_I^{(1:N)}|\mathcal{V}_I^{(0)})}\right] \geq \mathbb{E}_{q(\mathcal{V}_I^{(0)})}[-\log(p_\theta(\mathcal{V}_I^{(0)}))]. \tag{A3}$$

The objective can be further reformulated as a combination of KL divergences, denoted as $D_{\mathrm{KL}}(\cdot \parallel \cdot)$, plus an additional entropy term [36,64]:

$$\begin{aligned} L = \mathbb{E}_{q(\mathcal{V}_I^{(0)})}\Bigg[&\underbrace{D_{\mathrm{KL}}(p(\mathcal{V}_I^{(N)}|\mathcal{V}_I^{(0)}) \parallel p_\theta(\mathcal{V}_I^{(N)}))}_{L_N} \\ &+ \sum_{n>1}^{N}\underbrace{D_{\mathrm{KL}}(p(\mathcal{V}_I^{(n-1)}|\mathcal{V}_I^{(n)},\mathcal{V}_I^{(0)}) \parallel p_\theta(\mathcal{V}_I^{(n-1)}|\mathcal{V}_I^{(n)}))}_{L_{n-1}}\underbrace{-\log p_\theta(\mathcal{V}_I^{(0)}|\mathcal{V}_I^{(1)})}_{L_0}\Bigg] \end{aligned} \tag{A4}$$

The first term, L_N, has no learnable parameters as $p_\theta(\mathcal{V}_I^{(N)})$ is a Gaussian distribution, and can, therefore, be ignored during training. The terms within the second part of the summation, L_{n-1}, represent the KL divergence between $p_\theta(\mathcal{V}_I^{(n-1)}|\mathcal{V}_I^{(n)})$ and the posteriors of the forward process conditioned on $\mathcal{V}_I^{(0)}$, which are tractable using Bayes' theorem [43,76]:

$$p(\mathcal{V}_I^{(n-1)}|\mathcal{V}_I^{(n)},\mathcal{V}_I^{(0)}) \rightarrow \mathcal{V}_I^{(n-1)} \sim \mathcal{N}(\tilde{\mu}(\mathcal{V}_I^{(n)},\mathcal{V}_I^{(0)}), \tilde{\beta}_n\boldsymbol{I}), \tag{A5}$$

where

$$\tilde{\mu}_n(\mathcal{V}_I^{(n)},\mathcal{V}_I^{(0)}) := \frac{\sqrt{\bar{\alpha}_{n-1}}\beta_n}{1-\bar{\alpha}_n}\mathcal{V}_0 + \frac{\sqrt{\alpha_n}(1-\bar{\alpha}_{n-1})}{1-\bar{\alpha}_n}\mathcal{V}_I^{(n)} \tag{A6}$$

and

$$\tilde{\beta}_n := \frac{1-\bar{\alpha}_{n-1}}{1-\bar{\alpha}_n}\beta_n. \tag{A7}$$

By setting $\Sigma_\theta = \sigma_n^2\boldsymbol{I}$ to untrained constants, where σ_n^2 can be either β_n or $\tilde{\beta}_n$, as discussed in [36], the KL divergence between the two Gaussians in Equations (3) and (5) can be expressed as

$$L_{n-1} = \mathbb{E}_{q(\mathcal{V}_I^{(0)})}\left[\frac{1}{2\sigma_n^2}\|\tilde{\mu}_n(\mathcal{V}_I^{(n)},\mathcal{V}_I^{(0)}) - \mu_\theta(\mathcal{V}_I^{(n)},n)\|^2\right]. \tag{A8}$$

Given the Gaussian form of $p_\theta(\mathcal{V}_I^{(0)}|\mathcal{V}_I^{(1)})$ as presented in Equation (5), the term L_0 also results in the same form as Equation (A8). Substituting Equation (A2) into Equation (A6), we can express the mean of the conditioned posteriors as

$$\tilde{\mu}(\mathcal{V}_I^{(n)},\mathcal{V}_I^{(0)}) = \frac{1}{\sqrt{\alpha_n}}\left(\mathcal{V}_I^{(n)} - \frac{\beta_n}{\sqrt{1-\bar{\alpha}_n}}\epsilon\right). \tag{A9}$$

Given that $\mathcal{V}_I^{(n)}$ is available as input to the model, the parameterization can be chosen as

$$\mu_\theta(\mathcal{V}_I^{(n)}, n) = \frac{1}{\sqrt{\alpha_n}}\left(\mathcal{V}_I^{(n)} - \frac{\beta_n}{\sqrt{1-\bar{\alpha}_n}}\epsilon_\theta(\mathcal{V}_I^{(n)}, n)\right), \tag{A10}$$

where ϵ_θ is the predicted cumulative noise added to the current intermediate $\mathcal{V}_I^{(n)}$. This re-parameterization simplifies Equation (A8) as

$$L_{n-1} = \mathbb{E}_{q(\mathcal{V}_I^{(0)}),\epsilon}\left[\frac{\beta_n^2}{2\sigma_n^2\alpha_n(1-\bar{\alpha}_n)}\|\epsilon - \epsilon_\theta\left(\mathcal{V}_I^{(n)}(\mathcal{V}_I^{(0)}, \epsilon), n\right)\|^2\right]. \tag{A11}$$

In practice, we ignore the weighting term and optimize the following simplified variant of the variational bound:

$$L_{\text{simple}} = \mathbb{E}_{n, q(\mathcal{V}_I^{(0)}),\epsilon}\left[\|\epsilon - \epsilon_\theta\left(\mathcal{V}_I^{(n)}(\mathcal{V}_I^{(0)}, \epsilon), n\right)\|^2\right], \tag{A12}$$

where n is uniformly distributed between 1 and N. As demonstrated in [36], this approach improves the sample quality and simplifies implementation.

Appendix B. Implementation Details of DMs for Flow Field Reconstruction

During the training of both the RePaint and Palette models, we set the total number of diffusion steps $N = 2000$. A linear variance schedule is used, where the variances increase linearly from $\beta_1 = 10^{-6}$ to $\beta_N = 0.01$. Each model employs a U-Net architecture [62] characterized by two primary components: a downsampling stack and an upsampling stack, as shown in Figures 3a and 4a. The configuration of the upsampling stack mirrors that of the downsampling stack, creating a symmetrical structure. Each stack performs four steps of downsampling or upsampling, respectively. These steps consist of several residual blocks; some steps also include attention blocks. The two stacks are connected by an intermediate module, which consists of two residual blocks sandwiching an attention block [75]. Both DMs are trained with a batch size of 256 on four NVIDIA A100 GPUs for approximately 24 h.

For the RePaint model, the U-Net stages from the highest to lowest resolution (64×64 to 8×8) are configured with $[C, 2C, 3C, 4C]$ channels, where C equals 128. Three residual blocks are used at each stage. Attention mechanisms, specifically multi-head attention with four heads, are implemented after each residual block at the 16×16 and 8×8 resolution stages, and also within the intermediate module (Figure 3a). The model is trained using the AdamW optimizer [77] with a learning rate of 10^{-4} over 2×10^5 iterations. In addition, an exponential moving average (EMA) strategy with a decay rate of 0.999 is applied over the model parameters. During the reconstruction phase with a total of $N = 2000$ diffusion steps, the resampling technique is initiated at $n = 990$ and continues down to $n = 0$. In this approach, resampling is applied at every 10th step within this range, resulting in its application at 100 different points. At each point, the resampling involves a jump size of $j = 10$ and this procedure is iterated 9 times for each resampling point.

For the Palette model, the U-Net configuration uses $[C, 2C, 4C, 8C]$ channels across its stages, with C set to 64. Each stage has two residual blocks. Attention mechanisms are uniquely implemented in the intermediate module, with multi-head attention using 32 channels per head, as shown in Figure 4b. The model also incorporates a dropout rate of 0.2 for regularization. Following the approach in [39,53,54], we train Palette by conditioning the model on the continuous noise level $\bar{\alpha}$, instead of the discrete step index n. As a result, the loss function originally formulated in Equation (A12) is modified to

$$L_{\text{simple}} = \mathbb{E}_{\bar{\alpha}, q(\mathcal{V}_S, \mathcal{V}_G^{(0)}),\epsilon}\left[\|\epsilon - \epsilon_\theta\left(\mathcal{V}_S, \mathcal{V}_G^{(n)}(\mathcal{V}_G^{(0)}, \epsilon), \bar{\alpha}\right)\|^2\right]. \tag{A13}$$

In this process, we first uniformly sample n from 1 to N, and then uniformly sample $\bar{\alpha}$ in the range from $\bar{\alpha}_{n-1}$ to $\bar{\alpha}_n$. This approach allows Palette to use different noise schedules and total backward steps during inference. In fact during reconstruction, we use a total of 1000 backward steps with a linear noise schedule ranging from $\beta_1 = 10^{-4}$ to $\beta_N = 0.09$. The Adam optimizer [78] is used with a learning rate of 5×10^{-5}, training the model for approximately 720 to 750 epochs.

References

1. Le Dimet, F.X.; Talagrand, O. Variational algorithms for analysis and assimilation of meteorological observations: Theoretical aspects. *Tellus A Dyn. Meteorol. Oceanogr.* **1986**, *38*, 97–110. [CrossRef]
2. BEll, M.J.; Lefèbvre, M.; Le Traon, P.Y.; Smith, N.; Wilmer-Becker, K. GODAE: The global ocean data assimilation experiment. *Oceanography* **2009**, *22*, 14–21. [CrossRef]
3. Edwards, C.A.; Moore, A.M.; Hoteit, I.; Cornuelle, B.D. Regional ocean data assimilation. *Annu. Rev. Mar. Sci.* **2015**, *7*, 21–42. [CrossRef] [PubMed]
4. Wang, M.; Zaki, T.A. State estimation in turbulent channel flow from limited observations. *J. Fluid Mech.* **2021**, *917*, A9. [CrossRef]
5. Storer, B.A.; Buzzicotti, M.; Khatri, H.; Griffies, S.M.; Aluie, H. Global energy spectrum of the general oceanic circulation. *Nat. Commun.* **2022**, *13*, 5314. [CrossRef] [PubMed]
6. Shen, H.; Li, X.; Cheng, Q.; Zeng, C.; Yang, G.; Li, H.; Zhang, L. Missing information reconstruction of remote sensing data: A technical review. *IEEE Geosci. Remote Sens. Mag.* **2015**, *3*, 61–85. [CrossRef]
7. Zhang, Q.; Yuan, Q.; Zeng, C.; Li, X.; Wei, Y. Missing data reconstruction in remote sensing image with a unified spatial–temporal–spectral deep convolutional neural network. *IEEE Trans. Geosci. Remote Sens.* **2018**, *56*, 4274–4288. [CrossRef]
8. Merchant, C.J.; Embury, O.; Bulgin, C.E.; Block, T.; Corlett, G.K.; Fiedler, E.; Good, S.A.; Mittaz, J.; Rayner, N.A.; Berry, D.; et al. Satellite-based time-series of sea-surface temperature since 1981 for climate applications. *Sci. Data* **2019**, *6*, 223. [CrossRef]
9. Wang, Y.; Zhou, X.; Ao, Z.; Xiao, K.; Yan, C.; Xin, Q. Gap-Filling and Missing Information Recovery for Time Series of MODIS Data Using Deep Learning-Based Methods. *Remote Sens.* **2022**, *14*, 4692. [CrossRef]
10. Sammartino, M.; Buongiorno Nardelli, B.; Marullo, S.; Santoleri, R. An Artificial Neural Network to Infer the Mediterranean 3D Chlorophyll-a and Temperature Fields from Remote Sensing Observations. *Remote Sens.* **2020**, *12*, 4123. [CrossRef]
11. Courtier, P.; Thépaut, J.N.; Hollingsworth, A. A strategy for operational implementation of 4D-Var, using an incremental approach. *Q. J. R. Meteorol. Soc.* **1994**, *120*, 1367–1387.
12. Yuan, Z.; Wang, Y.; Wang, X.; Wang, J. Adjoint-based variational optimal mixed models for large-eddy simulation of turbulence. *arXiv* **2023**, arXiv:2301.08423.
13. Houtekamer, P.L.; Mitchell, H.L. A sequential ensemble Kalman filter for atmospheric data assimilation. *Mon. Weather Rev.* **2001**, *129*, 123–137. [CrossRef]
14. Mons, V.; Du, Y.; Zaki, T.A. Ensemble-variational assimilation of statistical data in large-eddy simulation. *Phys. Rev. Fluids* **2021**, *6*, 104607. [CrossRef]
15. Everson, R.; Sirovich, L. Karhunen–Loeve procedure for gappy data. *JOSA A* **1995**, *12*, 1657–1664. [CrossRef]
16. Borée, J. Extended proper orthogonal decomposition: A tool to analyse correlated events in turbulent flows. *Exp. Fluids* **2003**, *35*, 188–192. [CrossRef]
17. Venturi, D.; Karniadakis, G.E. Gappy data and reconstruction procedures for flow past a cylinder. *J. Fluid Mech.* **2004**, *519*, 315–336. [CrossRef]
18. Tinney, C.; Ukeiley, L.; Glauser, M.N. Low-dimensional characteristics of a transonic jet. Part 2. Estimate and far-field prediction. *J. Fluid Mech.* **2008**, *615*, 53–92. [CrossRef]
19. Discetti, S.; Bellani, G.; Örlü, R.; Serpieri, J.; Vila, C.S.; Raiola, M.; Zheng, X.; Mascotelli, L.; Talamelli, A.; Ianiro, A. Characterization of very-large-scale motions in high-Re pipe flows. *Exp. Therm. Fluid Sci.* **2019**, *104*, 1–8. [CrossRef]
20. Yildirim, B.; Chryssostomidis, C.; Karniadakis, G. Efficient sensor placement for ocean measurements using low-dimensional concepts. *Ocean Model.* **2009**, *27*, 160–173. [CrossRef]
21. Güemes, A.; Sanmiguel Vila, C.; Discetti, S. Super-resolution generative adversarial networks of randomly-seeded fields. *Nat. Mach. Intell.* **2022**, *4*, 1165–1173. [CrossRef]
22. Li, T.; Buzzicotti, M.; Biferale, L.; Bonaccorso, F. Generative adversarial networks to infer velocity components in rotating turbulent flows. *Eur. Phys. J. E* **2023**, *46*, 31. [CrossRef] [PubMed]
23. Li, T.; Buzzicotti, M.; Biferale, L.; Bonaccorso, F.; Chen, S.; Wan, M. Multi-scale reconstruction of turbulent rotating flows with proper orthogonal decomposition and generative adversarial networks. *J. Fluid Mech.* **2023**, *971*, A3. [CrossRef]
24. Buzzicotti, M. Data reconstruction for complex flows using AI: Recent progress, obstacles, and perspectives. *Europhys. Lett.* **2023**, *142*, 23001. [CrossRef]
25. Fukami, K.; Fukagata, K.; Taira, K. Super-resolution reconstruction of turbulent flows with machine learning. *J. Fluid Mech.* **2019**, *870*, 106–120. [CrossRef]
26. Liu, B.; Tang, J.; Huang, H.; Lu, X.Y. Deep learning methods for super-resolution reconstruction of turbulent flows. *Phys. Fluids* **2020**, *32*, 025105. [CrossRef]

27. Kim, H.; Kim, J.; Won, S.; Lee, C. Unsupervised deep learning for super-resolution reconstruction of turbulence. *J. Fluid Mech.* **2021**, *910*, A29. [CrossRef]
28. Buzzicotti, M.; Bonaccorso, F.; Di Leoni, P.C.; Biferale, L. Reconstruction of turbulent data with deep generative models for semantic inpainting from TURB-Rot database. *Phys. Rev. Fluids* **2021**, *6*, 050503. [CrossRef]
29. Guastoni, L.; Güemes, A.; Ianiro, A.; Discetti, S.; Schlatter, P.; Azizpour, H.; Vinuesa, R. Convolutional-network models to predict wall-bounded turbulence from wall quantities. *J. Fluid Mech.* **2021**, *928*, A27. [CrossRef]
30. Matsuo, M.; Nakamura, T.; Morimoto, M.; Fukami, K.; Fukagata, K. Supervised convolutional network for three-dimensional fluid data reconstruction from sectional flow fields with adaptive super-resolution assistance. *arXiv* **2021**, arXiv:2103.09020.
31. Yousif, M.Z.; Yu, L.; Hoyas, S.; Vinuesa, R.; Lim, H. A deep-learning approach for reconstructing 3D turbulent flows from 2D observation data. *Sci. Rep.* **2023**, *13*, 2529. [CrossRef] [PubMed]
32. Goodfellow, I.; Pouget-Abadie, J.; Mirza, M.; Xu, B.; Warde-Farley, D.; Ozair, S.; Courville, A.; Bengio, Y. Generative adversarial nets. *Adv. Neural Inf. Process. Syst.* **2014**, *27*.
33. Smith, R.C. *Uncertainty Quantification: Theory, Implementation, and Applications*; Siam: Philadelphia, PA, USA, 2013; Volume 12.
34. Hatanaka, Y.; Glaser, Y.; Galgon, G.; Torri, G.; Sadowski, P. Diffusion Models for High-Resolution Solar Forecasts. *arXiv* **2023**, arXiv:2302.00170.
35. Asahi, Y.; Hasegawa, Y.; Onodera, N.; Shimokawabe, T.; Shiba, H.; Idomura, Y. Generating observation guided ensembles for data assimilation with denoising diffusion probabilistic model. *arXiv* **2023**, arXiv:2308.06708.
36. Ho, J.; Jain, A.; Abbeel, P. Denoising diffusion probabilistic models. *Adv. Neural Inf. Process. Syst.* **2020**, *33*, 6840–6851.
37. Nichol, A.Q.; Dhariwal, P. Improved denoising diffusion probabilistic models. In Proceedings of the International Conference on Machine Learning, PMLR, Virtual, 18–24 July 2021; pp. 8162–8171.
38. Dhariwal, P.; Nichol, A. Diffusion models beat gans on image synthesis. *Adv. Neural Inf. Process. Syst.* **2021**, *34*, 8780–8794.
39. Chen, N.; Zhang, Y.; Zen, H.; Weiss, R.J.; Norouzi, M.; Chan, W. Wavegrad: Estimating gradients for waveform generation. *arXiv* **2020**, arXiv:2009.00713.
40. Brown, T.; Mann, B.; Ryder, N.; Subbiah, M.; Kaplan, J.D.; Dhariwal, P.; Neelakantan, A.; Shyam, P.; Sastry, G.; Askell, A.; et al. Language models are few-shot learners. *Adv. Neural Inf. Process. Syst.* **2020**, *33*, 1877–1901.
41. Shu, D.; Li, Z.; Farimani, A.B. A physics-informed diffusion model for high-fidelity flow field reconstruction. *J. Comput. Phys.* **2023**, *478*, 111972. [CrossRef]
42. Yang, G.; Sommer, S. A Denoising Diffusion Model for Fluid Field Prediction. *arXiv* **2023**, arXiv:2301.11661.
43. Li, T.; Biferale, L.; Bonaccorso, F.; Scarpolini, M.A.; Buzzicotti, M. Synthetic lagrangian turbulence by generative diffusion models. *arXiv* **2023**, arXiv:2307.08529.
44. Pouquet, A.; Sen, A.; Rosenberg, D.; Mininni, P.D.; Baerenzung, J. Inverse cascades in turbulence and the case of rotating flows. *Phys. Scr.* **2013**, *2013*, 014032. [CrossRef]
45. Oks, D.; Mininni, P.D.; Marino, R.; Pouquet, A. Inverse cascades and resonant triads in rotating and stratified turbulence. *Phys. Fluids* **2017**, *29*, 111109. [CrossRef]
46. Alexakis, A.; Biferale, L. Cascades and transitions in turbulent flows. *Phys. Rep.* **2018**, *767*, 1–101. [CrossRef]
47. Buzzicotti, M.; Aluie, H.; Biferale, L.; Linkmann, M. Energy transfer in turbulence under rotation. *Phys. Rev. Fluids* **2018**, *3*, 034802. [CrossRef]
48. Li, T.; Wan, M.; Wang, J.; Chen, S. Flow structures and kinetic-potential exchange in forced rotating stratified turbulence. *Phys. Rev. Fluids* **2020**, *5*, 014802. [CrossRef]
49. Herring, J.R. Statistical theory of quasi-geostrophic turbulence. *J. Atmos. Sci.* **1980**, *37*, 969–977. [CrossRef]
50. Herring, J. The inverse cascade range of quasi-geostrophic turbulence. *Meteorol. Atmos. Phys.* **1988**, *38*, 106–115. [CrossRef]
51. Herring, J.R.; Métais, O. Numerical experiments in forced stably stratified turbulence. *J. Fluid Mech.* **1989**, *202*, 97–115. [CrossRef]
52. Lugmayr, A.; Danelljan, M.; Romero, A.; Yu, F.; Timofte, R.; Van Gool, L. Repaint: Inpainting using denoising diffusion probabilistic models. In Proceedings of the IEEE/CVF Conference on Computer Vision and Pattern Recognition, New Orleans, LA, USA, 18–24 June 2022; pp. 11461–11471.
53. Saharia, C.; Ho, J.; Chan, W.; Salimans, T.; Fleet, D.J.; Norouzi, M. Image super-resolution via iterative refinement. *IEEE Trans. Pattern Anal. Mach. Intell.* **2022**, *45*, 4713–4726. [CrossRef]
54. Saharia, C.; Chan, W.; Chang, H.; Lee, C.; Ho, J.; Salimans, T.; Fleet, D.; Norouzi, M. Palette: Image-to-image diffusion models. In Proceedings of the ACM SIGGRAPH 2022 Conference Proceedings, Vancouver, BC, Canada, 7–11 August 2022; pp. 1–10.
55. Cambon, C.; Mansour, N.N.; Goderferd, F. Energy transfer in rotating turbulence. *J. Fluid Mech.* **1997**, *337*, 303–332. [CrossRef]
56. Goderferd, F.; Moisy, F. Structure and Dynamics of Rotating Turbulence: A Review of Recent Experimental and Numerical Results. *Appl. Mech. Rev.* **2015**, *67*, 030802. [CrossRef]
57. McWilliams, J.C. *Fundamentals of Geophysical Fluid Dynamics*; Cambridge University Press: Cambridge, UK, 2006.
58. Brandenburg, A.; Svedin, A.; Vasil, G.M. Turbulent diffusion with rotation or magnetic fields. *Mon. Not. R. Astron. Soc.* **2009**, *395*, 1599. [CrossRef]
59. Biferale, L.; Bonaccorso, F.; Buzzicotti, M.; Di Leoni, P.C. TURB-Rot. A large database of 3D and 2D snapshots from turbulent rotating flows. *arXiv* **2020**, arXiv:2006.07469.
60. Sawford, B. Reynolds number effects in Lagrangian stochastic models of turbulent dispersion. *Phys. Fluids A Fluid Dyn.* **1991**, *3*, 1577–1586. [CrossRef]

61. Buzzicotti, M.; Bhatnagar, A.; Biferale, L.; Lanotte, A.S.; Ray, S.S. Lagrangian statistics for Navier–Stokes turbulence under Fourier-mode reduction: Fractal and homogeneous decimations. *New J. Phys.* **2016**, *18*, 113047. [CrossRef]
62. Ronneberger, O.; Fischer, P.; Brox, T. U-net: Convolutional networks for biomedical image segmentation. In *Medical Image Computing and Computer-Assisted Intervention–MICCAI 2015, Proceedings of the 18th International Conference, Munich, Germany, 5–9 October 2015;* Proceedings, Part III 18; Springer: Berlin/Heidelberg, Germany, 2015; pp. 234–241.
63. Feller, W. On the theory of stochastic processes, with particular reference to applications. In *Selected Papers I;* Springer: Berlin/Heidelberg, Germany, 2015; pp. 769–798.
64. Sohl-Dickstein, J.; Weiss, E.; Maheswaranathan, N.; Ganguli, S. Deep unsupervised learning using nonequilibrium thermodynamics. In Proceedings of the International Conference on Machine Learning, PMLR, Lille, France, 7–9 July 2015; pp. 2256–2265.
65. Richardson, E.; Alaluf, Y.; Patashnik, O.; Nitzan, Y.; Azar, Y.; Shapiro, S.; Cohen-Or, D. Encoding in style: A stylegan encoder for image-to-image translation. In Proceedings of the IEEE/CVF Conference on Computer Vision and Pattern Recognition, Virtual, 19–25 June 2021; pp. 2287–2296.
66. Chung, H.; Kim, J.; Mccann, M.T.; Klasky, M.L.; Ye, J.C. Diffusion posterior sampling for general noisy inverse problems. *arXiv* **2022**, arXiv:2209.14687.
67. Zhang, G.; Ji, J.; Zhang, Y.; Yu, M.; Jaakkola, T.S.; Chang, S. Towards Coherent Image Inpainting Using Denoising Diffusion Implicit Models. In Proceedings of the 40th International Conference on Machine Learning, Honolulu, HI, USA, 23–29 July 2023.
68. Yu, J.; Lin, Z.; Yang, J.; Shen, X.; Lu, X.; Huang, T.S. Free-form image inpainting with gated convolution. In Proceedings of the IEEE/CVF International Conference on Computer Vision, Seoul, Republic of Korea, 27 October–2 November 2019; pp. 4471–4480.
69. Song, J.; Meng, C.; Ermon, S. Denoising diffusion implicit models. *arXiv* **2020**, arXiv:2010.02502.
70. Salimans, T.; Ho, J. Progressive distillation for fast sampling of diffusion models. *arXiv* **2022**, arXiv:2202.00512.
71. Stevens, R.J.; Meneveau, C. Flow structure and turbulence in wind farms. *Annu. Rev. Fluid Mech.* **2017**, *49*, 311–339. [CrossRef]
72. Gharaati, M.; Xiao, S.; Wei, N.J.; Martínez-Tossas, L.A.; Dabiri, J.O.; Yang, D. Large-eddy simulation of helical-and straight-bladed vertical-axis wind turbines in boundary layer turbulence. *J. Renew. Sustain. Energy* **2022**, *14*, 053301. [CrossRef]
73. Potisomporn, P.; Vogel, C.R. Spatial and temporal variability characteristics of offshore wind energy in the United Kingdom. *Wind Energy* **2022**, *25*, 537–552. [CrossRef]
74. Hochreiter, S.; Schmidhuber, J. Long short-term memory. *Neural Comput.* **1997**, *9*, 1735–1780. [CrossRef] [PubMed]
75. Vaswani, A.; Shazeer, N.; Parmar, N.; Uszkoreit, J.; Jones, L.; Gomez, A.N.; Kaiser, Ł.; Polosukhin, I. Attention is all you need. In *Advances in Neural Information Processing Systems;* Curran Associates, Inc.: Red Hook, NY, USA, 2017; Volume 30.
76. Weng, L. What Are Diffusion Models? 2021. Available online: lilianweng.github.io (accessed on 15 November 2023).
77. Loshchilov, I.; Hutter, F. Decoupled weight decay regularization. *arXiv* **2017**, arXiv:1711.05101.
78. Kingma, D.P.; Ba, J. Adam: A method for stochastic optimization. *arXiv* **2014**, arXiv:1412.6980.

atmosphere

MDPI

Article

2D and 3D Properties of Stably Stratified Turbulence

Yoshifumi Kimura [1,*] and Peter P. Sullivan [2]

1 Graduate School of Mathematics, Nagoya University, Furo-cho, Chikusa-ku, Nagoya 464-8602, Japan
2 Meoscale and Microscale Meteorology, National Center for Atmospheric Research, Boulder, CO 80301, USA; pps@ucar.edu
* Correspondence: kimura@math.nagoya-u.ac.jp; Tel.: +81-52-789-2819

Abstract: Through interactions between the modes of "waves" and "vortices", stably stratified turbulence exhibits characteristic features both in 2D and 3D. Using DNS of the Navier–Stokes equations coupled to an equation for temperature (the Bousinessq system), we investigate dynamical properties of stably stratified turbulence. After reviewing the importance of three characteristic length scales and their relations in stably stratified turbulence, we present numerical results showing how structures and spectra develop with the growth of the length scales. It is demonstrated that the temperature fluctuations make sharp fronts vertically which result in non-symmetric PDFs of the vertical derivative $\partial_z\theta$.

Keywords: stratified turbulence; buoyancy scale; Ozmidov scale; Kolmogorov scale; Craya-Herring decomposition; temperature front; cliff–ramp structures

1. Introduction

The first time YK met Jack was at the IUTAM Symposium on "Turbulence and Chaotic Phenomena in Fluids" which professor T. Tatsumi organized in Kyoto in 1983. That symposium was really a big event, and YK also meet many researchers whom he would re-encounter later in the US and elsewhere, for example, Annick Pouquet also attended the symposium. The next time he crossed paths with Jack was at the APS meeting at SUNY, Buffalo in 1988 where Jack organized a specialty conference on Multipoint Turbulent Closure. The speakers of the conference were Bill George, Marcel Lesieur, Bill Dannevik, Chuck Leith, Akira Yoshizawa, Jean-Pierre Bertoglio, and Jack. At that time YK had just begun his career as a postdoc at UCSD and subsequently at Los Alamos National Laboratory (LANL). At Los Alamos, he was affiliated with the Center for Nonlinear Studies (CNLS), and worked with Bob Kraichnan. (His official host at CNLS was Darryl Holm since Bob Kraichnan was by then a consultant at LANL). After Los Alamos, YK was awarded an Advanced Study Program (ASP) postdoc scholarship at NCAR and came to Boulder in 1991, and started full time collaborations with Jack.

The first project we conducted was "diffusion in stably stratified turbulence". Using decaying turbulence simulations without external forcing in a 2π-periodic box, we showed that randomly scattered pancake vortices are characteristic vortical structures in homogeneous turbulence [1]. This paper also considered the effect of stratification on eddy diffusion, which was stimulated by previous papers [2–6]. Then, we included a narrow band forcing term in the momentum equation in Fourier space to produce stationary turbulence, and investigated the power transition of the horizontal kinetic energy spectra in stably stratified turbulence [7]. In that paper, we confirmed the atmospheric observations of Nastrom and Gage (1985) that the horizontal spectrum shows a transition from $k_\perp^{-3}$ at the large scales to $k_\perp^{-5/3}$ at small scales where $k_\perp$ is the horizontal wave number [8].

Adding a narrow band forcing to the momentum equation is a simple but an artificial method to produce stationary turbulence, and we need to have a specific experimental design to choose a proper type of forcing. So far, various kinds of forcing are used for

Citation: Kimura, Y.; Sullivan, P.P. 2D and 3D Properties of Stably Stratified Turbulence. *Atmosphere* **2024**, *15*, 82. https://doi.org/10.3390/atmos15010082

Academic Editor: Doris Folini

Received: 20 November 2023
Revised: 30 December 2023
Accepted: 31 December 2023
Published: 9 January 2024

different purposes by many researchers [9–14]. The purpose of the present paper is to investigate how two and three-dimensional properties of stably stratified turbulence are developed by the interaction of "waves" and "vortices" in the turbulence. For this purpose, we first decouple the temperature equation from the momentum equation, and develop two-dimensional turbulence in the x, y coordinates which is uniform in the z coordinate in a 3D periodic cubic box. After two-dimensional turbulence is fully developed, the temperature equation is turned on, and we observe how the two-dimensional turbulence adjusts to the structures in stably stratified turbulence.

Here we explain the system of equations of the velocity and the temperature fluctuations and the method of the simulation [7]. Under the Boussinesq approximation, the equation of motion of stably stratified flow for the velocity vector, $\mathbf{u} = (u, v, w)$ and the temperature fluctuation from the linear (stable) mean, θ are written

$$\frac{\partial \mathbf{u}}{\partial t} + (\mathbf{u} \cdot \boldsymbol{\nabla})\mathbf{u} = -\boldsymbol{\nabla} p + \nu \nabla^2 \mathbf{u} + \theta \hat{\mathbf{z}} + \mathbf{f} \quad (1)$$

$$\frac{\partial \theta}{\partial t} + (\mathbf{u} \cdot \boldsymbol{\nabla})\theta = \kappa \nabla^2 \theta - N^2 w \quad (2)$$

$$\boldsymbol{\nabla} \cdot \mathbf{u} = 0 \quad (3)$$

where p is pressure, $\hat{\mathbf{z}}$ is the unit vector in the vertical direction, $\mathbf{f}$ is an external force, and ν and κ are kinematic viscosity and temperature diffusivity, respectively. N is the Brunt–Väisälä frequency, defined as $N = \sqrt{g\alpha(\partial \overline{T}/\partial z)/T_0}$, where α is the coefficient of thermal expansion. Here we assume $\overline{T}(z)$ satisfies $d\overline{T}/dz \equiv -N^2$. Periodic boundary conditions are used in the calculation of $\mathbf{u}$ and θ by assuming a linear ambient density profile. We set the Prandtl number $\nu/\kappa = 1$. For the time advancement, the low-storage third-order Runge–Kutta method is used [15]. The simulation code is decomposed into horizontal slabs and parallelized using the Message Passing Interface (MPI) system.

2. Vortical Structures in Stratified Turbulence

In stably stratified flows, waves produced by buoyancy and vortices coexist with turbulence, and their interaction produces many length scales. In general, at large scales the effect of buoyancy is assumed to be dominant, and by regarding the Brunt–Väisälä frequency as a frequency of stable oscillation, we define the buoyancy scale as L_B $(= U/N)$, where U is a characteristic horizontal scale.

At smaller scales, the turbulence characteristics become clearer. If we focus on the energy cascade, the energy dissipation rate ε is given by the velocity and the length scales, u and l, of the Kolmogorov inertial range as $\varepsilon \sim u^3/l$. If $l \sim L_B$ and $u \sim U$ are substituted in the dissipation expression, and U is eliminated, we obtain the Ozmidov scale L_O $(= \sqrt{\varepsilon/N^3})$ [16,17]. The Ozmidov scale is at the boundary between the buoyancy and the inertial range; scales smaller than L_O are unstable and tend to overturn. At even smaller scales, the Kolmogorov scale L_K $(= (\nu^3/\varepsilon)^{1/4})$ appears, and eddies smaller than L_K dissipate the kinetic energy by viscosity.

Thus we have three scales, L_B, L_O, L_K, as the characteristic length scales for stably stratified turbulence, and combinations of these describe different features of turbulence. The ratio of L_O and L_K is

$$\frac{L_O}{L_K} = \frac{(\varepsilon/N^3)^{1/2}}{\nu^3/\varepsilon)^{1/4}} = R_B^{3/4} \quad (4)$$

where R_B $\left(= \varepsilon/(\nu N^2)\right)$ is called the buoyancy Reynolds number [7,14,18]. When $R_B \gg 1$, the effect of viscosity is confined to scales much smaller than L_O, with the scale separation expected at the boundary L_O.

Figure 1 is the celebrated mesoscale horizontal spectra of Nastrom and Gage collected from aircraft flights [8] of the zonal (east–west) wind, the meridional (north–south) wind, and the potential temperature. In this plot, the abscissa denotes the magnitude of the horizontal wave number $k_\perp = \sqrt{x_x^2 + k_y^2}$.

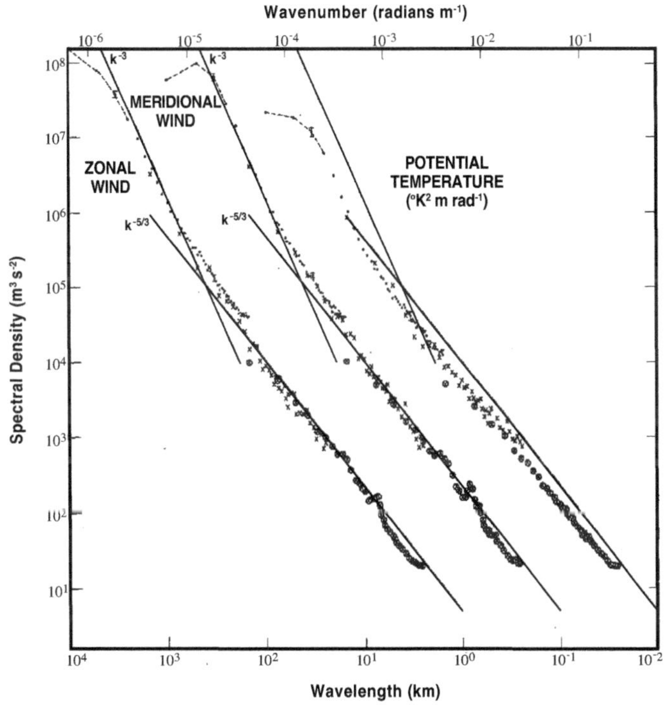

Figure 1. Nastrom-Gage spectrum ([8]) From the left, the zonal (east-west) wind, the meridional (north-south) wind, and the potential temperature. Each has the functional form of $k_\perp^\alpha$, and the power α is between -2 and -3 at large scales and near $-5/3$ at small scales.

Each spectrum has the functional form of $k_\perp^\alpha$, and the power α is between -2 and -3 at large scales and is close to $-5/3$ at small scales. Historically, this $-5/3$ range was first suspected to be the result of a 2D inverse energy cascade, but later was corroborated by analysis of the 3rd order horizontal velocity structure functions; the direction of energy cascade is regular, i.e., large to small scales [19,20]. The boundary wavenumber between the different powers corresponds to $k_O = 2\pi/L_O$, and for small scales below L_O the energy cascade is enhanced by eddy overturning.

When $R_B \gg 1$, since L_O is the smallest scale in the buoyancy dominated region, it can be used as a metric to characterize other length scales. If we assume that the horizontal length scale L_h is associated with the turbulent energy cascade process it can be expressed in terms of the characteristic horizontal velocity U and the rate of energy dissipation ε as $L_h \sim U^3/\varepsilon$ [21]. We further assume that the vertical length scale L_v corresponds to the buoyancy scale $L_B = U/N$. Then, by taking the ratio with L_O, and by expressing the results by the horizontal Froude number, $Fr_h \equiv U/(NL_h) = \varepsilon/(NU^2)$, we have the following relations:

$$\frac{L_h}{L_O} = Fr_h^{-3/2}, \qquad \frac{L_v}{L_O} = Fr_h^{-1/2}, \tag{5}$$

which indicates that $L_h/L_O \sim (L_v/L_O)^3$ holds in the buoyancy dominated region [14].

It was shown that randomly scattered horizontal pancake vortices are the characteristic vortical structures in stably stratified turbulence [7]. Based on the above relations, the ratio of horizontal and vertical length scales of the vortices is given as $L_h/L_O \sim (L_v/L_O)^3$, $L_h/L_v \sim Fr_h^{-1}$.

3. Development of 2dim and 3dim Energy and Spectrum

3.1. Craya–Herring Decomposition

The basic idea of the Craya–Herring decomposition is the following [22]. The Fourier transform of the velocity incompressibility condition provides $\mathbf{k} \cdot \tilde{\mathbf{u}}(\mathbf{k}) = 0$, where $\mathbf{k} = (k_x, k_y, k_z)^T$ is the wavenumber vector, directly shows that the Fourier transform of the velocity vector, $\tilde{\mathbf{u}}(\mathbf{k})$ spans two independent basis vectors $\mathbf{e}_1(\mathbf{k})$, $\mathbf{e}_2(\mathbf{k})$ in the plane perpendicular to $\mathbf{k}$ as $\tilde{\mathbf{u}}(\mathbf{k}) = \phi_1\mathbf{e}_1(\mathbf{k}) + \phi_2\mathbf{e}_2(\mathbf{k})$. As the orthonomal basis vectors, the following expressions follow:

$$\mathbf{e}_1(\mathbf{k}) = \frac{\mathbf{k} \times \hat{\mathbf{z}}}{||\mathbf{k} \times \hat{\mathbf{z}}||} = \frac{1}{\sqrt{k_x^2 + k_y^2}} \begin{pmatrix} k_y \\ -k_x \\ 0 \end{pmatrix}, \tag{6}$$

$$\mathbf{e}_2(\mathbf{k}) = \frac{\mathbf{k} \times (\mathbf{k} \times \hat{\mathbf{z}})}{||\mathbf{k} \times (\mathbf{k} \times \hat{\mathbf{z}})||} = \frac{1}{\sqrt{k_x^2 + k_y^2 + k_z^2}\sqrt{k_x^2 + k_y^2}} \begin{pmatrix} k_z k_x \\ k_z k_x \\ -(k_x^2 + k_y^2) \end{pmatrix}, \tag{7}$$

$$\mathbf{e}_3(\mathbf{k}) = \frac{\mathbf{k}}{||\mathbf{k}||} = \frac{1}{\sqrt{k_x^2 + k_y^2 + k_z^2}} \begin{pmatrix} k_x \\ k_y \\ k_z \end{pmatrix}. \tag{8}$$

Based on these expressions, ϕ_1 and ϕ_2 can be calculated explicitly as

$$\phi_1 = \tilde{\mathbf{u}} \cdot \mathbf{e}_1 = \frac{1}{\sqrt{k_x^2 + k_y^2}}(k_y\tilde{u} - k_x\tilde{v}) = \frac{i}{\sqrt{k_x^2 + k_y^2}}\tilde{\omega}_z \tag{9}$$

$$\phi_2 = \tilde{\mathbf{u}} \cdot \mathbf{e}_2 = -\frac{\sqrt{k_x^2 + k_y^2 + k_z^2}}{\sqrt{k_x^2 + k_y^2}}\tilde{w}, \tag{10}$$

where $\tilde{\omega}_z$ is the Fourier transform of the z-component of vorticity. These expressions define the energy densities $\Phi_1(\mathbf{k})$ and $\Phi_2(\mathbf{k})$ respectively as

$$\Phi_1(\mathbf{k}) = \frac{1}{2}|\phi_1|^2 = \frac{1}{2}\frac{|k_y\tilde{u} - k_x\tilde{v}|^2}{k_x^2 + k_y^2}, \quad \Phi_2(\mathbf{k}) = \frac{1}{2}|\phi_2|^2 = \frac{1}{2}\frac{(k_x^2 + k_y^2 + k_z^2)|\tilde{w}|^2}{k_x^2 + k_y^2}. \tag{11}$$

From the above equations; we see that ϕ_1 is related to the horizontal vorticity while ϕ_2 is related to the vertical velocity. Notice that the vertical velocity is coupled with the temperature fluctuation in an oscillatory manner, such as waves, we see that ϕ_1 is horizontal and vortical and ϕ_2 is vertical and wavy [7].

3.2. Numerical Results

We Fourier transform Equations (1)–(3) and integrate them pseudo-spectrally in a 2π periodic box in the x, y, z dimensions. Here, the results with 1024^3 grid points are presented. As the external force, we add an integrated white-noise to the horizontal velocity components, $\tilde{u}(k_x, k_y, k_z)$ and $\tilde{v}(k_x, k_y, k_z)$ in a narrow waveband at the horizontal wave number $k_f = \sqrt{k_x^2 + k_y^2} = 4,\ 10$.

$$\tilde{\mathbf{f}} = (f_x(k_x, k_y, 0), f_y(k_x, k_y, 0), 0)^T \tag{12}$$

Starting with zero kinetic energy, we first decouple the equation for the temperature fluctuation from the momentum equations for some time from $t = 0$, thus allowing the forcing to build two-dimensional turbulence uniformly in all horizontal layers in the z

direction; $\phi_2 = 0$ is imposed in this first stage. After two-dimensional turbulence is developed, the temperature equation is turned on to see how the energy in ϕ_2 grows.

Figure 2 shows the time development of the energy in ϕ_1 and ϕ_2, $\frac{1}{2}|\phi_1|^2$, and $\frac{1}{2}|\phi_2|^2$, for four different cases, $(N^2, k_f) = (10,4), (100,4), (10,10), (100,10)$, respectively. For all cases, the development of the features are summarized as follows:

1. The ϕ_1 energy grows rapidly in a fraction of time after $t = 0$, and continues to increase gradually. Then the equation for the temperature fluctuations is turned on. After a short period of time, the ϕ_2 energy begins to grow dramatically at an exponential growth, $e^{\alpha t}$. The growth rate α is larger for $k_f = 10$ than $k_f = 4$. As the energy in ϕ_2 grows comparable to ϕ_1, they begin to interact and the speed and the growth of ϕ_2 decreases (and slightly that of ϕ_1 as well).
2. As the interaction of ϕ_1 and ϕ_2 grows, and around the time when ϕ_2 energy becomes almost 10% of that of ϕ_1, the ϕ_1 energy reaches its maximum value, and then begins to decrease. Slightly after that time, ϕ_2 reaches its maximum value, followed by a gradual decrease in ϕ_1 and ϕ_2.
3. At long times, the interaction between ϕ_1 and ϕ_2 becomes almost stationary and produces small oscillations. The fluctuations in the forcing seems to be the cause of these oscillations. ϕ_1 and ϕ_2 oscillate almost in phase, which is presumably due to the incompressibility that the Boussinesq approximation assumes.

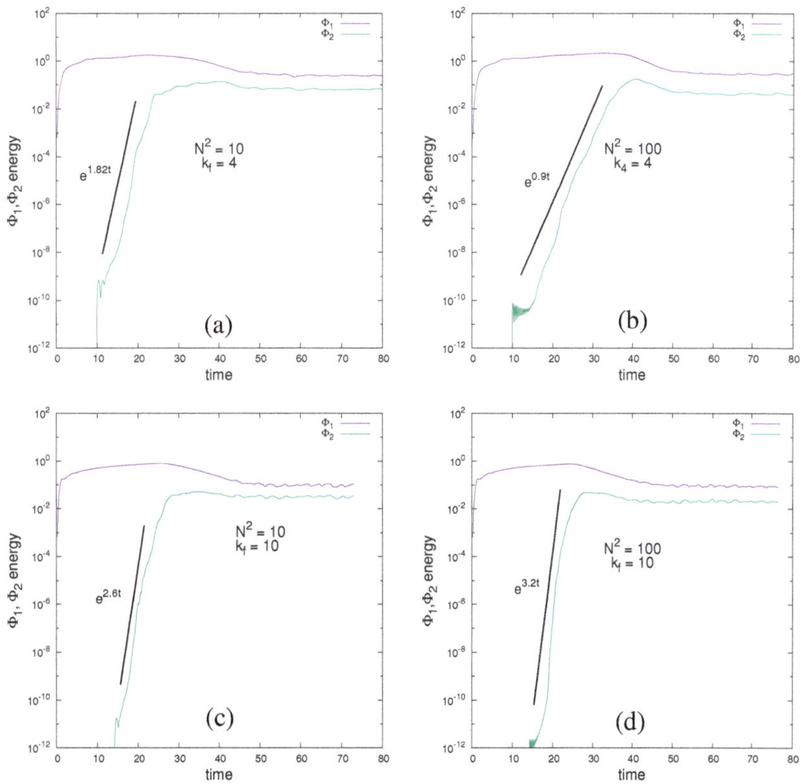

Figure 2. Development of ϕ_1 and ϕ_2 energy as a function of time: (**a**) $N^2 = 10, k_f = 4$, (**b**) $N^2 = 100$, $k_f = 4$, (**c**) $N^2 = 10, k_f = 10$, and (**d**) $N^2 = 100, k_f = 10$.

Next, we examine the development of statistical quantities in Figure 2d; the case $N^2 = 100, k_f = 10$ is an example. Figure 3 spans three time slots marked by orange solid lines, (i) $(0 \leq t \leq 22.79)$, (ii) $(22.84 \leq t \leq 45.04)$, and (iii) $(45.12 \leq t \leq 76.43)$, corresponding to the above three characteristic features (1)∼(3). Also in Figure 3, the dashed lines show the times $(t_1 \sim t_5)$ at which the flow features are analyzed in the later analysis.

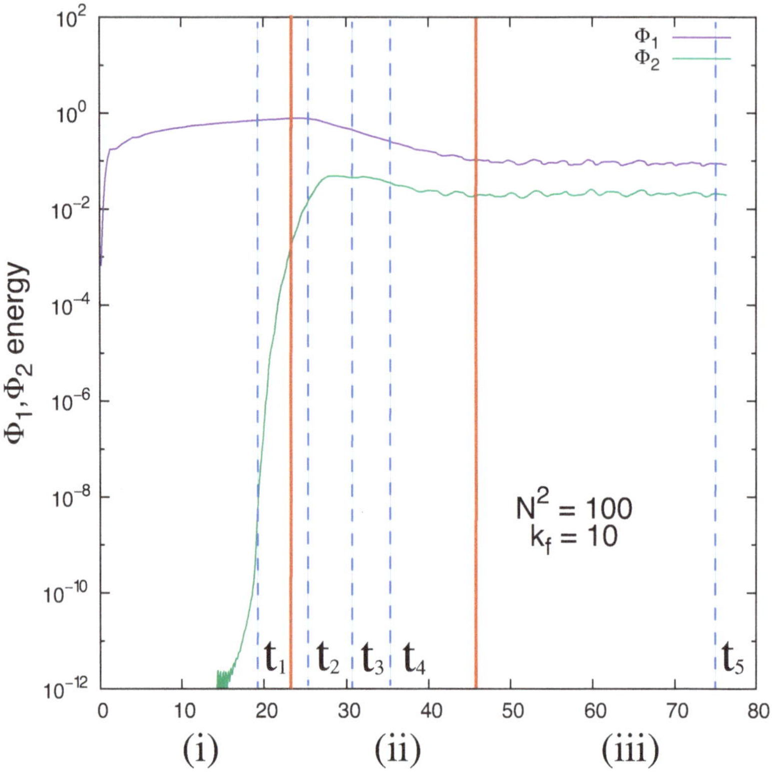

Figure 3. Division of the time intervals along Figure 2d: (i) $(0 \leq t \leq 22.79)$, (ii) $(22.84 \leq t \leq 45.04)$, and (iii) $(45.12 \leq t \leq 76.43)$. $t_1 \sim t_5$ (the boundary of these intervals are given as red lines) are the times at which statistics are shown in Figures 6 and 7.

Figure 4i shows the development of the $\phi_1(k_\perp)$ spectra during the (i) time slot in Figure 3. This time interval corresponds to the process of forced two-dimensional turbulence with $-5/3$ spectra at scales smaller than the forcing wavenumbers; this corresponds to an inverse cascade of energy. At larger scales, spectra with $-3 \sim -4$ slopes appear as typical spectra indicating a forward cascade of kinetic energy.

After the equation of the temperature fluctuations is turned on and ϕ_2 grows rapidly, the large scale part (small wavenumber part) of the ϕ_1 spectra tends to be flat as the small scale part (large wavenumber part) is elevated to be shallow spectra with an exponent approaching -1 (Figure 4ii). It seems obvious that the large scale energy is transferred significantly because of the interaction between ϕ_1 and ϕ_2.

When the interaction between ϕ_1 and ϕ_2 becomes almost stationary, the ϕ_1 spectra shows small changes, but tends to show a clear $-5/3$ slope at the high $k_\perp$ regions; at the same time the low $k_\perp$ spectra become flatter (Figure 4iii).

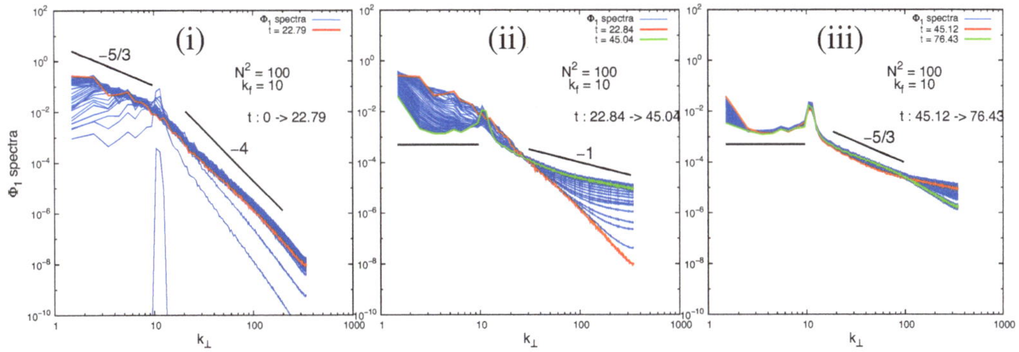

Figure 4. Development of ϕ_1 spectra in the time interval (**i–iii**) in Figure 3. Red and green colors correspond to the initial and the last time in each interval.

To understand the change in behavior in the ϕ_1 spectra, Figure 5 shows the change in the buoyancy, the Ozmidov and the Kolmogorov scales, explained in the previous section. A noteworthy phenomenon is that the Ozmidov, and the Kolmogorov scales cross around $t = 23$ at which the buoyancy scale reaches its maximum. In the previous figures, this time corresponds to the time when the 2D state begins to change to 3D by the interaction between waves and vortices. Another interesting point is that the development of the buoyancy scale follows a similar path as ϕ_1 in Figures 2 and 3. Throughout the simulation, the buoyancy scale is the largest scale in the flow. The similarity between the buoyancy scale and ϕ_1 implies that the horizontal flow continues to dominant, but we have to be a little careful to reach this conclusion—an explanation follows.

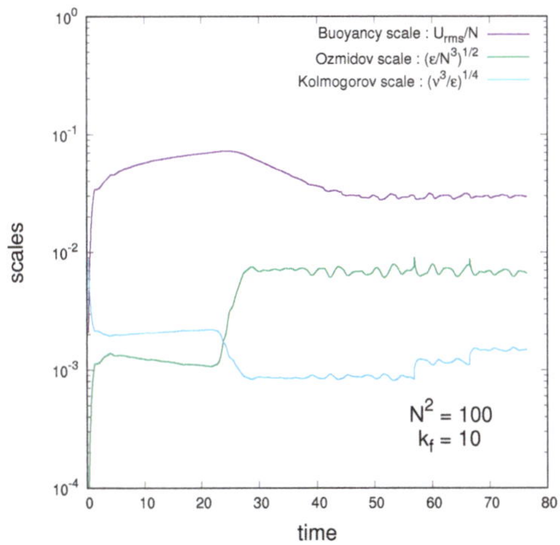

Figure 5. Growth of the tree scales, buoyancy, Ozmidov, and Kolmogorov scales for $N^2 = 100$, $k_f = 10$.

Figure 6 shows the development of the horizontal average of the variances, $\langle u^2 \rangle/2$, $\langle v^2 \rangle/2$, $\langle w^2 \rangle/2$, and the total energy $(\langle u^2 \rangle + \langle v^2 \rangle + \langle w^2 \rangle)/2$ for varying z. Results are shown at $t_1 \sim t_4$ in Figure 3 (upper row), and the horizontal average of the horizontal energy, $(\langle u^2 \rangle + \langle v^2 \rangle)/2$, and monitored also at $t_1 \sim t_4$ are the horizontal average of the square of the vertical derivative of temperature fluctuations, $\langle (\partial\theta/\partial z)^2 \rangle$ (lower row). The following observations can be made:

(1) At t_1, the energy in all directions is uniform in z. The vertical gradient of temperature fluctuations $\langle(\partial\theta/\partial z)^2\rangle$ shows small oscillation due to the initial small growth of ϕ_2. The total flow is mainly two-dimensional at this time.

(2) At t_2, fluctuations in the horizontal energy develop, but the vertical energy is still very close to zero. The oscillation of the energy is not uniform, and small deviations in frequency and amplitude are seen. However, the structures in the x and y directions are similar. More striking is the vertical gradient of the temperature fluctuations that shows a few periodic humps with the tops showing stronger oscillations.

(3) At $t = t_3$, the energy in the x and y directions, $\langle u^2\rangle/2$, $\langle v^2\rangle/2$, are in a similar range of magnitude with sharp peaks often at different locations in z. This implies that the velocity field is directionally intermittent, with the kinetic energy sometimes highly concentrated in one direction. Without looking at the coherence of the energy, we cannot readily identify the flow structure, but one possibility is there may at times be uni-directional wind flows.

(4) At $t = t_4$, the intermittency in both the energy and the temperature derivative distributions grows. In particular, the peaks of the latter become sharply outstanding from the signals of low but rapid oscillations.

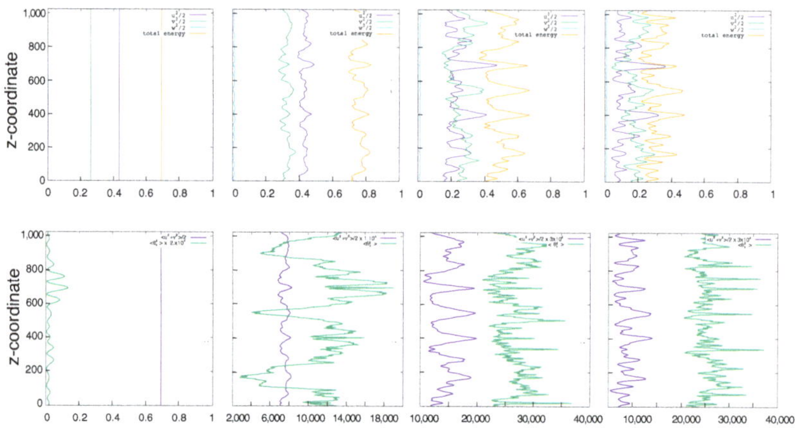

Figure 6. Upper row: The horizontal average of the energy, $\langle u^2\rangle/2$ (violet), $\langle v^2\rangle/2$ (green), $\langle w^2\rangle/2$ (blue) and the total energy $(\langle u^2\rangle + \langle v^2\rangle + \langle w^2\rangle)/2$ (orange) as a function of z. **Lower** row: The horizontal average of the horizontal energy, $(\langle u^2\rangle + \langle v^2\rangle)/2$ (violet) and the horizontal average of the square of the vertical derivative of temperature fluctuations, $\langle(\partial\theta/\partial z)^2\rangle$ (green). All statistics are monitored at $t_1 \sim t_4$ in Figure 3.

The above monitoring times, $t_1 \sim t_4$, are all in time slot (ii) in Figures 3 and 4. The characteristic strong intermittent growth during this time slot is attributed to the phenomenon that the 2D energy in the large scales is destroyed with the energy transferred into small scales producing spectra with a nearly -1 slope, much shallower than the Kolmogorov's $-5/3$ spectrum.

The dramatic growth in intermittency in time interval (ii) in Figures 3 and 4 is moderated in time interval (iii). In latter time range, the change in the large scale spectrum and the subsequent energy outflow towards small scales weakens (Figure 4iii), and the small scale spectrum has fallen to the Kolmogorov $-5/3$ spectrum. The end part of the spectra at the smallest scales has become steeper than the $-5/3$ power, showing a hint of dissipation spectrum. This observation has support if we look at the growth of the Kolmogorov scale at late time in Figure 5. It is interesting that the Kolmogorov scale increases in a stepped manner at $t \sim 56$ and $t \sim 67$.

The horizontal average of the horizontal energy, $(\langle u^2\rangle + \langle v^2\rangle)/2$, and the horizontal average of the square of the vertical derivative of temperature fluctuations, $\langle(\partial\theta/\partial z)^2\rangle$, at the time t_5 in Figure 3 are shown in Figure 7. After a long time, the oscillatory interaction of ϕ_1 and ϕ_2 is presumably near equilibrium. The intervals of the peaks are almost equally periodic, which means that the high intermittency of the temperature gradient is softened. The characteristic frequency of the peaks can be compared with the value of $2\pi U/N$ in Figure 7. However, the frequency should be related with that of the forcing, and the true reason for the interval is still an open question.

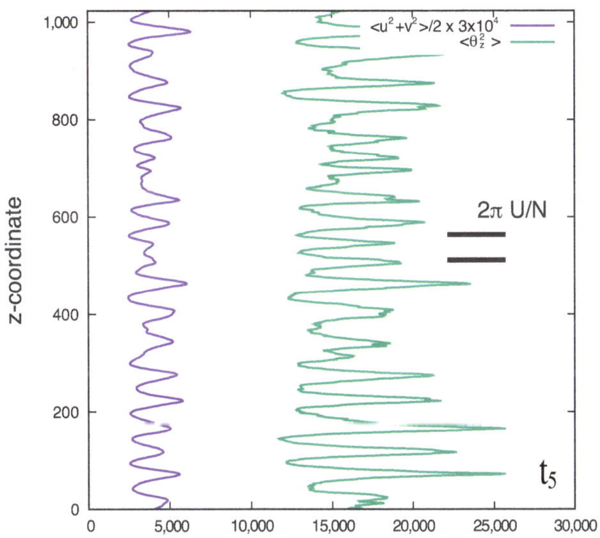

Figure 7. The horizontal average of the horizontal energy, $(\langle u^2\rangle + \langle v^2\rangle)/2$, and the horizontal average of the square of the vertical derivative of temperature fluctuations, $\langle(\partial\theta/\partial z)^2\rangle$, at about the final time t_5 in Figure 3. The characteristic frequency of the peaks is compared with the value of $2\pi U/N$.

Now we look at the contour plot of the temperature fluctuations in an $x - z$ plane at $y = \pi$ (grid number 512) for $N^2 = 10, k_f = 4$ at $t = 64.7340$ (Figure 8). (To see the structures clearly, a grey scale is used. The grid number is depicted in the figure as the x and z coordinates).

Many horizontally extended layers are observed which have sharp interfaces between black and white regions. The existence of such interfaces implies that the temperature fluctuations jump at the interfaces. To observe such jumps in temperature fluctuations, its value at $x = 134$ is monitored along the white line in Figure 8.

Figure 9a shows the graph of the temperature fluctuation θ as a function of z (in terms of grid number) along $x = 134$. Horizontally flat and straight lines are observed periodically which are bridged by finely oscillatory stairs.

In Figure 9b, the saw-tooth pattern of Figure 9a is superposed with the mean temperature, $\langle\theta\rangle$, corresponding to $N^2 = 10$. The overall saw-tooth structures in Figure 9a produce cliff–ramp structures in the total temperature field. (Major cliffs are marked with red circles). Temperature jumps at the height of the cliffs. In other words, the cliffs form the interface between cold and warm fluid masses. The most characteristic feature of the cliff–ramp structures is that ramps are mostly leftward (i.e., cooling as it goes up) with jitters. Similar temperature front structures are observed in simulations of homogeneous shear flows [23,24] as well as in large-eddy simulations of the stably stratified atmospheric boundary layer [25].

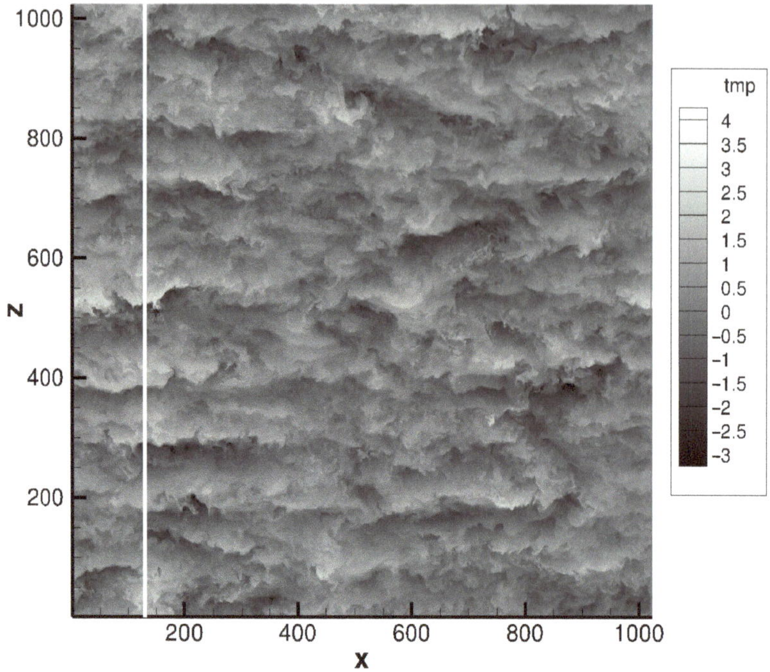

Figure 8. Temperature fluctuations in an xz plane at $y = \pi$ (grid number 512) for $N^2 = 10, k_f = 4$ at $t = 64.7340$.

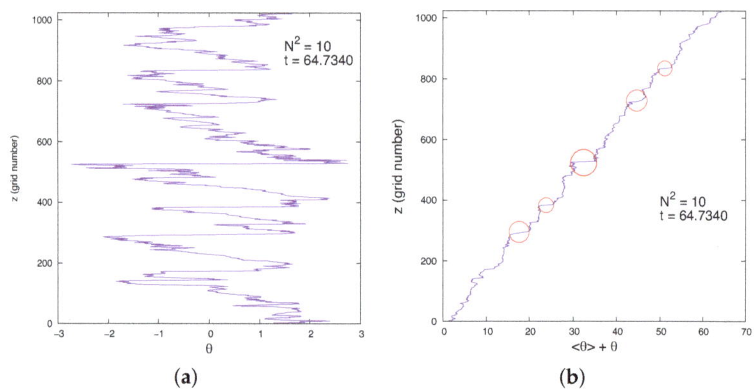

Figure 9. (**a**) Temperature fluctuation θ as a function of z (in grid number) along $x = 134$ in Figure 8 (white line). (**b**) The saw-tooth graph of (**a**) is superposed with the mean temperature, $\langle\theta\rangle$, corresponding to $N^2 = 10$. Cliff–ramp structures are seen, and major cliffs are marked with red circles.

To compensate for the cooling ramps, cliffs should be warming as it goes up. Therefore, there is an asymmetry in the slope of the temperature distribution. Such asymmetry is clearly seen statistically if we look at the PDFs of temperature fluctuation and its derivatives in the whole domain (Figure 10).

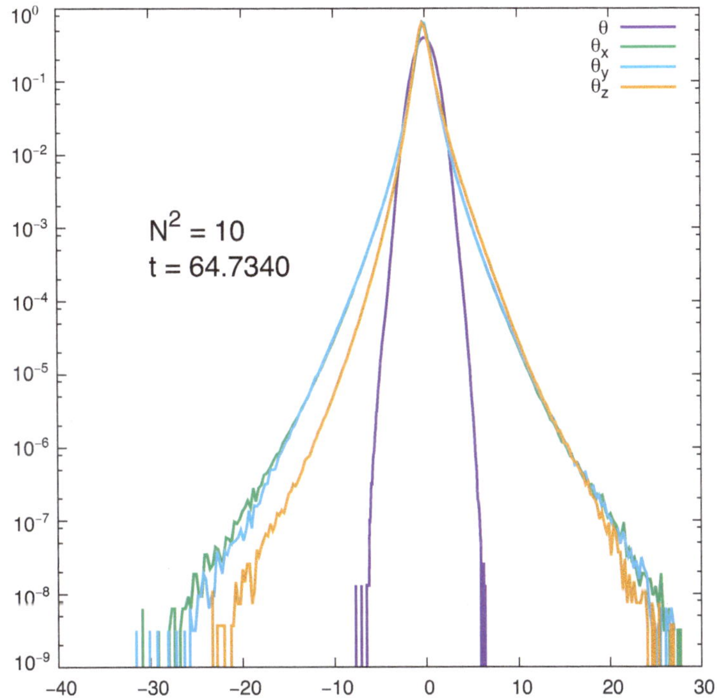

Figure 10. PDFs of temperature fluctuation and its derivatives in the whole domain. The PDF of temperature fluctuation is close to Gaussian, while those of its derivatives show exponential tails. The derivatives in x- and y- are symmetric, while the z- derivative is skewed.

In Figure 10, the PDF of temperature fluctuation is close to Gaussian, while those of its derivatives show exponential tails. Among the three derivatives, the derivatives in x- and y- are symmetric, while the z- derivative is skewed with the positive values being higher probability. This statistics supports the phenomenon that the cliffs are associated more with warming.

4. Discussion

Apparently it appears odd that the integration of the Navier–Stokes equations with periodic boundary conditions produces non-symmetric statistics of the derivative of the temperature fluctuations. However, we can check that the integration of θ_z in z is zero which is compatible with the periodic boundary condition. Even under this circumstance, we suspect that the intermittent velocity components (Figure 6) acting as a strong horizontal "wind" may produce a non-symmetric gradient of temperature fluctuations. Nevertheless, clarification of the true mechanism is still an open question.

The present research on the development of structures and spectra in stably stratified turbulence is largely related to the broad study of mixing and transport in turbulence. This old and important subject has been investigated by many researchers. The reader is referred to some of the basic reviews and the references therein for the historic and recent developments [18,26–29].

Author Contributions: Conceptualization and methodology including software, Y.K. and P.P.S.; validation, P.P.S.; writing—original draft preparation, Y.K.; writing—review and editing, P.P.S.; visualization, Y.K.; supervision, P.P.S.; funding acquisition, Y.K. All authors have read and agreed to the published version of the manuscript.

Funding: Y.K. is funded by JSPS (Japan Society for the Promotion of Science) KAKENHI (Grants-in-Aid for Scientific Research), grant number 19H00641 to conduct this reseach.

Institutional Review Board Statement: Not applicable.

Informed Consent Statement: Not applicable.

Data Availability Statement: The data presented in this study are available on request from the corresponding author due to privacy.

Acknowledgments: The final part of this paper is based on the talk presented in the VIIIth International Symposium on Stratified Flows, (29 August–1 September 2016, San Diego, CA, USA) [30]. Jack was one of the authors of this presentation. We were preparing for publicizing the result, but we could not finish it. This paper is dedicated to the memory of Jack Herring.

Conflicts of Interest: The authors declare no conflict of interest.

References

1. Kimura, Y.; Herring, J.R. Diffusion in stable stratified turbulence. *J. Fluid Mech.* **1996**, *328*, 253–269. [CrossRef]
2. Csanady, G.T. Turbulent diffusion in a stratified fluid. *J. Atmos. Sci.* **1964**, *21*, 439–447. [CrossRef]
3. Cox, C.Y.; Nagata, Y.; Osborn, T. Oceanic fine structure and internal waves. Papers in dedication to Prof. Michitaka Uda. *Bull. Jpn. Soc. Oceanogr.* **1969**, *1*, 67–71.
4. Lilly, D.K.; Walko, D.E.; Adelfang, S.I. Stratospheric Mixing estimated from high-altitude turbulence measurements. *J. Appl Met.* **1974**, *13*, 488–493. [CrossRef]
5. Weinstock, J. Vertical turbulent diffusion in a stably stratified fluid. *J. Atmos. Sci.* **1978**, *35*, 1022–1027. [CrossRef]
6. Pearson, H.J.; Puttock, J.S.; Hunt, J.C.R. A Statistical model of fluid<lement motions and vertical diffusion in a homogeneous stratified turbulent flow. *J. Fluid Mech.* **1983**, *129*, 219–249.
7. Kimura, Y.; Herring, J.R. Energy spectra of stably stratified turbulence. *J. Fluid Mech.* **2012**, *698*, 19–50. [CrossRef]
8. Nastrom, G.D.; Gage, K.S. A climatology of atmospheric wavenumber spectra of wind and temperature observed by commercial aircraft. *J. Atmos. Sci.* **1985**, *42*, 950–960. [CrossRef]
9. Riley, J.J.; de Bruyn Kops, S.M. Dynamics of turbulence strongly influenced by buoyancy. *Phys. Fluids* **2003**, *15*, 2047–2059. [CrossRef]
10. Waite, M.L.; Bartello, P. Stratified turbulence dominated by vortical motion. *J. Fluid Mech.* **2004**, *517*, 281–308. [CrossRef]
11. Bartello, P.; Tobias, S.M. Sensitivity of stratified turbulence to the buoyancy Reynolds number. *J. Fluid Mech.* **2013**, *725*, 1–22. [CrossRef]
12. Portwood, G.D.; de Bruyn Kops, S.M.; Taylor, J.R.; Salehipour, H.; Caulfield, C.P. Robust identification of dynamically distinct regions in stratified turbulence. *J. Fluid Mech.* **2016**, *807*, R2. [CrossRef]
13. Lucas, D.; Caulfield, C.P.; Kerswell, R.R. Layer formation in horizontally forced stratified turbulence: Connecting exact coherent structures to linear instabilities. *J. Fluid Mech.* **2017**, *832*, 409–437. [CrossRef]
14. Brethouwer, G.; Billant, P.; Lindborg, E.; Chomaz, J.-M. Scaling analysis and simulation of strongly stratified turbulent flows. *J. Fluid Mech.* **2007**, *585*, 343–368. [CrossRef]
15. Spalart, P.R.; Moser, R.D.; Rogers, M.M. Spectral methods for the Navier–Stokes equations with one infinite and two periodic directions. *J. Comput. Phys* **1991**, *96*, 297–324. [CrossRef]
16. Doughtery, J.P. The anisotropy of turbulence at the meteor level. *J. Atmos. Sci.* **1961**, *21*, 210–213.
17. Ozmidov, R.V. On the turbulent exchange in a stably stratified ocean. *Izv. Atmos. Ocean. Phys. Ser.* **1965**, *21*, 210–213.
18. Caulfield, C.P. Layering, Instabilities, and Mixing in Turbulent Stratified Flows. *Annu. Rev. Fluid Mech.* **2021**, *53*, 113–145. [CrossRef]
19. Cho, J.Y.N.; Lindborg, E. Horizontal velocity structure functions in the upper troposphere and lower stratosphere 1. Observations. *J. Geophys. Res.* **2001**, *106*, 10223–10232. [CrossRef]
20. Lindborg, E.; Cho, J.Y.N. Horizontal velocity structure functions in the upper troposphere and lower stratosphere 2. Theoretical considerations. *J. Geophys. Res.* **2001**, *106*, 10233–10241. [CrossRef]
21. Maffioli, A.; Davidson, P.A. Dynamics of stratified turbulence decaying from a high buoyancy Reynolds number. *J. Fluid Mech.* **2016**, *786*, 210–233. [CrossRef]
22. Sagaut, P.; Cambon, C. *Homogeneous Turbulence Dynamics*; Cambridge University Press: Cambridge, UK, 2008.
23. Gerz, T.; Howell, J.; Mahrt, L. Vortex structures and microfronts. *Phys. Fluids* **1994**, *6*, 1242–1251. [CrossRef]
24. Chung, D.; Matheou, G. Direct numerical simulation of stationary homogenous stratified sheared turbulence. *J. Fluid Mech.* **2012**, *696*, 434–467. [CrossRef]
25. Sullivan, P.P.; Weil, J.C.; Patton, E.G.; Jonker, H.J.J.; Mironov, D.V. Turbulent winds and temperature fronts in large eddy simulations of the stable atmospheric boundary layer. *J. Atmos. Sci.* **2016**, *73*, 1815–1840. [CrossRef]
26. Fernando, H.J.S. Turbulent mixing in stratified fluids. *Annu. Rev. Fluid Mech.* **1991**, *23*, 455–493. [CrossRef]
27. Peltier, W.R.; Caulfield, C.P. Mixing efficiency in stratified shear flows. *Annu. Rev. Fluid Mech.* **2003**, *35*, 135–167. [CrossRef]

28. Ivey, G.N.; Winters, K.B.; Koseff, J.R. Density stratification, turbulence, but how much mixing. *Annu. Rev. Fluid Mech.* **2008**, *40*, 169–184. [CrossRef]
29. Gregg, M.C.; D'Asaro, E.A.; Riley, J.J.; Kunze, E. Mixing efficiency in the ocean. *Annu. Rev. Mar. Sci.* **2018**, *10*, 443–473. [CrossRef]
30. Kimura, Y.; Sullivan, P.P.; Herring, J.R. Temperature front formation in stably stratified turbulence. In Proceedings of the VIIIth International Symposium on Stratified Flows, San Diego, CA, USA, 29 August–1 September 2016.

atmosphere

MDPI

Article

The Intermittency of Turbulence in Magneto-Hydrodynamical Simulations and in the Cosmos

Pierre Lesaffre [1,†], Edith Falgarone [1,*,†] and Pierre Hily-Blant [2,†]

1 Laboratoire de Physique de l'Ecole Normale Supérieure, ENS, Université Paris Sciences et Lettres, Centre National de la Recherche Scientifique, Sorbonne Université, 75005 Paris, France; pierre.lesaffre@ens.fr
2 Institut de Planétologie et d'Astrophysique de Grenoble, Centre National de la Recherche Scientifique, Université Grenoble Alpes, 38000 Grenoble, France; pierre.hily-blant@univ-grenoble-alpes.fr
* Correspondence: edith.falgarone@ens.fr
† These authors contributed equally to this work.

Abstract: Turbulent dissipation is a central issue in the star and galaxy formation process. Its fundamental property of space–time intermittency, well characterised in incompressible laboratory experiments, remains elusive in cosmic turbulence. Progress requires the combination of state-of-the-art modelling, numerical simulations and observations. The power of such a combination is illustrated here, where the statistical method intended to locate the extrema of velocity shears in a turbulent field, which are the signposts of intense dissipation extrema, is applied to numerical simulations of compressible magneto-hydrodynamical (MHD) turbulence dedicated to dissipation scales and to observations of a turbulent molecular cloud. We demonstrate that increments of several observables computed at the smallest lags can detect coherent structures of intense dissipation. We apply this statistical method to the observations of a turbulent molecular cloud close to the Sun in our galaxy and disclose a remarkable structure of extremely large velocity shear. At the location of the largest velocity shear, this structure is found to foster 10× more carbon monoxide molecules than standard diffuse molecular gas, an enrichment supported by models of non-equilibrium warm chemistry triggered by turbulent dissipation. In our simulations, we also compute structure functions of various synthetic observables and show that they verify Extended Self-Similarity. This allows us to compute their intermittency exponents, and we show how they constrain some properties of the underlying three-dimensional turbulence. The power of the combination of modelling and observations is also illustrated by the observations of the CH^+ cation that provide unique quantitative information on the kinetic energy trail in the massive, multi-phase and turbulent circum-galactic medium of a galaxy group at redshift $z = 2.8$.

Keywords: turbulence; intermittency; dissipation; coherent structures

Citation: Lesaffre, P.; Falgarone, E.; Hily-Blant P. The Intermittency of Turbulence in Magneto-Hydrodynamical Simulations and in the Cosmos. *Atmosphere* **2024**, *15*, 211. https://doi.org/10.3390/atmos15020211

Academic Editor: Annick Pouquet

Received: 11 January 2024
Revised: 26 January 2024
Accepted: 31 January 2024
Published: 8 February 2024

Correction Statement: This article has been republished with a minor change. The change does not affect the scientific content of the article and further details are available within the backmatter of the website version of this article.

1. Introduction

The unprecedented development of observational capabilities has revolutionised our knowledge of the cosmic history of star formation [1]. A consensus is emerging, on observational grounds, that the growth of galaxies in the early universe is largely governed by the conversion of massive gas reservoirs into stars over timescales as long as several Gyr, in apparent contradiction with the fast hierarchical merging imposed by dark matter (DM) structures [2]. This is one facet of the overcooling problem, also present in the local universe: the gas cools and forms stars too fast in simulations in comparison to observations. Theoretically, this problem is somewhat alleviated by stellar- or Active Galaxy Nuclei (AGN)-driven feedback [3].

Molecules, traditionally seen as the tracers of the truly cold universe in which stars form at temperatures considerably lower than the high temperatures reached by cosmological simulations, have unveiled a clue to solve the overcooling problem. An additional heating source—the dissipation of turbulence—is needed in cold interstellar matter (ISM) to explain its observed emission in the pure rotational lines of H_2 and the presence of molecules with highly endoenergetic formation routes, such as CH^+ (see the review by Hennebelle and Falgarone [4]). These observations, performed in diverse environments and at many scales, show that a major fraction of the gas internal energy is not dissipated at a high temperature but enters a multi-phasic turbulent cascade that dissipates radiatively at the low temperatures of the molecular gas. This unknown fraction is an uncharted link to the energetics of galaxy and star formation which must be explored to access the "relative importance of purely gravitational effects and of gas-dynamical effects involving dissipation and radiative cooling" stressed long ago by White and Rees [5]. This is why the intermittent dissipation of turbulent energy is key in the unravelling of the star and galaxy formation processes.

1.1. The Abyss between Cosmic Turbulence and Theory and Laboratory Experiments ...

Theoretical approaches and experiments on isotropic, homogeneous, incompressible turbulence remain a powerful source of inspiration and guidance to the studies of the complex turbulence pervading space from the terrestrial atmosphere and solar wind to galaxies and galaxy clusters. To name a few in the scope of this paper: Chandrasekhar and Fermi [6] for the expression of the turbulent pressure and viscosity, Méneveau and Sreenivasan [7] and She and Lévêque [8] on space–time intermittency, Alexakis et al. [9,10] on non-local interactions in MHD turbulence, Moffatt et al. [11] for qualifying the coherent structures of vorticity as the sinews of turbulence, Uritsky et al. [12] for the similarity of the statistical properties of clusters of coherent structures of vorticity and current in MHD simulations, Kimura et al. [13] and Kimura and Sullivan [14] for the formation of strong temperature fronts in stably stratified turbulence, Cadot et al. [15] for imaging intense small-scale vortices in turbulent water seeded with gas bubbles, Tabeling et al. [16] for experiments in superfluid Helium, Politano and Pouquet [17] for their predictions of scaling laws in incompressible magnetised turbulence and Schekochihin [18] for clarifying the links between MHD and kinetic turbulence.

The applicability to cosmic turbulence of theoretical results obtained in idealised conditions is unexpected because cosmic turbulence is highly compressible, magnetised and multi-phasic with Reynolds numbers $Re = Lu_{\mathrm{r.m.s.}}/\nu$ larger than 10^8, where $L > 100$ pc – two practical units for lengths/distances in astrophysics are the astronomical unit (au), the Earth–Sun distance, 1.5×10^{13} cm and the parsec (pc), 3×10^{18} cm, adapted to proto-planetary disks and galactic scales, respectively – and $u_{\mathrm{r.m.s.}} > 10$ km s^{-1} are the characteristic length and velocity of the integral scales of turbulence in galaxies and $\nu = \frac{1}{3}\lambda v_{\mathrm{th}} \sim 10^{18}$ cm^2 s^{-1}, with v_{th} the thermal velocity, is the molecular viscosity. The colossal range of densities and temperatures experienced by the gas along its cycle between stars and the interstellar medium (ISM) is illustrated in Figure 1.

The CNM and WNM (Figure 1) are two thermally stable phases [19] that build up the so-called "cold" ISM, with respect to the hot space-filling phase. Their characteristics are given in Table 1. They are in thermal pressure equilibrium and their average densities (and temperatures) differ by about 2 orders of magnitude. The ISM therefore harbours gas structures about 100× denser and colder than their environment called "clouds". The mass fraction in the dense phase depends on the ambient thermal pressure, and therefore on the distance from the stellar disks in galaxies. The ISM is fully or partially ionised depending on the ambient density of ionising ultra-violet (UV) photons and relativistic particles (cosmic rays, CR). The gas is therefore more or less coupled to the magnetic fields. The ISM also comprises sub-micron-size dust particles that contribute only 1% of the gas mass but play a key role in the gas coupling to the magnetic fields because these dust particles are charged and have a large collisional cross-section with the gas. They bear predominantly positive charges, except the smallest, which are predominantly negative [20], but their charge varies with the time and place depending on their environment (UV radiation and gas density).

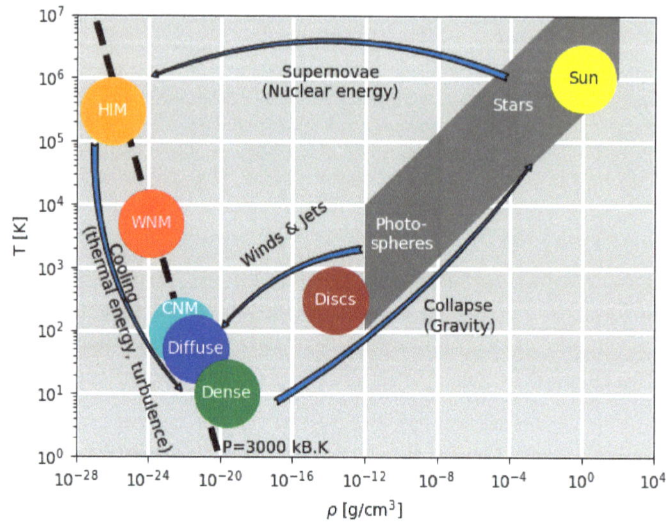

Figure 1. Temperature–density cycle of baryonic matter from the various thermal phases of the interstellar medium (ISM), the stars themselves and its ejection back to the ISM through winds, jets and supernova explosions. The different thermal phases are the hot ionised medium (HIM), the warm neutral medium (WNM), the cold (atomic) neutral medium (CNM) and denser molecular phases, noted as diffuse and dense (see Table 1 for the characteristics of each of these thermal phases). The energy and processes driving the evolution along each branch of the cycle are indicated. Note that turbulence is an actor along the cooling branch on the left of the cycle and that all the thermal phases of the ISM, except the densest, which are gravitationally bound, are in thermal pressure equilibrium. (Figure adapted from Lesaffre [21]).

Table 1. Characteristic scales and dimensionless numbers for various thermal phases in the galaxy, from the most dilute to the densest. HIM: hot ionised medium, WNM: warm neutral medium, CNM: cold neutral medium, diffuse and dense molecular gas. These characteristic values are orders of magnitude from Draine [22], (p.6, Table 1.3). Dimensionless numbers are defined as $\mathcal{M} = u_{\rm r.m.s.}/v_{\rm th}$, the Mach number; $Re = Lu_{\rm r.m.s.}/\nu$ (see text), the Reynolds number; $Rm = Lu_{\rm r.m.s.}/\eta$, the magnetic Reynolds number (the resistivity η is computed as in Equation (12) of Balbus and Terquem [23]); and $R_{AD} = L/(u_{\rm r.m.s.}t_a)$ is a dimensionless number based on the ambipolar diffusion process (see Equation (3) of Momferratos et al. [24]), where t_a is the ion-neutral momentum exchange time scale.

	HIM	WNM	CNM	Diffuse H_2	Dense H_2
Density n [cm^{-3}]	0.004	0.6	30	200	10^4
Temperature T [K]	3.10^5	5000	100	50	10
Length scale L [pc]	100	50	10	3	0.1
Velocity $u_{\rm r.m.s.}$ [km.s^{-1}]	10	10	10	3	0.1
$\mathcal{M}$	0.2	2	13	7	0.5
Re	10^2	10^5	10^7	10^7	10^6
Rm	10^{21}	10^{20}	10^{18}	10^{17}	10^{15}
R_{AD}	10^3	10^3	10^2	10^3	10^4
Ionisation fraction	1	10^{-2}	10^{-4}	10^{-4}	10^{-7}

Turbulence is ubiquitous throughout this cycle. Its origins are multiple, including the differential rotation of galaxies, the gravitational energy in gas accretion processes and the stellar feedback in the form of supernova explosions but also jets and winds all along the stellar lifetime that span several Gyr for solar-type stars. The role of turbulence in star formation, in particular, its link with the observed very low efficiency of star formation, is a highly debated issue because turbulence can both trigger and hamper star formation (see the review by Hennebelle and Falgarone [4]).

The large scales in cosmic turbulence concern rotating and shearing, stratified by the gravitation field, with an additional anisotropy due to the ubiquitous magnetic fields. The associated dynamical timescales are all different and span a broad range from hundreds of Myr to $\sim 10^3$ yr. The latter are shorter than collisional times, providing the ISM with facets of collisionless plasma (see the review by Ferrière [25]). A fraction of the cold medium lies in the thermally unstable phase because turbulence continuously drives the phase transition from the WNM to CNM [26]: the hydrodynamic pressure fluctuations of the turbulence destabilise the WNM and cause its fragmentation into CNM clouds, harboring droplets as dense as 10^4 cm^{-3}. Buoyancy in the galactic gravitational field is also inevitable given the gas equation of state and the multiplicity of energy and ionisation sources.

Finally, the weakly ionised ISM phases are not fully coupled to the magnetic fields. There is a whole range of high frequencies of magnetic waves which are not perceived by the plasma because the collisions of neutrals with ions and charged dust grains, closely coupled to the magnetic fields, are too rare. There is also a domain, at low frequency, where the waves no longer propagate because all their energy is dissipated in ion-neutral collisions. The frequencies that limit these regimes depend on the collision cross-sections between the ions and the neutrals and their respective densities [27]. Between these two extremes, the magnetic fields drift through the medium. This is an important phenomenon in the diffuse ISM because it is a major source of energy dissipation but it is also the reason that C-shocks (i.e., continuous shocks) exist, because, unlike J-shocks, they develop magnetic precursors [28–30]. This is also why they are so rich in molecules.

1.2. ...and yet

In spite of all the above, the electron density of the ISM and interplanetary ionised medium exhibits a Kolmogorov power spectrum spanning more than 10 orders of magnitude in scales down to $\sim 10^8$ cm (known as "The big power law in the sky") [31]. The slope of this spectrum suggests a link with turbulence.

Cold atomic gas too exhibits a structure down to very small scales of 10 to 10^4 au [32]. A few power spectra have been observed in the CNM: that of the continuum emission of dust in a high-galactic-latitude cloud extending from 20 pc down to 2000 au [33], that of light scattered by dust in a similar field extending over 3 orders of magnitude down to 10 mpc [34]. Both are consistent with the Kolmogorov spectrum, given the projection effects in observations (see next section).

The molecular component of the CNM (hereafter molecular clouds) also exhibits power law scalings between the velocity dispersion and size scale, the well-known linewidth-size relations of Larson [35], and is now observed in many environments including external galaxies (e.g., [4,36]) with a range of slope values close, but not equal, to the Kolmogorov spectrum. An extension of the Kolmogorov cascade, taking into account compressibility, has been proposed by Kritsuk et al. [37] using a density-weighted velocity $\mathbf{v} = \rho^{1/3}\mathbf{u}$ with ρ as the mass density; this preserves the Kolmogorov scaling of the power spectra and the third-order structure-function behaviour. But the respective roles of self-gravity and turbulence in these linewidth-size scalings remain an outstanding issue (e.g., [38]).

Diffuse molecular gas (Figure 1) is responsible for the very first steps of chemistry in space with the formation of light hydrides, such as CH, OH, CH^+, OH^+, ..., the most simple molecules that are the building blocks of the complex species found in dense proto-stellar cores or proto-planetary discs. It is in that diffuse medium of density $n_H \sim 10 - 100$ cm^{-3} that the turbulent dissipation scale, estimated as $l_{diss} = (\nu^3/\epsilon)^{1/4}$ with the spatially averaged kinetic energy transfer rate $\epsilon \approx 2 \times 10^{-25}$ erg cm^{-3} s^{-1} inferred from observations [4,39] is of the same order as the mean free path of atoms, or $\lambda \sim l_{diss} \sim 2$–10 au. It is therefore predictable that turbulent dissipation plays a key role in the very first steps of chemistry in space.

1.3. Specific Molecules As Tracers of Turbulent Dissipation

The emergence of molecules in the diffuse ISM raises outstanding issues: the observed abundances of the CO molecule in the diffuse molecular gas are an order of magnitude larger than what the chemistry driven by the UV photons and cosmic rays is able to produce (e.g., [40]). This problem emerged in fact in the 1940s with the discovery of the first molecules in diffuse medium, CH, CH^+ and CN, the formation of CH^+ being extremely endothermic. Energy inputs from stellar UV photons and cosmic rays are not sufficient to reproduce the observed column densities of these species. The same problem appeared later for HCO^+ (e.g., [41]) and for CO. Another major power source was needed. Several processes have been proposed, such as thermal conduction and turbulent transport [42] at the WNM/CNM interfaces or extreme dissipative events in interstellar turbulence, such as C-shocks [12,24,28,29,43–55], magnetised Burgers vortices [40]. The former two inject the thermal energy of the WNM into the CNM, while the latter feeds the chemistry with the turbulent energy of the CNM, which is an order of magnitude larger than its thermal energy, but is comparable to the thermal energy of the WNM. Ion-neutral drift in Alfvén waves have also been considered [44]. It is likely that they all contribute to exemplifying the tight coupling between the thermal and turbulent facets of ISM physics.

In situ measurements in the ISM are impossible and the measured quantities suffer projection effects. Velocities are measured via the Doppler effect of spectral lines so that only line-of-sight (*los*) velocities are accessible. Furthermore, the detected signal (i.e., a spectral line) is the line emission integrated along the whole line of sight: radiative transfer is therefore involved in the data interpretation. Displacements are inferred from distances projected in the plane of the sky (*pos*), so that only the *pos* variations of the *los*-integrated velocity are accessible, providing a proxy for the *pos* vorticity projection. Kinetic helicity cannot be directly measured. Only the intensity of the magnetic field *los* projection is accessible via the Zeeman effect [45] and the field direction, *pos* projected, is provided by polarisation measurements in different wavelength ranges (see the review by Ferrière [25]).

This is why, like luminescent plankton that shines in the spots of intense velocity shear in the breaking waves of tropical seas, molecules with highly endothermic formation rates are used as the tracers of dissipation bursts in cosmic turbulence. These specific molecules in diffuse molecular gas are therefore not only the tracers of extreme dissipation events in cosmic turbulence but also of the signposts of coherent structures that may be the seeds of more massive structures in molecular clouds, grown through braiding and coalescing [47]. Note that the extrema of current and vorticity in MHD turbulence are not randomly distributed in space and time but cluster in structures qualified as coherent because, following Moisy and Jiménez [46], their "inertial-range extent, implies a large-scale organization of the small-scale intermittent structures".

This paper reports on numerical simulations of compressible MHD turbulence, dedicated to dissipation scales, performed to characterise the nature of the coherent structures in ISM turbulence where dissipation is concentrated [48] and to illustrate the statistics performed on synthetic observations to disclose such Coherent Structures of Intense Dissipation Extrema—hereafter called CSIDE—in diffuse molecular gas. Then, the first example of such a statistical study in a nearby diffuse molecular cloud is presented. A long and thin parsec-scale coherent structure of intense velocity shear is identified [49,50]. High-angular resolution observations disclose an even more intense velocity shear within it, on linear scales of only a few mpc. Last, the power of the so-called "warm chemistry" triggered by turbulent dissipation [40] is illustrated by observations in the circum-galactic medium of a starburst galaxy group at redshift $z = 2.8$ [51].

2. Intermittency in Simulations of Magnetised Turbulence

2.1. Numerical Dissipation

Coherent structures that are the signposts of intermittency [11,12,15,18,46] are hard to characterise in controlled laboratory experiments because measurements can only be made at a small finite number of points. On the contrary, a numerical simulation allows its user to access every state variable at every position. Unfortunately, that advantage comes with distortion from the unavoidable discretisation artefacts, and the inability to perform simulations at Reynolds numbers as large as in the experiments.

Simulations of incompressible gases allow one to compute gradients in Fourier space. Such techniques ("pseudo-spectral") can obtain exponentially good accuracy with respect to the number of resolution elements. This greatly helps in trusting the dissipation terms when the resolution is large enough (i.e., when a bottleneck effect is absent). If the maximum wavenumber times the Kolmogorov dissipation scale is too small (lower than order unity, for example), Gibbs phenomena generate an excessive pile up of energy at small scales (i.e., the bottleneck effect). An exponential decay of power spectra at small scales is a good indication that numerical convergence has been obtained.The results of such incompressible simulations for hydrodynamical (HD) turbulence [46] show that dissipation is localised around vortices, and for MHD turbulence [12,24], they show that viscous and resistive dissipation are well separated on sheets which can therefore easily be identified as shearing or current sheets.

Compressible simulations are much more difficult to interpret. The finite size of the mesh and the necessary reconstruction of variables from the center to the edges of each pixel introduces a spurious diffusion length scale for every conservation equation (including mass conservation). To this effect, we built a method to estimate locally the total dissipation including the one inherent to the numerical scheme [24,48]. The method consists of replaying a simulation step while integrating a redundant conservation equation for a variant of the energy (total mechanical energy, for example). In isothermal simulations, that quantity is best taken as the generalised isothermal total mechanical energy $E = \frac{1}{2}\rho u^2 + \frac{1}{8\pi}B + p\log\rho$, where ρ, u, B and $p = \rho c_s^2$ are, respectively, the mass density, the velocity, the magnetic field intensity and the thermal pressure (with c_s as the isothermal sound speed). An estimation of the flux $\mathcal{F}_E$ of this quantity and its time derivative then allows one to recover the local total dissipation of energy as $\varepsilon = \partial_t E + \nabla.\mathcal{F}_E$.

Once the dissipation rate is known everywhere, we can study the regions of most intense dissipation. For instance, we can plot the *los*-integrated dissipation as on Figure 2: the regions of highest dissipation are not randomly distributed in space, they display large-scale and thin ridges of intense integrated dissipation. These ridges appear as a caustic effect from the projection of sheets seen almost edge-on where dissipation is intense (see [48]). They are coherent large-scale structures, hereafter called the CSIDE, as opposed to quasi-thermalized small-scale eddies.

2.2. The Nature of Coherent Structures in MHD Turbulence

Intense dissipation in MHD turbulence is therefore seen to lie on thin sheets whether it is incompressible [12,24] or compressible [48]. In incompressible MHD turbulence, viscous and ohmic dissipation always appear disconnected (e.g., [12,24]), and the same probably holds for reduced MHD, as in [52]. In these works, it makes it easy to differentiate shearing sheets and current sheets as connected sets of intense viscous or resistive dissipation.

In compressible MHD turbulence, this is much less obvious, as hinted at in Figure 2, where different natures of dissipation heating can be intermixed. Lehmann et al. [53], with the SHOCKFIND algorithm, and Richard et al. [48] designed several criteria which allow one to carefully select strong dissipation regions and characterise their physical nature as fast shocks, slow shocks, rotational discontinuities or Parker sheets. Some of these criteria are based on Rankine–Hugoniot relations [54], which seem to hold strikingly well far from their domain of application (i.e., in a non-stationary medium, inhomogeneous, with curved working surfaces, etc.). Some of these criteria are based on decomposition along free travelling waves, which characterise suprisingly well most of these structures. Gradients of fast, intermediate and slow waves form an orthogonal basis of gradients of MHD variables. It turns out that slow shocks decompose almost purely on gradients of slow waves, and similarly, fast shocks on fast waves.

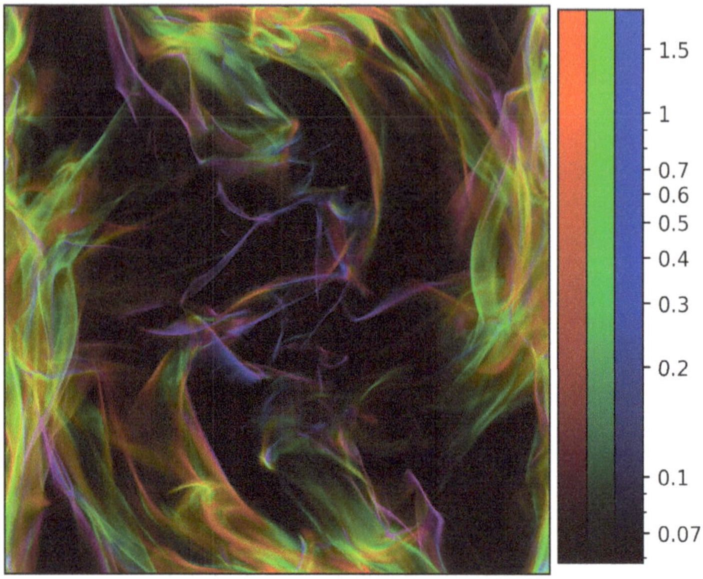

Figure 2. Integrated dissipation $\int \varepsilon dz$ along the line of sight coordinate z near dissipation peak for an initial r.m.s. Mach 4 simulation of decaying compressible MHD turbulence, starting from a perturbed Orszag–Tang initial configuration with a resolution of 1024^3 pixels (see [48]). The values of $\int \varepsilon dz$ are normalised by $< \rho > u^3_{\mathrm{r.m.s.}}$, where $< \rho >$ is the average mass density and $u_{\mathrm{r.m.s.}}$ is the initial r.m.s. velocity in the computational domain. The total intensity of pixels is coded according to the total dissipation $\int \varepsilon dz$, while red, green and blue color fractions of pixels scale according to the line-of-sight integrated relative fractions of Ohmic dissipation $\eta(\nabla \times B)^2$ (*red*), viscous shear dissipation $\rho\nu(\nabla \times u)^2$ (*green*) and compressible dissipation $\frac{4}{3}\rho\nu\ (\nabla.u)^2$ (*blue*), where η and ν are the resistive and viscous coefficients.

A systematic investigation of these CSIDE was performed by Richard et al. [48] for three-dimensional (3D) MHD turbulence and in Lesaffre et al. [55] for two-dimensional (2D) HD turbulence. For 2D HD, we were able to show that almost 80% of the dissipation was accounted for near the shock fronts, with the remainder as diffuse viscous shear in the wakes of these shocks [55]. In this setup, the effect of dissipation on the chemistry of the medium was to enhance the formation of molecules such as CH^+ or CO and collisionally excite H_2 molecules. For 3D MHD turbulence [48], we could not clearly prove whether there was a significant contribution from dissipation outside the detected CSIDE, given the technical difficulties introduced by the 3D geometry, but we computed that up to 30% of the dissipation occurs in less than 1% of the volume. We were surprised as most of the dissipation was accounted for by weakly compressive structures. This is illustrated by the distribution of the dissipation rate according to velocity convergence in Figure 3. Another surprise we found in Richard et al. [48] is that the entrance parameters in the Rankine–Hugoniot fronts we extract from the simulations do not span the whole parameter space available to them. Indeed, geometry and symmetries alone allow one to reduce to three the number of free parameters characterising a Rankine–Hugoniot discontinuity [54]. However, the dynamics of these dissipative structures seem to constrain them to lie on a two-dimensional parameter space. To be more precise, slow shocks and Alfvénic discontinuities have entrance magnetic fields nearly orthogonal to the discontinuity while fast shocks have their entrance magnetic fields nearly transverse. It will be the object of future investigations to understand the underlying reasons as to why this is.

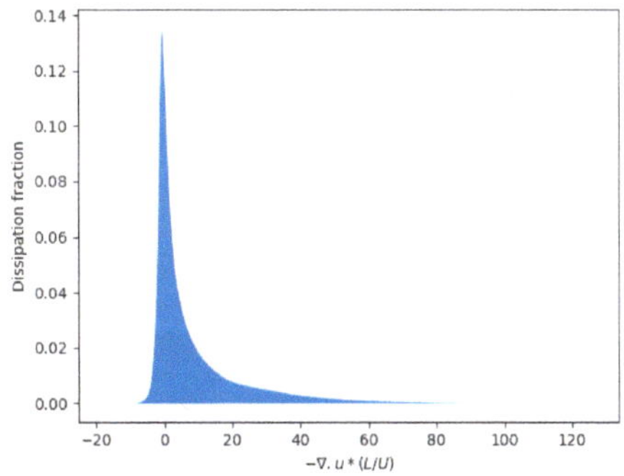

Figure 3. Fraction of energy dissipation for normalised velocity convergence (the opposite of velocity divergence). A normalised convergence value around unity indicates that $-\nabla.u{\sim}u_{\text{r.m.s.}}/L$, where $u_{\text{r.m.s.}}$ and L are the initial r.m.s. velocity and the size of the periodic domain, respectively. Most of the energy is dissipated at compression levels lower than this, although large values of the convergence exist (for example, in strongly compressive shocks). More than half of the energy is dissipated for normalised convergence values below 1. This graph is at a time near dissipation peak, when the compressive motions are maximal, and for initial Orszag–Tang conditions, which are known to generate large-scale shocks.

2.3. Synthetic Observables and the CSIDE

Laboratory measurements or in situ probes borne by satellites in the Solar Wind (e.g., [56]) allow one to probe only a small fraction of the volume of interest. In experiments, the progress of particle image velocimetry (PIV) techniques now authorise somewhat more widespread measurements, and Cadot et al. [15] cleverly used cavitation bubbles to reveal the most intense coherent structures in their experiments. On the other hand, astrophysical images of the nearby interstellar medium offer a global view, however they are projected on the *pos*: we lose one dimension of interest, and radiative transfer effects as well as chemical and line excitation issues conspire to distort the *los* integration process. Nevertheless, some distinctive features survive the projection. Figure 2 suggests that when viewed sideways, CSIDE can still be seen as prominent features in the integrated image.

Here, we compute some proxies for *los*-integrated observable variables in our isothermal simulations of decaying MHD turbulence (see [48] for a detailed presentation of the simulations), and we investigate where they vary strongly. The first obvious quantity we have access to is the column density $I = \int_0^L \rho dz$ (where L is the box length and z is the line of sight coordinate), probed indirectly by the intensity of the dust thermal continuum emission. Thanks to line profile measurements (see Section 4), one has access to a proxy of a density-weighted average *los* velocity $v_z = \int_0^L \rho\, u_z\, dz/I$, provided one finds a trustable tracer of density. In practice, temperature and chemical effects due to local heating as well as radiative transfer effects strongly hinder this measurement (see Sections 3 and 4). Finally, as the medium is magnetised, dust thermal emission is polarised. Modern astronomical instruments are able to measure the polarisation state of this emission, which is characterised through the Stokes parameters Q and U. The observed Stokes parameters are thus probes, however rather convoluted because polarisation is a pseudo-vector, of the *los*-integrated magnetic field direction. The polarisation of the dust thermal emission is due to the fact that dust grains are elongated and spin with their longer axis perpendicular to the magnetic field direction, thus orienting the polarisation vector of their thermal emission orthogonal to the magnetic field direction. The resulting Stokes parameters of dust emission, as computed in the numerical simulations, are therefore $Q + iU = \int_0^L dz \rho \hat{B}_\perp^2 e^{2i\phi}$, where ϕ is the position angle of the magnetic fields on the plane of sky and $\hat{B}_\perp$ is its relative norm compared to the local total fields.

In order to emphasise the locations where these observables change abruptly, we compute their one-pixel increments on the plane of the sky: i.e., around each pixel, we compute the standard deviation of the difference with its nearest neighbours. We then overlay 2-σ contours of these increments onto our integrated dissipation map (see Figure 4). In most cases, such contours delimit a region of strong dissipation. Indeed, whenever a physical quantity varies strongly, it is most often associated with dissipation. However, not all dissipation regions are detected, and various observables highlight different parts of the regions, depending on which component of the velocity or the magnetic fields is probed by the observable considered. Indeed, some dissipation regions can involve a jump in magnetic fields or velocity components which cannot be probed by astrophysical observables because of the projections discussed in Section 1. Unfortunately, we are not yet in a capacity to link the nature of the observable jumps to the nature of the underlying dissipative structure, except for the simple fact that strongly compressive shocks (slow shocks) should appear as a ridge of strong δI when their working surface contains the line of sight. But since the increments of synthetic observables provide a method to pinpoint and map CSIDE, these structures will be located in cosmic turbulence following this method, and further observed (e.g., at higher angular resolution, or with different observables) and the results of these additional observations will help us in understanding the nature of cosmic turbulence.

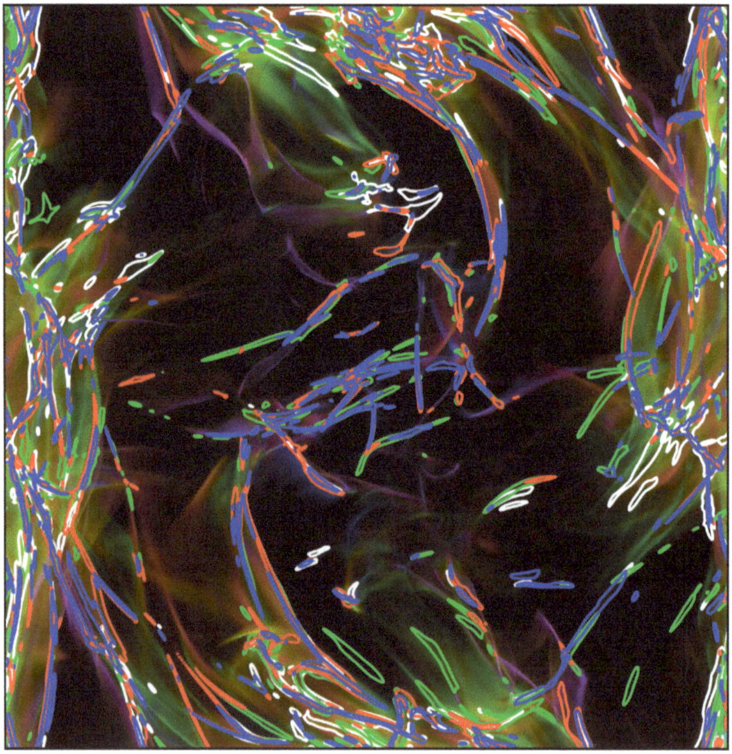

Figure 4. Same as Figure 2 for the background, overlaid with 2-σ contours of 1-pixel increments of integrated observables δI (*white*, column density), δv_z (*green*, centroid velocity), $\delta(Q/I)$ (*blue*, relative Stokes Q parameter) and $\delta(U/I)$ (*red*, relative Stokes U parameter). See text for more detailed definitions of I, Q, U, v_z.

2.4. Intermittency Statistics from Increments of Observables

Since the early works of Kolmogorov [57] (K41), who predicted it in the case of incompressible hydrodynamics, one of the first characteristics of turbulence that researchers have strived to predict is the power law slope of the velocity power spectrum. In the case of incompressible MHD turbulence, the exact scaling laws of the velocity and magnetic fields are still a matter of debate [18,58–60]. In compressible turbulence, theoretical discussions are scarce, but suggest that one could obtain the K41 scaling for $\rho^{1/3}u$ in isothermal turbulence [61], while numerical simulations [62] display different scalings for subsonic (K41) and supersonic scales (Burgers). Banerjee and Galtier [63] also managed to derive exact expressions for convoluted transfer functions in isothermal MHD, from which it is, however, hard to derive scaling laws for any simply defined quantity. The power spectrum of the density and the column density have also been discussed and either proven to be K41 in the Gaussian case [64], or shown to be close to K41 thanks to numerical simulations [65,66].

Twenty years after K41, Kolmogorov [67] realised that the spatial intermittency of the dissipation rate would introduce corrections to the structure function power law scalings with lag (the so-called anomalous scalings, characterised by intermittency exponents or multifractal spectra [68]). These anomalous scalings were linked to the fractal dimension of the dissipative structures, and generic models such as that of She and Leveque [8] were put in place, with shape parameters for the fractal geometry of the dissipative structures. Based on the observation that these were sheets in the case of MHD turbulence, this led to independent predictions from Grauer et al. [69] and Politano and Pouquet [70] for the Iroshnikov–Kraichnan scaling and later from Boldyrev et al. [71] for the Goldreich–Sridhar scaling (similar to K41).

Note that the causal link between the geometry of the coherent structures and the form of the intermittency exponents goes only one way. For instance, random fields built with controlled anomalous scalings close to those of hydrodynamics [72] or MHD [73] do not display any coherent structures. It is only in the case where a statistical collection of well-defined structures of strong dissipation is presupposed that the anomalous scaling coefficients can constrain the fractal dimension and geometry of these structures.

In the previous section, we demonstrated a spatial link between the increments of the synthetic observables and the CSIDE. We now investigate these increments at various lags, and look at classical structure functions built from these increments. Even though these indicators are very indirect and probably have not much to do with the underlying physics of the above intermittent theoretical models, we proceed to compute them as they will presumably still be good probes of the relative orientation and size of the salient features in the astrophysical images (they will be sensitive to their fractal structure, for example). As such, they provide a quantitative estimate to link the texture of the simulated images to the observed ones. In addition, the link to physics is blurred by the projections, and this analysis should rather be regarded as a quantitative characterisation of the evolution of the observable increments statistics with scale.

We now define $\delta_{\ell} X(\boldsymbol{r}) = X(\boldsymbol{r} + \boldsymbol{\ell}) - X(\boldsymbol{r})$ as the increment of an integrated observable X over a lag $\boldsymbol{\ell}$ at position $\boldsymbol{r}$ in the plane of sky and we investigate the statistics of these increments for lags where the norm is comprised between ℓ and $\ell + 1$ pixels (hence, we compute all increments in a corona of a width of 1 pixel around each pixel).

Figure 5, for example, shows how the probability distribution function (PDF) of the increment of the *los*-integrated velocity (v_z) varies from having strong non-Gaussian wings at small scales to being nearly Gaussian at large scales. As an illustration, the interior of the green contours of Figure 4 corresponds to the wings of the PDF for the smallest lag (dark blue on Figure 3), where the normalised increment $\delta v_z / \sigma(\delta v_z)$ exceeds 2 in its absolute value.

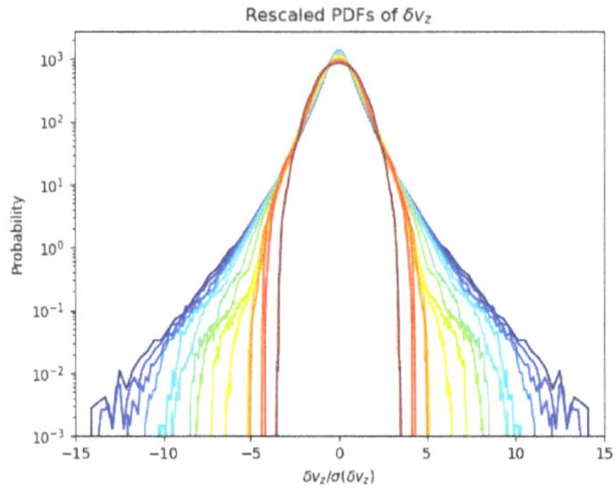

Figure 5. Probability distribution functions of the increments of the radial velocity v_z at dissipation peak (Orszag–Tang initial conditions) for a collection of lags ranging from small (1 pixel, *blue*) to large (256 pixels, *red*, or one quarter of the computational domain).

We then define "structure functions" in a natural way similarly to three-dimensional dynamically relevant structure functions: $S(p,\ell) =< |\delta_\ell X|^p >_{\ell\leq|\boldsymbol{\ell}|<\ell+1}$, where the average is presented over the image. These are displayed in Figure 6, left panel. We now measure scaling exponents ζ_p by adjusting a power law $S(p,\ell)\sim\ell^{\zeta_p}$ in a range of scales $\ell_{\rm min} \leq \ell \leq \ell_{\rm max}$. We adopt here the range of scales for the lags between 12 and 48 pixels (over a periodic computational domain of 1024 pixels aside) as this seems to minimise the dispersion (see Figure 6 left panel).

The values of the scalings, displayed on the left panel of Figure 7, usually fall way above K41: even if we would fit a slope to the values of ζ_p vs p, we would find slopes much larger than K41 or Iroshnikov–Kraichnan: we are clearly not in the domain of the application of these theories. Furthermore, all exponents measured on different variables (column density, projected velocity, U and Q Stokes parameters) display anomalous scalings in the sense that ζ_p is not proportional to p (see Figure 7). The uncertainty of the fit is displayed as error bars of one standard deviation around these intermittency exponents. It is a quantitative measurement of whether the scaling is indeed good or not, i.e., whether intermittency exponents are a good representation of the data. Note that these error bars do not reflect the temporal variation in the simulation, as they are displayed here only for one selected snapshot (at a time close to the dissipation peak, about a third of the non-linear initial turnover time). The temporal variability of these coefficients is in fact a few times larger than the displayed error bars.

Figure 6 demonstrates that when we plot the structure functions against $S(3,\ell)$ instead of the lag (right panel vs left panel), they behave much more like power laws. This was first discovered by Benzi et al. [74] for three-dimensional velocity field increments in hydrodynamics experiments. Because this allowed one to significantly extend the range of the scaling, this property was named "Extended Self-Similarity" (ESS). Similarly, we define ESS exponents as the exponents in the power law scalings $S(p,\ell)\sim S(3,\ell)^{\rm ESS\zeta_p}$. This significantly improves the quality of the fit, hence, our error bars (see right panel of Figure 7), even though we now use the full range of scales instead of the small restricted range of scales used in the left panel. This ESS property has been tested and verified in many independent data but has yet to find an underlying physical interpretation or proof. In addition, in our astrophysical case, for the observables integrated on the line of sight, the notion of an inertial range is hard to characterise. Indeed, projection brings larger scales onto smaller ones, and in doing so mixes dissipation, injection and inertial scales: this ESS property comes even more as a surprise.

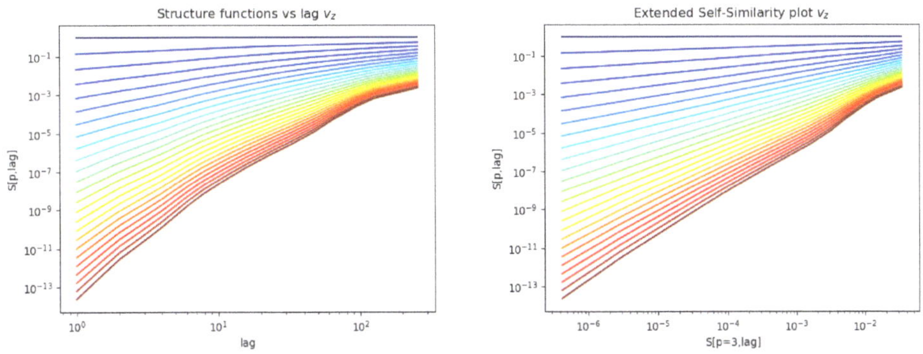

Figure 6. Dependence of structure functions for the radial velocity v_z at dissipation peak (Orszag–Tang initial conditions) versus lag ℓ in pixels (**left**) and versus $S(3, \ell)$ (**right**), where logarithmic scaling is seen to be extended to a larger range of scales. The order p ranges from 0 (*blue*) to 8 (*red*) in steps of 1/3.

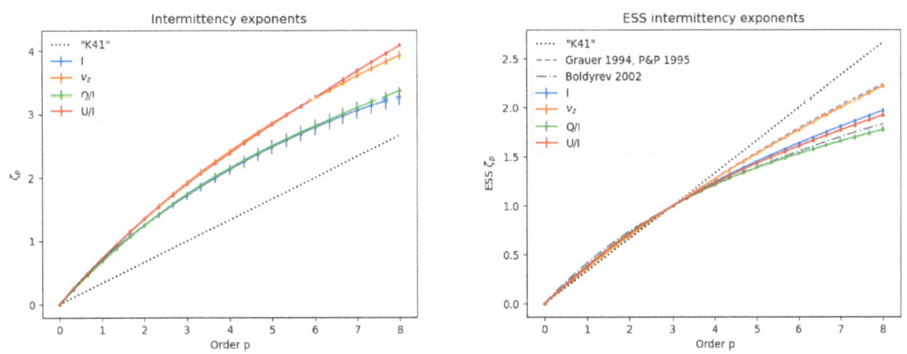

Figure 7. On the left panel, we show intermittency exponents measured for four variables, column density I, projected velocity v_z, U and Q Stokes parameters, probing scales within the range of lags 12 to 48 pixels for a simulation of 1024 pixels of side. ESS (see text) intermittency exponents (computed for the whole range of lags between 1 and 256 pixels) are displayed on the right panel. Error bars show the 1-σ standard deviation of the fit residuals over the selected range of scales. Error bars are significantly reduced when using ESS even though the lag range of the fit is much larger. These exponents are computed on a snapshot of a compressible MHD simulation of decaying turbulence (Orszag–Tang initial conditions), at a time near the dissipation peak, about a third of the initial non-linear turnover time [69–71].

We have investigated two types of initial conditions (an Arnold–Beltrami–Childress (ABC) flow, with large magnetic helicity and small cross helicity and an Orzag–Tang (OT) vortex, with zero magnetic helicity and large cross helicity, see [48]) and we have selected two different snapshots for each of these initial conditions. The earliest time is around dissipation peak, when stationarity is most likely to be realised (the second derivative in time of the total energy is zero), which also corresponds to the time when the CSIDE have just been born. The later time is after one turnover time, as a compromise between a time when turbulence has been well established and the initial conditions washed out, but nevertheless, the turbulence has not yet decayed too much and remains strong.

We summarise the results of our intermittency exponents measurements for all four cases in Figure 8. There is clear anomalous scaling (i.e., a departure from the linearity of ζ_p) in all four cases we investigated. However, the intermittency exponents are highly variable depending on the type of variable, the time and the initial conditions. Although the intermittency exponents differ slightly between the different observables, the use of ESS brings them closer to one another (see Figure 7 for example), spanning values between Grauer et al. [69] or Politano and Pouquet [70] and Boldyrev et al. [71]. In all cases, magnetic fields appear more intermittent than the velocity or column density, as already discussed in Schekochihin [18]. As in Figure 7, the error bars in Figure 8 characterise the goodness of the ESS fit for the whole lag range between 1 and 256 pixels. The exponents for the line of sight integrated velocity, v_z, are always close to the models of Grauer et al. [69] and Politano and Pouquet [70] and also close to the observed values measured in Polaris and Taurus by Hily-Blant et al. [49]. Although the variability of the anomalous scalings is strong, ESSζ_p for the column density scalings stays within one error bar of the scaling by Boldyrev et al. [71], except for the ABC flow at one turnover time. Provided we could neglect projection effects, and if we accept that dissipation is mainly occurring in sheets, this could mean that the correct ESS scaling for the velocity is closer to the Iroshnikov–Kraichnan scaling, while that of the mass density is closer to the Goldreich–Sridhar scaling.

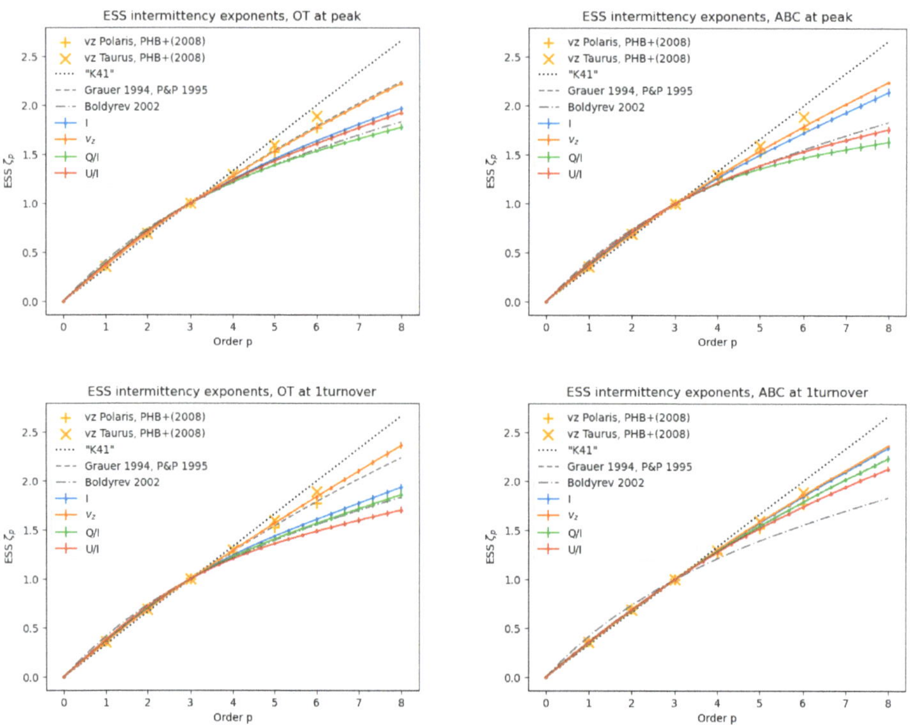

Figure 8. Comparison of our intermittency exponents to the model of Grauer et al. [69] or Politano and Pouquet [70] (P&P) and Boldyrev et al. [71] and to the observed ESS coefficients by Hily-Blant et al. [49] (PHB+(2008)) in the Polaris and Taurus regions. OT initial conditions for left panels, ABC flow for the right ones. At dissipation peak for top panels (at about 1/3 turnover time), after one initial turnover time for bottom ones.

3. Intermittency in Cosmic Turbulence

3.1. Extrema of Turbulent Dissipation in A Nearby Diffuse Molecular Cloud: A Source of CO molecules

A statistical analysis similar to that conducted on the simulations described in the previous section was performed in a nearby diffuse molecular cloud in the Polaris Flare, on the centroid velocities of ^{12}CO(J = 2-1) lines, defined as $\int vT(v)dv/\int T(v)dv$ where $T(v)$ is the line intensity as a function of velocity. The map of ^{12}CO(2-1) emission shown in Figure 9 (left) comprises almost 10^5 independent spectra obtained with the IRAM-30m telescope at an angular resolution of 11arcsec corresponding to ∼20 mpc at the distance of the cloud ($d = 350$ pc). The map of the thermal dust emission of the same field, Figure 9 (second panel), obtained at 250 µm with the *Herschel* satellite, displays the projected distribution of matter in the field, including bright and dense regions in the bottom of the map that harbour pre-stellar dense cores. The rectangular box encompasses three very faint dust filaments connected to the main massive structure.

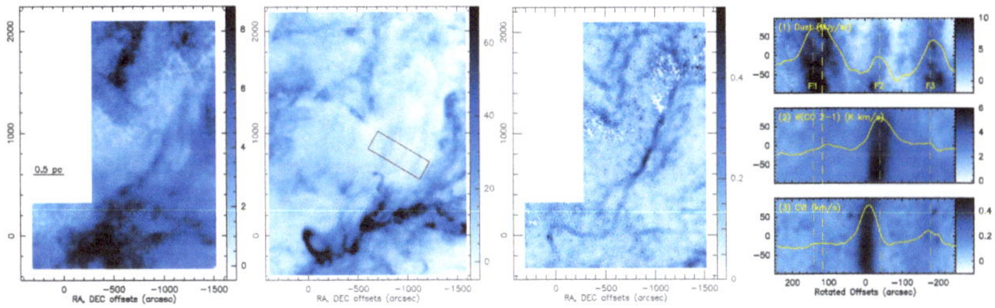

Figure 9. From left to right: Parsec-scale maps in the Polaris Flare of (1) the integrated ^{12}CO(2-1) line emission (expressed in K km s^{-1}) [50], (2) the dust continuum emission (in MJy/sr) measured at 250 µm by the SPIRE bolometers aboard the *Herschel* satellite [33], (3) the ^{12}CO(2-1) line centroid velocity increments (CVI, in km s^{-1}) measured at a lag of 60 arcsec (or 0.1 pc at d = 350 pc) Rightmost panel: Blow-up of the same quantities within the box drawn on the dust emission map and rotated by 30 deg: it encompasses three dust filaments among the weakest detected by *Herschel*/SPIRE. The yellow curves provide the quantities averaged along the filament directions: they show that the central filament, F2, barely detected in the dust emission, is the brightest in the CO(2-1) and CVI maps.

The probability distribution functions of the CO(2-1) line centroid velocity increments (CVI) between two positions separated by a lag ℓ, develop non-Gaussian wings, the wing intensity increasing with decreasing lag [49,50]. The locus of the positions contributing to these non-Gaussian wings (hereafter, the ECVI structure for extreme CVI structure) is shown in Figure 9 (third panel); as is evident, these large increments are not randomly distributed but form coherent structures, the most prominent of which is a parsec-long, thin filament (in the projection) that resembles coherent structures found in incompressible and compressible turbulence [37,75–77]. The velocity gradient reaches 40 km s^{-1} pc^{-1}, about 40× larger than the average velocity shear in molecular gas [50]. Furthermore, the coherent structures are themselves not random and seem to cluster around the most prominent one [46].

Interestingly, the behavior of the high-order structure functions of the CO centroid velocities with order p, measured in that Polaris field and in the Taurus molecular cloud, closely follows that found for the velocity in the numerical simulations of compressible MHD turbulence of Richard et al. [48] (Figure 8).

It is clear from Figure 9 (right) that the ECVI structure is bright in terms of the CO line emission, but almost undetected in the dust continuum. This ECVI structure is 10× more CO rich, given its total gas column density inferred from its dust sub mm brightness, than the two other nearby filaments that have the properties of diffuse molecular gas with an H_2 fraction of 0.3, inferred from the CO line width a standard CO-to-H_2 ratio.

High-angular resolution observations with the IRAM-NOEMA interferometer unveil similarly straight and elongated structures at the mpc scale, embedded within the ECVI structure and parallel to its parsec-scale orientation (Hily-Blant et al., in prep.). Not only the orientation of the large- and small-scale structures is preserved over three orders of magnitude in scale, but also the magnitude of the velocity differences. The remarkable coincidence at the 3 arcsec NOEMA resolution (in projection) of two elongated structures separated by 3 $\mathrm{km\,s^{-1}}$ reveals a local velocity shear larger than 600 $\mathrm{km\,s^{-1}\,pc^{-1}}$, or a dynamical timescale shorter than 1000 yr. Sharp variations across the ECVI structure of both the CO abundance and excitation are also derived from the $^{13}CO/^{12}CO$ and CO(2-1)/CO(1-0) line ratios.

The emergent scenario of CO-enriched gas produced by the intense velocity shear is inspired by the models of warm chemistry in turbulent dissipation regions of Godard et al. [40] and the numerical results of Richard et al. [48]. This CO-rich filament, associated with a velocity shear 600$\times$ larger than the average value in local molecular clouds, may be the first direct detection of the outcome of warm chemistry locally driven by a turbulent dissipation burst. The mpc scale is still far larger than the viscous dissipation scale, but these results unveil the possible role of intermittent turbulent dissipation in seeding the growth of molecular structures in diffuse gas, as suggested in Falgarone et al. [47].

The detailed modellings of the chemical outcome of turbulent dissipation are hard to reconcile with a coherent description of the energy cascade from the large scales of turbulence down to the dissipation scales, including intermittency. It has been attempted in two dimensions by Lesaffre et al. [55] and through the chemical post-processing of state-of-the-art numerical simulations of MHD turbulence, including ion-neutral drift (e.g., Moseley et al. [78]). These results are promising, but nevertheless, the former simulations are not magnetised, and the latter simulations are far from resolving the dissipation scales.

3.2. *Turbulent Dissipation in the Circum-Galactic Medium of A Galaxy Group at Redshift $z = 2.8$*

Another illustration of the power of chemistry fed by turbulent dissipation bursts is the analysis of the dynamics of kpc-scale highly turbulent environments of starburst galaxies at redshift $z = 2$ to 6 (Vidal-García et al. [51], Falgarone et al. [79]).

In numerical simulations, galaxies grow by accreting cold streams, modulo ejection of matter by stars and Active Galactic Nuclei (AGN) (Madau and Dickinson [1]). While the feedback from AGNs and star formation is observed through powerful outflows [3], cold-stream accretion is still elusive. The momentum exchange between the streams and a growing galaxy is so violent that a large turbulent region is generated around the galaxy. An indirect way of detecting cold stream accretion is therefore the detection of large turbulent reservoirs around galaxies. This is what has been achieved with the detection of the CH^+(1-0) line in emission and/or in absorption in all the gravitationally lensed starburst galaxies observed so far with ALMA at redshifts $z{\sim}2$ to 4 [79]. The unique spectroscopic and chemical properties of CH^+ allow for its $J = 1$-0 transition to highlight the sites of dissipation of mechanical energy. Moreover, it is such a fragile species that it cannot be transported and is observed where it forms. Absorption lines reveal massive (a few 10^{10} $M_\odot$), highly turbulent reservoirs of low density extending far out of the galaxies. Broad emission lines, with widths up to a few thousands of $\mathrm{km\,s^{-1}}$, arise in myriad low-velocity molecular shocks (Lehmann et al. [30]) powered by the feedback of star formation and AGNs.

4. Conclusions and Perspectives

We have used numerical simulations to build synthetic observables projected on the plane of the sky. From these synthetic maps, we have for the first time computed the plane-of-sky increments of these variables and measured their structure functions at various scales (or lags). We have shown that the large deviations in the smallest lags correlate spatially with the line-of-sight integrated dissipation: we discovered that the statistics at the shortest lags may be a way to detect the regions of strong dissipation on the sky plane. We hence provide a method to potentially detect dissipation structures from observables, and in addition, we suggest a new characterisation of the intermittency of cosmic turbulence through the statistical properties of the structure functions based on the increments of these observables.

Very suprisingly, we find that ESS applies to these structure functions. The "E" in "ESS" originally stands for "extended". This means that, when scales are remapped onto the third-order structure function, the range of scales where the self-similar scalings of the structure functions applies is considerably extended outside the inertial range. In this work, we have now shown that this ESS property can be furthermore extended. Indeed, it seems to apply to the following:

- In a situation outside stationary driven turbulence (we are in a case of decaying turbulence).
- For variables other than the proper transfer functions predicted by theory (such as Galtier and Banerjee [61] or Banerjee and Galtier [63] who nail the scaling for the order $p = 3$ in the inertial range and for very specific transfer functions).
- For projected variables and plane-of-sky increments instead of the actual 3D increments.

For all these reasons, we feel that there might be something more fundamental behind the ESS, with a lot still to be understood in the statistics of turbulence which may explain such behaviour of the structure functions.

Because ESS applies to our data, it makes sense to measure intermittency exponents. These exponents vary in time and according to the initial conditions. The velocity exponents lie close to the models of Grauer et al. [69] and Politano and Pouquet [70], designed for incompressible MHD and increments of Elsasser functions assuming an Iroshnikov–Kraichnan scaling. The exponents for the column density lie rather close to the model of Boldyrev et al. [71], a similar model based on the K41 scaling. Nevertheless, since we are far from the domain of the application of both of these theories, it is still premature to interpret these results as good matches. Meanwhile, we prefer to use the measurements of these intermittency exponents as a way to constrain quantitatively the geometry of the structures as they are observed on the plane of the sky.

The mere existence of large-scale and thin coherent structures in the observations of cosmic turbulence provide unique information on their stability. Thorough investigations of the predicted anisotropy of magnetised turbulence (i.e., structures linked to the intermittency of turbulence develop in the non-linear cascade perpendicular to the magnetic fields while coherence is driven by waves propagating along the fields, as in Schekochihin [18]), its balanced or imbalanced characteristics and the corresponding scaling laws (see the review by Schekochihin [18]) are a challenging opening for future observations.

Last, isothermal simulations are still very idealised compared to the complexity of the interstellar medium. The cooling, heating, chemistry and time-dependent excitation of molecules have to be included in order to bridge the gap between the simulations and observations. This has been done only in 2D without magnetic fields [55]: there is yet a lot more to be achieved in order to interpret observations. However, detailed computations in 1D for a well-chosen statistical collection of dissipative structures may be used to access the complexity of ISM turbulence. Simplified simulations in a multi-dimensional framework can help us pinpoint these statistics and hopefully understand what the interstellar medium has to tell us on magnetised turbulence. Another extension of this work will be the inclusion of idealised noise on our synthetic images to control the biases introduced by the instrumental defects or even by the unavoidable photon noise in weak signal data. Future studies should also focus on the distortions introduced by the radiative transfer which may differ from a simple projection in cases where the emission of the tracer molecules is not optically thin.

Finally, the observational study we propose here could be applied to turbulence in external galaxies provided they are not too far (i.e., they are well resolved by the observations). Any turbulent region is indeed amenable to this kind of study (statistics of increments measurements from observables), provided that (1) it is well separated spatially from any background that, if present, would blur the study, and (2) it comprises a large enough number of independent data points to allow for robust statistics. Possible targets are high-latitude galactic clouds, star formation regions, nearby galaxies but also massive circum-galactic environments and intra-cluster media.

Author Contributions: Conceptualisation and methodology, P.L., E.F. and P.H.-B.; software, P.L.; astronomical data acquisition, P.H.-B. and E.F., supervision, and project administration, E.F. All authors have read and agreed to the published version of the manuscript.

Funding: The research leading to these results has received funding from the European Research Council, under the European Community's Seventh framework Programme, through the Advanced Grant MIST (FP7/2017–2024, No. 742719).

Institutional Review Board Statement: Not applicable.

Informed Consent Statement: Not applicable.

Data Availability Statement: The data presented in this study are available on request to the corresponding author. The data are not publicly available due to privacy.

Acknowledgments: We thank our two anonymous referees for remarks which helped enlarge the scope of our results.

Conflicts of Interest: The authors declare no conflict of interest.

References

1. Madau, P.; Dickinson, M. Cosmic Star-Formation History. *ARA&A* **2014**, *52*, 415–486. [CrossRef]
2. Tacconi, L.J.; Genzel, R.; Sternberg, A. The Evolution of the Star-Forming Interstellar Medium Across Cosmic Time. *Annu. Rev. Astron. Astrophys.* **2020**, *58*, 157–203. [CrossRef]
3. Veilleux, S.; Maiolino, R.; Bolatto, A.D.; Aalto, S. Cool outflows in galaxies and their implications. *A&A Rev.* **2020**, *28*, 2. [CrossRef]
4. Hennebelle, P.; Falgarone, E. Turbulent molecular clouds. *A&ARv* **2012**, *20*, 55. [CrossRef]
5. White, S.D.M.; Rees, M.J. Core condensation in heavy halos: A two-stage theory for galaxy formation and clustering. *Mon. Not. R. Astron. Soc.* **1978**, *183*, 341–358. [CrossRef]
6. Chandrasekhar, S.; Fermi, E. Problems of Gravitational Stability in the Presence of a Magnetic Field. *Astrophys. J.* **1953**, *118*, 116. [CrossRef]
7. Meneveau, C.; Sreenivasan, K.R. The multifractal nature of turbulent energy dissipation. *J. Fluid Mech.* **1991**, *224*, 429–484. [CrossRef]
8. She, Z.S.; Leveque, E. Universal scaling laws in fully developed turbulence. *Phys. Rev. Lett.* **1994**, *72*, 336–339. [CrossRef]
9. Alexakis, A.; Mininni, P.D.; Pouquet, A. Shell-to-shell energy transfer in magnetohydrodynamics. I. Steady state turbulence. *Phys. Rev. E* **2005**, *72*, 046301. [CrossRef]
10. Alexakis, A.; Mininni, P.D.; Pouquet, A. Imprint of Large-Scale Flows on Turbulence. *Phys. Rev. Lett.* **2005**, *95*, 264503. [CrossRef] [PubMed]
11. Moffatt, H.K.; Kida, S.; Ohkitani, K. Stretched vortices—The sinews of turbulence; large-Reynolds-number asymptotics. *J. Fluid Mech.* **1994**, *259*, 241–264. [CrossRef]
12. Uritsky, V.M.; Pouquet, A.; Rosenberg, D.; Mininni, P.D.; Donovan, E.F. Structures in magnetohydrodynamic turbulence: Detection and scaling. *Phys. Rev. Stat. Nonlinear Soft Matter Phys.* **2010**, *82*, 1–15. [CrossRef] [PubMed]
13. Kimura, Y.; Sullivan, P.; Herring, J. Formation of temperature front in stably stratified turbulence. In Proceedings of the APS Division of Fluid Dynamics Meeting Abstracts, 2016, APS Meeting Abstracts, Portland, Oregonp, 20–22 November 2016; p. D35.005.
14. Kimura, Y.; Sullivan, P.P. 2D and 3D Properties of Stably Stratified Turbulence. *Atmosphere* **2024**, *15*, 82. [CrossRef]
15. Cadot, O.; Douady, S.; Couder, Y. Characterization of the low-pressure filaments in a three-dimensional turbulent shear flow. *Phys. Fluids* **1995**, *7*, 630–646. [CrossRef]
16. Tabeling, P.; Zocchi, G.; Belin, F.; Maurer, J.; Willaime, H. Probability density functions, skewness, and flatness in large Reynolds number turbulence. *Phys. Rev. E* **1996**, *53*, 1613–1621. [CrossRef] [PubMed]
17. Politano, H.; Pouquet, A. Dynamical length scales for turbulent magnetized flows. *Geophys. Res. Lett.* **1998**, *25*, 273–276. [CrossRef]
18. Schekochihin, A.A. MHD turbulence: A biased review. *J. Plasma Phys.* **2022**, *88*, 155880501. [CrossRef]
19. Field, G.B. Thermal Instability. *Astrophys. J.* **1965**, *142*, 531. [CrossRef]
20. Ibáñez-Mejía, J.C.; Walch, S.; Ivlev, A.V.; Clarke, S.; Caselli, P.; Joshi, P.R. Dust charge distribution in the interstellar medium. *MNRAS* **2019**, *485*, 1220–1247. [CrossRef]
21. Lesaffre, P. *Dynamics of the Galactic Matter Cycle*; Habilitation à Diriger des Recherches, Observatoire de Paris —PSL: Paris, France, 2018.
22. Draine, B.T. *Physics of the Interstellar and Intergalactic Medium*; Princeton University Press: Princeton, NJ, USA, 2011.
23. Balbus, S.A.; Terquem, C. Linear Analysis of the Hall Effect in Protostellar Disks. *Astrophys. J.* **2001**, *552*, 235–247. [CrossRef]
24. Momferratos, G.; Lesaffre, P.; Falgarone, E.; Pineau des Forêts, G. Turbulent energy dissipation and intermittency in ambipolar diffusion magnetohydrodynamics. *Mon. Not. R. Astron. Soc.* **2014**, *443*, 86–101. [CrossRef]
25. Ferrière, K. Magnetic fields and UHECR propagation. *Eur. Phys. J. Web Confs.* **2023**, *283*, 03001. [CrossRef]

26. Audit, E.; Hennebelle, P. On the structure of the turbulent interstellar clouds . Influence of the equation of state on the dynamics of 3D compressible flows. *A&A* **2010**, *511*, A76. [CrossRef]
27. Kulsrud, R.; Pearce, W.P. The Effect of Wave-Particle Interactions on the Propagation of Cosmic Rays. *Astrophys. J.* **1969**, *156*, 445. [CrossRef]
28. Draine, B.T.; Katz, N. Magnetohydrodynamic Shocks in Diffuse Clouds. II. Production of CH +, OH, CH, and Other Species. *Astrophys. J.* **1986**, *310*, 392. [CrossRef]
29. Flower, D.R.; Pineau des Forets, G. C-type shocks in the interstellar medium: Profiles of CH$\hat{+}$ and CH absorption lines. *MNRAS* **1998**, *297*, 1182–1188. [CrossRef]
30. Lehmann, A.; Godard, B.; Pineau des Forêts, G.; Vidal-García, A.; Falgarone, E. Self-generated ultraviolet radiation in molecular shock waves. II. CH^+ and the interpretation of emission from shock ensembles. *A&A* **2022**, *658*, A165. [CrossRef]
31. Armstrong, J.W.; Rickett, B.J.; Spangler, S.R. Electron Density Power Spectrum in the Local Interstellar Medium. *Astrophys. J.* **1995**, *443*, 209. [CrossRef]
32. Stanimirović, S.; Zweibel, E.G. Atomic and Ionized Microstructures in the Diffuse Interstellar Medium. *ARA&A* **2018**, *56*, 489–540. [CrossRef]
33. Miville-Deschênes, M.A.; Martin, P.G.; Abergel, A.; Bernard, J.P.; Boulanger, F.; Lagache, G.; Anderson, L.D.; André, P.; Arab, H.; Baluteau, J.P.; et al. Herschel-SPIRE observations of the Polaris flare: Structure of the diffuse interstellar medium at the sub-parsec scale. *A&A* **2010**, *518*, L104. [CrossRef]
34. Miville-Deschênes, M.A.; Duc, P.A.; Marleau, F.; Cuillandre, J.C.; Didelon, P.; Gwyn, S.; Karabal, E. Probing interstellar turbulence in cirrus with deep optical imaging: no sign of energy dissipation at 0.01 pc scale. *A&A* **2016**, *593*, A4. [CrossRef]
35. Larson, R.B. Turbulence and star formation in molecular clouds. *MNRAS* **1981**, *194*, 809–826. [CrossRef]
36. Heyer, M.; Krawczyk, C.; Duval, J.; Jackson, J.M. Re-Examining Larson's Scaling Relationships in Galactic Molecular Clouds. *Astrophys. J.* **2009**, *699*, 1092–1103. [CrossRef]
37. Kritsuk, A.G.; Norman, M.L.; Padoan, P.; Wagner, R. The Statistics of Supersonic Isothermal Turbulence. *Astrophys. J.* **2007**, *665*, 416–431. [CrossRef]
38. Field, G.B.; Blackman, E.G.; Keto, E.R. Does external pressure explain recent results for molecular clouds? *MNRAS* **2011**, *416*, 710–714. [CrossRef]
39. Miville-Deschênes, M.A.; Murray, N.; Lee, E.J. Physical Properties of Molecular Clouds for the Entire Milky Way Disk. *Astrophys. J.* **2017**, *834*, 57. [CrossRef]
40. Godard, B.; Falgarone, E.; Pineau des Forêts, G. Chemical probes of turbulence in the diffuse medium: The TDR model. *A&A* **2014**, *570*, A27. [CrossRef]
41. Lucas, R.; Liszt, H. The Plateau de Bure survey of galactic λ3mm HCO$\hat{+}$âbsorption toward compact extragalactic continuum sources. *A&A* **1996**, *307*, 237.
42. Godard, B.; Pineau des Forêts, G.; Hennebelle, P.; Bellomi, E.; Valdivia, V. 3D chemical structure of the diffuse turbulent interstellar medium. II. The origin of CH^+: A new solution to an 80-year mystery. *A&A* **2023**, *669*, A74. [CrossRef]
43. Flower, D.R.; Pineau des Forêts, G.; Hartquist, T.W. Theoretical studies of interstellar molecular shocks. I - General formulation and effects of the ion-molecule chemistry. *MNRAS* **1985**, *216*, 775–794. [CrossRef]
44. Federman, S.R.; Rawlings, J.M.C.; Taylor, S.D.; Williams, D.A. Synthesis of interstellar CH$\hat{+}$ without OH. *MNRAS* **1996**, *279*, L41–L46. [CrossRef]
45. Crutcher, R.M. Magnetic Fields in Molecular Clouds. *ARA&A* **2012**, *50*, 29–63. [CrossRef]
46. Moisy, F.; Jiménez, J. Geometry and clustering of intense structures in isotropic turbulence. *J. Fluid Mech.* **2004**, *513*, 111–133. [CrossRef]
47. Falgarone, E.; Pety, J.; Hily-Blant, P. Intermittency of interstellar turbulence: Extreme velocity-shears and CO emission on milliparsec scale. *A&A* **2009**, *507*, 355–368. [CrossRef]
48. Richard, T.; Lesaffre, P.; Falgarone, E.; Lehmann, A. Probing the nature of dissipation in compressible MHD turbulence. *A&A* **2022**, *664*, A193. [CrossRef]
49. Hily-Blant, P.; Falgarone, E.; Pety, J. Dissipative structures of diffuse molecular gas. III. Small-scale intermittency of intense velocity-shears. *A&A* **2008**, *481*, 367–380. [CrossRef]
50. Hily-Blant, P.; Falgarone, E. Intermittency of interstellar turbulence: Parsec-scale coherent structure of intense, velocity shear. *A&A* **2009**, *500*, L29–L32. [CrossRef]
51. Vidal-García, A.; Falgarone, E.; Arrigoni Battaia, F.; Godard, B.; Ivison, R.J.; Zwaan, M.A.; Herrera, C.; Frayer, D.; Andreani, P.; Li, Q.; et al. Where infall meets outflows: Turbulent dissipation probed by CH^+ and Lyα in the starburst/AGN galaxy group SMM J02399-0136 at z 2.8. *MNRAS* **2021**, *506*, 2551–2573. [CrossRef]
52. Zhdankin, V.; Uzdensky, D.A.; Perez, J.C.; Boldyrev, S. Statistical analysis of current sheets in three-dimensional magnetohydrodynamic turbulence. *Astrophys. J.* **2013**, *771*, 124. [CrossRef]
53. Lehmann, A.; Federrath, C.; Wardle, M. SHOCKFIND—An algorithm to identify magnetohydrodynamic shock waves in turbulent clouds. *Mon. Not. R. Astron. Soc.* **2016**, *463*, 1026–1039. [CrossRef]
54. Macquorn Rankine, W.J. On the Thermodynamic Theory of Waves of Finite Longitudinal Disturbance. *Philos. Trans. R. Soc. Lond. Ser. I* **1870**, *160*, 277–288.

55. Lesaffre, P.; Todorov, P.; Levrier, F.; Valdivia, V.; Dzyurkevich, N.; Godard, B.; Tram, L.N.; Gusdorf, A.; Lehmann, A.; Falgarone, E. Production and excitation of molecules by dissipation of two-dimensional turbulence. *Mon. Not. R. Astron. Soc.* **2020**, *495*, 816–834. [CrossRef]
56. Marino, R.; Sorriso-Valvo, L. Scaling laws for the energy transfer in space plasma turbulence. *Phys. Rep.* **2023**, *1006*, 1–144. [CrossRef]
57. Kolmogorov, A. The Local Structure of Turbulence in Incompressible Viscous Fluid for Very Large Reynolds' Numbers. *Akad. Nauk. Sssr Dokl.* **1941**, *30*, 301–305.
58. Iroshnikov, P.S. Turbulence of a Conducting Fluid in a Strong Magnetic Field. *Astron. Zhurnal* **1963**, *40*, 742.
59. Kraichnan, R.H. Inertial-Range Spectrum of Hydromagnetic Turbulence. *Phys. Fluids* **1965**, *8*, 1385–1387. [CrossRef]
60. Goldreich, P.; Sridhar, S. Toward a Theory of Interstellar Turbulence. II. Strong Alfvenic Turbulence. *Astrophys. J.* **1995**, *438*, 763. [CrossRef]
61. Galtier, S.; Banerjee, S. Exact Relation for Correlation Functions in Compressible Isothermal Turbulence. *Phys. Rev. Lett.* **2011**, *107*, 134501. [CrossRef]
62. Federrath, C.; Klessen, R.S.; Iapichino, L.; Beattie, J.R. The sonic scale of interstellar turbulence. *Nat. Astron.* **2021**, *5*, 365–371. [CrossRef]
63. Banerjee, S.; Galtier, S. Exact relation with two-point correlation functions and phenomenological approach for compressible magnetohydrodynamic turbulence. *Phys. Rev. E* **2013**, *87*, 013019. [CrossRef]
64. Lazarian, A.; Pogosyan, D. Velocity Modification of H I Power Spectrum. *Astrophys. J.* **2000**, *537*, 720–748. [CrossRef]
65. Kim, J.; Ryu, D. Density Power Spectrum of Compressible Hydrodynamic Turbulent Flows. *Astrophys. J. Lett.* **2005**, *630*, L45–L48. [CrossRef]
66. Saury, E.; Miville-Deschênes, M.A.; Hennebelle, P.; Audit, E.; Schmidt, W. The structure of the thermally bistable and turbulent atomic gas in the local interstellar medium. *A&A* **2014**, *567*, A16. [CrossRef]
67. Kolmogorov, A.N. A refinement of previous hypotheses concerning the local structure of turbulence in a viscous incompressible fluid at high Reynolds number. *J. Fluid Mech.* **1962**, *13*, 82–85. [CrossRef]
68. Frisch, U. *Turbulence*; Springer: Berlin/Heidelberg, Germany, 1996; p. 310.
69. Grauer, R.; Krug, J.; Marliani, C. Scaling of high-order structure functions in magnetohydrodynamic turbulence. *Phys. Lett. A* **1994**, *195*, 335–338. [CrossRef]
70. Politano, H.; Pouquet, A. Model of intermittency in magnetohydrodynamic turbulence. *Phys. Rev. E* **1995**, *52*, 636–641. [CrossRef] [PubMed]
71. Boldyrev, S.; Nordlund, Å.; Padoan, P. Scaling Relations of Supersonic Turbulence in Star-forming Molecular Clouds. *Astrophys. J.* **2002**, *573*, 678–684. [CrossRef]
72. Chevillard, L.; Robert, R.; Vargas, V. Random vectorial fields representing the local structure of turbulence. *J. Phys. Conf. Ser.* **2011**, *318*, 042002. [CrossRef]
73. Durrive, J.B.; Lesaffre, P.; Ferrière, K. Magnetic fields from multiplicative chaos. *MNRAS* **2020**, *496*, 3015–3034. [CrossRef]
74. Benzi, R.; Biferale, L.; Ciliberto, S.; Struglia, M.V.; Tripiccione, R. Scaling property of turbulent flows. *Phys. Rev. E* **1996**, *53*, R3025–R3027. [CrossRef]
75. Douady, S.; Couder, Y.; Brachet, M.E. Direct observation of the intermittency of intense vorticity filaments in turbulence. *Phys. Rev. Lett.* **1991**, *67*, 983–986. [CrossRef] [PubMed]
76. Ishihara, T.; Hunt, J.C.R.; Kaneda, Y. Intense dissipative mechanisms of strong thin shear layers in high Reynolds number turbulence. In Proceedings of the APS Division of Fluid Dynamics Meeting Abstracts, 2012, APS Meeting Abstracts, San Diego, CA, USA, 18–20 November 2012; p. A23.004.
77. Buaria, D.; Pumir, A.; Bodenschatz, E.; Yeung, P.K. Extreme velocity gradients in turbulent flows. *New J. Phys.* **2019**, *21*, 04300. [CrossRef]
78. Moseley, E.R.; Draine, B.T.; Tomida, K.; Stone, J.M. Turbulent dissipation, CH^+ abundance, H_2 line luminosities, and polarization in the cold neutral medium. *MNRAS* **2021**, *500*, 3290–3308. [CrossRef]
79. Falgarone, E.; Zwaan, M.A.; Godard, B.; Bergin, E.; Ivison, R.J.; Andreani, P.M.; Bournaud, F.; Bussmann, R.S.; Elbaz, D.; Omont, A.; et al. Large turbulent reservoirs of cold molecular gas around high-redshift starburst galaxies. *Nature* **2017**, *548*, 430–433. [CrossRef] [PubMed]

atmosphere

Article

Zonons Are Solitons Produced by Rossby Wave Ringing

Nimrod Cohen [1], Boris Galperin [2,*] and Semion Sukoriansky [1,*]

1 Department of Mechanical Engineering, Ben-Gurion University of the Negev, Beer-Sheva 84105, Israel; conim@post.bgu.ac.il
2 College of Marine Science, University of South Florida, St. Petersburg, FL 33701, USA
* Correspondence: bgalperin@usf.edu (B.G.); semion@bgu.ac.il (S.S.); Tel.: +1-727-553-1249 (B.G.)

Abstract: Along with the familiar Rossby–Haurwitz waves, two-dimensional flows on the surface of a rotating sphere in the regime of zonostrophic turbulence harbor another class of waves known as zonons. Zonons are wave packets produced by energetic large-scale Rossby–Haurwitz wave modes 'enslaving' other wave modes. They propagate westward with the phase speed of the enslaving modes. Zonons can be visualized as enslaving modes' 'ringing' in the enslaved ones with the frequencies of the former, the property that renders zonons non-dispersive. Zonons reside in high-shear regions confined between the opposing zonal jets yet they are mainly attached to westward jets and sustained by the ensuing barotropic instability. They exchange energy with the mean flow while preserving their identity in a fully turbulent environment, a feature characteristic of solitary waves. The goal of this study is to deepen our understanding of zonons' physics using direct numerical simulations, a weakly non-linear theory, and asymptotic analysis, and ascertain that zonons are indeed isomorphic to solitary waves in the Korteweg–de Vries framework. Having this isomorphism established, the analysis is extended to eddies detected in the atmospheres of Jupiter and Saturn based upon the observed mean zonal velocity profiles and earlier findings that circulations on both planets obey the regime of zonostrophic macroturbulence. Not only the analysis confirms that many eddies and eddy trains on both giant planets indeed possess properties of zonons, but the theory also correctly predicts latitudinal bands that confine zonal trajectories of the eddies.

Keywords: zonostrophic turbulence; eddies; solitary waves; giant planets

Citation: Cohen, N.; Galperin, B.; Sukoriansky, S. Zonons Are Solitons Produced by Rossby Wave Ringing. *Atmosphere* **2024**, *15*, 711. https://doi.org/10.3390/atmos15060711

Academic Editor: Leonardo Primavera

Received: 25 April 2024
Revised: 21 May 2024
Accepted: 26 May 2024
Published: 14 June 2024

1. Introduction

The paper by Salmon and Pizzo in this Special Issue [1] refers to fundamental contributions of Jack Herring to the theories of two-dimensional (2D) and quasigeostrophic turbulence [2–4]. The authors notice that much of the theoretical developments in these areas were accomplished in planar geometry in the hope that the β-plane approximation would capture the most important features of flows in the spherical geometry. Their primary conclusion, however, was that this is not the case, and that 2D incompressible flows on the sphere are significantly different from 2D flows on a β-plane, even in the case of vanishing rotation.

The situation is different for fast rotating flows with a β-effect, however, as on the largest scales, β-effect becomes predominant, and the interaction of the Rossby waves, or Rossby-Haurwitz waves (RHWs) in the spherical geometry, and turbulence leads to strong turbulence anisotropization, establishment of the regime of zonostrophic turbulence, see e.g., [5,6], and formation of zonal jets. As was demonstrated in [7,8], various features of zonostrophic turbulence, including its spectral properties, are amicable to representation in both planar and spherical geometries. Many facets of this dynamics are detailed in the recent book [9].

An important feature of the zonostrophic regime is the excitation of non-dispersive nonlinear waves termed zonons first described in [10]. These waves coexist with the linear RHWs and their physics is at the focus of the present study. In [10] it was established that zonons are forced oscillations induced by RHWs in other modes via nonlinear interactions, and their dispersion relation differs from that of the RHWs. Zonons reside in regions with a strong shear between opposing jets acting as wave guides for the long-living coherent wave packets. In the physical space, as shown in [11], zonons form westward propagating coherent eddies with zonal velocity equal to the phase speed of the most energetic RHWs.

Zonons can be viewed as a product of the interaction between energetic, or "ringing" RHWs and their overtones excited in other modes. This phenomenon is reminiscent of enslaved vibrations excited by forced, ringing strings of a musical instrument in quiescent, echoing strings. Once excited, the sound waves from the echoing strings blend with those from the ringing strings and form a specific chromatic signal slowly emanating from the body of the musical instrument as a virtually self-preserving "acoustic soliton", see e.g., [12]. The underlying mechanism behind the integration of ringing and echoing vibrations is the nonlinear interaction between the energetic and enslaved modes.

Is there anything in common between zonons and another type of ringing waves, recently discovered in [13]? They found that randomly excited global-scale atmospheric modes may come to a resonance with the lunar and solar atmospheric tidal waves described by the Laplace equation. In their words, these resonant interactions can make the atmosphere "ring like a bell".

While the resonant wave interactions are linear, the processes underlying the formation and maintenance of zonons are not. As they emerge from the typical for solitary waves nonlinear interaction between wave packets and the shear, one may hypothesize that zonons are indeed solitary waves, such as those studied in [14,15]. The latter papers focused on long waves in a homogeneous atmosphere with a uniform zonal flow and small velocity shear and arrived at the Korteweg–de Vries (KdV) equation. Following up studies [16,17] and others developed a general theory of solitary waves in a planetary zonal shear flow and demonstrated that the amplitude of long Rossby waves is governed by either the KdV equation or the modified KdV (mKdV) equation, dependent on the density stratification, see e.g., [18]. Later, a multi-scale method utilized in [19,20] showed that the primitive governing equation can be reduced to the KdV equation controlling the evolution of long, small-amplitude Rossby waves.

In [11], the connection between the solitary Rossby waves (SRWs) and zonons was outlined in the framework of the 2D barotropic vorticity equation on a β-plane. That study was akin to [16] where SRWs were investigated on a β-plane using the vorticity equation expanded around latitude θ_0 in the absence of forcing and damping. A direct mathematical link between zonons and solitary Rossby waves was not established, however. Substantiating such link is one of the main goals of this paper.

Other studies were concerned with eddies' energetics. In [21], this issue was considered through the prism of the zonal asymmetry of the Rossby wave propagation. The asymmetry prompts a sharp difference between the eastward and westward jets. The eastward jets are stable and sharp, while the westward ones are unstable and blunt. For Jupiter and Saturn, this asymmetry is illuminated in Figures 8 and 9 in Section 4 below, as well as in [22]. Westward jets are prone to barotropic and baroclinic instabilities that lead to eddy shedding [23,24], mixing of the potential vorticity (PV), and flattening of its profile across the jets [25]. The instabilities not only evoke eddies but also sustain them. The coexistence of eddies and jets can be viewed as eddy-jet symbiosis. The continuous two-way energy exchange between jets and eddies is facilitated by upscale and downscale turbulent cascades [11,26–28].

The dynamics of giant planets' atmospheres is strongly affected by the eddy–jet flow interaction [29–32]. For instance, almost all westward jets on Jupiter and Saturn are barotropically unstable [33–35] and shed cyclonic/anticyclonic eddies of various shapes and sizes.

Similar features are typical of the westward south-equatorial currents in all terrestrial oceans [21], where under the action of a seasonal forcing, the currents periodically become unstable and radiate the Tropical Instability Waves (TIWs), e.g., [36–38].

Do zonons and the SRWs [solitary Rossby-Haurwitz waves (SRHWs) on a sphere] represent different aspects of the same mathematical objects? This is an important non-trivial question as RHWs are spectral characteristics of a flow, while zonons construe it in the physical space.

The purpose of the present study is to answer this question for flows in the zonostrophic regime and thereby extend the results in [11]. The RHWs in a zonal flow on a surface of a rotating sphere will be scrutinized using the asymptotic multi-scale analysis and the assumption of weak nonlinearity. This technique yields the KdV equation in the spherical geometry, the hallmark of the solitary Rossby waves. For a given zonal velocity profile, the SRHWs phase speeds will be estimated by solving the generalized eigenvalue problem pertinent to the KdV equation. The RHWs velocities for barotropic flows can be computed analytically. The quantitative agreement between the two will be taken as an indication that the SRHWs and zonons are indeed two sides of the same coin.

An important question arises regarding the consistency of a mathematical framework describing forced and dissipative regime of zonostrophic turbulence harboring the inverse energy cascade and unforced, non-dissipative regime represented by the KdV equation featuring no cascades. We note that the backbone of both regimes is a quasi-steady zonal velocity profile that, while remaining intact, may develop shear instabilities.

In the KdV framework, the origin and the maintenance of the profile are immaterial yet the shear instabilities provide a waveguide for solitary waves [11]. In the zonostrophic regime, the profile is a product of a balance between the small-scale forcing and the large-scale energy withdrawal. As was noticed earlier, the profile may become barotropically unstable and instigate eddy shedding. As shown in [39], this process is sustained by the Available Rotational Kinetic Energy, or ARKE, which is a rotational equivalent of the available potential energy in flows with stable stratification. This energy, as well as the rate of the inverse cascade, can be estimated by a process of PV monotonizing in the horizontal plane [40] that is analogous to Thorpe's vertical reshuffling of fluid particles in stably stratified flows [41]. In [42] it was shown that the method of PV monotonizing allows one to estimate the inverse cascade rate on Jupiter and Saturn using a limited amount of data and with no information about details of the forcing. The mean zonal velocity profile not only contains information on the rate of the inverse cascade, but, as was shown in [21] and in analogy to the KdV framework, it sustains the energy and guides the propagation of coherent eddies. As will be shown in Section 3, the profile also reveals potential locations of the SRHWs and their trains.

Establishing the duality between SRHWs and zonons has important practical implications. For instance, it was found in [43] that the KdV equation describes a flow adjacent to the African easterly jet and helps to understand the evolution of a proto-vortex [44] growing into tropical storm Debby over West Africa in August 2006. That study hints that zonons may help to explain the formation of tropical cyclones and shed light upon their dynamics and evolution, at least during the initial stages.

The most far reaching application of the theory of Rossby solitons goes back to [45] where it was suggested that the SRWs may explain some facets of the dynamics of the Great Red Spot (GRS) and other features of the Jovian atmosphere. The application of the soliton solutions to the GRS was revisited in [46] while an extensive review of possible manifestations of SRWs in various planetary and geophysical environments was given in [47]. We shall see later that using the observed zonal velocity profiles, the KdV equation can quite accurately predict the latitudinal locations of the major eddies on both Jupiter and Saturn, including the GRS.

The investigation of the similarity between RHWs and zonons will proceed in the following fashion. The next section provides description of direct numerical simulations of the regime of zonostrophic turbulence which will be followed by the derivation of the KdV equation in a 2D flow on the surface of a rotating sphere. The following up section elaborates the existence and the features of the SRHWs. Section 3 elucidates the connections between the zonons and the SRHWs, Section 4 investigates the relationship between zonons and large eddies on Jupiter and Saturn, and Section 5 is discussion and conclusions.

2. The Regime of Zonostrophic Turbulence in Barotropic Simulations in Spherical Coordinate System

The barotropization of planetary atmospheres, i.e., accumulation of a large portion of their kinetic energy in the barotropic mode, was discussed in [6,26]. Both the barotropization and flow two-dimensionalization escalate with the increasing rate of rotation. Major features of the emerging flow regime can be represented by a mathematical model emulating barotropic, externally forced, dissipative 2D turbulence on the surface of a rotating sphere e.g., [6]. In a certain range of parameters, the flow attains the regime of zonostrophic turbulence with its inherent zonal jets and conjoined zonons. The model is based upon the vorticity evolution equation,

$$\frac{\partial \zeta}{\partial t} + J(\psi, q) = (-1)^{p+1} A_{2p} \nabla^{2p} \zeta - \lambda \zeta + \xi, \tag{1}$$

where ζ is the absolute vorticity, ψ is the stream function; $\nabla^2 \psi = \zeta$, and A_{2p} is the hyperviscosity coefficient in which p was set to 4. Furthermore, $q = \zeta + f$ is the PV, $f = 2\Omega \sin\theta$ is the Coriolis parameter; Ω is the angular velocity of sphere's rotation, and $\theta \in [-\frac{\pi}{2}, \frac{\pi}{2}]$ is the latitude. Since f is time-independent, Equation (1) could also be written in terms of the PV conservation. The Jacobian, $J(\psi, q)$, defined as $J(A, B) = (R^2 \cos\theta)^{-1}(A_\varphi B_\theta - A_\theta B_\varphi)$, with R and $\varphi \in [0, 2\pi]$ being sphere's radius and longitude, respectively, represents the nonlinear term in (1).

The flow governed by (1) features a small-scale forcing, ξ, with a zero mean, acting upon scales of the order of n_ξ^{-1}, and a large-scale energy withdrawal represented by a linear friction with the coefficient λ. The energy injected into the system by forcing drives the inverse cascade at a rate Π_ϵ. This cascade is terminated on large scales, of the order of n_{fr}^{-1}, by the linear friction that dominates on scales for which $n < n_{\mathrm{fr}}$. Under the action of forcing and friction, the flow eventually attains a steady state.

The stream function is decomposed into a series of spherical harmonics, $Y_n^m(\sin\theta, \varphi)$,

$$\psi(\theta, \varphi, t) = \sum_{n=1}^{N} \sum_{m=-n}^{n} \psi_n^m(t) Y_n^m(\sin\theta, \varphi). \tag{2}$$

They are eigenfunctions of the horizontal Laplacian operator, $\nabla_h^2 Y_n^m = -n(n+1) Y_n^m / R^2$, and depend on non-dimensional total and zonal indices, n and m, respectively.

Linearized Equation (1) that excludes forcing and dissipation admits a RHW solution whose dispersion relation is [6,48–50]

$$\omega_{RHW}(n, m) = -2\,\beta\,\frac{m}{n(n+1)} \tag{3}$$

yielding the phase speed

$$c_{RHW}(n) = \frac{\omega_{RHW}(n, m)}{m} = -\frac{2\beta}{n(n+1)}, \tag{4}$$

where $\beta = \Omega / R$. Equation (3) is consistent with a β-plane approximation format [49,51–53].

Equation (1) with the decomposition (2) was used for numerical investigation of various aspects of the nonlinear dynamics. Details of the numerical scheme and the simulations are given in [54].

Employing a unit radius, $R = 1$, eliminates the difference between the spherical harmonics' indices and the wave numbers. The energy spectrum can be computed following [55,56],

$$E(n) = \sum_{m=-n}^{n} \mathcal{E}(n,m) = \frac{n(n+1)}{4} \sum_{m=-n}^{n} \langle |\psi_n^m|^2 \rangle, \tag{5}$$

where the modal spectrum, $\mathcal{E}(n,m)$, is the spectral energy density per mode (n,m), and the angular brackets stand for ensemble or time average. The spectrum, $E(n)$, can be decomposed into a sum of zonal and nonzonal, i.e., residual components, $E(n) = E_Z(n) + E_R(n)$, where the zonal component is $E_Z(n) = \mathcal{E}(n,0)$.

In the regime of zonostrophic turbulence, the zonal and residual spectra acquire universal distributions [57] given by

$$E_Z(n) = C_Z \beta^2 n^{-5}, \qquad C_Z \approx 0.5, \tag{6a}$$

$$E_R(n) = C_R \Pi_\epsilon^{2/3} n^{-5/3}, \qquad C_R \approx 6. \tag{6b}$$

While the zonal spectrum (6a) develops a very steep slope, its residual counterpart (6b) closely follows the classical theory of isotropic 2D turbulence [58]. The intersection of (6a) and (6b) defines the transitional index, $n_\beta = (\beta^3/\Pi_\epsilon)^{1/5}$. The strength of anisotropy is characterized by the zonostrophy index, $R_\beta = n_\beta/n_R$ [54,59], where n_R is the Rhines index, $n_R = (\beta/2U)^{1/2}$, and U is the RMS zonal velocity. The threshold of the zonostrophic flow regime corresponds to $R_\beta \gtrsim 2.5$ [6,26,54,59,60].

Figure 1 shows the zonal and residual kinetic energy spectra obtained in 2D simulation using Equation (1). The parameters in this simulation were $\Omega = 1.4$, $R = 1$, and $R_\beta = 2.65$. The energy spectra in Figure 1 are in good agreement with Equations (6a) and (6b) clearly demonstrating the signature of the zonostrophic regime. Both spectra attain their maxima at a scale close to n_R^{-1}. On larger scales, friction becomes a dominant factor that causes the spectral amplitudes to rapidly decrease.

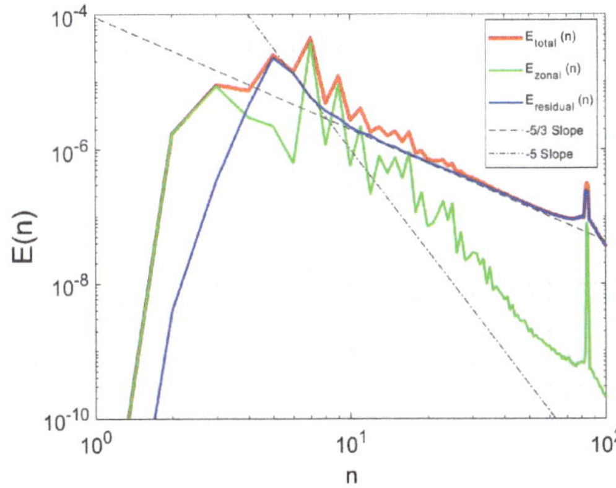

Figure 1. The zonal, $E_Z(n)$ (green line), residual, $E_R(n)$ (blue line), and total, $E(n)$ (red line) kinetic energy spectra obtained in numerical simulations of 2D turbulence on the surface of a rotating sphere in the zonostrophic regime ($R_\beta = 2.65$).

3. Rossby-Haurwitz Waves, Zonons, and Solitons

The zonal mode with the maximum energy, approximately quantified by $E_Z(n_R)$, dominates the zonal velocity profile while n_R divulges the corresponding number of alternating zonal jets [61]. Some of the westward jets feature a strong shear and are barotropically unstable [23,30,33,62]. Energized by incipient instabilities, nonlinear interactions between the RHWs lead to formation of nondispersive, westward propagating wave packets. In the physical space, these packets appear as coherent eddies termed zonons [10]. Zonons appear as eddies or trains of eddies riding on the most energetic RHWs [21,63], their speed being $c_{RHW}(n_R)$ [10,11]. Later it will be shown that zonons' dynamics is amenable to a description within the framework of the KdV equation, i.e., they possess properties of the solitary Rossby-Haurwitz waves (SRHWs). The congruity of RHWs, SRHWs and zonons will be demonstrated by exposing the consistency of their representations in the physical and Fourier spaces and closeness of their characteristic velocities. These issues will be addressed in the remainder of this section.

3.1. Flows on the Surface of a Rotating Sphere, the Mercator Projection, and the KdV Equation

As explained in the Introduction, the analysis of zonons is based upon consideration of a 2D flow on the surface of a rotating sphere absent external forcing, large-scale friction, and small-scale dissipation. The flow is governed by the barotropic vorticity Equation (1) modified to express the PV conservation [52]:

$$q_t + \frac{\psi_\varphi q_\theta - \psi_\theta q_\varphi}{R^2 \cos\theta} = 0, \tag{7a}$$

$$q = \frac{\cos\theta\left(\psi_\theta \cos\theta\right)_\theta + \psi_{\theta\theta}}{R^2 \cos^2\theta} + f. \tag{7b}$$

The variables in (7a), (7b) are nondimensionalized and marked by asterisks,

$$t = \Omega^{-1} t^*, \quad f = \Omega f^*, \quad \psi = \Omega R^2 \psi^*, \quad r = R r^*, \quad q = \Omega q^*. \tag{8}$$

In the nondimensional form, Equations (7a) and (7b) become

$$q^*_{t^*} + \frac{\psi^*_\varphi q^*_\theta - \psi^*_\theta q^*_\varphi}{\cos\theta} = 0, \tag{9a}$$

$$q^* = \frac{\cos\theta\left(\psi^*_\theta \cos\theta\right)_\theta + \psi^*_{\theta\theta}}{\cos^2\theta} + f^*. \tag{9b}$$

Further derivations are performed in the Mercator projection developed by the Flemish geographer and cartographer Gerardus Mercator in 1569. In this approach, the geographic map is projected from the globe to a tangential cylindrical surface touching the globe at the equator. The cylindrical surface is then unrolled into a planar map.

Another way of mapping a spherical surface onto a plane is by using a stereographic projection as done in [1]. A detailed comparison between the Mercator and stereographic projections is elaborated in [64]. Briefly, both techniques belong in a family of conformal mappings of a sphere onto a plane that only involve elementary calculus and trigonometry. In both cases, the equations governing 2D turbulence on a sphere become quite similar to the corresponding equations in the Cartesian coordinate system on a β-plane, e.g., [43].

The nondimensional Mercator projection variables, x^* and y^*, are defined by

$$x^* = r^* \varphi, \quad y^* = r^* \operatorname{arctanh}(\sin\theta) = r^* \operatorname{gd}^{-1}(\theta), \tag{10}$$

where $\operatorname{gd}^{-1}(\theta)$ is the inverse Gudermannian, and $r^* = 1$ on a sphere with the radius R.

The inverse Gudermannian in (10) maps $\theta \in [-\pi/2, \pi/2]$ onto $y^* \in (-\infty, \infty)$. The Gudermannian function was introduced by Johann Heinrich Lambert circa 1760 and later named after Christoph Gudermann who derived the relationship between circular and hyperbolic functions in 1830 [65]. The Gudermannian is also called the hyperbolic amplitude as it describes the limiting case of the Jacobi elliptic amplitude.

Using the new variables and dropping the asterisks, Equations (9a) and (9b) become

$$S^2 q_t + \psi_x q_y - \psi_y q_x = 0, \tag{11a}$$

$$q = \frac{1}{S^2}\left(\psi_{yy} + \psi_{xx}\right) + 2\Omega \tanh y, \tag{11b}$$

where $S \equiv \cos\theta = \operatorname{sech} y$.

Motivated by zonons' properties, we employ the asymptotic multiscale method that focuses upon large-scale, long-time solutions of Equations (11a) and (11b) using a long spatial and a slow temporal scales, X and T, respectively,

$$X = \varepsilon^{1/2}(x - c_0 t), \quad T = \varepsilon^{3/2} t. \tag{12}$$

Here, c_0 is a free parameter representing the speed of a coherent structure, and ε is a small parameter, $0 < \varepsilon \ll 1$, measuring the ratio of the fast and slow time scales. Using these notations, Equations (11a) and (11b) can be written in terms of X, y and T, and the stream function is then expanded in a series in powers of ε [11],

$$\psi(X, y, T) = \sum_{i=0}^{\infty} \varepsilon^i \psi^{(i)}. \tag{13}$$

Retaining a small $O(\varepsilon)$ term in (13), ψ can be written as a sum of the basic mean stream function, $\psi^{(0)}$, and a stream function of the coherent disturbance, $\psi^{(1)}$,

$$\psi(X, y, T) = \psi^{(0)} + \varepsilon\psi^{(1)} + O\left(\varepsilon^2\right). \tag{14}$$

A 2D velocity field on a sphere can be represented by its poloidal (meridional; $\hat{\boldsymbol{\theta}}$) and toroidal (azimuthal, zonal; $\hat{\boldsymbol{\varphi}}$) components,

$$\mathbf{u} = \nabla_\perp \psi = \frac{1}{R\cos\theta}\frac{\partial\psi}{\partial\varphi}\boldsymbol{\theta} - \frac{1}{R}\frac{\partial\psi}{\partial\theta}\boldsymbol{\varphi}. \tag{15}$$

Numerical simulations presented in [26,54,59] indicate that the zonal flow is comprised of alternating zonal jets with the amplitude $U_{jet}(\theta)$, where the mean velocity is $\langle \mathbf{u}(\theta) \rangle = U_{jet}(\theta)\hat{\boldsymbol{\varphi}}$, and so the zero-order stream function can be expressed via

$$U_{jet}(\theta) = -\frac{1}{R}\frac{d\psi^{(0)}}{d\theta}. \tag{16}$$

From (16) one infers that $\psi^{(0)}$ depends on θ only and can be computed by straightforward integration,

$$\psi^{(0)}(\theta) = -R\int_0^\theta U_{jet}(\theta')d\theta'. \tag{17}$$

In the Mercator projection, the zonal flow is $V(y)$, $U_{jet}(y) = SV(y)$, and

$$\psi^{(0)}(\tilde{y}) = -\int_0^{\tilde{y}} S^2 V(y')dy', \tag{18}$$

where $\tilde{y} = \arcsin(\tanh y) = \arctan(\sinh y) = \operatorname{gd}(y)$.

We drop the tilde over y and seek a solution to $\psi^{(i)}$, $i > 0$, using the method of separation of variables,

$$\psi^{(i)}(X, y, T) = A^{(i)}(X, T)\Phi^{(i)}(y), \tag{19}$$

which yields a power-series representation of (13) in the form

$$\psi(X, y, T) = -\int_0^y S^2 V dy' + \sum_{i=1}^{\infty} \varepsilon^i A^{(i)}(X, T)\Phi^{(i)}(y). \tag{20}$$

By rewriting Equations (11a) and (11b) in the new variables, X, y, T, substituting (13), and equating the terms with the same powers of the small parameter ε, we arrive at a system of coupled differential equations. In the leading order, $O(\varepsilon^{3/2})$, a second-order ordinary differential equation for $\Phi^{(1)}$ is obtained,

$$\Phi_{yy}^{(1)} + \frac{\Gamma_y}{V - c_0}\Phi^{(1)} = 0, \tag{21a}$$

$$\Gamma = f - \frac{(S^2 V)_y}{S^2}. \tag{21b}$$

Equations (21a) and (21b) have non-trivial solutions for a discrete set of values $c_0(i)$, $i = 0, 1, 2, \ldots$, and so they form a generalized eigenvalue problem. The solution of this problem will be discussed later.

Equation (21a) underscores the usefulness of the Mercator projection as it is similar to Equation (17) derived in [11] in a planar geometry using a β-plane approximation in the vicinity of a latitude θ_0. The differences appear, firstly, in the numerator, where the potential vorticity on a β-plane, $f - U_y$, is replaced by the meridional gradient of (21b) and, secondly, in the denominator, where U_{jet} is replaced by V. It is assumed that $V - c_0 \neq 0$ in (21a), i.e., that the normal modes are non-singular.

The stationary Equations (21a) and (21b) along with Equation (19) describe the variation of $\psi^{(1)}$ in the latitudinal direction. The evolution of the amplitude, $A^{(1)}(X,T)$, is determined by invoking the next-order approximation, $O(\varepsilon^{5/2})$, that yields the following equations:

$$\Phi_{yy}^{(2)} + \frac{\Gamma_y}{(V - c_0)}\Phi^{(2)} = -\frac{\mathcal{F}}{A_X^{(2)}(V - c_0)}, \tag{22a}$$

$$\mathcal{F} = A_T^{(1)}\Phi_{yy}^{(1)} + \kappa A^{(1)} A_X^{(1)} + A_{XXX}^{(1)}(V - c_0), \tag{22b}$$

$$\kappa = \Phi^{(1)}\left[\frac{\Phi_{yy}^{(1)}}{S^2}\right]_y - \Phi_y^{(1)}\frac{\Phi_{yy}^{(1)}}{S^2}. \tag{22c}$$

Equation (22b) shows that the effect of a longitudinal dispersion represented by the term $A_{XXX}^{(1)}$ and that of the nonlinearity, $A^{(1)}A_X^{(1)}$, appear only in the order $O(\varepsilon^{5/2})$.

After multiplying (21a) by $\Phi^{(2)}$ and (22a) by $\Phi^{(1)}$, subtracting the second product from the first one, and integrating the result over y in the interval $y \in (-\infty, \infty)$ using the integration by parts while utilizing the boundary conditions $\Phi^{(1)}(\pm\infty) = \Phi^{(2)}(\pm\infty) = 0$ (vanishing velocity component in the direction $\hat{y}$ at the poles), one finds:

$$-\frac{1}{A_X^{(2)}}\int_{-\infty}^{\infty} \frac{\mathcal{F}\Phi^{(1)}}{S^2(V - c_0)} dy = 0. \tag{23}$$

This equation implies that either $A_X^{(2)}$ is infinite, or $A^{(1)}(X,T)$ satisfies the KdV equation,

$$A_T^{(1)} + \mu A^{(1)} A_X^{(1)} + \delta A_{XXX}^{(1)} = 0, \tag{24}$$

where

$$I\mu = \int_{-\infty}^{\infty} \left[\frac{\Gamma_y}{S^2(V - c_0)} \right]_y \frac{\left[\Phi^{(1)}\right]^3}{V - c_0} dy, \tag{25a}$$

$$I\delta = -\int_{-\infty}^{\infty} \left[\Phi^{(1)}\right]^2 dy, \tag{25b}$$

$$I = \int_{-\infty}^{\infty} \frac{\Gamma_y}{(V - c_0)^2} \left[\Phi^{(1)}\right]^2 dy. \tag{25c}$$

This KdV equation is fully analogous to the one derived in the β-plane approximation see e.g., [11] if $\Gamma_y = \beta - U_{yy}$ is replaced by $\Gamma_y = f_y - [(S^2V)_y/S^2]_y$, U_{jet} is replaced by V, and the metric coefficient, S^2, appears in the expression for μ.

A well-known soliton solution of the KdV equation is

$$A^{(1)}(X,T) = A_* \operatorname{sech}^2\left(\frac{X - c_1 T}{\Delta_*}\right), \tag{26a}$$

$$A_* = 3\frac{c_1}{\mu}, \tag{26b}$$

$$\Delta_* = 2\left(\frac{\delta}{c_1}\right)^{1/2}. \tag{26c}$$

This solution describes a solitary wave with a characteristic width Δ_* that travels with a constant speed c_1, the latter being a free parameter that represents the speed in the moving frame (X,T).

3.2. Solitary Rossby Waves in Spherical Coordinates

Computing the perturbation stream function, $\psi^{(1)}(\theta,\varphi,t)$, in spherical coordinates starts with its factorization,

$$\psi^{(1)}(\theta,\varphi,t) = A^{(1)}(\varphi,t)\Phi^{(1)}(\theta), \tag{27}$$

and follows up by evaluation of $\Phi^{(1)}(\theta)$ and $A^{(1)}(\varphi,t)$ using Equations (21a), (21b) and (26a)–(26c). These equations are cast in terms of the slow variables. Inverting (12) restores them back to the unscaled coordinates (x,t). Next, by inverting (10), one restores the original spherical coordinates, (θ,φ), and arrives at a boundary-value/eigenvalue problem for $\Phi^{(1)}(\theta)$,

$$\left[\Phi_\theta^{(1)} \cos\theta\right]_\theta + \frac{R\Gamma_\theta}{V - c_0}\Phi^{(1)} = 0, \tag{28a}$$

$$\Gamma = 2\Omega \sin\theta - \frac{(V\cos^2\theta)_\theta}{R\cos\theta}, \tag{28b}$$

$$\text{N.B.C.} \quad \left.\frac{\partial\Phi^{(1)}}{\partial\theta}\right|_{\theta=\pm\pi/2} = 0. \tag{28c}$$

Here, $V(\theta)$ is computed from a given zonal velocity profile, $U_{jet}(\theta)$, and the Neumann boundary conditions, denoted N.B.C., follow from the requirement that both mean and coherent parts of the zonal velocity vanish at the poles.

The boundary value problem (28a)–(28c) possesses an infinite set of the eigenvalues, $\{c_0\}_i, i = 1,2,...$ They define a set of the eigenfunctions, $\{\Phi^{(1)}\}_i, i = 1,2,...$ To simplify the notations, we shall use c_i^0 and $\Phi_i^{(1)}$ instead of $\{c_0\}_i$ and $\{\Phi^{(1)}\}_i$ while keeping in mind that the lower indexes are introduced for the bookkeeping and not to denote differentiation.

In a weakly-nonlinear approximation,

$$\psi \cong \psi^{(0)} + \varepsilon\psi^{(1)} = \psi^{(0)} + \varepsilon A^{(1)}\Phi^{(1)}. \tag{29}$$

By multiplying (26a)–(26c) by ε, reverting the variables back to the unscaled coordinates (x,t), and then to the original coordinates (θ,φ) using (12) and (10), one derives

$$\varepsilon A^{(1)}(\varphi,t) = A_0\,\mathrm{sech}^2\left(\frac{R\varphi - ct}{\Delta}\right), \tag{30a}$$

$$A_0 = 3\frac{\varepsilon c_1}{\mu}, \tag{30b}$$

$$\Delta = 2\left(\frac{\delta}{\varepsilon c_1}\right)^{1/2}, \tag{30c}$$

$$c = c_0 + \varepsilon c_1, \tag{30d}$$

where Δ is soliton's characteristic width, εc_1 is the next order correction to c_0 that will be neglected, and the coefficients μ and δ are defined by

$$I\frac{\mu}{R} = \frac{1}{R^2}\int_{-\frac{\pi}{2}}^{\frac{\pi}{2}}\left[\frac{\Gamma_\theta}{(V-c_0)\cos\theta}\right]_\theta \frac{\left[\Phi^{(1)}\right]^3}{V-c_0}d\theta, \tag{31a}$$

$$I\frac{\delta}{R^3} = -\int_{-\frac{\pi}{2}}^{\frac{\pi}{2}}\frac{\left[\Phi^{(1)}\right]^2}{\cos\theta}d\theta, \tag{31b}$$

$$I = \int_{-\frac{\pi}{2}}^{\frac{\pi}{2}}\frac{\Gamma_\theta}{(V-c_0)^2}\left[\Phi^{(1)}\right]^2 d\theta. \tag{31c}$$

The parameters μ and δ depend on $V(\theta)$ and, thus, on the zonal velocity profile, $U_{jet}(\theta)$. The other two parameters, ε and c_1, can be varied in certain ranges such as to explore properties of the solution of the system (30a)–(31c).

A solitary wave appears stationary in a coordinate system moving along with it with the velocity c_0. The solitary waves' amplitudes, (30a), attain their maxima along the lines for which $R\varphi - c_0 t = 0$. Figure 2 shows the normalized soliton solution, $\varepsilon A^{(1)}(\varphi,t)/A_0$, as given by (30a), for different values of c_1 with $\varepsilon = 0.01$. The wave propagates in the zonal direction along a latitude isoline (θ = const) while preserving meridional symmetry about that isoline. For small c_1, $\varepsilon A^{(1)}(\varphi,t)/A_0$ is nearly constant but with increasing c_1, it gradually evolves into a sharp pulse. The limit $c_1 \to 0$ brings on $\Delta \to \infty$, and the soliton morphs into evanescent wave. Increase in c_1 raises wave's amplitude, A_0, while its characteristic width, Δ, decreases. Since the analytical solutions obtained on a rotating sphere and on a β-plane are practically identical, see e.g., [11] and [43], both panels in Figure 2 plotted here for a sphere are quite similar to Figures 1 and 2 drawn in [43] for a β-plane.

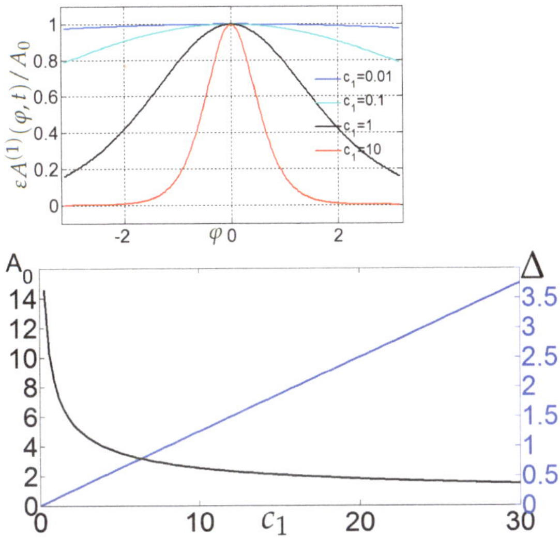

Figure 2. Top panel: normalized soliton solution $\varepsilon A^{(1)}(\varphi, t)/A_0$ in the propagation direction for different values of c_1 at $t = 0$. The values of other parameters are: $\varepsilon = 0.01$, $c_0 = -0.14$, and $\delta = 0.01$. **Bottom panel**: soliton's amplitude, A_0, (solid black line) and width, Δ, (solid blue line) as functions of c_1.

A system of Equations (30a)–(31c) can be used to diagnose whether or not a specific flow pattern gives rise to a solitary wave. The analysis is guided by the mean zonal velocity profile, $U_{jet}(\theta)$, that produces the pattern at hand, and the pattern's propagation speed, c_0. The latitudinal structure of the SRHW is determined by the eigenfunction $\Phi^{(1)}(\theta)$ given by (28a). For $\varepsilon\psi^{(1)}$, one derives a system of equations describing a wave with an amplitude $\mathcal{A}_0(\theta)$ traveling in the zonal direction with a constant speed c_0,

$$\varepsilon\psi^{(1)} = \varepsilon A^{(1)}\Phi^{(1)}(\theta) = \mathcal{A}_0(\theta)\,\mathrm{sech}^2\left(\frac{R\varphi - c_0 t}{\Delta}\right), \tag{32a}$$

$$\mathcal{A}_0(\theta) = 3\frac{\varepsilon c_1}{\mu}\Phi^{(1)}(\theta), \tag{32b}$$

$$\Delta = 2\left(\frac{\delta}{\varepsilon c_1}\right)^{1/2}. \tag{32c}$$

The initial system of equations can now be reduced to a single equation for the stream function $\psi(\theta, \varphi, t)$ [66] that yields an analytical expression for a SRHW,

$$\psi(\theta, \varphi, t) = -R\int_0^{\theta} U_{jet}(\theta')d\theta' + \mathcal{A}_0(\theta)\,\mathrm{sech}^2\left(\frac{R\varphi - c_0 t}{\Delta}\right). \tag{33}$$

In [11] it was established that zonons are long-living coherent eddies that can survive in a flow even after the forcing and friction are turned off. In the vicinity of eddy's center, the perturbation velocity inferred from $\psi^{(1)}$ approaches zero in both forced and unforced flows, or, equivalently, both $\psi^{(1)}$ and $\Phi^{(1)}(\theta)$ attain their extrema. These requirements imply that $\left[\Phi^{(1)}(\theta)\right]_\theta = 0$ somewhere inside an eddy. Thus, the behaviors of $\Phi^{(1)}(\theta)$ and its derivative divulge not only zonon's presence but also its latitudinal position.

The next subsection demonstrates how these results can be utilized to identify zonons in the physical space in the regime of zonostrophic turbulence as presented in Section 2 following [10].

3.3. Mean Zonal Velocity, Eigenfunctions, and Eigenvalues

The zonal velocity profile, $U_{jet}(\theta)$, obtained in simulations described in Section 2, is used to solve the eigenvalue problem (28a)–(28c). A numerical solution of this problem, using Matlab Version R2020a, yields discrete sets of eigenvalues, c_i^0, and eigenfunctions, $\Phi_i^{(1)}(\theta)$, $i = 1, 2, 3, \ldots$. The normalized eigenfunctions, $\Phi_i^{(1)}(\theta)/\Phi_i^{(1)}(\theta = 0)$, $i = 1, \ldots, 11$, are shown in Figure 3.

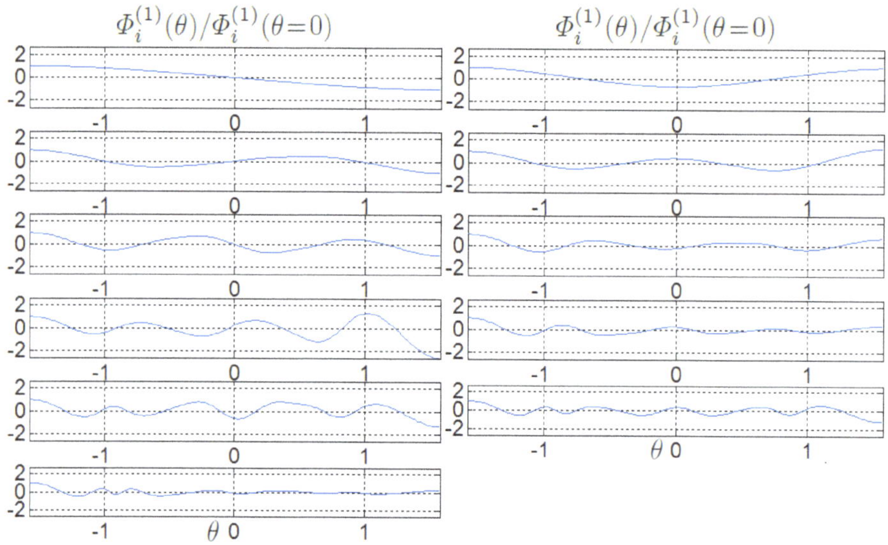

Figure 3. The first 11 normalized eigenfunctions, $\Phi_i^{(1)}(\theta)/\Phi_i^{(1)}(\theta = 0)$, shown from left to right and from top to bottom, from top left (i = 1) to bottom left (i = 11) corners.

Only $\Phi_i^{(1)}(\theta)$ with $i = 7$ exhibits 6 extrema; those can be collocated with eddies emerging along both sides of the three westward jets shown in Figure 4. Figure 1 demonstrates that $n = 7$ is the most energetic zonal mode while the maximum energy RHW is at $n = 5$. One infers that the closeness of the modes with most energy in either RHWs (residual spectrum) or zonal jets (zonal spectrum) may be a general feature of zonostrophic turbulence and so, in some situations where data is limited, one may be taken as a surrogate of the other.

Figure 4 also confirms that the extrema of $\Phi_7^{(1)}(\theta)$ as well as the zeros of $\left[\Phi_7^{(1)}(\theta)\right]_\theta$ coincide with the centers of the eddies as seen in the color maps of the potential vorticity perturbation, $q^{(1)}$.

In addition, Figure 4 illuminates the fact that the vortices reside in the regions of maximum shear between zonal jets. These locations correspond to those at which the potential vorticity gradient crosses zero and which, according to the Charney-Stern stability criterion, are the epicenters of the barotropic instability.

Using a detailed analysis, such locations were associated with the hot spots of eddy formation in [21] where the propagation speed of the traveling structures inferred from the stream function was compared with that derived from the centroids of different types of clusters construed via the Okubo–Weiss analysis. The results were compared with the theoretical phase speed of Rossby waves. The congruity of these two characteristics at the locations of maximum shear was quite remarkable. An important conclusion drawn from this analysis was that "the Rossby waves and eddies are inseparable as the former maintain the instability that sustains the latter. This symbiosis visually resembles the Rossby soliton". A similar idea about the coexistence of eddies and Rossby waves was discussed in [63].

The hypothesis about the unity of Rossby waves and Rossby solitons, formulated in [21], has not yet been substantiated within the framework of the KdV equation. The validation of this hypothesis within the KdV framework and its various applications are presented in the forthcoming sections.

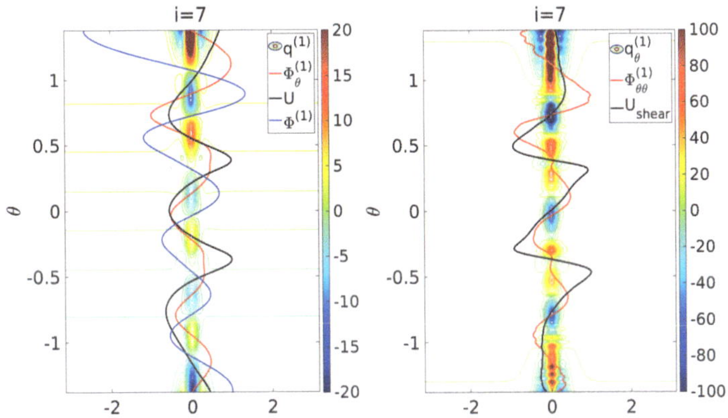

Figure 4. Left panel shows normalized perturbation velocity eigenfunction, $\Phi_i^{(1)}(\theta)$, along with its gradient, $[\Phi_i^{(1)}(\theta)]_\theta$, and normalized mean zonal velocity profile, $U_{jet}(\theta)$, vis-à-vis contours of the potential vorticity perturbation, $q^{(1)}$. **Right panel** shows normalized second derivative of the perturbation velocity eigenfunction, $[\Phi_i^{(1)}(\theta)]_{\theta\theta}$, and the normalized shear of the mean zonal velocity, $U_{shear}(\theta)$, vis-à-vis contours of the perturbation potential vorticity gradient, $q_\theta^{(1)}$. For both panels, $i = 7$.

3.4. *Numerical Simulations: Zonons and Eddies*

Table 1 compares a set of the first 11 eigenvalues, c_i^0, $i = 1, \ldots, 11$, of the boundary value problem (28a)–(28c) with a set of zonal phase speeds of the first 11 RHWs, $c_{RHW}(n), n = 1, \ldots, 11$, given by (4). The two sets are in close agreement as the phase speeds of the possible solitary waves, c_i^0, are close to $c_{RHW}(n)$ for $i = n$. For the most energetic RHW, $n = 5$, the difference between c_{RHW} and c_i^0 is about 10%, and for the most energetic zonal mode, $i = 7$, the difference is even smaller, only within 2%, blurring the contrast between waves and eddies. The eddy-wave duality was discussed in [21,44,63]. Proximity of c_{RHW} and c_i^0 for other wave numbers indicates that virtually all RHWs may generate secondary, i.e., enslaved waves, see [10], but their energies are much lower than those of the predominant zonons, and so they are less amenable to detection and observation. In addition, turbulence intensity raises with increasing n causing turbulence-induced scrambling to overwhelm wave processes.

Table 1. The zonal phase speeds of the RHWs vis-à-vis the zonal phase speeds of the SRHWs on a rotating sphere. The highlighted row with index $i = 7$ corresponds to the eigenfunction $\Phi_7^{(1)}(\theta)$ that divulges latitudinal locations of zonons.

i or n	Rotating Sphere		
	$c_{RHW}(n)$	c_i^0	Relative Difference [%]
1	−1.4	−1.445	3.2
2	−0.466	−0.465	0.2
3	−0.233	−0.2	14.1
4	−0.14	−0.144	2.8
5	−0.0933	−0.083	11
6	−0.066	−0.059	10.6
7	−0.05	−0.049	2
8	−0.038	−0.036	5.2
9	−0.031	−0.031	0
10	−0.025	−0.025	0
11	−0.02121	−0.023	8.4

The eigenvalues obtained from the solution of the boundary value problem are used to evaluate the parameters I_i, μ_i and δ_i needed to compute the coefficients of the KdV equations, and, ultimately, the amplitudes of the solitary waves.

3.5. Hovmöller Diagrams

The Hovmöller diagram technique [67] highlights propagation of RHWs and SRHWs in the physical space. The diagram axes are usually chosen as longitude or latitude and time and the variable plotted is the velocity stream function relative to the zonal flow; its values are represented by a color table. The slopes of lines of the same color reveal the speed of propagating waves and structures, i.e., packets of RHWs and SRHWs, relative to the zonal flow.

Figure 5 reproduces Hovmöller diagrams for zonostrophic regime ($R_\beta = 2.58$) investigated in Sukoriansky et al. [11]. Figure 5a reveals westward propagation of zonons that were identified by their spectral properties and confirms that their speed along the latitude of the maximum shear coincides with the phase speed of the most energetic RHW with $n = 5$. The same result was obtained for other latitudes thereby reiterating that the westward propagating eddies can indeed be identified with zonons moving with a constant speed relative to the zonal flow. Moreover, Figure 5b underscores that zonons' velocity in the meridional direction is zero. Figure 5b–d combined indicates that zonons reside in areas where the zonal velocity exhibits strong shear. Figure 5e–h displays similar results for the same flow 22,000 days after the forcing and the large-scale drag were switched off. In this case, the energy of RHWs decreases while zonons preserve their energy and become the dominant feature of the flow.

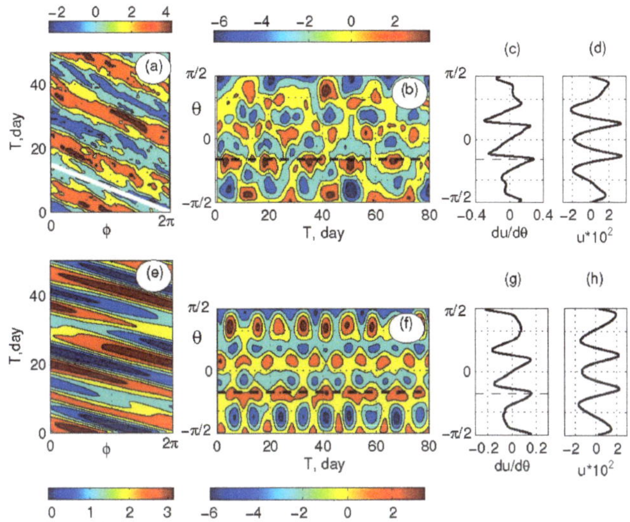

Figure 5. Latitudinal (**a**,**e**) and longitudinal (**b**,**f**) Hovmöller diagrams for a flow in zonostrophic regime ($R_\beta = 2.58$) analyzed in Sukoriansky et al. [11]. The top and bottom rows represent forced and unforced flows, respectively. The white line on plate (**a**) corresponds to the phase speed of RHW with $n = 5$. Plates (**d**,**h**) depict zonal velocity profiles while (**c**,**g**) show meridional shear of respective zonal velocities. The horizontal dashed lines on plates (**b**,**c**,**f**,**g**) correspond to the latitude of maximum shear that encompasses energetic zonons. After the forcing is switched off, the energy of velocity fluctuations decreases yet zonons retain their energy levels and integrity and become more apparent than RHWs.

In summary, zonons form long-living wave packets located between zonal jets. The shear exerted by opposing jets acts to tear the eddies apart while the nonlinearity opposes shear's action and blocks eddies' dispersion. As follows from the analysis employing the Hovmöller diagrams, whether the eddies are sustained by the shear instability or not, the summary action of two adjacent opposing jets results in their becoming a waveguide for the coherent wave packets. Such interaction between shear, nonlinearity, and wave propagation is typical of solitary waves, see e.g., [16], that include zonons.

How realistically a model based upon Equation (1) reflects basic feature of zonal jets and eddies on gaseous giant planets, such as Jupiter and Saturn? Detailed comparisons of jets' spectral properties in modeled and observed flows were presented in [9,26]. These comparisons confirmed the existence of the steep spectrum (6a) unique to giant planets' circulation. Studies of the energy exchange between the eddies and jets in [26,68] pointed to forcing of the jets by the eddies, at least in the extratropics. The energy source of these eddies is not well understood. Due to a strong barotropization pertinent to zonostrophic turbulence [26], the formation of zonal jets from eddy mixing and stirring is concurrent with a deep, quasi-barotropic response to the forcing occurring anywhere within the fluid layer. Rapid barotropization of localized forcing was demonstrated in laboratory experiments and in model simulations [68]. As a result, although the jets on Jupiter and Saturn appear spanning the entire depth of the weather layer, the locations of the eddy energy sources have not been identified as of yet. Multitude of processes can produce this energy, including deep convection, baroclinic instability, moist convection, and even gravitational tidal resonances induced by the Galilean moons. Another source, mentioned in [21], is the barotropic/baroclinic instability sustained by the eddies themselves via dynamical modifications of the surrounding flow field.

Even though the exact locations of the eddy energy sources on Jupiter and Saturn remain poorly known, the rapid barotropization of the injected energy renders the large-scale planetary dynamics amicable to analysis in the framework of the KdV theory in the zonostrophic regime. This analysis will be performed in the next section.

4. Zonons in the Atmospheres of Jupiter and Saturn

There exist numerous accounts of the forcing and dynamics of long-lived vortices on Jupiter and Saturn and their interaction with zonal jets, e.g., [9,28,69–73]. Read [68] provided the latest review that includes new data from *Cassini* and *Juno* missions. He emphasized two features of planetary vortices that have been at the focus of recent investigations: the depth of vortex penetration into planetary atmospheres and the mechanisms that sustain the vortices over a long time in spite of a strong horizontal shear due to zonal jets and disturbances caused by interactions with other eddies. An additional difficulty stems from the observed spontaneous eddy birth which is not readily reproduced in numerical models.

While the KdV framework developed in the previous sections for the zonostrophic regime does not address the issues related to the forcing and depth of the vortices, it does shed light upon their longevity and location. The theory applies to the atmospheres of Jupiter and Saturn as they both feature the regime of zonostrophic macroturbulence, with the values of the zonostrophy index, R_β, being about 5 for Jupiter [26] and about 4 for Saturn [42]. As mentioned in [68], early discussions of the physics of planetary vortices, such as the Great Red Spot, were based upon the quasi-geostrophic theory, see e.g., [46]. In this study, the vortices are analyzed in the framework of zonostrophic turbulence and associated KdV equation for the first time.

The velocity profiles needed for our investigation were published in [74] for Jupiter and [75] for Saturn. The analysis is based upon the boundary value problem, Equations (28a)–(28c), that yields the eigenvalues and eigenfunctions pertinent to the respective planetary atmospheres. These results, along with the spectra in [26,61], were used to estimate likely locations of planetary eddies. The outcome of these estimates is summarized in Sections 4.1–4.4.

4.1. A Brief Taxonomy of Jupiter's Eddies

The zonal velocity profile for Jupiter was obtained from the upper cloud level data collected by Voyager in 1980–1981 [74]. As noted in [24], Jupiter's mean zonal velocity, $U_{jet}(\theta)$, has not significantly changed at least since 1979.

Numerous studies of Jovian vortices, e.g., [23,24,32,76] reveal that they reside at many latitudes outside the near-equatorial belt $\pm 15^\circ$. In line with the mentioned in the Introduction tendency, most vortices are aligned in rows bound to the westward jets [24]. The eddies are either anticyclonic or cyclonic when they reside, respectively, on the poleward or equatorial flanks of a jet, as seen in Figure 6. Longitudinal interspersing of cyclones and anticyclones gives rise to a configuration termed 'Jovian vortex street' (JVS) [24,77]. A JVS may corrugate a westward jet and force it to meander between the cyclonic and anticyclonic eddies, just like a system of vortices wrinkles a zonal jet in the experiments on zonostrophic turbulence [21,39].

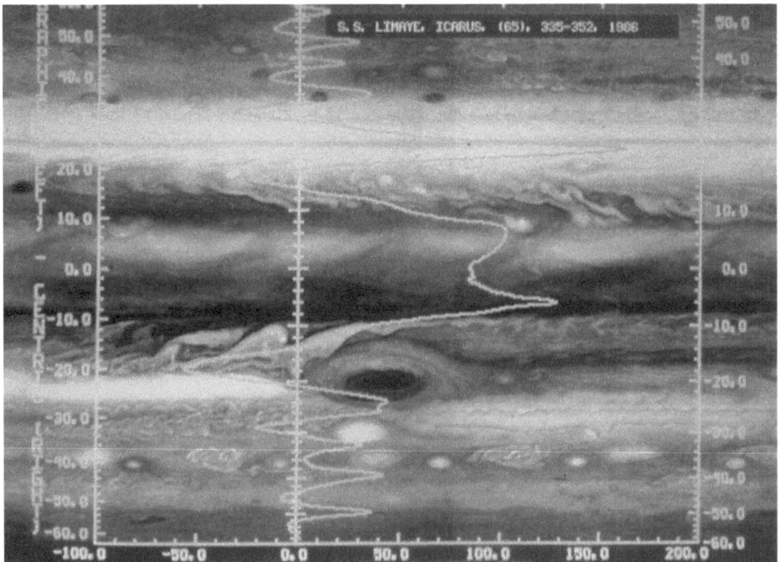

Figure 6. Voyager mosaic used in [23], Figure 2 (reprinted with permission from Elsevier), and in [24], Figure 1 (© American Meteorological Society. Used with permission). The mean zonal velocity profile, $U_{jet}(\theta)$, is represented by the white line. Rows of cyclones and anticyclones ride on the westward jets with the cyclones or anticyclones attached, respectively, to the equatorial or poleward sides. While the anticyclones are compact and bright, the cyclones appear as diffused filaments.

In [24], the JVSs were considered to be more fundamental coherent features of the Jovian atmosphere than the individual vortices. The dynamics of the JVS vortices implied quite elaborate yet bogus forcing and dissipation and it was unclear how they could affect the structure of turbulence. It was suggested that the JVS excludes the GRS because, due to its close proximity to the equator, the GRS does not possess a companion cyclonic vortex on its northern side while other anticyclonic vortices do.

After the GRS, the White Ovals (WOS) comprise the second in magnitude Jupiter's vortex system. It is located at about $-33°$, south of the GRS. The WOS formed in 1938 and then pinched off at six different longitudes labeled, alphabetically, as A, B, C, D, E, and F [76,78]. The vortices started moving towards each other while exhibiting signs of recirculation and eventually assumed oval shapes known today as BC, DE and FA. In 1998, the ovals BC and DE merged to a single storm, oval BE [79]. The new storm continued drifting along with oval FA and in the year 2000, the two storms merged into a single storm BA. This merger was described in [80] and, later on, in [76]. The oval BA is now the dominant vortex in its latitude. It has experienced various episodes of coloration changes.

The third longest-lived oval in Jupiter's atmosphere is a large anticyclone formed in the year 2006 at the planetographic latitude of 19° N [81]. As most large vortices, it is also bound to a westward jet and located in a zone characterized by a strong latitudinal shear that is prone to formation of closely interspersed cyclones and anticyclones.

Yet another major Jupiter's eddy is the so-called Jovian Ribbon. Located near 30° N, it is dynamically associated with a westward jet and exhibits irregular undulations in latitude [32], as seen in Figure 6.

4.2. *Jupiter: Eddies or Zonons?*

We now probe a congruence of Jupiter's eddies and zonons. The boundary-value problem (28a)–(28c) is solved for the observed mean zonal velocity profile reported in [74]. Table 2 compares the RHWs' phase speeds with the propagation velocities of the solitary waves. Similarly to the results of numerical simulations, there seems to be a good agreement between the propagation speeds of the corresponding zonons and RHWs (i.e., those for which $i = n$), they differ by less than 10% in the range of i between 17 and 23 although for several i from that range there was no convergence.

Table 2. Zonal phase speeds of the RHWs vis-à-vis zonal phase speeds of the SRWs on Jupiter.

	Jupiter		
i or n	**$c_{RHW}(n)$ ms^{-1}**	**c_i^0 ms^{-1}**	**Relative Difference [%]**
17	−80.4	−80.3	0.2
18	−72.0		
19	−64.8	−69.2	6.8
20	−58.6		
21	−53.3		
22	−48.7	−47.7	2.0
23	−44.6	−46.6	4.4
24	−41.0		
25	−37.9		

As explained earlier, the best visible zonons are produced by the most energetic RHWs, and to select proper modes, we need to scrutinize the residual energy spectrum. Figure 7 shows the zonal and residual energy spectra for Jupiter as estimated by the *Cassini* spacecraft during flyby in December 2000 [26]. The observed residual spectrum shows a peak at about $n = 23$. This wave number coincides with the energy peak in the zonal spectrum and is also indicative of the number of zonal jets [61]. The closeness of the wave numbers of maximum zonal and residual kinetic energies as observed in numerical simulations was mentioned in Section 3.3.

The mode with $n = 23$ is a good candidate for producing energetic secondary waves that can form zonons. If this possibility indeed materializes, then the extrema of $\Phi_{23}^{(1)}(\theta)$ would coincide with the locations of large coherent eddies observed on Jupiter's disk.

Figure 8 collocates the observed large Jupiter's eddies (left panel) and extrema of $\Phi_{23}^{(1)}(\theta)$ (right panel). The figure corroborates the premise that the majority of the eddies are associated with the westward jets and are tightly connected to the locations of maximum shear and energetic instabilities. The figure also demonstrates close correspondence between the locations of the observed and anticipated eddies, including the GRS. Given that some properties of Jupiter's eddies can be attributed to zonons, eddies' appearance can be aptly characterized as 'zonon streets'. Zonon streets are inclusive of the GRS.

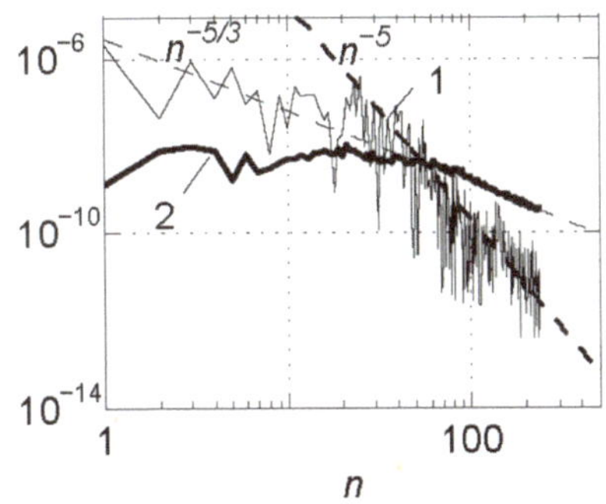

Figure 7. Jupiter's zonal (line 1) and residual (line 2) spectra estimated from the *Cassini* data [26]. The thick and thin dashed lines represent the zonostrophic n^{-5} and Kolmogorov $n^{-5/3}$ spectra, Equations (6a) and (6b), respectively. At $n = 23$, both spectra exhibit peaks whose coincidence was discussed earlier. The residual energy peak discloses the most energetic Rossby wave that generates zonons. These zonons may be identifiable with the observed eddies shown in Figure 8. The zonostrophy index for Jupiter is $R_\beta \gtrsim 5$ such that its atmospheric circulation is in the zonostrophic regime [26].

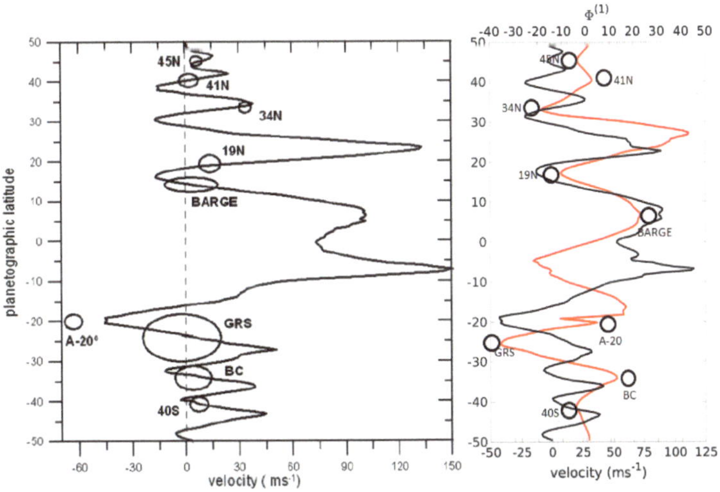

Figure 8. Left panel: Jupiter's zonal wind profile as measured by the HST [74] and used in [31] to trace large vortices (reprinted from [31], with permission from Elsevier). These vortices are represented by ovals and are scaled to their latitudinal size. **Right panel**: Jupiter's zonal velocity profile from the Voyager data [74] (black line), the eigenfunction $\Phi_{23}^{(1)}(\theta)$ (red line), and estimated locations of eddies, marked by small circles, at the extrema of the eigenfunction.

The equatorial belt between the latitudes $\pm 10^\circ$ features different dynamics [24] and was excluded from the analysis. For instance, new data obtained from the James Webb Space Telescope (JWST) observations of Jupiter reveals an intense narrow equatorial atmospheric jet located close to the tropopause [82]. The jet was extracted from the fast motions of equatorial hazes and is presumed to be a deep counterpart of equatorial oscillations in the stratosphere of Jupiter that are unrelated to the eddies.

4.3. A Brief Taxonomy of Saturn's Eddies

The regime of zonostrophic macroturbulence on Saturn is distinguished by vortices and cloud structures forming alternating zonal jets that encircle the planet with their beautiful mottled textures [9]. The jets' edges are wave-like but may become unstable and exhibit corrugation by turbulent eddies that may be evoked by either dynamic instabilities or moist convection e.g., [83–86].

Similarly to the alternating jet pattern on Jupiter, jets on Saturn also form an east-west stencil that is important for identification of instabilities and locations of vortices and storms. This stencil appears asymmetric with a bias towards the eastward jets. The use of the reference frame termed System IIIw [87] renders the east-west pattern more symmetric and similar to that on Jupiter.

Observations of Saturnine southern hemisphere reported in [88] have revealed that, just like on Jupiter, its zonal velocity remains nearly time-independent outside the equatorial belt. In addition, also in similarity to Jupiter's dynamics, eddies transfer momentum to eastward-going jets and remove momentum from their westward counterpart. On both planets, eddies are supplying the kinetic energy to the mean flow, i.e., the planets exhibit the inverse energy cascade. This cascade is terminated by a large-scale energy withdrawal due to the action of some large-scale drag. As elaborated in [61], the wave number associated with this drag on Saturn is about one half of that on Jupiter, and so the inverse cascade penetrates to wave numbers twice as small as those on Jupiter. As a result, in agreement with the tenets of the zonostrophic regime [26,61], the number of the zonal jets on Saturn is about one half of those on Jupiter, Saturnine total kinetic energy is about an order of magnitude larger than its Jupiter counterpart, and the velocity of its equatorial jet is about 4 times that of Jupiter. Fewer and broader zonal jets result in larger and more energetic storms on Saturn compared to Jupiter. However, while Jupiter's storms are always present, storms on Saturn are intermittent. Similarly to Jupiter, many of Saturn's vortices originate and propagate in close vicinity to its westward zonal jets. Although cloud structures on Jupiter and Saturn are similar, the cooler environment of Saturn (likely responsible for the smaller large-scale drag) causes its clouds to be seated deeper in its atmosphere.

Intense studies of Saturnine vortices and storms became possible with the emergence of satellite observations. Early evidence of its intermittent thunderstorms came from the Voyager flyby in 1980–1981. In 2004, the *Cassini* orbiter detected a series of storms at the narrow latitudinal band in the vicinity of 35° S known as the "Storm Alley" for its powerful lightning and thunderstorm activities [89]. Starting as localized bursts and cloud formations, the storms eventually developed into Dark Ovals. Similar dark spots were detected by *Cassini* in 2006 [90]. Most ovals resided in the shear zone, just to the south of the jet, and moved westward at speeds of about 25 m s^{-1}. Other spots, detected several degrees to the south, were moving eastward at comparable speeds. They rotated counterclockwise (anticyclonic) and resembled ball bearings between conveyor belts while preserving the sense of rotation of the ambient shear flow. The westward- and eastward-moving vortices at adjacent latitudinal bands could interact and sometimes produced a merger or a near-merger, suddenly emitting bright material forming in the close-by area [89]. This system of vortices moving in opposite directions is quite different from the Jovian vortex street described in Section 4.1 in which the vortices are straddled on both sides of the same jet. The difference may be due to the more active role of the moist convection on Saturn as it injects large-scale vorticity at the cloud level [89].

A large, long-lived cyclonic vortex centered at 45.5° S was reported in [91]. This vortex was associated with a westward jet in the System IIIw coordinate frame. Observations and modeling efforts suggested that the vortex was sustained by the meridional shear of the zonal wind rather than the energy flux from the surrounding smaller-scale eddies.

Another large cyclonic vortex, located at 50° N and associated with a westward jet in the System IIIw frame, was reported in [92]. Between 2006 and 2010, this large typical Saturnine cyclone was surrounded by smaller-scale convective clouds. Eight smaller-scale convective storms erupted within the vortex during this period, the seventh one bringing significant disruption and the eighth one quenching it. The storms were quite similar to the storm observed at about the same latitude in the southern hemisphere and reported in [93]. The latter storm was also amalgamated with a westward jet, as well as a jet at 48° S if considered in the System IIIw coordinate frame. That jet exhibited phenomena similar to those in the Storm Alley, with anticyclonic vortices rolling in the anticyclonic shear zone to the south of the jet [89].

Saturn spotlights a train of vortices known as the String of Pearls (SoP). Observations on 29 March 2008 [94] showed the 'pearls' as a chain of 17 dark, featureless cyclonic spots occupying latitudes between, approximately, 32.5° N and 34.5° N and coinciding with a westward jet. The bright appearance of the SoPs has earned them the interpretation of being thermal radiation escape outlets from deep layers. The SoPs may be a precursor of the Great White Spots (GWS) [95], the largest observed vortex system on Saturn. As was shown in [35], the initial positions of the two storms coincided, and a new storm erupted from the SoPs. The SoPs was last detected on 23 December 2010, after which the new storm has filled the entire latitudinal belt and the SoPs became indistinguishable within it.

The GWSs appear in intervals of about 30 years whereas in the last 140 years, 6 cases were documented [95]. These giant storms occurred in the northern hemisphere at different latitudes: at the equatorial region, 15° S–25° N; at mid-latitudes, 25° N–47° N, and at subpolar latitudes, 48° N–78° N. The heads, i.e., the leading bright spots of the storms, were found at 5° N, 12° N, 40° N, and 57° N. Aside from the near-equatorial storms, the other ones were located within the westward-propagating jets. Numerous studies converge on the idea that the onset stage of the GWS is driven by vigorous outbursts of moist convection [94,96].

Several weeks after formation of the GWS head, a chain of Dark Ovals (DOs) gradually formed in the storm's tail. The ovals were centered at the latitude of 25° N and moved eastward at +65 m s^{-1}. The SoPs moved westward, along the latitude 33° N, at -22.42 m s^{-1}. At that time, the DOs and the SoPs coexisted while differing dynamically and morphologically. Their colors also had different appearances [94]. All DOs were extinguished by 25 April.

As stated earlier, similar dark spots were detected in the southern hemisphere, around 35° S [89,90]. We note that most dark spots found on Saturn were previously located in anticyclonic shear zones. The spots in the vicinity of 25° N were formed in their cyclonic counterpart.

4.4. Saturn: Eddies or Zonons?

To ascertain whether or not Saturn's vortices are congruent with zonons, we solve the eigenvalue problem (28a)–(28c) using the observed mean zonal velocity profile given in [75]. Our analysis extends between, approximately, the latitudes ±60°. Since the planetary angular velocity in System IIIw differs from that in System III, it is important to verify the consistency of the eigenvalues and the RHWs' phase speeds in both systems. For the System IIIw, we used the mean zonal velocity profile presented in [87]. Table 3 demonstrates that, in similarity to numerical simulations and computations for Jupiter, $c_{RHW}(n) = c_i^0$ for $n = i$. The values of $c_{RHW}(n)$ and c_i^0 are mostly within 10% of each other in the range of i between 10 and 15 for both systems. For several values of i from that interval convergence in the System III was not achieved. The eigenfunctions in the System IIIw pointed to the same vortices as those found in the System III.

To chose the wave number that may correspond to zonons, one needs to locate the wave number n at which a RHW has maximum energy. In principle, this wave number could be identified by scrutinizing the residual spectrum. For Saturn, however, this spectrum has not been established yet, and so the zonal spectrum computed in [61] was used as its surrogate. The zonal spectrum indicates that the mode $n = 15$ appears most likely candidate to contain the maximum energy. The corresponding eigenfunction, $\Phi_{15}^{(1)}(\theta)$, is shown in Figure 9. Evidently, its extrema provide a reasonably accurate representation of the solitary waves' positions in the entire range of latitudes considered in this investigation.

Table 3. Zonal phase speeds of the RHWs vis-à-vis zonal phase speeds of the SRWs on Saturn in Systems III and IIIw.

i or *n*	System III			System IIIw		
	$c_{RHW}(n)$ ms^{-1}	c_i^0 ms^{-1}	Relative Difference [%]	$c_{RHW}(n)$ ms^{-1}	c_i^0 ms^{-1}	Relative Difference [%]
9	−232.6	−235.3	1.1	−221.3	−233.9	5.4
10	−190.3	−182.5	4.1	−181.0	−156.1	13.8
11	−158.6	−143.9	9.3	−150.9	−140.6	6.8
12	−134.2			−127.6	−124.3	2.6
13	−115.0			−109.4	−105.7	3.4
14	−99.7	−101.8	2.1	−94.8	−94.1	0.7
15	−87.2	−90.2	3.4	−83.0	−87.3	4.9
16	−77.0			−73.2	−75.3	2.8
17	−68.4			−65.1	−63.8	2.0
18	−61.2			−58.2	−58.9	1.2
19	−55.1			−52.4	−54.2	3.3

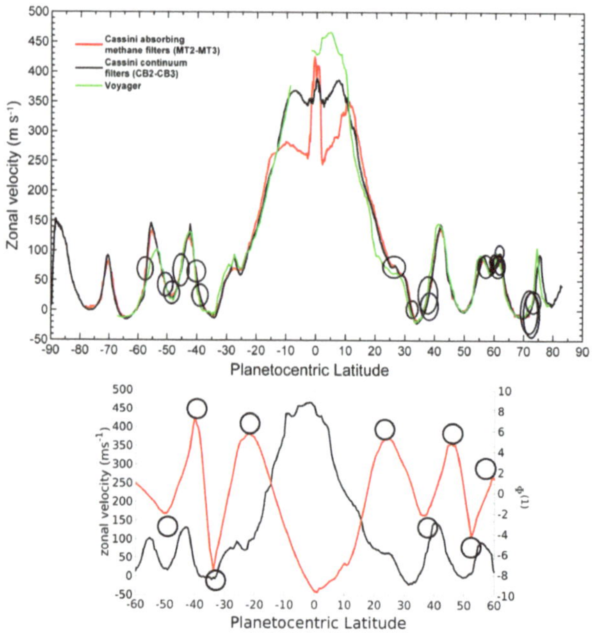

Figure 9. Top panel: Three Saturn's zonal wind profiles for two different epochs superimposed on each other. They reflect velocity fluctuations over the years [97]. The first epoch [75] is represented by the data for the upper cloud level collected by Voyager in 1980–1981. The second epoch [98] embodies the data collected by *Cassini* spacecraft in 2004–2009 and encompassing the upper cloud level (CB-data) and upper haze (MT-data). There is a notable variability in the equatorial region compounded with the vertical wind shear likely coupled with the semiannual oscillation (SAO) [99,100] in the temperature and wind fields that occur in the stratosphere in the latitudinal belt ±15° and affect Saturn's equatorial dynamics [97]. This belt was not considered in the analysis. The observed vortices are marked by ovals drawn by Agustin Sánchez-Lavega. Most of the vortices are described in Section 4.3. **Bottom panel**: Saturn's zonal velocity profile from the Voyager data [75] (black line), the eigenfunction $\Phi_{15}^{(1)}(\theta)$ (red line), and estimated latitudinal locations of eddies, marked by small circles placed at the extrema of the eigenfunction.

In analogy to Jupiter, vortices on Saturn are usually coming in trains although their structure may differ from that on Jupiter as in addition to the Charney-Stern instability, the moist convection processes are playing significant role in vortex formation and maintenance. Our analysis indicates, however, that similarly to Jupiter, the nonlinear dynamical processes underscore the physics of vortex evolution on Saturn and so its vortices and their trains can also be characterized as 'zonon streets'.

5. Discussion and Conclusions

This study outlines the connection between zonons and solitary Rossby-Haurwitz waves in the regime of zonostrophic turbulence. Zonons are 'slave' waves excited by RHWs in other wave modes via non-linear interactions. One can also view zonons as a product of RHWs ringing in other wave modes. Zonons' dispersion relation is different from that of the RHWs. They form long-lived, non-dispersive wave packets located between zonal jets that guide them around planets while replenishing their energy via shear instabilities. In the physical space, zonons form westward propagating coherent eddies whose zonal velocity is equal to the phase speed of the most energetic RHW.

Despite the dispersive action of the shear, the wave packets are being coagulated by the nonlinearity and preserve their identity. The balance between the effects of the shear and nonlinearity can be quantified within the KdV equation. The constructive interference between the effects of the shear and nonlinearity is typical of solitary waves, see e.g., [16], and so one infers that zonons are indeed solitary Rossby-Haurwitz waves.

In the preceding study of zonons in [11], the solitary wave framework was established using the KdV equation in the Cartesian coordinate system. This framework was not validated against data, however. The present study goes farther and develops the KdV foundation for zonons in a spherical coordinate system. The derivations are performed in the Mercator projection that allows for smooth two-way transitions of the governing equations between spectral and physical space representations. This approach leads to a boundary value problem whose nontrivial solutions yield eigenvalues and eigenfunctions. The eigenvalues correspond to zonal velocities of zonons, while the respective eigenfunctions disclose their latitudinal locations.

This framework was tested and validated against the results of direct numerical simulations of two-dimensional turbulence on the surface of a rotating sphere. The main outcome of this initial study was that large coherent eddies emerging in the regime of zonostrophic turbulence, or zonons, can indeed be identified with the solitary Rossby-Haurwitz waves bound to westward zonal jets and propagating inside the rails formed by the opposing jets.

The KdV framework was then applied to probe coherent vortices on Jupiter and Saturn, the two giant planets that boast zonostrophic regime. These comparisons involved observed mean zonal winds and available kinetic energy spectra. Guided by the results deduced from direct numerical simulations, it was established that the zonal velocities of large coherent eddies on the planets are well represented by the eigenvalues of the boundary value problem and their latitudinal locations are disclosed by the extrema of the corresponding eigenfunctions.

One of the hallmarks of this theory is that much information on the planetary eddies can be deduced from the zonal velocity profile which is much easier to establish than other characteristics such as, for instance, the residual energy spectrum. Another hallmark of the theory is in the offering an analytical framework providing latitudinal locations of nearly all major eddies on a rotating planet given that its atmosphere obeys the zonostrophic regime. This framework may apply not only to our neighboring giant planets but also some of their satellites as the simulations presented in [101] point to the possibility that the zonostrophic regime may dominate the circulation of subsurface oceans of Ganymede, Europa, Enceladus, and Titan. The KdV frame may apply to eddies on some exoplanets as well. The finding that zonons may include the Great Red Spot elevates the possibility that they may also be relevant to great spots on other planets [102,103] provided that those obey the zonostrophic regime.

These results may also be applicable in the terrestrial atmospheric environments where the zonostrophic regime is marginal. In Section 1, for example, it was noted that the solitary Rossby waves may help to understand genesis and early evolution of the proto-vortices developing into tropical storms. Moreover, zonons may be helpful in clarifying the role of eddies and spirals in oceanic circulation [104–107] given that the oceanic turbulence is marginally zonostrophic [6].

Author Contributions: Conceptualization, S.S.; Methodology, N.C. and B.G.; Software, N.C.; Validation, N.C., B.G. and S.S.; Formal analysis, N.C., B.G. and S.S.; Investigation, B.G. and S.S.; Data curation, N.C.; Writing—original draft, N.C. and B.G.; Supervision, B.G. and S.S. All authors have read and agreed to the published version of the manuscript.

Funding: This study was supported by the Israel Science Foundation governed by Israeli Academy of Sciences (Grant No. 134/03 and Grant No. 408/15).

Institutional Review Board Statement: Not applicable.

Informed Consent Statement: Not applicable.

Data Availability Statement: The raw data supporting the conclusions of this article will be made available by the authors on request.

Acknowledgments: Tov Elperin, who untimely passed away, played an important role in initial advising of Nimrod Cohen in his doctoral dissertation. Roger H. Grimshaw was instrumental in the development of the KdV framework in spherical geometry. Digitized profiles of the mean zonal velocities on Jupiter and Saturn used in Section 4 were made available to us by Agustin Sánchez-Lavega and Peter Read; they also provided editorial assistance.

Conflicts of Interest: The authors declare no conflicts of interest.

References

1. Salmon, R.; Pizzo, N. Two-Dimensional Flow on the Sphere. *Atmosphere* **2023**, *14*, 747. [CrossRef]
2. Herring, J.; Orszag, S.; Kraichnan, R.; Fox, D. Decay of two-dimensional homogeneous turbulence. *J. Fluid Mech.* **1974**, *66*, 417–444. [CrossRef]
3. Herring, J.R. On the Statistical Theory of Two-Dimensional Topographic Turbulence. *J. Atmos. Sci.* **1977**, *34*, 1731–1750. [CrossRef]
4. Herring, J.R. Statistical theory of quasi-geostrophic turbulence. *J. Atmos. Sci.* **1980**, *37*, 969–977. [CrossRef]
5. Galperin, B.; Sukoriansky, S. Zonostrophic turbulence. *Phys. Scr.* **2008**, *T132*, 014034. [CrossRef]
6. Galperin, B.; Sukoriansky, S.; Young, R.M.B.; Chemke, R.; Kaspi, Y.; Read, P.L.; Dikovskaya, N. Barotropic and Zonostrophic Turbulence. In *Zonal Jets: Phenomenology, Genesis, and Physics*; Galperin, B., Read, P.L., Eds.; Cambridge University Press: Cambridge, UK, 2019; pp. 220–237. [CrossRef]
7. Chekhlov, A.; Orszag, S.; Sukoriansky, S.; Galperin, B.; Staroselsky, I. The effect of small-scale forcing on large-scale structures in two-dimensional flows. *Phys. D* **1996**, *98*, 321–334. [CrossRef]
8. Huang, H.P.; Robinson, W. Two-dimensional turbulence and persistent zonal jets in a global barotropic model. *J. Atmos. Sci.* **1998**, *55*, 611–632. [CrossRef]
9. Galperin, B.; Read, P.L. (Eds.) *Zonal Jets: Phenomenology, Genesis, and Physics*; Cambridge University Press: Cambridge, UK, 2019; p. 504. [CrossRef]
10. Sukoriansky, S.; Dikovskaya, N.; Galperin, B. Nonlinear waves in zonostrophic turbulence. *Phys. Rev. Lett.* **2008**, *101*, 178501. [CrossRef]
11. Sukoriansky, S.; Dikovskaya, N.; Grimshaw, R.; Galperin, B. Rossby waves and zonons in zonostrophic turbulence. *AIP Conf. Proc.* **2012**, *1439*, 111–122. [CrossRef]
12. Péronne, E.; Chuecos, N.; Thevenard, L.; Perrin, B. Acoustic solitons: A robust tool to investigate the generation and detection of ultrafast acoustic waves. *Phys. Rev. B* **2017**, *95*, 064306. [CrossRef]
13. Sakazaki, T.; Hamilton, K. An Array of Ringing Global Free Modes Discovered in Tropical Surface Pressure Data. *J. Atmos. Sci.* **2020**, *77*, 2519–2539. [CrossRef]
14. Long, R. Solitary waves in the westerlies. *J. Atmos. Sci.* **1964**, *21*, 197–200. [CrossRef]
15. Benney, D. Long nonlinear waves in fluid flows. *J. Math. Phys.* **1966**, *45*, 52–63. [CrossRef]
16. Redekopp, L. On the theory of solitary Rossby waves. *J. Fluid Mech.* **1977**, *82*, 725–745. [CrossRef]
17. Redekopp, L.; Weidman, P. Solitary Rossby waves in zonal shear flows and their interactions. *J. Atmos. Sci.* **1978**, *35*, 790–804. [CrossRef]

18. Wadati, M. The modified Korteweg–de Vries equation. *J. Phys. Soc. Jpn.* **1973**, *34*, 1289–1296. [CrossRef]
19. Boyd, J. Equatorial soliton waves, Part 1: Rossby solitons. *J. Phys. Oceanogr.* **1980**, *10*, 1699–1717. [CrossRef]
20. Boyd, J. The nonlinear equatorial Kelvin wave. *J. Phys. Oceanogr.* **1980**, *10*, 1–11. [CrossRef]
21. Espa, S.; Cabanes, S.; King, G.P.; Nitto, G.D.; Galperin, B. Eddy–wave duality in a rotating flow. *Phys. Fluids* **2020**, *32*, 076604. [CrossRef]
22. Galperin, B.; Sukoriansky, S.; Espa, S.; Lacorata, G.; Dikovskaya, N.; Hoemann, J. Turbulence, Diffusion and Mixing Barriers in Flows with Zonal Jets. In *Zonal Jets: Phenomenology, Genesis, and Physics*; Galperin, B.; Read, P.L., Eds.; Cambridge University Press: Cambridge, UK, 2019; pp. 450–460. [CrossRef]
23. Limaye, S.S. Jupiter: New estimates of the mean zonal flow at the cloud level. *Icarus* **1986**, *65*, 335–352. [CrossRef]
24. Humphreys, T.; Marcus, P.S. Vortex street dynamics: The selection mechanism for the areas and locations of Jupiter's vortices. *J. Atmos. Sci.* **2007**, *64*, 1318–1333. [CrossRef]
25. Dritschel, G.; McIntyre, M.E. Multiple jets as PV staircases: The Phillips effect and the resilience of eddy-transport barriers. *J. Atmos. Sci.* **2008**, *65*, 855–874. [CrossRef]
26. Galperin, B.; Young, R.; Sukoriansky, S.; Dikovskaya, N.; Read, P.; Lancaster, A.; Armstrong, D. Cassini observations reveal a regime of zonostrophic macroturbulence on Jupiter. *Icarus* **2014**, *229*, 295–320. [CrossRef]
27. Alexakis, A.; Biferale, L. Cascades and transitions in turbulent flows. *Phys. Rep.* **2018**, *767*, 1–101.
28. Ingersoll, A.P. Cassini exploration of the planet Saturn: A comprehensive review. *Space Sci. Rev.* **2020**, *216*. 122. [CrossRef] [PubMed]
29. Trammell, J.; Li, L.; Jiang, X.; Smith, M.; Hörst, S.; Vasavada, A. The global vortex analysis of Jupiter and Saturn based on Cassini imaging science subsystem. *Icarus* **2014**, *242*, 122–129. [CrossRef]
30. Li, L.; Ingersoll, A.P.; Vasavada, A.R.; Porco, C.C.; Del Genio, A.D.; Ewald, S.P. Life cycles of spots on Jupiter from Cassini images. *Icarus* **2004**, *172*, 9–23. [CrossRef]
31. Legarreta, J.; Sánchez-Lavega, A. Vertical structure of Jupiter's troposphere from nonlinear simulations of long-lived vortices. *Icarus* **2008**, *196*, 184–201. [CrossRef]
32. Cosentino, G.; Simon, A.; Morales-Juberias, R.; Sayanagi, K.M. Observations and numerical modeling of the Jovian Ribbon. *Astrophys. J.* **2015**, *810*, L10. [CrossRef]
33. Ingersoll, A.P.; Beebe, R.F.; Mitchell, J.L.; Garneau, G.W.; Yagi, G.M.; Müller, J.P. Interaction of eddies and mean zonal flow on Jupiter as inferred from Voyager 1 and 2 images. *J. Geophys. Res.* **1981**, *A86*, 8733–8743. [CrossRef]
34. Vasavada, A.R.; Hörst, S.M.; Kennedy, M.R.; Ingersoll, A.P.; Porco, C.C.; Del Genio, A.D.; West, R.A. Cassini imaging of Saturn: Southern hemisphere winds and vortices. *J. Geophys. Res.* **2006**, *111*, E05004. [CrossRef]
35. Sayanagi, K.M.; Dyudina, U.A.; Ewald, S.P.; Muro, G.D.; Ingersoll, A.P. Cassini ISS observation of Saturn's string of pearls. *Icarus* **2014**, *229*, 170–180. [CrossRef]
36. Qiao, L.; Weisberg, R.H. Tropical instability wave kinematics: Observations from the tropical instability wave experiment. *J. Geophys. Res. Ocean.* **1995**, *100*, 8677–8693. [CrossRef]
37. Chelton, D.B.; Wentz, F.J.; Gentemann, C.L.; de Szoeke, R.A.; Schla, M.G. Satellite microwave SST observations of transequatorial tropical instability waves. *Geophys. Res. Lett.* **2000**, *27*, 1239–1242. [CrossRef]
38. Contreras, F. Long-term observations of tropical instability waves. *J. Phys. Oceanogr.* **2002**, *32*, 2715–2722. [CrossRef]
39. Galperin, B.; Hoemann, J.; Espa, S.; DiNitto, G.; Lacorata, G. Anisotropic macroturbulence and diffusion associated with a westward zonal jet—From laboratory to planetary atmospheres and oceans. *Phys. Rev. E* **2016**, *94*, 063102. [CrossRef] [PubMed]
40. Galperin, B.; Hoemann, J.; Espa, S.; Di Nitto, G. Anisotropic turbulence and Rossby waves in an easterly jet—An experimental study. *Geophys. Res. Lett.* **2014**, *41*, 6237–6243. [CrossRef]
41. Thorpe, S. *The Turbulent Ocean*; Cambridge University Press: Cambridge, UK, 2005.
42. Cabanes, S.; Espa, S.; Galperin, B.; Young, R.M.B.; Read, P.L. Revealing the intensity of turbulent energy transfer in planetary atmospheres. *Geophys. Res. Lett.* **2020**, *47*, e2020GL088685. [CrossRef]
43. Lenouo, A.; Nkankam, F.K. Solitary Rossby Waves in the Lower Tropical Troposphere. *ISRN Atmos. Sci.* **2013**, *2013*, 124965. [CrossRef]
44. Dunkerton, T.J.; Montgomery, M.T.; Wang, Z. Tropical cyclogenesis in a tropical wave critical layer: Easterly waves. *Atmos. Chem. Phys.* **2009**, *9*, 5587–5646. [CrossRef]
45. Maxworthy, T.; Redekopp, L. A solitary wave theory of the Great Red Spot and other features in the Jovian atmosphere. *Icarus* **1976**, *29*, 261–271. [CrossRef]
46. Marcus, P. Jupiter's Great Red Spot and other vortices. *Annu. Rev. Astron. Astrophys.* **1993**, *31*, 523–573. [CrossRef]
47. Boyd, J.P. Chapter 6. Planetary solitary waves. In *WIT Transactions on State of the Art in Science and Engineering*; WIT Press: Southampton, UK, 2007; Volume 9. [CrossRef]
48. Holton, J. *Dynamic Meteorology*, 4th ed.; Elsevier: Amsterdam, The Netherlands, 2004.
49. Sánchez-Lavega, A. *An Introduction to Planetary Atmospheres*; CRC Press: Boca Raton, FL, USA, 2011.
50. Zaqarashvili, T.V.; Albekioni, M.; Ballester, J.L.; Bekki, Y.; Biancofiore, L.; Birch, A.C.; Dikpati, M.; Gizon, L.; Gurgenashvili, E.; Heifetz, E.; et al. Rossby Waves in Astrophysics. *Space Sci. Rev.* **2021**, *217*, 1–93. [CrossRef]
51. Pedlosky, J. *Geophysical Fluid Dynamics*, 2nd ed.; Springer: Berlin/Heidelberg, Germany, 1987.

52. Huang, H.P.; Galperin, B.; Sukoriansky, S. Anisotropic spectra in two-dimensional turbulence on the surface of a rotating sphere. *Phys. Fluids* **2001**, *13*, 225–240. [CrossRef]
53. Vallis, G. *Atmospheric and Oceanic Fluid Dynamics*; Cambridge University Press: Cambridge, UK, 2006.
54. Sukoriansky, S.; Dikovskaya, N.; Galperin, B. On the 'arrest' of the inverse energy cascade and the Rhines scale. *J. Atmos. Sci.* **2007**, *64*, 3312–3327. [CrossRef]
55. Boer, G. Homogeneous and isotropic turbulence on the sphere. *J. Atmos. Sci.* **1983**, *40*, 154–163. [CrossRef]
56. Boer, G.; Shepherd, T. Large-scale two-dimensional turbulence in the atmosphere. *J. Atmos. Sci.* **1983**, *40*, 164–184. [CrossRef]
57. Sukoriansky, S.; Galperin, B.; Dikovskaya, N. Universal spectrum of two-dimensional turbulence on a rotating sphere and some basic features of atmospheric circulation on giant planets. *Phys. Rev. Lett.* **2002**, *89*, 124501. [CrossRef] [PubMed]
58. Kraichnan, R. Inertial-range transfer in two- and three-dimensional turbulence. *J. Fluid Mech.* **1971**, *47*, 525–535. [CrossRef]
59. Galperin, B.; Sukoriansky, S.; Dikovskaya, N.; Read, P.; Yamazaki, Y.; Wordsworth, R. Anisotropic turbulence and zonal jets in rotating flows with a β-effect. *Nonlinear Proc. Geophys.* **2006**, *13*, 83–98. [CrossRef]
60. Lemasquerier, D.; Favier, B.; Le Bars, M. Zonal jets experiments in the gas giants' zonostrophic regime. *Icarus* **2023**, *390*, 115292. [CrossRef]
61. Galperin, B.; Sukoriansky, S.; Huang, H.P. Universal n^{-5} spectrum of zonal flows on giant planets. *Phys. Fluids* **2001**, *13*, 1545–1548. [CrossRef]
62. Li, L.; Ingersoll, A.; Huang, X. Interaction of moist convection with zonal jets on Jupiter and Saturn. *Icarus* **2006**, *180*, 113–123. [CrossRef]
63. Polito, S.; Sato, O.T. Do eddies ride on Rossby waves? *J. Geophys. Res. Ocean.* **2015**, *120*, 5417–5435. [CrossRef]
64. Daners, D. The Mercator and Stereographic Projections, and Many in Between. *Amer. Math. Mon.* **2012**, *119*, 199–210. [CrossRef]
65. Gudermannian Function. In *CRC Standard Mathematical Tables and Formulae*, 31st ed.; Zwillinger, D., Ed.; CRC Press: Boca Raton, FL, USA, 1995; pp. 530–532.
66. Crighton, D. Applications of KdV. *Acta Applic. Math.* **1995**, *39*, 39–67. [CrossRef]
67. Hovmöller, E. The Trough-and-Ridge diagram. *Tellus* **1949**, *1*, 62–66. [CrossRef]
68. Read, P.L. The Dynamics of Jupiter's and Saturn's Weather Layers: A Synthesis After Cassini and Juno. *Annu. Rev. Fluid Mech.* **2024**, *56*, 271–293. [CrossRef]
69. Ingersoll, A.P.; Dowling, T.E.; Gierasch, P.J.; Orton, G.S.; Read, P.L.; Sánchez-Lavega, A.; Showman, A.P.; Simon-Miller, A.A.; Vasavada, A.R. Dynamics of Jupiter's atmosphere. In *Jupiter: The Planet, Satellites and Magnetosphere*; Bagenal, F., Dowling, T.E., McKinnon, W.B., Eds.; Cambridge University Press: Cambridge, UK, 2004; pp. 105–128.
70. Vasavada, A.R.; Showman, A.P. Jovian atmospheric dynamics: An update after Galileo and Cassini. *Rep. Prog. Phys.* **2005**, *68*, 1935–1996. [CrossRef]
71. Del Genio, A.D.; Achterberg, R.K.; Baines, K.H.; Flasar, F.M.; Read, P.L.; Sánchez-Lavega, A.; Showman, A.P. Saturn atmospheric structure and dynamics. In *Saturn from Cassini–Huygens*; Dougherty, M.K., Esposito, L.W., Krimigis, S.M., Eds.; Springer: London, UK, 2009; pp. 113–159. [CrossRef]
72. Sayanagi, K.M.; Baines, K.H.; Dyudina, U.A.; Fletcher, L.N.; Sánchez-Lavega, A.; West, R.A. Saturn's polar atmosphere. In *Saturn in the 21st Century*; Baines, K.H., Flasar, F.M., Krupp, N., Stallard, T.E., Eds.; Cambridge University Press: Cambridge, UK, 2018; pp. 295–336.
73. Showman, A.P.; Ingersoll, A.P.; Achterberg, R.; Kaspi, Y. The global atmospheric circulation of Saturn. In *Saturn in the 21st Century*; Baines, K.H., Flasar, F.M., Krupp, N., Stallard, T.E., Eds.; Cambridge University Press: Cambridge, UK, 2018; pp. 337–376.
74. García-Melendo, E.; Sánchez-Lavega, A. A study of the stability of Jovian zonal winds from HST Images: 1995–2000. *Icarus* **2001**, *152*, 316–330. [CrossRef]
75. Sánchez-Lavega, A.; Rojas, J.F.; Sada, P.V. Saturn's zonal winds at cloud level. *Icarus* **2000**, *147*, 405–420. [CrossRef]
76. Morales-Juberías, R.; Simon, A.A.; Cosentino, R.G. Analysis of the long-term drift rates and oscillations of Jupiter's largest vortices. *Icarus* **2022**, *372*, 114732. [CrossRef]
77. Marcus, P.S. Prediction of a global climate change on Jupiter. *Nature* **2004**, *428*, 828–831. [CrossRef] [PubMed]
78. Rogers, J.H. *The giant planet Jupiter*; Cambridge University Press: Cambridge, UK, 1995; Volume 6.
79. Sánchez-Lavega, A.; Rojas, J.F.; Hueso, R.; Lecacheux, J.; Colas, F.; Acarreta, J.R.; Miyazaki, I.; Parker, D. Interaction of Jovian white ovals BC and DE in 1998 from Earth-based observations in the visual range. *Icarus* **1999**, *142*, 116–124. [CrossRef]
80. Sánchez-Lavega, A.; Orton, G.S.; Morales, R.; Lecacheux, J.; Colas, F.; Fisher, B.; Fukumura-Sawada, P.; Golisch, W.; Griep, D.; Kaminski, C.; et al. The merger of two giant anticyclones in the atmosphere of Jupiter. *Icarus* **2001**, *149*, 491–495. [CrossRef]
81. Barrado-Izagirre, N.; Legarreta, J.; Sánchez-Lavega, A.; Pérez-Hoyos, S.; Hueso, R.; Iñurrigarro, P.; Rojas, J.F.; Mendikoa, I.; Ordonez-Etxeberria, I.; the IOPW Team. Jupiter's third largest and longest-lived oval: Color changes and dynamics. *Icarus* **2021**, *361*, 114394. [CrossRef]
82. Hueso, R.; Sánchez-Lavega, A.; Fouchet, T.; de Pater, I.; Nano, A.A.; Fletcher, L.N.; Wong, M.H.; Rodríguez-Ovalle, P.; Sromovsky, L.A.; Fry, P.M.; et al. An intense narrow equatorial jet in Jupiter's lower stratosphere observed by JWST. *Nat. Astron.* **2023**. [CrossRef]
83. Aurnou, J.M.; Heimpel, M.H. Zonal jets in rotating convection with mixed mechanical boundary conditions. *Icarus* **2004**, *169*, 492–498. [CrossRef]
84. Sayanagi, K.M.; Showman, A.P. Effects of a large convective storm on Saturn's equatorial jet. *Icarus* **2007**, *187*, 520–539. [CrossRef]

85. Showman, A.P. Numerical simulations of forced shallow-water turbulence: Effects of moist convection on the large-scale circulation of Jupiter and Saturn. *J. Atmos. Sci.* **2007**, *64*, 3132–3157. [CrossRef]
86. Palotai, C.; Brueshaber, S.; Sankar, R.; Sayanagi, K. Moist convection in the giant planet atmospheres. *Remote Sens.* **2023**, *15*, 219. [CrossRef]
87. Read, P.L.; Dowling, T.E.; Schubert, G. Saturn's rotation period from its atmospheric planetary-wave configuration. *Nature* **2009**, *460*, 608–610. [CrossRef]
88. Del Genio, A.D.; Barbara, J.M.; Ferrier, J.; Ingersoll, A.P.; West, R.A.; Vasavada, A.R.; Spitale, J.; Porco, C.C. Saturn eddy momentum fluxes and convection: First estimates from Cassini images. *Icarus* **2007**, *189*, 479–492. [CrossRef]
89. Porco, C.C.; Baker, E.; Barbara, J.; Beurle, K.; Brahic, A.; Burns, J.A.; Charnoz, S.; Cooper, N.; Dawson, D.D.; Del Genio, A.D.; et al. Cassini imaging science: Initial results on Saturn's atmosphere. *Science* **2005**, *307*, 1243–1247. [CrossRef] [PubMed]
90. Dyudina, U.A.; Ingersoll, A.P.; Ewald, S.P.; Porco, C.C.; Fischer, G.; Kurth, W.; Desch, M.; Del Genio, A.; Barbara, J.; Ferrier, J. Lightning storms on Saturn observed by Cassini ISS and RPWS during 2004–2006. *Icarus* **2007**, *190*, 545–555. [CrossRef]
91. del Río-Gaztelurrutia, T.; Legarreta, J.; Hueso, R.; Pérez-Hoyos, S.; Sánchez-Lavega, A. A long-lived cyclone in Saturn's atmosphere: Observations and models. *Icarus* **2010**, *209*, 665–681. [CrossRef]
92. Gunnarson, J.L.; Sayanagi, K.M.; Fischer, G.; Barry, T.; Wesley, A.; Dyudina, U.A.; Ewald, S.P.; Ingersoll, A.P. Multiple convective storms within a single cyclone on Saturn. *Icarus* **2023**, *389*, 115228. [CrossRef]
93. Sánchez-Lavega, A.; Lecacheux, J.; Gomez, J.M.; Colas, F.; Laques, P.; Noll, K.; Gilmore, D.; Miyazaki, I.; Parker, D. Large-Scale Storms in Saturn's Atmosphere During 1994. *Science* **1996**, *271*, 631–634. [CrossRef]
94. Sayanagi, K.M.; Dyudina, U.A.; Ewald, S.P.; Fischer, G.; Ingersoll, A.P.; Kurth, W.S.; Muro, G.D.; Porco, C.C.; West, R.A. Dynamics of Saturn's great storm of 2010–2011 from Cassini ISS and RPWS. *Icarus* **2013**, *223*, 460–478. [CrossRef]
95. García-Melendo, E.; Sánchez-Lavega, A. Shallow water simulations of Saturn's giant storms at different latitudes. *Icarus* **2017**, *286*, 241–260. [CrossRef]
96. Sánchez-Lavega, A. Saturn's Great White Spots. *Chaos* **1994**, *4*, 341–353. [CrossRef]
97. Sánchez-Lavega, A.; García-Melendo, E.; Pérez-Hoyos, S.; Hueso, R.; Wong, M.H.; Simon, A.; Sanz-Requena, J.F.; Antuñano, A.; Barrado-Izagirre, N.; Garate-Lopez, I.; et al. An enduring rapidly moving storm as a guide to Saturn's Equatorial jet's complex structure. *Nat. Commun.* **2016**, *7*, 13262. [CrossRef] [PubMed]
98. García-Melendo, E.; Pérez-Hoyos, S.; Sánchez-Lavega, A.; Hueso, R. Saturn's zonal wind profile in 2004–2009 from Cassini ISS images and its long-term variability. *Icarus* **2011**, *215*, 62–74. [CrossRef]
99. Orton, G.S.; Yanamandra-Fisher, P.A.; Fisher, B.M.; Friedson, A.J.; Parrish, P.D.; Nelson, J.F.; Bauermeister, A.S.; Fletcher, L.; Gezari, D.Y.; Varosi, F.; et al. Semi-annual oscillations in Saturn's low-latitude stratospheric temperatures. *Nature* **2008**, *453*, 197–199. [CrossRef] [PubMed]
100. Fouchet, T.; Guerlet, S.; Strobel, D.F.; Simon-Miller, A.A.; Bézard, B.; Flasar, F.M. An equatorial oscillation in Saturn's middle atmosphere. *Nature* **2008**, *453*, 200–202. [CrossRef] [PubMed]
101. Cabanes, S.; Gastine, T.; Fournier, A. Zonostrophic turbulence in the subsurface oceans of the Jovian and Saturnian moons. *Icarus* **2024**, *415*, 116047. [CrossRef]
102. Kundt, W.; Lüttgens, G. Rings around planets, atmospheric super-rotation, and their great spots. *Astrophys. Space Sci.* **1998**, *257*, 33–47. [CrossRef]
103. Imamura, T.; Mitchell, J.; Lebonnois, S.; Kaspi, Y.; Showman, A.P.; Korablev, O. Superrotation in Planetary Atmospheres. *Space Sci. Rev.* **2020**, *216*, 87. [CrossRef]
104. Chelton, D.; Schlax, M.; Samelson, R.; de Szoeke, R. Global observation of large oceanic eddies. *Geophys. Res. Lett.* **2007**, *34*. [CrossRef]
105. Chelton, D.; Schlax, M.; Samelson, R. Global observations of nonlinear mesoscale eddies. *Progr. Oceanogr.* **2011**, *91*, 167–216. [CrossRef]
106. Early, J.; Samelson, R.; Chelton, D. The evolution and propagation of quasigeostrophic ocean eddies. *J. Phys. Oceanogr.* **2011**, *41*, 1535–1555. [CrossRef]
107. Wunsch, C. A time-average ocean: Thermal wind and flow spirals. *Prog. Oceanogr.* **2024**, *221*, 103206. [CrossRef]

MDPI AG
Grosspeteranlage 5
4052 Basel
Switzerland
Tel.: +41 61 683 77 34

Atmosphere Editorial Office
E-mail: atmosphere@mdpi.com
www.mdpi.com/journal/atmosphere

www.ingramcontent.com/pod-product-compliance
Lightning Source LLC
Chambersburg PA
CBHW042014260926
51850CB00020B/3408